Studies in Logic
Volume 112

Implicative-groups vs. Groups and Generalizations
Second Edition

Volume 102
Bounded Reasoning Volume 1: Classical Propositional Logic
Marcello D'Agostino, Dov Gabbay, Costanza Larese, Sanjay Modgil

Volume 103
The Fertile Debate. Affective Exploration of a Controversy
Claire Polo

Volume 104
Argument, Sex and Logic
Dov Gabbay, Gadi Rozenberg and Lydia Rivlin

Volume 105
Logic as a Tool. A Guide to Formal Logical Reasoning
Valentin Goranko

Volume 106
New Directions in Term Logic
George Englebretsen, ed

Volume 107
Non-commutative Algebras. Pseudo-BCK Algebreas versus m-pseudo-BCK Algebras
Afrodita Iorgulescu

Volume 108
Semitopology: decentralised collaborative action via topology, algebra, and logic
Murdoch J. Gabbay

Volume 109
The Cognitive Dimension of Social Argumentation. Proceedings of the 4th European Conference on Argumentation, Volume I. Fabio Paglieri, Alessandro Ansani and Marco Marini, eds.

Volume 110
The Cognitive Dimension of Social Argumentation. Proceedings of the 4th European Conference on Argumentation, Volume II. Fabio Paglieri, Alessandro Ansani and Marco Marini, eds.

Volume 111
The Cognitive Dimension of Social Argumentation. Proceedings of the 4th European Conference on Argumentation, Volume III. Fabio Paglieri, Alessandro Ansani and Marco Marini, eds.

Volume 112
Implicative-groups vs. Groups and Generalizations. Second Edition
Afrodita Iorgulescu

Studies in Logic Series Editor
Dov Gabbay dov.gabbay@kcl.ac.uk

Implicative-groups vs. Groups and Generalizations

Second Edition

Afrodita Iorgulescu

ISBN 978-1-84890-486-6

College Publications
Scientific Director: Dov Gabbay
Managing Director: Jane Spurr

http://www.collegepublications.co.uk

Cover prepared by Debbie Hunt

To the Royal House of Romania

Preface to the second edition

The first edition of this monograph was launched in February 2018, by Matrix Rom, Bucharest, in Romania. It had 730 pages, 22 chapters divided into two parts: Part I (Implicative-groups vs. groups and generalizations, 17 chapters) and Part II (Examples by PASCAL programs, 5 chapters). Its back cover contained the kind appreciation of Francesco Paoli, Università degli Studi di Cagliari, Italy, presented also on the back cover of this second edition.

This second edition contains the revised text of the first edition, i.e. the text where Part II is eliminated and the small errors, mostly typing errors, observed in time, are corrected. We have eliminated Part II, containing over 100 Pascal programs with examples, in order to shorten the book and taking into consideration the appearance of the automated theorem prover PROVER9-MACE4, developed by William W. McCune (1953-2011), which provides proofs but also counter-examples/examples.

We have added to the Bibliography our new papers and the two monographs written after 2018.

The book is supposed to be of interest mostly to the researchers in the topics of algebras of logic and group theory, but it may be also of interest to the students in mathematics/computer science attracted by these topics, because there are many things to develop further.

Afrodita Iorgulescu
afrodita.iorgulescu@ase.ro

Bucharest, March 2025

Preface to the first edition

The present monograph treats mainly our results obtained in the last nine years, many of them obtained in the last two years and presented here for the first time.

To specify and understand the context, we make first an overview of some algebras of logic (with additional operation(s)) vs. an overview of some unital magmas (with additional operation(s)), in the non-commutative case. On this occasion, we introduced the pseudo-M algebra (and some subclasses) as the most general algebra of logic studied in this work.

We then generalize the pseudo-M algebras to the quasi-pseudo-M algebras and we generalize the unital magmas to the quasi-m unital magmas; it is the beginning of two parallel theories connected to quantum computation.

The core of the monograph is made by the presentation of the implicative-groups (in the "world" of pseudo-M algebras) vs. the groups (in the "world" of unital magmas); we present the implicative-group and the partially-ordered (lattice-ordered) implicative-group as a term equivalent definition of the group and the partially-ordered (lattice-ordered) group, respectively. The lattice-ordered implicative-group is the great piece which missed from the puzzle showing the connections between lattice-ordered groups and some algebras of logic.

We then present new generalizations of the implicative-groups vs. new generalizations of the groups, with connections.

We also introduce the quasi-implicative groups, and some generalizations of them, vs. the quasi-m groups, and some generalizations of them, with connections.

We present connections at group level and connections between the group level (unital magmas level) and the algebras of logic level. We discuss filters/ideals and the deductive systems of the implicative-groups/groups. We study the normal filters/ideals and the compatible deductive systems, the representability of some of the involved algebras and we present other connections between the group level and the algebras of logic level. Finally, we present the implicative-states and the Bosbach-states on l-groups with strong unit.

The whole picture is completed by the examples and the informations provided by over 100 programs in the PASCAL programming language.

We believe that the new notions and the new connections, presented in a unifying way, open new perspectives in the study of the algebras of logic (of the pseudo-M algebras, in particular) vs. the unital magmas.

Since there are many new notions and results (the content of chapters 5, 10, 11, 12, 13 is entirely new), there can be some errors; we kindly ask the reader to point out any error found to the e-mail address below.

Afrodita Iorgulescu
afrodita.iorgulescu@ase.ro

Bucharest, January 2018

Acknowledgements

All my gratitude to Sergiu Rudeanu, to Paul Flondor and to Francesco Paoli, for their valuable and very useful remarks and suggestions.

I am specially grateful to George Georgescu for his encouragements in doing this research.

All my gratitude to the following mathematicians who helped me, by their remarks on the final draft, to improve the quality of the monograph: Cristian Calude, George Georgescu, Francesco Paoli, Mircea Sularia, Dragoş Vaida.

I wanted to dedicate this monograph to my dear mother, Elisabeta Iorgulescu, former teacher of mathematics, now of age 96. She was by my side all my life.

I wanted to dedicate this monograph to my dear colleague (we graduated in the same year, 1969, the Faculty of mathematics and mechanics of the University of Bucharest) and friend George Georgescu, professor at the University of Bucharest. We work in the same area of mathematics, the algebra of logic. He was my collaborator in many papers, he encouraged me in writing my books, we disscuss every day everything.

I wanted to dedicate this monograph to my dear professor Sergiu Rudeanu, now of age 83. He was my advisor for the PhD Thesis, at the University of Bucharest, and my friend and advisor since then.

I wanted finally to dedicate this monograph to the memory of Franco Montagna. I have met him at some conferences in Europe and I was very impressed by his sensibility as a human being. Once, a conference was organized at the University of Sienna, his place, and I wanted to present my research, just made, on implicative-groups; the conference chair rejected my application and Franco wrote me an e-mail telling he was very sorry for the decision and that he would invite me to the University later, for some days - invitation that I have declined very politely. We never met again and, very unfortunately, he died in 2016.

I love all these people.

I dedicate this monograph to the Royal House of Romania, which has celebrated 150 years in 2016. The Royal period, 1866-1947, was the most beneficial for Romania, a republic since 1948.

Bucharest, January 2018

Contents

Chapter 1

Introduction

This monograph is based mainly on author's papers [99], [100], [104], [101], on preprints [97], [98], [103] and on author's book [94], but many new notions and results are also presented (in Chapters 5, 10, 11, 12, 13).

The work on the monograph was begun in January 2016 and was finished in January 2018.

The core of this monograph is composed mainly by the *implicative-group* (a term equivalent definition of the group), the *po-implicative-group* (a term equivalent definition of the po-group) and the l-implicative-group (a term equivalent definition of the l-group), all coming from algebras of logic, notions introduced and studied in [99]. Other related topics concerning these notions are taken from [100].

In order to create the frame-work and to help to better understand the definitions, the results and the connections, we start by presenting an overview of some algebras of logic and an overview of some unital magmas [94], in the non-commutative case, and some connections existing between the two "worlds". We introduce the *pseudo-M algebras* and some new subclasses of them, which complete the area of those algebras of logic involved in the monograph.

We extend the notions of quasi-algebra and quasi-structure (generalizations of algebras of logic) introduced and studied in [101] to the non-commutative case, introducing the notions of quasi-pseudo-algebra and quasi-pseudo-structure. We then introduce and study the quasi-pseudo-M algebras and the quasi-pseudo-BCI algebras, as generalizations of pseudo-M algebras and pseudo-BCI algebras, respectively. On the other hand, we introduce the new notion of *quasi-m algebra*. In particular, we introduce and study the quasi-m unital magmas and the quasi-m monoids, as generalizations of the unital magmas and monoids, respectively, and we establish connections.

We then introduce new generalizations of the implicative-groups vs. new generalizations of the groups, and their corresponding quasi-pseudo-algebras vs. the corresponding quasi-m algebras, respectively.

The goal of the monograph is to present a unifying treatement of the subjects. For this, the basic properties used in the monograph have

unique names.

1.1 Introduction

Pseudo-MV algebras, the non-commutative generalizations of Chang's MV algebras [26], were introduced in 1999 [58] and developed in [61]. Pseudo-MV algebras are intervals $[0, u]$ in l-groups [47] and pseudo-MV algebras are termwise equivalent [24], [25] to pseudo-Wajsberg algebras. Hence, the pseudo-Wajsberg algebras must be connected to a notion that is termwise equivalent to the l-group.

$$\begin{array}{ccc} ? & \Leftrightarrow & \mathbf{l-group} \\ \Downarrow & & \Downarrow \\ \mathbf{pseudo-Wajsberg\ algebra} & \Leftrightarrow & \mathbf{pseudo-MV\ algebra} \end{array}$$

Pseudo-product algebras, the non-commutative generalizations of Hájek's product algebras [72], were introduced in 2002 [42]. Pseudo-product algebras are obtained as bounded sets $-\{\infty\} \cup G^{-}$ in l-groups and pseudo-product algebras are equivalent to pseudo-Hájek(pRP) algebras verifying some properties. Hence, the pseudo-Hájek(pRP) algebras verifying some properties must be connected also to a notion that is termwise equivalent to the l-group.

$$\begin{array}{ccc} ? & \Leftrightarrow & \mathbf{l-group} \\ \Downarrow & & \Downarrow \\ \mathbf{pseudo-Hajek(pRP)\ algebra} & \Leftrightarrow & \mathbf{pseudo-product\ algebra} \\ \mathbf{with\ some\ properties} & & \end{array}$$

The notion termwise equivalent to the l-group is the *l-implicative-group*; it is the great piece which had missed from the puzzle showing the connections between algebras of logic and unital magmas.

The research concerning the implicative-group, the po-implicative-group and the l-implicative-group was begun in 2009's; it was presented first in the preprints [97] and [98], then in the papers [99] and [100]. The core of the monograph is based on these papers.

In order to create the frame-work and to better understand the definitions, the results and the connections, we start by presenting a survey of some algebras of logic in the non-commutative case (see [104] for the commutative case); we introduce the *pseudo-M algebras* and some new subclasses of them, which complete the area of algebras of logic involved in the study. We also present a survey of some unital magmas [94]. Some connections between the "world" of algebras of logic (of pseudo-M algebras) and the "world" of unital magmas are then presented.

We then extend the notions of quasi-algebra and quasi-structure (generalizations of algebras of logic comming from quantum computing, introduced and studied in

[101]) to the non-commutative case, introducing the quasi-pseudo-algebra and the quasi-pseudo-structure. As examples, we introduce the quasi-pseudo-M algebras and the quasi-pseudo-BCI algebras, as generalizations of the pseudo-M, pseudo-BCI, respectively. On the other hand, we introduce the new notion of *quasi-m algebra*. As examples, we introduce and study the quasi-m unital magmas and the quasi-m monoids, as generalizations of the unital magmas and monoids, respectively.

Then, after recalling useful definitions and results concerning the groups, we present the *implicative-groups*, the *po-implicative-groups* and the *l-implicative-groups*, and prove their termwise equivalences with the groups, the po-groups and the l-groups, respectively.

We then introduce some generalizations of the group (the hub, the moon, the goop and also the qm-group, the qm-hub, the qm-moon, the qm-goop, respectively) and some corresponding generalizations of the implicative-group (the implicative-hub, the implicative-moon, the implicative-goop and also the q-implicative-group, the q-implicative-hub, the q-implicative-moon, the q-implicative-goop, respectively). We establish some connections.

G. Dymek [50] (see [138] for the commutative case) made in 2012 the connection between the pseudo-BCI algebras and the groups, by introducing the subclass of p-semisimple pseudo-BCI algebras and proving that they are termwise equivalent with the groups. We then have concluded that the p-semisimple pseudo-BCI algebras are termwise equivalent with the implicative-groups. We also define and study other p-semisimple algebras of logic. We then make the connections between the pseudo-BCI algebras and the pseudo-*aRM** algebras, on one hand, and the implicative-groups and the implicative-hubs, respectively, on the other hand, by using the new notion of *p-semisimple* pseudo-algebra, and the connections between the quasi-pseudo-BCI algebras and the quasi-pseudo-*aRM** algebras, on one hand, and the qi-implicative-groups and the q-implicative-hubs, respectively, on the other hand, by using the new notion of *quasi-p-semisimple* quasi-pseudo-algebra.

The announced connections between the l-implicative-groups and the pseudo-Wajsberg algebras and the pseudo-Hájek(pRP) algebras verifying some properties are then presented.

We study the normal filters/ideals and the compatible deductive systems, the representability of some of the involved algebras and we establish other connections between the group (unital magmas) level and the algebras of logic level. We study the implicative-states and the Bosbach-states on l-groups with strong unit.

Note that most of the algebras recalled or introduced in this monograph have essentially a partial order relation (an order relation, for short) (i.e. a binary relation that is reflexive, antisymmetrique and transitive), denoted by $\leq$, which usually does not appear explicitely in the definitions. The pair $(A, \leq)$ is called a partially-ordered set, or poset (po-set) for short.

The presence of the *order* relation $\leq$ implies the presence of the *duality principle*. Thus, each such algebra has a dual algebra, with a *dual order* relation denoted by $\geq$. We have given names to the dual algebras: left-algebras and right-algebras [52], [86], [87], [94], namely:

- *left-algebras* are those algebras having a pseudo-t-norm $\odot$ and/or a (reversed) pseudo-residuum ($\to^L$, $\rightsquigarrow^L$) and an element 1 as principal primitive operations on a *poset with* 1 *as greatest element* and, dually,

- *right-algebras* are those algebras having a pseudo-t-conorm $\oplus$ and/or a (reversed) pseudo-coresiduum ($\to^R$, $\rightsquigarrow^R$) and an element 0 as principal primitive operations on a *poset with* 0 *as smallest element.*

Hence, the notions of left-algebra and right-algebra are dual; they are connected with the left-continuity of a pseudo-t-norm and with the right-continuity of a pseudo-t-conorm, respectively, or with the negative (left) cone and the positive (right) cone of a partially-ordered group, respectively, etc. One can find more about left- and right- algebras in [94].

In order to address also the algebras based on posets containing an element 1 (an element 0, dually), as the pseudo-BCI algebras and the po-groups are, we need, in this monograph, to extend the above definitions of left-algebras and right-algebras by considering just posets containing an element 1 (0, dually), instead of posets with greatest element 1 (posets with smallest element 0, dually). Moreover, any generalization of a left-algebra (right-algebra) is also named a left-algebra (right-algebra).

Consequently, more generally,

- *left-algebras* will be those algebras having a pseudo-t-norm $\odot$ and/or a (reversed) pseudo-residuum ($\to^L$, $\rightsquigarrow^L$) and an element 1 as principal primitive operations on a *poset containing* 1; any generalization of a left-algebra is also a left-algebra;

dually,

- *right-algebras* will be those algebras having a pseudo-t-conorm $\oplus$ and/or a (reversed) pseudo-coresiduum ($\to^R$, $\rightsquigarrow^R$) and an element 0 as principal primitive operations on a *poset containing* 0; any generalization of a right-algebra is also a right-algebra.

We obtain then the following generalized definitions:

Definition 1.1.1 (See [52] and see [68], [128] for t-norms on $[0,1]$)

(i) A *pseudo-t-norm* on the poset $(A^L, \leq)$ containing an element 1 is a binary operation $\odot$ (called *pseudo-product*) verifying: for all $x, y, z, a \in A^L$,

(Ass) $x \odot (y \odot z) = (x \odot y) \odot z$,

(pU) $x \odot 1 = x = 1 \odot x$,

(pCp) $x \leq y \Longrightarrow a \odot x \leq a \odot y$ and $x \odot a \leq y \odot a$.

(ii) Dually, a *pseudo-t-conorm* on the poset $(A^R, \leq)$ containing an element 0 is a binary operation $\oplus$ (called *pseudo-sum*) verifying: for all $x, y, z, a \in A^R$,

(Ass) $x \oplus (y \oplus z) = (x \oplus y) \oplus z$,

(pU) $x \oplus 0 = x = 0 \oplus x$,

(pCp) $x \leq y \Longrightarrow a \oplus x \leq a \oplus y$ and $x \oplus a \leq y \oplus a$.

A pseudo-t-norm (pseudo-t-conorm) $\otimes$ is *commutative*, if it verifies: $x \otimes y = y \otimes x$, for all x, y. A commutative pseudo-t-norm (pseudo-t-conorm) is called a *t-norm* (*t-conorm*, respectively) and a commutative pseudo-product (pseudo-sum) becomes a *product* (*sum*, respectively).

Remark 1.1.2 The sets A^L and A^R may coincide in some cases.

Definition 1.1.3 [86], [87]

(i) A *pseudo-residuum* on the poset $(A^L, \leq)$ containing an element 1 is a pair of binary operations $(\rightarrow^L, \rightsquigarrow^L)$ (called *pseudo-implication*) verifying: for all $x, y, z \in A^L$,
(pBB^L) $y \rightarrow^L z \leq (z \rightarrow^L x) \rightsquigarrow^L (y \rightarrow^L x)$, $y \rightsquigarrow^L z \leq (z \rightsquigarrow^L x) \rightarrow^L (y \rightsquigarrow^L x)$,
(pM^L) $1 \rightarrow^L x = x = 1 \rightsquigarrow^L x$,
(IdEq^L) $x \rightarrow^L y = 1 \Longleftrightarrow x \rightsquigarrow^L y = 1$,
(pEq^L) $x \leq y \Longleftrightarrow x \rightarrow^L y = 1 \overset{(IdEq^L)}{\Longleftrightarrow} x \rightsquigarrow^L y = 1$.

(ii) A *pseudo-coresiduum* on the poset $(A^R, \leq)$ containing an element 0 is a pair of binary operations $(\rightarrow^R, \rightsquigarrow^R)$ (called *pseudo-coimplication*) verifying: for all $x, y, z \in A^R$,
(pBB^R) $y \rightarrow^R z \geq (z \rightarrow^R x) \rightsquigarrow^R (y \rightarrow^R x)$, $y \rightsquigarrow^R z \geq (z \rightsquigarrow^R x) \rightarrow^R (y \rightsquigarrow^R x)$,
(pM^R) $0 \rightarrow^R x = x = 0 \rightsquigarrow^R x$,
(IdEq^R) $x \rightarrow^R y = 0 \Longleftrightarrow x \rightsquigarrow^R y = 0$,
(pEq^R) $x \geq y \Longleftrightarrow x \rightarrow^R y = 0 \overset{(IdEq^R)}{\Longleftrightarrow} x \rightsquigarrow^R y = 0$.

Remark 1.1.4 Note that, in [86], [87], we have considered in the definition of pseudo-residuum four axioms: (pBB^L), (pM^L), (pEq^L) and (p^{*L}), where:
(p^{*L}) $x \leq y \Longrightarrow z \rightarrow^L x \leq z \rightarrow^L y$, $z \rightsquigarrow^L x \leq z \rightsquigarrow^L y$,
but (p^{*L}) can be obtained from (pBB^L), (pM^L), (pEq^L).

A pseudo-residuum (pseudo-coresiduum) $(\rightarrow, \rightsquigarrow)$ is *commutative*, if it verifies $\rightarrow = \rightsquigarrow$. A commutative pseudo-residuum (pseudo-coresiduum) is called a *residuum* (*coresiduum*, respectively) and it is denoted by $\rightarrow$; a commutative pseudo-implication (pseudo-coimplication) is called an *implication* (*coimplication*, respectively).

A *reversed* pseudo-residuum is a pair of binary operations $(\square, \#)$, where $x \square y = y \rightarrow^L x$ and $x \# y = y \rightsquigarrow^L x$, with the corresponding properties.

Dually, a *reversed* pseudo-coresiduum is a pair of binary operations $(\star, \circ)$, where $x \star y = y \rightarrow^R x$ and $x \circ y = y \rightsquigarrow^R x$, with the corresponding properties.

Remark 1.1.5 The most important continuous t-norms on $[0, 1]$ are the following three: Łukasiewicz t-norm, Product t-norm, Gödel t-norm. These three t-norms and their associated residua are the following:

(1) Łukasiewicz:

$$x \odot_L y = \max(0, x+y-1), \quad x \rightarrow_L y = \begin{cases} 1, & \text{if } x \leq y \\ 1-x+y, & \text{if } x > y \end{cases} = \min(1, 1-x+y);$$

(2) Product (Gaines):

$$x \odot_P y = xy, \quad x \rightarrow_P y = \begin{cases} 1, & \text{if } x \leq y \\ y/x, & \text{if } x > y, \end{cases} \quad \text{(Goguen implication)}$$

(3) Gödel (Brouwer):

$$x \odot_G y = \min(x, y), \quad x \rightarrow_G y = \begin{cases} 1, & \text{if } x \leq y \\ y, & \text{if } x > y, \end{cases} \quad \text{(Gödel implication).}$$

They correspond to the most significant fuzzy logics: Łukasiewicz logic, Product logic and Gödel logic, respectively. The MV algebras, the Product algebras and the Gödel algebras constitute the algebraic models for these three types of logics. The class of BL algebras contains the MV algebras [26], [32], the Product algebras [73], [141], [72] and the Gödel algebras [72] (or linear Heyting algebras (A. Monteiro; cf. L. Monteiro [149]) or L-algebras (Horn [80]); cf. [11]).

Note that a left-algebra (dually for a right-algebra) belongs:
- either to "the world of $\odot$, 1", called here "the world of unital magmas", when $\odot$ and 1 are the principal primitive operations,
- or to "the world of $(\rightarrow^L, \rightsquigarrow^L)$, 1", called "the world of algebras of logic (of pseudo-M algebras, more precisely)", when $(\rightarrow^L, \rightsquigarrow^L)$ and 1 are the principal primitive operations.

Additional operations (secondary primitive and derivate operations) can appear in these worlds.

Note that the algebraists work usually with the additive groups and with the positive (right) cone of a partially-ordered group $(G, \leq, +, -, 0)$, where there are essentially a pseudo-sum $\oplus = +$ and an element 0. Sometimes, the negative (left) cone is needed also, where there are essentially a pseudo-product $\odot = +$ and an element $1 = 0$. Hence, the algebraists usually work with the *right-unital magmas.*

By contrary, note that the logicians work with the logic of *truth*, where the *truth* is represented by 1, and there is essentially one implication (two implications, in the non-commutative case); we could name this logic "left-logic". One can imagine also a "right-logic", as a logic of *false*, where the *false* is represented by 0. Hence, the logicians (usually) work with the *left-algebras of logic* (or the *algebras of left-logic*).

Summarizing, for algebraists, the appropriate algebras are the unital magmas, not the algebras of logic, and among the unital magmas, the appropriate algebras are the right-algebras. For logicians, by contrary, the appropriate algebras are the algebras of logic, not the unital magmas, and among the algebras of logic, the appropriate algebras are the left-algebras. This explains why, for examples, the MV algebras (and hence the pseudo-MV algebras), were initially introduced as right-unital magmas, while the Wajsberg algebras (and hence the pseudo-Wajsberg algebras) were initially introduced as left-algebras of logic.

Therefore, in this monograph we shall work as much possible both with unital magmas and with algebras of logic (pseudo-M algebras, more precisely) and both with left-algebras and with right-algebras. But, **since we are comming from algebras of logic side, we shall put on first line the algebras of logic and the left-algebras**; therefore, the group (and the unital magmas) will be defined multiplicatively.

The monograph is organized in 17 chapters, as follows.

In **Chapter 2 (Algebras of logic (pM algebras (with additional operations)))**, we introduce the equivalent notions of pseudo-algebra and pseudo-structure (of logic) and the p-semisimple property. We introduce the (strong) pM, pML, pME, pMEL algebras, in order to complete the hierarchies from [104], generalized here to the non-commutative case. We recall the definitions of left- and right- pseudo-BCI algebras, (bounded) pseudo-BCK algebras, (bounded) pseudo-BCK(pP) algebras (lattices), (bounded) pseudo-BCK(pRP) algebras (lattices), bounded pseudo-BCK algebras with property (pDN), pseudo-Hájek(pP) algebras, pseudo-Hájek(pRP) algebras, pseudo-Wajsberg algebras and implicative-Boolean algebras. We introduce and study the pseudo-BCI(pP) algebras (lattices) and pseudo-BCI(pRP) algebras (lattices). We present some results, including the equivalences (EL1), (iEL1), (EL2), (iEL2), (iEL3) - (iEL10) between classes of algebras of logic, and hierarchies of some of the involved algebras of logic.

In **Chapter 3 (Unital magmas (with additional operations))**, we recall the definitions of the unital magmas, monoids, loops, groups, posets, (bounded) lattices and of the left- and right- po-ms, po-ims, l-ms, l-ims, po-ims(pR), po-ims(pPR), l-ims(pR), l-ims(pPR), pseudo-residuated lattices, pseudo-BL algebras, pseudo-product algebras, pseudo-MV algebras and Boolean algebras. We present some results, including the equivalences (EM1), (iEM1), (EM2), (iEM2), (iEM3) - (iEM9) betwwen classes of unital magmas, and hierarchies of some of the involved unital magmas [94].

In **Chapter 4 (Connections between the two kinds of algebras)**, we show connections existing between the pseudo-algebras of logic (pM algebras) and the unital magmas presented in the previous Chapters 2 and 3, respectively [94]. Namely, in Section 1, we present the implications (ILM1) and (ILM2) and the equivalences (iELM1) - (iELM9) existing between the intermediary notions from the world of pM algebras and the intermediary notions from the world of unital magmas. In Section 2, based on the previous connections, we present the implications (I1) and (I2) and the equivalences (iE1) - (iE9), (iE9'), (iE10), (iE11') existing between the four types of notions from the world of pM algebras and from the world of unital magmas.

In **Chapter 5 (Quasi-pM algebras vs. quasi-m unital magmas)**, we are dealing with generalizations of pseudo-algebras (pseudo-structures) of logic (of pM algebras), called quasi-pseudo-algebras (quasi-pseudo-structures) vs. generalizations of unital magmas, called quasi-m algebras. In Section 1, we introduce the quasi-pseudo-algebras (quasi-pseudo-structures) (as non-commutative generalizations of the quasi-algebras (quasi-structures) comming from quantum computing [101]) and some results in their theory. We introduce the quasi-pM and the quasi-pBCI algebras, as generalizations of the pM and the pBCI algebras, respectively, as first examples. We generalize the p-semisimple property to the quasi-p-semisimple property. We generalize to the non-commutative case the notion of quasi-algebra (quasi-structure) with quasi-negation from [103]. In Section 2, we introduce the quasi-m algebras and some results in their theory. We introduce the quasi-m unital magmas and the quasi-m monoids, as generalizations of the unital magmas and of the monoids, respectively, as first examples. We introduce the quasi-m algebras with quasi-m negations. In Section 3, we establish a connection between quasi-pseudo-algebras with quasi-negations and quasi-m algebras with quasi-m negations.

In **Chapter 6 (Examples of the two kinds of algebras)**, we present some examples of (quasi-) pseudo-M algebras and of (quasi-m) unital magmas. We also present some programs in the PASCAL programming language to check some properties.

In **Chapter 7 (Groups, po-groups, l-groups)**, we recall [97], [100] the group, the po-group and the l-group, the intermediary notions of *X-po-group*, *X-l-group* and *X-group*, respectively, and prove the equivalences (EG1), (EG2), (EG3) between these notions, respectively.

In **Chapter 8 (Implicative-groups, po-(l-)implicative-groups)**, we recall and study [97], [100] the implicative-group, the po-implicative-group, the l-implicative-group and the intermediary notions of X-implicative-group, X-po-implicative-group, X-l-implicative-group, respectively, and prove the equivalences (EI1), (EI2), (EI3) between these notions, respectively.

In **Chapter 9 (Connections between groups and implicative-groups)**, we present [97], [100] the equivalences (EIG1), (EIG2), (EIG3) between the intermediary notions, in Section 1, and the equivalences (Eq1), (Eq2), (Eq3) between the corresponding four notions, in Section 2. In Section 3, we present an example of non-commutative group (implicative-group) and some examples of l-groups (l-implicative-groups).

In **Chapter 10 (Generalizations of groups)**, we introduce and study the new notions of hub, moon and goop and we show their connections with the two definitions of the groups. We also introduce and study the new notions of qm-hub, qm-moon, qm-group and qm-goop and we show the connections between all these generalizations of the groups. Final connections of some generalizations of the groups involved in this monograph are presented.

In **Chapter 11 (Generalizations of implicative-groups)**, we introduce and study the analogous new notions of those from Chapter 10, namely, the implicative-hub, the implicative-moon and the implicative-goop, and we show their connections with the implicative-group (in Section 1); the q-implicative-hub, the q-implicative-moon, the q-implicative-group and the q-implicative-goop, and we show the connections between all these generalizations of the implicative-goups (in Section 3). We also introduce the (strong) quasi-pseudo-*aRM** algebras (in Section 2). Final connections of some generalizations of the implicative-groups are also presented.

In **Chapter 12 (Connections between generalizations)**, we establish the connections between the goops and the implicative-goops, the qm-groups and the q-implicative-groups, the (super) qm-goops and the (super) q-implicative-goops. We prove final connections and we present resuming connections.

In **Chapter 13 ((Quasi-) p-semisimple (quasi-) pseudo-algebras)**, Section 1, we present some p-semisimple pseudo-algebras, as for example the p-semisimple pBCI, pBCI(pP), pBCI(pRP) algebras (which are the implicative-groups) and the p-semisimple p*aRM** algebras (which are the implicative-hubs). In Section 2, we present some quasi-p-semisimple quasi-pseudo-algebras, as the quasi-p-semisimple q-pBCI (which are the exchanged q-i-hubs) and the quasi-p-semisimple q-p*aRM** algebras (which are the super q-i-hubs). In Section 3, we present resuming connections.

In **Chapter 14 (Connections at lattice-order level)**, we present notions and results from [97], [98], [99], [100]. In Section 1, we present connections between the l-implicative-groups and the pseudo-BCK(pP) lattices which provide the pseudo-Wajsberg algebras and the equivalent of pseudo-product algebras. Namely, we prove that the l-implicative-group operations $\vee, \wedge, \rightarrow, \rightsquigarrow, 1$ and $\cdot$, restricted to the negative (positive) cone, determine a structure of left- (right-) pseudo-BCK lattice with pseudo-product (pseudo-sum, respectively) verifying some properties; we prove, consequently, that by bounding in two different ways the above mentioned left- (right-) pseudo-BCK lattices with pseudo-product (pseudo-sum, respectively) verifying some properties, we obtain a left- (right-) pseudo-Wajsberg algebra or a left-pseudo-Hájek(pP) algebra with the properties (pP1^L), (pP2^L) (right-pseudo-Hájek(pS) algebra with the properties (pP1^R), (pP2^R), dually). In Section 2, we obtain the corresponding connections between the l-groups and the po-ims(pR) which provide the pseudo-MV algebras and the pseudo-product algebras. The analysis of three important examples ends the chapter.

In **Chapter 15 (Normal filters/ideals and compatible deductive systems)**, we present notions and results from [98], [99]. We recall the notion of *deductive system* of a po-implicative-group and prove that the deductive systems of a po-implicative-group coincide with the convex subgroups of the termwise equivalent po-group. We introduce the notion of *compatible* deductive system of a po-implicative-group vs. the old notion of *normal* convex po-subgroup of a po-group and prove the equivalence "*compatible* if and only if *normal*". Then, we mainly prove that the normality (compatibility) at l-group (l-implicative-group) G level is inherited by the algebras obtained by restricting the l-group (l-implicative-group) operations to G^- and to G^+, to $[u', 1]$ and to $[1, u]$, and finally to $G^-_{-\infty} = \{-\infty\} \cup G^-$ and to $G^+_{+\infty} = G^+ \cup \{+\infty\}$ and, also, that the equivalence "*compatible* if and only if *normal*", existing at l-group (l-implicative-group) level, is preserved by the algebras obtained by restricting the l-group (l-implicative-group) operations to G^- and to G^+, to $[u', 1]$ and to $[1, u]$, and finally to $G^-_{-\infty}$ and to $G^+_{+\infty}$.

In **Chapter 16 (Representability)**, we study the representability [98], [99]. First, we present equivalent conditions for an l-implicative-group to be representable. Then, we prove that representability at l-implicative-group G level is inherited by the algebras obtained by restricting the operations from G to G^- and to G^+. The research here must be continued: connections between the representability at l-group (l-implicative-group) G level and the representability at $[u', 1] \subset G^-$, $[1, u] \subset G^+$ level and at $G^-_{-\infty}$, $G^+_{+\infty}$ level must be found.

In **Chapter 17 (States and implicative-states)**, we present the states [98], [99]. We introduce the notions of *additive-state*, or *state* for short, on a po-group with strong unit and *implicative-state* on a po-implicative-group with strong unit and prove they coincide. Next, we introduce the notions of *state morphism* on an l-group with strong unit and *implicative-state morphism* on an l-implicative-group with strong unit and prove they coincide. Finally, we define the distance functions d_1^L, d_2^L and d_1^R, d_2^R on an l-group and prove some properties, following the ideas in the pseudo-BL algebras case. Then, we introduce the notion of *Bosbach-state* on an l-group with strong unit, prove some properties and prove that any state is a Bosbach-state, following the ideas from [57].

1.2 Basic (regular) properties used in the monograph

1.2.1 Basic (regular) properties used in pM algebras (implicative-groups, implicative-hubs)

The lists I and II of the basic (regular) properties used in the monograph is divided into two parts: the properties in Part 1 are those that will be generalized when considering the quasi-algebras (quasi-structures) or the quasi-m algebras (quasi-m structures). Here is the list (L comes from "Left", R comes from "Right"):

List I, Part 1

(pM^L) $1 \to^L x = x = 1 \rightsquigarrow^L x$,
(pM^R) $0 \to^R x = x = 0 \rightsquigarrow^R x$;
(pN^L) $1 \le x \Longrightarrow x = 1$,
(pN^R) $0 \ge x \Longrightarrow x = 0$;
(An) (Antisymmetry) $x \le y$ and $y \le x \Longrightarrow x = y$,
(pAn^L') (Antisymmetry) $x \to^L y = 1 = y \to^L x \Longrightarrow x = y$,
(pAn^R') (Antisymmetry) $x \to^R y = 0 = y \to^R x \Longrightarrow x = y$,
(pEq^L) $x \le y \Longleftrightarrow x \to^L y = 1 \; (\overset{(IdEq^L)}{\Leftrightarrow} \; x \rightsquigarrow^L y = 1)$,
(pEq^R) $x \ge y \Longleftrightarrow x \to^R y = 0 \; (\overset{(IdEq^R)}{\Leftrightarrow} \; x \rightsquigarrow^R y = 0)$;
(p-s^L) (p-semisimple) $x \le y \Longrightarrow x = y$,
(p-s^R) (p-semisimple) $x \ge y \Longrightarrow x = y$;
(pNeg^L) $x^- = x \to^L 0$, $x^\sim = x \rightsquigarrow^L 0$, where $0 = 1^- = 1^\sim$,
(pNeg^R) $x^- = x \to^R 1$, $x^\sim = x \rightsquigarrow^R 1$, where $1 = 0^- = 0^\sim$;
($\overline{pM^L}$) $1 \to^L x^- = x^-$, $1 \to^L x^\sim = x^\sim$, $1 \rightsquigarrow^L x^- = x^-$, $1 \rightsquigarrow^L x^\sim = x^\sim$,
($\overline{pM^R}$) $0 \to^R x^- = x^-$, $0 \to^R x^\sim = x^\sim$, $0 \rightsquigarrow^R x^- = x^-$, $0 \rightsquigarrow^R x^\sim = x^\sim$;
==========particular cases: ========================
(pM) $1 \to x = x = 1 \rightsquigarrow x$;
(pEq=) $x = y \Longleftrightarrow x \to y = 1 \overset{(IdEq)}{\Leftrightarrow} x \rightsquigarrow y = 1$;
(pD=)=(pC^L) $(y \to x) \rightsquigarrow x = y = (y \rightsquigarrow x) \to x$;
(pNeg) $x^- = x \to 1$, $x^\sim = x \rightsquigarrow 1$, where $1 = 1^- = 1^\sim$;
($\overline{pM}$) $1 \to x^{-1} = x^{-1}$, $1 \rightsquigarrow x^{-1} = x^{-1}$, with $x^{-1} \overset{def.}{=} x^- = x^\sim$;

List I, Part 2

(pM1^L) $1 \to^L 1 = 1 \; (\Leftrightarrow 1 \rightsquigarrow^L 1 = 1)$,
(pM0^R) $0 \to^R 0 = 0 \; (\Leftrightarrow 0 \rightsquigarrow^R 0 = 0)$;
(IdEq^L) $x \to^L y = 1 \Leftrightarrow x \rightsquigarrow^L y = 1$,
(IdEq^R) $x \to^R y = 0 \Leftrightarrow x \rightsquigarrow^R y = 0$;
(pEx^L) $x \to^L (y \rightsquigarrow^L z) = y \rightsquigarrow^L (x \to^L z)$,
(pEx^R) $x \to^R (y \rightsquigarrow^R z) = y \rightsquigarrow^R (x \to^R z)$;
(pEx1^L) $x \to^L (y \rightsquigarrow^L 1) = y \rightsquigarrow^L (x \to^L 1)$,

(pEx0^R) $x \to^R (y \leadsto^R 0) = y \leadsto^R (x \to^R 0)$;
(F) (First element) $x \geq 0$,
(pF') (First element) $x \to^R 0 = 0$;
(L) (Last element) $x \leq 1$,
(pL') (Last element) $x \to^L 1 = 1$;
(Id^L) $x \to^L 1 = x \leadsto^L 1$,
(Id^R) $x \to^R 0 = x \leadsto^R 0$;
(pBB^L) $y \to^L z \leq (z \to^L x) \leadsto^L (y \to^L x)$, $y \leadsto^L z \leq (z \leadsto^L x) \to^L (y \leadsto^L x)$,
(pBB^L') $(y \to^L z) \to^L ((z \to^L x) \leadsto^L (y \to^L x)) = 1$,
$(y \leadsto^L z) \to^L ((z \leadsto^L x) \to^L (y \leadsto^L x) = 1)$,
(pBB^R) $y \to^R z \geq (z \to^R x) \leadsto^R (y \to^R x)$, $y \leadsto^R z \geq (z \leadsto^R x) \to^R (y \leadsto^R x)$;
(pBB^R') $(y \to^R z) \to^R ((z \to^R x) \leadsto^R (y \to^R x)) = 0$,
$(y \leadsto^R z) \to^R ((z \leadsto^R x) \to^R (y \leadsto^R x)) = 0$;
(pBB1^L) $y \to^L z \leq (z \to^L 1) \leadsto^L (y \to^L 1)$, $y \leadsto^L z \leq (z \leadsto^L 1) \to^L (y \leadsto^L 1)$,
(pBB0^R) $y \to^R z \geq (z \to^R 0) \leadsto^R (y \to^R 0)$, $y \leadsto^R z \geq (z \leadsto^R 0) \to^R (y \leadsto^R 0)$;
(pB^L) $z \to^L x \leq (y \to^L z) \to^L (y \to^L x)$, $z \leadsto^L x \leq (y \leadsto^L z) \leadsto^L (y \leadsto^L x)$,
(pB^L') $(z \to^L x) \to^L ((y \to^L z) \to^L (y \to^L x)) = 1$,
$(z \leadsto^L x) \to^L ((y \leadsto^L z) \leadsto^L (y \leadsto^L x)) = 1$,
(pB^R) $z \to^R x \geq (y \to^R z) \to^R (y \to^R x)$, $z \leadsto^R x \geq (y \leadsto^R z) \leadsto^R (y \leadsto^R x)$;
(pB^R') $(z \to^R x) \to^R ((y \to^R z) \to^R (y \to^R x)) = 0$,
$(z \leadsto^R x) \to^R ((y \leadsto^R z) \leadsto^R (y \leadsto^R x)) = 0$;
(pC^L) $x \to^L (y \leadsto^L z) \leq y \leadsto^L (x \to^L z)$, $x \leadsto^L (y \to^L z) \leq y \to^L (x \leadsto^L z)$,
(pC^R) $x \to^R (y \leadsto^R z) \geq y \leadsto^R (x \to^R z)$, $x \leadsto^R (y \to^R z) \geq y \to^R (x \leadsto^R z)$;
(pC1^L) $x \to^L (y \leadsto^L 1) \leq y \leadsto^L (x \to^L 1)$, $x \leadsto^L (y \to^L 1) \leq y \to^L (x \leadsto^L 1)$,
(pC0^R) $x \to^R (y \leadsto^R 0) \geq y \leadsto^R (x \to^R 0)$, $x \leadsto^R (y \to^R 0) \geq y \to^R (x \leadsto^R 0)$;
(pD^L) $y \leq (y \to^L x) \leadsto^L x$, $y \leq (y \leadsto^L x) \to^L x$,
(pD^L') $y \leadsto^L ((y \to^L x) \leadsto^L x) = 1$, $y \to^L ((y \leadsto^L x) \to^L x) = 1$;
(pD^R) $y \geq (y \to^R x) \leadsto^R x$, $y \geq (y \leadsto^R x) \to^R x$,
(pD^R') $y \leadsto^R ((y \to^R x) \leadsto^R x) = 0$, $y \to^R ((y \leadsto^R x) \to^R x) = 0$;
(pD1^L) $y \leq (y \to^L 1) \leadsto^L 1$, $y \leq (y \leadsto^L 1) \to^L 1$,
(pD0^R) $y \geq (y \to^R 0) \leadsto^R 0$, $y \geq (y \leadsto^R 0) \to^R 0$;
(pEq#^{L}) $x \leq y \to^L z \Leftrightarrow y \leq x \leadsto^L z$,
(pEq#^{R}) $x \geq y \to^R z \Leftrightarrow y \geq x \leadsto^R z$;
(pK^L) $x \leq y \to^L x$ and $x \leq y \leadsto^L x$,
(pK^R) $x \geq y \to^R x$ and $x \geq y \leadsto^R x$;
(Re) (Reflexivity) $x \leq x$,
(pRe^L) (Reflexivity) $x \to^L x = 1$ ($\Leftrightarrow x \leadsto^L x = 1$),
(pRe^R) (Reflexivity) $x \to^R x = 0$ ($\Leftrightarrow x \leadsto^R x = 0$);
(Tr) (Transitivity) $x \leq y$ and $y \leq z \Longrightarrow x \leq z$,
(pTr^L') (Transitivity) $x \to^L y = 1 = y \to^L z \Longrightarrow x \to^L z = 1$,
(pTr^R') (Transitivity) $x \to^R y = 0 = y \to^R z \Longrightarrow x \to^R z = 0$;
(p*^{L}) $x \leq y \Longrightarrow (z \to^L x \leq z \to^L y$ and $z \leadsto^L x \leq z \leadsto^L y)$,
(p*^{L}') $x \to^L y = 1 \Longrightarrow$
$((z \to^L x) \to^L (z \to^L y) = 1$ and $(z \leadsto^L x) \to^L (z \leadsto^L y) = 1)$,
(p*^{R}) $x \geq y \Longrightarrow (z \to^R x \geq z \to^R y$ and $z \leadsto^R x \geq z \leadsto^R y)$;
(p**^{L}) $x \leq y \Longrightarrow (y \to^L z \leq x \to^L z$ and $y \leadsto^L z \leq x \leadsto^L z)$,

(p**^{L}') $x \to^L y = 1 \Longrightarrow$
$((y \to^L z) \to^L (x \to^L z) = 1$ and $(y \rightsquigarrow^L z) \to^L (x \rightsquigarrow^L z) = 1)$,
(p**^{R}) $x \geq y \Longrightarrow (y \to^R z \geq x \to^R z$ and $y \rightsquigarrow^R z \geq x \rightsquigarrow^R z)$;
(p@rel^L) $x \to^L y \leq (x \odot z) \to^L (y \odot z),\ x \rightsquigarrow^L y \leq (z \odot x) \rightsquigarrow^L (z \odot y)$,
(p@rel^R) $x \to^R y \geq (x \oplus z) \to^R (y \oplus z),\ x \rightsquigarrow^R y \geq (z \oplus x) \rightsquigarrow^R (z \oplus y)$;
(p@^{L}) $x \to^L y = (x \odot z) \to^L (y \odot z),\ x \rightsquigarrow^L y = (z \odot x) \rightsquigarrow^L (z \odot y)$,
(p@^{R}) $x \to^R y = (x \oplus z) \to^R (y \oplus z),\ x \rightsquigarrow^R y = (z \oplus x) \rightsquigarrow^R (z \oplus y)$;
(pP) (pseudo-product)
$\exists\, x \odot y \overset{not.}{=} \min\{z \mid x \leq y \to^L z\} = \min\{z \mid y \leq x \rightsquigarrow^L z\}$,
(pS) (pseudo-sum)
$\exists\, x \oplus y \overset{not.}{=} \max\{z \mid x \geq y \to^R z\} = \max\{z \mid y \geq x \rightsquigarrow^R z\}$;
(pPP) $x \leq y \to^L (x \odot y),\ x \leq y \rightsquigarrow^L (y \odot x)$,
(pSS) $x \geq y \to^R (x \oplus y),\ x \geq y \rightsquigarrow^R (y \oplus x)$;
(pRR) $(y \to^L z) \odot y \leq z,\ y \odot (y \rightsquigarrow^L z) \leq z$,
(pcoRR) $(y \to^R z) \oplus y \geq z,\ y \oplus (y \rightsquigarrow^R z) \geq z$;
(pRP) $x \leq y \to^L z \Longleftrightarrow y \leq x \rightsquigarrow^L z \Longleftrightarrow x \odot y \leq z$, (pRP)=(pPR)
(pcoRS) $x \geq y \to^R z \Longleftrightarrow y \geq x \rightsquigarrow^R z \Longleftrightarrow x \oplus y \geq z$ (pcoRS)=(pScoR);

(pprel^L) (pseudo-prelinearity) $(x \to^L y) \vee (y \to^L x) = 1 = (x \rightsquigarrow^L y) \vee (y \rightsquigarrow^L x)$,
(pprel^R) (pseudo-prelinearity) $(x \to^R y) \wedge (y \to^R x) = 0 = (x \rightsquigarrow^R y) \wedge (y \rightsquigarrow^R x)$;
(pdiv^L) (pseudo-divisibility) $x \wedge y = (x \to^L y) \odot x = x \odot (x \rightsquigarrow^L y)$,
(pdiv^R) (pseudo-divisibility) $x \vee y = (x \to^R y) \oplus x = x \oplus (x \rightsquigarrow^R y)$;
(pCC^L) (C.C. Chang) $x \vee y = (x \to^L y) \rightsquigarrow^L y = (x \rightsquigarrow^L y) \to^L y$,
(pCC^R) (C.C. Chang) $x \wedge y = (x \to^R y) \rightsquigarrow^R y = (x \rightsquigarrow^R y) \to^R y$;
(pcomm^L) (sup-commutativity)
$(x \to^L y) \rightsquigarrow^L y = (y \to^L x) \rightsquigarrow^L x, \quad (x \rightsquigarrow^L y) \to^L y = (y \rightsquigarrow^L x) \to^L x$,
(pcomm^R) (inf-commutativity)
$(x \to^R y) \rightsquigarrow^R y = (y \to^R x) \rightsquigarrow^R x, \quad (x \rightsquigarrow^R y) \to^R y = (y \rightsquigarrow^R x) \to^R x$;
(pDN^L) (pseudo-Double Negation) $(x^{-^L})^{\sim^L} = x = (x^{\sim^L})^{-^L}$,
where $x^{-^L} = x \to^L 0$ and $x^{\sim^L} = x \rightsquigarrow^L 0$,
(pDN^R) (pseudo-Double Negation) $(x^{-^R})^{\sim^R} = x = (x^{\sim^R})^{-^R}$,
where $x^{-^R} = x \to^R 1$ and $x^{\sim^R} = x \rightsquigarrow^R 1$;
(pP)$_{(pDN)}$ $x \odot y = (x \to^L y^-)^\sim = (y \rightsquigarrow^L x^\sim)^-$,
(pS)$_{(pDN)}$ $x \oplus y = (x \to^R y^-)^\sim = (y \rightsquigarrow^R x^\sim)^-$;
(pR)$_{(pDN)}$ $x \to^L y = (x \odot y^\sim)^-, \quad x \rightsquigarrow^L y = (y^- \odot x)^\sim$,
(pcoR)$_{(pDN)}$ $x \to^R y = (x \oplus y^\sim)^-, \quad x \rightsquigarrow^R y = (y^- \oplus x)^\sim$;
(pNg^{Lb}) $x \leq y \Longrightarrow y^{-^L} \leq x^{-^L}, \quad y^{\sim^L} \leq x^{\sim^L}$,
(pNg^{Rb}) $x \leq y \Longrightarrow y^{-^R} \leq x^{-^R}, \quad y^{\sim^R} \leq x^{\sim^R}$;
(pEqNg^{Lb}) $x \leq y \Leftrightarrow y^{-^L} \leq x^{-^L} \Leftrightarrow y^{\sim^L} \leq x^{\sim^L}$,
(pEqNg^{Rb}) $x \leq y \Leftrightarrow y^{-^R} \leq x^{-^R} \Leftrightarrow y^{\sim^R} \leq x^{\sim^R}$;
(N1^L) $1^{-1} = 1$,
(N0^R) $0^{-1} = 0$;
(pN1^{Lb}) $0 = 1^- = 1^\sim,\ 1 = 0^- = 0^\sim$,
(pN0^{Rb}) $1 = 0^- = 0^\sim,\ 0 = 1^- = 1^\sim$;

(pDN$^{\leq}$) $x \leq (x^{-1})^{-1}$,
(pDN$^{\geq}$) $x \geq (x^{-1})^{-1}$;
(pNg1^L) $x \to^L y^{-1} = y \rightsquigarrow^L x^{-1}$,
(pNg1^R) $x \to^R y^{-1} = y \rightsquigarrow^R x^{-1}$;
(pNg1^{Lb}) $x \to^L y^{\sim} = y \rightsquigarrow^L x^{-}$,
(pNg1^{Rb}) $x \to^R y^{\sim} = y \rightsquigarrow^R x^{-}$;
(pDNeg1^{Lb}) $x \to^L y = y^{-} \rightsquigarrow^L x^{-}$, $x \rightsquigarrow^L y = y^{\sim} \to^L x^{\sim}$,
(pDNeg1^{Rb}) $x \to^R y = y^{-} \rightsquigarrow^R x^{-}$, $x \rightsquigarrow^R y = y^{\sim} \to^R x^{\sim}$;
(pDNeg2^{Lb}) $x^{\sim} \to^L y = y^{-} \rightsquigarrow^L x$,
(pDNeg2^{Rb}) $x^{\sim} \to^R y = y^{-} \rightsquigarrow^R x$;
(pDNeg3^{Lb}) $(x \to^L y^{-})^{\sim} = (y \rightsquigarrow^L x^{\sim})^{-}$,
(pDNeg3^{Rb}) $(x \to^R y^{-})^{\sim} = (y \rightsquigarrow^R x^{\sim})^{-}$;
(pP1^L) $x \wedge x^{-^L} = 0 = x \wedge x^{\sim^L}$,
(pP1^R) $x \vee x^{-^R} = 1 = x \vee x^{\sim^R}$;
(pP2^L) $(z^{-^L})^{-^L} \odot [(x \odot z) \to^L (y \odot z)] \leq x \to^L y$,
$(z^{\sim^L})^{\sim^L} \odot [(z \odot x) \rightsquigarrow^L (z \odot y)] \leq x \rightsquigarrow^L y$,
(pP2^R) $(z^{-^R})^{-^R} \oplus [(x \oplus z) \to^R (y \oplus z)] \geq x \to^R y$,
$(z^{\sim^R})^{\sim^R} \oplus [(z \oplus x) \rightsquigarrow^R (z \oplus y)] \geq x \rightsquigarrow^R y$;

===========particular cases: ==========================

(pM1) $1 \to 1 = 1 = 1 \rightsquigarrow 1$;
(IdEq) $x \to y = 1 \Leftrightarrow x \rightsquigarrow y = 1$,
(Id) $x \to 1 = x \rightsquigarrow 1$;
(pBB=) $y \to z = (z \to x) \rightsquigarrow (y \to x)$, $y \rightsquigarrow z = (z \rightsquigarrow x) \to (y \rightsquigarrow x)$,
(pBB1=) $y \to z = (z \to 1) \rightsquigarrow (y \to 1)$, $y \rightsquigarrow z = (z \rightsquigarrow 1) \to (y \rightsquigarrow 1)$,
(pB=) $z \to x = (y \to z) \to (y \to x)$, $z \rightsquigarrow x = (y \rightsquigarrow z) \rightsquigarrow (y \rightsquigarrow x)$,
(pD1=) $x = (x \to 1) \rightsquigarrow 1)$, $x = (x \rightsquigarrow 1) \to 1)$,
(pEq#$^=$) $x = y \to z \Leftrightarrow y = x \rightsquigarrow z$,
(p*Eq) $x \leq y \Leftrightarrow z \to x \leq z \to y \Leftrightarrow z \rightsquigarrow x \leq z \rightsquigarrow y$;
(pEx) $x \to (y \rightsquigarrow z) = y \rightsquigarrow (x \to z)$,
(pEx1) $x \to (y \rightsquigarrow 1) = y \rightsquigarrow (x \to 1)$,
(pEx2) $y^{-1} \to (x^{-1} \rightsquigarrow 1) = y^{-1} \to x$, $x^{-1} \rightsquigarrow (y^{-1} \to 1) = x^{-1} \rightsquigarrow y$;
(pRe') $x \to x = 1 = x \rightsquigarrow x = 1$;
(p@) $x \to y = (x \cdot z) \to (y \cdot z)$, $x \rightsquigarrow y = (z \cdot x) \rightsquigarrow (z \cdot y)$;
(pP=) $\exists\, y \cdot x = z$ such that $x \to z = y$ and $x \cdot y = u$ such that $x \rightsquigarrow u = y$;
(α) $(y \to z) \cdot y = z = y \cdot (y \rightsquigarrow z)$
(it is the analogous of (pRR) and (pcoRR));
(β) $y \to (x \cdot y) = x = y \rightsquigarrow (y \cdot x)$
(it is the analogous of (pPP) and (pSS));
(pGa$^=$) $x \cdot y = z \Longleftrightarrow x = y \to z \overset{(pEq\#^=)}{\Leftrightarrow} y = x \rightsquigarrow z$
(it is the analogous of (pPR)=(pRP) and (pScoR)=(pcoRS)),
(pGa$^{\leq}$) $x \cdot y \leq z \Longleftrightarrow x \leq y \to z \overset{(pEq\#^{\leq})}{\Leftrightarrow} y \leq x \rightsquigarrow z$
(it is the analogous of (pPR) = (pRP)),
(pGa$^{\geq}$) $x \cdot y \geq z \Longleftrightarrow x \geq y \to z \overset{(pEq\#^{\geq})}{\Leftrightarrow} y \geq x \rightsquigarrow z$

(it is the analogous of (pScoR) = (pcoRS));
(DN) (Double Negation) $(x^{-1})^{-1} = x$,
(N1) $1^{-1} = 1$,
(pNg) $(x \to y)^{-1} = y \to x$, $(x \rightsquigarrow y)^{-1} = y \rightsquigarrow x$,
(pNg1) $x \to y^{-1} = y \rightsquigarrow x^{-1}$;
(pDNeg1) $x \to y = y^{-1} \rightsquigarrow x^{-1}$, $x \rightsquigarrow y = y^{-1} \to x^{-1}$,
(pDNeg2) $x^{-1} \to y = y^{-1} \rightsquigarrow x (= y \cdot x)$,
(pDNeg3) $(x \to y^{-1})^{-1} = (y \rightsquigarrow x^{-1})^{-1}$.

1.2.2 Basic (regular-m) properties used in unital magmas (groups, hubs)

List II, Part 1

(pU) (pseudo-Unit element) $x \odot 1 = x = 1 \odot x$ (multiplicative, or left notation);
(pEqP^{Lb}) $x \leq y \Longleftrightarrow x \odot y^{\sim} = 0 \ (\stackrel{(IdEqP^{Lb})}{\Longleftrightarrow} y^{-} \odot x = 0)$,
(pEqS^{Rb}) $x \geq y \Longleftrightarrow x \oplus y^{\sim} = 1 \ (\stackrel{(IdEqS^{Rb})}{\Longleftrightarrow} y^{-} \oplus x = 1)$;
(pIv^L) $x \odot x^{\sim} = 0 = x^{-} \odot x$, where $0 = 1^{-} = 1^{\sim}$,
(pIv^R) $x \oplus x^{\sim} = 1 = x^{-} \oplus x$, where $1 = 0^{-} = 0^{\sim}$;
($\overline{pU^L}$) $1 \odot x^{-} = x^{-} = x^{-} \odot 1$, $1 \odot x^{\sim} = x^{\sim} = x^{\sim} \odot 1$;
==========particular cases: ==========================
(pEqP=) $x = y \Longleftrightarrow x \cdot y^{-1} = 1 \ (\stackrel{(IdEqP)}{\Longleftrightarrow} y^{-1} \cdot x = 1)$;
(pIv) (pseudo-Inverse) $x \cdot x^{-1} = 1 = x^{-1} \cdot x$, where $1 = 1^{-1} = 1^{-} = 1^{\sim}$;
($\overline{pU}$) $1 \cdot x^{-1} = x^{-1} = x^{-1} \cdot 1$;

List II, Part 2

(U1) $1 \odot 1 = 1$ (multiplicative, or left notation);
(Ass) (Associativity) $x \odot (y \odot z) = (x \odot y) \odot z$ (multiplicative, or left notation);
(L) (Last element) $x \leq 1$,
(F) (First element) $x \geq 0$;
(pR) (pseudo-residuum)

$\exists\, y \to^L z \stackrel{not.}{=} \max\{x \mid x \odot y \leq z\}$, $\exists\, y \rightsquigarrow^L z \stackrel{not.}{=} \max\{x \mid y \odot x \leq z\}$,

(pcoR) (pseudo-coresiduum)

$\exists\, y \to^R z \stackrel{not.}{=} \min\{x \mid x \oplus y \geq z\}$, $\exists\, y \rightsquigarrow^R z \stackrel{not.}{=} \min\{x \mid y \oplus x \geq z\}$;

(pPR) $x \odot y \leq z \Longleftrightarrow x \leq y \to^L z \Longleftrightarrow y \leq x \rightsquigarrow^L z$, (pPR)=(pRP),
(pScoR) $x \oplus y \geq z \Longleftrightarrow x \geq y \to^R z \Longleftrightarrow y \geq x \rightsquigarrow^R z$, (pScoR)=(pcoRS;
(pRR) $(y \to^L z) \odot y \leq z$, $y \odot (y \rightsquigarrow^L z) \leq z$,
(pcoRR) $(y \to^R z) \oplus y \geq z$, $y \oplus (y \rightsquigarrow^R z) \geq z$;
(pPP) $x \leq y \to^L (x \odot y)$, $x \leq y \rightsquigarrow^L (y \odot x)$,
(pSS) $x \geq y \to^R (x \oplus y)$, $x \geq y \rightsquigarrow^R (y \oplus x)$;
(IdEqP^{Lb}) $x \odot y^{\sim} = 0 \Longleftrightarrow y^{-} \odot x = 0$, with $0 = 1^{-} = 1^{\sim}$,
(IdEqS^{Rb}) $x \oplus y^{\sim} = 1 \Longleftrightarrow y^{-} \oplus x = 1$, with $1 = 0^{-} = 0^{\sim}$;
(pDNP1) $(x \odot y^{\sim})^{-} = ((x^{-})^{-} \odot y^{-})^{\sim}$, $(y^{-} \odot x)^{\sim} = (y^{\sim} \odot (x^{\sim})^{\sim})^{-}$,

(pDNP2) $(x^- \odot y^-)^\sim = (x^\sim \odot y^\sim)^-$;

===========particular cases: ==========================

(pIv') (pseudo-Inverse) for all x, there exists y, such that $x \cdot y = 1 = y \cdot x$,

(pIv") (pseudo-Inverse) for all x, there exists a unique y, such that $x \cdot y = 1 = y \cdot x$,

(pIv"') for all $x \in A$, there exists a unique $y \in A$, such that $x \cdot y = 1$;

(pCp) (pseudo-Compatibility) $x \leq y \Longrightarrow (a \cdot x \leq a \cdot y,\ \ x \cdot a \leq y \cdot a)$,

(pEqCp) (pseudo-Compatibility) $x \leq y \Leftrightarrow a \cdot x \leq a \cdot y \Leftrightarrow x \cdot a \leq y \cdot a$;

(pIvP) $x \cdot y = 1 = y \cdot x \Longleftrightarrow y = x^{-1}$,

(pIvG) $(x \cdot y^{-1})^{-1} = y \cdot x^{-1}$, $(y^{-1} \cdot x)^{-1} = x^{-1} \cdot y$;

(DN) (Double Negation) $(x^{-1})^{-1} = x$,

(N1) $1^{-1} = 1$;

(ImNeg) $x \leq y \Longrightarrow y^{-1} \leq x^{-1}$,

(EqNeg) $x \leq y \Longleftrightarrow y^{-1} \leq x^{-1}$;

(NegP) $(x \cdot y)^{-1} = y^{-1} \cdot x^{-1}$;

(IdEqP) $x \cdot y^{-1} = 1 \Longleftrightarrow y^{-1} \cdot x = 1$;

(pR=) $\exists\ x \to y = z$ such that $z \cdot x = y$ and $\exists\ x \rightsquigarrow y = u$ such that $x \cdot u = y$;

(α) $(y \to z) \cdot y = z = y \cdot (y \rightsquigarrow z)$
(it is the analogous of (pRR) and (pcoRR));

(β) $y \to (x \cdot y) = x = y \rightsquigarrow (y \cdot x)$
(it is the analogous of (pPP) and (pSS)).

1.3 Basic quasi-properties (quasi-m properties) used in the monograph

1.3.1 Basic quasi-properties

The List qI of the basic quasi-properties used in the monograph is divided into three parts: the properties in Part 1 are the proper quasi-properties, the properties in Part 2 are the same as the regular properies in List I, Part 2, and the properties in Part 3 are those always verified in regular case.

List qI, Part 1

(q-pM^L) $1 \to^L (x \rightsquigarrow^L y) = x \rightsquigarrow^L y$, $1 \rightsquigarrow^L (x \to^L y) = x \to^L y$,

(q-pM^R) $0 \to^R (x \rightsquigarrow^R y) = x \rightsquigarrow^R y$, $0 \rightsquigarrow^R (x \to^R y) = x \to^R y$;

(q-pM$_d^L$) $1 \to^L (x \to^L y) = x \to^L y$, $1 \rightsquigarrow^L (x \rightsquigarrow^L y) = x \rightsquigarrow^L y$,

(q-pM$_d^R$) $0 \to^R (x \to^R y) = x \to^R y$, $0 \rightsquigarrow^R (x \rightsquigarrow^R y) = x \rightsquigarrow^R y$;

(q-pN^L) $1 \leq x \to^L y \Longrightarrow x \to^L y = 1$, $1 \leq x \rightsquigarrow^L y \Longrightarrow x \rightsquigarrow^L y = 1$,

(q-pN^R) $0 \geq x \to^R y \Longrightarrow x \to^R y = 0$, $0 \geq x \rightsquigarrow^R y \Longrightarrow x \rightsquigarrow^R y = 0$;

(q-pAn^L) (quasi-pseudo-Antisymmetry)
$(x \leq y$ (and $y \leq x)) \Longrightarrow (1 \to^L x = 1 \to^L y$ and $1 \rightsquigarrow^L x = 1 \rightsquigarrow^L y)$;

(q-pAn^R) (quasi-pseudo-Antisymmetry)
$(x \leq y$ (and $y \leq x)) \Longrightarrow (0 \to^R x = 0 \to^R y$ and $0 \rightsquigarrow^R x = 0 \rightsquigarrow^R y)$;

(q-p-s^L) (quasi-p-semisimple)
$x \leq y \Longleftrightarrow (1 \to^L x = 1 \to^L y$ and $1 \rightsquigarrow^L x = 1 \rightsquigarrow^L y)$,

(q-p-s^R) (quasi-p-semisimple)
$x \geq y \Longleftrightarrow (0 \to^R x = 0 \to^R y$ and $0 \rightsquigarrow^R x = 0 \rightsquigarrow^R y)$;
(q-pNeg^L) $x \to^L 0 = 1 \to^L x^- = (1 \rightsquigarrow^L x)^-$, $\quad x \rightsquigarrow^L 0 = 1 \rightsquigarrow^L x^\sim = (1 \to^L x)^\sim$,
with $0 = 1^- = 1^\sim$,
(q-pNeg^R) $x \to^R 1 = 0 \to^R x^- = (0 \rightsquigarrow^R x)^-$, $x \rightsquigarrow^R 1 = 0 \rightsquigarrow^R x^\sim = (0 \to^R x)^\sim$,
with $1 = 0^- = 0^\sim$;
(q-$\overline{pM^L}$) $1 \to^L (x \to^L y)^- = (x \to^L y)^-$, $1 \to^L (x \rightsquigarrow^L y)^- = (x \rightsquigarrow^L y)^-$,
$1 \rightsquigarrow^L (x \rightsquigarrow^L y)^\sim = (x \rightsquigarrow^L y)^\sim$, $1 \rightsquigarrow^L (x \to^L y)^\sim = (x \to^L y)^\sim$,
(q-$\overline{pM^R}$) $0 \to^R (x \to^R y)^- = (x \to^R y)^-$, $0 \to^R (x \rightsquigarrow^R y)^- = (x \rightsquigarrow^R y)^-$,
$0 \rightsquigarrow^R (x \rightsquigarrow^R y)^\sim = (x \rightsquigarrow^R y)^\sim$, $0 \rightsquigarrow^R (x \to^R y)^\sim = (x \to^R y)^\sim$;

(q-$\overline{pM(1 \to^L x, 1 \rightsquigarrow^L x)}$) $1 \to^L (1 \to^L x)^- = (1 \to^L x)^-$, $1 \to^L (1 \rightsquigarrow^L x)^- = (1 \rightsquigarrow^L x)^-$, $1 \rightsquigarrow^L (1 \rightsquigarrow^L x)^\sim = (1 \rightsquigarrow^L x)^\sim$, $1 \rightsquigarrow^L (1 \to^L x)^\sim = (1 \to^L x)^\sim$;
==========particular cases: =========================
(q-pM) $\quad 1 \to (x \rightsquigarrow y) = x \rightsquigarrow y, \quad 1 \rightsquigarrow (x \to y) = x \to y$,
(q-pM$_d$) $\quad 1 \to (x \to y) = x \to y, \quad 1 \rightsquigarrow (x \rightsquigarrow y) = x \rightsquigarrow y$,
(q-pM($\cdot$)) $\quad 1 \to (x \cdot y) = x \cdot y = 1 \rightsquigarrow (x \cdot y)$,
(q-pM($1 \to x, 1 \rightsquigarrow x$)) $\quad 1 \to (1 \rightsquigarrow x) = 1 \rightsquigarrow x, \quad 1 \rightsquigarrow (1 \to x) = 1 \to x$;
(q-pN) $\quad 1 \to (x \rightsquigarrow y) = 1 \Longrightarrow x \rightsquigarrow y = 1, \quad 1 \rightsquigarrow (x \to y) = 1 \Longrightarrow x \to y = 1$;
(q-pEq=) $\quad (1 \to x = 1 \to y$ and $1 \rightsquigarrow x = 1 \rightsquigarrow y) \Longleftrightarrow x \to y = 1 \overset{(IdEq)}{\Longleftrightarrow} x \rightsquigarrow y = 1$,
(q-pD=) $\quad 1 \to y = (y \to x) \rightsquigarrow x, \quad 1 \rightsquigarrow y = (y \rightsquigarrow x) \to x$;
(q-pNeg) $\quad x \to 1 = 1 \to x^- = (1 \rightsquigarrow x)^- = x \rightsquigarrow 1 = 1 \rightsquigarrow x^- = (1 \to x)^-$,
with $1 = 1^-$;
(q-$\overline{pM}$) $\quad 1 \to (x \rightsquigarrow y)^- = (x \rightsquigarrow y)^-$, $1 \rightsquigarrow (x \to y)^- = (x \to y)^-$,

(q-$\overline{pM(1 \to x, 1 \rightsquigarrow x)}$) $\quad 1 \to (1 \rightsquigarrow x)^- = (1 \rightsquigarrow x)^-$, $1 \rightsquigarrow (1 \to x)^- = (1 \to x)^-$;

List qI (= List I), Part 2

(pM1^L), ..., (pDNeg3);

List qI, Part 3

(IdR^L) $\quad 1 \to^L x = 1 \rightsquigarrow^L x$,
(IdcoR^R) $\quad 0 \to^R x = 0 \rightsquigarrow^R x$;
(q-pRe^L($1 \to^L x, 1 \rightsquigarrow^L x$))
$(1 \to^L x) \to^L (1 \to^L x) = 1, \ (1 \rightsquigarrow^L x) \to^L (1 \rightsquigarrow^L x) = 1$,
(q-pRe^R($0 \to^R x, 0 \rightsquigarrow^R x$))
$(0 \to^R x) \to^R (0 \to^R x) = 0, \ (0 \rightsquigarrow^R x) \to^R (0 \rightsquigarrow^R x) = 0$;
(q-pR1^L) $\quad (1 \rightsquigarrow^L x) \to^L x = 1 = (1 \to^L x) \rightsquigarrow^L x$,
(q-pR1^R) $\quad (0 \rightsquigarrow^R x) \to^R x = 0 = (0 \to^R x) \rightsquigarrow^R x$;
(q-pR2^L) $\quad x \to^L (1 \rightsquigarrow^L x) = 1 = x \rightsquigarrow^L (1 \to^L x)$,
(q-pR2^R) $\quad x \to^R (0 \rightsquigarrow^R x) = 0 = x \rightsquigarrow^R (0 \to^R x)$;
(q-pI1^L) $\quad (1 \to^L x) \rightsquigarrow^L 1 = (1 \rightsquigarrow^L x) \to^L 1$,
(q-pI1^R) $\quad (0 \to^R x) \rightsquigarrow^R 0 = (0 \rightsquigarrow^R x) \to^R 0$;

(q-pI2^L) $x \to^L y = x \to^L (1 \rightsquigarrow^L y)$, $x \rightsquigarrow^L y = x \rightsquigarrow^L (1 \to^L y)$,
(q-pI2^R) $x \to^R y = x \to^R (0 \rightsquigarrow^R y)$, $x \rightsquigarrow^R y = x \rightsquigarrow^R (0 \to^R y)$;
===========particular cases: ==========================
(IdR) $1 \to x = 1 \rightsquigarrow x$;
(q-pRe($1 \to x, 1 \rightsquigarrow x$)) $(1 \to x) \to (1 \to x) = 1$, $(1 \rightsquigarrow x) \to (1 \rightsquigarrow x) = 1$;
(q-pR1) $(1 \rightsquigarrow x) \to x = 1 = (1 \to x) \rightsquigarrow x$;
(q-pR2) $x \to (1 \rightsquigarrow x) = 1 = x \rightsquigarrow (1 \to x)$;
(q-pII) $x \to y = (1 \to x) \to (1 \to y)$, $x \rightsquigarrow y = (1 \rightsquigarrow x) \rightsquigarrow (1 \rightsquigarrow y)$,
(q-pI1) $(1 \to x) \rightsquigarrow 1 = (1 \rightsquigarrow x) \to 1$;
(q-pI2) $x \to y = x \to (1 \rightsquigarrow y)$, $x \rightsquigarrow y = x \rightsquigarrow (1 \to y)$.

1.3.2 Basic quasi-m properties

The list qmII of the basic quasi-m properties used in the monograph is divided into three parts: the properties in Part 1 are the proper quasi-m properties, the properties in Part 2 are the same as the regular-m properies in List II, Part 2, and the properties in Part 3 are those always verified in regular-m case.

List qmII, Part 1

(qm-pU) $1 \odot (x \odot y) = x \odot y = (x \odot y) \odot 1$,
(qm-An) (qm-Antisymmetry)
$(x \le y$ and $y \le x) \Longrightarrow (1 \odot x = 1 \odot y$ and $x \odot 1 = y \odot 1)$;
(qm-NegP^L) $(x \odot 1)^- = x^- \odot 1 = 1 \odot x^- = (1 \odot x)^-$,
$(x \odot 1)^\sim = x^\sim \odot 1 = 1 \odot x^\sim = (1 \odot x)^\sim$;
(qm-$\overline{pU^L}$) $1 \odot (x \odot y)^- = (x \odot y)^- = (x \odot y)^- \odot 1$;
$1 \odot (x \odot y)^\sim = (x \odot y)^\sim = (x \odot y)^\sim \odot 1$,
(qm-$\overline{pU(1 \odot x, x \odot 1)^L}$) $1 \odot (1 \odot x)^- = (1 \odot x)^- = (1 \odot x)^- \odot 1$,
$1 \odot (1 \odot x)^\sim = (1 \odot x)^\sim = (1 \odot x)^\sim \odot 1$, $1 \odot (x \odot 1)^- = (x \odot 1)^- = (x \odot 1)^- \odot 1$,
$1 \odot (x \odot 1)^\sim = (x \odot 1)^\sim = (x \odot 1)^\sim \odot 1$;
===========particular cases: ==========================
(qm-pU($\to, \rightsquigarrow$)) $1 \cdot (x \to y) = x \to y = (x \to y) \cdot 1$,
$1 \cdot (x \rightsquigarrow y) = x \rightsquigarrow y = (x \rightsquigarrow y) \cdot 1$;
(qm-pD=) $(y \to x) \rightsquigarrow x = y \cdot 1 \overset{(IdP)}{=} 1 \cdot y = (y \rightsquigarrow x) \to x$;
(qm-pEq=) $(1 \cdot x = 1 \cdot y$ and $x \cdot 1 = y \cdot 1) \Longleftrightarrow x \to y = 1 \overset{(IdEq)}{\Leftrightarrow} x \rightsquigarrow y = 1$;
(qm-pEqP=) $(1 \cdot x = 1 \cdot y$ and $x \cdot 1 = y \cdot 1) \Longleftrightarrow y \cdot x^- = 1 \overset{(IdEqP)}{\Longleftrightarrow} x^- \cdot y = 1$;
(qm-pIvP) $(u \cdot v) \cdot (x \cdot y) = 1 (\Leftrightarrow (x \cdot y) \cdot (u \cdot v) = 1) \Longleftrightarrow x \cdot y = (u \cdot v)^-$;
(qm-NegP) $(x \cdot 1)^- = x^- \cdot 1 = 1 \cdot x^- = (1 \cdot x)^-$;
(qm-$\overline{pU}$) $1 \cdot (x \cdot y)^{-1} = (x \cdot y)^{-1} = (x \cdot y)^{-1} \cdot 1$, with $x^{-1} = x^- = x^\sim$,

(qm-$\overline{pU(1 \cdot x, x \cdot 1)}$) $1 \cdot (1 \cdot x)^- = (1 \cdot x)^- = (1 \cdot x)^- \cdot 1$,
$1 \cdot (x \cdot 1)^- = (x \cdot 1)^- = (x \cdot 1)^- \cdot 1$;

List qmII (= List II), Part 2

(L), ..., (β),

List qmII, Part 3

(IdP) $1 \cdot x = x \cdot 1$,
(Ass1) $x \cdot (1 \cdot y) = (x \cdot 1) \cdot y$,
(Ass2) $x \cdot (1 \cdot y) = x \cdot y$ and $(x \cdot 1) \cdot y = x \cdot y$,
(qm-PI) $x \cdot y = (1 \cdot x) \cdot (1 \cdot y)$, $x \cdot y = (x \cdot 1) \cdot (y \cdot 1)$,
(qm-PI1) $x \cdot y = (1 \cdot x) \cdot y$,
(qm-PI2) $x \cdot y = x \cdot (y \cdot 1)$.

Chapter 2

Algebras of logic (pM algebras (with additional operations))

The "world" of (commutative and non-commutative) algebras of logic is very large. We overview here only those algebras of logic connected to (commutative) implicative-groups (groups): this is the "world" of pseudo-M algebras (and its descendents), which is a "world" of algebras having an *internal*, or *natural*, binary relation, $\leq$ or, dually, $\geq$. Those algebras are briefly the following.

The *BCK* and *BCI algebras* were introduced in 1966 by K. Iséki [121], as algebraic models of BCK-logic and of BCI-logic, respectively; the BCK algebras are particular cases of the BCI algebras. The axioms of the propositional calculus of the BCK-logic are the following:
(B) $(\psi \to \chi) \to ((\varphi \to \psi) \to (\varphi \to \chi))$
(C) $(\varphi \to (\psi \to \chi)) \to (\psi \to (\varphi \to \chi))$
(K) $\varphi \to (\psi \to \varphi)$.
The axioms of the propositional calculus of the BCI-logic are the above (B), (C) and the following (I):
(I) $\varphi \to \varphi$.
Hilbert algebras are particular cases of BCK algebras [94]. *Hilbert algebras* were introduced in 1950, in a dual form, by Henkin [75], under the name "implicative model", as a model of positive implicative propositional calculus - an important fragment of classical propositional calculus introduced by Hilbert [77], [79]. Cf. A. Diego [40], it was Antonio Monteiro who has given the name "Hilbert algebras" to the dual algebras of Henkin's implicative models.

Hundreds of papers were writen on BCK and BCI algebras, and the books [143] and [94] on BCK algebras and the book [83] on BCI algebras.

Most of the commutative algebras of logic (such as residuated lattices, Boolean algebras, MV algebras, Wajsberg algebras, BL algebras, Gödel algebras, product algebras, Hilbert algebras, Heyting algebras, NM algebras, MTL algebras, IMTL algebras, R_0 algebras, weak-R_0 algebras etc.) can be expressed as particular cases of BCK algebras (more precisely, of reversed left-BCK algebras) (see [94]; see also [155], [139], [36]).

The commutative groups can be expressed as particular cases of BCI algebras [138], [142].

Several generalizations of BCI or of BCK algebras were introduced in time, namely:

BCH algebras were introduced in 1983 by Q.P. Hu and X. Li [81]; the examples of proper BCH algebras given in [82] are in fact BCI algebras. The exact connection between BCH algebras and BCI algebras was established in [104], Part I.

BCC algebras, also called *BIK$^+$ algebras*, were introduced in 1984 by Y. Komori [129], [130] (see [163]).

BZ algebras, also called *weak-BCC algebras*, were introduced in 1995 by X.H. Zhang and R. Ye [164].

BE algebras were introduced in 2006 by H.S. Kim and Y.H. Kim [126].

Pre-BCK algebras were introduced in 2010 by D. Buşneag and S. Rudeanu [23]. In pre-BCK algebras, the binary relation $\leq$ is only a pre-order (i.e. reflexive and transitive). A BCK algebra is just a pre-BCK algebra verifying also the antisymmetry. And, as it can be noticed from [23], a pre-BCK algebra is a BE algebra verifying an additional property.

Note that in April 2014 we found about the existence of another notion of *pre-BCK algebra*, introduced and studied in 2003 by Matthew Spinks in his PhD Thesis [159]. Spinks's pre-BCK algebras are more general than Buşneag and Rudeanu's pre-BCK algebras: more precisely, Spinks's *pre-BCK-algebras* are just our *pre-BBBCC algebras*, introduced and studied in the papers [104].

In the papers [104], starting with a list of properties of BCK algebras and with the algebras BCH, BCC, BZ, BE, pre-BCK, old generalizations of BCI algebras or of BCK algebras, we have introduced new generalizations of BCI or of BCK algebras, up to RM and RML algebras, respectively, and, consequently, new generalizations of Hilbert algebras. Namely, we have found **thirty one** new distinct generalizations of BCI or of BCK algebras and **twenty** new distinct generalizations of Hilbert algebras. We have presented the hierarchies existing between all these algebras, old or new ones. We have presented proper examples for each old or new algebra.

We found later, in 2016, that our RME algebras were introduced by B.L. Meng [144] in 2010, under the name of *CI-algebras*, as a generalization of BE algebras and of "dual BCK/BCI/BCH-algebras".

We found later also, in 2017, that our aRM algebras were introduced in 1998 by Y.B. Jun, E.H. Roh and H.S. Kim [125], under the name of *BH algebras*, as a generalization of BCH and BZ algebras.

Pseudo-BCK algebras were introduced by G. Georgescu and A. Iorgulescu in 2001 [60], as a non-commutative generalization of Iseki's BCK algebras. *Pseudo-BCI algebras* were introduced by W.A. Dudek and Y.B. Jun in 2008 [44], as a non-commutative generalization of Iseki's BCI algebras. *Pseudo-BE algebras* were introduced by R.A. Borzooei et al. in 2013 [13], as a non-commutative generalization of BE algebras. *Pseudo-BCH algebras* were introduced by A. Walendziak in 2015 [161], as a non-commutative generalization of BCH algebras. *Pseudo-CI algebras* were introduced by A. Rezaei et al. in 2016 [157], as a non-commutative generalization of RME=CI algebras.

In this chapter, we introduce the equivalent notions of *pseudo-algebra of logic* and *pseudo-structure of logic* and we are looking for what happens when the internal, binary relation $\leq$, or $\geq$, turns into equality $=$, i.e. the *p-semisimple* transformation - generalized here to all pseudo-algebras (pseudo-structures) of logic. We introduce the most general non-commutative algebra of logic, the *pseudo-M algebra* and its new descendents: pseudo-ML, pseudo-ME and pseudo-MEL algebras; their hierarchy is presented as Hierarchy No.1.

We also generalize to the non-commutative case some of the commutative algebras presented in [104], up to pseudo-RM and pseudo-RML algebras, and we present their hierarchies (of ascendents of pseudo-BCI and pseudo-BCK algebras, respectively) as Hierarchies No.2 and No. 4. A global hierarchy, including almost all the ascendents of pseudo-BCI and pseudo-BCK algebras - and descendents of pseudo-M algebras in the same time - is presented as Hierarchy No. 3.

Moreover, we introduce the notion of *strong* pseudo-M algebra.

We then recall the definitions of left and right pseudo-BCI algebras, (bounded) pseudo-BCK algebras, (bounded) pseudo-BCK(pP) algebras (lattices), (bounded) pseudo-BCK(pRP) algebras (lattices), bounded pseudo-BCK algebras with property (pDN), pseudo-Hájek(pP) algebras, pseudo-Hájek(pRP) algebras, pseudo-Wajsberg algebras and implicative-Boolean algebras. We introduce and study the pseudo-BCI(pP) algebras (lattices) and the pseudo-BCI(pRP) algebras (lattices). We present some results, including the equivalences (EL1), (iEL1), (EL2), (iEL2), (iEL3) - (iEL10) between these classes of algebras of logic. We present finally a simplified hierarchy of some descendents of pseudo-BCI and pseudo-BCK algebras as Hierarchy No. 5.

Remark 2.0.1 The reader will note through this monograph the crucial role played by the dual properties (pM^L), (pM^R) and (pEx^L), (pEx^R) in the "world" of pseudo-M algebras (of logic):
- (pM^L), (pM^R) correspond to the property (pU) in the "world" of non-commutative unital magmas (presented in Chapter 3); (pM^L), (pM^R) will be generalized to (q-pM^L), (q-pM^R) and (pU) will be generalized to (qm-pU), in Chapter 5;
- (pEx^L), (pEx^R) correspond (see Chapter 12) to the property (Ass) in the "world" of non-commutative unital magmas (Chapter 3).

The chapter has 14 sections.

2.1 Pseudo-M algebras and some descendents

We introduce in this section:
- the equivalent new notions of pseudo-algebra and pseudo-structure (of logic) and the p-semisimple property,
- the new notions of pseudo-M, pseudo-ML, pseudo-ME and pseudo-MEL algebras, directly in the non-commutative case,

- generalizations to the non-commutative case of some of the commutative algebras presented in [104],
- the new notion of *strong* pseudo-M algebra.

We study these algebras in some details and we present the Hasse and the Hasse-type diagrams.

2.1.1 Pseudo-algebra vs. pseudo-structure. The p-semisimple property

First, we introduce the general equivalent notions of *pseudo-algebra* and *pseudo-structure* of logic, which are the non-commutative notions corresponding to *algebra* and *structure*, respectively, in the commutative case. Following the context, we shall use one or the other equivalent notion, in the work with the non-commutative algebras of logic.

Definitions 2.1.1 ("L" comes from "Left")

(i) A *left-pseudo-algebra of logic*, or a *left-pseudo-algebra* for short, is an algebra $\mathcal{A}^L = (A^L, \to^L, \rightsquigarrow^L, 1)$ of type $(2,2,0)$ verifying the property: for all $x, y \in A^L$,
(IdEq^L) $\quad x \to^L y = 1 \Longleftrightarrow x \rightsquigarrow^L y = 1$.

An *internal* or *natural* binary relation $\leq$ can always be defined on A^L by: for all $x, y \in A^L$,

$$x \leq y \stackrel{def.}{\Longleftrightarrow} x \to^L y = 1 \; (\stackrel{(IdEq^L)}{\Longleftrightarrow} x \rightsquigarrow^L y = 1). \tag{2.1}$$

(i') Any algebra $\mathcal{A}' = (A, \sigma)$ whose signature σ contains $\to^L, \rightsquigarrow^L, 1$ and (IdEq^L) is verified is also called *left-pseudo-algebra*.

(i") Any algebra $\mathcal{A}'' = (A, \tau)$ which is term equivalent to a left-pseudo-algebra $\mathcal{A}' = (A, \sigma)$ is also called *left-pseudo-algebra*.

Equivalently,

(j) a *left-pseudo-structure of logic*, or a *left-pseudo-structure* for short, is a structure $\mathcal{A}^{\mathcal{L}} = (A^L, \leq, \to^L, \rightsquigarrow^L, 1)$, with a binary relation $\leq$ on A^L, two binary operations on A^L, $\to^L$ and $\rightsquigarrow^L$, and an element $1 \in A^L$ such that the above property (IdEq^L) holds and $\leq$ and $\to^L$, $\rightsquigarrow^L$, 1 are connected by:
(pEq^L) $x \leq y \Longleftrightarrow x \to^L y = 1 \; (\stackrel{(IdEq^L)}{\Leftrightarrow} x \rightsquigarrow^L y = 1)$.

(j') Any structure $\mathcal{A}' = (A, \leq, \sigma)$ whose signature σ contains $\to^L, \rightsquigarrow^L$, 1 and (IdEq^L) and (pEq^L) are verified is also called *left-pseudo-structure*.

(j") Any structure $\mathcal{A}'' = (A, \leq, \tau)$ which is term equivalent to a left-pseudo-structure $\mathcal{A}' = (A, \leq, \sigma)$ is also called *left-pseudo-structure*.

Remark 2.1.2 In Definitions 2.1.1, the *equivalence*:

$$x \leq y \Longleftrightarrow x \to^L y = 1 \; (\stackrel{(IdEq^L)}{\Longleftrightarrow} x \rightsquigarrow^L y = 1)$$

is used either
- as the *definition* (2.1) of the binary relation $\leq$, in the algebra $(A, \to^L, \rightsquigarrow^L, 1)$, or
- as the *connection* (pEq^L) between the binary relation $\leq$ and the operations $\to^L, \rightsquigarrow^L, 1$, in the structure $(A, \leq, \to^L, \rightsquigarrow^L, 1)$.

The dual definitions of *right-pseudo-algebra* and *right-pseudo-structure* are obvious and are omitted; we have ("R" comes from "Right"):
(IdEq^R) $x \to^R y = 0 \Longleftrightarrow x \rightsquigarrow^R y = 0$ and

$$x \geq y \overset{def.}{\Longleftrightarrow} x \to^R y = 0 \; (\overset{(IdEq^R)}{\Longleftrightarrow} x \rightsquigarrow^R y = 0). \tag{2.2}$$

or
(pEq^R) $x \geq y \Longleftrightarrow x \to^R y = 0 \; (\overset{(IdEq^R)}{\Leftrightarrow} x \rightsquigarrow^R y = 0)$.

The left-pseudo-algebra (left-pseudo-structure) is *commutative* if $\to^L = \rightsquigarrow^L$, and in this case: the unique binary operation is denoted by $\to^L$, (IdEq^L) becomes superfluous, (pEq^L) becomes (Eq^L) and we say that we have a *left-algebra* (*left-structure*, respectively).

The right-pseudo-algebra (right-pseudo-structure) is *commutative* if $\to^R = \rightsquigarrow^R$, and in this case: the unique binary operation is denoted by $\to^R$, (IdEq^R) becomes superfluous, (pEq^R) becomes (Eq^R) and we say that we have a *right-algebra* (*right-structure*, respectively).

We introduce now the *p-semisimple property*, starting from an equivalent notion of *p-semisimple pseudo-BCI algebra* used in [50] (see [138] for the commutative case) and recalled in Chapter 13.

Definition 2.1.3

(i) We say that a left-pseudo-algebra (or left-pseudo-structure) is *p-semisimple*, if the following property holds: for all $x, y \in A^L$,
(p-s^L) $x \leq y \implies x = y$.

(i') We say that a right-pseudo-algebra (or right-pseudo-structure) is *p-semisimple*, if the following property holds: for all $x, y \in A^R$,
(p-s^R) $x \geq y \implies x = y$.

2.1.2 pM, pML, pME, pMEL algebras

Let $\mathcal{A}^L = (A^L, \leq, \to^L, \rightsquigarrow^L, 1)$ be a left-pseudo-structure. Consider the following properties that can be verified by $\mathcal{A}^L$ (properties that are verified by any left-pseudo-BCK algebra): for all $x, y, z \in A^L$

(An) (Antisymmetry) $x \leq y$ and $y \leq x \Longrightarrow x = y$;
(pBB^L) $y \to^L z \leq (z \to^L x) \rightsquigarrow^L (y \to^L x)$, $y \rightsquigarrow^L z \leq (z \rightsquigarrow^L x) \to^L (y \rightsquigarrow^L x)$,
(pB^L) $z \to^L x \leq (y \to^L z) \to^L (y \to^L x)$, $z \rightsquigarrow^L x \leq (y \rightsquigarrow^L z) \rightsquigarrow^L (y \rightsquigarrow^L x)$,
(pC^L) $x \to^L (y \rightsquigarrow^L z) \leq y \rightsquigarrow^L (x \to^L z)$, $x \rightsquigarrow^L (y \to^L z) \leq y \to^L (x \rightsquigarrow^L z)$,
(pD^L) $y \leq (y \to^L x) \rightsquigarrow^L x$, $y \leq (y \rightsquigarrow^L x) \to^L x$,
(pEx^L) $x \to^L (y \rightsquigarrow^L z) = y \rightsquigarrow^L (x \to^L z)$,
(Id^L) $x \to^L 1 = x \rightsquigarrow^L 1$,
(pK^L) $x \leq y \to^L x$ and $x \leq y \rightsquigarrow^L x$;
(L) (Last element) $x \leq 1$,
(pM^L) $1 \to^L x = x = 1 \rightsquigarrow^L x$,
(pN^L) $1 \leq x \Longrightarrow x = 1$,
(Re) (Reflexivity) $x \leq x$,

(p*^{L}) $x \leq y \Longrightarrow (z \to^L x \leq z \to^L y,\ z \rightsquigarrow^L x \leq z \rightsquigarrow^L y)$,
(p**^{L}) $x \leq y \Longrightarrow (y \to^L z \leq x \to^L z,\ y \rightsquigarrow^L z \leq x \rightsquigarrow^L z)$;
(Tr) (Transitivity) $x \leq y$ and $y \leq z \Longrightarrow x \leq z$.

We introduce the following new notions as left-pseudo-structures (the dual notions are omitted):

Definition 2.1.4 A left-pseudo-structure $\mathcal{A}^L = (A^L, \leq, \to^L, \rightsquigarrow^L, 1)$ is called:

i. *left-pseudo-M algebra*, or *left-pM algebra* for short, if it verifies the axiom: (pM^L).

ii. *left-pseudo-ML algebra*, or *left-pML algebra* for short, if it verifies the axioms: (pM^L), (L).

iii. *left-pseudo-ME algebra*, or *left-pME algebra* for short, if it verifies the axioms: (pM^L), (pEx^L).

iv. *left-pseudo-MEL algebra*, or *left-pMEL algebra* for short, if it verifies the axioms: (pM^L), (pEx^L), (L).

Denote by $\mathbf{pM}^L$, $\mathbf{pML}^L$, $\mathbf{pME}^L$, $\mathbf{pMEL}^L$ the classes of left- pM, pML, pME, pMEL algebras, respectively.

Examples of pM, pML, pME, pMEL algebras will be presented in Chapter 6, Section 6.1.

The pM and pML algebras have a special property:

Proposition 2.1.5

(1) Let $\mathcal{A}^L = (A^L, \leq, \to^L, \rightsquigarrow^L, 1)$ be a left-pM algebra (left-pML algebra). Then, $(A^L, \leq, \to^L, 1)$ and $(A^L, \leq, \rightsquigarrow^L, 1)$ are left-M algebras (left-ML algebras, respectively).

(1') Conversely, let $(A^L, \leq, \to^L, 1)$ and $(A^L, \leq, \rightsquigarrow^L, 1)$ be any two left-M algebras (two left-ML algebras) such that (IdEq^L) holds. Then, $\mathcal{A}^L = (A^L, \leq, \to^L, \rightsquigarrow^L, 1)$ is a left-pM algebra (left-pML algebra, respectively).

Proof. Obviously. □

2.1.3 20 new or old pseudo-structures and results

We recall now some old notions or we introduce some new notions (see [104] for the commutative case) as left-pseudo-structures (the dual notions are omitted):

Definitions 2.1.6 A left-pseudo-structure $\mathcal{A}^L = (A^L, \leq, \to^L, \rightsquigarrow^L, 1)$ is called:

1. *left-pseudo-RM algebra*, or *left-pRM algebra* for short, if it verifies the axioms: (Re), (pM^L).

2. *left-pseudo-RML algebra*, or *left-pRML algebra* for short, if it verifies the axioms: (Re), (pM^L), (L).

3. *left-pseudo-RME algebra*, or *left-pRME algebra* for short (or *left-pseudo-CI algebra*, or *left-pCI algebra* for short), if it verifies the axioms: (Re), (pM^L), (pEx^L).

4. *left-pseudo-BE algebra*, or *left-pBE algebra* for short, if it verifies the axioms (Re), (pM^L), (pEx^L), (L).

5. *left-pre-pseudo-BZ algebra*, or *left-pre-pBZ algebra* for short, if it verifies the axioms: (Re), (pM^L), (pB^L).

6. *left-pre-pseudo-BCC algebra*, or *left-pre-pBCC algebra* for short, if it verifies the axioms: (Re), (pM^L), (pB^L), (L).

7. *left-pre-pseudo-BCI algebra*, or *left-pre-pBCI algebra* for short, if it verifies the axioms: (Re), (pM^L), (pEx^L), (pB^L).

8. *left-pre-pseudo-BCK algebra*, or *left-pre-pBCK algebra* for short, if it verifies the axioms (Re), (pM^L), (pEx^L), (p*^{L}), (L).

9. *left-pseudo-aRM algebra*, or *left-paRM algebra* for short (or *left-pseudo-BH algebra*, or *left-pBH algebra* for short), if it verifies the axioms: (Re), (pM^L), (An).

10. *left-pseudo-aRML algebra*, or *left-paRML algebra* for short, if it verifies the axioms: (Re), (pM^L), (An), (L).

11. *left-pseudo-BCH algebra*, or *left-pBCH algebra* for short, if it verifies the axioms (Re), (pEx^L), (An), hence (pM^L).

12. *left-pseudo-aBE algebra*, or *left-paBE algebra* for short, if it verifies the axioms: (Re), (pM^L), (pEx^L), (An), (L).

13. *left-pseudo-BZ algebra*, or *left-pBZ algebra* for short, if it verifies the axioms (Re), (pM^L), (pB^L), (An).

14. *left-pseudo-BCC algebra*, or *left-pBCC algebra* for short, if it verifies the axioms (Re), (pM^L), (pB^L), (An), (L).

15. *left-pseudo-BCI algebra*, or *left-pBCI algebra* for short, if it verifies the axioms (pBB^L), (pD^L), (Re), (An), hence the axioms (Re), (pM^L), (pEx^L), (pBB^L), (An).

16. *left-pseudo-BCK algebra*, or *left-pBCK algebra* for short, if it verifies the axioms (pBB^L), (pD^L), (Re), (L), (An), hence the axioms (Re), (pM^L), (pEx^L), (pBB^L), (An), (L).

Denote by **pRM**^{L}, **pRML**^{L}, **pRME**^{L} =**pCI**^{L}, **pBE**^{L}, **pre-pBZ**^{L}, **pre-pBCC**^{L}, **pre-pBCI**^{L}, **pre-pBCK**^{L}, **paRM**^{L} =**pBH**^{L}, **paRML**^{L}, **pBCH**^{L}, **paBE**^{L}, **pBZ**^{L}, **pBCC**^{L}, **pBCI**^{L}, **pBCK**^{L} the classes of left- pRM, pRML, pRME = pCI, pBE, pre-pBZ, pre-pBCC, pre-pBCI, pre-pBCK, paRM, paRML, pBCH, paBE, pBZ, pBCC, pBCI, pBCK algebras, respectively.

Note that we could define another twenty two new pseudo-RM (or pseudo-RML) algebras, generalizations of pseudo-BCI algebras (or of pseudo-BCK algebras, respectively), following [104], in the commutative case. We shall define here only the following four:

Definitions 2.1.7

17. A *left-pseudo-oRM algebra*, or a *left-poRM algebra* for short, is a left-pseudo-aRM algebra verifying (Tr) ("o" comes from "ordered", because $\leq$ is an order).

18. A *left-pseudo-*aRM algebra*, or a *left-p*aRM algebra* for short, is a left-pseudo-aRM algebra verifying (p*^{L}).

19. A *left-pseudo-aRM** algebra*, or a *left-paRM** algebra* for short, is a left-paRM algebra verifying (p**^{L}).

20. A *left-pseudo-*aRM** algebra*, or a *left-p*aRM** algebra* for short, is a left-aRM algebra verifying (p^{*L}), (p^{**L}).

Denote by **poRM**^{L}, **p*aRM**^{L}, **paRM****^{L}, **p*aRM****^{L} the classes of left-poRM algebras, left-p*aRM algebras, left-paRM** algebras, left-p*aRM** algebras, respectively.

Examples of proper commutative such left-algebras are presented in [104].

The pBCI and the pBCK algebras and some of their subclasses are recalled in some details further in this chapter.

- **General properties**

In order to establish the connections between the above classes of algebras of logic, we have the following results (the dual case is omitted).

Proposition 2.1.8 *(See [104] for the commutative case and for more results)*
Let $(A, \leq, \to^L, \rightsquigarrow^L, 1)$ be a left-pseudo-structure. Then, we have:
(1) (pM^L) + (pEq^L) $\Longrightarrow$ (pN^L),
(1') (L) + (An) $\Longrightarrow$ (pN^L),
(1") (pK^L) + (An) + (pEq^L) $\Longrightarrow$ (pN^L),
(1"') (pD^L) + (Re) + (An) + (pEq^L) $\Longrightarrow$ (pN^L);
(2) (pK^L) + (Re) + (pEq^L) $\Longrightarrow$ (L),
(3) (pK^L) + (pN^L) + (pEq^L) $\Longrightarrow$ (L),
(3') (pK^L) + (pM^L) + (pEq^L) $\Longrightarrow$ (L),
(4) (L) + (pEx^L) + (Re) + (pEq^L) $\Longrightarrow$ (pK^L);
(5) (pC^L) + (An) $\Longrightarrow$ (pEx^L),
(6) (pEx^L) + (pEq^L) + (pB^L) $\Longrightarrow$ (pBB^L),
(6') (pEx^L) + (pEq^L) + (pBB^L) $\Longrightarrow$ (pB^L),
(6") (pEx^L) + (pEq^L) $\Longrightarrow$ (pBB^L) $\Leftrightarrow$ (pB^L);
(7) (Re) + (pEx^L) + (pEq^L) $\Longrightarrow$ (pD^L),
(7') (pBB^L) + (pM^L) $\Longrightarrow$ (pD^L);
(8) (pK^L) + (pC^L) + (An) + (pEq^L) $\Longrightarrow$ (Re),
(8') (pM^L) + (pBB^L) + (pEq^L) $\Longrightarrow$ (Re);
(9) (Re) + (pEx^L) + (pEq^L) + (p^{*L}) $\Longrightarrow$ (pBB^L),
(10) (pM^L) + (pEq^L) + (pB^L) $\Longrightarrow$ (p^{*L}),
(11) (pN^L) + (pBB^L) + (pEq^L) $\Longrightarrow$ (p^{**L});
(12) (Re) + (An) + (pEx^L) + (pEq^L) $\Longrightarrow$ (pM^L),
(13) (pBB^L) + (pD^L) + (pN^L) $\Longrightarrow$ (pC^L),
(14) (pM^L) + (pBB^L) + (An) + (pEq^L) $\Longrightarrow$ (pEx^L),
(15) (pN^L) + (pEq^L) + (p^{*L}) $\Longrightarrow$ (Tr),
(15') (pN^L) + (pEq^L) + (p^{**L}) $\Longrightarrow$ (Tr).

Proof. (1): If $1 \leq x$, i.e. $1 \to^L x = 1 = 1 \rightsquigarrow^L x$, by ($pEq^L$), then $x = 1$, by (pM^L); thus (pN^L) holds.

(1'): Suppose $1 \leq x$; by (L), we also have $x \leq 1$; hence, $x = 1$, by (An). Thus, (pN^L) holds.

(1”): Suppose $1 \leq x$, i.e. $1 \to^L x = 1$, by (pEq^L); by (pK^L), $x \leq 1 \to^L x = 1$; then, $x = 1$, by (An). Thus, (pN^L) holds.

(1”’): Suppose $1 \leq x$, i.e. $1 \to^L x = 1$, by (pEq^L); by (pD^L), $x \leq (x \rightsquigarrow^L x) \to^L x \overset{(Re)+(pEq^L)}{=} 1 \to^L x = 1$; hence, $x = 1$, by (An).

(2): Take $y = x$ in (pK^L); we obtain: $x \leq x \to^L x = 1$, by (Re) and (pEq^L); thus, (L) holds.

(3): By (pK^L), for all x, $1 \leq x \to^L 1$; then, by (pN^L), $x \to^L 1 = 1$, i.e. $x \leq 1$, by (pEq^L); thus, (L) holds.

(3’): By above (1), (pM^L) + (pEq^L) $\Longrightarrow$ (pN^L), and by above (3), (pK^L) + (pN^L) + (pEq^L) $\Longrightarrow$ (L); thus, (L) holds.

(4): In (pEx^L) $(x \to^L (y \rightsquigarrow^L z) = y \rightsquigarrow^L (x \to^L z))$, take $z = x$; we obtain: $x \to^L (y \rightsquigarrow^L x) = y \rightsquigarrow^L (x \to^L x) = y \rightsquigarrow^L 1 = 1$, by (Re), (pEq^L), (L); hence $x \leq y \rightsquigarrow^L x$, by (pEq^L). Similarly, in (pEx^L) $(x \rightsquigarrow^L (y \to^L z) = y \to^L (x \rightsquigarrow^L z))$, take $z = x$; we obtain: $x \rightsquigarrow^L (y \to^L x) = y \to^L (x \rightsquigarrow^L x) = y \to^L 1 = 1$, hence $x \leq y \to x$, by (pEq^L). Thus, (pK^L) holds.

(5): By (pC^L),
$x \to^L (y \rightsquigarrow^L z) \leq y \rightsquigarrow^L (x \to^L z)$ and $y \rightsquigarrow^L (x \to^L z) \leq x \to^L (y \rightsquigarrow^L z)$;
hence, by (An), $x \to^L (y \rightsquigarrow^L z) = y \rightsquigarrow^L (x \to^L z)$; thus, (pEx^L) holds.

(6): By (pB^L),
$z \to^L x \leq (y \to^L z) \to^L (y \to^L x)$ and $z \rightsquigarrow^L x \leq (y \rightsquigarrow^L z) \rightsquigarrow^L (y \rightsquigarrow^L x)$;
hence, by (pEq^L),
$(z \to^L x) \rightsquigarrow^L ((y \to^L z) \to^L (y \to^L x)) = 1$ and
$(z \rightsquigarrow^L x) \to^L ((y \rightsquigarrow^L z) \rightsquigarrow^L (y \rightsquigarrow^L x)) = 1$;
then, by (pEx^L), we obtain:
$(y \to^L z) \to^L ((z \to^L x) \rightsquigarrow^L (y \to^L x)) = 1$ and
$(y \rightsquigarrow^L z) \rightsquigarrow^L ((z \rightsquigarrow^L x) \to^L (y \rightsquigarrow^L x)) = 1$,
i.e., by (pEq^L),
$y \to^L z \leq (z \to^L x) \rightsquigarrow^L (y \to^L x)$ and $y \rightsquigarrow^L z \leq (z \rightsquigarrow^L x) \to^L (y \rightsquigarrow^L x)$; thus, (pBB^L) holds.

(6’): By (pBB^L),
$y \to^L z \leq (z \to^L x) \rightsquigarrow^L (y \to^L x)$ and $y \rightsquigarrow^L z \leq (z \rightsquigarrow^L x) \to^L (y \rightsquigarrow^L x)$;
hence, by (pEq^L),
$(y \to^L z) \to^L ((z \to^L x) \rightsquigarrow^L (y \to^L x)) = 1$ and
$(y \rightsquigarrow^L z) \rightsquigarrow^L ((z \rightsquigarrow^L x) \to^L (y \rightsquigarrow^L x)) = 1$;
then, by (pEx^L), we obtain:
$(z \to^L x) \rightsquigarrow^L ((y \to^L z) \to^L (y \to^L x)) = 1$ and
$(z \rightsquigarrow^L x) \to^L ((y \rightsquigarrow^L z) \rightsquigarrow^L (y \rightsquigarrow^L x)) = 1$, i.e., by (pEq^L),
$z \to^L x \leq (y \to^L z) \to^L (y \to^L x)$ and $z \rightsquigarrow^L x \leq (y \rightsquigarrow^L z) \rightsquigarrow^L (y \rightsquigarrow^L x)$; thus, (pB^L) holds.

(6”): By (6) and (6’).

(7): $y \to^L [(y \to^L x) \rightsquigarrow^L x] \overset{(pEx^L)}{=} (y \to^L x) \rightsquigarrow^L (y \to^L x) = 1$, by (Re) and (pEq^L), hence $y \leq (y \to^L x) \rightsquigarrow^L x$, by (pEq^L).

Similarly, $y \rightsquigarrow^L [(y \rightsquigarrow^L x) \to^L x] \overset{(pEx^L)}{=} (y \rightsquigarrow^L x) \to^L (y \rightsquigarrow^L x) = 1$, hence $y \leq (y \rightsquigarrow^L x) \to^L x$. Thus, (pD^L) holds.

(7'): In (pBB^L), take $y = 1$; we obtain:
$z \overset{(pM^L)}{=} 1 \to^L z \leq (z \to^L x) \rightsquigarrow^L (1 \to^L x) \overset{(pM^L)}{=} (z \to^L x) \rightsquigarrow^L x$ and
$z \overset{(pM^L)}{=} 1 \rightsquigarrow^L z \leq (z \rightsquigarrow^L x) \to^L (1 \rightsquigarrow^L x) \overset{(pM^L)}{=} (z \rightsquigarrow^L x) \to^L x$.

(8): By above (1"), (pK^L) + (An) + (pEq^L) $\Longrightarrow$ (pN^L); by above (5), (pC^L) + (An) $\Longrightarrow$ (pEx^L); by (pK^L), $1 \leq [(x \to^L x) \to^L 1]$; by (pN^L), $(x \to^L x) \to^L 1 = 1$, i.e. $x \to^L x \leq 1$, by (pEq^L).

On the other hand, $1 \rightsquigarrow^L (x \to^L x) \overset{(pEx^L)}{=} x \to^L (1 \rightsquigarrow^L x) = 1$, by (pK^L) and (pEq^L); hence, $1 \leq x \to^L x$.
Now, by (An), $x \to^L x = 1$, i.e. $x \leq x$, by (pEq^L). Thus, (Re) holds.

(8'): First, by (1), (pM^L) + (pEq^L) $\Longrightarrow$ (pN^L). Then, take $y = z = 1$ in (pBB^L):
$y \to^L z \leq (z \to^L x) \rightsquigarrow^L (y \to^L x)$ and $y \rightsquigarrow^L z \leq (z \rightsquigarrow^L x) \to^L (y \rightsquigarrow^L x)$;
by (pM^L), we then obtain:
$1 \leq (1 \to^L x) \rightsquigarrow^L (1 \to^L x) = x \rightsquigarrow^L x$ and $1 \leq (1 \rightsquigarrow^L x) \to^L (1 \rightsquigarrow^L x) = x \to^L x$;
hence, by (pN^L), $x \rightsquigarrow^L x = 1$ and $x \to^L x = 1$, i.e. $x \leq x$, by (pEq^L).

(9): By (7), (Re) + (pEx^L) + (pEq^L) $\Longrightarrow$ (pD^L), thus (pD^L) holds. Hence, $z \leq (z \to x) \rightsquigarrow x$ and $z \leq (z \rightsquigarrow x) \to x$. Then, by (p*^{L}), we obtain:
$y \to^L z \leq y \to^L [(z \to^L x) \rightsquigarrow^L x] \overset{(pEx^L)}{=} (z \to^L x) \rightsquigarrow^L (y \to^L x)$ and
$y \rightsquigarrow^L z \leq y \rightsquigarrow^L [(z \rightsquigarrow^L x) \to^L x] \overset{(pEx^L)}{=} (z \rightsquigarrow^L x) \to^L (y \rightsquigarrow^L x)$; thus, (pBB^L) holds.

(10): By (1), (pM^L) + (pEq^L) $\Longrightarrow$ (pN^L). Suppose $y \leq z$, i.e. $y \to^L z = 1 = y \rightsquigarrow^L z$, by (pEq^L). Since, by (pB^L),
$y \to^L z \leq (x \to^L y) \to^L (x \to^L z)$ and $y \rightsquigarrow^L z \leq (x \rightsquigarrow^L y) \rightsquigarrow^L (x \rightsquigarrow^L z)$,
then, by (pN^L, $(x \to^L y) \to^L (x \to^L z) = 1$ and $(x \rightsquigarrow^L y) \rightsquigarrow^L (x \rightsquigarrow^L z) = 1$, i.e. $x \to^L y \leq x \to^L z$ and $x \rightsquigarrow^L y \leq x \rightsquigarrow^L z$; thus, (p*^{L}) holds.

(11): By (pBB^L),
$y \to^L z \leq (z \to^L x) \rightsquigarrow^L (y \to^L x)$ and $y \rightsquigarrow^L z \leq (z \rightsquigarrow^L x) \to^L (y \rightsquigarrow^L x)$.
Suppose $y \leq z$, i.e. $y \to^L z = 1 = y \rightsquigarrow^L z$, by (pEq^L); then,
$1 \leq (z \to^L x) \rightsquigarrow^L (y \to^L x)$ and $1 \leq (z \rightsquigarrow^L x) \to^L (y \rightsquigarrow^L x)$; hence, by (pN^L),
$1 = (z \to^L x) \rightsquigarrow^L (y \to^L x)$ and $1 = (z \rightsquigarrow^L x) \to^L (y \rightsquigarrow^L x)$, i.e. by (pEq^L),
$z \to^L x \leq y \to^L x$ and $z \rightsquigarrow^L x \leq y \rightsquigarrow^L x$; thus, (p**^{L}) holds.

(12): By (7), (Re) + (pEx^L) + (pEq^L) $\Longrightarrow$ (pD^L) and,
by (1"'), (pD^L) + (Re) + (An) + (pEq^L) $\Longrightarrow$ (pN^L).

$x \rightsquigarrow^L (1 \to^L x) \overset{(pEx^L)}{=} 1 \to^L (x \rightsquigarrow^L x) = 1 \to^L 1 = 1$, by (Re) and (pEq^L); hence, $x \leq 1 \to^L x$.
On the other hand, $(1 \to^L x) \rightsquigarrow^L x = 1$; indeed, by (pD^L), $1 \leq (1 \to^L x) \rightsquigarrow^L x$, hence by (pN^L), $(1 \to^L x) \rightsquigarrow^L x = 1$; hence, $1 \to^L x \leq x$. And $x \leq 1 \to^L x$ and $1 \to^L x \leq x$ imply $1 \to^L x = x$, by (An). Similarly, $1 \rightsquigarrow^L x = x$. Thus, (pM^L) holds.

(13): First, by (11), (pBB^L) + (pN^L) + (pEq^L) imply (p**^{L}).
Then, by (pBB^L): $(Y \to^L Z \leq (Z \to^L X) \rightsquigarrow^L (Y \to^L X))$,
for $X = u \rightsquigarrow^L x$, $Y = y$, $Z = z \rightsquigarrow^L x$, we obtain:
$y \to^L (z \rightsquigarrow^L x) \leq ((z \rightsquigarrow^L x) \to^L (u \rightsquigarrow^L x)) \rightsquigarrow^L (y \to^L (u \rightsquigarrow^L x))$.

Then, by (p^{**L}), we obtain:
$V_1 \overset{notation}{=} [((z \rightsquigarrow^L x) \to^L (u \rightsquigarrow^L x)) \rightsquigarrow^L (y \to^L (u \rightsquigarrow^L x))] \to^L$
$[(u \rightsquigarrow^L z) \rightsquigarrow^L (y \to^L (u \rightsquigarrow^L x))] \leq$
$(y \to^L (z \rightsquigarrow^L x)) \to^L [(u \rightsquigarrow^L z) \rightsquigarrow^L (y \to^L (u \rightsquigarrow^L x))] \overset{notation}{=} W_1$.
But, the left side $V_1 = 1$; indeed, by (pBB^L), we have:
$u \rightsquigarrow^L z \leq (z \rightsquigarrow^L x) \to^L (u \rightsquigarrow^L x)$;
then, by (p^{**L}), we obtain:
$[(z \rightsquigarrow^L x) \to^L (u \rightsquigarrow^L x)) \rightsquigarrow^L (y \to^L (u \rightsquigarrow^L x)) \leq (u \rightsquigarrow^L z) \rightsquigarrow^L (y \to^L (u \rightsquigarrow^L x))$,
i.e. $V_1 = 1$.
Then, by (pN^L), $W_1 = 1$, i.e.
$y \to^L (z \rightsquigarrow^L x) \leq (u \rightsquigarrow^L z) \rightsquigarrow^L (y \to^L (u \rightsquigarrow^L x))$,
which for $z = y \to^L x$ and $u = z$ gives:
$y \to^L ((y \to^L x) \rightsquigarrow^L x) \leq (z \rightsquigarrow^L (y \to^L x)) \rightsquigarrow^L (y \to^L (z \rightsquigarrow^L x))$;
but, by (pD^L), the left side $y \to^L ((y \to^L x) \rightsquigarrow^L x) = 1$; hence, by ($pN^L$),
$(z \rightsquigarrow^L (y \to^L x)) \rightsquigarrow^L (y \to^L (z \rightsquigarrow^L x)) = 1$,
i.e. $z \rightsquigarrow^L (y \to^L x) \leq y \to^L (z \rightsquigarrow^L x)$, by ($pEq^L$).

Similarly, by (pBB^L) again: $(Y \rightsquigarrow^L Z \leq (Z \rightsquigarrow^L X) \to^L (Y \rightsquigarrow^L X))$,
for $X = u \to^L x$, $Y = y$, $Z = z \to^L x$, we obtain:
$y \rightsquigarrow^L (z \to^L x) \leq ((z \to^L x) \rightsquigarrow^L (u \to^L x)) \to^L (y \rightsquigarrow^L (u \to^L x))$.
Then, by (p^{**L}), we obtain:
$V_2 \overset{notation}{=} [((z \to^L x) \rightsquigarrow^L (u \to^L x)) \to^L (y \rightsquigarrow^L (u \to^L x))] \rightsquigarrow^L$
$[(u \to^L z) \to^L (y \rightsquigarrow^L (u \to^L x))] \leq$
$(y \rightsquigarrow^L (z \to^L x)) \rightsquigarrow^L [(u \to^L z) \to^L (y \rightsquigarrow^L (u \to^L x))] \overset{notation}{=} W_2$.
But, the left side $V_2 = 1$; indeed, by (pBB^L), we have:
$u \to^L z \leq (z \to^L x) \rightsquigarrow^L (u \to^L x)$;
then, by (p^{**L}), we obtain:
$[(z \to^L x) \rightsquigarrow^L (u \to^L x)] \to^L (y \rightsquigarrow^L (u \to^L x)) \leq (u \to^L z) \to^L (y \rightsquigarrow^L (u \to^L x))$,
i.e. $V_2 = 1$.
Then, by (pN^L), $W_2 = 1$, i.e.
$y \rightsquigarrow^L (z \to^L x) \leq (u \to^L z) \to^L (y \rightsquigarrow^L (u \to^L x))$,
which for $z = y \rightsquigarrow^L x$ and $u = z$ gives:
$y \rightsquigarrow^L ((y \rightsquigarrow^L x) \to^L x) \leq (z \to^L (y \rightsquigarrow^L x)) \to^L (y \rightsquigarrow^L (z \to^L x))$;
but, by (pD^L), the left side $y \rightsquigarrow^L ((y \rightsquigarrow^L x) \to^L x) = 1$; hence, by ($pN^L$),
$(z \to^L (y \rightsquigarrow^L x)) \to^L (y \rightsquigarrow^L (z \to^L x)) = 1$, i.e $z \to^L (y \rightsquigarrow^L x) \leq y \rightsquigarrow^L (z \to^L x)$;
hence, $y \to^L (z \rightsquigarrow^L x) \leq z \rightsquigarrow^L (y \to^L x)$. Thus, ($pC^L$) holds.

(14): By (1), (pM^L) + (pEq^L) $\Longrightarrow$ (pN^L),
by (7'), (pBB^L) + (pM^L) $\Longrightarrow$ (pD^L),
by (13), (pBB^L) + (pD^L) + (pN^L) $\Longrightarrow$ (pC^L) and
by (5), (pC^L) + (An) $\Longrightarrow$ (pEx^L). Thus, (pEx^L) holds.

(15): Suppose $x \leq y$ and $y \leq z$, i.e. $x \to^L y = 1$, by (pEq^L); then, by (p^{*L}), $x \to^L y \leq x \to^L z$, hence $1 \leq x \to^L z$; then, by (pN^L), $x \to^L z = 1$, i.e. $x \leq z$, by (pEq^L) again. Thus, (Tr) holds.

(15'): Suppose $x \leq y$ and $y \leq z$, i.e. $y \to^L z = 1$, by (pEq^L); then, by (p^{**L}), $y \to^L z \leq x \to^L z$, hence $1 \leq x \to^L z$; then, by (pN^L), $x \to^L z = 1$, i.e. $x \leq z$, by (pEq^L) again. $\square$

Theorem 2.1.9 *(See [104], Part I, Theorem 1, for the commutative case)*
Let $\mathcal{A}^L = (A^L, \leq, \to^L, \rightsquigarrow^L, 1)$ be a left-pseudo-structure such that ((IdEq^L), (pEq^L)) (Re), (pM^L), (pEx^L) hold, i.e. $\mathcal{A}^L$ is a left-pRME algebra (= left-pCI algebra). Then, we have:

$$(pBB^L) \Leftrightarrow (pB^L) \Leftrightarrow (p*^L).$$

Proof. By Proposition 2.1.8 (6"), (pEx^L) + (pEq^L) $\Longrightarrow$ (pBB^L) $\Leftrightarrow$ (pB^L).
By (9), (Re) + (pEx^L) + (pEq^L) + (p*^{L}) $\Longrightarrow$ (pBB^L).
By (10), (pM^L) + (pEq^L) + (pB^L) $\Longrightarrow$ (p*^{L}). Hence, we have:

$$(p*^L) \Longrightarrow (pBB^L) \Leftrightarrow (pB^L) \Longrightarrow (p*^L),$$

so (pBB^L) $\Leftrightarrow$ (pB^L) $\Leftrightarrow$ (p*^{L}). □

Theorem 2.1.10 *(See [104], Part I, Theorem 3, for the commutative case)*
Let $\mathcal{A}^L = (A^L, \leq, \to^L, \rightsquigarrow^L, 1)$ be a left-pseudo-structure such that ((IdEq^L), (pEq^L)) (An), (pM^L), (pB^L) hold. Then, we have:

$$(pEx^L) \Leftrightarrow (pBB^L).$$

Proof. By Proposition 2.1.8 (6), (pEx^L) + (pEq^L) + (pB^L) $\Longrightarrow$ (pBB^L).
By Proposition 2.1.8 (14), (pM^L) + (pBB^L) + (An) + (pEq^L) $\Longrightarrow$ (pEx^L). □

The next theorem was proved first in the commutative case and communicated us by Michael Kinyon, following our two open problems announced in the initial preprint of [104] on December 2013 on: http://arxiv.org/abs/1312.2494; he proved (i) first by using the automated theorem proving tool PROVER9.

Theorem 2.1.11 *(See [104], Part I, Theorem 4 (Michael Kinyon))*
In any pseudo-algebra $(A, \to = \to^L, \rightsquigarrow = \rightsquigarrow^L, 1)$ we have:
(i) (pM^L) + (pBB^L) imply (pB^L),
*(ii) (pM^L) + (pB^L) imply (p**^{L}).*

Proof. (i): By Proposition 2.1.8 (7'), we have (pM^L) + (pBB^L) imply (pD^L).
Suppose that (pBB^L) holds, i.e.

$$(1)\ x \to y \leq (y \to z) \rightsquigarrow (x \to z),\ (2)\ x \rightsquigarrow y \leq (y \rightsquigarrow z) \to (x \rightsquigarrow z).$$

- First, in (pBB^L) (1), set $x = u$ and $y = (u \to v) \rightsquigarrow v$, to get:
$(u \to [(u \to v) \rightsquigarrow v]) \to [(((u \to v) \rightsquigarrow v) \to z) \rightsquigarrow (u \to z)] \overset{(pD^L)}{=}$
$1 \to [(((u \to v) \rightsquigarrow v) \to z) \rightsquigarrow (u \to z)] \overset{(pM^L)}{=} (((u \to v) \rightsquigarrow v) \to z) \rightsquigarrow (u \to z) = 1.$
After renaming variables, we get:

$$(a1) \qquad (((x \to y) \rightsquigarrow y) \to z) \rightsquigarrow (x \to z) = 1.$$

Next, in (pBB^L) (2), set $x = u \rightsquigarrow v$ and $y = (v \rightsquigarrow w) \to (u \rightsquigarrow w)$, to get:
$((u \rightsquigarrow v) \rightsquigarrow [(v \rightsquigarrow w) \to (u \rightsquigarrow w)]) \to$

$[(((v \rightsquigarrow w) \to (u \rightsquigarrow w)) \rightsquigarrow z) \to ((u \rightsquigarrow v) \rightsquigarrow z)] \overset{(pBB^L)(2)}{=}$
$1 \to [(((v \rightsquigarrow w) \to (u \rightsquigarrow w)) \rightsquigarrow z) \to ((u \rightsquigarrow v) \rightsquigarrow z)] \overset{(pM^L)}{=}$
$(((v \rightsquigarrow w) \to (u \rightsquigarrow w)) \rightsquigarrow z) \to ((u \rightsquigarrow v) \rightsquigarrow z) = 1.$
After renaming variables, we get:

$$(b1) \qquad (((x \rightsquigarrow y) \to (u \rightsquigarrow y)) \rightsquigarrow z) \to ((u \rightsquigarrow x) \rightsquigarrow z) = 1.$$

Taking $z = u \rightsquigarrow y$ in (b1), we get:

$$(c1) \qquad (((x \rightsquigarrow y) \to (u \rightsquigarrow y)) \rightsquigarrow (u \rightsquigarrow y)) \to ((u \rightsquigarrow x) \rightsquigarrow (u \rightsquigarrow y)) = 1.$$

Now, in (a1), set $x = v \rightsquigarrow w,\ \ y = t \rightsquigarrow w,\ \ z = (t \rightsquigarrow v) \rightsquigarrow (t \rightsquigarrow w)$ to get:
$[(((v \rightsquigarrow w) \to (t \rightsquigarrow w)) \rightsquigarrow (t \rightsquigarrow w)) \to ((t \rightsquigarrow v) \rightsquigarrow (t \rightsquigarrow w))] \rightsquigarrow$
$((v \rightsquigarrow w) \to ((t \rightsquigarrow v) \rightsquigarrow (t \rightsquigarrow w))) \overset{(c1)}{=}$
$1 \rightsquigarrow ((v \rightsquigarrow w) \to ((t \rightsquigarrow v) \rightsquigarrow (t \rightsquigarrow w))) \overset{(pM)^L}{=}$
$(v \rightsquigarrow w) \to ((t \rightsquigarrow v) \rightsquigarrow (t \rightsquigarrow w)) = 1,$
i.e. (pBL) (2) $(v \rightsquigarrow w \leq (t \rightsquigarrow v) \rightsquigarrow (t \rightsquigarrow w))$ holds.

- Now, "dually", in (pBB^L) (2), set $x = u$ and $y = (u \rightsquigarrow v) \to v$, to get:
$(u \rightsquigarrow [(u \rightsquigarrow v) \to v]) \to [(((u \rightsquigarrow v) \to v) \rightsquigarrow z) \to (u \rightsquigarrow z)] \overset{(pD^L)}{=}$
$1 \to [(((u \rightsquigarrow v) \to v) \rightsquigarrow z) \to (u \rightsquigarrow z)] \overset{(pM^L)}{=} (((u \rightsquigarrow v) \to v) \rightsquigarrow z) \to (u \rightsquigarrow z) = 1.$
After renaming variables, we get:

$$(a2) \qquad (((x \rightsquigarrow y) \to y) \rightsquigarrow z) \to (x \rightsquigarrow z) = 1.$$

Next, in (pBB^L) (1), set $x = u \to v$ and $y = (v \to w) \rightsquigarrow (u \to w)$, to get:
$((u \to v) \to [(v \to w) \rightsquigarrow (u \to w)]) \to$
$[(((v \to w) \rightsquigarrow (u \to w)) \to z) \rightsquigarrow ((u \to v) \to z)] \overset{(pBB^L)(1)}{=}$
$1 \to [(((v \to w) \rightsquigarrow (u \to w)) \to z) \rightsquigarrow ((u \to v) \to z)] \overset{(pM^L)}{=}$
$(((v \to w) \rightsquigarrow (u \to w)) \to z) \rightsquigarrow ((u \to v) \to z) = 1.$
After renaming variables, we get:

$$(b2) \qquad (((x \to y) \rightsquigarrow (u \to y)) \to z) \rightsquigarrow ((u \to x) \to z) = 1.$$

Taking $z = u \to y$ in (b2), we get:

$$(c2) \qquad (((x \to y) \rightsquigarrow (u \to y)) \to (u \to y)) \rightsquigarrow ((u \to x) \to (u \to y)) = 1.$$

Now, in (a2), set $x = v \to w,\ \ y = t \to w,\ \ z = (t \to v) \to (t \to w)$ to get:
$[(((v \to w) \rightsquigarrow (t \to w)) \to (t \to w)) \rightsquigarrow ((t \to v) \to (t \to w))] \to$
$((v \to w) \rightsquigarrow ((t \to v) \to (t \to w))) \overset{(c2)}{=}$
$1 \to ((v \to w) \rightsquigarrow ((t \to v) \to (t \to w))) \overset{(pM^L)}{=}$
$(v \to w) \rightsquigarrow ((t \to v) \to (t \to w)) = 1,$
i.e. (pB^L) (1) $(v \to w \leq (t \to v) \to (t \to w))$ holds.

(ii): Suppose (pB^L) is

$$(1)\ y \to z \leq (x \to y) \to (x \to z), (2)\ y \rightsquigarrow z \leq (x \rightsquigarrow y) \rightsquigarrow (x \rightsquigarrow z).$$

If $x \leq y$, i.e. $x \to y = 1 = x \rightsquigarrow y$, then we get, from (pB^L):
$y \to z \leq 1 \to (x \to z) \overset{(pM^L)}{=} x \to z$ and $y \rightsquigarrow z \leq 1 \rightsquigarrow (x \rightsquigarrow z) \overset{(pM^L)}{=} x \rightsquigarrow z$, i.e. (**^{L}) holds. □

By Theorem 2.1.11 (i) and Proposition 2.1.8 (10), we obtain immediately that:

Corollary 2.1.12 (See [104], Part I, Corollary 2, for the commutative case)
(pM^L) + (pEq^L) + (pBB^L) imply (p*^{L}).

Concluding, by above Theorem 2.1.11 and Proposition 2.1.8 (1), (10), (11), (15), (15'), we immediately obtain:

Corollary 2.1.13 (See [104], Part I, Corollary 3)
In any pseudo-algebra $(A, \to^L, \rightsquigarrow^L, 1)$ verifying (pM^L) and (pEq^L), we have:

$$(pBB^L) \implies (pB^L) \implies (p*^L), (p**^L) \implies (Tr).$$

• **Pseudo-RME (= pseudo-CI) algebras**

First, we present the following Proposition.

Proposition 2.1.14 *Let $\mathcal{A}^L = (A^L, \leq, \to^L, \rightsquigarrow^L, 1)$ be a left-pRME (= left-pCI) algebra. Then we have: for all $x, y \in A^L$,*
(Id^L) $x \to^L 1 = x \rightsquigarrow^L 1$,
(pD^L) $y \leq (y \to^L x) \rightsquigarrow^L x,\ \ y \leq (y \rightsquigarrow^L x) \to^L x$.

Proof. (Id^L): $x \to^L 1 \overset{(pRe^L)}{=} x \to^L (x \rightsquigarrow^L x) \overset{(pEx)}{=} x \rightsquigarrow^L (x \to^L x) \overset{(pRe^L)}{=} x \rightsquigarrow^L 1$.

(pD^L): $y \to^L [(y \to^L x) \rightsquigarrow^L x] \overset{(pEx)}{=} [(y \to^L x) \rightsquigarrow^L (y \to^L x) \overset{(pRe^L)}{=} 1$, hence $y \leq (y \to^L x) \rightsquigarrow^L x$, by (pEq^L), and
$y \rightsquigarrow^L [(y \rightsquigarrow^L x) \to^L x] \overset{(pEx)}{=} [(y \rightsquigarrow^L x) \to^L (y \rightsquigarrow^L x) \overset{(pRe^L)}{=} 1$, hence $y \leq (y \rightsquigarrow^L x) \to^L x$, by (pEq^L). □

Let $\mathcal{A}^L = (A^L, \leq, \to^L, \rightsquigarrow^L, 1)$ be a left-pRME (= left-pCI) algebra. We can define two negations $^- = ^{-^L}$ and $^\sim = ^{\sim^L}$ by: for all $x \in A^L$,

$$x^- \overset{def.}{=} x \to^L 1, \quad x^\sim \overset{def.}{=} = x \rightsquigarrow^L 1.$$

But, since $x \to^L 1 \overset{(Id^L)}{=} x \rightsquigarrow^L 1$, it follows that $^- = ^\sim$, i.e. there exists in fact only one negation, denoted by $^{-1}$:

$$x^{-1} \overset{def.}{=} x \to^L 1 \overset{(Id^L)}{=} x \rightsquigarrow^L 1.$$

Then we obtain the following results.

Proposition 2.1.15 *(See Proposition 2.1.17)*
Let $\mathcal{A}^L = (A^L, \leq, \to^L, \leadsto^L, 1)$ be a left-pRME (= left-pCI) algebra. Then we have:
(pDN$^\leq$) $x \leq (x^{-1})^{-1}$,
(N1^L) $1^{-1} = 1$,
(pNg1^L) $x \to^L y^{-1} = y \leadsto^L x^{-1}$.

Proof. (pDN$^\leq$): $x \overset{(pD^L)}{\leq} (x \to^L 1) \leadsto^L 1 \overset{def.}{=} (x^{-1})^{-1}$.

(N1^L): $1^{-1} \overset{def.}{=} 1 \to^L 1 \overset{(pM^L)}{=} 1$.

(pNg1^L): $x \to^L y^{-1} \overset{def.}{=} x \to^L (y \leadsto^L 1) \overset{(pEx^L)}{=} y \leadsto^L (x \to^L 1) \overset{def.}{=} y \leadsto^L x^{-1}$.
□

2.1.4 Strong pM algebras

We introduce here a subclass of the class of pseudo-M algebras (the dual case is omitted):

Definition 2.1.16 Let $\mathcal{A}^L = (A^L, \leq, \to^L, \leadsto^L, 1)$ be a left-pM algebra. We say that $\mathcal{A}^L$ is *strong*, if it verifies: for all $x, y \in A^L$,
(Id^L) $x \to^L 1 = x \leadsto^L 1$,
(pEx1^L) $x \to^L (y \leadsto^L 1) = y \leadsto^L (x \to^L 1)$, i.e. a weaker pseudo-Exchange property.

We can define in general two negations $^- = ^{-^L}$ and $^\sim = ^{\sim^L}$ by: for all $x \in A^L$,

$$x^- \overset{def.}{=} x \to^L 1, \quad x^\sim \overset{def.}{=} x \leadsto^L 1.$$

But, since $x \to^L 1 \overset{(Id^L)}{=} x \leadsto^L 1$, it follows that $^- = ^\sim$, i.e. there exists in fact only one negation, denoted by $^{-1}$:

$$x^{-1} \overset{def.}{=} x \to^L 1 \overset{(Id^L)}{=} x \leadsto^L 1.$$

Then we obtain the following results.

Proposition 2.1.17 *Let $\mathcal{A}^L = (A^L, \leq, \to^L, \leadsto^L, 1)$ be a strong left-pRM (left-paRM) algebra. Then, we have:*
(N1^L) $1^{-1} = 1$,
(pD1^L) $y \leq (y \to^L 1) \leadsto^L 1$, $y \leq (y \leadsto^L 1) \to^L 1$,
(pDN$^\leq$) $x \leq (x^{-1})^{-1}$,
(pNg1^L) $x \to^L y^{-1} = y \leadsto^L x^{-1}$.

Proof. (N1^L): $1^{-1} \overset{def.}{=} 1 \to^L 1 \overset{(pM^L)}{=} 1$.

(pD1^L): $y \to^L [(y \to^L 1) \leadsto^L 1] \overset{(pEx1^L)}{=} (y \to^L 1) \leadsto^L (y \to^L 1) \overset{(pRe^L)}{=} 1$,
hence $y \leq (y \to^L 1) \leadsto^L 1$, by (pEq^L), and
$y \leadsto^L [(y \leadsto^L 1) \to^L 1] \overset{(pEx1^L)}{=} (y \leadsto^L 1) \to^L (y \leadsto^L 1) \overset{(pRe^L)}{=} 1$,
hence $y \leq (y \leadsto^L 1) \to^L 1$, by (pEq^L).

(pDN$^{\leq}$): $x \overset{(pD1^L)}{\leq} (x \to^L 1) \rightsquigarrow^L 1 \overset{def.}{=} (x^{-1})^{-1}$.

(pNg1^L): $x \to^L y^{-1} \overset{def.}{=} x \to^L (y \rightsquigarrow^L 1) \overset{(pEx1^L)}{=} y \rightsquigarrow^L (x \to^L 1) \overset{def.}{=} y \rightsquigarrow^L x^{-1}$. □

Consider the following properties:
(pBB1^L) $y \to^L z \leq (z \to^L 1) \rightsquigarrow^L (y \to^L 1)$, $y \rightsquigarrow^L z \leq (z \rightsquigarrow^L 1) \to^L (y \rightsquigarrow^L 1)$,
(pC1^L) $x \to^L (y \rightsquigarrow^L 1) \leq y \rightsquigarrow^L (x \to^L 1)$, $x \rightsquigarrow^L (y \to^L 1) \leq y \to^L (x \rightsquigarrow^L 1)$.

Proposition 2.1.18 *Let $\mathcal{A}^L = (A^L, \leq, \to^L, \rightsquigarrow^L, 1)$ be a pseudo-structure. Then,*
(a) (pEx^L) $\Longrightarrow$ (pEx1^L),
(b) (pD^L) $\Longrightarrow$ (pD1^L),
(c) (pBB^L) $\Longrightarrow$ (pBB1^L),
(d) (pC^L) $\Longrightarrow$ (pC1^L),
(e) (pEx1^L) + (pD1^L) + (p^L) $\Longrightarrow$ (pBB1^L),*
*(f) (pBB1^L) + (pD1^L) + (p**^{L}) + (Tr) $\Longrightarrow$ (pC1^L),*
*(g) (pBB1^L) + (pD1^L) + (p**^{L}) + (Tr) + (An) $\Longrightarrow$ (pEx1^L).*

Proof. (a), (b), (c), (d): obviously.

(e): By (pD1^L), $z \leq (z \to^L 1) \rightsquigarrow^L 1$; then, by (p*^{L}),
$y \to^L z \leq y \to^L [(z \to^L 1) \rightsquigarrow^L 1] \overset{(pEx1^L)}{=} (z \to^L 1) \rightsquigarrow^L (y \to^L 1)$.
Similarly, by (pD1^L), $z \leq (z \rightsquigarrow^L 1) \to^L 1$; then, by (p*^{L}),
$y \rightsquigarrow^L z \leq y \rightsquigarrow^L [(z \rightsquigarrow^L 1) \to^L 1] \overset{(pEx1^L)}{=} (z \rightsquigarrow^L 1) \to^L (y \rightsquigarrow^L 1)$. Thus, (pBB1^L) holds.

(f): $x \to^L (y \rightsquigarrow^L 1) \overset{(pBB1^L)}{\leq} ((y \rightsquigarrow^L 1) \to^L 1) \rightsquigarrow^L (x \to^L 1)$;
by (pD1^L), $y \leq (y \rightsquigarrow^L 1) \to^L 1$; then, by (p**^{L}),
$((y \rightsquigarrow^L 1) \to^L 1) \rightsquigarrow^L (x \to^L 1) \leq y \rightsquigarrow^L (x \to^L 1)$;
then, by (Tr), $x \to^L (y \rightsquigarrow^L 1) \leq y \rightsquigarrow^L (x \to^L 1)$.
Similarly, $y \rightsquigarrow^L (x \to^L 1) \overset{(pBB1^L)}{\leq} ((x \to^L 1) \rightsquigarrow^L 1) \to^L (y \rightsquigarrow^L 1)$;
by (pD1^L), $x \leq (x \to^L 1) \rightsquigarrow^L 1$; then, by (p**^{L}),
$((x \to^L 1) \rightsquigarrow^L 1) \to^L (y \rightsquigarrow^L 1) \leq x \to^L (y \rightsquigarrow^L 1)$;
then, by (Tr), $y \rightsquigarrow^L (x \to^L 1) \leq x \to^L (y \rightsquigarrow^L 1)$; thus, (pC1^L) holds.

(g): By above (f). □

Note that, by above Propositions 2.1.14 and 2.1.18, **any left-pRME=pCI algebra is strong.**

By above Proposition 2.1.18, we obtain:

Corollary 2.1.19 *Let $\mathcal{A}^L = (A^L, \leq, \to^L, \rightsquigarrow^L, 1)$ be a strong left-paRM=left-pBH algebra. If (p*^{L}) and (p**^{L}) hold, then we have:*

$$(pEx1^L) \iff (pBB1^L).$$

Proof. By the definition of a strong left-paRM algebra, the properties (pM^L), (pRe^L), (IdEq^L), (pEq^L), (An), (Id^L) and (pEx1^L) hold. By Proposition 2.1.17, (pD1^L) holds too. By Proposition 2.1.18 (e), (pEx1^L) + (pD1^L) + (p*^{L}) $\Longrightarrow$ (pBB1^L); thus, (pBB1^L) holds.

Conversely, by Proposition 2.1.8 (1), (pM^L) + (pEq^L) $\Longrightarrow$ (pN^L) and by Proposition 2.1.8 (15'), (pN^L) + (pEq^L) + (p**^{L}) $\Longrightarrow$ (Tr). Thus, (Tr) holds. Hence, by Proposition 2.1.18 (g), (pBB1^L) + (pD1^L) + (p**^{L}) + (Tr) + (An) $\Longrightarrow$ (pEx1^L). Thus, if (p*^{L}) and (p**^{L}) hold, then we have: (pEx1^L) $\Longleftrightarrow$ (pBB1^L). □

Note that, by Corollary 2.1.19, in a strong left-p*aRM** algebra, we have: (pEx1^L) $\Longleftrightarrow$ (pBB1^L).

The dual results are omitted.

We shall see in Chapter 13 that p-semisimple pM, pRM, paRM and p*aRM** algebras coincide with the implicative-hubs, while the p-semisimple strong pM, pRM, paRM and p*aRM** algebras coincide with the strong implicative-hubs.

2.1.5 Hasse diagrams and Hasse-type diagrams

We introduce now the following definition (see [104] for the commutative case).

Definition 2.1.20 Let $\mathcal{A} = (A, \leq, \rightarrow, \rightsquigarrow, 1)$ be a left-(right-) pseudo-structure.

1) We shall say that $\mathcal{A}$ is *reflexive* if $\leq$ is reflexive (i.e. it satisfies property (Re)).

2) We shall say that $\mathcal{A}$ is *antisymmetric* if $\leq$ is antisymmetric (i.e. it satisfies property (An)).

3) We shall say that $\mathcal{A}$ is *transitive* if $\leq$ is transitive (i.e. it satisfies property (Tr)).

4) We shall say that $\mathcal{A}$ is *pre-ordered* if $\leq$ is a pre-order relation (i.e. it is reflexive and transitive).

5) We shall say that $\mathcal{A}$ is *ordered* if $\leq$ is a partial-order relation (i.e. it is reflexive, antisymmetrique and transitive).

6) We shall say that $\mathcal{A}$ is a *lattice* if $\leq$ is a lattice-order relation (i.e. it is a partial-order such that there exists sup(x,y) and inf(x,y) for each $x, y \in A$); we shall use the notation $x \vee y$ for sup(x,y) and $x \wedge y$ for inf(x,y), with $x \leq y \Leftrightarrow x \vee y = y \Leftrightarrow x \wedge y = x$.

Then, the natural binary relation $\leq$ is an *order* relation in left- pBZ, pBCC, pBCI and pBCK algebras and is only a *pre-order* in left- pre-pBZ, pre-pBCC, pre-pBCI and pre-pBCK algebras (see [104] for the commutative case).

Remark 2.1.21

(1) If the pseudo-algebra (pseudo-structure) is not ordered (i.e. it is neither reflexive nor antisymmetric nor transitive, or it is only reflexive, or reflexive and transitive, or reflexive and antisymmetric), then a *Hasse-type diagram* is used, where each element is represented by a circ $\circ$, and if $x \leq y$ and $y \leq x$ and $x \neq y$

(i.e. x and y have the *same height*, or are *parallel*)), then a horizontal line will connect them.

(2) If the pseudo-algebra (pseudo-structure) is ordered, then the usual *Hasse diagram* is used, where each element is represented by a bullet $\bullet$, and if $x \leq y$ and there is no z such that $x < z < y$, then x is reprezented below y and a line will connect them.

2.2 The hierarchies Nos. 1, 2, 3, 4 of pseudo-algebras of logic

A hierarchy of classes of pseudo-algebras will be represented by a kind of Hasse-type diagram, where the pseudo-algebras are represented as follows:
- a class of *reflexive* pseudo-algebras, by $\bigcirc$
- a class of *antisymmetrique* pseudo-algebras, by $\circ$
- a class of *transitive* pseudo-algebras, by $\bullet$
- a class of pseudo-algebras which does not verify (Re), (An), (Tr), by $\Box$.

Consequently, we shall represent:
- a class of pseudo-algebras where $\leq$ is *reflexive and antisymetrique*,
- a class of *pre-ordered* pseudo-algebras (i.e. $\leq$ is reflexive and transitive),
- a class of *ordered* pseudo-algebras, respectively by:

$$\circledcirc \qquad \odot \qquad \bullet$$

The left-pseudo-algebras of logic (left-PM algebras) defined in the previous section are connected as in the following hierarchies (without proofs) (see [104], Part I, for the commutative case) in Figures 2.1, 2.2, 2.3 and 2.4.

2.3 Pseudo-BCI and pseudo-BCK algebras

As the BCK algebras, the pseudo-BCK algebras were introduced as right-algebras with 0, i.e. as structures $(A^R, \leq, \star, \circ, 0)$.

The *reversed right-pseudo-BCK algebra* is obtained by reversing the operations $\star$ and $\circ$, i.e. by replacing $x \star y$ by $y \to^R x$ and $x \circ y$ by $y \rightsquigarrow^R x$, for all $x, y \in A^R$ [88]; hence, a reversed right-pseudo-BCK algebra is a structure $(A^R, \leq, \to^R, \rightsquigarrow^R, 0)$.

Dually, the left-pseudo-BCK algebra is obtained from the right-pseudo-BCK algebra by replacing the relation $\leq$ by the dual relation, $\geq$, $\star$ by $\Box$, $\circ$ by $\#$ and 0 by 1; hence, a left-pseudo-BCK algebra is a structure $(A^L, \geq, \Box, \#, 1)$.

The *reversed left-pseudo-BCK algebra* is obtained by reversing the operations $\Box$ and $\#$, i.e. by replacing $x \Box y$ by $y \to^L x$ and $x \# y$ by $y \rightsquigarrow^L x$, for all $x, y \in A^L$. Hence, a reversed left-pseudo-BCK algebra is a structure $(A^L, \geq, \to^L, \rightsquigarrow^L, 1)$.

As the pseudo-BCK algebras, the pseudo-BCI algebras were also introduced as right-algebras with 0.

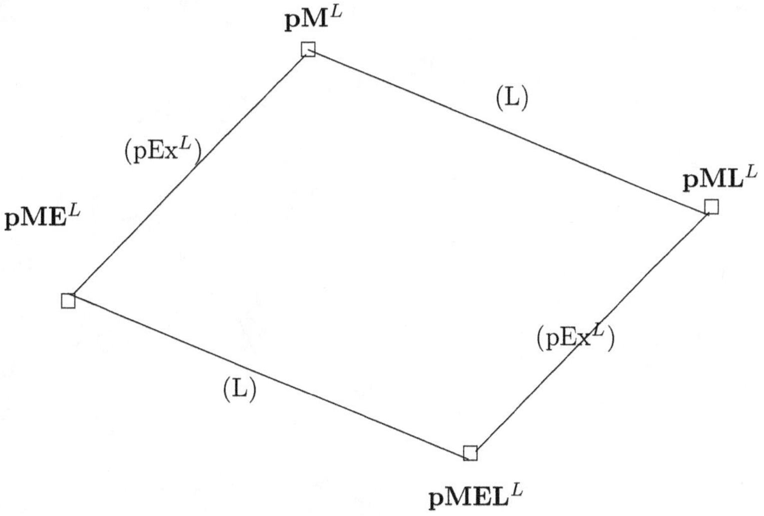

Figure 2.1: The hierarchy No. 1 of left-pseudo-algebras of logic

We shall use the following definitions as pseudo-structures.

Definition 2.3.1 (Definition 1 of pBCI algebras)

(i) A *reversed left-pseudo-BCI algebra*, or a *left-pBCI algebra* for short, is a structure

$$\mathcal{A}^L = (A^L, \leq, \to^L, \leadsto^L, 1),$$

where $\leq$ is a binary relation on A^L, $(\to^L, \leadsto^L)$ is a pair of binary operations on A^L (called *pseudo-implication*) and 1 is an element of A^L verifying the axioms: for all $x, y, z \in A^L$,

(pBB^L) $y \to^L z \leq (z \to^L x) \leadsto^L (y \to^L x)$,
$y \leadsto^L z \leq (z \leadsto^L x) \to^L (y \leadsto^L x)$,
(pD^L) $y \leq (y \to^L x) \leadsto^L x$, $y \leq (y \leadsto^L x) \to^L x$,
(Re) $x \leq x$ (Reflexivity),
(An) $x \leq y$ and $y \leq x \Longrightarrow x = y$ (Antisymmetry),
(IdEq^L) $x \to^L y = 1 \Longleftrightarrow x \leadsto^L y = 1$,
(pEq^L) $x \leq y \Longleftrightarrow x \to^L y = 1$ $(\overset{(IdEq^L)}{\Longleftrightarrow}$ $x \leadsto^L y = 1)$.

(i') Dually, a *reversed right-pseudo-BCI algebra*, or a *right-pBCI algebra* for short, is a structure

$$\mathcal{A}^R = (A^R, \geq, \to^R, \leadsto^R, 0),$$

where $\geq$ is a binary relation on A^R, $(\to^R, \leadsto^R)$ is a pair of binary operations on A^R (called *pseudo-coimplication*) and 0 is an element of A^R verifying the axioms: for all $x, y, z \in A^R$,

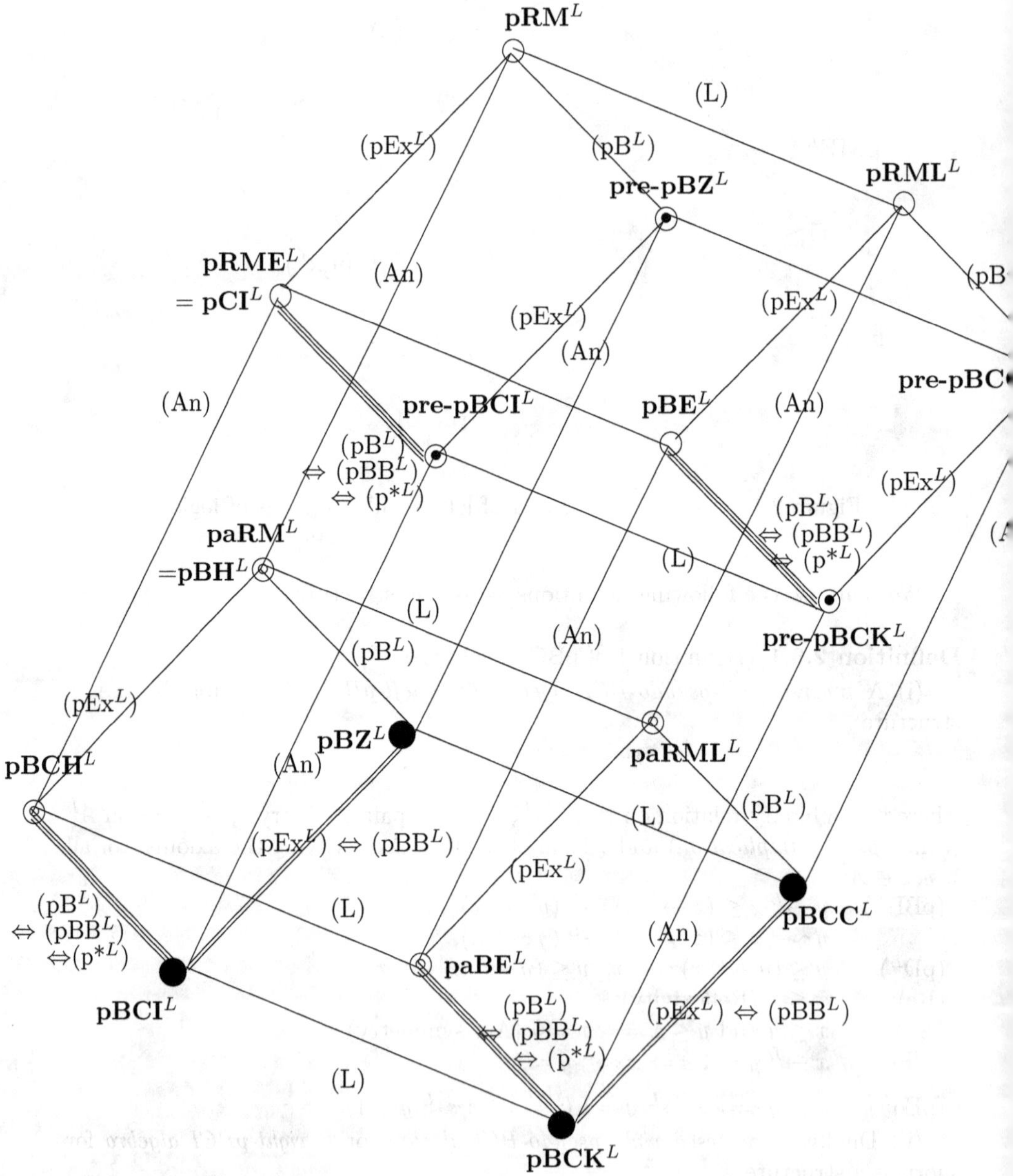

Figure 2.2: The hierarchy No. 2 of left-pseudo-algebras of logic

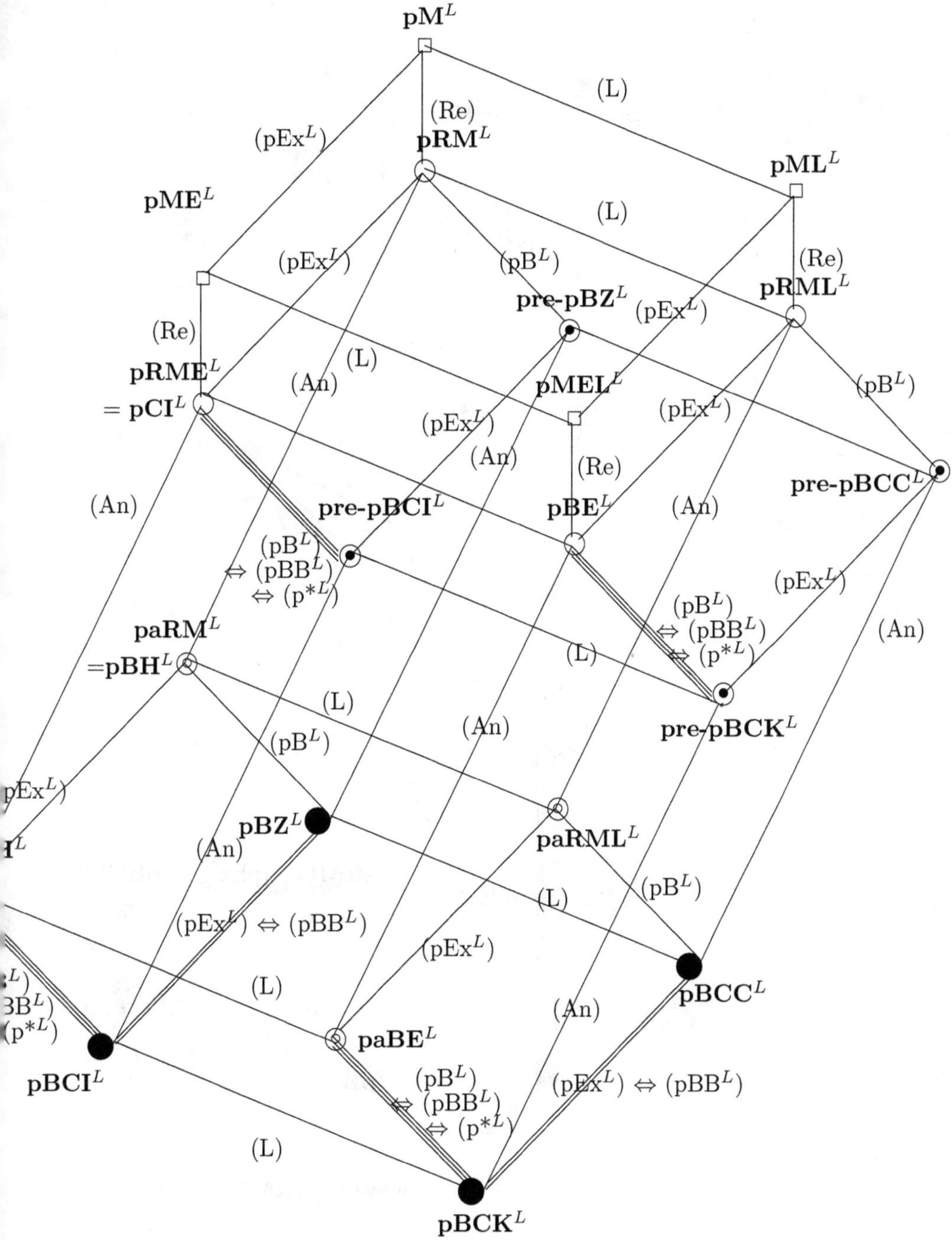

Figure 2.3: The hierarchy No. 3 (see No. 1 and No. 2) of left-pseudo-algebras of logic

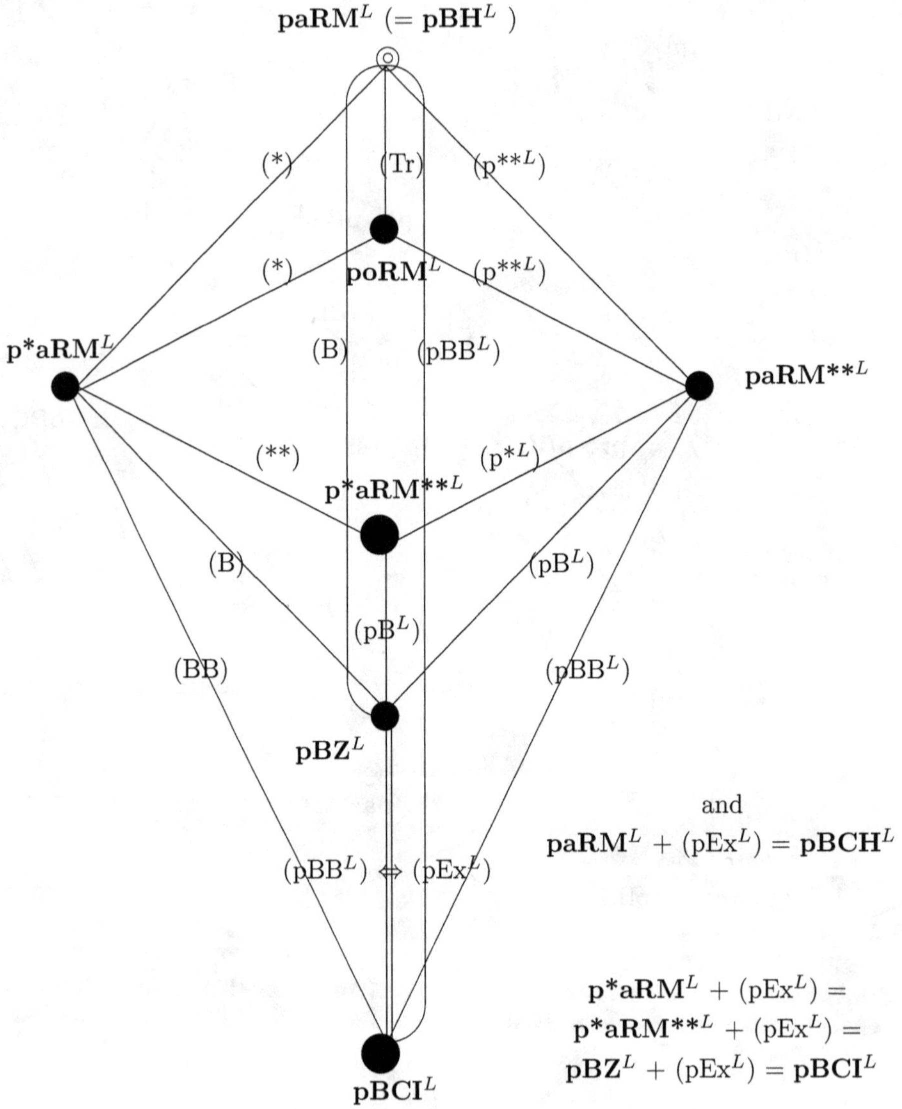

Figure 2.4: The detailed hierarchy No. 4 between **paRM**^{L} and **pBCI**^{L}

(pBB^R) $y \to^R z \geq (z \to^R x) \rightsquigarrow^R (y \to^R x)$,
$y \rightsquigarrow^R z \geq (z \rightsquigarrow^R x) \to^R (y \rightsquigarrow^R x)$,
(pD^R) $y \geq (y \to^R x) \rightsquigarrow^R x$, $y \geq (y \rightsquigarrow^R x) \to^R x$,
(Re) $x \geq x$ (Reflexivity),
(An) $x \geq y$ and $y \geq x \Longrightarrow x = y$ (Antisymmetry),
(IdEq^R) $x \to^R y = 0 \Longleftrightarrow x \rightsquigarrow^R y = 0$,
(pEq^R) $x \geq y \Longleftrightarrow x \to^R y = 0$ ($\overset{(IdEq^R)}{\Longleftrightarrow}$ $x \rightsquigarrow^R y = 0$).

Denote by $\mathbf{pBCI}^L$ the class of left-pBCI algebras and by $\mathbf{pBCI}^R$ the class of right-pBCI algebras.

Definition 2.3.2 (Definition 1 of pBCK algebras)
(i) A *reversed left-pseudo-BCK algebra*, or a *left-pBCK algebra* for short, is a structure [88], [94]

$$\mathcal{A}^L = (A^L, \leq, \to^L, \rightsquigarrow^L, 1),$$

where $\leq$ is a binary relation on A^L, $(\to^L, \rightsquigarrow^L)$ is a pair of binary operations on A^L (called *pseudo-implication*) and 1 is an element of A^L verifying (pBB^L), (pD^L), (Re), (An), (IdEq^L), (pEq^L) and
(L) $x \leq 1$, for all $x \in A^L$ (Last element).
(i') Dually, a *reversed right-pseudo-BCK algebra*, or a *right-pBCK algebra* for short, is a structure [60]

$$\mathcal{A}^R = (A^R, \geq, \to^R, \rightsquigarrow^R, 0)$$

verifying (pBB^R), (pD^R), (Re), (An), (IdEq^R), (pEq^R) and
(F) $x \geq 0$, for all $x \in A^R$ (First element).

Denote by $\mathbf{pBCK}^L$ the class of all left-pBCK algebras and by $\mathbf{pBCK}^R$ the class of all right-pBCK algebras.

Hence, a left-pBCK algebra is a left-pBCI algebra verifying (L) and, dually, a right-pBCK algebra is a right-pBCI algebra verifying (F), i.e.:

$\mathbf{pBCK}^L = \mathbf{pBCI}^L$ + (L) and, dually, $\mathbf{pBCK}^R = \mathbf{pBCI}^R$ + (F).

Note that we could say that a pBCK algebra is an "integral" pBCI algebra.
Note that we can define alternatively a left-pBCI (left-pBCK) algebra as pseudo-algebra, as follows.

Definition 2.3.3 (Definition 1' of pBCI (pBCK) algebras)
(i) A *left-pBCI (left-pBCK) algebra* for short is an algebra

$$\mathcal{A}^L = (A^L, \to^L, \rightsquigarrow^L, 1)$$

verifying: for all $x, y, z \in A^L$,

(pBB^{L}') $(y \to^L z) \rightsquigarrow^L ((z \to^L x) \rightsquigarrow^L (y \to^L x)) = 1$,
$(y \rightsquigarrow^L z) \to^L ((z \rightsquigarrow^L x) \to^L (y \rightsquigarrow^L x)) = 1$,
(pD^{L}') $y \rightsquigarrow^L ((y \to^L x) \rightsquigarrow^L x) = 1,\ y \to^L ((y \rightsquigarrow^L x) \to^L x) = 1$,
(pRe^{L}') $x \to^L x = 1$ ($\overset{(IdEq^L)}{\Longleftrightarrow}$ $x \rightsquigarrow^L x = 1$) (Reflexivity),
(pAn^{L}') $x \to^L y = 1$ ($= x \rightsquigarrow^L y$) and $y \to^L x = 1$ ($= y \rightsquigarrow^L x$) $\Longrightarrow x = y$ (Antisymmetry),
(IdEq^{L}) $x \to^L y = 1 \Longleftrightarrow x \rightsquigarrow^L y = 1$,
(and (pL') $x \to^L 1 = 1$ ($\overset{(IdEq^L)}{\Longleftrightarrow}$ $x \rightsquigarrow^L 1$) (Last element), respectively),
where a binary relation $\leq$ is then defined by: for all $x, y \in A^L$,

$$x \leq y \overset{def.}{\Longleftrightarrow} x \to^L y = 1 \ (\overset{(IdEq^L)}{\Longleftrightarrow} x \rightsquigarrow^L y = 1).$$

(i') Dually, a *right-pBCI (right-pBCK) algebra* for short is an algebra

$$\mathcal{A}^R = (A^R, \to^R, \rightsquigarrow^R, 0)$$

verifying the dual (pBB^{R}'), (pD^{R}'), (pRe^{R}'), (pAn^{R}'), (IdEq^{R}) (and (pF'), respectively), where a binary relation $\geq$ is then defined by: for all $x, y \in A^R$,

$$x \geq y \overset{def.}{\Longleftrightarrow} x \to^R y = 0 \ (\overset{(IdEq^R)}{\Longleftrightarrow} x \rightsquigarrow^R y = 0).$$

A left-pBCI (left-pBCK) algebra $\mathcal{A}^L = (A^L, \leq, \to^L, \rightsquigarrow^L, 1)$ is *commutative* if $\to^L = \rightsquigarrow^L$. A commutative left-pBCI (left-pBCK) algebra is a left-BCI (left-BCK) algebra.

Dually, a right-pBCI (right-pBCK) algebra $\mathcal{A}^R = (A^R, \geq, \to^R, \rightsquigarrow^R, 0)$ is *commutative* if $\to^R = \rightsquigarrow^R$. A commutative right-pBCI (right-pBCK) algebra is a right-BCI (right-BCK) algebra.

We have the following result (the dual case is omitted).

Proposition 2.3.4 *(see [94], Proposition 9.5.4, for the integral case)*
The following properties hold in a left-pBCI algebra (hence in a left-pBCK algebra): for all x, y, z,
(pM^{L}) $1 \to^L x = x = 1 \rightsquigarrow^L x$;
(pN^{L}) $1 \leq x \Longrightarrow x = 1$;
(Tr) (Transitivity) $x \leq y$ *and* $y \leq z \Longrightarrow x \leq z$;
(pEx^{L}) (pseudo-Exchange) $z \rightsquigarrow^L (y \to^L x) = y \to^L (z \rightsquigarrow^L x)$;
(pC^{L}) $x \to^L (y \rightsquigarrow^L z) \leq y \rightsquigarrow^L (x \to^L z)$ *and*
$y \rightsquigarrow^L (x \to^L z) \leq x \to^L (y \rightsquigarrow^L z)$,
(pEq#^{L}) $x \leq y \to^L z \Leftrightarrow y \leq x \rightsquigarrow^L z$;
(pB^{L}) $z \to^L x \leq (y \to^L z) \to^L (y \to^L x)$,
$z \rightsquigarrow^L x \leq (y \rightsquigarrow^L z) \rightsquigarrow^L (y \rightsquigarrow^L x)$;
(p^{L})* $x \leq y \Longrightarrow z \to^L x \leq z \to^L y,\quad z \rightsquigarrow^L x \leq z \rightsquigarrow^L y$;
*(p**^{L})* $x \leq y \Longrightarrow y \to^L z \leq x \to^L z,\quad y \rightsquigarrow^L z \leq x \rightsquigarrow^L z$.

Note that, in a left-pBCI algebra (left-pBCK algebra)
$\mathcal{A}^L = (A^L, \leq, \to^L, \rightsquigarrow^L, 1)$, the binary relation $\leq$ is a partial order relation, since

the properties (Re), (An), (Tr) hold.

Consider the following property: for all x, y,
(pK^L) $x \le y \to^L x$ and $x \le y \rightsquigarrow^L x$.

Remark 2.3.5 Since the property (L) holds in a left-pBCK algebra, but does not hold in a left-pBCI algebra, it follows, by Proposition 2.1.8, that the property (pK^L) holds in a left-pBCK algebra, but does not hold in a left-pBCI algebra. Similarly in the dual case.

Remarks 2.3.6 (See [104]) Consider the above property (pN^L) ($1 \le x \Longrightarrow x = 1$). Note that this property says that 1 **is the maximal element** of the left-pBCI algebra. Hence,
- if $\mathcal{A}^L$ is a left-pBCI algebra, then $(A^L, \le)$ is a **poset with maximal element** 1, while
- if $\mathcal{A}^L$ is a left-pBCK algebra, then $(A^L, \le, 1)$ is a **poset with greatest element** 1.

The following theorem is presented in [94] for pseudo-BCK algebras. We shall prove that it is valid for pseudo-BCI algebras too. It gives us an equivalent definition of a left-pBCI algebra (and hence of a left-pBCK algebra) (the right-case is omitted).

Theorem 2.3.7 *(See [94], Theorem 9.5.5, for the integral case)*
(1) Let $\mathcal{A}^L = (A^L, \le, \to^L, \rightsquigarrow^L, 1)$ be a structure such that $(A^L, \le)$ is a poset (i.e. (Re), (An), (Tr) hold) and the following properties hold: (pBB^L), (pM^L), (IdEq^L), (pEq^L) (i.e. the pair $(\to^L, \rightsquigarrow^L)$ is a pseudo-residuum). Then, $\mathcal{A}^L$ is a left-pBCI algebra.
(1') Conversely, every left-pBCI algebra $(A^L, \le, \to^L, \rightsquigarrow^L, 1)$ is a poset satisfying (pBB^L), (pM^L), (IdEq^L), (pEq^L).

Proof. (1): It remains to prove that (pD^L) holds; indeed, by (pBB^L), for all $x, y, z \in A$, $z \to^l y \le (y \to^L x) \rightsquigarrow^L (z \to^L x)$ and $z \rightsquigarrow^L y \le (y \rightsquigarrow^L x) \to^L (z \rightsquigarrow^L x)$; take then $z = 1$ and apply (pM^L); we get $y \le (y \to^L x) \rightsquigarrow^L x$ and $y \le (y \rightsquigarrow^L x) \to^L x$. Thus, (pD^L) holds.
(1'): follows by Definition 2.3.1 and by Proposition 2.3.4. □

By above theorem, we obtain the following equivalent definitions (the dual case is omitted).

Definition 2.3.8 (Definition 2 of pBCI (pBCK) algebras)
(1) A *left-pBCI algebra* is a structure $\mathcal{A}^L = (A^L, \le, \to^L, \rightsquigarrow^L, 1)$ s.t. $(A^L, \le)$ is a poset (with maximal element 1) and the properties (pBB^L), (pM^L), (IdEq^L), (pEq^L) hold.
(2) A *left-pBCK algebra* is a structure $\mathcal{A}^L = (A^L, \le, \to^L, \rightsquigarrow^L, 1)$ such that $(A^L, \le, 1)$ is a poset with greatest element 1 and the properties (pBB^L), (pM^L), (IdEq^L), (pEq^L) hold.

Remark 2.3.9

The statement: "The structure $(A^L, \leq, \rightarrow^L, \rightsquigarrow^L, 1)$ is a left-pBCI algebra" is equivalent to the statement: "The pseudo-implication $(\rightarrow^L, \rightsquigarrow^L)$ is a pseudo-residuum on the poset $(A^L, \leq)$ (with maximal element 1)", by Definition 1.1.3.

Dually, the statement: "The structure $(A^R, \geq, \rightarrow^R, \rightsquigarrow^R, 0)$ is a right-pBCI algebra" is equivalent to the statement: "The pseudo-coimplication $(\rightarrow^R, \rightsquigarrow^R)$ is a pseudo-coresiduum on the poset $(A^R, \leq)$ (with minimal element 0)", by Definition 1.1.3.

The following theorem gives us another equivalent definition of a left-pBCI algebra (and hence of a left-pBCK algebra) (the right-case is omitted).

Theorem 2.3.10

(1) Let $\mathcal{A}^L = (A^L, \leq, \rightarrow^L, \rightsquigarrow^L, 1)$ be a structure such that $(A^L, \leq)$ is a poset (i.e. (Re), (An), (Tr) hold) and the following properties hold: (pM^L), (pEx^L), (p^L), (IdEq^L), (pEq^L). Then, $\mathcal{A}^L$ is a left-pBCI algebra.*

(1') Conversely, every left-pBCI algebra $(A^L, \leq, \rightarrow^L, \rightsquigarrow^L, 1)$ is a poset satisfying (pM^L), (pEx^L), (p^L), (IdEq^L), (pEq^L).*

Proof. (1): $\mathcal{A}$ is a left-pME algebra verifying (Re), (An), hence is a left-pBCH algebra verifying (p*^{L}), hence is a left-pBCI algebra, by Theorem 2.1.9.

(1'): Every left-pBCI algebra verifies (pM^L), (pEx^L), (p*^{L}), (IdEq^L), (pEq^L) and (Re), (An), (Tr), by Definition 2.3.1 and Proposition 2.3.4. □

By above Theorem 2.3.10, we obtain the following equivalent definitions, that will be used in the sequel (the dual case is omitted).

Definition 2.3.11 (Definition 3 of pBCI (pBCK) algebras)

(1) A *left-pBCI algebra* is a structure $\mathcal{A}^L = (A^L, \leq, \rightarrow^L, \rightsquigarrow^L, 1)$ s.t. $(A^L, \leq)$ is a poset (with maximal element 1) and the properties (pM^L), (pEx^L), (p*^{L}), (IdEq^L), (pEq^L) hold.

(2) A *left-pBCK algebra* is a structure $\mathcal{A}^L = (A^L, \leq, \rightarrow^L, \rightsquigarrow^L, 1)$ such that $(A^L, \leq, 1)$ is a poset with greatest element 1 and the properties (pM^L), (pEx^L), (p*^{L}), (IdEq^L), (pEq^L) hold.

To introduce the following equivalent definition of pBCI algebras, comming from the non-commutative BCI logic, we need the following result.

Theorem 2.3.12 *(See [104], Propositions 9, 10, Theorem 5 - in commutative case)*

(1) Let $\mathcal{A}^L = (A^L, \leq, \rightarrow^L, \rightsquigarrow^L, 1)$ be a structure such that (pB^L), (pC^L), (Re) and (An), (IdEq^L) and (pEq^L) hold. Then, $\mathcal{A}^L$ is a left-pBCI algebra.

(1') Conversely, every left-pBCI algebra verifies (pB^L), (pC^L), (Re) and (An), (IdEq^L), (pEq^L).

Proof. (1): Following the Definition 1 of left-pBCI algebras, it remains to prove that (pBB^L) and (pD^L) hold. Indeed, by Proposition 2.1.8, (pC^L) + (An) $\Longrightarrow$

(pEx^L), (pEx^L) + (pB^L) + (pEq^L) $\Longrightarrow$ (pBB^L) and (Re) + (pEx^L) + (pEq^L) $\Longrightarrow$ (pD^L).

(1'): It remains to prove that every left-pBCI verifies (pB^L). Indeed, by Proposition 2.3.4, (pB^L) holds. □

By above theorem, we obtain the following equivalent definition of left-pBCI algebras comming from the non-commutative BCI logic. Note that the name "BCI" for pBCI (BCI) algebras comes from the names ("B", "C", "I") given to the logical axioms of BCI logic, where the name (Re) (comming from "Reflexivity") is used in the monograph, instead of (I) (Identity), for the corresponding axiom.

Definition 2.3.13 (Definition 4 of pBCI algebras)
A *left-pBCI algebra* is a structure $\mathcal{A}^L = (A^L, \leq, \rightarrow^L, \rightsquigarrow^L, 1)$ such that the properties (pB^L), (pC^L), (Re) and (An), (IdEq^L), (pEq^L) hold.

To introduce the following equivalent definition of pBCK algebras, comming from the non-commutative BCK logic, we need the following result.

Theorem 2.3.14 *(See [104], Propositions 11, 12, Theorem 6 - in commutative case)*
(1) Let $\mathcal{A}^L = (A^L, \leq, \rightarrow^L, \rightsquigarrow^L, 1)$ be a structure such that (pB^L), (pC^L), (pK^L) and (An), (IdEq^L) and (pEq^L) hold. Then, $\mathcal{A}^L$ is a left-pBCK algebra.
(1') Conversely, every left-pBCK algebra verifies (pB^L), (pC^L), (pK^L) and (An), (IdEq^L), (pEq^L).

Proof. (1): Following the Definition 1 of left-pBCK algebras, the properties (pBB^L), (pD^L), (Re), (L), (An), (IdEq^L), (pEq^L) hold; hence, it remains to prove that (pBB^L), (pD^L), (Re), (L) hold. Indeed, by Proposition 2.1.8:
- (pC^L) + (An) $\Longrightarrow$ (pEx^L) and (pEx^L) + (pB^L) + (pEq^L) $\Longrightarrow$ (pBB^L); thus, (pBB^L) holds;
- (pK^L) + (An) + (pEq^L) $\Longrightarrow$ (pN^L) and (pK^L) + (pN^L) + (pEq^L) $\Longrightarrow$ (L); thus, (L) holds;
- (pK^L) + (pC^L) + (An) + (pEq^L) $\Longrightarrow$ (Re); thus (Re) holds;
- (Re) + (pEx^L) + (pEq^L) $\Longrightarrow$ (pD^L); thus, (pD^L) holds.

(1'): It remains to prove that every left-pBCK verifies (pB^L), (pC^L) and (pK^L). Indeed, by Proposition 2.3.4, (pB^L) and (pC^L) hold, and by Proposition 2.1.8, (L) + (pEx^L) + (Re) + (pEq^L) $\Longrightarrow$ (pK^L). □

By above theorem, we obtain the following equivalent definition of left-pBCK algebras comming from the non-commutative BCK logic. Note that the name "BCK" for pBCK (BCK) algebras comes from the names ("B", "C", "K") given to the logical axioms of BCK logic.

Definition 2.3.15 (Definition 4 of pBCK algebras)
A *left-pBCK algebra* is a structure $\mathcal{A}^L = (A^L, \leq, \rightarrow^L, \rightsquigarrow^L, 1)$ such that the properties (pB^L), (pC^L), (pK^L) and (An), (IdEq^L), (pEq^L) hold.

Theorem 2.3.16 *[51] Any pseudo-BCK algebra can be extended to a pseudo-BCI algebra.*

Proof. Let $(A, \leq, \to, \rightsquigarrow, 1)$ be a pseudo-BCK algebra and $\delta \notin A$. Define on $A' = A \cup \{\delta\}$ the operations: for all $x, y \in A'$,

$$x \to' y = \begin{cases} x \to y, & \text{if } x, y \in A, \\ \delta, & \text{if } x = \delta,\ y \in A \text{ or } x \in A,\ y = \delta, \\ 1, & \text{if } x = y = \delta, \end{cases}$$

$$x \rightsquigarrow' y = \begin{cases} x \rightsquigarrow y, & \text{if } x, y \in A, \\ \delta, & \text{if } x = \delta,\ y \in A \text{ or } x \in A,\ y = \delta, \\ 1, & \text{if } x = y = \delta. \end{cases}$$

Then, by routine calculations, $(A', \leq, \to', \rightsquigarrow', 1)$ is only a pseudo-BCI algebra, since $\delta \not\leq 1$. □

2.3.1 The negative and positive cones of pBCI algebras

Let $\mathcal{A}^L = (A^L, \leq, \to^L, \rightsquigarrow^L, 1)$ be a left-pBCI algebra. Define the "negative (left) cone" and the "positive (right) cone" of $\mathcal{A}^L$ as follows, respectively:

$$A^{-L} \stackrel{def.}{=} \{x \in A^L \mid x \leq 1\} \quad \textit{and} \quad A^{+L} \stackrel{def.}{=} \{x \in A^L \mid x \geq 1\}.$$

Then, we obtain:

Lemma 2.3.17 *We have:*
(1) A^{-L} is closed under $\to^L$, $\rightsquigarrow^L$ and
(2) $A^{+L} = \{1\}$.

Proof. (1): Let $x, y \in A^{-L}$, i.e. $x \leq 1$ and $y \leq 1$; then, $x \to^L y \stackrel{(p*^L)}{\leq} x \to^L 1 = 1$ and $x \rightsquigarrow^L y \stackrel{(p*^L)}{\leq} x \rightsquigarrow^L 1 = 1$, by (pEq^L); hence, $x \to^L y$, $x \rightsquigarrow^L y \in A^{-L}$.
(2): Obviously, by (pN^L). □

Dually, let $\mathcal{A}^R = (A^R, \geq, \to^R, \rightsquigarrow^R, 0)$ be a right-pBCI algebra. Define the "positive (right) cone" and the "negative (left) cone" of $\mathcal{A}^R$ as follows, respectively:

$$A^{+R} \stackrel{def.}{=} \{x \in A^R \mid x \geq 0\} \quad \textit{and} \quad A^{-R} \stackrel{def.}{=} \{x \in A^R \mid x \leq 0\}.$$

Then, we obtain, dually:

Lemma 2.3.18 *We have:*
(1') A^{+R} is closed under $\to^R$, $\rightsquigarrow^R$ and
(2') $A^{-R} = \{0\}$.

Then, by Lemmas 2.3.17 and 2.3.18, we obtain immediately the following result.

Proposition 2.3.19
(1) Let $\mathcal{A}^L = (A^L, \leq, \to^L, \rightsquigarrow^L, 1)$ be a left-pBCI algebra.
Then

$$\mathcal{A}^{-L} = (A^{-L}, \leq, \to^L, \rightsquigarrow^L, 1)$$

is a left-pBCK algebra.

(1') Dually, let $\mathcal{A}^R = (A^R, \geq, \to^R, \rightsquigarrow^R, 0)$ be a right-pBCI algebra. Then

$$\mathcal{A}^{+R} = (A^{+R}, \geq, \to^R, \rightsquigarrow^R, 0)$$

is a right-pBCK algebra.

2.3.2 One negation on pBCI algebras

First, we present the following Proposition.

Proposition 2.3.20 *Let $\mathcal{A}^L = (A^L, \leq, \to^L, \rightsquigarrow^L, 1)$ be a left-pBCI algebra. Then we have: for all $x, y \in A^L$,*
(Id^L) $x \to^L 1 = x \rightsquigarrow^L 1$ [50],
(1^L) $x \to^L y \leq (y \to^L x) \rightsquigarrow^L 1$, $x \rightsquigarrow^L y \leq (y \rightsquigarrow^L x) \to^L 1$ [50],
(2^L) $(y \to^L x) \to^L 1 = (y \to^L 1) \rightsquigarrow^L (x \rightsquigarrow^L 1)$,
$(y \rightsquigarrow^L x) \rightsquigarrow^L 1 = (y \rightsquigarrow^L 1) \to^L (x \to^L 1)$ [50],
(3^L) $y \to^L x \leq (x \to^L 1) \rightsquigarrow^L (y \to^L 1)$, $y \rightsquigarrow^L x \leq (x \rightsquigarrow^L 1) \to^L (y \rightsquigarrow^L 1)$,
(4^L) $x \leq (x \to^L 1) \rightsquigarrow^L 1$, $x \leq (x \rightsquigarrow^L 1) \to^L 1$.

Let $\mathcal{A}^L = (A^L, \leq, \to^L, \rightsquigarrow^L, 1)$ be a left-pBCI algebra. We can define two negations $^- = ^{-^L}$ and $^\sim = ^{\sim^L}$ by: for all $x \in A^L$,

$$x^- \stackrel{def.}{=} x \to^L 1, \quad x^\sim \stackrel{def.}{=} = x \rightsquigarrow^L 1.$$

But, since $x \to^L 1 \stackrel{(Id^L)}{=} x \rightsquigarrow^L 1$, it follows that $^- = ^\sim$, i.e. there exists in fact only one negation, denoted by $^{-1}$:

$$x^{-1} \stackrel{def.}{=} x \to^L 1 \stackrel{(Id^L)}{=} x \rightsquigarrow^L 1. \tag{2.3}$$

Then we obtain the following properties:

Proposition 2.3.21 *(See Proposition 2.1.17)*
($1'^L$) $x \to^L y \leq (y \to^L x)^{-1}$, $x \rightsquigarrow^L y \leq (y \rightsquigarrow^L x)^{-1}$,
($2'^L$) $(y \to^L x)^{-1} = y^{-1} \rightsquigarrow^L x^{-1}$, $(y \rightsquigarrow^L x)^{-1} = y^{-1} \to^L x^{-1}$,
($3'^L$) $y \to^L x \leq x^{-1} \rightsquigarrow^L y^{-1}$, $y \rightsquigarrow^L x \leq x^{-1} \to^L y^{-1}$;
($pDN^{\leq}$) $x \leq (x^{-1})^{-1}$,
($N1^L$) $1^{-1} = 1$,
($pNg1^L$) $x \to^L y^{-1} = y \rightsquigarrow^L x^{-1}$.

Proof. ($1'^L$) - ($3'^L$), ($pDN^{\leq}$): By Proposition 2.3.20.

($N1^L$): $1^{-1} \stackrel{def.}{=} 1 \to^L 1 \stackrel{(pM^L)}{=} 1$.

($pNg1^L$): $x \to^L y^{-1} \stackrel{def.}{=} x \to^L (y \rightsquigarrow^L 1) \stackrel{(pEx^L)}{=} y \rightsquigarrow^L (x \to^L 1) \stackrel{def.}{=} y \rightsquigarrow^L x^{-1}$.
□

Note that we could define alternatively a left-pBCI algebra as a structure

$$(A^L, \leq, \to^L, \rightsquigarrow^L, ^{-1}, 1),$$

where the negation $^{-1}$ would be expressed in terms of the other operations, $(\to^L, \leadsto^L)$, 1: for all $x \in A^L$, $x^{-1} = x \to^L 1 \overset{(Id^L)}{=} x \leadsto^L 1$.

Remark 2.3.22 Let $\mathcal{A}^L = (A^L, \leq, \to^L, \leadsto^L, 1)$ be a left-pBCK algebra. Then, for all $x \in A^L$, $x^{-1} = 1$; it follows that this type of negation has no sense in left-pBCK algebras.

The dual of above Propositions 2.3.20, 2.3.21 are omitted.

2.3.3 Bounded pBCK algebras. Two negations

Definition 2.3.23 (Definition 1 of bounded pBCK algebras)

(i) Let $\mathcal{A}^L = (A^L, \leq, \to^L, \leadsto^L, 1)$ be a left-pBCK algebra. If there is an element $0 \in A^L$ satisfying (F) ($0 \leq x$, for all $x \in A^L$), then 0 is called the *first element* of $\mathcal{A}^L$. A left-pBCK algebra with 0 is called to be *bounded* and it is denoted by:

$$\mathcal{A}^{Lb} = (A^L, \leq, \to^L, \leadsto^L, 0, 1).$$

(i') Dually, let $\mathcal{A}^R = (A^R, \geq, \to^R, \leadsto^R, 0)$ be a right-pBCK algebra. If there is an element $1 \in A^R$ satisfying (L) ($1 \geq x$, for all $x \in A^R$), then 1 is called the *last element* of $\mathcal{A}^R$. A right-pBCK algebra with 1 is called to be *bounded* and it is denoted by:

$$\mathcal{A}^{Rb} = (A^R, \geq, \to^R, \leadsto^R, 0, 1).$$

Denote by $\mathbf{pBCK}^{Lb}$ the class of all bounded left-pBCK algebras and, dually, denote by $\mathbf{pBCK}^{Rb}$ the class of all bounded right-pBCK algebras.

Note that, by Definition 2.3.23, we have:

$\mathbf{pBCK}^{Lb} \subseteq \mathbf{pBCK}^L$ and $\mathbf{pBCK}^{Rb} \subseteq \mathbf{pBCK}^R$.

Note that another definition of bounded pBCK algebras is the following.

Definition 2.3.24 (Definition 2 of bounded pBCK algebras)

(i) A *bounded left-pBCK algebra* is a structure $(A^L, \leq, \to^L, \leadsto^L, 0, 1)$ such that $(A^L, \leq, \to^L, \leadsto^L, 1)$ is a left-pBCK algebra and $0 \in A^L$ verifies (F) ($0 \leq x$, for all $x \in A^L$).

(i') Dually, a *bounded right-pBCK algebra* is a structure $(A^R, \geq, \to^R, \leadsto^R, 0, 1)$ such that $(A^R, \geq, \to^R, \leadsto^R, 0)$ is a right-pBCK algebra and $1 \in A^R$ satisfies (L) ($1 \geq x$, for all $x \in A^R$).

Let $\mathcal{A}^{Lb} = (A^L, \leq, \to^L, \leadsto^L, 0, 1)$ be a bounded left-pBCK algebra. Define, for all $x \in A^L$, two *negations*, $^- = ^{-^L}$ and $^\sim = ^{\sim^L}$, by: for all $x \in A^L$,

$$x^- \overset{def.}{=} x \to^L 0, \quad x^\sim \overset{def.}{=} x \leadsto^L 0.$$

Dually, let $\mathcal{A}^{Rb} = (A^R, \geq, \to^R, \rightsquigarrow^R, 0, 1)$ be a bounded right-pBCK algebra. Define, for all $x \in A^R$, two *negations*, $^- = {^-}^R$ and $^\sim = {^\sim}^R$, by: for all $x \in A^R$,

$$x^- \stackrel{def.}{=} x \to^R 1, \quad x^\sim \stackrel{def.}{=} x \rightsquigarrow^R 1.$$

Then, the following properties hold (the dual case is omitted).

Proposition 2.3.25 *[88] In a bounded left-pBCK algebra $\mathcal{A}^L$ we have: for all $x, y \in A$,*

(pN1^{Lb}) $\quad 0 = 1^- = 1^\sim,\ 1 = 0^- = 0^\sim,$

(pNg^{Lb}) $\quad x \leq y \Rightarrow y^- \leq x^-,\ y^\sim \leq x^\sim,$

(2.4) $\quad x \leq (x^-)^\sim, \quad x \leq (x^\sim)^-$

(2.5) $\quad ((x^-)^\sim)^- = x^-, \quad ((x^\sim)^-)^\sim = x^\sim.$

(2.6) $\quad x \to^L y \leq y^- \rightsquigarrow^L x^-, \quad x \rightsquigarrow^L y \leq y^\sim \to^L x^\sim,$

(pNg1^{Lb}) $\quad x \to^L y^\sim = y \rightsquigarrow^L x^-.$

2.4 Pseudo-BCI and pseudo-BCK lattices

If the order relation $\leq$ is a lattice order, we denote by $\wedge$ and $\vee$ the Dedekind lattice operations: for all x, y, $x \leq y \Leftrightarrow x \wedge y = x \Leftrightarrow x \vee y = y$.

Note that a lattice is self-dual. We shall say that $(A, \wedge, \vee)$ is the *left-lattice*, while $(A, \vee, \wedge)$ is the *right-lattice.*

Definition 2.4.1

(i) A *left-pBCI lattice* (*left-pBCK lattice*) is a left-pBCI algebra (left-pBCK algebra)

$$\mathcal{A}^L = (A^L, \leq, \to^L, \rightsquigarrow^L, 1)$$

where $\leq$ is a lattice order relation. It will be denoted by

$$(A^L, \wedge, \vee, \to^L, \rightsquigarrow^L, 1).$$

(i') Dually, a *right-pBCI lattice* (*right-pBCK lattice*) is a right-pBCI algebra (right-pBCK algebra)

$$\mathcal{A}^R = (A^R, \geq, \to^R, \rightsquigarrow^R, 0)$$

where $\geq$ is a lattice order relation. It will be denoted by

$$(A^R, \vee, \wedge, \to^R, \rightsquigarrow^R, 0).$$

Denote by **pBCI-L**^{L} (**pBCK-L**^{L}) the class of all left-pBCI lattices (left-pBCK lattices, respectively) and, dually, denote by **pBCI-L**^{R} (**pBCK-L**^{R}) the class of all right-pBCI lattices (right-pBCK lattices, respectively).

Lemma 2.4.2 *([94], Lemma 10.1.13)*
Let $(A^L, \wedge, \vee, \to^L, \rightsquigarrow^L, 1)$ *be a left-pBCK lattice. Then, for any* $x, y, z \in A^L$, *we have:*
(i) $z \to^L (x \wedge y) \leq (z \to^L x) \wedge (z \to^L y)$, $z \rightsquigarrow^L (x \wedge y) \leq (z \rightsquigarrow^L x) \wedge (z \rightsquigarrow^L y)$;
(ii) $z \to^L (x \vee y) \geq (z \to^L x) \vee (z \to^L y)$, $z \rightsquigarrow^L (x \vee y) \geq (z \rightsquigarrow^L x) \vee (z \rightsquigarrow^L y)$;
(iii) $(x \wedge y) \to^L z \geq (x \to^L z) \vee (y \to^L z)$, $(x \wedge y) \rightsquigarrow^L z \geq (x \rightsquigarrow^L z) \vee (y \rightsquigarrow^L z)$.

2.4.1 The negative and positive cones of pBCI lattices

By Proposition 2.3.19, we obtain the following result.

Proposition 2.4.3
(1) Let $\mathcal{A}^L = (A^L, \wedge, \vee, \to^L, \rightsquigarrow^L, 1)$ *be a left-pBCI lattice and* $A^{-L} = \{x \in A^L \mid x \leq 1\}$.
Then

$$\mathcal{A}^{-L} = (A^{-L}, \wedge, \vee, \to^L, \rightsquigarrow^L, 1)$$

is a left-pBCK lattice.
(1') Dually, let $\mathcal{A}^R = (A^R, \vee, \wedge, \to^R, \rightsquigarrow^R, 0)$ *be a right-pBCI lattice and* $A^{+R} = \{x \in A^R \mid x \geq 0\}$.
Then

$$\mathcal{A}^{+R} = (A^{+R}, \vee, \wedge, \to^R, \rightsquigarrow^R, 0)$$

is a right-pBCK lattice.

2.4.2 Bounded pBCK lattices. Two negations

Definition 2.4.4
(i) A left-pBCK lattice $\mathcal{A}^L = (A^L, \wedge, \vee, \to^L, \rightsquigarrow^L, 1)$ is said to be *bounded*, if the left-pBCK algebra $(A^L, \leq, \to^L, \rightsquigarrow^L, 1)$ is bounded ($x \leq y \Leftrightarrow x \wedge y = x \Leftrightarrow x \vee y = y$), with 0 as the first (smallest) element. A *bounded left-pBCK lattice* is denoted by

$$\mathcal{A}^{Lb} = (A^L, \wedge, \vee, \to^L, \rightsquigarrow^L, 0, 1).$$

(i') Dually, a right-pBCK lattice $\mathcal{A}^R = (A^R, \vee, \wedge, \to^R, \rightsquigarrow^R, 0)$ is said to be *bounded*, if the right-pBCK algebra $(A^R, \geq, \to^R, \rightsquigarrow^R, 0)$ is bounded ($x \leq y \Leftrightarrow x \vee y = y \Leftrightarrow x \wedge y = x$, $x \leq y \Leftrightarrow y \geq x$), with 1 as the last (greatest) element. A *bounded right-pBCK lattice* is denoted by:

$$\mathcal{A}^{Rb} = (A^R, \vee, \wedge, \to^R, \rightsquigarrow^R, 0, 1).$$

Denote by **pBCK-L**^{Lb} the class of all bounded left-pBCK lattices and, dually, denote by **pBCK-L**^{Rb} the class of all bounded right-pBCK lattices.

Let $\mathcal{A}^{Lb} = (A^L, \wedge, \vee, \rightarrow^L, \rightsquigarrow^L, 0, 1)$ be a bounded left-pBCK lattice. Define, for all $x \in A^L$, two *negations* by: for all $x \in A^L$,

$$x^- = x^{-^L} \stackrel{def.}{=} x \rightarrow^L 0, \quad x^\sim = x^{\sim^L} \stackrel{def.}{=} x \rightsquigarrow^L 0.$$

Dually, let $\mathcal{A}^{Rb} = (A^R, \vee, \wedge, \rightarrow^R, \rightsquigarrow^R, 0, 1)$ be a bounded right-pBCK lattice. Define, for all $x \in A^R$, two *negations* by: for all $x \in A^R$,

$$x^- = x^{-^R} \stackrel{def.}{=} x \rightarrow^R 1, \quad x^\sim = x^{\sim^R} \stackrel{def.}{=} x \rightsquigarrow^R 1.$$

2.5 Pseudo-BCI(pP) and pseudo-BCK(pP) algebras

Definition 2.5.1

(i) A *left-pBCI algebra (left-pBCK algebra) with property (pP) (i.e. with pseudo-product)*, or a *pBCI(pP) algebra (pBCK(pP) algebra)* for short, is a left-pBCI algebra (left-pBCK algebra, respectively)

$$\mathcal{A}^L = (A^L, \leq, \rightarrow^L, \rightsquigarrow^L, 1)$$

verifying additionally the property (pP) holds: for all $x, y \in A^L$, there exist $\min\{z \in A^L \mid x \leq y \rightarrow^L z\}$ and $\min\{z \in A^L \mid y \leq x \rightsquigarrow^L z\}$, they are equal, and the element so obtained is denoted $x \odot y$; for short,

(pP) (pseudo-product) $\exists\, x \odot y \stackrel{notation}{=} \min\{z \mid x \leq y \rightarrow^L z\} = \min\{z \mid y \leq x \rightsquigarrow^L z\}.$

(i') Dually, a *right-pBCI algebra (right-pBCK algebra) with property (pS) (i.e. with pseudo-sum)*, or a *pBCI(pS) algebra (pBCK(pS) algebra)* for short, is a right-pBCI algebra (right-pBCK algebra, respectively)

$$\mathcal{A}^R = (A^R, \geq, \rightarrow^R, \rightsquigarrow^R, 0)$$

verifying additionally the property (pS) holds: for all $x, y \in A^R$,

(pS) (pseudo-sum) $\exists\, x \oplus y \stackrel{notation}{=} \max\{z \mid x \geq y \rightarrow^R z\} = \max\{z \mid y \geq x \rightsquigarrow^R z\}.$

Denote by **pBCI(pP)** the class of all pBCI(pP) algebras and by **pBCI(pS)** the class of all pBCI(pS) algebras.

Denote by **pBCK(pP)** the class of all pBCK(pP) algebras and by **pBCK(pS)** the class of all pBCK(pS) algebras.

Hence, we have:

pBCK(pP) = **pBCI(pP)** + (L) and, dually, **pBCK(pS)** = **pBCI(pS)** + (F).

Note that, by Definition 2.5.1 (i), in a pBCI(pP) algebra (pBCK(pP) algebra) $\mathcal{A}^L$ we have, for all $x, y \in A^L$,

$$x \odot y \in \{z \in A^L \mid x \leq y \rightarrow^L z\}, \quad x \odot y \in \{z \in A^L \mid y \leq x \rightsquigarrow^L z\},$$

hence we have the following property, for all $x, y \in A^L$,
(pPP) $\quad x \leq y \rightarrow^L (x \odot y)$, $y \leq x \rightsquigarrow^L (x \odot y)$ (or $x \leq y \rightsquigarrow^L (y \odot x)$).

Dually, by Definition 2.5.1 (i'), in a pBCI(pS) algebra (pBCK(pS) algebra) $\mathcal{A}^R$ we have, for all $x, y \in A^R$,

$$x \oplus y \in \{z \in A^R \mid x \geq y \rightarrow^R z\}, \quad x \oplus y \in \{z \in A^R \mid y \geq x \rightsquigarrow^R z\},$$

hence we have the following property, for all $x, y \in A^R$,
(pSS) $\quad x \geq y \rightarrow^R (x \oplus y)$, $y \geq x \rightsquigarrow^R (x \oplus y)$ (or $x \geq y \rightsquigarrow^R (y \oplus x)$).

Remark 2.5.2 A pBCI(pP) algebra (pBCK(pP) algebra) can be defined equivalently as a structure

$$\mathcal{A}^L = (A^L, \leq, \rightarrow^L, \rightsquigarrow^L, \odot, 1)$$

such that $(A^L, \leq, \rightarrow^L, \rightsquigarrow^L, 1)$ is a left-pBCI algebra (left-pBCK algebra, respectively) where, for all $x, y \in A^L$, there exist $\min\{z \mid x \leq y \rightarrow^L z\}$ and $\min\{z \mid y \leq x \rightsquigarrow^L z\}$ and they are equal, and the pseudo-product $\odot$ can be expressed in terms of the pseudo-implication $(\rightarrow^L, \rightsquigarrow^L)$ as follows:

$$x \odot y = \min\{z \mid x \leq y \rightarrow^L z\} = \min\{z \mid y \leq x \rightsquigarrow^L z\}.$$

Similarly in the dual case.

2.6 Pseudo-BCI(pRP) and pseudo-BCK(pRP) algebras - two intermediary notions

In this section, we present the two intermediary notions between pBCI(pP) algebras, pBCK(pP) algebras (as algebras of logic) and the unital magmas: the pseudo-BCI(pRP) algebras and the pseudo-BCK(pRP) algebras, respectively.

Definition 2.6.1
(i) A *left-pBCI algebra with the property (pRP)*, or a *pBCI(pRP) algebra* for short, is a structure

$$\mathcal{A}^L = (A^L, \leq, \rightarrow^L, \rightsquigarrow^L, \odot, 1)$$

such that $(A^L, \leq)$ is a poset, $(A^L, \leq, \rightarrow^L, \rightsquigarrow^L, 1)$ is a left-pME algebra and the pseudo-product $\odot$ is connected to the pseudo-implication $(\rightarrow^L, \rightsquigarrow^L)$ by the following property (pRP): for all $x, y, z \in A^L$,
(pRP) $\quad x \leq y \rightarrow^L z \Longleftrightarrow y \leq x \rightsquigarrow^L z \Longleftrightarrow x \odot y \leq z$.

(i') Dually, a *right-pBCI algebra with the property (pcoRS)*, or a *pBCI(pcoRS) algebra* for short, is a structure

$$\mathcal{A}^R = (A^R, \geq, \rightarrow^R, \rightsquigarrow^R, \oplus, 0)$$

such that $(A^R, \geq)$ is a poset, $(A^R, \geq, \to^R, \rightsquigarrow^R, 0)$ is a right-pME algebra and the pseudo-sum $\oplus$ is connected to the pseudo-coimplication $(\to^R, \rightsquigarrow^R)$ by the following property (pcoRS): for all $x, y, z \in A^R$,
(pcoRS) $\quad x \geq y \to^R z \Longleftrightarrow y \geq x \rightsquigarrow^R z \Longleftrightarrow x \oplus y \geq z$.

Denote by **pBCI(pRP)** the class of all pBCI(RP) algebras and by **pBCI(pcoRS)** the class of all pBCI(pcoRS) algebras.

Note that in (pRP), the equivalence $x \leq y \to^L z \Longleftrightarrow y \leq x \rightsquigarrow^L z$ is just (pEq$\#^L$), that is always true in any left-pBCI algebra, by Proposition 2.3.4.

Dually, note that in (pcoRS), the equivalence $x \geq y \to^R z \Longleftrightarrow y \geq x \rightsquigarrow^R z$ is just (pEq$\#^R$), that is always true in any right-pBCI algebra, by the dual of Proposition 2.3.4.

Definition 2.6.2

(i) A *left-pseudo-BCK algebra with the property (pRP)*, or a *pBCK(pRP) algebra* for short, is a structure

$$\mathcal{A}^L = (A^L, \leq, \to^L, \rightsquigarrow^L, \odot, 1)$$

such that $(A^L, \leq, 1)$ is a poset with last element 1, $(A^L, \leq, \to^L, \rightsquigarrow^L, 1)$ is a left-pME algebra and the pseudo-product $\odot$ is connected to the pseudo-implication $(\to^L, \rightsquigarrow^L)$ by the property (pRP): for all $x, y, z \in A^L$,
(pRP) $\quad x \leq y \to^L z \Longleftrightarrow y \leq x \rightsquigarrow^L z \Longleftrightarrow x \odot y \leq z$.

(i') Dually, a *right-pseudo-BCK algebra with the property (pcoRS)*, or a *pBCK(pcoRS) algebra* for short, is a structure

$$\mathcal{A}^R = (A^R, \geq, \to^R, \rightsquigarrow^R, \oplus, 0)$$

such that $(A^R, \geq, 0)$ is a poset with first element 0, $(A^R, \geq, \to^R, \rightsquigarrow^R, 0)$ is a right-pME algebra and the pseudo-sum $\oplus$ is connected to the pseudo-coimplication $(\to^R, \rightsquigarrow^R)$ by the property (pcoRS): for all $x, y, z \in A^R$,
(pcoRS) $\quad x \geq y \to^R z \Longleftrightarrow y \geq x \rightsquigarrow^R z \Longleftrightarrow x \oplus y \geq z$.

Denote by **pBCK(pRP)** the class of all pBCK(RP) algebras and by **pBCK(pcoRS)** the class of all pBCK(pcoRS) algebras.

Hence, we have:

pBCK(pRP) = **pBCI(pRP)** + (L) and, dually,
pBCK(pcoRS) = **pBCI(pcoRS)** + (F).

Note that the dual equivalences (pRP) and (pcoRS) are non-commutative *Galois connections.*

An important lemma is the following:

Lemma 2.6.3 *(See [94], Lemma 9.5.11, for posets with* 1*)*

Let $\mathcal{A}^L = (A^L, \leq, \rightarrow^L, \rightsquigarrow^L, \odot)$ (or $\mathcal{A}^L = (A^L, \leq, \odot, \rightarrow^L, \rightsquigarrow^L)$) be a structure such that:
(a) $(A, \leq)$ is a poset,
(b) the equivalences (pRP) hold: for all $x, y, z \in A^L$,
(pRP) $x \leq y \rightarrow^L z \Longleftrightarrow y \leq x \rightsquigarrow^L z \Longleftrightarrow x \odot y \leq z.$

Then, we have the following properties: for all $x, y, z \in A^L$,
(pPP) $y \leq x \rightarrow^L (y \odot x), \quad x \leq y \rightsquigarrow^L (y \odot x),$
(pRR) $(y \rightarrow^L x) \odot y \leq x, \quad y \odot (y \rightsquigarrow^L x) \leq x,$
(pCp) $x \leq y \Longrightarrow x \odot z \leq y \odot z, \quad z \odot x \leq z \odot y,$
(p^{*L}) $x \leq y \Longrightarrow z \rightarrow^L x \leq z \rightarrow^L y, \quad z \rightsquigarrow^L x \leq z \rightsquigarrow^L y,$
(p^{**L}) $x \leq y \Longrightarrow y \rightarrow^L z \leq x \rightarrow^L z, \quad y \rightsquigarrow^L z \leq x \rightsquigarrow^L z.$

Proof. (pPP): $y \leq x \rightarrow^L (y \odot x) \overset{(pRP)}{\Leftrightarrow} y \odot x \leq y \odot x$, which is true by (a). $x \leq y \rightsquigarrow^L (y \odot x) \overset{(pRP)}{\Leftrightarrow} y \odot x \leq y \odot x$, which is true by (a).

(pRR): $(y \rightarrow^L x) \odot y \leq x \overset{(pRP)}{\Leftrightarrow} y \rightarrow^L x \leq y \rightarrow^L x$, which is true by (a). $y \odot (y \rightsquigarrow^L x) \leq x \overset{(pRP)}{\Leftrightarrow} y \rightsquigarrow^L x \leq y \rightsquigarrow^L x$, which is true by (a).

(pCp): By (pPP), $y \leq z \rightarrow^L (y \odot z)$ and if $x \leq y$, it follows that $x \leq z \rightarrow^L (y \odot z)$; hence, $x \odot z \leq y \odot z$, by (pRP). By (pPP) also, $y \leq z \rightsquigarrow^L (z \odot y)$ and if $x \leq y$, it follows that $x \leq z \rightsquigarrow^L (z \odot y)$; hence, $z \odot x \leq z \odot y$, by (pRP).

(p^{*L}): By (pRR), $(z \rightarrow^L x) \odot z \leq x$ and if $x \leq y$, it follows that $(z \rightarrow^L x) \odot z \leq y$; hence, $z \rightarrow^L x \leq z \rightarrow^L y$, by (pRP). By (pRR) also, $z \odot (z \rightsquigarrow^L x) \leq x$ and if $x \leq y$, it follows that $z \odot (z \rightsquigarrow^L x) \leq y$; hence, $z \rightsquigarrow^L x \leq z \rightsquigarrow^L y$, by (pRP).

(p^{**L}): Let $x \leq y$. By (pCp), we obtain $(y \rightarrow^L z) \odot x \leq (y \rightarrow^L z) \odot y \leq z$, by (pRR); hence, $(y \rightarrow^L z) \odot x \leq z$, which implies, by (pRP), that $y \rightarrow^L z \leq x \rightarrow^L z$. By (pCp) again, we obtain $x \odot (y \rightsquigarrow^L z) \leq y \odot (y \rightsquigarrow^L z) \leq z$, by (pRR); hence, $x \odot (y \rightsquigarrow^L z) \leq z$, which implies, by (pRP), that $y \rightsquigarrow^L z \leq x \rightsquigarrow^L z$. □

Proposition 2.6.4 *Let $\mathcal{A}^L = (A^L, \leq, \rightarrow^L, \rightsquigarrow^L, \odot, 1)$ be a pBCI(pRP) algebra or a pBCK(pRP) algebra. Then, the properties (pPP), (pRR), (pCp), (p^{*L}), (p^{**L}) hold.*

Proof. Obviously, by Lemma 2.6.3. □

Recall that a *monoid* is an algebra $(A, \odot, 1)$ of type $(2, 0)$ verifying: for all $x, y, z \in A$,
(pU) (Unit element) $1 \odot x = x = x \odot 1$,
(Ass) (Associativity) $x \odot (y \odot z) = (x \odot y) \odot z$.

Then, we have the following result.

Proposition 2.6.5 *(See [94], Proposition 9.5.15, for pBCK algebras)*
Let $\mathcal{A}^L = (A^L, \leq, \rightarrow^L, \rightsquigarrow^L, \odot, 1)$ be a pBCI(pRP) algebra. Then, the algebra $(A^L, \odot, 1)$ is a monoid.

Proof. (Ass): $(x \odot y) \odot z \leq a \overset{(pRP)}{\Leftrightarrow} x \odot y \leq z \rightarrow^L a \overset{(pRP)}{\Leftrightarrow} x \leq y \rightarrow^L (z \rightarrow^L a) \overset{(pRP)}{\Leftrightarrow} y \leq x \rightsquigarrow^L (z \rightarrow^L a) \overset{(pEx)}{\Leftrightarrow} y \leq z \rightarrow^L (x \rightsquigarrow^L a)$ and

$x \odot (y \odot z) \leq a \overset{(pRP)}{\Leftrightarrow} x \leq (y \odot z) \to^L a \overset{(pRP)}{\Leftrightarrow} y \odot z \leq x \rightsquigarrow^L a \overset{(pRP)}{\Leftrightarrow} y \leq z \to^L (x \rightsquigarrow^L a)$. Thus, $(x \odot y) \odot z = x \odot (y \odot z)$, by the reflexivity and antisymmetry of $\leq$.

(pU): $x \odot 1 \leq a \overset{(pRP)}{\Leftrightarrow} x \leq 1 \to^L a \overset{(pM^L)}{\Leftrightarrow} x \leq a$; thus, $x \odot 1 = x$, by the reflexivity and antisymmetry of $\leq$.

$1 \odot x \leq a \overset{(pRP)}{\Leftrightarrow} x \leq 1 \rightsquigarrow^L a \overset{(pM^L)}{\Leftrightarrow} x \leq a$; thus, $1 \odot x = x$. □

Remark 2.6.6 In the non-integral case, the pseudo-product $\odot$ verifies the properties (pU), (Ass), by Proposition 2.6.5, and (pCp), by Lemma 2.6.3, hence is a pseudo-t-norm on the poset $(A^L, \leq)$, by Definition 1.1.1.

Proposition 2.6.7 *Let $\mathcal{A}^L = (A^L, \leq, \to^L, \rightsquigarrow^L, \odot, 1)$ be a pBCI(pRP) algebra. Then, the following properties hold: for all $x, y, z \in A^L$,*
(pD^L) $y \leq (y \to^L x) \rightsquigarrow^L x,\ y \leq (y \rightsquigarrow^L x) \to^L x,$
(pBB^L) $y \to^L z \leq (z \to^L x) \rightsquigarrow^L (y \to^L x),$
$y \rightsquigarrow^L z \leq (z \rightsquigarrow^L x) \to^L (y \rightsquigarrow^L x),$

Proof. (pD^L): $y \leq (y \to^L x) \rightsquigarrow^L x \overset{(pRP)}{\Leftrightarrow} y \to^L x \leq y \to^L x$, which is true by the reflexivity of $\leq$.

(pBB^L): $(z \to^L x) \rightsquigarrow^L (y \to^L x) \overset{(pEx^L)}{=} y \to^L ((z \to^L x) \rightsquigarrow^L x)$; but, $z \leq (z \to^L x) \rightsquigarrow^L x$, by (pD^L), hence, by (p*^{L}), $y \to^L z \leq y \to^L ((z \to^L x) \rightsquigarrow^L x)$, hence $y \to^L z \leq (z \to^L x) \rightsquigarrow^L (y \to^L x)$ and

$(z \rightsquigarrow^L x) \to^L (y \rightsquigarrow^L x) \overset{(pEx^L)}{=} y \rightsquigarrow^L ((z \rightsquigarrow^L x) \to^L x)$; but, $z \leq (z \rightsquigarrow^L x) \to^L x$, by (pD^L), hence, by (p*^{L}), $y \rightsquigarrow^L z \leq y \rightsquigarrow^L ((z \rightsquigarrow^L x) \to^L x)$, hence $y \rightsquigarrow^L z \leq (z \rightsquigarrow^L x) \to^L (y \rightsquigarrow^L x)$. Thus, (pBB^L) holds. □

Proposition 2.6.8 *Let $\mathcal{A}^L = (A^L, \leq, \to^L, \rightsquigarrow^L, \odot, 1)$ be a pBCK(pRP) algebra. Then, the following property holds: for all $x, y, z \in A^L$,*
(pN^L) $1 \leq x \Longrightarrow x = 1$.

Proof. (pN^L): By (L), $x \leq 1$, for all $x \in A^L$; hence, if $1 \leq x$, then $x = 1$, by the antisymmetry of $\leq$. □

Proposition 2.6.9 *(See [88], ([94], Proposition 10.1.3))*
Let $\mathcal{A}^L = (A^L, \leq, \to^L, \rightsquigarrow^L, \odot, 1)$ be a pBCK(pRP) algebra. Then, we have: for all $x, y, z \in A^L$,

(2.7) $$x \odot y \leq x,\ y,$$

(2.8) $$(x \to^L y) \odot x \leq x, y, \quad x \odot (x \rightsquigarrow^L y) \leq x,\ y,$$

(p@rel^L) $$x \to^L y \leq (x \odot z) \to^L (y \odot z), \quad x \rightsquigarrow^L y \leq (z \odot x) \rightsquigarrow^L (z \odot y),$$

(2.9) $$x \odot (y \to^L z) \leq y \to^L (x \odot z), \quad (y \rightsquigarrow^L z) \odot x \leq y \rightsquigarrow^L (z \odot x),$$

(2.10) $$(y \to^L z) \odot (x \to^L y) \leq x \to^L z, \quad (x \rightsquigarrow^L y) \odot (y \rightsquigarrow^L z) \leq x \rightsquigarrow^L z,$$

(2.11) $$x \to^L (y \to^L z) = (x \odot y) \to^L z, \quad x \rightsquigarrow^L (y \rightsquigarrow^L z) = (y \odot x) \rightsquigarrow^L z,$$

(2.12) $$(x \odot z) \to^L (y \odot z) \leq x \to^L (z \to^L y),$$
$$(z \odot x) \rightsquigarrow^L (z \odot y) \leq x \rightsquigarrow^L (z \rightsquigarrow^L y),$$

(2.13) $$x \to^L y \leq (x \odot z) \to^L (y \odot z) \leq x \to^L (z \to^L y),$$
$$x \rightsquigarrow^L y \leq (z \odot x) \rightsquigarrow^L (z \odot y) \leq x \rightsquigarrow^L (z \rightsquigarrow^y).$$

2.7 The basic equivalences (EL1) and (iEL1)

In this section, we present the basic equivalences:
- (EL1), between pBCI(pP) and pBCI(pRP) algebras, and between their duals,
- (iEL1), between pBCK(pP) and pBCK(pRP) algebras, and between their duals.

• **In the non-integral case**, we have the following theorem (the theorem in right-case is omitted).

Theorem 2.7.1 *(See [94], Theorem 9.5.13, for pseudo-BCK algebras)*
(1) Let $\mathcal{A}^L = (A^L, \leq, \to^L, \rightsquigarrow^L, 1)$ be a pBCI(pP) algebra, where for all $x, y \in A^L$:
(pP) $\exists x \odot y \overset{notation}{=} \min\{z \mid x \leq y \to^L z\} = \min\{z \mid y \leq x \rightsquigarrow^L z\}$.
Define $\pi(\mathcal{A}^L) = (A^L, \leq, \to^L, \rightsquigarrow^L, \odot, 1)$.
Then, $\pi(\mathcal{A}^L)$ is a pBCI(pRP) algebra.
(1') Conversely, let $\mathcal{A}^L = (A^L, \leq, \to^L, \rightsquigarrow^L, \odot, 1)$ be a pBCI(pRP) algebra. Define $\pi'(\mathcal{A}^L) = (A^L, \leq, \to^L, \rightsquigarrow^L, 1)$.
Then, $\pi'(\mathcal{A}^L)$ is a pBCI(pP) algebra, where for all $x, y \in A^L$:
$$\min\{z \mid x \leq y \to^L z\} = \min\{z \mid y \leq x \rightsquigarrow^L z\} = x \odot y.$$
(2) The above defined mappings π and π' are mutually inverse.

Proof. (1): We must prove that (pRP) holds. Indeed, if $x \leq y \to^L z$, then by (pP), $x \odot y \leq z$; by Proposition 2.3.4, (p^{*L}) holds, so if $x \odot y \leq z$, then, by (p^{*L}), $y \to^L (x \odot y) \leq y \to^L z$ and since by (pP), (pPP) also holds, then $x \leq y \to^L (x \odot y)$, hence $x \leq y \to^L z$, by transitivity.
Also, if $y \leq x \rightsquigarrow^L z$, then by (pP), $x \odot y \leq z$; if $x \odot y \leq z$, then, by (p^{*L}), $x \rightsquigarrow^L (x \odot y) \leq x \rightsquigarrow^L z$ and since, by (pPP), $y \leq x \rightsquigarrow^L (x \odot y)$, then $y \leq x \rightsquigarrow^L z$. Thus, (pRP) holds. Then, apply Theorem 2.3.10.
(1'): We must prove that (pP) holds. Indeed, by Proposition 2.6.4, the property (pPP) holds, hence $x \leq y \to^L (x \odot y)$, i.e. $x \odot y \in \{z \mid x \leq y \to^L z\}$; if z verifies $x \leq y \to^L z$, then by (pRP), $x \odot y \leq z$; so there exists $\min\{z \mid x \leq y \to^L z\} = x \odot y$.

Also, by (pPP) again, $y \leq x \rightsquigarrow^L (x \odot y)$, i.e. $x \odot y \in \{z \mid y \leq x \rightsquigarrow^L z\}$; if z verifies $y \leq x \rightsquigarrow^L z$, then by (pRP), $x \odot y \leq z$; so there exists $\min\{z \mid y \leq x \rightsquigarrow^L z\} = x \odot y$ too. Thus, (pP) holds.

We must also prove that $(A^L, \leq, \rightarrow^L, \rightsquigarrow^L, 1)$ is a left-pBCI algebra (Definition 3). Indeed, by Lemma 2.6.3, (p$*^L$) holds; then, apply Theorem 2.3.10.

(2): Obviously. □

Remark 2.7.2 By above proof, we remark that for pBCI algebras:

$$(pP) + (p*^L) \Leftrightarrow (pRP).$$

We then have the following dual basic equivalences:

(EL1) **pBCI(pP)** $\Leftrightarrow$ **pBCI(pRP)** and, dually,
pBCI(pS) $\Leftrightarrow$ **pBCI(pcoRS)**.

• **In the integral case**, we have the following theorem (the theorem in right-case is omitted).

Theorem 2.7.3

(1) Let $\mathcal{A}^L = (A^L, \leq, \rightarrow^L, \rightsquigarrow^L, 1)$ be a pBCK(pP) algebra, where for all $x, y \in A^L$:

(pP) $\exists x \odot y \overset{notation}{=} \min\{z \mid x \leq y \rightarrow^L z\} = \min\{z \mid y \leq x \rightsquigarrow^L z\}$.

Define $\pi(\mathcal{A}^L) = (A^L, \leq, \rightarrow^L, \rightsquigarrow^L, \odot, 1)$.

Then, $\pi(\mathcal{A}^L)$ is a pBCK(pRP) algebra.

(1') Conversely, let $\mathcal{A}^L = (A^L, \leq, \rightarrow^L, \rightsquigarrow^L, \odot, 1)$ be a pBCK(pRP) algebra. Define $\pi'(\mathcal{A}^L) = (A^L, \leq, \rightarrow^L, \rightsquigarrow^L, 1)$.

Then, $\pi'(\mathcal{A}^L)$ is a pBCK(pP) algebra, where for all $x, y \in A^L$:

$$\min\{z \mid x \leq y \rightarrow^L z\} = \min\{z \mid y \leq x \rightsquigarrow^L z\} = x \odot y.$$

(2) The above defined mappings π and π' are mutually inverse.

Proof. The same proof as in above Theorem 2.7.1. □

Consequently, we have the following dual equivalences:

(iEL1) **pBCK(pP)** $\Leftrightarrow$ **pBCK(pRP)** and, dually,
pBCK(pS) $\Leftrightarrow$ **pBCK(pcoRS)**.

2.8 Pseudo-BCI(pP) and pseudo-BCK(pP) lattices

Definition 2.8.1

(i) A *pBCI(pP) lattice (pBCK(pP) lattice)* is an algebra

$$\mathcal{A}^L = (A^L, \wedge, \vee, \rightarrow^L, \rightsquigarrow^L, 1)$$

that is simultaneously a pBCI(pP) algebra (pBCK(pP) algebra) and a left-pBCI lattice (left-pBCK lattice, respectively).

(i') Dually, a *pBCI(pS) lattice (pBCK(pS) lattice)* is an algebra

$$\mathcal{A}^R = (A^R, \vee, \wedge, \to^R, \rightsquigarrow^R, 0)$$

that is simultaneously a pBCI(pS) algebra (pBCK(pS) algebra) and a right-pBCK lattice (right-pBCK lattice, respectively).

Denote by **pBCI(pP)-L** the class of all pBCI(pP) lattices and by **pBCI(pS)-L** the class of all pBCI(pS) lattices.

Denote by **pBCK(pP)-L** the class of all pBCK(pP) lattices and by **pBCK(pS)-L** the class of all pBCK(pS) lattices.

Proposition 2.8.2 *(See [94], Proposition 10.2.2)*

Let $\mathcal{A}^L$ be a pBCK(pP) lattice and I be an arbitrary set.. Then the following properties hold, for all $x, y, z, x_i, y_i \in A^L$:

$$x \odot (\vee_{i\in I} y_i) = \vee_{i\in I}(x \odot y_i), \qquad (\vee_{i\in I} y_i) \odot x = \vee_{i\in I}(y_i \odot x), \tag{2.14}$$

$$x \vee (y \odot z) \geq (x \vee y) \odot (x \vee z), \tag{2.15}$$

$$(\vee_{i\in I} x_i) \to^L y = \wedge_{i\in I}(x_i \to^L y), \quad (\vee_{i\in I} x_i) \rightsquigarrow^L y = \wedge_{i\in I}(x_i \rightsquigarrow^L y), \tag{2.16}$$

$$y \to^L (\wedge_{i\in I} x_i) = \wedge_{i\in I}(y \to^L x_i), \quad y \rightsquigarrow^L (\wedge_{i\in I} x_i) = \wedge_{i\in I}(y \rightsquigarrow^L x_i). \tag{2.17}$$

whenever the arbitrary unions and meets exist.

Proposition 2.8.3 *(See [94], Proposition 10.2.3)*

Let $\mathcal{A}^L$ be a pBCK(pP) lattice. Then we have: for all $x, y \in A^L$,

$$y \odot x \leq (x \to^L y) \odot x \leq x \wedge y, \quad x \odot y \leq x \odot (x \rightsquigarrow^L y) \leq x \wedge y, \tag{2.18}$$

$$x \to^L (x \wedge y) = x \to^L y, \quad x \rightsquigarrow^L (x \wedge y) = x \rightsquigarrow^L y. \tag{2.19}$$

Proposition 2.8.4 *(See [94], Proposition 10.3.3)*

Let $\mathcal{A}^L = (A, \wedge, \vee, \to^L, \rightsquigarrow^L 1)$ be a pBCK(pP) lattice and I an arbitrary set. Then,

$$(pdiv^L) \Longrightarrow a \wedge (\vee_{i\in I} b_i) = \vee_{i\in I}(a \wedge b_i),$$

whenever the arbitrary unions exist, where: for all $x, y \in A^L$,

$(pdiv^L)$ (pseudo-divisibility) $x \wedge y = (x \to^L y) \odot x = x \odot (x \rightsquigarrow^L y)$.

By this proposition, we immediately get the following result.

Corollary 2.8.5 *(See [94], Corollary 10.3.4)*

Let $\mathcal{A}^L = (A^L, \wedge, \vee, \to^L, \rightsquigarrow^L, 1)$ be a pBCK(pP) lattice verifying $(pdiv^L)$. Then, the reduct $(A^L, \wedge, \vee)$ is a distributive lattice.

2.9 Pseudo-BCI(pRP) and pseudo-BCK(pRP) lattices - two intermediary notions

In this section, we present the two intermediary notions between pBCI(pP) lattices and pBCK(pP) lattices (as algebras of logic) and the unital magmas: the pseudo-BCI(pRP) lattices and the pseudo-BCK(pRP) lattices, respectively. We present the equivalences (EL2) and (iEL2).

Definition 2.9.1
(i) A *pBCI(pRP) lattice (pBCK(pRP) lattice)* is an algebra

$$\mathcal{A}^L = (A^L, \wedge, \vee, \to^L, \leadsto^L, \odot, 1)$$

that is simultaneously a pBCI(pRP) algebra (pBCK(pRP) algebra) and $\leq$ is a lattice order relation.

(i') Dually, a *pBCI(pcoRS) lattice (pBCK(pcoRS) lattice)* is an algebra

$$\mathcal{A}^R = (A^R, \vee, \wedge, \to^R, \leadsto^R, \oplus, 0)$$

that is simultaneously a pBCI(pcoRS) algebra (pBCK(pcoRS) algebra) and $\geq$ is a lattice order relation.

Denote by **pBCI(pRP)-L** the class of all pBCI(pRP) lattices and denote by **pBCI(pcoRS)-L**^{R} the class of all pBCI(pcoRS) lattices.

Denote by **pBCK(pRP)-L** the class of all pBCK(pRP) lattices and denote by **pBCK(pcoRS)-L**^{R} the class of all pBCK(pcoRS) lattices.

• **In the non-integral case**, by (EL1), we obtain the dual equivalences:

(EL2) **pBCI(pP)-L** $\Leftrightarrow$ **pBCI(pRP)-L** and, dually,
pBCI(pS)-L $\Leftrightarrow$ **pBCI(pcoRS)-L**.

• **In the integral case**, by (iEL1), we obtain consequently, the dual equivalences:

(iEL2) **pBCK(pP)-L** $\Leftrightarrow$ **pBCK(pRP)-L** and, dually,
pBCK(pS)-L $\Leftrightarrow$ **pBCK(pcoRS)-L**.

2.10 Bounded t-normed algebras of logic. Two negations

We have bounded algebras only in the integral case. In this section, we recall the definitions of bounded algebras of logic having a pseudo-t-norm, $\odot$ (pseudo-t-conorm, $\oplus$) (see Remark 2.6.6) - in a word, that are "t-normed". Thus, we recall [94] the definitions of bounded pBCK(pP) and pBCK(pRP) algebras, of bounded pBCK(pP) and pBCK(pRP) lattices, and of their duals.

We present some results, including the equivalences (iEL3) and (iEL4) (in lattice case).

Definition 2.10.1
(i) A pBCK(pP) algebra $\mathcal{A}^L = (A^L, \leq, \to^L, \leadsto^L, 1)$ is said to be *bounded*, if the left-pBCK algebra $(A^L, \leq, \to^L, \leadsto^L, 1)$ is bounded, with 0 as the first element. A *bounded pBCK(pP) algebra* is denoted by:

$$\mathcal{A}^{Lb} = (A^L, \leq, \to^L, \leadsto^L, 0, 1).$$

(ii) A pBCK(pRP) algebra $\mathcal{A}^L = (A^L, \leq, \to^L, \leadsto^L, \odot, 1)$ is said to be *bounded*, if the poset with 1 $(A^L, \leq, 1)$ is bounded, with 0 as the first (smallest) element. A *bounded pBCK(pRP) algebra* is denoted by:

$$\mathcal{A}^{Lb} = (A^L, \leq, \to^L, \leadsto^L, \odot, 0, 1).$$

(i') Dually, a pBCK(pS) algebra $\mathcal{A}^R = (A^R, \geq, \to^R, \leadsto^R, 0)$ is said to be *bounded*, if the right-pBCK algebra $(A^R, \geq, \to^R, \leadsto^R, 0)$ is bounded, with 1 as the last (greatest) element. A *bounded pBCK(pS) algebra* is denoted by:

$$\mathcal{A}^{Rb} = (A^R, \geq, \to^R, \leadsto^R, 0, 1).$$

(ii') A pBCK(pcoRS) algebra $\mathcal{A}^R = (A^R, \geq, \to^R, \leadsto^R, \oplus, 0)$ is said to be bounded, if the poset with 0 $(A^R, \leq, 0)$ is bounded, with 1 as the last (greatest) element. A *bounded pBCK(pcoRS) algebra* is denoted by:

$$\mathcal{A}^{Rb} = (A^R, \geq, \to^R, \leadsto^R, \oplus, 0, 1).$$

Denote by $\mathbf{pBCK(pP)}^b$ the class of all bounded pBCK(pP) algebras and, dually, denote by $\mathbf{pBCK(pS)}^b$ the class of all bounded pBCK(pS) algebras.

Denote by $\mathbf{pBCK(pRP)}^b$ the class of all bounded pBCK(pRP) algebras and, dually, denote by $\mathbf{pBCK(pcoRS)}^b$ the class of all bounded pBCK(pcoRS) algebras.

We then have, by (iEL1), the following dual equivalences:

(iEL3) $\mathbf{pBCK(pP)}^b \Leftrightarrow \mathbf{pBCK(pRP)}^b$ and, dually,
$\mathbf{pBCK(pS)}^b \Leftrightarrow \mathbf{pBCK(pcoRS)}^b$.

Proposition 2.10.2 *(See [88], ([94], Proposition 10.1.10), ([94], Proposition 10.1.11))*
Let $\mathcal{A}^{Lb} = (A^L, \leq, \to^L, \leadsto^L, 0, 1)$ be a bounded pBCK(pP) algebra. Then, we have: for all $x, y \in A^L$,

(2.20) $$0 \odot x = x \odot 0 = 0,$$

(2.21) $$x \to^L y^- = (x \odot y)^-, \quad x \leadsto^L y^\sim = (y \odot x)^\sim.$$

(pIv^L) $x \odot x^\sim = 0 = x^- \odot x$,
where $x^- = x \to^L 0$, $x^\sim = x \leadsto^L 0$.

In lattice case, we have the following definitions.

Definition 2.10.3
(i) A pBCK(pP) lattice $\mathcal{A}^L = (A^L, \wedge, \vee, \rightarrow^L, \rightsquigarrow^L, 1)$ is said to be *bounded*, if the left-pBCK lattice $(A^L, \wedge, \vee, \rightarrow^L, \rightsquigarrow^L, 1)$ is bounded, with 0 as the first (smallest) element. A *bounded pBCK(pP) lattice* is denoted by:

$$\mathcal{A}^{Lb} = (A^L, \wedge, \vee, \rightarrow^L, \rightsquigarrow^L, 0, 1).$$

(ii) A pBCK(pRP) lattice $\mathcal{A}^L = (A^L, \wedge, \vee, \rightarrow^L, \rightsquigarrow^L, \odot, 1)$ is said to be *bounded*, if the lattice with 1 $(A^L, \wedge, \vee, 1)$ is bounded, with 0 as the first (smallest) element. A *bounded pBCK(pRP) lattice* is denoted by:

$$\mathcal{A}^{Lb} = (A^L, \wedge, \vee, \rightarrow^L, \rightsquigarrow^L, \odot, 0, 1).$$

(i') Dually, a pBCK(pS) lattice $\mathcal{A}^R = (A^R, \vee, \wedge, \rightarrow^R, \rightsquigarrow^R, 0)$ is said to be *bounded*, if the right-pBCK lattice $(A^R, \vee, \wedge, \rightarrow^R, \rightsquigarrow^R, 0)$ is bounded, with 1 as the last (greatest) element. A *bounded pBCK(pS) lattice* is denoted by:

$$\mathcal{A}^{Rb} = (A^R, \vee, \wedge, \rightarrow^R, \rightsquigarrow^R, 0, 1).$$

(ii') A pBCK(pcoRS) lattice $\mathcal{A}^R = (A^R, \vee, \wedge, \rightarrow^R, \rightsquigarrow^R, \oplus, 0)$ is said to be *bounded*, if the lattice with 0 $(A^R, \vee, \wedge, 0)$ is bounded, with 1 as the last (greatest) element. A *bounded pBCK(pcoRS) lattice* is denoted by:

$$\mathcal{A}^{Rb} = (A^R, \vee, \wedge, \rightarrow^R, \rightsquigarrow^R, \oplus, 0, 1).$$

Denote by **pBCK(pP)-L**^{b} the class of all bounded pBCK(pP) lattices and, dually, denote by **pBCK(pS)-L**^{b} the class of all bounded pBCK(pS) lattices.
Denote by **pBCK(pRP)-L**^{b} the class of all bounded pBCK(pRP) lattices and, dually, denote by **pBCK(pcoRS)-L**^{b} the class of all bounded pBCK(pcoRS) lattices.

We then have, by (iEL2), the following dual equivalences, in lattice case:

(iEL4) **pBCK(pP)-L**^{b} $\Leftrightarrow$ **pBCK(pRP)-L**^{b} and, dually,
pBCK(pS)-L^b $\Leftrightarrow$ **pBCK(pcoRS)-L**^{b}.

Note that the two negations, $^- = ^{-^L}$ and $^\sim = ^{\sim^L}$, are those defined earlier for left-pBCK algebras or left-pBCK lattices and, dually, the two negations, $^- = ^{-^R}$ and $^\sim = ^{\sim^R}$, are those defined for right-pBCK algebras or right-pBCK lattices.

Proposition 2.10.4 *(See [94], Proposition 10.2.4)*
Let $\mathcal{A}^{Lb} = (A^L, \wedge, \vee, \rightarrow^L, \rightsquigarrow^L, 0, 1)$ be a bounded pBCK(pP) lattice. Then, we have: for all $x, y \in A^L$,

(2.22) $$(x \vee y)^\sim = x^\sim \wedge y^\sim, \quad (x \vee y)^- = x^- \wedge y^-,$$

(2.23) $$x^- \vee y^- \leq (x \wedge y)^-, \quad x^\sim \vee y^\sim \leq (x \wedge y)^\sim.$$

2.11 The property (pDN). The equivalences (iEL5), (iEL6)

In this section, we recall the definitions of pseudo-BCK algebras having the property (pDN) and we recall some results, including the important Theorem 2.11.4 and some important remarks.

We present the important equivalences (iEL5) and (iEL6) for the sequel.

Definition 2.11.1

(i) A bounded left-pBCK algebra $\mathcal{A}^{Lb} = (A^L, \leq, \to^L, \rightsquigarrow^L, 0, 1)$ is said to be *with property (pDN^L)*, or *involutive*, if it verifies the property: for all $x \in A^L$,
(pDN^L) (pseudo-Double Negation) $\quad (x^\sim)^- = x = (x^-)^\sim$,
where $x^- = x \to^L 0$, $x^\sim = x \rightsquigarrow^L 0$.

(i') Dually, a bounded right-pBCK algebra $\mathcal{A}^{Rb} = (A^R, \geq, \to^R, \rightsquigarrow^R, 0, 1)$ is said to be *with property (pDN^R)*, or *involutive*, if it verifies the property: for all $x \in A^R$,
(pDN^R) (pseudo-Double Negation) $\quad (x^\sim)^- = x = (x^-)^\sim$,
where $x^- = x \to^R 1$, $x^\sim = x \rightsquigarrow^R 1$.

Denote by $\mathbf{pBCK}^L_{(pDN)}$ the class of involutive bounded left-pBCK algebras and, dually, denote by $\mathbf{pBCK}^R_{(pDN)}$ the class of involutive bounded right-pBCK algebras.

Lemma 2.11.2 *(See [88], ([94], Lemma 10.1.18))*

Let $\mathcal{A}^{Lb}$ be a bounded left-pBCK algebra with property (pDN^L). Then, we have: for all $x, y \in A^L$,
(pEqNg^{Lb}) $\quad x \leq y \Leftrightarrow y^- \leq x^- \Leftrightarrow y^\sim \leq x^\sim$,
(pDNeg1^{Lb}) $\quad x \to^L y = y^- \rightsquigarrow^L x^-, \quad x \rightsquigarrow^L y = y^\sim \to^L x^\sim$,
(pDNeg2^{Lb}) $\quad y^- \rightsquigarrow^L x = x^\sim \to^L y$.

Proposition 2.11.3 *(See [60], ([94], Proposition 10.1.20))*

Let $\mathcal{A}^{Lb}$ be a bounded left-pBCK algebra with property (pDN^L). Then, we have: for all $x, y \in A^L$,
(pDNeg3^{Lb}) $\quad (x \to^L y^-)^\sim = (y \rightsquigarrow^L x^\sim)^-$.

We have then the following important dual results.

Theorem 2.11.4 *(See [60], Theorem 10.1.26)*

(1) Let $\mathcal{A}^{Lb} = (A^L, \leq, \to^L, \rightsquigarrow^L, 0, 1)$ be a bounded left-pBCK algebra with property (pDN^L). Then, $\mathcal{A}^{Lb}$ is with property (pP) and we have: for all $x, y \in A^L$,
(2.24)
$$x \odot y \overset{not.}{=} \min\{z \mid x \leq y \to^L z\} = \min\{z \mid y \leq x \rightsquigarrow^L z\} = (x \to^L y^-)^\sim = (y \rightsquigarrow^L x^\sim)^-,$$

(2.25) $$x \to^L y = (x \odot y^\sim)^-, \quad x \rightsquigarrow^L y = (y^- \odot x)^\sim.$$

(1') Dually, let $\mathcal{A}^{Rb} = (A^R, \geq, \to^R, \rightsquigarrow^R, 0, 1)$ be a bounded right-pBCK algebra with property (pDN^R) (involutive). Then, $\mathcal{A}^{Rb}$ is with property (pS) and we have:

for all $x, y \in A^R$,
(2.26)
$x \oplus y \overset{not.}{=} \max\{z \mid x \geq y \to^R z\} = \max\{z \mid y \geq x \leadsto^R z\} = (x \to^R y^-)^\sim = (y \leadsto^R x^\sim)^-$,

$$x \to^R y = (x \oplus y^\sim)^-, \quad x \leadsto^R y = (y^- \oplus x)^\sim. \tag{2.27}$$

Remark 2.11.5

(1) By above Theorem 2.11.4 (1), if $\mathcal{A}^{Lb} = (A^L, \leq, \to^L, \leadsto^L, 0, 1)$ is a bounded pBCK(pP) algebra with property (pDN^L), then we have the properties:

(pP)$_{(pDN)}$ $\quad x \odot y = (x \to^L y^-)^\sim \overset{(pDNeg3^{Lb})}{=} (y \leadsto^L x^\sim)^-$,
(pR)$_{(pDN)}$ $\quad x \to^L y = (x \odot y^\sim)^-, \quad x \leadsto^L y = (y^- \odot x)^\sim$.

(1') Dually, by above Theorem 2.11.4 (1'), if $\mathcal{A}^{Rb} = (A^R, \geq, \to^R, \leadsto^R, 0, 1)$ is a bounded pBCK(pS) algebra with property (pDN^R), then we have the properties:

(pS)$_{(pDN)}$ $\quad x \oplus y = (x \to^R y^-)^\sim \overset{(pDNeg3^{Rb})}{=} (y \leadsto^R x^\sim)^-$,
(pcoR)$_{(pDN)}$ $\quad x \to^R y = (x \oplus y^\sim)^-, \quad x \leadsto^R y = (y^- \oplus x)^\sim$.

Corollary 2.11.6 *Let $\mathcal{A}^{Lb} = (A^L, \leq, \to^L, \leadsto^L, 0, 1)$ be a bounded left-pBCK algebra with property (pDN^L). Then, we have: for all $x, y \in A^L$,*
(IdEqP^{Lb}) $\quad x \odot y^\sim = 0 \Longleftrightarrow y^- \odot x = 0$, *with* $0 = 1^- = 1^\sim$,
(pDNP1) $\quad (x \odot y^\sim)^- = ((x^-)^- \odot y^-)^\sim, \quad (y^- \odot x)^\sim = (y^\sim \odot (x^\sim)^\sim)^-$,
(pDNP2) $\quad (x^- \odot y^-)^\sim = (x^\sim \odot y^\sim)^-$.

Proof. (IdEqP^{Lb}): $x \odot y^\sim = 0 \overset{(pP)_{(pDN)}}{\Leftrightarrow} (x \to^L (y^\sim)^-)^\sim = 0 \overset{(pDN^L)}{\Leftrightarrow} (x \to^L y)^\sim = 0 \Leftrightarrow x \to^L y = 1$ and
$y^- \odot x = 0 \overset{(pP)_{(pDN)}}{\Leftrightarrow} (x \leadsto^L (y^-)^\sim)^- = 0 \overset{(pDN^L)}{\Leftrightarrow} (x \leadsto^L y)^- = 0 \Leftrightarrow x \leadsto^L y = 1$;
then, by (IdEq^L), (IdEqP^{Lb}) holds.

(pDNP1): By (pR)$_{(pDN)}$, $(x \odot y^\sim)^- = x \to^L y$, $((x^-)^- \odot y^-)^\sim = y^- \leadsto^L x^-$ and $(y^- \odot x)^\sim = x \leadsto^L y$, $(y^\sim \odot (x^\sim)^\sim)^- = y^\sim \to^L x^\sim$; then, by (pDNeg1^{Lb}), (pDNP1) holds.

(pDNP2): By (pR)$_{(pDN)}$, $(x^- \odot y^-)^\sim = y^- \leadsto^L x$, $(x^\sim \odot y^\sim)^- = x^\sim \to^L y$; then, by (pDNeg2^{Lb}), (pDNP2) holds. □

Remarks 2.11.7 (See [93])

• Let $\mathcal{A}^{Lb} = (A^L, \leq, \to^L, \leadsto^L, 0, 1)$ be a bounded pBCK(pP) algebra (with the pseudo-product $\odot$) with the property (pDN^L) (i.e. involutive). Then, by Theorem 2.11.4 (1), we have: for all $x, y \in A^L$,
$x \odot y = (x \to^L y^-)^\sim \overset{(pDNeg3^{Lb})}{=} (y \leadsto^L x^\sim)^-$ and $x \to^L y = (x \odot y^\sim)^-$, $x \leadsto^L y = (y^- \odot x)^\sim$, where $x^- = x \to^L 0$ and $x^\sim = x \leadsto^L 0$.

We can define additional "right" operations: $\oplus^L$ and $\Rightarrow^L$, $\approx>^L$, in the following way:

$$x \oplus^L y \overset{def.}{=} (y^- \odot x^-)^\sim \overset{(pDNP2)}{=} (y^\sim \odot x^\sim)^- \quad \textit{and} \tag{2.28}$$

$$x \Rightarrow^L y \stackrel{def.}{=} (x \oplus^L y^\sim)^-, \quad x \approx>^L y \stackrel{def.}{=} (y^- \oplus^L x)^\sim, \tag{2.29}$$

by (2.27).

Note that this definition (2.28) was initially used to define $\oplus$ in a left-pseudo-MV algebra $\mathcal{A}^L = (A^L, \odot, ^-, ^\sim, 0, 1)$; hence there are also the implications $\Rightarrow^L$, $\approx>^L$ in $\mathcal{A}^L$ (i.e. there are four implications).

Then, we have:

(i) $x^-(= x \to^L 0) = x \Rightarrow^L 1$ and $x^\sim(= x \rightsquigarrow^L 0) = x \approx>^L 1$.
Indeed,
$x \Rightarrow^L 1 \stackrel{def.}{=} (x \oplus^L 1^\sim)^- = (((1^\sim)^- \odot x^-)^\sim)^- = 1 \odot x^- = x^-$ and
$x \approx>^L 1 \stackrel{def.}{=} (1^- \oplus^L x)^\sim = ((x^\sim \odot (1^-)^\sim)^-)^\sim = x^\sim \odot 1 = x^\sim$.

(ii) The connections between the "right" operations $\oplus^L$, $\Rightarrow^L$, $\approx>^L$ are:

$$x \oplus^L y = (x \Rightarrow^L y^-)^\sim = (y \approx>^L x^\sim)^-. \tag{2.30}$$

Indeed,
$(x \Rightarrow^L y^-)^\sim \stackrel{def.}{=} ((x \oplus^L (y^-)^\sim)^-)^\sim = x \oplus^L y$ and
$(y \approx>^L x^\sim)^- \stackrel{def.}{=} (((x^\sim)^- \oplus^L y)^\sim)^- = x \oplus^L y$.

(iii) The "left" operations ($\odot$, $\to^L$, $\rightsquigarrow^L$) expressed in terms of the "right" operations ($\oplus^L$, $\Rightarrow^L$, $\approx>^L$) are:

$$x \odot y = (y^- \oplus^L x^-)^\sim = (y^\sim \oplus^L x^\sim)^- \quad and \tag{2.31}$$

$$x \to^L y = (x^- \approx>^L y^\sim)^-, \quad x \rightsquigarrow^L y = (x^\sim \Rightarrow^L y^-)^\sim. \tag{2.32}$$

Indeed,
$(y^- \oplus^L x^-)^\sim \stackrel{def.}{=} (((x^-)^\sim \odot (y^-)^\sim)^-)^\sim \stackrel{(pDN^L)}{=} x \odot y$ and
$(y^\sim \oplus^L x^\sim)^- \stackrel{def.}{=} (((x^\sim)^- \odot (y^\sim)^-)^\sim)^- \stackrel{(pDN^L)}{=} x \odot y$;
$(x^- \approx>^L y^\sim)^- \stackrel{def.}{=} (((y^\sim)^- \oplus^L x^-)^\sim)^- = y \oplus^L x^- \stackrel{def.}{=} ((x^-)^\sim \odot y^\sim)^- = (x \odot y^\sim)^-$
$= x \to^L y$,
$(x^\sim \Rightarrow^L y^-)^\sim \stackrel{def.}{=} ((x^\sim \oplus^L (y^-)^\sim)^-)^\sim = x^\sim \oplus^L y \stackrel{def.}{=} (y^- \odot (x^\sim)^-)^\sim = (y^- \odot x)^\sim$
$= x \rightsquigarrow^L y$.

(iv) The "right" operations ($\oplus^L$, $\Rightarrow^L$, $\approx>^L$) expressed in terms of the "left" operations ($\odot$, $\to^L$, $\rightsquigarrow^L$) are:

$$x \Rightarrow^L y = (x^- \rightsquigarrow^L y^\sim)^-, \quad x \approx>^L y = (x^\sim \to^L y^-)^\sim. \tag{2.33}$$

Indeed,
$(x^- \rightsquigarrow^L y^\sim)^- = (((y^\sim)^- \odot x^-)^\sim)^- = y \odot x^- = (((x^-)^\sim \oplus^L y^\sim)^-$
$= (x \oplus^L y^\sim)^- \stackrel{def.}{=} x \Rightarrow^L y$,
$(x^\sim \to^L y^-)^\sim = ((x^\sim \odot (y^-)^\sim)^-)^\sim = x^\sim \odot y = (y^- \oplus^L (x^\sim)^-)^\sim$
$= (y^- \oplus^L x)^\sim \stackrel{def.}{=} x \approx>^L y$.

Consequently, the algebra $\mathcal{A}^{bLR} = (A^L, \leq, \Rightarrow^L, \approx>^L, 0, 1)$ is a bounded pBCK(pS) algebra (with the pseudo-sum $\oplus^L$) with property (pDN^R) that is termwise equivalent with $\mathcal{A}^{Lb}$. Hence, $\mathcal{A}^{Lb}$ is in the same time a (left-) pBCK(pP)$_{(pDN)}$ algebra and a (right-) pBCK(pS))$_{(pDN)}$ algebra. We say that $\mathcal{A}^{Lb}$ is *selfdual.*

• Dually, let $\mathcal{A}^{Rb} = (A^R, \geq, \rightarrow^R, \rightsquigarrow^R, 0, 1)$ be a bounded pBCK(pS) algebra (with the pseudo-sum $\oplus$) with property (pDN^R). Then, by Theorem 2.11.4 (1'), we have: for all $x, y \in A^R$,
$x \oplus y = (x \rightarrow^R y^-)^\sim = (y \rightsquigarrow^R x^\sim)^-$ and $x \rightarrow^R y = (x \oplus y^\sim)^-$, $x \rightsquigarrow^R y = (y^- \oplus x)^\sim$, where now $x^- = x \rightarrow^R 1$ and $x^\sim = x \rightsquigarrow^R 1$.

We can define additional "left" operations: $\odot^R$ and $\Rightarrow^R$, $\approx>^R$, in the following way (see the initial definition of (right-) pseudo-MV algebras in [61], [42]):

$$x \odot^R y \overset{def.}{=} (y^- \oplus x^-)^\sim = (y^\sim \oplus x^\sim)^- \quad \textit{and} \tag{2.34}$$

$$x \Rightarrow^R y \overset{def.}{=} (x \odot^R y^\sim)^-, \quad x \approx>^R y \overset{def.}{=} ((y^- \odot^R x)^\sim. \tag{2.35}$$

Note that this definition (2.34) was initially used in [58], [61] to define $\odot$ in a right-pseudo-MV algebra $\mathcal{A}^R = (A^R, \oplus, ^-, ^\sim, 0, 1)$, hence there are also the implications $\Rightarrow^R$, $\approx>^R$ (i.e. there are four implications in $\mathcal{A}^{Rb}$).

Then, we have:

(i') $x^-(= x \rightarrow^R 1) = x \Rightarrow^R 0$ and $x^\sim(= x \rightsquigarrow^R 1) = x \approx>^R 0$.

(ii') The connections between the "left" operations $\odot^R$, $\Rightarrow^R$, $\approx>^R$ are:

$$x \odot^R y = (x \Rightarrow^R y^-)^\sim = (y \approx>^R x^\sim)^-. \tag{2.36}$$

(iii') The "right" operations ($\oplus$, $\rightarrow^R$, $\rightsquigarrow^R$) expressed in terms of the "left" operations ($\odot^R$, $\Rightarrow^R$, $\approx>^R$) are:

$$x \oplus y = (y^- \odot^R x^-)^\sim = (y^\sim \odot^R x^\sim)^- \quad \textit{and} \tag{2.37}$$

$$x \rightarrow^R y = (x^- \approx>^R y^\sim)^-, \quad x \rightsquigarrow^R y = (x^\sim \Rightarrow^R y^-)^\sim. \tag{2.38}$$

(iv') The "left" operations ($\odot^R$, $\Rightarrow^R$, $\approx>^R$) expressed in terms of the "right" operations ($\oplus$, $\rightarrow^R$, $\rightsquigarrow^R$) are:

$$x \Rightarrow^R y = (x^- \rightsquigarrow^R y^\sim)^-, \quad x \approx>^R y = (x^\sim \rightarrow^R y^-)^\sim. \tag{2.39}$$

Consequently, the algebra $\mathcal{A}^{bRL} = (A^R, \geq, \Rightarrow^R, \approx>^R, 0, 1)$ is a bounded pBCK(pP) algebra (with the pseudo-product $\odot^R$) with property (pDN) (involutive) that is termwise equivalent with $\mathcal{A}^{Rb}$. We say that $\mathcal{A}^{Rb}$ is *selfdual.*

Note that the definitions (2.28) and (2.34) given in above Remarks 2.11.7 of the additional operations $\oplus^L$ and $\odot^R$, respectively, are not unique. The alternative definitions (2.40) and (2.48), respectively, are presented in the next similar Remarks 2.11.8.

Remarks 2.11.8

• Let $\mathcal{A}^{Lb} = (A^L, \leq, \rightarrow^L, \rightsquigarrow^L, 0, 1)$ be a bounded pBCK(pP) algebra (with the pseudo-product $\odot$) with property (pDN^L) (i.e. involutive), where: for all $x, y \in A^L$, $x \odot y = (x \rightarrow^L y^-)^\sim \overset{(pDNeg3^{Lb})}{=} (y \rightsquigarrow^L x^\sim)^-$ and $x \rightarrow^L y = (x \odot y^\sim)^-$, $x \rightsquigarrow^L y = (y^- \odot x)^\sim$, where $x^- = x \rightarrow^L 0$ and $x^\sim = x \rightsquigarrow^L 0$.

We can define the additional "right" operations: $\oplus^L$ and $\Rightarrow^L$, $\approx>^L$, in the following alternative way (see Definition 3.12.3):

$$(2.40) \qquad x \oplus^L y \overset{def.}{=} (x^- \odot y^-)^\sim \overset{(pDNP2)}{=} (x^\sim \odot y^\sim)^- \quad \textit{and hence}$$

$$(2.41) \qquad x \Rightarrow^L y \overset{def.}{=} (x \oplus^L y^\sim)^-, \quad x \approx>^L y \overset{def.}{=} (y^- \oplus^L x)^\sim.$$

Then, we have:

(j) $x^-(= x \rightarrow^L 0) = x \Rightarrow^L 1$ and $x^\sim(= x \rightsquigarrow^L 0) = x \approx>^L 1$.

Indeed,
$x \Rightarrow^L 1 \overset{def.}{=} (x \oplus^L 1^\sim)^- = ((x^- \odot (1^\sim)^-)^\sim)^- = x^- \odot 1 = x^-$ and
$x \approx>^L 1 \overset{def.}{=} (1^- \oplus^L x)^\sim = ((1^-)^\sim)^- \odot (x^\sim)^\sim = 1 \odot x^\sim = x^\sim$.

(jj) The connections between the "right" operations $\oplus^L$, $\Rightarrow^L$, $\approx>^L$ are:

$$(2.42) \qquad x \oplus^L y = (x \Rightarrow^L y^-)^\sim = (y \approx>^L x^\sim)^-.$$

Indeed,
$(x \Rightarrow^L y^-)^\sim \overset{def.}{=} ((x \oplus^L (y^-)^\sim)^-)^\sim = x \oplus^L y$ and
$(y \approx>^L x^\sim)^- \overset{def.}{=} (((x^\sim)^- \oplus^L y)^\sim)^- = x \oplus^L y$.

(jjj) The "left" operations ($\odot$, $\rightarrow^L$, $\rightsquigarrow^L$) expressed in terms of the "right" operations ($\oplus^L$, $\Rightarrow^L$, $\approx>^L$) are:

$$(2.43) \qquad x \odot y = (x^- \oplus^L y^-)^\sim = (x^\sim \oplus^L y^\sim)^- \quad \textit{and}$$

$$(2.44) \qquad x \rightarrow^L y = (x^- \Rightarrow^L y^-)^\sim = (x^\sim \Rightarrow^L y^\sim)^-,$$

$$(2.45) \qquad x \rightsquigarrow^L y = (x^- \approx>^L y^-)^\sim = (x^\sim \approx>^L y^\sim)^-.$$

Indeed,
$(x^- \oplus^L y^-)^\sim \overset{def.}{=} (((x^-)^\sim \odot (y^-)^\sim)^-)^\sim \overset{(pDN^L)}{=} x \odot y$ and
$(x^\sim \oplus^L y^\sim)^- \overset{def.}{=} (((x^\sim)^- \odot (y^\sim)^-)^\sim)^- \overset{(pDN^L)}{=} x \odot y$;
$(x^- \Rightarrow^L y^-)^\sim \overset{def.}{=} ((x^- \oplus^L (y^-)^\sim)^-)^\sim = x^- \oplus^L y = ((x^-)^\sim \odot y^\sim)^-$
$= (x \odot y^\sim)^- = x \rightarrow^L y$ and
$(x^\sim \Rightarrow^L y^\sim)^- \overset{def.}{=} ((x^\sim \oplus^L (y^\sim)^\sim)^-)^- = ((((x^\sim)^- \odot ((y^\sim)^\sim)^-)^\sim)^-)^-$
$= (x \odot y^\sim)^- = x \rightarrow^L y$;
$(x^- \approx>^L y^-)^\sim \overset{def.}{=} (((y^-)^- \oplus^L x^-)^\sim)^\sim = (((((y^-)^-)^\sim \odot (x^-)^\sim)^-)^\sim)^\sim$

$= (y^- \odot x)^\sim = x \rightsquigarrow^L y$ and
$(x^\sim \approx>^L y^\sim)^- \stackrel{def.}{=} (((y^\sim)^- \oplus^L x^\sim)^\sim)^- = y \oplus^L x^\sim = (y^- \odot (x^\sim)^-)^\sim$
$= (y^- \odot x)^\sim = x \rightsquigarrow^L y$.

(jv) The "right" operations ($\oplus^L$, $\Rightarrow^L$, $\approx>^L$) expressed in terms of the "left" operations ($\odot$, $\rightarrow^L$, $\rightsquigarrow^L$) are:

(2.46) $$x \Rightarrow^L y = (x^- \rightarrow^L y^-)^\sim = (x^\sim \rightarrow^L y^\sim)^-,$$

(2.47) $$x \approx>^L y = (x^- \rightsquigarrow^L y^-)^\sim = (x^\sim \rightsquigarrow^L y^\sim)^-.$$

Indeed,
$(x^- \rightarrow^L y^-)^\sim = ((x^- \odot (y^-)^\sim)^-)^\sim = x^- \odot y = ((x^-)^\sim \oplus^L y^\sim)^-$
$= (x \oplus^L y^\sim)^- \stackrel{def.}{=} x \Rightarrow^L y$ and
$(x^\sim \rightarrow^L y^\sim)^- = ((x^\sim \odot (y^\sim)^\sim)^-)^- = ((((x^\sim)^- \oplus^L ((y^\sim)^\sim)^-)^\sim)^-)^-$
$= (x \oplus^L y^\sim)^- \stackrel{def.}{=} x \Rightarrow^L y$;
$(x^- \rightsquigarrow^L y^-)^\sim = (((y^-)^- \odot x^-)^\sim)^\sim = (((((y^-)^-)^\sim \oplus^L (x^-)^\sim)^-)^\sim)^\sim$
$= (y^- \oplus^L x)^\sim \stackrel{def.}{=} x \approx>^L y$ and
$(x^\sim \rightsquigarrow^L y^\sim)^- = (((y^\sim)^- \odot x^\sim)^\sim)^- = y \odot x^\sim = (y^- \oplus^L (x^\sim)^-)^\sim$
$= (y^- \oplus^L x)^\sim \stackrel{def.}{=} x \approx>^L y$.

Consequently, the algebra $\mathcal{A}^{bLR} = (A^L, \leq, \Rightarrow^L, \approx>^L, 0, 1)$ is a bounded pBCK(pS) algebra (with the pseudo-sum $\oplus^L$) with property (pDN^R) that is termwise equivalent with $\mathcal{A}^{Lb}$. We say that $\mathcal{A}^{Lb}$ is *selfdual.*

• Dually, let $\mathcal{A}^{Rb} = (A^R, \geq, \rightarrow^R, \rightsquigarrow^R, 0, 1)$ be a bounded right-pBCK(pS) algebra (with the pseudo-sum $\oplus$) with property (pDN^R) . We have: for all $x, y \in A^R$, $x \oplus y = (x \rightarrow^R y^-)^\sim = (y \rightsquigarrow^R x^\sim)^-$ and $x \rightarrow^R y = (x \oplus y^\sim)^-$, $x \rightsquigarrow^R y = (y^- \oplus x)^\sim$, where now $x^- = x \rightarrow^R 1$ and $x^\sim = x \rightsquigarrow^R 1$.

We can define additional "left" operations: $\odot^R$ and $\Rightarrow^R$, $\approx>^R$, in the following alternative way (see Definition 3.12.3):

(2.48) $$x \odot^R y \stackrel{def.}{=} (x^- \oplus y^-)^\sim = (x^\sim \oplus y^\sim)^- \quad \textit{and}$$

(2.49) $$x \Rightarrow^R y \stackrel{def.}{=} (x \odot^R y^\sim)^-, \quad x \approx>^R y \stackrel{def.}{=} (y^- \odot^R x)^\sim.$$

Then, we have:

(j') $x^-(= x \rightarrow^R 1) = x \Rightarrow^R 0$ and $x^\sim(= x \rightsquigarrow^R 1) = x \approx>^R 0$. (jj') The connections between the "left" operations $\odot^R$, $\Rightarrow^R$, $\approx>^R$ are:

(2.50) $$x \odot^R y = (x \Rightarrow^R y^-)^\sim = (y \approx>^R x^\sim)^-.$$

(jjj') The "right" operations ($\oplus$, $\rightarrow^R$, $\rightsquigarrow^R$) expressed in terms of the "left" operations ($\odot^R$, $\Rightarrow^R$, $\approx>^R$) are:

(2.51) $$x \oplus y = (x^- \odot^R y^-)^\sim = (x^\sim \odot^R y^\sim)^- \quad \textit{and}$$

(2.52) $$x \rightarrow^R y = (x^- \Rightarrow^R y^-)^\sim = (x^\sim \Rightarrow^R y^\sim)^-,$$

(2.53) $$x \rightsquigarrow^R y = (x^\sim \approx>^R y^\sim)^- = (x^- \approx>^R y^-)^\sim.$$

(jv') The "left" operations ($\odot^R$, $\Rightarrow^R$, $\approx>^R$) expressed in terms of the "right" operations ($\oplus$, $\rightarrow^R$, $\rightsquigarrow^R$) are:

(2.54) $$x \Rightarrow^R y = (x^- \rightarrow^R y^-)^\sim = (x^\sim \rightarrow^R y^\sim)^-,$$

(2.55) $$x \approx>^R y = (x^- \rightsquigarrow^R y^-)^\sim = (x^\sim \rightsquigarrow^R y^\sim)^-.$$

Consequently, the algebra $\mathcal{A}^{bRL} = (A^R, \geq, \Rightarrow^R, \approx>^R, 0, 1)$ is a bounded pBCK(pP) algebra (with the pseudo-product $\odot^R$) with property (pDN^L) that is termwise equivalent with $\mathcal{A}^{Rb}$. We say that $\mathcal{A}^{Rb}$ is *selfdual*.

Remark 2.11.9 Looking at the starting formula: (2.40) from Remarks 2.11.8, vs. (2.28) from Remarks 2.11.7, we remark that the derivated formulas: (2.44), (2.45) from Remarks 2.11.8, vs. (2.32) from Remarks 2.11.7, and the derivated formulas: (2.46), (2.47) from Remarks 2.11.8, vs. (2.33) from Remarks 2.11.7, are better, by their symmetry. **Hence, the formula (2.40) from Remarks 2.11.8 is better than (2.28) from Remarks 2.11.7 and therefore it will be used in this monograph** when defining left- and right- pseudo-MV algebras (see Definition 3.12.3), for examples.

By (iEL3), we obtain the dual equivalences:

(iEL5) $\mathbf{pBCK(pP)}_{(pDN)} \Leftrightarrow \mathbf{pBCK(pRP)}_{(pDN)}$ and, dually,
$\mathbf{pBCK(pS)}_{(pDN)} \Leftrightarrow \mathbf{pBCK(pcoRS)}_{(pDN)}$

and, by (iEL4), we obtain the dual equivalences, in lattice case:

(iEL6) $\mathbf{pBCK(pP)\text{-}L}_{(pDN)} \Leftrightarrow \mathbf{pBCK(pRP)\text{-}L}_{(pDN)}$ and, dually,
$\mathbf{pBCK(pS)\text{-}L}_{(pDN)} \Leftrightarrow \mathbf{pBCK(pcoRS)\text{-}L}_{(pDN)}$.

Proposition 2.11.10 *(See [94], Proposition 10.2.9)*
Let $\mathcal{A}^{Lb} = (A, \leq, \rightarrow^L, \rightsquigarrow^L, 0, 1)$ be a pBCK(pP)$_{(pDN)}$ lattice. Then, we have: for all $x, y \in A^L$,

(2.56) $$(x \wedge y)^- = x^- \vee y^-, \quad (x \wedge y)^\sim = x^\sim \vee y^\sim,$$

(2.57) $$x \wedge y = (x^- \vee y^-)^\sim, \quad x \wedge y = (x^\sim \vee y^\sim)^-.$$

Proposition 2.11.11 *(See [94], Proposition 10.2.10)*
Let $\mathcal{A}^{Lb}$ be a pBCK(pP)$_{(pDN)}$ lattice which satisfies the property: for all $x \in A^L$,
(pP1^L) $x \wedge x^- = 0 = x \wedge x^\sim = 0$.
Then, $\mathcal{A}^{Lb}$ is a Boolean algebra.

2.12 Pseudo-Hájek(pP), pseudo-Hájek(pRP), pseudo-Wajsberg algebras

In this section, we recall the definitions of pseudo-Hájek(pP), pseudo-Hájek(pRP) and of pseudo-Wajsberg algebras and some results.

We present the equivalences (iEL7), (iEL8), (iEL9), (iEL10).

The pseudo-Hájek(pP) and the pseudo-Hájek(pRP) algebras were introduced in [87] as a non-commutative generalization of the Hájek(P) and Hájek(RP) algebras introduced in [86] (see also [89], [90], [94]).

Definition 2.12.1

(i) A *pseudo-Hájek(pP) algebra*, or a *pHájek(pP) algebra*, is a bounded pBCK(pP) lattice

$$\mathcal{A}^{Lb} = (A^L, \wedge, \vee, \to^L, \rightsquigarrow^L, 0, 1)$$

verifying (pprel^L) and (pdiv^L): for all $x, y \in A^L$,
(pprel^L) (pseudo-prelinearity) $(x \to^L y) \vee (y \to^L x) = 1 = (x \rightsquigarrow^L y) \vee (y \rightsquigarrow^L x)$,
(pdiv^L) (pseudo-divisibility) $x \wedge y = (x \to^L y) \odot x = x \odot (x \rightsquigarrow^L y)$.

(i') Dually, a *pseudo-Hájek(pS) algebra*, or a *pHájek(pS) algebra*, is a bounded pBCK(pS) lattice

$$\mathcal{A}^{Rb} = (A^R, \vee, \wedge, \to^R, \rightsquigarrow^R, 0, 1)$$

verifying (pprel^R) and (pdiv^R): for all $x, y \in A^R$,
(pprel^R) (pseudo-prelinearity) $(x \to^R y) \wedge (y \to^R x) = 0 = (x \rightsquigarrow^R y) \wedge (y \rightsquigarrow^R x)$,
(pdiv^R) (pseudo-divisibility) $x \vee y = (x \to^R y) \oplus x = x \oplus (x \rightsquigarrow^R y)$.

Denote by **pHa(pP)** the class of all pHájek(pP) algebras and by **pHa(pS)** the class of all pHájek(pS) algebras. Hence, we have:

pHa(pP) = **pBCK(pP)-L**^{b} + (pprel^L) + (pdiv^L) and, dually,
pHa(pS) = **pBCK(pS)-L**^{b} + (pprel^R) + (pdiv^R).

Definition 2.12.2

(i) A *pseudo-Hájek(pRP) algebra*, or a *pHájek(pRP) algebra*, is a bounded pBCK(pRP) lattice

$$\mathcal{A}^{Lb} = (A^L, \wedge, \vee, \to^L, \rightsquigarrow^L, \odot, 0, 1)$$

verifying (pprel^L) and (pdiv^L).

(i') Dually, a *pseudo-Hájek(pcoRS) algebra*, or a *pHájek(pcoRS) algebra*, is a bounded pBCK(pcoRS) lattice

$$\mathcal{A}^{Rb} = (A^R, \vee, \wedge, \to^R, \rightsquigarrow^R, \oplus, 0, 1)$$

verifying (pprel^R) and (pdiv^R).

Denote by **pHa(pRP)** the class of all pHájek(pRP) algebras and by **pHa(pcoRS)** the class of all pHájek(pcoRS) algebras. Hence, we have:

pHa(pRP) = **pBCK(pRP)-L**^{b} + (pprel^L) + (pdiv^L) and, dually,
pHa(pcoRS) = **pBCK(pcoRS)-L**^{b} + (pprel^R) + (pdiv^R).

We then have, by (iEL4), the following dual equivalences:

(iEL7) **pHa(pP)** $\Leftrightarrow$ **pHa(pRP)** and, dually,
pHa(pS) $\Leftrightarrow$ **pHa(pcoRS)**

and, by (iEL7), we have the following dual equivalences:

(iEL8) **pHa(pP)** +(pP1^L)+(pP2^L) $\Leftrightarrow$ **pHa(pRP)** +(pP1^L)+(pP2^L)
and, dually,
pHa(pS) +(pP1^R)+(pP2^R) $\Leftrightarrow$ **pHa(pcoRS)**+(pP1^R)+(pP2^R),

where:

(pP1^L) $x \wedge x^{-^L} = 0 = x \wedge x^{\sim^L}$,
(pP2^L) $(z^{-^L})^{-^L} \odot [(x \odot z) \to^L (y \odot z)] \leq x \to^L y$,
$(z^{\sim^L})^{\sim^L} \odot [(z \odot x) \rightsquigarrow^L (z \odot y)] \leq x \rightsquigarrow^L y$;
(pP1^R) $x \vee x^{-^R} = 1 = x \vee x^{\sim^R}$,
(pP2^R) $(z^{-^R})^{-^R} \oplus [(x \oplus z) \to^R (y \oplus z)] \geq x \to^R y$,
$(z^{\sim^R})^{\sim^R} \oplus [(z \oplus x) \rightsquigarrow^R (z \oplus y)] \geq x \rightsquigarrow^R y$.

Definition 2.12.3
(i) We say that a left-pBCK lattice

$$\mathcal{A}^L = (A^L, \wedge, \vee, \to^L, \rightsquigarrow^L, 1)$$

is *with property (pCC^L)*, if we have: for all $x, y \in A^L$,
(pCC^L) $x \vee y = (x \rightsquigarrow^L y) \to^L y = (x \to^L y) \rightsquigarrow^L y$.
(i') Dually, we say that a right-pBCK lattice

$$\mathcal{A}^R = (A^R, \vee, \wedge, \to^R, \rightsquigarrow^R, 0)$$

is *with property (pCC^R)*, if we have: for all $x, y \in A^R$,
(pCC^R) $x \wedge y = (x \to^R y) \rightsquigarrow^R y = (x \rightsquigarrow^R y) \to^R y$.

Theorem 2.12.4 *([94], Theorem 11.1.8)*
(1) Let $\mathcal{A}^L = (A^L, \wedge, \vee, \to^L, \rightsquigarrow^L, 1)$ be a pBCK(pP) lattice. Then,

$$(pC^L) \implies (pprel^L) + (pdiv^L).$$

(1') Dually, let $\mathcal{A}^R = (A^R, \vee, \wedge, \to^R, \rightsquigarrow^R, 0)$ be a pBCK(pS) lattice. Then,

$$(pC^R) \implies (pprel^R) + (pdiv^R).$$

Corollary 2.12.5 *([94], Corollary 10.2.15)*

(1) Let $\mathcal{A}^{Lb} = (A^L, \wedge, \vee, \to^L, \rightsquigarrow^L, 0, 1)$ *be a bounded pBCK(pP) lattice with* (pCC^L) *property. Then,* $\mathcal{A}^{Lb}$ *is with* (pDN^L) *property.*

(1') Dually, let $\mathcal{A}^{Rb} = (A^R, \vee, \wedge, \to^R, \rightsquigarrow^R, 0, 1)$ *be a bounded pBCK(pS) lattice with* (pCC^R) *property. Then,* $\mathcal{A}^{Rb}$ *is with* (pDN^R) *property.*

Pseudo-Wajsberg algebras were introduced in 1999 [24] by R. Ceterchi, as a non-commutative generalization of Wajsberg algebras introduced by Font, Rodriguez and Torrens [53]. They are termwise equivalent to pseudo-MV algebras.

Definition 2.12.6

(i) A *left-pseudo-Wajsberg algebra*, or a *left-pWajsberg algebra* for short, is an algebra

$$\mathcal{A}^L = (A^L, \to^L, \rightsquigarrow^L, ^-, ^\sim, 1)$$

of type $(2,2,1,1,0)$, where $^- = ^{-^L}$ and $^\sim = ^{\sim^L}$, verifying: for all $x, y, z \in A^L$,

(pWL1) $1 \to^L x = x$, $1 \rightsquigarrow^L x = x$ (it is (pM^L)),

(pWL2) $(x \rightsquigarrow^L y) \to^L y = (y \rightsquigarrow^L x) \to^L x = (x \to^L y) \rightsquigarrow^L y = (y \to^L x) \rightsquigarrow^L x$,

(pWL3) $(x \to^L y) \to^L [(y \to^L z) \rightsquigarrow^L (x \to^L z)] = 1$,
$(x \rightsquigarrow^L y) \rightsquigarrow^L [(y \rightsquigarrow^L z) \to^L (x \rightsquigarrow^L z)] = 1$,

(pWL4) $1^- = 1^\sim$ (and let 0 denote this element),

(pWL5) $(x^- \rightsquigarrow^L y^-) \to^L (y \to^L x) = 1$, $(x^\sim \to^L y^\sim) \to^L (y \rightsquigarrow^L x) = 1$,

(pWL6) $(x \to^L y^-)^\sim = (y \rightsquigarrow^L x^\sim)^-$.

(i') Dually, a *right-pseudo-Wajsberg algebra*, or a *right-pWajsberg algebra* for short, is an algebra

$$\mathcal{A}^R = (A^R, \to^R, \rightsquigarrow^R, ^-, ^\sim, 0)$$

of type $(2,2,1,1,0)$, where $^- = ^{-^R}$ and $^\sim = ^{\sim^R}$, verifying: for all $x, y, z \in A^R$,

(pWR1) $0 \to^R x = x$, $0 \rightsquigarrow^R x = x$ (it is (pM^R)),

(pWR2) $(x \rightsquigarrow^R y) \to^R y = (y \rightsquigarrow^R x) \to^R x = (x \to^R y) \rightsquigarrow^R y = (y \to^R x) \rightsquigarrow^R x$,

(pWR3) $(x \to^R y) \to^R [(y \to^R z) \rightsquigarrow^R (x \to^R z)] = 0$,
$(x \rightsquigarrow^R y) \rightsquigarrow^R [(y \rightsquigarrow^R z) \to^R (x \rightsquigarrow^R z)] = 0$,

(pWR4) $0^- = 0^\sim$ (and let 1 denote this element),

(pWR5) $(x^- \rightsquigarrow^R y^-) \to^R (y \to^R x) = 0$, $(x^\sim \to^R y^\sim) \to^R (y \rightsquigarrow x) = 0$,

(pWR6) $(x \to^R y^-)^\sim = (y \rightsquigarrow^R x^\sim)^-$.

Denote by $\mathbf{pW}^L$ the class of all left-pWajsberg algebras and by $\mathbf{pW}^R$ the class of all right-pWajseberg algebras.

A commutative left-pWajsberg algebra is called *left-Wajsberg algebra* and a commutative right-pWajsberg algebra is called *right-Wajsberg algebra.*

The pWajsberg algebras are connected to the pMV algebras in Chapter 4.

Note that a left-pWajsberg algebra verifies (pDN^L) (it is *involutive*), hence it is self-dual, i.e. the dual of $(A^L, \to^L, \rightsquigarrow^L, ^-, ^\sim, 1)$ is $(A^L, \Rightarrow^L, \approx>^L, ^-, ^\sim, 0)$ and vice-versa (see Remarks 2.11.7, 2.11.8, 2.11.9).

Dually, a right-pWajsberg algebra verifies (pDN^R) (it is *involutive*), hence it is self-dual, i.e. the dual of $(A^R, \to^R, \rightsquigarrow^R, ^-, ^\sim, 0)$ is $(A^R, \Rightarrow^R, \approx>^R, ^-, ^\sim, 1)$ and vice-versa (see Remarks 2.11.7, 2.11.8, 2.11.9).

Recall that a left-pseudo-BCK algebra $(A, \leq, \rightarrow^L, \rightsquigarrow^L, 1)$ is *sup-commutative*, if the following property holds: for all $x, y \in A$,
(pcomm^L) $(x \rightarrow^L y) \rightsquigarrow^L y = (y \rightarrow^L x) \rightsquigarrow^L x$,
$(x \rightsquigarrow^L y) \rightarrow^L y = (y \rightsquigarrow^L x) \rightarrow^L x$.

Dually, a right-pseudo-BCK algebra $(A, \geq, \rightarrow^R, \rightsquigarrow^R, 0)$ is *inf-commutative*, if the following property holds: for all, $x, y \in A$,
(pcomm^R) $(x \rightarrow^R y) \rightsquigarrow^R y = (y \rightarrow^R x) \rightsquigarrow^R x$,
$(x \rightsquigarrow^R y) \rightarrow^R y = (y \rightsquigarrow^R x) \rightarrow^R x$.

Then, we have the following important result.

Theorem 2.12.7 *([94], Theorem 10.1.39) (The dual case is omitted)*
The left-pWajsberg algebras are categorically equivalent to the bounded sup-commutative left-pBCK algebras.

Remark 2.12.8 By above theorem, we can define equivalently the left-pWajsberg algebras as *bounded sup-commutative left-pBCK algebras.* Thus, we obtain examples of left-pWajsberg algebras from the examples of bounded sup-commutative left-pBCK algebras.

We have then the dual equivalences (iEL9):

(iEL9) $\mathbf{pHa(pRP)}_{(pDN)} \iff \mathbf{pW}^L$ and, dually,
$\mathbf{pHa(pcoRS)}_{(pDN)} \iff \mathbf{pW}^R$,

where
$\mathbf{pHa(pRP)}_{(pDN)} = \mathbf{pBCK(pRP)\text{-}L}_{(pDN)}$ + (pprel^L) + (pdiv^L),
$\mathbf{pHa(pcoRS)}_{(pDN)} = \mathbf{pBCK(pcoRS)\text{-}L}_{(pDN)}$ + (pprel^R) + (pdiv^R),

namely we have the following theorem (the dual one is omitted).

Theorem 2.12.9 *(See Theorem 3.12.4)*
1) Let $\mathcal{A}^L = (A^L, \rightarrow^L, \rightsquigarrow^L, ^-, ^\sim, 1)$ be a left-pWajsberg algebra.
Define $\rho(\mathcal{A}^L) \stackrel{def.}{=} (A^L, \wedge, \vee, \rightarrow^L, \rightsquigarrow^L, \odot, 0, 1)$ by: for all $x, y \in A^L$,
$x \wedge y \stackrel{def.}{=} x \odot (x \rightsquigarrow^L y) = y \odot (y \rightsquigarrow^L x) = (y \rightarrow^L x) \odot y = (x \rightarrow^L y) \odot x$,
$x \vee y \stackrel{def.}{=} (x \rightsquigarrow^L y) \rightarrow^L y = (y \rightsquigarrow^L x) \rightarrow^L x = (x \rightarrow^L y) \rightsquigarrow^L y = (y \rightarrow^L x) \rightsquigarrow^L x$,
by (pWL2),
$x \odot y \stackrel{def.}{=} (x \rightarrow^L y^-)^\sim = (y \rightsquigarrow^L x^\sim)^-$, $0 \stackrel{def.}{=} 1^- = 1^\sim$.
Then, $\rho(\mathcal{A}^L)$ is a pHájek(pRP) algebra with (pDN^L).
1') Conversely, let $\mathcal{A}^{Lb} = (A^L, \wedge, \vee, \rightarrow^L, \rightsquigarrow^L, \odot, 0, 1)$ be a pHájek(pRP) algebra with (pDN^L), where $x^- \stackrel{def.}{=} x \rightarrow^L 0$ and $x^\sim \stackrel{def.}{=} x \rightsquigarrow^L 0$.
Define $\rho'(\mathcal{A}^{Lb}) \stackrel{def.}{=} (A^L, \rightarrow^L, \rightsquigarrow^L, ^-, ^\sim, 1)$.
Then, $\rho'(\mathcal{A}^{Lb})$ is a left-pWajsberg algebra.
2) The above defined mappings ρ and ρ' are mutually inverse.

The proof is routine. □

We also have the following important result:

Theorem 2.12.10 *([94], Theorem 10.2.16)*

(1) The bounded pBCK(pP) lattice with (pCC^L) property is an equivalent definition of the left-pWajsberg algebra.

(1') Dually, the bounded pBCK(pS) lattice with (pCC^R) property is an equivalent definition of the right-pWajsberg algebra.

Theorem 2.12.10 says that we have the special equivalences (iEL10):

(iEL10) $\mathbf{pBCK(pP)\text{-}L}^b$ + (pCC^L) $\Longleftrightarrow$ $\mathbf{pW}^L$ and, dually,
$\mathbf{pBCK(pS)\text{-}L}^b$ + (pCC^R) $\Longleftrightarrow$ $\mathbf{pW}^R$.

2.13 Implicative-Boolean algebras

The implicative-Boolean algebra, a term equivalent definition of Boolean algebra, was introduced in [94] and studied in [95], [62].

Definition 2.13.1

(i) A *left-implicative-Boolean algebra* is an algebra $\mathcal{A}^L = (A^L, \rightarrow^L, ^-, 1)$ of type $(2, 1, 0)$, where $^- = ^{-^L}$, verifying the following axioms: for all $x, y, z \in A^L$,
(G1-L) $x \rightarrow^L (y \rightarrow^L x) = 1$,
(G2-L) $[x \rightarrow^L (y \rightarrow^L z)] \rightarrow^L [(x \rightarrow^L y) \rightarrow^L (x \rightarrow^L z)] = 1$,
(G3-L) $(y^- \rightarrow^L x^-) \rightarrow^L (x \rightarrow^L y) = 1$,
(G4-L)=(An^L) $x \rightarrow^L y = 1$ and $y \rightarrow^L x = 1$ implies $x = y$.

(i') Dually, a *right-implicative-Boolean algebra* is an algebra $\mathcal{A}^R = (A^R, \rightarrow^R, ^-, 0)$ of type $(2, 1, 0)$, where $^- = ^{-^R}$, verifying the following axioms: for all $x, y, z \in A^R$,
(G1-R) $x \rightarrow^R (y \rightarrow^R x) = 0$,
(G2-R) $[x \rightarrow^R (y \rightarrow^R z)] \rightarrow^R [(x \rightarrow^R y) \rightarrow^R (x \rightarrow^R z)] = 0$,
(G3-R) $(y^- \rightarrow^R x^-) \rightarrow^R (x \rightarrow^R y) = 0$,
(G4-R)=(An^R) $x \rightarrow^R y = 0$ and $y \rightarrow^R x = 0$ implies $x = y$.

Denote by **implicative-Boole**^{L} the class of left-implicative-Boolean algebras and by **implicative-Boole**^{R} the class of right-implicative-Boolean algebras.

The implicative-Boolean algebras are connected to the Boolean algebras in Chapter 4.

Note that the left-implicative-Boolean algebra verifies (pDN^L) (it is *involutive*), hence it is self-dual, i.e. the dual of $(A^L, \rightarrow^L, ^-, 1)$ is $(A^L, \Rightarrow^L, ^-, 0)$ and vice-versa (see Remarks 2.11.7, 2.11.8).

Dually, the right-implicative-Boolean algebra verifies (pDN^R) (it is *involutive*), hence it is self-dual, i.e. the dual of $(A^R, \rightarrow^R, ^-, 0)$ is $(A^R, \Rightarrow^R, ^-, 1)$ and vice-versa (see Remarks 2.11.7, 2.11.8).

2.14 The hierarchy No. 5 of pseudo-algebras of logic

We present a simplified hierarchy (we consider the (bounded) pseudo-M algebras with additional operation(s) as particular cases of pseudo-M algebras) of some of the pseudo-algebras of logic (pM algebras) (that are particular cases of pBCI algebras and pBCK algebras) recalled in this chapter, in Figure 2.5.

The six classes of algebras: **pBCI**^{L}, **pBCI(pP)**, **pBCI-L**^{L}, **pBCI(pP)-L**, **pBCK**^{L}, **pBCK-L**^{L} which have no equivalent in the world of unital magmas are represented by a smaller bullet.

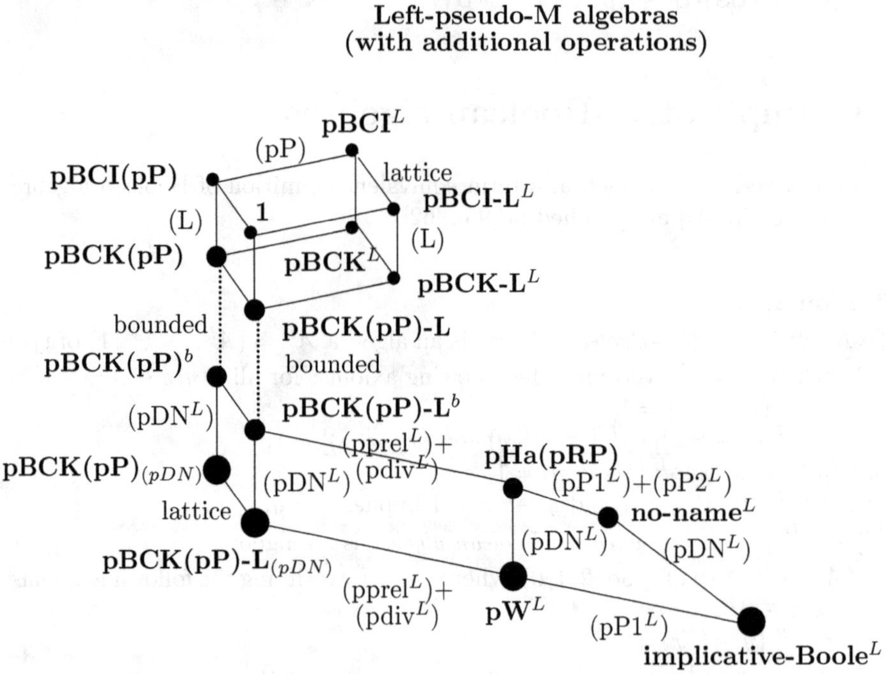

Figure 2.5: The hierarchy No. 5 of some left-pseudo-algebras of logic, where **1** is **pBCI(pP)-L** and **no-name**^{L} is **pHa(pRP)** + (pP1^L) + (pP2^L)

Chapter 3

Unital magmas (with additional operations)

In this chapter, we recall the definitions of *unital magmas*, monoids, loops and groups and their hierarchy; the groups, the po-groups and the l-groups will be recalled and analized in some details in Chapter 7.

We shall also recall in this chapter the definitions of posets, (bounded) lattices and of the left and right po-ms, po-ims, l-ms, l-ims, po-ims(pR), po-ims(pPR), l-ims(pR), l-ims(pPR), pseudo-residuated lattices, pseudo-BL algebras, pseudo-product algebras, pseudo-MV algebras and Boolean algebras.

We present some results, including the equivalences (EM1), (iEM1), (EM2), (iEM2), (iEM3) - (iEM9) between some classes of unital magmas, and some hierarchies of some of the overviewed unital magmas [94].

Remark 3.0.1 (see Remark 2.0.1)

The reader will note through this monograph the crucial role played by the properties (pU) and (Ass) in the "world" of non-commutative unital magmas:
- (pU) corresponds to the dual properties (pM^L), (pM^R) in the "world" of pseudo-M algebras (of logic) (presented in Chapter 2); (pU) will be generalized to (qm-pU) and (pM^L), (pM^R) will be generalized to (q-pM^L), (q-pM^R), in Chapter 5;
- (Ass) corresponds to the dual properties (pEx^L), (pEx^R) in the "world" of pseudo-M algebras (of logic) (Chapter 2).

The chapter has 14 sections.

3.1 Unital magmas. Monoids, loops, groups. The hierarchy No. 1

We recall the definition of *unital magma*, the analogous, in the "world" of unital magmas, of the *pseudo-M algebra*, from the "world" of pseudo-M algebras (of logic).

Definition 3.1.1 A *unital magma* is an algebra $(A, \otimes, e)$ of type $(2, 0)$ verifying the axiom: for all $x \in A$,
(pU) $x \otimes e = x = e \otimes x$, i.e. e is the *unit* element of A.

Note that a unital magma can be denoted:
- multiplicatively (or left), by $(A, \odot, 1)$ or $(A, \cdot, 1)$, in which case the unit 1 is called the *identity* element, or
- additively (or right), by $(A, \oplus, 0)$ or $(A, +, 0)$, in which case the unit 0 is called the *neutral* element.

Denote by **unital magma** the class of all unital magmas.

An unital magma is *commutative*, or *abelian*, if $\otimes$ is commutative, i.e. $x \otimes y = y \otimes x$, for all x, y; in this, case (pU) becomes (U): $x \otimes e = x$, for all x.

Definitions 3.1.2

A *monoid* is an unital magma $(A, \otimes, e)$ verifying the axiom:
(Ass) (Associativity) $x \otimes (y \otimes z) = (x \otimes y) \otimes z$, for all $x, y, z \in A$.

A *loop* is a quasigroup with a unit element [22], i.e. is an unital magma $(A, \otimes, e)$ such that:
(qg) in the equation $x \otimes y = z$, if any two of the symbols x, y, z belong to A, then the third is uniquely determined as an element of A.

A *group* is an unital magma $(A, \otimes, e)$ verifying (Ass) and
(pIv') (pseudo-Inverse) for all $x \in A$, there exists $y \in A$, such that $x \otimes y = e = y \otimes x$.

Denote by **monoid** the class of all monoids, by **loop** the class of all loops and by **group** the class of all groups.

Hence we have the equivalences:

(UMM) **unital magma** + (Ass) $\Longleftrightarrow$ **monoid**,
(MG) **monoid** + (pIv') $\Longleftrightarrow$ **group**,
(UML) **unital magma** + (qg) $\Longleftrightarrow$ **loop**,
(LG) **loop** + (Ass) $\Longleftrightarrow$ **group**.

Hence, we have the hierarchy No. 1 from Figure 3.1.

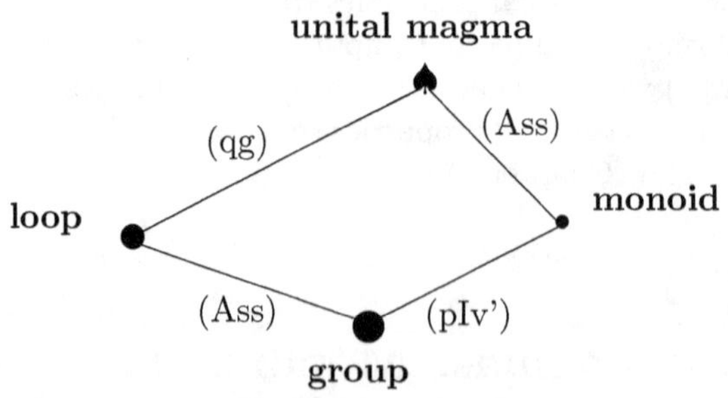

Figure 3.1: The hierarchy No. 1 of unital magmas

Note that in Chapter 10 we shall introduce a new notion, the *hub*, different of loop, as a special unital magma and a generalization of the group.

3.2 Posets and (bounded) lattices

A *partially-ordered set*, or *poset* or *po-set* for short, is a structure $(A, \leq)$, where $\leq$ is an external partially-ordered binary relation, or an order relation for short, on A, i.e. $\leq$ is a binary relation that is reflexive (the property (Re): $x \leq x$, for all $x \in A$), antisymmetric (the property (An): $x \leq y,\ y \leq x \Longrightarrow x = y$, for all $x, y \in A$) and transitive (the property (Tr): $x \leq y,\ y \leq z \Longrightarrow x \leq z$, for all $x, y, z \in A$). We say that the poset $(A, \leq)$ is a *chain*, if the order relation $\leq$ is total, i.e. $x \leq y$ or $y \leq x$, for all $x, y \in A$.

The presence of the *order* relation $\leq$ implies the presence of the *duality principle*. Thus, each poset $(A^L, \leq)$, with the order relation $\leq$, has a dual poset, $(A^R, \geq)$, with the *dual order* relation denoted by $\geq$. Note that we may have $A^L = A^R = A$, in which case we say that $(A, \leq)$ is self-dual, with $x \geq y \Leftrightarrow y \leq x$, for all $x, y \in A$.

A poset with *maximal element,* 1 (*minimal element,* 0) is a structure $(A, \leq, 1)$ $((A, \leq, 0))$ such that $1 \in A$ $(0 \in A)$ and $1 \leq x \Longrightarrow x = 1$ ($x \leq 0 \Longrightarrow x = 0$, respectively), for all $x \in A$.

A poset with *greatest element,* 1 (*smallest element,* 0) is a structure $(A, \leq, 1)$ $((A, \leq, 0))$ such that $1 \in A$ $(0 \in A)$ and $x \leq 1$ ($0 \leq x$, respectively), for all $x \in A$.

A *bounded* poset is a poset $(A, \leq)$ having both greatest element, 1, and smallest element, 0, i.e. we have $0 \leq x \leq 1$, for all $x \in A$; it is denoted by $(A, \leq, 0, 1)$.

If the order relation $\leq$ on a set A is a lattice order (i.e. there exist $\inf(x, y)$ and $\sup(x, y)$ for all $x, y \in A$), then the poset $(A, \leq)$ is called *Ore lattice.*

If we denote by $\wedge$ and $\vee$ the Dedekind lattice operations on A (i.e. $\wedge$ and $\vee$ are idempotent ($x \wedge x = x$, $x \vee x = x$, for all $x \in A$), commutative ($x \wedge y = y \wedge x$, $x \vee y = y \vee x$, for all $x, y \in A$), associative ($x \wedge (y \wedge z) = (x \wedge y) \wedge z$, $x \vee (y \vee z) = (x \vee y) \vee z$, for all $x, y, z \in A$) and the absorption laws hold ($x \wedge (x \vee y) = x$, $x \vee (x \wedge y) = x$, for all $x, y \in A$)), where for all $x, y \in A$,

$$x \leq y \Leftrightarrow x \wedge y = x \Leftrightarrow x \vee y = y,$$

then the algebra $(A, \wedge, \vee)$ (or $(A, \vee, \wedge)$) is called *Dedekind lattice.*

The two definitions of the lattices (Ore and Dedekind) are definitionally equivalent, i.e. we have the following well known result.

Theorem 3.2.1

(1) Let $\mathcal{A} = (A, \leq)$ be an Ore lattice. Define $\Phi(\mathcal{A}) \stackrel{def.}{=} (A, \wedge, \vee)$, where, for all $x, y \in A$,

$$x \wedge y \stackrel{def.}{=} \inf(x, y), \quad x \vee y \stackrel{def.}{=} \sup(x, y). \tag{3.1}$$

Then, $\Phi(\mathcal{A})$ is a Dedekind lattice.

(1') Let $\mathcal{A} = (A, \wedge, \vee)$ be a Dedekind lattice. Define $\Psi(\mathcal{A}) \stackrel{def.}{=} (A, \leq)$, where, for all $x, y \in A$,

$$x \leq y \stackrel{def.}{\Leftrightarrow} x \vee y = y \quad (\text{or } x \wedge y = x). \tag{3.2}$$

Then, the binary relation $\leq$ is an order relation and the structure $\Psi(\mathcal{A})$ is an Ore lattice, where for all $x, y \in A$,

$$\inf(x, y) \overset{def.}{=} x \wedge y, \quad \sup(x, y) \overset{def.}{=} x \vee y. \tag{3.3}$$

(2) The two mappings, Φ and Ψ, are mutually inverse.

We shall freely use in the sequel both definitions of lattices.

Note that a lattice is self-dual, i.e. the dual of $(A, \leq)$ is $(A, \geq)$ and vice-versa, or the dual of $(A, \wedge, \vee)$ is $(A, \vee, \wedge)$ and vice-versa. We shall say that $(A, \wedge, \vee)$ is the *left-lattice*, while $(A, \vee, \wedge)$ is the *right-lattice*.

A lattice $(A, \leq, 0, 1)$ (or $(A, \wedge, \vee, 0, 1)$) is *bounded*, if the poset $(A, \leq, 0, 1)$ is bounded, in which case we have $0 \leq x \leq 1$ (or $0 \vee x = x = x \wedge 1$, respectively). A bounded lattice is denoted either by $(A, \leq, 0, 1)$ (dually, by $(A, \geq, 0, 1)$) or by $(A, \wedge, \vee, 0, 1)$ (dually, by $(A, \vee, \wedge, 0, 1)$).

Note that if $(A, \wedge, \vee, 0, 1)$ is a bounded lattice, then $(A, \wedge, 1)$ and $(A, \vee, 0)$ are commutative monoids.

• **Hasse diagrams and Hasse-type diagrams**

Let $\mathcal{A} = (A, \leq)$ be a structure, where $\leq$ is a binary relation on A.

(1) If $\leq$ is an order relation, then $\mathcal{A}$ it is represented by the usual *Hasse diagram*: each element of A is represented by a bullet $\bullet$ and if $x \leq y$ and there is no z such that $x < z < y$, then x is reprezented below y and a line will connect them.

(1) If $\leq$ is not an order relation (i.e. the binary relation $\leq$ is neither reflexive nor antisymmetric nor transitive, or it is only reflexive, or reflexive and transitive, or reflexive and antisymmetric), then it is represented by a *Hasse-type diagram*: each element of the algebra is represented by a circ $\circ$, and if $x \leq y$ and $y \leq x$ and $x \neq y$ (i.e. x and y have the *same height*, or are *parallel*)), then a horizontal line will connect them.

3.3 Po-ms and po-ims

Definition 3.3.1

(i) A *left-partially-ordered monoid*, or a *left-po-m* for short, is a structure

$$\mathcal{A}^L = (A^L, \leq, \odot, 1)$$

such that $(A^L, \leq)$ is a poset containing an element 1, $(A^L, \odot, 1)$ is a monoid and $\leq$ is compatible with $\odot$, i.e. the following property holds: for all $x, y, a \in A^L$,
(pCp) $x \leq y \Longrightarrow a \odot x \leq a \odot y,\ x \odot a \leq y \odot a$.

(i') Dually, a *right-partially-ordered monoid*, or a *right-po-m* for short, is a structure

$$\mathcal{A}^R = (A^R, \geq, \oplus, 0)$$

such that $(A^R, \leq, 0)$ is a poset containing an element 0 (where $x \leq y \Leftrightarrow y \geq x$, for all $x, y \in A^R$), $(A^R, \oplus, 0)$ is a monoid and $\geq$ is compatible with $\oplus$, i.e. the

following property holds: for all $x, y, a \in A^R$,
(pCp) $x \geq y \Longrightarrow a \oplus x \geq a \oplus y,\ x \oplus a \geq y \oplus a$.

A po-m is *commutative*, if the monoid operation $\odot$ is commutative. A commutative po-m is named *po-cm*.

Denote by $\mathbf{po\text{-}m}^L$ the class of all left-po-ms and by $\mathbf{po\text{-}m}^R$ the class of all right-po-ms.

Remark 3.3.2

The statement: "The structure $(A^L, \leq, \odot, 1)$ is a left-po-m" is equivalent to the statement: "The pseudo-product $\odot$ is a pseudo-t-norm on the poset $(A^L, \leq)$ containing the element 1", by Definition 1.1.1.

Dually, the statement: "The structure $(A^R, \geq, \oplus, 0)$ is a right-po-m" is equivalent to the statement: "The pseudo-sum $\oplus$ is a pseudo-t-conorm on the poset $(A^R, \leq)$ containing the element 0, by Definition 1.1.1".

Definition 3.3.3

(i) A *left-partially-ordered integral monoid*, or a *left-po-im* for short, is a left-po-m

$$\mathcal{A}^L = (A^L, \leq, \odot, 1)$$

such that $(A^L, \leq, 1)$ is a poset with 1 as greatest element, i.e. the additional property (L) holds: for all $x \in A^L$,
(L) (Last element) $\quad x \leq 1$;
integral means that the greatest element of the poset coincides with the unit element of the monoid.

(i') Dually, a *right-partially-ordered integral monoid*, or a *right-po-im* for short, is a right-po-m

$$\mathcal{A}^R = (A^R, \geq, \oplus, 0)$$

such that $(A^R, \leq, 0)$ is a poset with 0 as smallest element (where $x \leq y \Leftrightarrow y \geq x$, for all $x, y \in A^R$), i.e. the additional property (F) holds: for all $x \in A^R$,
(F) (First element) $\quad x \geq 0$;
integral means that the smallest element of the poset coincides with the unit element of the monoid.

A commutative po-im is named *po-cim*.

Denote by $\mathbf{po\text{-}im}^L$ the class of all left-po-ims and by $\mathbf{po\text{-}im}^R$ the class of all right-po-ims.

Hence, a left-po-im is a left-po-m verifying (L) and a right-po-im is a right-po-m verifying (F). We write:

$\mathbf{po\text{-}im}^L = \mathbf{po\text{-}m}^L + $ (L) and, dually, $\mathbf{po\text{-}im}^R = \mathbf{po\text{-}m}^R + $ (F).

3.3.1 The negative and positive cones of po-ms

Let $\mathcal{A} = (A, \leq, +, 0)$ be a po-m (left-po-m or right-po-m).
(i) Define the "negative (left) cone" of $\mathcal{A}$ as follows:

$$A^- \stackrel{def.}{=} \{x \in A \mid x \leq 0\}.$$

(i') Define the "positive (right) cone" of $\mathcal{A}$ as follows:

$$A^+ \stackrel{def.}{=} \{x \in A \mid x \geq 0\}.$$

Then, we obtain the following results.

Lemma 3.3.4 *A^- and A^+ are closed under $+$.*

Proof. Let $x, y \in A^-$, i.e. $x \leq 0$ and $y \leq 0$. Then, $x + y \stackrel{(pCp)}{\leq} 0 + y \stackrel{(pU)}{=} y \leq 0$; hence, $x + y \leq 0$, by transitivity, i.e. $x + y \in A^-$. Similarly, A^+ is closed under $+$.
□

Proposition 3.3.5
Let $\mathcal{A} = (A, \leq, +, 0)$ be a po-m. Then,
(1) $(A^-, \leq, \odot = +, \mathbf{1} = 0)$ is a left-po-im.
(1') $(A^+, \geq, \oplus = +, \mathbf{0} = 0)$ is a right-po-im.

Proof. By Lemma 3.3.4 □

3.3.2 Bounded po-ims

Definition 3.3.6
(i) A left-po-im $\mathcal{A}^L = (A^L, \leq, \odot, 1)$ is *bounded*, if the poset $(A^L, \leq, 1)$ is bounded, i.e. besides the last (greatest) element 1, there exists a first (smallest) element 0 too. A *bounded left-po-im* is denoted by

$$\mathcal{A}^{Lb} = (A^L, \leq, \odot, 0, 1).$$

(i') Dually, a right-po-im $\mathcal{A}^R = (A^R, \geq, \oplus, 0)$ is *bounded*, if the poset $(A^R, \leq, 0)$ is bounded, i.e. besides the first (smallest) element 0, there exists a last (greatest) element 1 too. A *bounded right-po-im* is denoted by

$$\mathcal{A}^{Rb} = (A^R, \geq, \oplus, 0, 1).$$

Denote by $\mathbf{po\text{-}im}^{Lb}$ ($\mathbf{po\text{-}im}^{Rb}$) the class of bounded left-po-ims (bounded right-po-ims, respectively).

Proposition 3.3.7 *Any finite po-im is bounded.*

Proof. Let $(A^L, \leq, \odot, 1)$ be a finite left-po-im. Suppose there exist $a, b \in A^L$ such that $a \neq b$ and a, b are incomparable. Since $a, b \leq 1$, it follows that $a \odot b \leq 1 \odot b = b$ and $a \odot b \leq a \odot 1 = a$, by (pCp) and (pU), i.e. there is $a \odot b \leq a, b$. Then, apply induction. □

Since any finite left-po-im is bounded, i.e. it has a smallest element too, we shall denote the smallest element by 0. Hence, a finite left-po-im $\mathcal{A}^L = (A^L, \leq, \odot, 1)$ becomes $\mathcal{A}^{Lb} = (A^L, \leq, \odot, 0, 1)$. Similarly in the dual case.

Remark 3.3.8 There exist infinite po-ims which are not bounded (see Example 5 in Chapter 5).

Theorem 3.3.9 *(See Theorem 2.3.16)*
Any bounded po-im can be extended to a po-m.

Proof. Let $(M, \leq, \odot, 0, 1)$ be a bounded left-po-im and $\delta \notin M$. Define on $M' = M \cup \{\delta\}$ the operation $\odot'$ by: for all $x, y \in M'$,

$$x \odot' y = \begin{cases} x \odot y, & \text{if } x, y \in M, \\ \delta, & \text{if } x = \delta,\ y \in M \text{ or } x \in M,\ y = \delta, \\ 0, & \text{if } x = y = \delta. \end{cases}$$

Then, obviously, $(M', \leq, \odot', 1)$ is only a po-m, because $\delta \not\leq 1$. □

3.4 *l*-ms and *l*-ims

In lattice case (i.e. $\leq$ is a lattice operation), we have the following definitions.

Definition 3.4.1 (i) A *left-l-m* (*left-l-im*) is a is a left-po-m (left-po-im, respectively)

$$\mathcal{A}^L = (A^L, \leq, \odot, 1)$$

where $\leq$ is a lattice order. A left-*l*-m (left-*l*-im) will be denoted by

$$(A^L, \wedge, \vee, \odot, 1).$$

(i') Dually, a *right-l-m* (*right-l-im*) is a is a right-po-m (right-po-im, respectively)

$$\mathcal{A}^R = (A^R, \geq, \oplus, 0)$$

where $\geq$ is a lattice order. A right-*l*-m (right-*l*-im) will be denoted by

$$(A^R, \vee, \wedge, \oplus, 0).$$

Denote by $\boldsymbol{l}$**-m**^{L} ($\boldsymbol{l}$**-im**^{L}) the class of all left-*l*-ms (left-*l*-ims, respectively) and denote by $\boldsymbol{l}$**-m**^{R} ($\boldsymbol{l}$**-im**^{R}) the class of all right-*l*-ms (right-*l*-ims, respectively).

3.4.1 The negative and positive cones of l-ms

Lemma 3.4.2 *Let $\mathcal{A} = (A, \wedge, \vee, +, 0)$ be a l-m (left-l-m or right-l-m). Then, A^- and A^+ are closed under $\wedge$ and $\vee$.*

Proof.
Let $x, y \in A^-$, i.e. $x \leq 0$ and $y \leq 0$. Then, 0 is an upper bound of $\{x, y\}$; hence, $0 \geq x \vee y$, i.e. $x \vee y \in A^-$; since $0 \geq x \vee y \geq x \wedge y$, it follows that $0 \geq x \wedge y$, i.e. $x \wedge y \in A^-$ too.
Let now $x, y \in A^+$, i.e. $x \geq 0$ and $y \geq 0$. Then, 0 is a lower bound of $\{x, y\}$; hence, $0 \leq x \wedge y$, i.e. $x \wedge y \in A^+$; since $0 \leq x \wedge y \leq x \vee y$, it follows that $0 \leq x \vee y$, i.e. $x \vee y \in A^+$ too. □

Then, by Proposition 3.3.5 and Lemma 11.1.41, we obtain obviously the following result.

Proposition 3.4.3 *Let $\mathcal{A} = (A, \wedge, \vee, +, 0)$ be a l-m. Then,*
(1) $(A^-, \wedge, \vee, \odot = +, \mathbf{1} = 0)$ is a left-l-im,
(1') $(A^+, \vee, \wedge, \oplus = +, \mathbf{0} = 0)$ is a right-l-im.

3.4.2 Bounded l-ims

Definition 3.4.4
(i) A left-l-im $\mathcal{A}^L = (A^L, \wedge, \vee, \odot, 1)$ is said to be *bounded*, if the left-po-im $(A^L, \leq, \odot, 1)$ is bounded ($x \leq y \Leftrightarrow x \wedge y = x \Leftrightarrow x \vee y = y$), with 0 as the first (smallest) element. A *bounded left-l-im* is denoted by

$$\mathcal{A}^{Lb} = (A^L, \wedge, \vee, \odot, 0, 1).$$

(i') Dually, a right-l-im $\mathcal{A}^R = (A^R, \vee, \wedge, \oplus, 0)$ is said to be *bounded*, if the right-po-im $(A^R, \geq, \oplus, 0)$ is bounded ($x \leq y \Leftrightarrow x \vee y = y \Leftrightarrow x \wedge y = x$, $x \leq y \Leftrightarrow y \geq x$), with 1 as the last (greatest) element. A *bounded right-l-im* is denoted by:

$$\mathcal{A}^{Rb} = (A^R, \vee, \wedge, \oplus, 0, 1).$$

Denote by l-$\mathbf{im}^{Lb}$ the class of all bounded left-l-ims and, dually, denote by l-$\mathbf{im}^{Rb}$ the class of all bounded right-l-ims.

3.5 Po-ms(pR) and po-ims(pR)

Definition 3.5.1
(i) A *left-po-m (left-po-im) with pseudo-residuum*, or a *po-m(pR) (po-im(pR))* for short, is a left-po-m (left-po-im, respectively)

$$\mathcal{A}^L = (A^L, \leq, \odot, 1)$$

verifying additionally the property (pR): for all $y, z \in A^L$, there exist $\max\{x \in A^L \mid x \odot y \leq z\}$ and $\max\{x \in A^L \mid y \odot x \leq z\}$ (that are unique by the antisymmetry of

$\leq$) and they are denoted by $y \rightarrow^L z$ and $y \rightsquigarrow^L z$, respectively; for short,
(pR) (pseudo-residuum) $\exists\, y \rightarrow^L z \overset{notation}{=} \max\{x \in A^L \mid x \odot y \leq z\}$,
$\exists\, y \rightsquigarrow^L z \overset{notation}{=} \max\{x \in A^L \mid y \odot x \leq z\}$.

(i') Dually, a *right-po-m (right-po-im) with pseudo-coresiduum*, or a *po-m(pcoR) (po-im(pcoR))* for short, is a right-po-m (right-po-im, respectively)

$$\mathcal{A}^R = (A^R, \geq, \oplus, 0)$$

verifying additionally the property (pcoR): for all $y, z \in A^R$,
(pcoR) (pseudo-coresiduum) $\exists\, y \rightarrow^R z \overset{notation}{=} \min\{x \in A^R \mid x \oplus y \geq z\}$,
$\exists\, y \rightsquigarrow^R z \overset{notation}{=} \min\{x \in A^R \mid y \oplus x \geq z\}$.

We shall see later (Theorem 4.2.1) that, in the case of po-ims(pR), the defined pseudo-implication $(\rightarrow^L, \rightsquigarrow^L)$ is indeed a pseudo-residuum and that the defined pseudo-coimplication $(\rightarrow^R, \rightsquigarrow^R)$ is indeed a pseudo-coresiduum, by Definition 1.1.3. But for po-ms, this is not always the case (see Example 5' in Chapter 6).

Denote by **po-m(pR)** the class of all po-ms(pR) and by **po-m(pcoR)** the class of all po-ms(pcoR).

Denote by **po-im(pR)** the class of all po-ims(pR) and by **po-im(pcoR)** the class of all po-ims(pcoR).

Note that, by Definition 3.5.1 (i), in a po-m(pR) (po-im(pR)) $\mathcal{A}^L$ we have, for all $y, z \in A^L$,

$$y \rightarrow^L z \in \{x \in A^L \mid x \odot y \leq z\}, \quad y \rightsquigarrow^L z \in \{x \in A^L \mid y \odot x \leq z\},$$

hence we have the following property: for all $y, z \in A^L$,
(pRR) $\quad (y \rightarrow^L z) \odot y \leq z, \quad y \odot (y \rightsquigarrow^L z) \leq z$.

Dually, by Definition 3.5.1 (i'), in a po-m(pcoR) (po-im(pcoR)) $\mathcal{A}^R$ we have, for all $y, z \in A^R$,

$$y \rightarrow^R z \in \{x \in A^R \mid x \oplus y \geq z\}, \quad y \rightsquigarrow^R z \in \{x \in A^R \mid y \oplus x \geq z\},$$

hence we have the following property: for all $y, z \in A^R$,
(pcoRR) $\quad (y \rightarrow^R z) \oplus y \geq z, \quad y \oplus (y \rightsquigarrow^R z) \geq z$.

Remark 3.5.2 We can define equivalently a po-m(pR) (po-im(pR)) as a structure

$$\mathcal{A}^L = (A^L, \leq, \odot, \rightarrow^L, \rightsquigarrow^L, 1)$$

such that $(A^L, \leq, \odot, 1)$ is a left-po-m (left-po-im, respectively) where, for all $y, z \in A^L$, there exist $\max\{x \in A^L \mid x \odot y \leq z\}$ and $\max\{x \in A^L \mid y \odot x \leq z\}$, and the pseudo-implication $(\rightarrow^L, \rightsquigarrow^L)$ can be expressed in terms of $\odot$: for all $y, z \in A^L$,

$$y \rightarrow^L z = \max\{x \in A^L \mid x \odot y \leq z\}, \quad y \rightsquigarrow^L z = \max\{x \in A^L \mid y \odot x \leq z\}.$$

Similarly in the dual case.

A commutative po-m(pR) is named *po-cm(pR)* and a commutative po-m(pcoR) is named *po-cm(pcoR)*. A commutative po-im(pR) is named *po-cim(pR)* and a commutative po-im(pcoR) is named *po-cim(pcoR)*.

3.6 Po-ms(pPR) and po-ims(pPR) - two intermediary notions

Definition 3.6.1

(i) A *left-po-m (left-po-im) with the property (pPR)*, or a *po-m(pPR) (po-im(pPR))* for short, is a structure

$$\mathcal{A}^L = (A^L, \leq, \odot, \to^L, \leadsto^L, 1)$$

such that $(A^L, \leq)$ is a poset ($(A^L, \leq, 1)$ is a poset with 1, respectively), $(A^L, \odot, 1)$ is a monoid and the pseudo-implication $(\to^L, \leadsto^L)$ is connected to the pseudo-product $\odot$ by the following property (pPR): for all $x, y, z \in A^L$,
(pPR) $\quad x \odot y \leq z \Longleftrightarrow x \leq y \to^L z \Leftrightarrow y \leq x \leadsto^L z.$

(i') Dually, a *right-po-m (right-po-im) with the property (pScoR)*, or a *po-m(pScoR) (po-im(pScoR))* for short, is a structure

$$\mathcal{A}^R = (A^R, \geq, \oplus, \to^R, \leadsto^R, 0)$$

such that $(A^R, \leq)$ is a poset ($(A^R, \leq, 0)$ is a poset with 0, respectively), $(A^R, \oplus, 0)$ is a monoid and the pseudo-coimplication $(\to^R, \leadsto^R)$ is connected to the pseudo-sum $\oplus$ by the following property (pScoR): for all $x, y, z \in A^R$,
(pScoR) $\quad x \oplus y \geq z \Longleftrightarrow x \geq y \to^R z \Leftrightarrow y \geq x \leadsto^R z.$

Denote by **po-m(pPR)** the class of all po-ms(pPR) and by **po-m(pScoR)** the class of all po-ms(pScoR).

Denote by **po-im(pPR)** the class of all po-ims(pPR) and by **po-im(pScoR)** the class of all po-ims(pScoR).

Note that the property (pPR) coincides with the property (pRP) and that the dual property (pScoR) coincides with the dual property (pcoRS) that appeared before in the study of algebras of logic (Chapter 2):

$$(pPR) = (pRP) \quad \textit{and, dually,} \quad (pScoR) = (pcoRS).$$

Note that the dual equivalences (pPR)=(pRP) and (pScoR)=(pcoRS) are non-commutative *Galois connections.*

Note, finally, that the po-im(pR), the po-im(pPR), the po-im(pcoR) and the po-im(pScoR) are all called in the literature under the unique name: *partially-ordered, residuated, integral monoid*, or *porim* for short - *pocrim*, in the commutative case.

Proposition 3.6.2 *Let $\mathcal{A}^L = (A^L, \leq, \odot, \to^L, \leadsto^L, 1)$ be a po-m(pPR). Then, the properties (pPP), (pRR), (pCp), (p$*^L$), (p$**^L$) and (pBB^L), (pM^L) hold.*

Proof. The properties (pPP), (pRR), (pCp), (p$*^L$), (p$**^L$) hold by Lemma 2.6.3, since (pPR)=(pRP).

(pBB^L): $y \to^L z \leq (z \to^L x) \leadsto^L (y \to^L x) \overset{(pPR)}{\Longleftrightarrow}$
$(z \to^L x) \odot (y \to^L z) \leq y \to^L x \overset{(pPR)}{\Longleftrightarrow}$
$[(z \to^L x) \odot (y \to^L z)] \odot y \leq x \overset{(Ass)}{\Longleftrightarrow}$
$(z \to^L x) \odot [(y \to^L z) \odot y] \leq x$,
which is always true; indeed, by (pRR), $(y \to^L z) \odot y \leq z$; then, by (pCp), $(z \to^L x) \odot [(y \to^L z) \odot y)] \leq (z \to^L x) \odot z$; but, $(z \to^L x) \odot z \leq x$, by (pRR) again; it follows that $(z \to^L x) \odot [(y \to^L z) \odot y] \leq x$, by transitivity.

Also, $y \leadsto^L z \leq (z \leadsto^L x) \to^L (y \leadsto^L x) \overset{(pPR)}{\Longleftrightarrow}$
$(y \leadsto^L z) \odot (z \leadsto^L x) \leq y \leadsto^L x \overset{(pPR)}{\Longleftrightarrow}$
$y \odot [(y \leadsto^L z) \odot (z \leadsto^L x)] \leq x \overset{(Ass)}{\Longleftrightarrow}$
$[y \odot (y \leadsto^L z)] \odot (z \leadsto^L x) \leq x$,
which is always true; indeed, by (pRR), $y \odot (y \leadsto^L z) \leq z$; then, by (pCp), $[y \odot (y \leadsto^L z)] \odot (z \leadsto^L x) \leq z \odot (z \leadsto^L x)$; but, $z \odot (z \leadsto^L x) \leq x$, by (pRR) again; it follows that $[y \odot (y \leadsto^L z)] \odot (z \leadsto^L x) \leq x$, by transitivity. Thus, (pBB^L) holds.

(pM^L): $1 \to^L x \overset{(pU)}{=} (1 \to^L x) \odot 1 \leq x \overset{(pPR)}{\Leftrightarrow} 1 \to^L x \leq 1 \to^L x$, which is true by the reflexivity of $\leq$.
$x \leq 1 \to^L x \overset{(pPR)}{\Leftrightarrow} x \overset{(pU)}{=} x \odot 1 \leq x$, which is true by the reflexivity of $\leq$ again. Hence, by the antisymmetry of $\leq$, we obtain that $1 \to^L x = x$.

Also, $1 \leadsto^L x \overset{(pU)}{=} 1 \odot (1 \to^L x) \leq x \overset{(pPR)}{\Leftrightarrow} 1 \leadsto^L x \leq 1 \leadsto^L x$, which is true.
$x \leq 1 \leadsto^L x \overset{(pPR)}{\Leftrightarrow} x \overset{(pU)}{=} 1 \odot x \leq x$, which is true. Hence, $1 \leadsto^L x = x$; thus, (pM^L) holds. □

Since (pCp) holds, then we obtain:

Corollary 3.6.3 *A po-m(pPR) (po-im(pPR)) is a po-m (po-im).*

Proposition 3.6.4 *Let $\mathcal{A}^L = (A^L, \leq, \odot, \to^L, \leadsto^L, 1)$ be a po-im(pPR). Then, the additional properties hold: for all $x, y \in A^L$,*
(pN^L) If $1 \leq x$, then $x = 1$;
(IdEq^L) $x \to^L y = 1 \Leftrightarrow x \leadsto^l y = 1$;
(pEq^L) $x \leq y \Longleftrightarrow x \to^L y = 1 \overset{(IdEq^L)}{\Leftrightarrow} x \leadsto^l y = 1$.

Proof.

(pN^L): By (L), $x \leq 1$, for all $x \in A^L$. Hence, if $1 \leq x$, then $x = 1$, by the antisymmetry of $\leq$.

(IdEq^L) : First, note that $x = 1 \Leftrightarrow (x \leq 1$ and $1 \leq x)$; but since $x \leq 1$ is always true, by (L), then we obtain that $x = 1 \Leftrightarrow 1 \leq x$.

$x \to^L y = 1 \Leftrightarrow 1 \leq x \to^L y \overset{(pPR)}{\Longleftrightarrow} 1 \odot x \leq y \overset{(pU)}{\Longleftrightarrow} x \leq y$ and
$x \leadsto^L y = 1 \Leftrightarrow 1 \leq x \leadsto^L y \overset{(pRP)}{\Longleftrightarrow} x \odot 1 \leq y \overset{(pU)}{\Longleftrightarrow} x \leq y$.
Thus (IdEq^L) holds.

(pEq^L) : By above proof. □

Remark 3.6.5 In po-ims(pPR), the pseudo-implication $(\to^L, \rightsquigarrow^L)$ verifies (pBB^L), (pM^L), by Proposition 3.6.2, (IdEq^L), (pEq^L), by Proposition 3.6.4, hence is a pseudo-residuum on the poset $(A^L, \leq, 1)$ with last element 1, by Definition 1.1.3.

3.7 The basic equivalences (EM1) and (iEM1)

In this section, we present the basic equivalences:
- (EM1), between po-ms(pR) and po-ms(pPR), and between their duals, and
- (iEM1), between po-ims(pR) and po-ims(pPR), and between their duals.

• **In the non-integral case**, we have the following theorem (the theorem in right-case is omitted).

Theorem 3.7.1 *(See [94], Theorem 9.5.23, for the integral case)*
(1) Let $\mathcal{A}^L = (A^L, \leq, \odot, 1)$ be a po-m(pR), where for all $y, z \in A^L$:

(pR) $\exists y \to^L z \overset{notation}{=} \max\{x \mid x \odot y \leq z\}, \quad \exists x \rightsquigarrow^L z \overset{notation}{=} \max\{y \mid x \odot y \leq z\}.$

Define $\rho(\mathcal{A}^L) \overset{def.}{=} (A^L, \leq, \odot, \to^L, \rightsquigarrow^L, 1)$.
Then, $\rho(\mathcal{A}^L)$ is a po-m(pPR).
(1') Conversely, let $\mathcal{A}^L = (A^L, \leq, \odot, \to^L, \rightsquigarrow^L, 1)$ be a po-m(pPR).
Define $\rho'(\mathcal{A}^L) \overset{def.}{=} (A^L, \leq, \odot, 1)$.
Then, $\rho'(\mathcal{A}^L)$ is a po-m(pR), where for all $y, z \in A^L$:
$\max\{x \mid x \odot y \leq z\} = y \to^L z, \quad \max\{x \mid y \odot x \leq z\} = y \rightsquigarrow^L z.$
(2) The above defined mappings ρ and ρ' are mutually inverse.

Proof.
(1): We must prove that (pPR) holds.
If $x \odot y \leq z$, then by (pR), $x \leq y \to^L z$; if $x \leq y \to^L z$, then by (pCp), $x \odot y \leq (y \to^L z) \odot y$ and since we also have, by (pR), that (pRR) holds, then $(y \to^L z) \odot y \leq z$, so $x \odot y \leq z$, by transitivity.
Also, if $x \odot y \leq z$, then by (pR) again, $y \leq x \rightsquigarrow^L z$; if $y \leq x \rightsquigarrow^L z$, then by (pCp), $x \odot y \leq x \odot (x \rightsquigarrow^L z)$ and since (pRR) holds, then $x \odot (x \rightsquigarrow^L z) \leq z$, so $x \odot y \leq z$. Thus, (pPR) holds.
(1'): We must prove that (pCp) and (pR) hold.
By above Proposition 3.6.2, (pCp) holds. It remains to prove that (pR) holds. Indeed, by above Proposition 3.6.2 again, (pRR) holds, hence
$(y \to^L z) \odot y \leq z$, i.e. $y \to^L z \in \{x \mid x \odot y \leq z\}$; if x verifies $x \odot y \leq z$, then, by (pPR), $x \leq y \to^L z$; hence, there exists $\max\{x \mid x \odot y \leq z\} = y \to^L z$.
Also, since (pRR) holds, then $x \odot (x \rightsquigarrow^L z) \leq z$, i.e. $x \rightsquigarrow^L z \in \{y \mid x \odot y \leq z\}$; if y verifies $x \odot y \leq z$, then, by (pPR), $y \leq x \rightsquigarrow^L z$; hence, there exists $\max\{y \mid x \odot y \leq z\} = x \rightsquigarrow^L z$. Thus, (pR) holds.
(2): Obviously. □

Remark 3.7.2 By above proof, we have for po-ms that:

$$(pR) + (pCp) \Leftrightarrow (pPR).$$

We have, consequently, the following dual basic equivalences:

(EM1) **po-m(pPR)** $\Leftrightarrow$ **po-m(pR)** and, dually,
po-m(pScoR) $\Leftrightarrow$ **po-m(pcoR)**.

• **In the integral case**, we have consequently, by (EM1), the following dual equivalences:

(iEM1) **po-im(pPR)** $\Leftrightarrow$ **po-im(pR)** and, dually,
po-im(pScoR) $\Leftrightarrow$ **po-im(pcoR)**.

3.8 l-ms(pR) and l-ims(pR)

In this section, we present the l-ms(pR) and the l-ims(pR) and their duals.

Definition 3.8.1
(i) A *lattice-ordered (integral) monoid with the property (pR)*, or a *l-m(pR) (l-im(pR))* for short, is a po-m(pR) (po-im(pR), respectively)

$$\mathcal{A}^L = (A^L, \leq, \odot, 1)$$

where $\leq$ is a lattice order. A l-m(pR) (l-im(pR)) will be denoted by $(A^L, \wedge, \vee, \odot, 1)$.
(i') Dually, a *lattice-ordered (integral) monoid with the property (pcoR)*, or a *l-m(pcoR) (l-im(pcoR))* for short, is a po-m(pcoR) (po-im(pcoR), respectively)

$$\mathcal{A}^R = (A^R, \geq, \oplus, 0)$$

where $\geq$ is a lattice order. A l-m(pcoR) (l-im(pcoR)) will be denoted by $(A^R, \vee, \wedge, \oplus, 0)$.

Denote by **l-m(pR)** the class of all l-ms(pR) and by **l-m(pcoR)** the class of all l-ms(pcoR). Denote by **l-im(pR)** the class of all l-ims(pR) and by **l-im(pcoR)** the class of all l-ims(pcoR).

3.9 l-ms(pPR) and l-ims(pPR) - two intermediary notions

In this section, we present the l-ms(pPR) (left-pseudo-residuated lattices) and the l-ims(pPR) (integral left-pseudo-residuated latticse) - intermediary notions between the algebras of logic and the l-ms(pR) and l-ims(pR), respectively.
We also present the equivalences (EM2) and (iEM2).

Definition 3.9.1
(i) A *lattice-ordered (integral) monoid with the property (pPR)*, or a *l-m(pPR) (l-im(pPR))* for short, is a po-m(pPR) (po-im(pPR), respectively)

$$\mathcal{A}^L = (A^L, \leq, \odot, \rightarrow^L, \rightsquigarrow^L, 1)$$

where $\leq$ is a lattice-order. A l-m(pPR) (l-im(pPR)) will be denoted

$$(A^L, \wedge, \vee, \odot, \rightarrow^L, \rightsquigarrow^L, 1).$$

(i') Dually, a *lattice-ordered (integral) monoid with the property (pScoR)*, or a *l-m(pScoR) (l-im(pScoR))* for short, is a po-m(pScoR) (po-im(pScoR), respectively)

$$\mathcal{A}^R = (A^R, \geq, \oplus, \rightarrow^R, \rightsquigarrow^R, 0)$$

where $\geq$ is a lattice-order. A l-m(pScoR) (l-im(pScoR)) will be denoted

$$(A^R, \vee, \wedge, \oplus, \rightarrow^R, \rightsquigarrow^R, 0).$$

Denote by l-**m(pPR)** the class of all l-ms(pPR) and by l-**m(pScoR)** the class of all l-ms(pScoR).

Denote by l-**im(pPR)** the class of all l-ims(pPR) and by l-**im(pScoR)** the class of all l-ims(pScoR).

Definition 3.9.2

(i) A *left-non-commutative residuated lattice*, or a *left-pseudo-residuated lattice*, is an algebra

$$\mathcal{A}^L = (A^L, \wedge, \vee, \odot, \rightarrow^L, \rightsquigarrow^L, 1)$$

of type $(2, 2, 2, 2, 2, 0)$ such that $(A^L, \wedge, \vee)$ is a lattice (under lattice order $\leq$), $(A^L, \odot, 1)$ is a monoid and the property (pPR) holds; consequently, it is just the l-m(pPR), because (pCp) also holds, by Lemma 2.6.3.

(ii) An *integral left-non-commutative residuated lattice*, or an *integral left-pseudo-residuated lattice*, is an algebra

$$\mathcal{A}^L = (A^L, \wedge, \vee, \odot, \rightarrow^L, \rightsquigarrow^L, 1)$$

of type $(2, 2, 2, 2, 2, 0)$ such that $(A^L, \wedge, \vee, 1)$ is a lattice with last element 1 (under lattice order $\leq$), $(A^L, \odot, 1)$ is a monoid and the property (pPR) holds; consequently, it is just the l-im(pPR), because (pCp) also holds, by Lemma 2.6.3.

(i') Dually, a *right-non-commutative residuated lattice*, or a *right-pseudo-residuated lattice*, is an algebra

$$\mathcal{A}^R = (A^R, \vee, \wedge, \oplus, \rightarrow^R, \rightsquigarrow^R, 0)$$

of type $(2, 2, 2, 2, 2, 0)$ such that $(A, \vee, \wedge)$ is a lattice (under lattice order $\geq$), $(A, \oplus, 0)$ is a monoid and the property (pScoR) holds; consequently, it is just the l-m(pScoR), because (pCp) also holds, by the dual of Lemma 2.6.3.

(ii') Dually, an *integral right-non-commutative residuated lattice*, or an *integral right-pseudo-residuated lattice*, is an algebra

$$\mathcal{A}^R = (A^R, \vee, \wedge, \oplus, \rightarrow^R, \rightsquigarrow^R, 0)$$

of type $(2, 2, 2, 2, 2, 0)$ such that $(A, \vee, \wedge, 0)$ is a lattice with first element 0 (under lattice order $\geq$), $(A, \oplus, 0)$ is a monoid and the property (pScoR) holds; consequently, it is just the l-im(pScoR), because (pCp) also holds, by the dual of Lemma 2.6.3.

Denote by $\mathbf{pR\text{-}L}^L$ the class of all left-pseudo-residuated lattices and by $\mathbf{pR\text{-}L}^R$ the class of all right-pseudo-residuated lattices.

Denote by $\mathbf{ipR\text{-}L}^L$ the class of all integral left-pseudo-residuated lattices and by $\mathbf{ipR\text{-}L}^R$ the class of all integral right-pseudo-residuated lattices.

Hence, we have:

$\mathbf{pR\text{-}L}^L = l\text{-}\mathbf{m(pPR)}$ and, dually, $\mathbf{pR\text{-}L}^R = l\text{-}\mathbf{m(pScoR)}$, and
$\mathbf{ipR\text{-}L}^L = l\text{-}\mathbf{im(pPR)}$ and, dually, $\mathbf{ipR\text{-}L}^R = l\text{-}\mathbf{im(pScoR)}$.

• **In the non-integral case,** we obtain, by (EM1), the dual equivalences (EM2) between l-ms(pR) and l-ms(pPR) (left-pseudo-residuated lattices):

(EM2) $l\text{-}\mathbf{m(pPR)} = \mathbf{pR\text{-}L}^L \Leftrightarrow l\text{-}\mathbf{m(pR)}$ and, dually,
$l\text{-}\mathbf{m(pScoR)} = \mathbf{pR\text{-}L}^R \Leftrightarrow l\text{-}\mathbf{m(pcoR)}$.

• **In the integral case**, we obtain, by (iEM1), the dual equivalences (iEM2) between l-ims(pR) and l-ims(pPR) (integral left-pseudo-residuated lattices):

(iEM2) $l\text{-}\mathbf{im(pPR)} = \mathbf{ipR\text{-}L}^L \Leftrightarrow l\text{-}\mathbf{im(pR)}$ and, dually,
$l\text{-}\mathbf{im(pScoR)} = \mathbf{ipR\text{-}L}^R \Leftrightarrow l\text{-}\mathbf{im(pcoR)}$.

3.10 Bounded residuated unital magmas. Two negations

We have bounded algebras only in the integral case. In this section, we recall the definitions of bounded unital magmas having a pseudo-residuum, $(\rightarrow^L, \rightsquigarrow^L)$ (pseudo-coresiduum, $(\rightarrow^R, \rightsquigarrow^R)$) (see Remark 3.6.5) - in a word, that are "residuated". Thus, we recall [94] the definitions of bounded po-ims(pR), po-ims(pPR), l-ims(pR), l-ims(pPR) and their duals.

We present the equivalences (iEM3) and (iEM4) (in the lattice case).

Definition 3.10.1

(i) A po-im(pR) $\mathcal{A}^L = (A^L, \leq, \odot, 1)$ is said to be *bounded*, if the left-po-im $(A^L, \leq, \odot, 1)$ is bounded, with 0 as the first element. A *bounded po-im(pR)* is denoted by:

$$\mathcal{A}^{Lb} = (A^L, \leq, \odot, 0, 1).$$

(ii) A po-im(pPR) $\mathcal{A}^L = (A^L, \leq, \odot, \rightarrow^L, \rightsquigarrow^L, 1)$ is said to be *bounded*, if the left-po-im $(A^L, \leq, \odot, 1)$ is bounded, with 0 as the first (smallest) element. A *bounded po-im(pPR)* is denoted by:

$$\mathcal{A}^{Lb} = (A^L, \leq, \odot, \rightarrow^L, \rightsquigarrow^L, 0, 1).$$

(i') Dually, a po-im(pcoR) $\mathcal{A}^R = (A^R, \geq, \oplus, 0)$ is said to be *bounded*, if the right-po-im $(A^R, \geq, \oplus, 0)$ is bounded, with 1 as the last (greatest) element. A *bounded po-im(pcoR)* is denoted by:

$$\mathcal{A}^{Rb} = (A^R, \geq, \oplus, 0, 1).$$

(ii') A po-im(pScoR) $\mathcal{A}^R = (A^R, \geq, \oplus, \to^R, \rightsquigarrow^R, 0)$ is said to be bounded, if the right-po-im $(A^R, \geq, \oplus, 0)$ is bounded, with 1 as the last (greatest) element. A *bounded po-im(pScoR)* is denoted by:

$$\mathcal{A}^{Rb} = (A^R, \geq, \oplus, \to^R, \rightsquigarrow^R, 0, 1).$$

Denote by **po-im(pR)**^{b} the class of all bounded po-ims(pR) and, dually, denote by **po-im(pcoR)**^{b} the class of all bounded po-ims(pcoR).

Denote by **po-im(pPR)**^{b} the class of all bounded po-ims(pPR) and, dually, denote by **po-im(pScoR)**^{b} the class of all bounded po-ims(pScoR).

Then, by (iEM1), we obtain the dual equivalences:

(iEM3) **po-im(pPR)**^{b} $\Leftrightarrow$ **po-im(pR)**^{b} and, dually,
po-im(pScoR)^b $\Leftrightarrow$ **po-im(pcoR)**^{b}.

In a bounded po-im(pR) $(A^L, \leq, \odot, 0, 1)$, or in a bounded po-im (pPR) $(A^L, \leq, \odot, \to^L, \rightsquigarrow^L, 0, 1)$, we can define two *negations* as follows: for all $x \in A^L$,

$$x^- = x^{-^L} \stackrel{def.}{=} x \to^L 0 = \max\{y \mid y \odot x = 0\} \quad \textit{and}$$

$$x^\sim = x^{\sim^L} \stackrel{def.}{=} x \rightsquigarrow^L 0 = \max\{y \mid x \odot y = 0\}.$$

Dually, in a bounded po-im(pcoR) $(A^R, \leq, \oplus, 0, 1)$, or in a bounded po-im(pScoR) $(A^R, \geq, \oplus, \to^R, \rightsquigarrow^R, 0, 1)$, we can define two *negations* as follows: for all $x \in A^R$,

$$x^- = x^{-^R} \stackrel{def.}{=} x \to^R 1 = \min\{y \mid y \oplus x = 1\} \quad \textit{and}$$

$$x^\sim = x^{\sim^R} \stackrel{def.}{=} x \rightsquigarrow^R 1 = \min\{y \mid x \oplus y = 1\}.$$

In lattice case, we have the following definitions.

Definition 3.10.2

(i) An l-im(pR) $\mathcal{A}^L = (A^L, \wedge, \vee, \odot, 1)$ is said to be *bounded*, if the left-l-im $(A^L, \wedge, \vee, \odot, 1)$ is bounded, with 0 as the first element. A *bounded l-im(pR)* is denoted by:

$$\mathcal{A}^{Lb} = (A^L, \wedge, \vee, \odot, 0, 1).$$

(ii) An l-im(pPR) $\mathcal{A}^L = (A^L, \wedge, \vee, \odot, \to^L, \rightsquigarrow^L, 1)$ is said to be *bounded*, if the left-l-im $(A^L, \wedge, \vee, \odot, 1)$ is bounded, with 0 as the first (smallest) element. A *bounded l-im(pPR)* is denoted by:

$$\mathcal{A}^{Lb} = (A^L, \wedge, \vee, \odot, \to^L, \rightsquigarrow^L, 0, 1).$$

(i') Dually, an l-im(pcoR) $\mathcal{A}^R = (A^R, \vee, \wedge, \oplus, 0)$ is said to be *bounded*, if the right-l-im $(A^R, \vee, \wedge, \oplus, 0)$ is bounded, with 1 as the last (greatest) element. A *bounded l-im(pcoR)* is denoted by:

$$\mathcal{A}^{Rb} = (A^R, \vee, \wedge, \oplus, 0, 1).$$

(ii') An l-im(pScoR) $\mathcal{A}^R = (A^R, \vee, \wedge, \oplus, \rightarrow^R, \rightsquigarrow^R, 0)$ is said to be bounded, if the right-l-im $(A^R, \vee, \wedge, \oplus, 0)$ is bounded, with 1 as the last (greatest) element. A *bounded l-im(pScoR)* is denoted by:

$$\mathcal{A}^{Rb} = (A^R, \vee, \wedge, \oplus, \rightarrow^R, \rightsquigarrow^R, 0, 1).$$

Denote by l-**im(pR)**^{b} the class of all bounded l-ims(pR) and, dually, denote by l-**im(pcoR)**^{b} the class of all bounded l-ims(pcoR).

Denote by l-**im(pPR)**^{b} the class of all bounded l-ims(pPR) and, dually, denote by l-**im(pScoR)**^{b} the class of all bounded lo-ims(pScoR).

Then, by (iEM2), we obtain the dual equivalences:

(iEM4) l-**im(pPR)**^{b} = **ipR-L**^{Lb} $\Leftrightarrow$ l-**im(pR)**^{b} and, dually,
l-**im(pScoR)**^{b} = **ipR-L**^{Rb} $\Leftrightarrow$ l-**im(pcoR)**^{b}.

In a bounded l-im(pR) $(A^L, \wedge, \vee, \odot, 0, 1)$, or in a bounded l-im (pPR) $(A^L, \wedge, \vee, \odot, \rightarrow^L, \rightsquigarrow^L, 1)$, we can define the two *negations* $^{-^L}$ and $^{\sim^L}$ as above.

Dually, in a bounded l-im(pcoR) $(A^R, \vee, \wedge, \oplus, 0, 1)$, or in a bounded l-im(pScoR) $(A^R, \vee, \wedge, \oplus, \rightarrow^R, \rightsquigarrow^R, 0)$ we can define the two *negations* $^{-^R}$ and $^{\sim^R}$ as above.

3.11 The property (pDN). The equivalences (iEM5), (iEM6)

In this section, we present a general definition of involutive unital magmas and the equivalences (iEM5) and (iEM6) (in the lattice case).

Other results can be obtain, in "mirror" to those from algebras of logic, by the the connections between the algebras of logic (pB algebras, more precisely) and the unital magmas, presented further in Chapter 4.

Definition 3.11.1

(i) A bounded left-algebra $\mathcal{A}^{Lb}$ is said to be *with property (pDN^L)*, or *involutive*, if it verifies (pDN^L): for all $x \in A^L$,
(pDN^L) (pseudo-Double Negation) $(x^-)^\sim = x = (x^\sim)^-$,
where $x^- = x \rightarrow^L 0$ and $x^\sim = x \rightsquigarrow^L 0$.

(i') Dually, a bounded right-algebra $\mathcal{A}^{Rb}$ is said to be *with property (pDN^R)*, or *involutive*, if it verifies (pDN^R): for all $x \in A^R$,
(pDN^R) (pseudo-Double Negation) $(x^-)^\sim = x = (x^\sim)^-$,
where $x^- = x \rightarrow^R 1$ and $x^\sim = x \rightsquigarrow^R 1$.

Denote by $\cdot^{L}_{(pDN)}$ the class of all involutive bounded left-algebras and by $\cdot^{R}_{(pDN)}$ the class of all involutive bounded right-algebras.

Then, by (iEM3), we obtain the dual equivalences:

(iEM5) $\mathbf{po\text{-}im(pPR)}_{(pDN)} \Leftrightarrow \mathbf{po\text{-}im(pR)}_{(pDN)}$ and, dually,
$\mathbf{po\text{-}im(pScoR)}_{(pDN)} \Leftrightarrow \mathbf{po\text{-}im(pcoR)}_{(pDN)}$

and by (iEM4), we obtain the dual equivalences, in lattice case:

(iEM6) $\mathbf{\mathit{l}\text{-}im(pPR)}_{(pDN)} = \mathbf{ipR\text{-}L}^{L}_{(pDN)} \Leftrightarrow \mathbf{\mathit{l}\text{-}im(pR)}_{(pDN)}$ and, dually,
$\mathbf{\mathit{l}\text{-}im(pScoR)}_{(pDN)} = \mathbf{ipR\text{-}L}^{R}_{(pDN)} \Leftrightarrow \mathbf{\mathit{l}\text{-}im(pcoR)}_{(pDN)}$.

3.12 Pseudo-BL, pseudo-product and pseudo-MV algebras

In this section, we present the pseudo-BL, pseudo-product and pseudo-MV algebras and the equivalences (iEM7), (iEM8), (iEM9).

Pseudo-BL algebras were introduced by G. Georgescu and A. Iorgulescu in 2000 [59], [42], [43], as a non-commutative generalization of P. Hajék's BL algebras, introduced in 1996 [71], [72].

Definition 3.12.1

(i) A *left-pseudo-BL algebra*, or a *left-pBL algebra* for short, is a bounded integral left-pseudo-residuated lattice

$$\mathcal{A}^{Lb} = (A^L, \wedge, \vee, \odot, \rightarrow^L, \rightsquigarrow^L, 0, 1)$$

verifying (pprel^L) and (pdiv^L): for all $x, y \in A^L$,
(pprel^L) (pseudo-prelinearity) $(x \rightarrow^L y) \vee (y \rightarrow^L x) = 1 = (x \rightsquigarrow^L y) \vee (y \rightsquigarrow^L x)$,
(pdiv^L) (pseudo-divisibility) $x \wedge y = (x \rightarrow^L y) \odot x = x \odot (x \rightsquigarrow^L y)$.

(i') Dually, a *right-pseudo-BL algebra*, or a *right-pBL algebra* for short, is a bounded integral right-pseudo-residuated lattice

$$\mathcal{A}^{Rb} = (A^R, \vee, \wedge, \oplus, \rightarrow^R, \rightsquigarrow^R, 0, 1)$$

verifying additionally (pprel^R) and (pdiv^R): for all $x, y \in A^R$,
(pprel^R) (pseudo-prelinearity) $(x \rightarrow^R y) \wedge (y \rightarrow^R x) = 0 = (x \rightsquigarrow^R y) \wedge (y \rightsquigarrow^R x)$,
(pdiv^R) (pseudo-divisibility) $x \vee y = (x \rightarrow^R y) \oplus x = x \oplus (x \rightsquigarrow^R y)$.

Denote by $\mathbf{pBL}^L$ the class of all left-pBL algebras and by $\mathbf{pBL}^R$ the class of all right-pBL algebras.

Hence, we have:
$\mathbf{pBL}^L = \mathbf{ipR\text{-}L}^{Lb} + (\text{pprel}^L) + (\text{pdiv}^L) = \mathbf{\mathit{l}\text{-}im(pPR)}^b + (\text{pprel}^L) + (\text{pdiv}^L)$ and, dually,

$\mathbf{pBL}^R = \mathbf{ipR\text{-}L}^{Rb} + (\mathrm{pprel}^R) + (\mathrm{pdiv}^R) = l\text{-}\mathbf{im(pScoR)}^b + (\mathrm{pprel}^R) + (\mathrm{pdiv}^R)$.

A commutative pBL algebra is called *BL algebra*.

Pseudo-product algebras were introduced in 2002 [42], as a non-commutative generaliztions of Hajék's product algebras [72].

Definition 3.12.2

(i) A *left-pseudo-product algebra*, or a *left-pproduct algebra* for short, is a left-pBL algebra

$$\mathcal{A}^{Lb} = (A^L, \wedge, \vee, \odot, \to^L, \rightsquigarrow^L, 0, 1)$$

verifying additionally (pP1^L) and (pP2^L): for all $x, y, z \in A^L$,
(pP1^L) $x \wedge x^{-^L} = 0 = x \wedge x^{\sim^L}$,
(pP2^L) $(z^{-^L})^{-^L} \odot [(x \odot z) \to^L (y \odot z)] \leq x \to^L y$,
$(z^{\sim^L})^{\sim^L} \odot [(z \odot x) \rightsquigarrow^L (z \odot y)] \leq x \rightsquigarrow^L y$.

(i') Dually, a *right-pseudo-product algebra*, or a *right-pproduct algebra* for short, is a right-pBL algebra

$$\mathcal{A}^{Rb} = (A^R, \vee, \wedge, \oplus, \to^R, \rightsquigarrow^R, 0, 1)$$

verifying additionally (pP1^R) and (pP2^R): for all $x, y, z \in A^R$,
(pP1^R) $x \vee x^{-^R} = 1 = x \vee x^{\sim^R}$,
(pP2^R) $(z^{-^R})^{-^R} \oplus [(x \oplus z) \to^R (y \oplus z)] \geq x \to^R y$,
$(z^{\sim^R})^{\sim^R} \oplus [(z \oplus x) \rightsquigarrow^R (z \oplus y)] \geq x \rightsquigarrow^R y$.

Denote by $\mathbf{pproduct}^L$ the class of all left-pproduct algebras and by $\mathbf{pproduct}^R$ the class of all right-pproduct algebras.

Hence, we have:
$\mathbf{pproduct}^L = \mathbf{pBL}^L + (\mathrm{pP1}^L) + (\mathrm{pP2}^L)$ and, dually,
$\mathbf{pproduct}^R = \mathbf{pBL}^R + (\mathrm{pP1}^R) + (\mathrm{pP2}^R)$.

A commutative pproduct algebra is called *product algebra*.

Note that both pseudo-BL algebras and pseudo-product algebras are intermediary notions.

Then, by (iEM4), we have the dual equivalences:

(iEM7) $\mathbf{pBL}^L \Leftrightarrow l\text{-}\mathbf{im(pR)}^b$+(pprel^L)+(pdiv^L) and, dually,
$\mathbf{pBL}^R \Leftrightarrow l\text{-}\mathbf{im(pcoR)}^b$+(pprel^R)+ (pdiv^R).

Consequently, we have the dual equivalences:

(iEM8) $\mathbf{pproduct}^L \Leftrightarrow l\text{-}\mathbf{im(pR)}^b$ +(pprel^L)+ (pdiv^L)+(pP1^L)+(pP2^L)
and, dually,
$\mathbf{pproduct}^R \Leftrightarrow l\text{-}\mathbf{im(pcoR)}^b$+(pprel^R)+ (pdiv^R)+(pP1^R)+(pP2^R).

Pseudo-MV algebras were introduced as a non-commutative generalization with eight axioms of Chang's MV algebras in about 1996 by George Georgescu, in a manuscript which circulated some years, in order to find someone interested to complete the results with ideal theory. In 1998, we completed the results with (normal) ideal theory and the draft of the paper was distributed to some colleagues to start their own reasearch on the new algebras; consequently, at the section of *Algebra of Logic and Applications* of the Fourth International Symposium on Economic Informatics held in 1999, May 6-9, in Bucharest, there were presented four communications on pseudo-MV algebras:
- [58], which is a resuming of the paper [61] appeared in 2001, where the definition with 8 axioms of pseudo-MV algebras is presented and many other results, including (normal) ideal theory, but also
- [24], where R. Ceterchi introduced pseudo-Wajsberg algebras and proved their term-equivalence with pseudo-MV algebras, which is a resuming of the paper [25] appeared in 2001,
- [45], where A. Dvurecenskij introduced a partial addition in pseudo-MV algebras and
- [85], where some results on pseudo-MV algebras as semigroups are presented.

Later on, J. Rachůnek introduced another non-commutative MV algebra with 24 axioms (the analogous of Chang's axioms for MV algebra), called "(non-commutative) MV algebra" and studied its connection with bounded DRl-monoids in a paper [156] appeared in 2002; Rachůnek's non-commutative MV algebra is now called *GMV algebra* (i.e. *generalized MV-algebra*) in the literature and was proved to be equivalent to pseudo-MV algebra. Note that a different structure with the same name *GMV algebra* was also introduced in 2005 by N. Galatos and C. Tsinakis [55] (see also [122]), as a pseudo-residuated lattice that satisfies $x \vee y = x/((x \vee y) \setminus x) = (x/(x \vee y)) \setminus x$ (cf. [56], pag. 186), i.e. a more general algebra.

Definition 3.12.3 (i) A *left-pseudo-MV algebra*, or a *left-pMV algebra* for short, is an algebra [61]

$$\mathcal{A}^L = (A^L, \odot, ^-, ^\sim, 0, 1)$$

of type $(2,1,1,0,0)$, where $^- = ^{-^L}$ and $^\sim = ^{\sim^L}$, such that the following axioms are satisfied: for all $x, y, z \in A^L$,
(pMVL1) $x \odot (y \odot z) = (x \odot y) \odot z$, which is (Ass),
(pMVL2) $x \odot 1 = 1 \odot x = x$, which is (pU),
(pMVL3) $x \odot 0 = 0 \odot x = 0$,
(pMVL4) $0^- = 1$, $0^\sim = 1$,
(pMVL5) $(x^- \odot y^-)^\sim = (x^\sim \odot y^\sim)^-$,
(pMVL6) $x \odot (y \oplus x^\sim) = y \odot (x \oplus y^\sim) = (y^- \oplus x) \odot y = (x^- \oplus y) \odot x$,
(pMVL7) $(x^- \odot y) \oplus x = y \oplus (x \odot y^\sim)$,
(pMVL8) $(x^-)^\sim = x$,
where $x \oplus y \overset{def.}{=} (x^- \odot y^-)^\sim = (x^\sim \odot y^\sim)^-$ (see Remarks 2.11.7, 2.11.8, 2.11.9).

(i') Dually, a *right-pseudo-MV algebra*, or a *right-pMV algebra* for short, [58],

[61] is an algebra

$$\mathcal{A}^R = (A^R, \oplus, ^-, ^\sim, 0, 1)$$

of type $(2, 1, 1, 0, 0)$, where $^- = ^{-^R}$ and $^\sim = ^{\sim^R}$, such that the following axioms are satisfied: for all $x, y, z \in A^R$,
(pMVR1) $x \oplus (y \oplus z) = (x \oplus y) \oplus z$, which is (Ass)
(pMVR2) $x \oplus 0 = 0 \oplus x = x$, which is (pU),
(pMVR3) $x \oplus 1 = 1 \oplus x = 1$,
(pMVR4) $1^\sim = 0; \quad 1^- = 0$,
(pMVR5) $(x^- \oplus y^-)^\sim = (x^\sim \oplus y^\sim)^-$,
(pMVR6) $x \oplus (y \odot x^\sim) = y \oplus (x \odot y^\sim) = (y^- \odot x) \oplus y = (x^- \odot y) \oplus x$,
(pMVR7) $(x^- \oplus y) \odot x = y \odot (x \oplus y^\sim)$,
(pMVR8) $(x^-)^\sim = x$,
where $x \odot y \stackrel{def.}{=} (x^- \oplus y^-)^\sim = (x^\sim \oplus y^\sim)^-$ (see Remarks 2.11.7, 2.11.8, 2.11.9).

Denote by $\mathbf{pMV}^L$ the class of all left-pMV algebras and by $\mathbf{pMV}^R$ the class of all right-pMV algebras.

A commutative pMV algebra is called *MV algebra.*

Note that a left-pMV algebra verifies the property (pDN^L) (it is *involutive*), and hence it is self dual, i.e. the dual of $(A^L, \odot, ^-, ^\sim, 0, 1)$ is $(A^L, \oplus, ^-, ^\sim, 0, 1)$ and vice-versa. Dually, a right-pMV algebra verifies the property (pDN^R) (it is *involutive*), and hence it is self dual, i.e. the dual of $(A^R, \oplus, ^-, ^\sim, 0, 1)$ is $(A^R, \odot, ^-, ^\sim, 0, 1)$ and vice-versa (see Remarks 2.11.7, 2.11.8, 2.11.9).

We have the following dual equivalences:

(iEM9) $\mathbf{pBL}^L_{(pDN)} \Leftrightarrow \mathbf{pMV}^L$ and, dually,
$\mathbf{pBL}^R_{(pDN)} \Leftrightarrow \mathbf{pMV}^R$,

where:
$\mathbf{pBL}^L_{(pDN)} = l\text{-}\mathbf{im(pPR)}_{(pDN)} + (\mathrm{pprel}^L) + (\mathrm{pdiv}^L)$ and
$\mathbf{pBL}^R_{(pDN)} = l\text{-}\mathbf{im(pScoR)}_{(pDN)} + (\mathrm{pprel}^R) + (\mathrm{pdiv}^R)$,

namely we have the following theorem (the theorem in the right-case is omitted).

Theorem 3.12.4 *(See [42], Corollary 3.29)*
1) Let $\mathcal{A}^L = (A^L, \odot, ^-, ^\sim, 0, 1)$ be a left-pMV algebra.
Define $\rho(\mathcal{A}^L) \stackrel{def.}{=} (A^L, \wedge, \vee, \odot, \to^L, \leadsto^L, 0, 1)$ by: for all $x, y \in A^L$,
$x \wedge y \stackrel{def.}{=} x \odot (x^\sim \oplus y) = y \odot (y^\sim \oplus x) = (x \oplus y^-) \odot y = (y \oplus x^-) \odot x$, *by (pMVL6),*
$x \vee y \stackrel{def.}{=} x \oplus (x^- \odot y) = (x \odot y^\sim) \oplus y$, *by (pMVL7),*
$x \to^L y \stackrel{def.}{=} (x \odot y^\sim)^- = x^- \oplus y$,
$x \leadsto^L y \stackrel{def.}{=} (y^- \odot x)^\sim = y \oplus x^\sim$.
Then, $\rho(\mathcal{A}^L)$ is a left-pBL algebra with (pDN^L).

1') Conversely, let $\mathcal{A}^{Lb} = (A^L, \wedge, \vee, \odot, \to^L, \leadsto^L, 0, 1)$ *be a left-pBL algebra with* (pDN^L)*, where* $x^- \stackrel{def.}{=} x \to^L 0$ *and* $x^\sim \stackrel{def.}{=} x \leadsto^L 0$*, for all* $x \in A^L$.
Define $\rho'(\mathcal{A}^{Lb}) \stackrel{def.}{=} (A^L, \odot, ^-, ^\sim, 0, 1)$.

Then, $\rho'(\mathcal{A}^{Lb})$ *is a left-pMV algebra.*

2) The above defined mappings ρ *and* ρ' *are mutually inverse.*

We conclude, by (iEM9), (iEM6), that:

$\mathbf{pMV}^L \Leftrightarrow$ ***l*-im(pR)**$_{(pDN)}$ + (pprel^L) + (pdiv^L) and, dually,
$\mathbf{pMV}^R \Leftrightarrow$ ***l*-im(pcoR)**$_{(pDN)}$ + (pprel^R) + (pdiv^R).

3.13 Boolean algebras

Boolean algebras were introduced in 1854 by George Boole. They are only commutative. The most used definition is as a complemented, distributive, bounded (Dedekind) lattice.

Definition 3.13.1

(i) A *left-Boolean algebra* is an algebra

$$\mathcal{A}^L = (A^L, \wedge, \vee, ^{-^L}, 0, 1)$$

verifying: for all $x, y, z \in A^L$,
(B1L) $x \wedge x = x$, $\quad x \vee x = x$ (idempotency of $\wedge$, $\vee$),
(B2L) $x \wedge y = y \wedge x$, $\quad x \vee y = y \vee x$ (commutativity of $\wedge$, $\vee$),
(B3L) $x \wedge (y \wedge z) = (x \wedge y) \wedge z$, $\quad x \vee (y \vee z) = (x \vee y) \vee z$ (associativity of $\wedge$, $\vee$),
(B4L) $x \vee (x \wedge y) = x$, $\quad x \wedge (x \vee y) = x$ (absorption laws),
(B5L) $x \vee (y \wedge z) = (x \vee y) \wedge (x \vee z)$, $x \wedge (y \vee z) = (x \wedge y) \vee (x \wedge z)$ (distributivity),
(B6L) $x \wedge 1 = x$, $\quad x \vee 0 = x$ (i.e. $0 \leq x \leq 1$),
(B7L) $x \wedge x^{-^L} = 0$, $\quad x \vee x^{-^L} = 1$.

(i') Dually, a *right-Boolean algebra* is an algebra

$$\mathcal{A}^R = (A^R, \vee, \wedge, ^{-^R}, 0, 1)$$

verifying the dual properties (B1R) - (B7R).

Denote by $\mathbf{Boole}^L$ the class of all left-Boolean algebras and by $\mathbf{Boole}^R$ the class of all right-Boolean algebras.

Recall that [94]:

$\mathbf{Boole}^L = \mathbf{pMV}^L$ + (pP1^L) and, dually, $\mathbf{Boole}^R = \mathbf{pMV}^R$ + (pP1^R) and also
$\mathbf{Boole}^L = \mathbf{pproduct}^L$ + (pDN^L) and, dually, $\mathbf{Boole}^R = \mathbf{pproduct}^R$ + (pDN^R).

Note that a left-Boolean algebra verifies (pDN^L) (it is *involutive*), hence it is self-dual, i.e. the dual of $(A^L, \wedge, \vee, ^{-^L}, 0, 1)$ is $(A^L, \vee, \wedge, ^{-^L}, 0, 1)$ and vice-versa.

Dually, a right-Boolean algebra verifies (pDN^R) (it is *involutive*), hence it is self-dual, i.e. the dual of $(A^R, \vee, \wedge, ^{-^R}, 0, 1)$ is $(A^R, \wedge, \vee, ^{-^R}, 0, 1)$ and vice-versa.

3.14 The hierarchy No. 2 of some unital magmas

We present a simplified hierarchy (we consider the (bounded) unital magmas with additional operation(s) as particular cases of unital magmas) of some of the left-unital magmas recalled in this chapter, in Figure 3.2. The six classes of algebras: **po-m**^{L}, **po-m(pR)**, l**-m**^{L}, l**-m(pR)**, **po-im**^{L}, l**-im**^{L}, which have no equivalent in the world of pseudo-algebras of logic, are represented by a smaller bullet.

Left-unital magmas
(with additional operations)

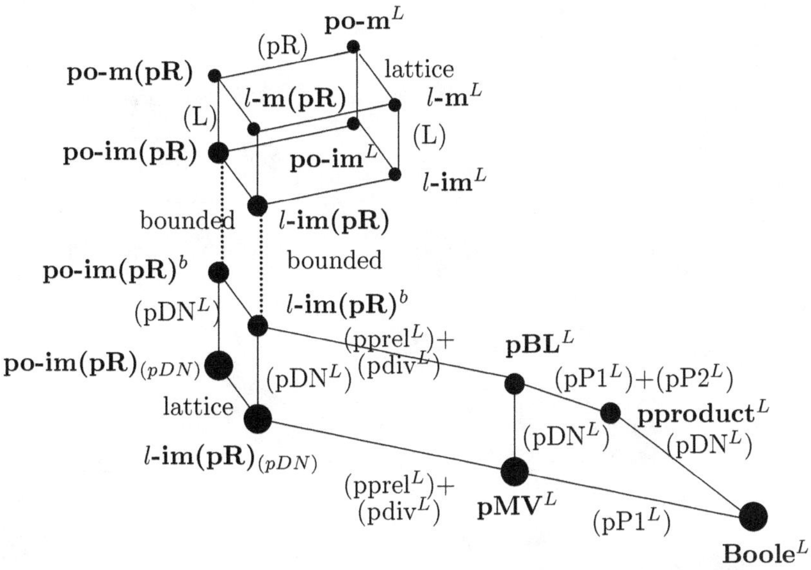

Figure 3.2: The hierarchy No. 2 of some left-unital magmas (with additional operations)

Chapter 4

Connections between the two kinds of algebras

In this chapter, we show connections existing between the pseudo-algebras of logic (pM algebras) and the unital magmas presented in the previous Chapters 2 and 3, respectively [94]. Namely, in Section 1, we present the implications (ILM1) and (ILM2) and the equivalences (iELM1) - (iELM9) existing between the intermediary notions from the world of pM algebras and the intermediary notions from the world of unital magmas. In Section 2, based on the previous connections, we present the implications (I1) and (I2) and the equivalences (iE1) - (iE9), (iE9'), (iE10), (iE11') existing between the four types of notions from the world of pM algebras and from the world of unital magmas.

The chapter has two sections.

4.1 Connections between the two intermediary notions

In this section, we present the implications (ILM1) and (ILM2) and the equivalences (ELM1), (ELM2) and (iELM1) - (iELM9) existing between the intermediary notions from the world of pseudo-algebras of logic (pM algebras) and the intermediary notions from the world of unital magmas presented in the previous Chapters 2 and 3, respectively.

4.1.1 Basic implications (ILM1) and equivalences (ELM1), (iELM1)

We shall see that the dual basic connections between the intermediary pM algebras and the intermediary unital magmas are the implications (ILM1) and the equivalences (ELM1) (in the non-integral case) and the equivalences (iELM1) (in the integral case).

• **In the non-integral case**, we have the following theorem (the theorem in the dual case is omitted), which states in what conditions we have an equivalence.

Theorem 4.1.1
(1) Let $\mathcal{A}^L = (A^L, \leq, \rightarrow^L, \rightsquigarrow^L, \odot, 1)$ *be a pBCI(pRP) algebra.*
Define $\Phi'(\mathcal{A}^L) \stackrel{def.}{=} (A^L, \leq, \odot, \rightarrow^L, \rightsquigarrow^L, 1)$.
Then, $\Phi'(\mathcal{A}^L)$ *is a po-m(pPR) verifying (pEq*^{L}*).*
(1') Conversely, let $\mathcal{A}^L = (A^L, \leq, \odot, \rightarrow^L, \rightsquigarrow^L, 1)$ *be a po-m(pPR) verifying (pEq*^{L}*).*
Define $\Psi'(\mathcal{A}^L) \stackrel{def.}{=} (A^L, \leq, \rightarrow^L, \rightsquigarrow^L, \odot, 1)$.
Then, $\Psi'(\mathcal{A}^L)$ *is a pBCI(pRP) algebra.*
(2) The above defined mappings Φ' *and* Ψ' *are mutually inverse.*

Proof.
(1): We have to prove that $(A^L, \odot, 1)$ is a monoid, which follows by Proposition 2.6.5.
(1'): Let $\mathcal{A}^L = (A^L, \leq, \odot, \rightarrow^L, \rightsquigarrow^L, 1)$ be a po-m(pPR) verifying (pEq^L). By Proposition 3.6.2, it also verifies (pBB^L) and (pM^L). Then, $(A^L, \leq, \rightarrow^L, \rightsquigarrow^L, 1)$ is a pBCI algebra. Hence, $\Psi'(\mathcal{A}^L)$ is a pBCI(pRP) algebra.
(2): Obviously. □

Hence, by above Theorem 4.1.1, we have the implications:

(ILM1) **pBCI(pRP)** $\Longrightarrow$ **po-m(pPR)** and, dually,
pBCI(pcoRS) $\Longrightarrow$ **po-m(pScoR)**,

and the equivalences:

(ELM1) **pBCI(pRP)** $\Leftrightarrow$ **po-m(pPR)** + (pEq^L) and, dually,
pBCI(pcoRS) $\Leftrightarrow$ **po-m(pScoR)** + (pEq^R).

• **In the integral case**, we have the following theorem (the theorem in the dual case is omitted).

Theorem 4.1.2 *(See [94], Theorem 9.5.29)*
(1) Let $\mathcal{A}^L = (A^L, \leq, \rightarrow^L, \rightsquigarrow^L, \odot, 1)$ *be a pBCK(pRP) algebra.*
Define $\Phi'(\mathcal{A}^L) \stackrel{def.}{=} (A^L, \leq, \odot, \rightarrow^L, \rightsquigarrow^L, 1)$.
Then, $\Phi'(\mathcal{A}^L)$ *is a po-im(pPR).*
(1') Conversely, let $\mathcal{A}^L = (A^L, \leq, \odot, \rightarrow^L, \rightsquigarrow^L, 1)$ *be a po-im(pPR).*
Define $\Psi'(\mathcal{A}^L) \stackrel{def.}{=} (A^L, \leq, \rightarrow^L, \rightsquigarrow^L, \odot, 1)$.
Then, $\Psi'(\mathcal{A}^L)$ *is a pBCK(pRP) algebra.*
(2) The above defined mappings Φ' *and* Ψ' *are mutually inverse.*

Proof.

(1): By above Theorem 4.1.1.

(1'): We have to prove that (pBB^L), (pM^L), (pEq^L) hold. Indeed, (pBB^L) and (pM^L) follow by Proposition 3.6.2, (pEq^L) follows by Proposition 3.6.4.

(2): Obviously. □

Hence, in the integral case, we have the equivalences:

(iELM1) **pBCK(pRP)** $\Leftrightarrow$ **po-im(pPR)** and, dually,
pBCK(pcoRS) $\Leftrightarrow$ **po-im(pScoR)**.

4.1.2 The lattice case

• **In the non-integral case**, based on the implications (ILM1) and the equivalences (ELM1), we obtain:

(ILM2) **pBCI(pRP)-L** $\Longrightarrow$ ***l*-m(pPR)** = **pR-L^L** and, dually,
pBCI(pcoRS)-L $\Longrightarrow$ ***l*-m(pScoR)** = **pR-L^R**,

(ELM2) **pBCI(pRP)-L** $\Leftrightarrow$ ***l*-m(pPR)** + (pEq^L) and, dually,
pBCI(pcoRS)-L $\Leftrightarrow$ ***l*-m(pScoR)** + (pEq^R).

• **In the integral case**, based on the equivalence (iELM1), we obtain:

(iELM2) **pBCK(pRP)-L** $\Leftrightarrow$ ***l*-im(pPR)** = **ipR-L^L** and, dually,
pBCK(pcoRS)-L $\Leftrightarrow$ ***l*-im(pScoR)** = **ipR-L^R**.

4.1.3 The bounded case

In bounded case (hence the integral case), based on the above equivalences (iELM1) and (iELM2) (in lattice case), we obtain the following dual connections between the intermediary pM algebras (of logic) and the intermediary unital magmas:

(iELM3) **pBCK(pRP)^b** $\Leftrightarrow$ **po-im(pPR)^b** and, dually,
pBCK(pcoRS)^b $\Leftrightarrow$ **po-im(pScoR)^b**;

(iELM4) **pBCK(pRP)-L^b** $\Leftrightarrow$ ***l*-im(pPR)^b** = **ipR-L^{Lb}** and, dually,
pBCK(pcoRS)-L^b $\Leftrightarrow$ ***l*-im(pScoR)^b** = **ipR-L^{Rb}**.

4.1.4 The property (pDN). Equivalences (iELM5), (iELM6)

In the involutive case, based on the above equivalences (iELM3) and (iELM4) (in lattice case), we obtain the following dual connections:

(iELM5) **pBCK(pRP)$_{(pDN)}$** $\Leftrightarrow$ **po-im(pPR)$_{(pDN)}$** and, dually,

pBCK(pcoRS)$_{(pDN)}$ ⇔ **po-im(pScoR)**$_{(pDN)}$;

(iELM6) **pBCK(pRP)-L**$_{(pDN)}$ ⇔ ***l*-im(pPR)**$_{(pDN)}$ and, dually,
pBCK(pcoRS)-L$_{(pDN)}$ ⇔ ***l*-im(pScoR)**$_{(pDN)}$.

4.1.5 Other connections

By (iELM4) and (iELM6), we obtain the following dual connections between the intermediary pM algebras and the intermediary unital magmas:

(iELM7) **pHa(pRP)** ⇔ **pBL**^{L} and, dually,
pHa(pcoRS) ⇔ **pBL**^{R};

(iELM8) **pHa(pRP)** +(pP1^L)+(pP2^L) ⇔ **pproduct**^{L} and, dually,
pHa(pcoRS) +(pP1^R)+(pP2^R) ⇔ **pproduct**^{R};

(iELM9) **pHa(pRP)**$_{(pDN)}$ ⇔ **pBL**$^L_{(pDN)}$ and, dually,
pHa(pcoRS)$_{(pDN)}$ ⇔ **pBL**$^R_{(pDN)}$.

4.2 Connections between the four notions

In this section, based on the previous connections, we present the implications (I1) and (I2) and the equivalences (E1), (E2) and (iE1) - (iE9), (iE9'), (iE10), (iE11') existing between the four notions from the world of pM algebras (of logic) and from the world of unital magmas presented in the previous Chapters 2 and 3, respectively.

4.2.1 Basic implication (I1) and equivalences (E1), (iE1)

• **In the non-integral case**, by equivalences (EL1), (EM1) and above implications (ILM1), we obtain the following dual basic connections:

(I1) **pBCI(pP)** ⇔ **pBCI(pRP)** ⟹ **po-m(pPR)** ⇔ **po-m(pR)** and, dually,
pBCI(pS) ⇔ **pBCI(pcoRS)** ⟹ **po-m(pScoR)** ⇔ **po-m(pcoR)**.

By equivalences (EL1), (EM1) and (ELM1), we obtain the equivalences:

(E1) **pBCI(pP)** ⇔ **pBCI(pRP)** ⇔ **po-m(pPR)**+(pEq^L) ⇔ **po-m(pR)**+(pEq^L) and, dually,
pBCI(pS) ⇔ **pBCI(pcoRS)** ⇔ **po-m(pScoR)**+(pEq^R) ⇔ **po-m(pcoR)**+ (pEq^R).

• **In the integral case**, by equivalences (iEL1), (iEM1) and above (iELM1), we obtain the following dual equivalences:

(iE1) **pBCK(pP)** $\Leftrightarrow$ **pBCK(pRP)** $\Leftrightarrow$ **po-im(pPR)** $\Leftrightarrow$ **po-im(pR)**
and, dually,
pBCK(pS) $\Leftrightarrow$ **pBCK(pcoRS)** $\Leftrightarrow$ **po-im(pScoR)** $\Leftrightarrow$ **po-im(pcoR)**.

Hence, by Theorems 3.7.1, 2.7.1, 4.1.2, we obtain obviously the following theorem, which states that **pBCK(pP)** $\Leftrightarrow$ **po-im(pR)**.

Theorem 4.2.1 *(See [94], Theorem 9.5.30)*
(1) Let $\mathcal{A}^L = (A^L, \leq, \to^L, \rightsquigarrow^L, 1)$ be a pBCK(pP) algebra, where for all $x, y \in A^L$:
(pP) $\exists x \odot y \overset{notation}{=} \min\{z \mid x \leq y \to^L z\} = \min\{z \mid y \leq x \rightsquigarrow^L z\}$.
Define $\Phi(\mathcal{A}^L) \overset{def.}{=} (A, \leq, \odot, 1)$.
Then, $\Phi(\mathcal{A}^L)$ is a po-im(pR), where for all $y, z \in A^L$:

$$\max\{x \mid x \odot y \leq z\} = y \to^L z, \quad \max\{x \mid y \odot x \leq z\} = y \rightsquigarrow^L z.$$

(1') Conversely, let $\mathcal{A}^L = (A^L, \leq, \odot, 1)$ be a po-im(pR), where for any $y, z \in A^L$:
(pR) $\exists y \to^L z \overset{notation}{=} \max\{x \mid x \odot y \leq z\}$, $\exists y \rightsquigarrow^L z \overset{notation}{=} \max\{x \mid y \odot x \leq z\}$.
Define $\Psi(\mathcal{A}^L) \overset{def.}{=} (A^L, \leq, \to^L, \rightsquigarrow^L, 1)$.
Then, $\Psi(\mathcal{A}^L)$ is a pBCK(pP) algebra, where for all $x, y \in A^L$:

$$\min\{z \mid x \leq y \to^L z\} = \min\{z \mid y \leq x \rightsquigarrow^L z\} = x \odot y.$$

(2) The above defined mappings Φ and Ψ are mutually inverse.

4.2.2 The lattice case

• **In the non-integral case**, by equivalences (EL2), (EM2) and above implications (ILM2), we obtain:

(I2) **pBCI(pP)-L** $\Leftrightarrow$ **pBCI(pRP)-L** $\Longrightarrow$ ***l*-m(pPR)** $=$ **pR-L**^{L} $\Leftrightarrow$ ***l*-m(pR)**
and, dually,
pBCI(pS)-L $\Leftrightarrow$ **pBCI(pcoRS)-L** $\Longrightarrow$ ***l*-m(pScoR)** $=$ **pR-L**^{R} $\Leftrightarrow$ ***l*-m(pcoR)**.

By equivalences (EL2), (EM2) and (ELM2), we obtain:

(E2) **pBCI(pP)-L** $\Leftrightarrow$ **pBCI(pRP)-L** $\Leftrightarrow$ ***l*-m(pPR)**+(pEq^L) $\Leftrightarrow$ ***l*-m(pR)**+(pEq^L)
and, dually,
pBCI(pS)-L $\Leftrightarrow$ **pBCI(pcoRS)-L** $\Leftrightarrow$ ***l*-m(pScoR)**+(pEq^R) $\Leftrightarrow$ ***l*-m(pcoR)**+ (pEq^R).

• **In the integral case**, by equivalences (iEL2), (iEM2) and above (iELM2), we obtain the equivalences:

(iE2) **pBCK(pP)-L** $\Leftrightarrow$ **pBCK(pRP)-L** $\Leftrightarrow$ l-**im(pPR)** = **ipR-L**^{L} $\Leftrightarrow$ l-**im(pR)**
and, dually,
pBCK(pS)-L $\Leftrightarrow$ **pBCK(pcoRS)-L** $\Leftrightarrow$ l-**im(pScoR)** = **ipR-L**^{R}
$\Leftrightarrow$ l-**im(pcoR)**.

4.2.3 The bounded case

In bounded case (hence the integral case), by (iEL3), (iEM3) and above (iELM3), we obtain:

(iE3) **pBCK(pP)**^{b} $\Leftrightarrow$ **pBCK(pRP)**^{b} $\Leftrightarrow$ **po-im(pPR)**^{b} $\Leftrightarrow$ **po-im(pR)**^{b}
and, dually,
pBCK(pS)^b $\Leftrightarrow$ **pBCK(pcoRS)**^{b} $\Leftrightarrow$ **po-im(pScoR)**^{b} $\Leftrightarrow$ **po-im(pcoR)**^{b}.

By (iEL4), (iEM4) and above (iELM4), we obtain:

(iE4) **pBCK(pP)-L**^{b} $\Leftrightarrow$ **pBCK(pRP)-L**^{b} $\Leftrightarrow$ l-**im(pPR)**^{b} = **ipR-L**^{Lb}
$\Leftrightarrow$ l-**im(pR)**^{b} and, dually,
pBCK(pS)-L^b $\Leftrightarrow$ **pBCK(pcoRS)-L**^{b} $\Leftrightarrow$ l-**im(pScoR)**^{b} = **ipR-L**^{Rb}
$\Leftrightarrow$ l-**im(pcoR)**^{b}.

4.2.4 The property (pDN). Equivalences (iE5), (iE6)

By (iEL5), (iEM5) and above (iELM5), we obtain:

(iE5) **pBCK(pP)**$_{(pDN)}$ $\Leftrightarrow$ **pBCK(pRP)**$_{(pDN)}$ $\Leftrightarrow$ **po-im(pPR)**$_{(pDN)}$
$\Leftrightarrow$ **po-im(pR)**$_{(pDN)}$ and, dually,
pBCK(pS)$_{(pDN)}$ $\Leftrightarrow$ **pBCK(pcoRS)**$_{(pDN)}$ $\Leftrightarrow$ **po-im(pScoR)**$_{(pDN)}$
$\Leftrightarrow$ **po-im(pcoR)**$_{(pDN)}$.

By (iEL6), (iEM6) and above (iELM6), we obtain:

(iE6) **pBCK(pP)-L**$_{(pDN)}$ $\Leftrightarrow$ **pBCK(pRP)-L**$_{(pDN)}$ $\Leftrightarrow$ l-**im(pPR)**$_{(pDN)}$
= **ipR-L**$^L_{(pDN)}$ $\Leftrightarrow$ l-**im(pR)**$_{(pDN)}$ and, dually,
pBCK(pS)-L$_{(pDN)}$ $\Leftrightarrow$ **pBCK(pcoRS)-L**$_{(pDN)}$ $\Leftrightarrow$ l-**im(pScoR)**$_{(pDN)}$
= **ipR-L**$^R_{(pDN)}$ $\Leftrightarrow$ l-**im(pcoR)**$_{(pDN)}$.

4.2.5 Other connections

By (iEL7), (iEM7) and above (iELM7), we obtain:

(iE7) **pHa(pP)** $\Leftrightarrow$ **pHa(pRP)** $\Leftrightarrow$ **pBL**^{L} $\Leftrightarrow$ l-**im(pR)**^{b}+(pprel^L)+(pdiv^L)
and, dually,
pHa(pS) $\Leftrightarrow$ **pHa(pcoRS)** $\Leftrightarrow$ **pBL**^{R} $\Leftrightarrow$ l-**im(pcoR)**^{b}+(pprel^R)+ (pdiv^R).

By (iEL8), (iEM8) and above (iELM8), we obtain:

(iE8) **pHa(pP)** +(pP1^L)+(pP2^L) $\Leftrightarrow$ **pHa(pRP)**+(pP1^L)+(pP2^L)
$\Leftrightarrow$ **pproduct**^{L} $\Leftrightarrow$ ***l*-im(pR)**^{b} +(pprel^L)+ (pdiv^L)+(pP1^L)+(pP2^L)
and, dually,
pHa(pS) +(pP1^R)+(pP2^R) $\Leftrightarrow$ **pHa(pcoRS)**+(pP1^R)+(pP2^R)
$\Leftrightarrow$ **pproduct**^{R} $\Leftrightarrow$ ***l*-im(pcoR)**^{b}+(pprel^R)+ (pdiv^R)+(pP1^R)+(pP2^R).

By (iEL9), (iEM9) and above (iELM9), we obtain:

(iE9) **pW**^{L} $\Leftrightarrow$ **pHa(pRP)**$_{(pDN)}$ $\Leftrightarrow$ **pBL**$^L_{(pDN)}$ $\Leftrightarrow$ **pMV**^{L} and, dually,
pW^R $\Leftrightarrow$ **pHa(pcoRS)**$_{(pDN)}$ $\Leftrightarrow$ **pBL**$^R_{(pDN)}$ $\Leftrightarrow$ **pMV**^{R}.

Hence, we have the equivalences:
(iE9') **pW**^{L} $\Leftrightarrow$ **pMV**^{L} and, dually, **pW**^{R} $\Leftrightarrow$ **pMV**^{R},

namely we have the following theorem (the dual case is omitted).

Theorem 4.2.2 *(See [24], [25], ([88], Theorem 6.7) and ([94], Theorem 9.2.11))*
(1) Let $\mathcal{A}^L = (A^L, \to^L, \rightsquigarrow^L, ^-, ^\sim, 1)$ *be a left-pWajsberg algebra, where* $x^- = x^{-^L}$ *and* $x^\sim = x^{\sim^L}$.
Define $\Phi(\mathcal{A}^L) \stackrel{def.}{=} (A^L, \odot, ^-, ^\sim, 0, 1)$ *as follows: for all* $x, y \in A^L$,
$x \odot y \stackrel{def.}{=} (x \to^L y^-)^\sim = (y \rightsquigarrow^L x^\sim)^-$, $0 \stackrel{def.}{=} 1^- = 1^\sim$.
Then, $\Phi(\mathcal{A}^L)$ *is a left-pMV algebra.*
(1') Conversely, let $\mathcal{A}^L = (A^L, \odot, ^-, ^\sim, 0, 1)$ *be a left-pMV algebra.*
Define $\Psi(\mathcal{A}^L) \stackrel{def.}{=} (A^L, \to^L, \rightsquigarrow^L, ^-, ^\sim, 1)$ *as follows: for all* $x, y \in A^L$,
$x \to^L y \stackrel{def.}{=} (x \odot y^\sim)^-$, $\quad x \rightsquigarrow^L y \stackrel{def.}{=} (y^- \odot x)^\sim$.
Then, $\Psi(\mathcal{A}^L)$ *is a left-pWajsberg algebra.*
(2) The above defined mappings Φ *and* Ψ *are mutually inverse.*

By (iEL10), (iE4) and (iE9'), we obtain the following equivalences:

(iE10) **pW**^{L} $\equiv$ **pBCK(pP)-L**^{b} + (pCC^L) $\Leftrightarrow$ **pBCK(pRP)-L**^{b} + (pCC^L) $\Leftrightarrow$
***l*-im(pPR)**^{b} + (pCC^L) $\Leftrightarrow$ ***l*-im(pR)**^{b} + (pCC^L) $\equiv$ **pMV**^{L}
and, dually,
pW^R $\equiv$ **pBCK(pS)-L**^{b} + (pCC^R) $\Leftrightarrow$ **pBCK(pcoRS)-L**^{b} + (pCC^R) $\Leftrightarrow$
***l*-im(pScoR)**^{b} + (pCC^R) $\Leftrightarrow$ ***l*-im(pcoR)**^{b} + (pCC^R) $\equiv$ **pMV**^{R}.

We have finally the following connections between the pM algebras and the unital magmas:

(iE11') **implicative-Boole**^{L} $\Leftrightarrow$ **Boole**^{L} and, dually,
implicative-Boole^R $\Leftrightarrow$ **Boole**^{R},

namely we have the following theorem (the dual, "right" one, is omitted).

Theorem 4.2.3 *[62]*
(1) Let $\mathcal{B}^L = (B^L, \to^L, ^-, 1)$ *be a left-implicative-Boolean algebra. Define* $\Phi(\mathcal{B}^L) \overset{def.}{=} (B^L, \wedge, \vee, ^-, 0, 1)$ *as follows: for every* $x, y \in B^L$, $x \wedge y \overset{def.}{=} (x \to^L y^-)^-$, $x \vee y \overset{def.}{=} (x^- \wedge y^-)^- = x^- \to^L y$, $0 \overset{def.}{=} 1^-$.
Then $\Phi(\mathcal{B}^L)$ *is a left-Boolean algebra.*
(1') Conversely, let $\mathcal{B}^L = (B^L, \wedge, \vee, ^-, 0, 1)$ *be a left-Boolean algebra. Define* $\Psi(\mathcal{B}^L) \overset{def.}{=} (B, \to^L, ^-, 1)$ *as follows: for every* $x, y \in B^L$, $x \to^L y \overset{def.}{=} (x \wedge y^-)^- = x^- \vee y$.
Then $\Psi(\mathcal{B}^L)$ *is a left-implicative-Boolean algebra.*
(2) The mappings Φ *and* Ψ *are mutual inverse.*

4.2.6 Resuming remarks

Remarks 4.2.4
(1) Resuming, we have the basic equivalences (iE1) and the corresponding definitions illustrated in Figure 4.1.

pBCK(pP)	$\Leftrightarrow$	**pBCK(pRP)**	$\Leftrightarrow$	**po-im(pPR)**	$\Leftrightarrow$	**po-im(pR)**
$(A^L, \leq,$		$(A^L, \leq,$		$(A^L, \leq,$		$(A^L, \leq,$
$\to, \rightsquigarrow, 1)$		$\to, \rightsquigarrow, \odot, 1)$		$\odot, \to, \rightsquigarrow, 1)$		$\odot, 1)$
poset with 1		poset with 1		poset with 1		poset with 1
(pBB^L),		(pBB^L),		(Ass),		(Ass),
(pM^L),		(pM^L),		(pU)		(pU),
(pEq^L)						(pCp)
(pP)		(pRP)		(pPR)		(pR)
(I.1)		(I.2)		(II.2)		(II.1)

Figure 4.1: The equivalences (iE1), where $\to = \to^L$, $\rightsquigarrow = \rightsquigarrow^L$

(2) By adding the property (pDN^L) to the bounded algebras from above (1), we obtain the definitions and the termwise-equivalences (iE5) illustrated in Figure 4.2.

Remarks 4.2.5 Thus, there exist four (two plus two) different equivalent notions (definitions) for a left-algebra having the operations $\odot$ and/or $(\to, \rightsquigarrow)$, and the element 1, that are connected by the non-commutative Galois connections (pRP)=(pPR) or, equivalently, by (pP) or (pR), denoted by:

$I.1\ (\to, \rightsquigarrow, 1)$,	$I.2\ (\to, \rightsquigarrow, \odot, 1)$;	$II.2\ (\odot, \to, \rightsquigarrow, 1)$,	$II.1\ (\odot, 1)$
direct	*intermediary*	*intermediary*	*direct*
notions,	*notions;*	*notions,*	*notions.*

pBCK(pP) with (pDN)	$\Leftrightarrow$	**pBCK(pRP) with (pDN)**	$\Leftrightarrow$	**po-im(pPR) with (pDN)**	$\Leftrightarrow$	**po-im(pR) with (pDN)**
$(A^L, \leq, \to, \rightsquigarrow, 0, 1)$		$(A^L, \leq, \to, \rightsquigarrow, \odot, 0, 1)$		$(A^L, \leq, \odot, \to, \rightsquigarrow, 0, 1)$		$(A^L, \leq, \odot, 0, 1)$
poset with $0, 1$		poset with $0, 1$		poset with $0, 1$		poset with $0, 1$
(pBB^L), (pM^L), (pEq^L)		(pBB^L), (pM^L),		(Ass), (pU)		(Ass), (pU), (pCp)
$x^- = x \to 0$,		$x^- = x \to 0$,		$x^- = x \to 0$,		$x^- = \max\{y \mid y \odot x = 0\}$,
$x^\sim = x \rightsquigarrow 0$,		$x^\sim = x \rightsquigarrow 0$,		$x^\sim = x \rightsquigarrow 0$,		$x^\sim = \max\{y \mid x \odot y = 0\}$
(pDN^L)		(pDN^L)		(pDN^L)		(pDN^L)
(pP)$_{(pDN)}$ $\exists\, x \odot y = (x \to y^-)^\sim = (y \rightsquigarrow x^\sim)^-$		(pRP)		(pPR)		(pR)$_{(pDN)}$ $\exists\, x \to y = (x \odot y^\sim)^-$ $\exists\, x \rightsquigarrow y = (y^- \odot x)^\sim$
(I.1)		(I.2)		(II.2)		(II.1)

Figure 4.2: The equivalence (iE5), where $\to = \to^L$, $\rightsquigarrow = \rightsquigarrow^L$

Note also that usually none, one or maximum two, among the four different types of notions (definitions) are used in the literature, for each algebra. For examples, for left-pBCK algebras, none of the four different definitions of left-algebras appears in the literature, because $\odot$ does not exist; for left-pBCK algebras with pseudo-product, two definitions appear in the literature: left-pBCK(pP) and left-pBCK(pRP) algebras; for left-pseudo-BL algebras, one definition appears in the literature.

Consequently, the four different equivalent notions (definitions) of algebras involved in a study determine a *left- "Mendeleev-type" table (matrix)* (Sergiu Rudeanu's remark) with 4 columns and as many rows as distinct (i.e. not equivalent) non-commutative algebras are involved in the study.

The non-commutative algebras given in the study will fill some cells of the table and the empty cells of each row will be filled with the "missing" algebras; the only problem is *the problem of names* for the "missing" algebras.

Consider, for example, the following six classes of bounded left-algebras (see [94] for the commutative case):
- the bounded pBCK(pP) algebras ($\mathbf{pBCK(pP)}^b$),
- the bounded po-ims(pR) ($\mathbf{po\text{-}im(pR)}^b$),
- the bounded integral left-pseudo-residuated lattices ($\mathbf{ipR\text{-}L}^{Lb}$),
- the left-pBL algebras ($\mathbf{pBL}^L$),
- the left-pWajsberg algebras ($\mathbf{pW}^L$) and
- the left-pMV algebras ($\mathbf{pMV}^L$).

They determine a "Mendeleev-type" table (matrix) with 4 columns and 4 rows, where only six cells are filled. One can easy see that the classes of algebras on the same row (in the four columns) are equivalent. The table of all $16 = 4 \times 4$ classes of bounded algebras is presented below (see [94]), where the initial six classes of bounded algebras are marked by a bullet.

The world of bounded left-algebras			
The general world of $\rightarrow, \rightsquigarrow$, $0, 1$ (The general world of pM algebras of logic) (I)		The general world of $\odot$, $0, 1$ (The general world of unital magmas) (II)	
(I.1) The world of $\rightarrow, \rightsquigarrow$, $0, 1$ (direct notions)	(I.2) The world of $\rightarrow, \rightsquigarrow$, $\odot$, $0, 1$ (intermediary notions)	(II.2) The world of $\odot$, $\rightarrow, \rightsquigarrow$, $0, 1$ (intermediary notions)	(II.1) The world of $\odot$, $0, 1$ (direct notions)
• **pBCK(pP)**^{b} $(A^L, \leq,$ $\rightarrow, \rightsquigarrow, 0, 1)$ poset with $0, 1$ (pBB^L),(pM^L), (pEq^L) (pP)	**pBCK(pRP)**^{b} $(A^L, \leq,$ $\rightarrow, \rightsquigarrow, \odot, 0, 1)$ poset with $0, 1$ (pBB^L),(pM^L), (pRP)	**po-im(pPR)**^{b} $(A^L, \leq,$ $\odot, \rightarrow, \rightsquigarrow, 0, 1)$ poset with $0, 1$ (Ass), (pU) (pPR)	• **po-im(pR)**^{b} $(A^L, \leq,$ $\odot, 0, 1)$ poset with $0, 1$ (Ass), (pU), (pCp) (pR)
pBCK(pP)-L^b $(A^L, \wedge, \vee,$ $\rightarrow, \rightsquigarrow, 0, 1)$ lattice with $0, 1$ (pBB^L),(pM^L), (pEq^L) (pP)	**pBCK(pRP)-L**^{b} $(A^L, \wedge, \vee,$ $\rightarrow, \rightsquigarrow, \odot, 0, 1)$ lattice with $0, 1$ (pBB^L),(pM^L), (pRP)	• **l-im(pPR)**^{b} = **ipR-L**^{Lb} $(A^L, \wedge, \vee,$ $\odot, \rightarrow, \rightsquigarrow, 0, 1)$ lattice with $0, 1$ (Ass), (pU) (pPR)	**l-im(pR)**^{b} $(A^L, \wedge, \vee,$ $\odot, 0, 1)$ lattice with $0, 1$ (Ass), (pU), (pCp) (pR)
pHa(pP) $(A^L, \wedge, \vee,$ $\rightarrow, \rightsquigarrow, 0, 1)$ = **pBCK(pP)-L**^{b} +(pprel) +(pdiv)	**pHa(pRP)** $(A^L, \wedge, \vee,$ $\rightarrow, \rightsquigarrow, \odot, 0, 1)$ = **pBCK(pRP)-L**^{b} +(pprel) +(pdiv)	• **pBL**^{L} $(A^L, \wedge, \vee,$ $\odot, \rightarrow, \rightsquigarrow, 0, 1)$ = **ipR-L**^{Lb} +(pprel)+(pdiv)	**no name** $(A^L, \wedge, \vee,$ $\odot, 0, 1)$ = **l-im(pR)**^{b} +(pprel)+(pdiv)
• **pW**^{L} $(A^L, \rightarrow, \rightsquigarrow,$ $^-, ^\sim, 1)$	**pHa(pRP)** + (pDN^L)	**pBL**^{L} + (pDN^L)	• **pMV**^{L} $(A^L, \odot,$ $^-, ^\sim, 0, 1)$

The table with four columns
corresponding to the four different equivalent notions (definitions)
of bounded non-commutative left-algebras, where ^{b} means "bounded",
and where $\rightarrow = \rightarrow^L$, $\rightsquigarrow = \rightsquigarrow^L$, (pprel)=(pprel^L), (pdiv)=(pdiv^L)

Remarks 4.2.6 We have developed in some details the theory of pseudo-algebras of logic (pM algebras, more precisely) and we did not developed at all the theory of unital magmas, because it can be obtained by the equivalences presented in this chapter.

For instance, by (iE5), we obtain the analogous, in the theory of unital magmas, of the important Theorem 2.11.4, namely we have:

Theorem 4.2.7

(1) Let $\mathcal{A}^{Lb} = (A^L, \leq, \odot, 0, 1)$ *be a bounded left-po-im(pR) with the property* (pDN^L) *(involutive). Then, we have: for all* $x, y \in A^L$,

$$x \to^L y = (x \odot y^{\sim})^{-}, \quad x \rightsquigarrow^L y = (y^{-} \odot x)^{\sim} \tag{4.1}$$

and

$$x \odot y = (x \to^L y^{-})^{\sim} = (y \rightsquigarrow^L x^{\sim})^{-}. \tag{4.2}$$

(1') Dually, let $\mathcal{A}^{Rb} = (A^R, \geq, \oplus, 0, 1)$ *be a bounded right-po-im(pcoR) with the property* (pDN^R) *(involutive). Then, we have: for all* $x, y \in A^R$,

$$x \to^R y = (x \oplus y^{\sim})^{-}, \quad x \rightsquigarrow^R y = (y^{-} \oplus x)^{\sim}. \tag{4.3}$$

and

$$x \oplus y = (x \to^R y^{-})^{\sim} = (y \rightsquigarrow^R x^{\sim})^{-}. \tag{4.4}$$

By (iE5) also, one can obtain the analogous of the important Remarks 2.11.7, 2.11.8 and 2.11.9.

Finally, we have the following remark.

Remark 4.2.8 The pM algebras presented in Figure 2.5, at the end of Chapter 2, and the unital magmas presented in Figure 3.2, at the end of Chapter 3, are connected as follows:

pBCK(pP)	⇔	**po-im(pR)**,	by (iE1);
pBCK(pP)-L	⇔	***l*-im(pR)**,	by (iE2);
pBCK(pP)^b	⇔	**po-im(pR)**^{b},	by (iE3);
pBCK(pP)-L^b	⇔	***l*-im(pR)**^{b},	by (iE4);
pBCK(pP)$_{(pDN)}$	⇔	**po-im(pR)**$_{(pDN)}$,	by (iE5);
pBCK(pP)-L$_{(pDN)}$	⇔	***l*-im(pR)**$_{(pDN)}$,	by (iE6);
pHa(pRP)	⇔	**pBL**^{L},	by (iELM7);
pW^L	⇔	**pMV**^{L},	by (iE9');
no-name^L	⇔	**pproduct**^{L},	by (iELM8);
implicative-Boole^L	⇔	**Boole**^{L},	by (iE11').

See also the general Theorem 5.3.1 in next chapter.

Chapter 5

Generalizations: quasi-pM algebras vs. quasi-m unital magmas (with additional operations)

In this chapter, we are dealing with generalizations of pseudo-algebras (pseudo-structures) of logic, called *quasi-pseudo-algebras* (*quasi-pseudo-structures*), vs. generalizations of algebras (from the "world" of unital magmas), called quasi-m algebras.

In Section 1, we introduce the quasi-pseudo-algebras (quasi-pseudo-structures) (as non-commutative generalizations of the quasi-algebras (quasi-structures) comming from quantum computing [101]) and we present some results. We introduce the quasi-pM and the quasi-pBCI algebras, as generalizations of the pM and of the pBCI algebras, respectively, as first examples. We generalize the p-semisimple property to the quasi-p-semisimple property. We generalize to the non-commutative case the notion of quasi-algebra (quasi-structure) with quasi-negation from [103].

In Section 2, we introduce the quasi-m algebras and we present some results. We introduce the quasi-m unital magmas and the quasi-m monoids, as generalizations of the unital magmas and of the monoids, respectively, as first examples. We introduce the quasi-m algebras with quasi-m negations.

In Section 3, we establish a connection between quasi-pseudo-algebras with quasi-negations and quasi-m algebras with quasi-m negations.

The departure point for the research presented in this chapter was:
- the introducing of the *quasi-MV algebras* in 2006 [136], as generalizations of MV algebras introduced in 1958 [26], following an investigation into the foundations of quantum computing (see [65]); since then, many papers investigated them [154], [20], [131], [123];
- the introducing of the *quasi-Wajsberg algebras* in 2010 [21], as generalizations of

Wajsberg algebras introduced in 1984 [53]; they are term-equivalent to quasi-MV algebras, just as Wajsberg algebras are term equivalent to MV algebras:

Definition 5.0.1 [136] A (right-) *quasi-MV algebra* is an algebra $(A, \oplus, ^-, 0, 1)$ of type $(2, 1, 0, 0)$ satisfying the following properties: for all $x, y, z \in A$,
(A1) $x \oplus (y \oplus z) = (x \oplus z) \oplus y$,
(A2)=(DN^R) $(x^-)^- = x$,
(A3) $x \oplus 1 = 1$,
(A4) $(x^- \oplus y)^- \oplus y = (y^- \oplus x)^- \oplus x$,
(A5) $(x \oplus 0)^- = x^- \oplus 0$,
(A6) $(x \oplus y) \oplus 0 = x \oplus y$,
(A7) $0^- = 1$.

Definition 5.0.2 [21] A (left-) *quasi-Wajsberg algebra* is an algebra $(A, \rightarrow, ^-, 1)$ of type $(2, 1, 0)$ satisfying the following properties: for all $x, y, z \in A$,
(q-M^L) $1 \rightarrow (x \rightarrow y) = x \rightarrow y$,
(BB^{L}') $(x \rightarrow y) \rightarrow [(y \rightarrow z) \rightarrow (x \rightarrow z)] = 1$,
(comm^L) $(x \rightarrow y) \rightarrow y = (y \rightarrow x) \rightarrow x$,
(DN1) $(x^- \rightarrow y^-) \rightarrow (y \rightarrow x) = 1$,
(DN^L) $(x^-)^- = x$,
$(q\overline{M}(1 \rightarrow x))$ $1 \rightarrow (1 \rightarrow x)^- = (1 \rightarrow x)^-$.

In paper [101], starting from the quasi-Wajsberg algebras, whose regular sets are the Wajsberg algebras, we have introduced a theory of quasi-algebras (of logic) vs. a theory of regular algebras (of logic), in the commutative case; the first part of the theory of quasi-algebras vs. the theory of regular algebras is there presented. We have introduced the quasi-RM and the quasi-RML algebras and we have presented two equivalent definitions of quasi-BCI and of quasi-BCK algebras. We then developed the theory, including the introduction of the *quasi-negation*, but the research is in the preprints [102], [103].

Very recently we found that also the *quasi-pseudo-MV algebras* were introduced, in 2016 by W.J. Chen and W.A. Dudek [28], as non-commutative generalizations of the quasi-MV algebras and of the pseudo-MV algebras; see also [29] and [30]:

Definition 5.0.3 [28] A (leftt-) *quasi-pseudo-MV algebra* is an algebra $(A, \odot, ^-, ^\sim, 1)$ of type $(2, 1, 1, 0)$ satisfying the following properties: for all $x, y, z \in A$,
(QP1)=(Ass) $x \odot (y \odot z) = (x \odot y) \odot z$,
(QP2) $1^- = 1^\sim$ (denoted by 0),
(QP3)=(IdP) $x \odot 1 = 1 \odot x$,
(QP4) $x \odot 1^- = 1^- = 1^- \odot x$,
(QP5)=(pDNP2) $(x^- \odot y^-)^\sim = (x^\sim \odot y^\sim)^-$,
(QP6) =(pDN^L) $(x^-)^\sim = x = (x^\sim)^-$,
(QP7) $y \odot (x^- \odot y)^\sim = (y \odot x^\sim)^- \odot y = x \odot (y^- \odot x)^\sim = (x \odot y^\sim)^- \odot x$,
(QP8) $(x \odot 1)^- = x^- \odot 1, \quad (x \odot 1)^\sim = x^\sim \odot 1$,
(QP9) $x \odot y \odot 1 = x \odot y$.

In this monograph, we have generalized some results from the paper [101] and from the preprint [103] to the non-commutative case, namely we have introduced the first part of the theory of quasi-pseudo-algebras vs. the theory of regular pseudo-algebras.

Writing this monograph, we have realized that in fact there are two kinds of "quasi"-generalizations: "quasi" and "quasi-m", corresponding to the two kinds of algebras: *pM algebras* (inside the algebras of logic) and *unital magmas*, respectively. Therefore, we present here an introduction to the new theory of *quasi-m algebras* also, directly in the non-commutative case.

Hence, in this chapter we generalize:
- the *pM algebras* to *quasi-pM algebras*, as the most general quasi-pseudo-algebra, by generalizing the principal, defining dual properties (pM^L), (pM^R) to (q-pM^L), (q-pM^R), respectively (in Section 1), and
- the *unital magmas* to *quasi-m unital magmas*, by generalizing the principal, defining property (pU) to (qm-pU) (in Section 2).

We introduce also the quasi-pseudo-algebras with (involutive) quasi-negations and the quasi-pseudo-BCI algebras (in Section 1) and the quasi-m algebras with (involutive) quasi-m negations and the quasi-m monoids (in Section 2).

Other quasi-pseudo-algebras will be introduced in Chapter 10 and other quasi-m algebras will be introduced in Chapter 11.

Some of the *quasi-pM algebras* and some of the *quasi-m unital magmas* will be connected by a Galois connection in Chapter 12 (as happens in the regular case, where their corresponding *pM algebras* and *unital magmas*, respectively, are connected by a Galois connection). In Section 3, we establish some general connections.

Thus, it will be rather obvious that the quasi-Wajsberg algebras are particular cases of *quasi-M algebras*, while the quasi-MV algebras are particular cases of commutative quasi-m unital magmas (and the quasi-pseudo-MV algebras are particular cases of quasi-m unital magmas).

The chapter has 3 sections.

5.1 Quasi-pM algebras vs. regular pM algebras

In this section, we generalize to the non-commutative case some definitions and results from [101] and [103]. We introduce the *quasi-pM algebra*, a generalization of the pM algebra.

Let $(A, \to^L, \rightsquigarrow^L, 1)$ be an algebra of type $(2,2,0)$ or let $(A, \leq, \to^L, \rightsquigarrow^L, 1)$ be a structure.

Let us consider the following properties: for all $x, y \in A$,
(IdEq^L) $\quad x \to^L y = 1 \Longleftrightarrow x \rightsquigarrow^L y = 1$,
(pEq^L) $\quad x \leq y \Longleftrightarrow x \to^L y = 1 \overset{(IdEq^L)}{\Leftrightarrow} x \rightsquigarrow^L y = 1$,
(pM^L) $\quad 1 \to^L x = x = 1 \rightsquigarrow^L x$,
(pEx^L) $\quad x \to^L (y \rightsquigarrow^L z) = y \rightsquigarrow^L (x \to^L z)$,
(IdR^L) $\quad 1 \to^L x = 1 \rightsquigarrow^L x \quad$ and
(q-pM^L) $\quad 1 \rightsquigarrow^L (x \to^L y) = x \to^L y, \quad 1 \to^L (x \rightsquigarrow^L y) = x \rightsquigarrow^L y$,
(q-pM^L($1 \to^L x, 1 \rightsquigarrow^L x$)) $\quad 1 \rightsquigarrow^L (1 \to^L x) = 1 \to^L x, \ 1 \to^L (1 \rightsquigarrow^L x) = 1 \rightsquigarrow^L x$,
(pM1^L) $\quad 1 \to^L 1 = 1 \ (\Leftrightarrow 1 \rightsquigarrow^L 1 = 1)$.

Note that we have, obviously:
(i) (pM^L) $\Longrightarrow$ (pM1^L),
(ii) (pM^L) $\Longrightarrow$ (q-pM^L) $\Longrightarrow$ (q-pM^L($1 \to^L x, 1 \leadsto^L x$)).
Besides the set A, let us define the following subset of A:

$$V_{pM^L} \stackrel{def.}{=} \{x \in A \mid x \stackrel{(pM^L)}{=} 1 \to^L x = 1 \leadsto^L x\}.$$

Note that, if property (pM^L) holds, then $A \subseteq V_{pM^L}$, hence $V_{pM^L} = A$.

5.1.1 Regular pseudo-algebras (of logic)

Recall that the pseudo-algebras (equivalently, the pseudo-structures) of logic were introduced by Definitions 2.1.1 in Chapter 2.

Definitions 5.1.1 (The dual definitions are omitted)
1. The left-pseudo-algebra $(A, \to^L, \leadsto^L, 1)$ (or, equivalently, the left-pseudo-structure $(A, \leq, \to^L, \leadsto^L, 1)$) is called *regular*, if it satisfies the property (pM^L).
1'. Any left-pseudo-algebra (left-pseudo-structure) $\mathcal{A}' = (A, \sigma)$ whose signature σ contains $\to^L, \leadsto^L$, 1 ($\leq$, $\to^L, \leadsto^L$, 1, respectively) is also called *regular*, if it satisfies the property (pM^L).
1". Any left-pseudo-algebra (left-pseudo-structure) $\mathcal{A}'' = (A, \tau)$ which is term equivalent to a regular left-pseudo-algebra (left-pseudo-structure) $\mathcal{A}' = (A, \sigma)$ is also called *regular*.
2. The pseudo-implication $(\to^L, \leadsto^L)$ from a regular left-pseudo-algebra (left-pseudo-structure) is called *regular pseudo-implication.*
3. The binary relation $\leq$ of a regular left-pseudo-algebra (left-pseudo-structure) is called *binary regular relation.*

Note that, by above Definitions 5.1.1 (1), the most general regular left-pseudo-structure is the left-pseudo-M algebra, introduced in Chapter 2. Other regular pseudo-algebras were recalled or presented also in Chapter 2.

Roughly speaking, the *regular* left-pseudo-algebras (left-pseudo-structures) are those classical left-pseudo-algebras (left-pseudo-structures) verifying the property (pM^L). All the properties verified by a regular left-pseudo-algebra (left-pseudo-structure), including the property (pM^L), are called *regular properties.* A *theory of regular left-pseudo-algebras (left-pseudo-structures)* establishes the connections existing between the property (pM^L) and the other regular properties (i.e. if a regular property is proved by (pM^L) or not), as for example the Proposition 2.1.8. The regular properties which depend on (pM^L) can be modified when we generalize (pM^L) to (q-pM^L), i.e. when we develop a theory of left-quasi-pseudo-algebras (left-quasi-pseudo-structures) - see further in this section.

Recall from Chapter 2 that the internal, natural binary regular relation $\leq$ defined in any regular pseudo-algebra is an *order* relation in left- pBZ, pBCC, pBCI and pBCK algebras and is only a *pre-order* in left- pre-pBZ, pre-pBCC, pre-pBCI and pre-pBCK algebras (see [104] for the commutative case).

The ordered regular pseudo-algebras are represented by Hasse diagrams, while all the others are represented by Hasse-type diagrams.

Remark 5.1.2 In any regular pseudo-algebra (pseudo-structure) $\mathcal{A}$ we have that: $V_{pM^L} = A$ and this is the basic, definable property of regular pseudo-algebras (pseudo-structures).

• **Bounded regular pseudo-algebras**

Let $\mathcal{A} = (A, \to^L, \rightsquigarrow^L, 1)$ be an ordered regular left-pseudo-algebra (or, equivalently, let $\mathcal{A} = (A, \leq, \to^L, \rightsquigarrow^L, 1)$ be an ordered regular left-pseudo-structure) and $\leq$ be the order of $\mathcal{A}$ (i.e. (pM^L), (Re), (An), (Tr) hold).

Definitions 5.1.3

i. An element $l \in A$ is called *last regular element* or *the greatest regular element* of $\mathcal{A}$, if $x \to^L l = 1 = x \rightsquigarrow^L l$ (or $x \leq l$), for every $x \in A$.

ii. An element $f \in A$ is called *first regular element* or *the smallest regular element* of $\mathcal{A}$, if $f \to^L x = 1 = f \rightsquigarrow^L x$ (or $f \leq x$), for every $x \in A$.

Hence, the notions of *first regular element* and *last regular element* are *dual to each other*. Note that both the first regular element and the last regular element of $\mathcal{A}$ are unique, by (An).

The special element $1 \in A$ can be the last regular element. In this case, we have the property:
(L') (Last) $x \to^L 1 = 1 = x \rightsquigarrow^L 1$, for all $x \in A$, or, equivalently,
(L) (Last) $x \leq 1$, for all $x \in A$.

An ordered regular pseudo-algebra with property (L) is an ordered pseudo-ML algebra.

The first regular element will be denoted by 0; hence we have the property:
(F') (First) $0 \to^L x = 1 = 0 \rightsquigarrow^L x$, for all $x \in A$, or, equivalently,
(F) (First) $0 \leq x$, for all $x \in A$.

Definition 5.1.4 An ordered regular left-pseudo-algebra (left-pseudo-structure) with first regular element 0 and last regular element 1 is called *bounded* and is denoted by $(A, \to^L, \rightsquigarrow^L, 0, 1)$ ($(A, \leq, \to^L, \rightsquigarrow^L, 0, 1)$, respectively).

Note that a bounded ordered regular left-pseudo-algebra is a bounded ordered left-pseudo-ML algebra.

Let $\mathcal{A} = (A, \to^L, \rightsquigarrow^L, 0, 1)$ be a bounded ordered regular left-pseudo-algebra (or, equivalently, let $\mathcal{A} = (A, \leq, \to^L, \rightsquigarrow^L, 0, 1)$ be a bounded ordered regular left-pseudo-structure) (i.e. properties (pM^L), (Re), (An), (Tr) and (F), (L) hold).

We define two *negations* $^- : A \longrightarrow A$ and $^\sim : A \longrightarrow A$ by: for all $x \in A$,

$$x^- \stackrel{def.}{=} x \to^L 0, \quad x^\sim \stackrel{def.}{=} x \rightsquigarrow^L 0.$$

Hence, the negations are *unique* and we have:
$1^- \stackrel{def.}{=} 1 \to^L 0 \stackrel{(pM^L)}{=} 0$ and $1^\sim \stackrel{def.}{=} 1 \rightsquigarrow^L 0 \stackrel{(pM^L)}{=} 0$ and
$0^- \stackrel{def.}{=} 0 \to^L 0 \stackrel{(pRe^L)}{=} 1$ and $0^\sim \stackrel{def.}{=} 0 \rightsquigarrow^L 0 \stackrel{(pRe^L)}{=} 1$,
i.e. the following property holds:
(pN1^{Lb}) $\quad 0 = 1^- = 1^\sim, \quad 1 = 0^- = 0^\sim$.

Remark 5.1.5 If the regular left-pseudo-algebra (left-pseudo-structure) is not ordered, then one can extend the above notions.

- **Regular pseudo-algebras with (involutive) regular negations**

As we know, in bounded left-pBCK algebras $(A, \leq, \rightarrow^L, \rightsquigarrow^L, 0, 1)$, we can *define* two negations $^-$ and $^\sim$ by: $x^- = x \rightarrow^L 0$, $x^\sim = x \rightsquigarrow^L 0$; they are not included in the signature. But, in the left-pseudo-Wajsberg algebras $(A, \rightarrow^L, \rightsquigarrow^L, ^-, ^\sim, 0, 1)$, there exists two negations ($^-$ and $^\sim$) included in the signature, and, if we denote $1^- = 1^\sim$ by 0, then the two negations are *connected* to $\rightarrow^L$, $\rightsquigarrow^L$ and 1 by:
(pNeg^L) $\quad x^- = x \rightarrow^L 0, \quad x^\sim = x \rightsquigarrow^L 0.$
Consequently, we cam make the following remark.

Remark 5.1.6 (See Remark 2.1.2)

Just as the *equivalence*: $\quad x \leq y \quad \Longleftrightarrow \quad x \rightarrow^L y = 1 \ (\overset{(IdEq^L)}{\Longleftrightarrow} \ x \rightsquigarrow^L y = 1)$
can be used:
- either as the *definition* of the binary regular relation $\leq$ in the regular left-pseudo-algebra $(A, \rightarrow^L, \rightsquigarrow^L, 1)$,
- or as the *connection* (pEq^L) between the binary regular relation $\leq$ and $\rightarrow^L, \rightsquigarrow^L$, 1 in the regular left-pseudo-structure $(A, \leq, \rightarrow^L, \rightsquigarrow^L, 1)$,
the same, the *equalities*: $\quad x^- = x \rightarrow^L 0, \quad x^\sim = x \rightsquigarrow^L 0$
can be used:
- either as the *definition* of the negations $^-$, $^\sim$ in the *bounded regular left-pseudo-algebra* $(A, \rightarrow^L, \rightsquigarrow^L, 0, 1)$ *(left-pseudo-structure* $(A, \leq, \rightarrow^L, \rightsquigarrow^L, 0, 1)$*)*, and in this case:
$1^- \overset{def.}{=} 1 \rightarrow^L 0 = 0$, $1^\sim \overset{def.}{=} 1 \rightsquigarrow^L 0 = 0$ by (pM^L), and
$0^- \overset{def.}{=} 0 \rightarrow^L 0 = 1$, $0^\sim \overset{def.}{=} 0 \rightsquigarrow^L 0 = 1$ by (Re) (see the sequel),
- or as the *connection* (pNeg^L) between the negations $^-$, $^\sim$ and $\rightarrow^L, \rightsquigarrow^L$, 1 in the *regular left-pseudo-algebra* $(A, \rightarrow^L, \rightsquigarrow^L, ^-, ^\sim, 1)$ *(left-pseudo-structure* $(A, \leq, \rightarrow^L, \rightsquigarrow^L, ^-, ^\sim, 1)$*) with two negations*, and in this case:
$0 \overset{notation}{=} 1^- = 1^\sim$ and 0^- must equal $0 \rightarrow^L 0$, $0^\sim$ must equal $0 \rightsquigarrow^L 0$.
Note that, in both cases, the negations are *unique*.

Definitions 5.1.7 (The dual case is omitted)

i. A *regular left-pseudo-algebra (left-pseudo-structure) with negations* is a regular left-pseudo-algebra $\mathcal{A} = (A, \rightarrow^L, \rightsquigarrow^L, 1)$ (left-pseudo-structure $\mathcal{A} = (A, \leq, \rightarrow^L, \rightsquigarrow^L, 1)$) (i.e. (pM^L) holds) with $^-$, $^\sim$ two unary operations on A, called *regular negations*, which are connected with $\rightarrow^L, \rightsquigarrow^L$, 1 by (pNeg^L), where 0 is the notation for the element $1^- = 1^\sim$.

i'. A *regular left-pseudo-algebra (left-pseudo-structure) with negations* is an algebra $\mathcal{A} = (A, \rightarrow^L, \rightsquigarrow^L, ^-, ^\sim, 1)$ (a structure $\mathcal{A} = (A, \leq, \rightarrow^L, \rightsquigarrow^L, ^-, ^\sim, 1)$) such that $(A, \rightarrow^L, \rightsquigarrow^L, 1)$ ($(A, \leq, \rightarrow^L, \rightsquigarrow^L, 1)$ respectively) is a regular left-pseudo-algebra (left-pseudo-structure) (i.e. (pM^L) holds) and $^-$, $^\sim$ are unary operations on A, called *regular negations*, which are connected with $\rightarrow^L, \rightsquigarrow^L$, 1 by (pNeg^L), where 0 is the notation for the element $1^- = 1^\sim$.

ii. If the regular negations verify (pDN^L), we shall say that $\mathcal{A}$ is a regular left-pseudo-algebra (left-pseudo-structure) with *involutive* regular negations.

Hence, the negations are *unique* and (pN1^{Lb}) holds:
$0 \overset{notation}{=} 1^- = 1^\sim$ and $0^- = 0 \to^L 0 = 1$, $0^\sim = 0 \rightsquigarrow^L 0 = 1$, if (Re) holds.

Remark 5.1.8 If $^- = ^\sim$, and we denote this by $^{-1}$, and if (Id) holds, then (pNeg^L) becomes:
(pNeg) $x^{-1} = x \to 1 \overset{(Id)}{=} x \rightsquigarrow 1$, with $1 = 1^{-1}$
and (pDN^L) becomes:
(DN) $(x^{-1})^{-1} = x$.

Hence, a *regular left-pseudo-algebra with involutive regular negation* $\mathcal{A}$ verifies (pM), (pNeg) and (DN).

Note that the Wajsberg algebra and the implicative-Boolean algebra are examples of regular algebras with involutive regular negation (see [103]) (in the "world" of M algebras with additional operations). Other examples are presented in Chapter 11.

5.1.2 Quasi-pseudo-algebras vs. regular pseudo-algebras. The equivalence (Q)

Let $\mathcal{A} = (A, \Rightarrow^L, \approx>^L, 1)$ be a *left-pseudo-algebra* or, equivalently,
let $\mathcal{A} = (A, \preceq, \Rightarrow^L, \approx>^L, 1)$ be a *left-pseudo-structure.*

Let us consider the following properties: for all $x, y \in A$,
(q-pM^L) $1 \approx>^L (x \Rightarrow^L y) = x \Rightarrow^L y$, $1 \Rightarrow^L (x \approx>^L y) = x \approx>^L y$,
(pM1^L) $1 \Rightarrow^L 1 = 1 (\Leftrightarrow 1 \approx>^L 1 = 1)$.

Definitions 5.1.9

(q1) The left-pseudo-algebra $(A, \Rightarrow^L, \approx>^L, 1)$ (or, equivalently, the left-pseudo-structure $(A, \preceq, \Rightarrow^L, \approx>^L, 1)$) is called *left-quasi-pseudo-algebra* (*left-quasi-pseudo-structure*, respectively), if it satisfies the properties (q-pM^L) and (pM1^L).

(q1') Any pseudo-algebra (pseudo-structure) $\mathcal{A}' = (A, \sigma)$ whose signature σ contains $\Rightarrow^L, \approx>^L, 1$ ($\preceq$, $\Rightarrow^L, \approx>^L$, 1, respectively) is also called *left-quasi-pseudo-algebra* (*left-quasi-pseudo-structure*), if it satisfies the properties (q-pM^L) and (pM1^L).

(q1") Any algebra (structure) $\mathcal{A}'' = (A, \tau)$ which is term equivalent to a left-quasi-pseudo-algebra (left-pseudo-structure) $\mathcal{A}' = (A, \sigma)$, is also called *left-quasi-pseudo-algebra* (*left-quasi-pseudo-structure*).

(q2) The pseudo-implication $(\Rightarrow^L, \approx>^L)$ from a left-quasi-pseudo-algebra (left-quasi-pseudo-structure) is called *quasi-pseudo-implication.*

(q3) The binary relation $\preceq$ of a left-quasi-pseudo-algebra (left-quasi-pseudo-structure) is called *binary quasi-relation.*

Roughly speaking, the *left-quasi-pseudo-algebras (left-quasi-pseudo-structures)* are those left-pseudo-algebras (left-pseudo-structures) verifying the properties (q-pM^L) and (pM1^L). All the properties verified by a left-quasi-pseudo-algebra (left-quasi-pseudo-structure), including the property (q-pM^L), are called *quasi-properties.*

A quasi-property is *proper*, if it is not a regular property. A *theory of left-quasi-pseudo-algebras (left-quasi-pseudo-structures)* establishes the connections existing between the property (q-pM^L) and the other quasi-properties (i.e. if a quasi-property is proved by (q-pM^L) or not), as for example the next Proposition 5.1.25.

Besides the set A, let us define now the following subset of A:

$$V_{pM^L} \stackrel{def.}{=} \{x \in A \mid x \stackrel{(pM^L)}{=} 1 \Rightarrow^L x = 1 \approx>^L x\}.$$

Remarks 5.1.10

(i) Note that:
- if (q-pM^L) holds, then $V_{pM^L} \neq A$, i.e. $V_{pM^L} \subset A$;
- if (pM^L) holds, then $V_{pM^L} = A$.

(ii) Since (pM^L) implies (q-pM^L), it follows that any quasi-pseudo-algebra (quasi-pseudo-structure) will be a generalization of the corresponding regular pseudo-algebra (pseudo-structure). For example, in the commutative case, the quasi-Wajsberg algebra introduced in [21] is a generalization of the Wajsberg algebra, the Wajsberg algebras being called in this context *regular algebras*.

(iii) The above remarks show the central role played by the property (pM^L); we obtain the general rule that, roughly speaking:

(Q) **quasi-pseudo-algebra (quasi-pseudo-structure) + (pM^L) $\Longleftrightarrow$ regular pseudo-algebra (pseudo-structure).**

In the view of the above Remarks 5.1.10, we introduce the following definitions:

Definitions 5.1.11 1. For every quasi-pseudo-algebra (quasi-pseudo-structure) $\mathcal{A}$, the subset V_{pM^L} of A will be called the *regular* set of $\mathcal{A}$ and will be denoted by $R(A)$:

$$R(A) \stackrel{def.}{=} V_{pM^L} \subseteq A.$$

The elements of $R(A)$ are called the *regular elements* of A.

2. The quasi-pseudo-algebra (quasi-pseudo-structure) $\mathcal{A}$ is called *q-proper*, if $R(A) \neq A$ (i.e. (pM^L) $\not\Longleftrightarrow$ (q-pM^L)); otherwise, $\mathcal{A}$ is a regular pseudo-algebra (pseudo-structure).

Definition 5.1.12 For every q-proper left-quasi-pseudo-algebra $\mathcal{A} = (A, \Rightarrow^L, \approx>^L, 1)$ (or, equivalently, left-quasi-pseudo-structure $\mathcal{A} = (A, \preceq, \Rightarrow^L, \approx>^L, 1)$), any subset $S \subseteq A$ closed under $\Rightarrow^L, \approx>^L$ and containing 1 is called a *quasi-pseudo-subalgebra* (*quasi-pseudo-substructure*) of $\mathcal{A}$.

We then have the following important result:

Theorem 5.1.13 *Let $\mathcal{A} = (A, \Rightarrow^L, \approx>^L, 1)$ be a q-proper quasi-pseudo-algebra (or, equivalently, let $\mathcal{A} = (A, \preceq, \Rightarrow^L, \approx>^L, 1)$ be a q-proper quasi-pseudo-structure). Then,*

$$\mathcal{R}(\mathcal{A}) = (R(A), \rightarrow^L, \rightsquigarrow^L, 1)$$

is a regular pseudo-algebra (or, equivalently, $\mathcal{R}(\mathcal{A}) = (R(A), \leq, \rightarrow^L, \rightsquigarrow^L, 1)$ is a regular pseudo-structure, respectively), where

$$\rightarrow^L = \Rightarrow^L|_{R(A)}, \quad \rightsquigarrow^L = \approx>^L|_{R(A)}, \qquad \leq = \preceq|_{R(A)}.$$

Proof: First, we prove that the regular set $R(A)$ is closed under $\Rightarrow^L, \approx>^L$ and that $1 \in R(A)$. Indeed, if $x, y \in R(A) \subset A$, then $x \Rightarrow^L y \overset{(q-pM^L)}{=} 1 \approx>^L (x \Rightarrow^L y)$ and $x \approx>^L y \overset{(q-pM^L)}{=} 1 \Rightarrow^L (x \approx>^L y)$, hence $x \Rightarrow^L y \in R(A)$ and $x \approx>^L y \in R(A)$; since $1 \Rightarrow^L 1 = 1$ (by (pM1^L)), it follows that $1 \in R(A)$. Consequently, $R(A)$ is a quasi-pseudo-subalgebra (quasi-pseudo-substructure) of $\mathcal{A}$. Hence, $\mathcal{R}(\mathcal{A}) = (R(A), \Rightarrow^L, \approx>^L, 1)$ is a quasi-pseudo-algebra ($\mathcal{R}(\mathcal{A}) = (R(A), \preceq, \Rightarrow^L, \approx>^L, 1)$ is a quasi-pseudo-structure). Moreover, (q-pM^L) coincides with (pM^L) on $R(A)$, i.e. (q-pM^L)$|_{R(A)} \Longleftrightarrow$ (pM^L), by the definition of $R(A)$. Consequently, $\mathcal{R}(\mathcal{A}) = (R(A), \rightarrow^L, \rightsquigarrow^L, 1)$ is a regular pseudo-algebra (or, equivalently, $\mathcal{R}(\mathcal{A}) = (R(A), \leq, \rightarrow^L, \rightsquigarrow^L, 1)$ is a regular pseudo-structure, respectively). □

Conventions: In order to simplify the writing,
- the *quasi-pseudo-implication* $(\Rightarrow^L, \approx>^L)$ of a quasi-pseudo-algebra (quasi-pseudo-structure) $\mathcal{A}$ and the corresponding (its restriction to $R(A)$) *regular pseudo-implication* $(\rightarrow^L, \rightsquigarrow^L)$ of the regular pseudo-algebra (pseudo-structure) $\mathcal{R}(\mathcal{A})$ will be denoted the same in the sequel, namely by $(\rightarrow^L, \rightsquigarrow^L)$.
- the *binary quasi-relation* $\preceq$ of a quasi-pseudo-algebra (quasi-pseudo-structure) $\mathcal{A}$ and the corresponding (its restriction to $R(A)$) *binary regular relation* $\leq$ of $\mathcal{R}(\mathcal{A})$ will be also denoted the same in the sequel, namely by $\leq$.

- **The quasi-pM algebras**

Following the above Definitions 5.1.9 (q1), we introduce the most general quasi-pseudo-algebra, namely (the dual case is omitted):

Definition 5.1.14 A *left-quasi-pseudo-M algebra*, or a *left-q-pM algebra* for short, is a left-pseudo-structure $\mathcal{A}^L = (A^L, \leq, \rightarrow^L, \rightsquigarrow^L, 1)$ verifying the properties (q-pM^L) and (pM1^L).

Denote by **q-pM**^{L} the class of all left-q-pM algebras and, dually, denote by **q-pM**^{R} the class of all right-q-pM algebras.

Now, one can define analogously the quasi-versions of all the (regular) pseudo-algebras defined or recalled in Chapter 2. For example,

Definition 5.1.15 A *left-quasi-pseudo-ME algebra*, or a *left-q-pME algebra* for short, is a left-pseudo-structure $\mathcal{A}^L = (A^L, \leq, \rightarrow^L, \rightsquigarrow^L, 1)$ verifying the properties (q-pM^L), (pM1^L) and (pEx^L).

We immediately obtain the following two Corollaries of Theorem 5.1.13:

Corollary 5.1.16 *Let* $\mathcal{A}^L = (A^L, \leq, \rightarrow^L, \rightsquigarrow^L, 1)$ *be a left-q-pM algebra. Then,* $\mathcal{R}(\mathcal{A}^L) = (R(A^L), \leq, \rightarrow^L, \rightsquigarrow^L, 1)$ *is a left-pM algebra.*

Corollary 5.1.17 *Let* $\mathcal{A}^L = (A^L, \leq, \rightarrow^L, \rightsquigarrow^L, 1)$ *be a left-q-pME algebra. Then,* $\mathcal{R}(\mathcal{A}^L) = (R(A^L), \leq, \rightarrow^L, \rightsquigarrow^L, 1)$ *is a left-pME algebra.*

Hence, we have the following equivalences, by (Q):
q-pM^L + (pM^L) $\Longleftrightarrow$ **pM**^{L} and, dually,

q-pM^R + (pM^R) $\Longleftrightarrow$ **pM**^{R};
q-pME^L + (pM^L) $\Longleftrightarrow$ **pME**^{L} and, dually,
q-pME^R + (pM^R) $\Longleftrightarrow$ **pME**^{R}.

An important result is the following.

Proposition 5.1.18 *Consider the property:*
(q-pM^L): $1 \to^L (x \rightsquigarrow^L y) = x \rightsquigarrow^L y$ *and* $1 \rightsquigarrow^L (x \to^L y) = x \to^L y$.
Then, the following property is derived:
(q-pM^L_d) $1 \to^L (x \to^L y) = x \to^L y$, $1 \rightsquigarrow^L (x \rightsquigarrow^L y) = x \rightsquigarrow^L y$.

Proof.
$$1 \to^L (x \to^L y) \overset{(pM^L)}{=} 1 \to^L [1 \rightsquigarrow^L (x \to^L y)] \overset{(pM^L)}{=} 1 \rightsquigarrow^L (x \to^L y) \overset{(pM^L)}{=} x \to^L y$$
$$1 \rightsquigarrow^L (x \rightsquigarrow^L y) = 1 \rightsquigarrow^L [1 \to^L (x \rightsquigarrow^L y)] \overset{(pM^L)}{=} 1 \to^L (x \rightsquigarrow^L y) = x \rightsquigarrow^L y. \quad \Box$$

Proposition 5.1.19 *Let $(A, \to^L, \rightsquigarrow^L, 1)$ be a quasi-pseudo-algebra (or $(A, \leq, \to^L, \rightsquigarrow^L, 1)$ be a quasi-pseudo-structure).*
If the property (pEx^L) holds, then the property (IdR^L) holds.

Proof. $1 \to^L x \overset{(q-pM^L)}{=} 1 \rightsquigarrow^L (1 \to^L x) \overset{(pEx^L)}{=} 1 \to^L (1 \rightsquigarrow^L x) \overset{(q-pM^L)}{=} 1 \rightsquigarrow^L x$. $\Box$

Hence, we obtain:

Corollary 5.1.20 *Let $\mathcal{A}^L = (A^L, \leq, \to^L, \rightsquigarrow^L, 1)$ be a left-q-pME algebra. Then, the property (IdR^L) holds.*

- **Quasi-order. The quasi-p-semisimple property**

Let $\mathcal{A}$ be a q-proper quasi-pseudo-algebra (quasi-pseudo-structure) (i.e. (q-pM^L), which differs from (pM^L), and (pM1^L) hold); then its subalgebra $\mathcal{R}(\mathcal{A})$ is a regular pseudo-algebra (pseudo-structure) (i.e. (pM^L) holds), by Theorem 5.1.13.

Consider the following properties of $(\to^L, \rightsquigarrow^L)$:

(pRe^L) $x \to^L x = 1 (\Leftrightarrow x \rightsquigarrow^L x = 1)$,
(pAn^L) $x \to^L y = 1 = y \to^L x \Longrightarrow x = y$,
$x \rightsquigarrow^L y = 1 = y \rightsquigarrow^L x \Longrightarrow x = y$,
(q-pAn^L') $x \to^L y = 1 = y \to^L x \Longrightarrow 1 \to^L x = 1 \to^L y$,
$x \rightsquigarrow^L y = 1 = y \rightsquigarrow^L x \Longrightarrow 1 \rightsquigarrow^L x = 1 \rightsquigarrow^L y$,
(pTr^L) $x \to^L y = 1 = y \to^L z \Longrightarrow x \to^L z = 1$,
$x \rightsquigarrow^L y = 1 = y \rightsquigarrow^L z \Longrightarrow x \rightsquigarrow^L z = 1$,

or the following properties of $\leq$:

(Re) $x \leq x$,
(An) $x \leq y$ and $y \leq x \Longrightarrow x = y$,
(q-pAn^L) $x \leq y$ and $y \leq x \Longrightarrow (1 \to^L x = 1 \to^L y$ and $1 \rightsquigarrow x^L = 1 \rightsquigarrow^L y)$,
(Tr) $x \leq y$ and $y \leq z \Longrightarrow x \leq z$.

Definitions 5.1.21 We shall say that $\mathcal{A}$ is:
- *reflexive*, if property (pRe^L) (or (Re)) is satisfied,

- *quasi-antisymmetric*, if property (q-pAn^L') (or (q-pAn^L)) is satisfied,
- *transitive*, if property (pTr^L) (or (Tr)) is satisfied,
- *quasi-pre-ordered*, and $\leq$ is a *quasi-pre-order*, if it is reflexive and transitive,
- *quasi-ordered*, and $\leq$ is a *quasi-order* (or a *q-order* for short), if it is reflexive, quasi-antisymmetric and transitive.

A quasi-ordered quasi-pseudo-algebra (quasi-pseudo-structure) will be simply called: "a quasi-ordered pseudo-algebra (pseudo-structure)".

Remarks 5.1.22

(i) (pM^L) + (q-pAn^L) $\Longrightarrow$ (pAn^L).
Indeed, suppose that $x \to^L y = 1 = y \to^L x$; then, by (q-pAn^L), $1 \to^L x = 1 \to^L y$, hence, by (pM^L), $x = y$. Similarly, suppose that $x \leadsto^L y = 1 = y \leadsto^L x$; then, by (q-pAn^L), $1 \leadsto^L x = 1 \leadsto^L y$, hence, by (pM^L), $x = y$. Consequently, in the presence of (pM^L), any q-order becomes a regular order.

(ii) The above remark (i) shows again the central role played by the property (pM^L): we obtain, by (Q), the general rule that, roughly speaking,

quasi-ordered pseudo-algebra (pseudo-structure) + (pM^L) $\Longleftrightarrow$
ordered regular pseudo-algebra (pseudo-structure).

Note that quasi-Wajsberg algebras are in fact *quasi-ordered algebras* and that Wajsberg algebras are in fact *ordered regular algebras.*

We now present an important remark:

Remark 5.1.23 Let $\mathcal{A} = (A, \to^L, \leadsto^L, 1)$ be a quasi-ordered pseudo-algebra (or $\mathcal{A} = (A, \leq, \to^L, \leadsto^L, 1)$ be a quasi-ordered pseudo-structure). Since (q-pM^L) coincides with (pM^L) on $R(A)$, by Theorem 5.1.13, it follows that (q-pAn^L) coincides with (pAn^L) on $R(A)$. Consequently,
- the q-order relation $\leq$ on A becomes an order relation on $R(A)$;
- $\mathcal{R}(\mathcal{A}) = (R(A), \to^L, \leadsto^L, 1)$ is an ordered regular pseudo-algebra (or, equivalently, $\mathcal{R}(\mathcal{A}) = (R(A), \leq, \to^L, \leadsto^L, 1)$ is an ordered regular pseudo-structure, respectively).

We introduce now the following definition.

Definition 5.1.24

i. We say that a left-quasi-pseudo-algebra (or, equivalently, a left-quasi-pseudo-structure) $\mathcal{A}^L$ is *quasi-p-semisimple*, if the following property is verified: for all $x, y \in A^L$,
(q-p-s^L) $x \leq y \Longleftrightarrow (1 \to^L x = 1 \to^L y$ and $1 \leadsto x^L = 1 \leadsto^L y)$.

i'. Dually, we say that a right-quasi-pseudo-algebra (or, equivalently, a right-quasi-pseudo-structure) $\mathcal{A}^R$ is *quasi-p-semisimple*, if the following property is verified: for all $x, y \in A^R$,
(q-p-s^R) $x \geq y \Longleftrightarrow (0 \to^R x = 0 \to^R y$ and $0 \leadsto x^R = 0 \leadsto^R y)$.

- **General properties**

Consider the following properties: for all $x, y \in A$,
(q-pN^L) $1 \leq x \to^L y \Longrightarrow x \to^L y = 1$, $1 \leq x \rightsquigarrow^L y \Longrightarrow x \rightsquigarrow^L y = 1$,
(q-pI2^L) $x \to^L y = x \to^L (1 \rightsquigarrow^L y)$, $x \rightsquigarrow^L y = x \rightsquigarrow^L (1 \to^L y)$,
(q-pR2^L) $x \to^L (1 \rightsquigarrow^L x) = 1$, $x \rightsquigarrow^L (1 \to^L x) = 1$.

Proposition 5.1.25 *(See [101] for the commutative case) (See Proposition 2.1.8 in the regular case)*
Let $(A, \to^L, \rightsquigarrow^L, 1)$ be a quasi-pseudo-algebra
(or $(A, \leq, \to^L, \rightsquigarrow^L, 1)$ be a quasi-pseudo-structure). The following properties hold:
(q-1) (q-pM^L) + (pEq^L) $\Longrightarrow$ (q-pN^L),
(q-2) (pC^L) + (q-pM^L) + (q-pAn^L) $\Longrightarrow$ (pEx^L),
*(q-3) (q-pN^L) + (pBB^L) + (pEq^L) $\Longrightarrow$ (p**^{L}),*
*(q-3') (q-pM^L) + (pBB^L) + (pEq^L) $\Longrightarrow$ (p**^{L}),*
(q-4) (q-pN^L) + (pB^L) + (pEq^L) $\Longrightarrow$ (p^L),*
(q-4') (q-pM^L) + (pB^L) + (pEq^L) $\Longrightarrow$ (p^L),*
(q-5) (pBB^L) + (pD^L) + (q-pN^L) + (pEq^L) $\Longrightarrow$ (pC^L),
(q-6) (pEx^L) + (q-pM^L) $\Longrightarrow$ (q-pI2^L),
(q-7) (q-pI2^L) + (pRe^L) $\Longrightarrow$ (q-pR2^L).

Proof. (q-1): If $1 \leq x \rightsquigarrow^L y$, i.e. $1 \to^L (x \rightsquigarrow^L y) = 1$, by (pEq^L), then $x \rightsquigarrow^L y = 1$, by (q-pM^L). If $1 \leq x \to^L y$, i.e. $1 \rightsquigarrow^L (x \to^L y) = 1$, by (pEq^L), then $x \to^L y{=}1$, by (q-pM^L). Thus, (q-pN^L) holds.

(q-2): By (pC^L),
$x \to^L (y \rightsquigarrow^L z) \leq y \rightsquigarrow^L (x \to^L z)$ and $y \rightsquigarrow^L (x \to^L z) \leq x \to^L (y \rightsquigarrow^L z)$.
Then, by (q-pAn^L), we obtain:
$1 \to^L [x \to^L (y \rightsquigarrow^L z)] = 1 \to^L [y \rightsquigarrow^L (x \to^L z)]$,
$1 \rightsquigarrow^L [x \to^L (y \rightsquigarrow^L z)] = 1 \rightsquigarrow^L [y \rightsquigarrow^L (x \to^L z)]$. Hence,

$x \to^L (y \rightsquigarrow^L z) \overset{(q-pM^L)}{=} 1 \rightsquigarrow^L [x \to^L (y \rightsquigarrow^L z)] = 1 \rightsquigarrow^L [y \rightsquigarrow^L (x \to^L z)]$
$\overset{(q-pM^L)}{=} 1 \rightsquigarrow^L (1 \to^L [y \rightsquigarrow^L (x \to^L z)]) \overset{(q-pM^L)}{=} 1 \to^L [y \rightsquigarrow^L (x \to^L z)] \overset{(q-pM^L)}{=}$
$y \rightsquigarrow^L (x \to^L z)$. Thus, (pEx^L) holds.

(q-3): Suppose $y \leq z$, i.e. $y \to^L z = 1$, by (pEq^L); then, by (pBB^L),
$1 \leq (z \to^L x) \rightsquigarrow^L (y \to^L x)$; hence, by (q-pN^L),
$(z \to^L x) \rightsquigarrow^L (y \to^L x) = 1$, i.e. $x \to^L x \leq y \to^L x$.
Suppose again $y \leq z$, i.e. $y \rightsquigarrow^L z = 1$; then, by (pBB^L),
$1 \leq (z \rightsquigarrow^L x) \to^L (y \rightsquigarrow^L x)$; hence, by (q-pN^L),
$(z \rightsquigarrow^L x) \to^L (y \rightsquigarrow^L x) = 1$, i.e. $z \rightsquigarrow^L x \leq y \rightsquigarrow^L x$. Thus, (p**^{L}) holds.

(q-3'): By (q-1), (q-pM^L) + (pEq^L) $\Longrightarrow$ (q-pN^L); then, apply above (q-3).

(q-4): Suppose $y \leq z$, i.e. $y \to^L z = 1$, by (pEq^L); then, by (pB^L),
$1 = y \to^L z \leq (x \to^L y) \to^L (x \to^L z)$; hence, by (q-pN^L),
$(x \to^L y) \to^L (x \to^L z) = 1$, i.e. $x \to^L y \leq x \to^L z$.
Suppose again $y \leq z$, i.e. $y \rightsquigarrow^L z = 1$, by (pEq^L); then, by (pB^L),
$1 = y \rightsquigarrow^L z \leq (x \rightsquigarrow^L y) \rightsquigarrow^L (x \rightsquigarrow^L z)$; hence, by (q-pN^L),
$(x \rightsquigarrow^L y) \rightsquigarrow^L (x \rightsquigarrow^L z) = 1$, i.e. $x \rightsquigarrow^L y \leq x \rightsquigarrow^L z$. Thus, (p*^{L}) holds.

(q-4'): By (q-1), (q-pM^L) + (pEq^L) $\Longrightarrow$ (q-pN^L); then, apply above (q-4).

(q-5): First, by (q-3), (pBB^L) + (q-pN^L) + (pEq^L) imply (p**^{L}).

Then, by (pBB^L) $(Y \to^L Z \leq (Z \to^L X) \rightsquigarrow^L (Y \to^L X))$,
for $X = u \rightsquigarrow^L x$, $Y = y$, $Z = z \rightsquigarrow^L x$, we obtain:
$y \to^L (z \rightsquigarrow^L x) \leq ((z \rightsquigarrow^L x) \to^L (u \rightsquigarrow^L x)) \rightsquigarrow^L (y \to^L (u \rightsquigarrow^L x))$.
Then, by (p**^{L}), we obtain:
$V_1 \overset{notation}{=} [((z \rightsquigarrow^L x) \to^L (u \rightsquigarrow^L x)) \rightsquigarrow^L (y \to^L (u \rightsquigarrow^L x))] \to^L$
$[(u \rightsquigarrow^L z) \rightsquigarrow^L (y \to^L (u \rightsquigarrow^L x))] \leq$
$(y \to^L (z \rightsquigarrow^L x)) \to^L [(u \rightsquigarrow^L z) \rightsquigarrow^L (y \to^L (u \rightsquigarrow^L x))] \overset{notation}{=} W_1$.
But, the left side $V_1 = 1$; indeed, by (pBB^L), we have:
$u \rightsquigarrow^L z \leq (z \rightsquigarrow^L x) \to^L (u \rightsquigarrow^L x)$;
then, by (p**^{L}), we obtain:
$[(z \rightsquigarrow^L x) \to^L (u \rightsquigarrow^L x)) \rightsquigarrow^L (y \to^L (u \rightsquigarrow^L x)) \leq (u \rightsquigarrow^L z) \rightsquigarrow^L (y \to^L (u \rightsquigarrow^L x))$,
i.e. $V_1 = 1$.
Then, by (q-pN^L), $W_1 = 1$, i.e.
$y \to^L (z \rightsquigarrow^L x) \leq (u \rightsquigarrow^L z) \rightsquigarrow^L (y \to^L (u \rightsquigarrow^L x))$,
which for $z = y \to^L x$ and $u = z$ gives:
$y \to^L ((y \to^L x) \rightsquigarrow^L x) \leq (z \rightsquigarrow^L (y \to^L x)) \rightsquigarrow^L (y \to^L (z \rightsquigarrow^L x))$;
but, by (pD^L), the left side $y \to^L ((y \to^L x) \rightsquigarrow^L x) = 1$; hence, by (q-pN^L),
$(z \rightsquigarrow^L (y \to^L x)) \rightsquigarrow^L (y \to^L (z \rightsquigarrow^L x)) = 1$,
i.e. $z \rightsquigarrow^L (y \to^L x) \leq y \to^L (z \rightsquigarrow^L x)$.

Similarly, by (pBB^L) again $(Y \rightsquigarrow^L Z \leq (Z \rightsquigarrow^L X) \to^L (Y \rightsquigarrow^L X))$,
for $X = u \to^L x$, $Y = y$, $Z = z \to^L x$, we obtain:
$y \rightsquigarrow^L (z \to^L x) \leq ((z \to^L x) \rightsquigarrow^L (u \to^L x)) \to^L (y \rightsquigarrow^L (u \to^L x))$.
Then, by (p**^{L}), we obtain:
$V_2 \overset{notation}{=} [((z \to^L x) \rightsquigarrow^L (u \to^L x)) \to^L (y \rightsquigarrow^L (u \to^L x))] \rightsquigarrow^L$
$[(u \to^L z) \to^L (y \rightsquigarrow^L (u \to^L x))] \leq$
$(y \rightsquigarrow^L (z \to^L x)) \rightsquigarrow^L [(u \to^L z) \to^L (y \rightsquigarrow^L (u \to^L x))] \overset{notation}{=} W_2$.
But, the left side $V_2 = 1$; indeed, by (pBB^L), we have:
$u \to^L z \leq (z \to^L x) \rightsquigarrow^L (u \to^L x)$;
then, by (p**^{L}), we obtain:
$[(z \to^L x) \rightsquigarrow^L (u \to^L x)] \to^L (y \rightsquigarrow^L (u \to^L x)) \leq (u \to^L z) \to^L (y \rightsquigarrow^L (u \to^L x))$,
i.e. $V_2 = 1$.
Then, by (q-pN^L), $W_2 = 1$, i.e.
$y \rightsquigarrow^L (z \to^L x) \leq (u \to^L z) \to^L (y \rightsquigarrow^L (u \to^L x))$,
which for $z = y \rightsquigarrow^L x$ and $u = z$ gives:
$y \rightsquigarrow^L ((y \rightsquigarrow^L x) \to^L x) \leq (z \to^L (y \rightsquigarrow^L x)) \to^L (y \rightsquigarrow^L (z \to^L x))$;
but, by (pD^L), the left side $y \rightsquigarrow^L ((y \rightsquigarrow^L x) \to^L x) = 1$; hence, by (q-pN^L),
$(z \to^L (y \rightsquigarrow^L x)) \to^L (y \rightsquigarrow^L (z \to^L x)) = 1$, i.e $z \to^L (y \rightsquigarrow^L x) \leq y \rightsquigarrow^L (z \to^L x)$;
hence, $y \to^L (z \rightsquigarrow^L x) \leq z \rightsquigarrow^L (y \to^L x)$. Thus, (pC^L) holds.

(q-6): $x \to^L y \overset{(q-pM^L)}{=} 1 \rightsquigarrow^L (x \to^L y) \overset{(pEx^L)}{=} x \to^L (1 \rightsquigarrow^L y)$ and
$x \rightsquigarrow^L y \overset{(q-pM^L)}{=} 1 \to^L (x \rightsquigarrow^L y) \overset{(pEx^L)}{=} x \rightsquigarrow^L (1 \to^L y)$. Thus, (q-pI2^L) holds.

(q-7): $x \to^L (1 \rightsquigarrow^L x) \overset{(q-pI2^L)}{=} x \to^L x \overset{(pRe^L)}{=} 1$ and
$x \rightsquigarrow^L (1 \to^L x) \overset{(q-pI2^L)}{=} x \rightsquigarrow^L x \overset{(pRe^L)}{=} 1$. Thus, (q-pR2^L) holds. □

- **The duality principle**

Let $\mathcal{A} = (A, \rightarrow^L, \rightsquigarrow^L, 1)$ be a quasi-ordered pseudo-algebra (or, equivalently, let $\mathcal{A} = (A, \leq, \rightarrow^L, \rightsquigarrow^L, 1)$ be a quasi-ordered pseudo-structure) and $\leq$ be the quasi-order of $\mathcal{A}$. The relation $\geq$, defined on A as follows: for every $x, y \in A$, $x \geq y \overset{def.}{\Longleftrightarrow} y \leq x$, is also a q-order, called the *inverse q-order relation* or the *dual q-order relation* of the q-order relation $\leq$.

The duality principle for quasi-ordered pseudo-algebras (pseudo-structures) is the following:
"every statement (definition of a notion, proposition, theorem, etc.) concerning the quasi-ordered pseudo-algebra $\mathcal{A} = (A, \rightarrow^L, \rightsquigarrow^L, 1)$ (quasi-ordered pseudo-structure $\mathcal{A} = (A, \leq, \rightarrow^L, \rightsquigarrow^L, 1)$) remains valid if we replace everywhere inside it the q-order relation $\leq$ with the inverse q-order relation, $\geq$".

The pseudo-algebra $(A, \rightarrow^L, \rightsquigarrow^L, 1)$ (the pseudo-structure $(A, \geq, \rightarrow^L, \rightsquigarrow^L, 1)$) obtained in this way is also a quasi-ordered pseudo-algebra (quasi-ordered pseudo-structure), called the *dual* of $\mathcal{A}$. The statement obtained in this way (definition of a notion, proposition, theorem, etc.) is the *dual* statement of the first statement (definition of a notion, proposition, theorem, etc.). We say also that the two quasi-ordered pseudo-algebras (pseudo-structures) or statements are *dual to each other* or simple *dual.*

- **Elements of the same height. The clouds**

Definition 5.1.26 Let $\mathcal{A} = (A, \rightarrow^L, \rightsquigarrow^L, 1)$ be a quasi-ordered pseudo-algebra (or, equivalently, let $\mathcal{A} = (A, \leq, \rightarrow^L, \rightsquigarrow^L, 1)$ be a quasi-ordered pseudo-structure).

We say that $a, b \in A$ have *the same height*, or are *parallel*, and we denote this by $a \parallel b$, if $a \rightarrow^L b = 1 = a \rightsquigarrow^L b$ and $b \rightarrow^L a = 1 = b \rightsquigarrow^L a$ (or, equivalently, $a \leq b$ and $b \leq a$).

Note that if (pM^L) holds, then $a \parallel b \Leftrightarrow a = b$, by (pAn^L).

Note also that if $a \parallel b$, then $1 \rightarrow^L a = 1 \rightarrow^L b$ and $1 \rightsquigarrow^L a = 1 \rightsquigarrow^L b$, by (q-pAn^L).

Corollary 5.1.27 *If (pEx^L) holds, then: for all $x, y \in A$,*

$$x \parallel y \Longleftrightarrow 1 \rightarrow^L x \overset{(IdR^L)}{(=} 1 \rightsquigarrow^L x) = 1 \rightarrow^L y \overset{(IdR^L)}{(=} 1 \rightsquigarrow^L y).$$

Proof. $\Longrightarrow$: By Proposition 5.1.19, $1 \rightarrow^L x = 1 \rightsquigarrow^L x$ and $1 \rightarrow^L y = 1 \rightsquigarrow^L y$.

$\Longleftarrow$: We have seen that (pEx^L) + (q-pM^L) $\Longrightarrow$ (q-pI2) and (q-pI2) + (pRe^L) $\Longrightarrow$ (q-pR2), hence (q-pI2) and (q-pR2) hold. Hence, if $1 \rightarrow^L x = 1 \rightarrow^L y$, then $x \rightsquigarrow^L y \overset{(q-pI2)}{=} x \rightsquigarrow^L (1 \rightarrow^L y) = x \rightsquigarrow^L (1 \rightarrow^L x) \overset{(q-pR2)}{=} 1$ and $y \rightsquigarrow^L x \overset{(q-pI2)}{=} y \rightsquigarrow^L (1 \rightarrow^L x) = y \rightsquigarrow^L (1 \rightarrow^L y) \overset{(q-pR2)}{=} 1$; thus, $x \leq y$ and $y \leq x$, i.e. $x \parallel y$. $\Box$

Proposition 5.1.28 *The relation $\parallel$ is an equivalence relation of $\mathcal{A}$.*

Proof. For all $x \in A$, $x \parallel x$ means $x \leq x$, which is true by (Re); thus, $\parallel$ is reflexive. For all $x, y \in A$, $x \parallel y$, i.e. $x \leq y$ and $y \leq x$, implies $y \leq x$ and $x \leq y$, i.e. $y \parallel x$; thus, $\parallel$ is symmetric. For all $x, y, z \in A$, $x \parallel y$ and $y \parallel z$, i.e. $x \leq y$, $y \leq x$ and $y \leq z$, $z \leq y$, imply $x \leq z$ and $z \leq x$, by (Tr), i.e. $x \parallel z$; thus, $\parallel$ is transitive. □

Proposition 5.1.29 *If properties (p^{*L}), (p^{**L}) also hold, then $\parallel$ is a congruence relation of $\mathcal{A}$.*

Proof. We must prove that $\parallel$ is compatible with $\to^L$ and $\rightsquigarrow^L$. Indeed, if $a \parallel b$ and $x \parallel y$, i.e. $a \leq b$, $b \leq a$ and $x \leq y$, $y \leq x$, then:
- $a \to^L x \overset{(p*^L)}{\leq} a \to^L y \overset{(p**^L)}{\leq} b \to^L y$ and $b \to^L y \overset{(p*^L)}{\leq} b \to^L x \overset{(p**^L)}{\leq} a \to^L x$; hence, $(a \to^L x) \parallel (b \to^L y)$;
- $a \rightsquigarrow^L x \overset{(p*^L)}{\leq} a \rightsquigarrow^L y \overset{(p**^L)}{\leq} b \rightsquigarrow^L y$ and $b \rightsquigarrow^L y \overset{(p*^L)}{\leq} b \rightsquigarrow^L x \overset{(p**^L)}{\leq} a \rightsquigarrow^L x$; hence, $(a \rightsquigarrow^L x) \parallel (b \rightsquigarrow^L y)$. □

Proposition 5.1.30 *If (p^{*L}), (p^{**L}) hold and if $b \parallel a$, then, in the tables of $\to^L$, $\rightsquigarrow^L$ we have:*
(i) the row of b coincides with the row of a;
(ii) the column of b coincides with the column of a.

Proof: Suppose $b \parallel a$, i.e. $b \leq a$ and $a \leq b$. Then,

(i): By (p^{**L}), we obtain: for each $z \in A$, $(a \to z \leq b \to z, b \to z \leq a \to z)$ and $(a \rightsquigarrow z \leq b \rightsquigarrow z, b \rightsquigarrow z \leq a \rightsquigarrow z)$. Then, by (q-pAn^L), $1 \rightsquigarrow (b \to z) = 1 \rightsquigarrow (a \to z)$ and $1 \to (b \rightsquigarrow z) = 1 \to (a \rightsquigarrow z)$, hence, by (q-pM), $b \to z = a \to z$ and $b \rightsquigarrow z = a \rightsquigarrow z$.

(ii): By (p^{*L}), we obtain: for each $z \in A$, $(z \to^L b \leq z \to^L a, z \to^L a \leq z \to^L b)$ and $(z \rightsquigarrow^L b \leq z \rightsquigarrow^L a, z \rightsquigarrow^L a \leq z \rightsquigarrow^L b)$. Then, by (q-pAn^L), $1 \rightsquigarrow^L (z \to^L b) = 1 \rightsquigarrow^L (z \to^L a)$ and $1 \to^L (z \rightsquigarrow^L b) = 1 \to^L (z \rightsquigarrow^L a)$, hence, by (q-pM^L), $z \to^L b = z \to^L a$ and $z \rightsquigarrow^L b = z \rightsquigarrow^L a$. □

For each $x \in A$, we denote its equivalence class by

$$| x | \overset{notation}{=} \{y \in A \mid y \parallel x\}$$

and we denote by $A/\parallel$ the quotient set (i.e. the set of all equivalence classes):

$$A/\parallel \overset{notation}{=} \{| x | \mid x \in A\}.$$

We shall call the equivalence classes determined by $\parallel$ *clouds* (as in [136]). Thus, we have, for each $x \in A$:

$$C(x) = | x | .$$

We have, more generally:

Lemma 5.1.31 *(See [101] for the commutative case)*
If property (pEx^L) holds, then every cloud in a quasi-ordered pseudo-algebra (structure) $\mathcal{A}$ contains exactly one regular element.

Proof. We have seen that (pEx^L) + (q-pM^L) $\Longrightarrow$ (q-pI2^L) and (q-pI2^L) + (pRe^L) $\Longrightarrow$ (q-pR2^L), hence (q-pI2^L) and (q-pR2^L) hold. Also, (IdR^L) holds.

Let C be a cloud, i.e. there exists $c \in A$ such that $C = C(c) = \mid c \mid$.

• If $c \in R(A)$, then $c = 1 \to^L c = 1 \rightsquigarrow^L c$. Let $a \in C$ such that $a \neq c$; then $a \parallel c$, i.e. $a \leq c$ and $c \leq a$, hence $1 \to^L a = 1 \to^L c$, by (q-pAn^L); we shall prove that $a \notin R(A)$. Indeed, if, by absurdum hypothesis, $a \in R(A)$, i.e. $a = 1 \to^L a = 1 \rightsquigarrow^L a$, then $a = 1 \to^L a = 1 \to^L c = c$, hence $a = c$: contradiction.

• If $c \notin R(A)$, we put $b_c = 1 \to^L c \overset{(IdR^L)}{=} 1 \rightsquigarrow^L c$. Then,

$$b_c = 1 \to^L c \overset{(q-pM^L)}{=} 1 \rightsquigarrow^L (1 \to^L c) = 1 \rightsquigarrow^L b_c \overset{(IdR^L)}{=} 1 \to^L b_c,$$

hence $b_c \in R(A)$.

We prove now that $b_c \in C$. Indeed,

$b_c \to^L c \overset{(q-pI2^L)}{=} b_c \to^L (1 \rightsquigarrow^L c) = b_c \to^L b_c \overset{(pRe^L)}{=} 1$ and

$c \to^L b_c \overset{(q-pI2^L)}{=} c \to^L (1 \rightsquigarrow^L b_c) = c \to^L (1 \rightsquigarrow^L c) \overset{(q-pR2)}{=} 1$,

hence $b_c \parallel c$, i.e. $b_c \in \mid c \mid= C$.

We prove now that any other element a of C (hence $a \parallel c$), s.t. $a \neq b_c$, is not a regular element. Indeed, if, by absurdum hypothesis, $a \in R(A)$, i.e. $a = 1 \to^L a = 1 \rightsquigarrow^L a$, then, since $a \parallel c$, it follows that:

$a \to^L b_c \overset{(q-pI2^L)}{=} a \to^L (1 \rightsquigarrow^L b_c) = a \to^L (1 \rightsquigarrow^L c) = a \to^L (1 \rightsquigarrow^L a) \overset{(q-pR2^L)}{=} 1$

and

$b_c \to^L a \overset{(q-pI2^L)}{=} b_c \to^L (1 \rightsquigarrow^L a) = b_c \to^L (1 \rightsquigarrow^L c) = b_c \to^L b_c \overset{(pRe^L)}{=} 1$;

hence, by (q-pAn^L), $1 \to^L a = 1 \to^L b_c$, i.e. $a = b_c$: contradiction. □

We define implications of clouds as follows: for all $x, y \in A$,

$$\mid x \mid \to^L \mid y \mid \overset{def.}{=} \mid x \to^L y \mid, \quad \mid x \mid \rightsquigarrow^L \mid y \mid \overset{def.}{=} \mid x \rightsquigarrow^L y \mid .$$

Then, we have the following expected result:

Proposition 5.1.32 *If (pEx^L) also hold, then the quotient pseudo-algebra $(A/\parallel, \to^L, \rightsquigarrow^L, \mid 1 \mid)$ is a regular ordered pseudo-algebra, isomorphic to $(R(A), \to^L, \rightsquigarrow^L, 1)$.*

Proof: Routine, by Lemma 5.1.31. □

Remark 5.1.33 Given a finite regular ordered pseudo-algebra $(X, \to^L, \rightsquigarrow^L, 1)$, we can obtain, **in general**, an infinity of finite quasi-ordered pseudo-algebras: $(A_1, \to^L, \rightsquigarrow^L, 1)$, $(A_2, \to^L, \rightsquigarrow^L, 1)$, ... such that $R(A_1) = R(A_2) = ... = X$, by adding one or more elements parallel with some (all) elements of X.

We can quickly draw the tables of $\to^L$, $\rightsquigarrow^L$ for such a finite quasi-ordered pseudo-algebra $(A, \to^L, \rightsquigarrow^L, 1)$ with $R(A) = X$ by using Proposition 5.1.30, if (p*^{L}), (p**^{L}) hold.

• **Quasi-Hasse diagrams and quasi-Hasse-type diagrams**

A quasi-order relation $\leq$ on A will be represented graphically by a *quasi-Hasse diagram*, i.e.:
- a regular element is represented by a bullet $\bullet$,
- an element parallel with a regular element is represented by a bigcirc $\bigcirc$,
- the fact that $x < y$ (i.e. $x \leq y$ and $x \neq y$) and there is no z with $x < z < y$ is represented by:
· a line connecting the two points, y being higher than x, if the elements x, y are regular,
· a horizontal line connecting the two points, if the elements x, y have the same height (are parallel).

Consequently, the regular order relation $\leq$ on $R(A)$ will be represented graphically by a Hasse diagram.

The quasi-Hasse diagram is useful for recognizing the properties of a quasi-order relation - just as the Hasse diagram is useful for recognizing the properties of a (regular) order relation.

If the binary quasi-relation $\leq$ is not a quasi-order relation, then the quasi-pseudo-algebra will be represented graphically by a *quasi-Hasse-type diagram*, i.e.:
- a regular element is represented by a small circle $\circ$,
- an element parallel with a regular element is represented by a bigcirc $\bigcirc$,
- the fact that $x < y$ (i.e. $x \leq y$ and $x \neq y$) and there is no z with $x < z < y$ is represented by:
· a line connecting the two points, y being higher than x, if the elements x, y are regular,
· a horizontal line connecting the two points, if the elements x, y have the same height (are parallel).

Consequently, the binary regular relation $\leq$ on $R(A)$ will be represented graphically by a Hasse-type diagram.

• Bounded quasi-algebras (see [103] for the commutative case)

Let $\mathcal{A} = (A, \rightarrow^L, 1)$ be a left-quasi-ordered algebra (or, equivalently, let $\mathcal{A} = (A, \leq, \rightarrow^L, 1)$ be a left-quasi-ordered structure) and $\leq$ be the quasi-order of $\mathcal{A}$ (i.e. (q-M^L), (Re) (hence (M1^L)), (q-An), (Tr) hold).

Then, by Theorem 5.1.13, $\mathcal{R}(\mathcal{A}) = (R(A), \rightarrow^L, 1)$ is a regular ordered left-algebra (or, equivalently, $\mathcal{R}(\mathcal{A}) = (R(A), \leq, \rightarrow^L, 1)$ is a regular ordered left-structure) and $\leq$ is a regular order of $R(A)$ (i.e. (M^L), (Re), (An), (Tr) hold).

Definitions 5.1.34

i. An element $l \in A$ is called *last quasi-element* or *the greatest quasi-element* of $\mathcal{A}$, if $(1 \rightarrow^L x) \rightarrow^L (1 \rightarrow^L l) = 1$ (or $1 \rightarrow^L x \leq 1 \rightarrow^L l$), for every $x \in A$.

ii. An element $f \in A$ is called *first quasi-element* or *the smallest quasi-element)* of $\mathcal{A}$, if $(1 \rightarrow^L f) \rightarrow^L (1 \rightarrow^L x) = 1$ (or $1 \rightarrow^L f \leq 1 \rightarrow^L x$), for every $x \in A$.

Hence, *the notions of first quasi-element and last quasi-element are dual to each other.*

Note that in a quasi-ordered left-algebra (left-structure) there can be more first quasi-elements and more last quasi-elements, i.e. the first quasi-element and the last quasi-element can be not unique (when they exist) - see next Examples 5.1.37. Anyway, they form clouds: the cloud of first elements and the cloud of last elements.

By Lemma 5.1.31, if the propery (Ex^L) holds, then every cloud in a quasi-ordered pseudo-algebra (pseudo-structure) $\mathcal{A}$ contains exactly one regular element.

The special element $1 \in A$ is a regular element (i.e. $1 \in R(A)$) and it can be the last quasi-element; in this case, we have the property: for every $x \in A$,
(L') (Last) $x \to^L 1 = 1$ or, equivalently, (L) (Last) $x \leq 1$.

Note that there can be other last quasi-elements too, but not as regular elements, by Lemma 5.1.31, if the property (Ex^L) holds; in this case, all the last quasi-elements of $\mathcal{A}$ form the cloud of 1, denoted by $C(1) =\mid 1 \mid$.

In the cloud of the first quasi-elements, only one is a regular element, by Lemma 5.1.31, if the property (Ex^L) holds; the first quasi-element which is a regular element also will be denoted by 0. Consequently,
(1) $0 \in R(A)$ and, consequently, $1 \to^L 0 = 0$, by (M^L);
(2) we have the property: for every $x \in A$,
(F') (First) $0 \to^L x = 1$ or, equivalently, (F) (First) $0 \leq x$;
(3) all the first quasi-elements of $\mathcal{A}$ form the cloud of 0, denoted by $C(0) =\mid 0 \mid$.

Definition 5.1.35 A quasi-ordered left-algebra (quasi-ordered left-structure) with first quasi-element 0 and last quasi-element 1 is called *bounded* and is denoted by $\mathcal{A} = (A, \to^L, 0, 1)$ ($\mathcal{A} = (A, \leq, \to^L, 0, 1)$, respectively).

Let us introduce the new property (10-0):
(10-0) $1 \to^L 0 = 0$.

Remarks 5.1.36 Let $\mathcal{A} = (A, \to^L, 0, 1)$ be a bounded left-quasi-algebra. Then:
(1) $\mathcal{R}(\mathcal{A}) = (R(A), \to^L, 0, 1)$ is a bounded regular left-algebra;
(2) the property (10-0) holds.

Examples 5.1.37 Let us consider the following two left-quasi-algebras $\mathcal{A}_1 = (A_1 = \{0, x, 1\}, \to^L, 1)$ and $\mathcal{A}_2 = (A_2 = \{0, x, y, 1\}, \to^L, 1)$ given by the following tables of $\to^L$:

$\mathcal{A}_1$

$\to^L$	**0**	x	**1**
0	**1**	1	**1**
x	1	1	1
1	**0**	0	**1**

$\mathcal{A}_2$

$\to^L$	**0**	x	y	**1**
0	**1**	1	1	**1**
x	1	1	1	1 .
y	0	0	1	1
1	**0**	0	1	**1**

Then, the quasi-Hasse diagrams from Figure 5.1 present for each of the left-quasi-algebras $\mathcal{A}_1$ and $\mathcal{A}_2$ two ways of drawing the diagrams.

In Figure 5.1, both the left-quasi-structures $\mathcal{A}_1$ and $\mathcal{A}_2$ are bounded: $\mathcal{A}_1$ has 1 as quasi-last element and 0 and x as quasi-first elements, i.e. $C(0) =\mid 0 \mid 0 = \{0, x\}$ and $C(1) =\mid 1 \mid= \{1\}$; $\mathcal{A}_2$ has 1 and y as quasi-last elements and 0 and x as quasi-first elements, i.e. $C(0) =\mid 0 \mid= \{0, x\}$ and $C(1) =\mid 1 \mid= \{1, y\}$.

Note that $R(A_1) = R(A_2) = \{0, 1\}$.

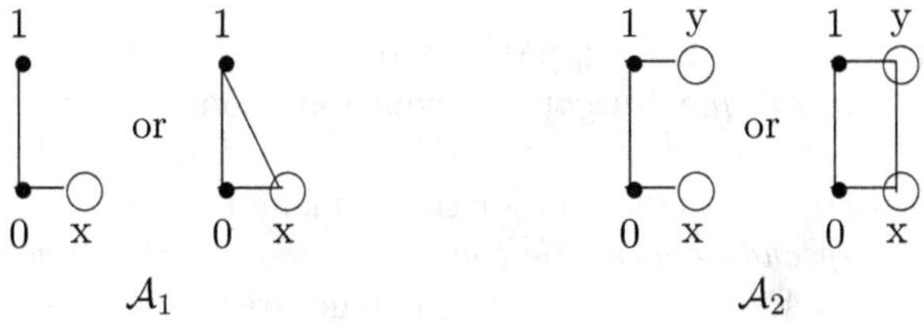

Figure 5.1: The quasi-Hasse diagrams of quasi-algebras $\mathcal{A}_1$ and $\mathcal{A}_2$

• Quasi-pseudo-structures with (involutive) quasi-negations (see [103] for the commutative case)

Let $\mathcal{A} = (A, \leq, \rightarrow^L, \rightsquigarrow^L, 0, 1)$ be a bounded left-quasi-pseudo-structure such that the associated bounded regular left-pseudo-structure $\mathcal{R}(\mathcal{A})$ contains the *involutive* regular negations $^{-'}$ (where $x^{-'} = x \rightarrow^L 0$) and $^{\sim'}$ (where $x^{\sim'} = x \rightsquigarrow^L 0$), i.e. (pDN^L) holds.

We extend the involutive regular negations $^{-'} : R(A) \longrightarrow R(A)$ and $^{\sim'} : R(A) \longrightarrow R(A)$ to the so called *quasi-negations* $^- : A \longrightarrow A$ and $^\sim : A \longrightarrow A$, respectively, such that the following property be verified: for all $x \in A$,

(q-pNeg^L) $x \rightarrow^L 0 = 1 \rightarrow^L x^- = (1 \rightsquigarrow^L x)^-$, $x \rightsquigarrow^L 0 = 1 \rightsquigarrow^L x^\sim = (1 \rightarrow^L x)^\sim$, where $0 = 1^- = 1^\sim$.

Proposition 5.1.38 *The quasi-negations $^-$ and $^\sim$ on A are indeed extensions of the regular negations $^{-'}$ and $^{\sim'}$ on $R(A)$, i.e.*

$$^- |_{R(A)} = {}^{-'}, \quad ^\sim |_{R(A)} = {}^{\sim'}.$$

Proof. For all $r \in R(A)$, by (q-pNeg^L),
$r \rightarrow^L 0 = 1 \rightarrow^L r^- = (1 \rightsquigarrow^L r)^-$, hence $r \rightarrow^L 0 = r^-$, by (pM^L) and
$r \rightsquigarrow^L 0 = 1 \rightsquigarrow^L r^\sim = (1 \rightarrow^L r)^\sim$, hence $r \rightsquigarrow^L 0 = r^\sim$, by (pM^L);
but, $r \rightarrow^L 0 = r^{-'}$ and $r \rightsquigarrow^L 0 = r^{\sim'}$. Hence, $r^- = r^{-'}$ and $r^\sim = r^{\sim'}$. □

Remark 5.1.39 For any $\mathcal{A}$, there can be more quasi-negations verifying (q-pNeg^L), i.e. the quasi-negations are not necessarily unique, when they exists.

If there exists quasi-negations, we then look among them for the quasi-negations $^-$, $^\sim$ that verify (pDN^L) ($(x^-)^\sim = x = (x^\sim)^-$, for all $x \in A$), which are called *involutive*. If such a pair of involutive quasi-negations exists, we then denote $\mathcal{A}$ by $(A, \leq, \rightarrow^L, \rightsquigarrow^L, ^-, ^\sim, 0, 1)$ and we say that it is a bounded left-quasi-pseudo-structure with *involutive* quasi-negations.

We then have the following alternative definitions.

Definition 5.1.40 (See Definition 5.1.7 in the regular case) (The dual case is omitted)

i. A *left-quasi-pseudo-algebra (left-quasi-pseudo-structure) with quasi-negations* is a left-quasi-pseudo-algebra $\mathcal{A} = (A, \rightarrow^L, \rightsquigarrow^L, 1)$ (a left-quasi-pseudo-structure

$\mathcal{A} = (A, \leq, \rightarrow^L, \rightsquigarrow^L, 1))$ (i.e. (q-pM^L) and (pM1^L) hold) with $^-$, $^\sim$ two unary operations on A, called *quasi-negations*, which are connected with $\rightarrow^L, \rightsquigarrow^L$, 1 by:
(q-pNeg^L) $x \rightarrow^L 0 = 1 \rightarrow^L x^- = (1 \rightsquigarrow^L x)^-$, $x \rightsquigarrow^L 0 = 1 \rightsquigarrow^L x^\sim = (1 \rightarrow^L x)^\sim$,
where 0 is the notation for the element $1^- = 1^\sim$.

i'. A *left-quasi-pseudo-algebra (left-quasi-pseudo-structure) with quasi-negations* is an algebra $\mathcal{A} = (A, \rightarrow^L, \rightsquigarrow^L, ^-, ^\sim, 1)$ (a structure $\mathcal{A} = (A, \leq, \rightarrow^L, \rightsquigarrow^L, ^-, ^\sim, 1))$ such that $(A, \rightarrow^L, \rightsquigarrow^L, 1)$ ($(A, \leq, \rightarrow^L, \rightsquigarrow^L, 1)$ respectively) is a left-quasi-pseudo-algebra (left-quasi-pseudo-structure) (i.e. (q-pM^L) and (pM1^L) hold) and $^-$, $^\sim$ are unary operations on A, called *quasi-negations*, which are connected with $\rightarrow^L, \rightsquigarrow^L$, 1 by:
(q-pNeg^L) $x \rightarrow^L 0 = 1 \rightarrow^L x^- = (1 \rightsquigarrow^L x)^-$, $x \rightsquigarrow^L 0 = 1 \rightsquigarrow^L x^\sim = (1 \rightarrow^L x)^\sim$,
where 0 is the notation for the element $1^- = 1^\sim$.

ii. If the quasi-negations verify (pDN^L), we say that $\mathcal{A}$ is a left-quasi-pseudo-algebra (left-quasi-pseudo-structure) with *involutive* quasi-negations.

Note that the quasi-Wajsberg algebra [21] is an example of quasi-algebra with involutive quasi-negation (see [103]) (in the "world" of quasi-M algebras with additional operations). Other examples are presented in Chapter 11.

Remark 5.1.41 For the same left-quasi-pseudo-algebra $\mathcal{A} = (A, \rightarrow^L, \rightsquigarrow^L, 1)$ (left-quasi-pseudo-structure $\mathcal{A} = (A, \leq, \rightarrow^L, \rightsquigarrow^L, 1)$ respectively), there can be more quasi-negations ($^-$, $^\sim$) verifying (q-pNeg^L), i.e. there can be more left-quasi-pseudo-algebras (left-quasi-pseudo-structures) with quasi-negations associated to $\mathcal{A}$.

Note that we have:
$0 \overset{notation}{=} 1^- = 1^\sim$ and $0^- = 0 \rightarrow^L 0 = 1$, $0^\sim = 0 \rightsquigarrow^L 0 = 1$, if (pRe^L) holds.

Remarks 5.1.42 (See the analogous Remark 5.2.33)

(1) If the two quasi-negations $^-$ and $^\sim$ exist and coincide, and if we denote the unique quasi-negation by $^-$, then it verifies the property (i.e. (q-pNeg^L) becomes):
(q-pNeg') $x \rightarrow 1 = 1 \rightarrow x^- = (1 \rightsquigarrow x)^-$, $x \rightsquigarrow 1 = 1 \rightsquigarrow x^- = (1 \rightarrow x)^-$,
with $1 = 1^-$.

If such an involutive quasi-negation exists, then we shall say that $\mathcal{A}$ is a *left-quasi-pseudo-algebra (left-pseudo-structure) with an involutive quasi-negation.*

(2) Moreover, if (Id) holds ($x \rightarrow 1 = x \rightsquigarrow 1$), then (q-pNeg') becomes:
(q-pNeg) $x \rightarrow 1 = 1 \rightarrow x^- = (1 \rightsquigarrow x)^- = x \rightsquigarrow 1 = 1 \rightsquigarrow x^- = (1 \rightarrow x)^-$,
with $1 = 1^-$.

We have the following remark.

Remark 5.1.43 (See Remarks 2.1.2, 5.1.6)

Just as the *equivalence:* $x \leq y \iff x \rightarrow^L y = 1$ ($\Leftrightarrow x \rightsquigarrow^L y = 1$) can be used either
- as the *definition* of the binary quasi-relation $\leq$ in the quasi-pseudo-algebra $(A, \rightarrow^L, \rightsquigarrow^L, 1)$ or
- as the *connection* between the binary quasi-relation $\leq$ and $\rightarrow^L, \rightsquigarrow^L$, 1 in the quasi-pseudo-structure $(A, \leq, \rightarrow^L, \rightsquigarrow^L, 1)$,

the same, the *equalities*:
(q-pNeg^L) $x \to^L 0 = 1 \to^L x^- = (1 \rightsquigarrow^L x)^-$, $x \rightsquigarrow^L 0 = 1 \rightsquigarrow^L x^\sim = (1 \to^L x)^\sim$
will be used either
- in the *definition* of the quasi-negations $^-$, $^\sim$, in the *bounded left-quasi-pseudo-algebra* $(A, \to^L, \rightsquigarrow^L, 1)$ *(left-quasi-pseudo-structure* $(A, \leq, \to^L, \rightsquigarrow^L, 1)$*)*, and in this case:
$1^- \overset{def.}{=} 1 \to^L 0 = 0$ and $1^\sim \overset{def.}{=} 1 \rightsquigarrow^L 0 = 0$ by (pM^L), and
$0^- \overset{def.}{=} 0 \to^L 0 = 1$ and $0^\sim \overset{def.}{=} 0 \rightsquigarrow^L 0 = 1$, by (pRe^L) or
- as the *connection* between the quasi-negations $^-$, $^\sim$ and $\to^L, \rightsquigarrow^L$, 1 in the *left-quasi-pseudo-algebra* $(A, \to^L, \rightsquigarrow^L, ^-, ^\sim, 1)$ *(left-quasi-pseudo-structure* $(A, \leq, \to^L, \rightsquigarrow^L, ^-, ^\sim, 1)$*) with quasi-negations*, and in this case:
$0 \overset{notation}{=} 1^- = 1^\sim$ and 0^- must equal $0 \to^L 0 = 1$ and $0^\sim$ must equal $0 \rightsquigarrow^L 0 = 1$, by (pRe^L).

Consider now the following properties (the dual case is omitted):
$(\overline{pM^L})$ $1 \to^L x^- = x^-$, $1 \to^L x^\sim = x^\sim$, $1 \rightsquigarrow^L x^- = x^-$, $1 \rightsquigarrow^L x^\sim = x^\sim$,
$(\text{q-}\overline{pM^L})$ $1 \to^L (x \to^L y)^- = (x \to^L y)^-$, $1 \to^L (x \rightsquigarrow^L y)^- = (x \rightsquigarrow^L y)^-$,
$1 \rightsquigarrow^L (x \rightsquigarrow^L y)^\sim = (x \rightsquigarrow^L y)^\sim$, $1 \rightsquigarrow^L (x \to^L y)^\sim = (x \to^L y)^\sim$,
$(\text{q-}\overline{pM^L_d})$ $1 \to^L (x \to^L y)^\sim = (x \to^L y)^\sim$, $1 \to^L (x \rightsquigarrow^L y)^\sim = (x \rightsquigarrow^L y)^\sim$,
$1 \rightsquigarrow^L (x \rightsquigarrow^L y)^- = (x \rightsquigarrow^L y)^-$, $1 \rightsquigarrow^L (x \to^L y)^- = (x \to^L y)^-$,

$(\text{q-}\overline{pM(1 \to^L x, 1 \rightsquigarrow^L x)})$ $1 \to^L (1 \to^L x)^- = (1 \to^L x)^-$,
$1 \to^L (1 \rightsquigarrow^L x)^- = (1 \rightsquigarrow^L x)^-$,
$1 \rightsquigarrow^L (1 \rightsquigarrow^L x)^\sim = (1 \rightsquigarrow^L x)^\sim$,
$1 \rightsquigarrow^L (1 \to^L x)^\sim = (1 \to^L x)^\sim$.

Proposition 5.1.44 *We have:*
(i) $(\text{q-}\overline{pM^L}) \Longrightarrow (\text{q-}\overline{pM^L_d})$.
(ii) $(\overline{pM^L}) \Longrightarrow (\text{q-}\overline{pM^L}) \Longrightarrow (\text{q-}\overline{pM(1 \to^L x, 1 \rightsquigarrow^L x)})$.

(iii) If (q-pNeg^L) *holds, then* $(\text{q-}\overline{pM(1 \to^L x, 1 \rightsquigarrow^L x)})$ *holds too.*

Proof. (i): $1 \to^L (x \to^L y)^\sim = 1 \to^L [1 \rightsquigarrow^L (x \to^L y)^\sim] \overset{(q-\overline{pM^L})}{=} 1 \rightsquigarrow^L (x \to^L y)^\sim = (x \to^L y)^\sim$. The other properties have similar proofs.
(ii): Obviously.
(iii): $1 \to^L (1 \to^L x)^- = (1 \rightsquigarrow^L (1 \to^L x))^- \overset{(q-\overline{pM^L})}{=} (1 \to^L x)^-$.
$1 \to^L (1 \rightsquigarrow^L x)^- = (1 \rightsquigarrow^L (1 \rightsquigarrow^L x))^- \overset{(q-\overline{pM^L_d})}{=} (1 \rightsquigarrow^L x)^-$. Similarly for the rest.
□

Remark 5.1.45 If $^- = ^\sim$, then the above properties become:
$(\overline{pM})$ $1 \to^L x^- = x^-$, $1 \rightsquigarrow^L x^- = x^-$,
$(\text{q-}\overline{pM})$ $1 \to (x \rightsquigarrow y)^- = (x \rightsquigarrow y)^-$, $1 \rightsquigarrow (x \to y)^- = (x \to y)^-$,
$(\text{q-}\overline{pM_d})$ $1 \to (x \to y)^- = (x \to y)^-$, $1 \rightsquigarrow (x \rightsquigarrow y)^- = (x \rightsquigarrow y)^-$,

$(\text{q-}\overline{pM(1 \to x, 1 \rightsquigarrow x)})$ $1 \to (1 \rightsquigarrow x)^- = (1 \rightsquigarrow x)^-$, $1 \rightsquigarrow (1 \to x)^- = (1 \to x)^-$.

Consequently, we obtain by above Proposition 5.1.44:

Proposition 5.1.46 *We have:*
(i) $(q\text{-}\overline{pM}) \Longrightarrow (q\text{-}\overline{pM_d})$.
(ii) $(\overline{pM}) \Longrightarrow (q\text{-}\overline{pM}) \Longrightarrow (q\text{-}\overline{pM(1 \to x, 1 \rightsquigarrow x)})$.

(iii) If (q-pNeg) holds, then $(q\text{-}\overline{pM(1 \to x, 1 \rightsquigarrow x)})$ *holds too.*

5.1.3 Quasi-pBCI algebras vs. pBCI algebras

We introduce here (see [101] for the commutative case) the following definition of quasi-pBCI algebra as a structure, following the Definition 1 of pBCI algebras (the dual case is omitted):

Definition 5.1.47 (Definition 1 of q-pBCI algebras)
A *left-quasi-pseudo-BCI algebra*, or a *left-q-pBCI algebra* for short, is a structure $\mathcal{A}^L = (A^L, \leq, \to^L, \rightsquigarrow^L, 1)$ where $\leq$ is a binary relation, $\to^L$ and $\rightsquigarrow^L$ are binary operations on A^L and $1 \in A^L$ verifying: for all $x, y, z \in A^L$,
(pBB^L) $y \to^L z \leq (z \to^L x) \rightsquigarrow^L (y \to^L x)$, $y \rightsquigarrow^L z \leq (z \rightsquigarrow^L x) \to^L (y \rightsquigarrow^L x)$,
(pD^L) $y \leq (y \to^L x) \rightsquigarrow^L x$, $y \leq (y \rightsquigarrow^L x) \to^L x$,
(Re) $x \leq x$ (Reflexivity),
(q-pM^L) $1 \to^L (x \rightsquigarrow^L y) = x \rightsquigarrow^L y$, $1 \rightsquigarrow^L (x \to^L y) = x \to^L y$,
(q-pAn^L) $(x \leq y$ and $y \leq x) \Longrightarrow (1 \to^L x = 1 \to^L y$ and $1 \rightsquigarrow^L x = 1 \rightsquigarrow^L y)$ (quasi-Antisymmetry),
(IdEq^L) $x \to^L y = 1 \Longleftrightarrow x \rightsquigarrow^L y = 1$,
(pEq^L) $x \leq y \Longleftrightarrow x \to^L y = 1 \overset{(IdEq^L)}{\Longleftrightarrow} x \rightsquigarrow^L y = 1$.

We can define equivalently a quasi-pBCI algebra as an algebra, following the Definition 1' of pBCI algebras:

Definition 5.1.48 (Definition 1' of q-pBCI algebras)
A *left-quasi-pseudo-BCI algebra*, or a *left-q-pBCI algebra* for short, is an algebra $\mathcal{A}^L = (A^L, \to^L, \rightsquigarrow^L, 1)$ of type $(2, 2, 0)$ verifying: for all $x, y, z \in A^L$,
(pBB^L') $(y \to^L z) \rightsquigarrow^L ((z \to^L x) \rightsquigarrow^L (y \to^L x)) = 1$,
$(y \rightsquigarrow^L z) \to^L ((z \rightsquigarrow^L x) \to^L (y \rightsquigarrow^L x)) = 1$,
(pD^L') $y \rightsquigarrow^L ((y \to^L x) \rightsquigarrow^L x) = 1$, $y \to^L ((y \rightsquigarrow^L x) \to^L x) = 1$,
(pRe^L') $x \to^L x = 1 = x \to^L x$ (Reflexivity),
(q-pM^L) $1 \to^L (x \rightsquigarrow^L y) = x \rightsquigarrow^L y)$, $1 \rightsquigarrow^L (x \to^L y) = x \to^L y$,
(q-pAn^L') $x \to^L y = 1$ $(\Leftrightarrow x \rightsquigarrow^L y = 1)$ and $y \to^L x = 1$ $(\Leftrightarrow y \rightsquigarrow^L x = 1) \Longrightarrow$ $(1 \to^L x = 1 \to^L y$ and $1 \rightsquigarrow^L x = 1 \rightsquigarrow^L y)$ (quasi-Antisymmetry)
(IdEq^L) $x \to^L y = 1 \Longleftrightarrow x \rightsquigarrow^L y = 1$,
where a binary relation $\leq$ is then defined by (IdEq^L): for all $x, y \in A^L$,

$$x \leq y \overset{def.}{\Longleftrightarrow} x \to^L y = 1 \overset{(IdEq^L)}{\Longleftrightarrow} x \rightsquigarrow^L y = 1.$$

A left-q-pBCI algebra is *commutative*, if $\to^L = \rightsquigarrow^L$ (dually, if $\to^R = \rightsquigarrow^R$). A commutative left-q-pBCI algebra is a left-q-BCI algebra (defined and studied in [101]).

Denote by **q-pBCI**^{L} the class of all left-q-pBCI algebras. Dually, denote by **q-pBCI**^{R} the class of all right-q-pBCI algebras.

Note that a left-q-pBCI algebra $\mathcal{A}^L$ is a left-quasi-pseudo-structure, by above Definition 1, or a left-quasi-pseudo-algebra, by above Definition 1', and $\mathcal{R}(\mathcal{A}^L)$ is a left-pBCI algebra (regular left-pseudo-structure or regular left-pseudo-algebra), i.e. we can write, by (Q), the special equivalences:

$$\mathbf{q\text{-}pBCI}^L + (\mathrm{pM}^L) \Longleftrightarrow \mathbf{pBCI}^L \text{ and, dually, } \mathbf{q\text{-}pBCI}^R + (\mathrm{pM}^R) \Longleftrightarrow \mathbf{pBCI}^R.$$

Proposition 5.1.49 *Let $\mathcal{A}^L = (A^L, \to^L, \leadsto^L, 1)$ be a left-q-pBCI algebra (Definition 1). Then, the properties (pEx^L), (Id^L), (p^{*L}), (p^{**L}) hold.*

Proof. (pEx^L): By Proposition 5.1.25 (q-1), (q-pM^L) + (pEq^L) $\Longrightarrow$ (q-pN^L), by Proposition 5.1.25 (q-5), (pBB^L) + (pD^L) + (q-pN^L) + (pEq^L) $\Longrightarrow$ (pC^L), by Proposition 5.1.25 (q-2), (pC^L) + (q-pM^L) + (q-pAn^L) $\Longrightarrow$ (pEx^L); thus, (pEx^L) holds.

(Id^L): $x \to^L 1 \overset{(Re)+(pEq^L)}{=} x \to^L (x \leadsto^L x) \overset{(pEx^L)}{=} x \leadsto^L (x \to^L x) = x \leadsto^L 1.$

(p^{*L}): By Proposition 2.1.8 (6'), (pEx^L) + (pBB^L) + (pEq^L) $\Longrightarrow$ (pB^L), by Proposition 5.1.25 (q-4'), (q-pM^L) + (pB^L) + (pEq^L) $\Longrightarrow$ (p^{*L}); thus, (p^{*L}) holds.

(p^{**L}): By Proposition 5.1.25 (q-3'), (q-pM^L) + (pBB^L) + (pEq^L) $\Longrightarrow$ (p^{**L}); thus, (p^{**L}) holds. □

Note that, since the properties (pEx^L), (p^{*L}), (p^{**L}) hold, by above Proposition 5.1.49, then all the results from Proposition 5.1.19 and Corollary 5.1.27 to Proposition 5.1.32 hold.

This section will be connected further to implicative-groups and their generalizations. Examples of these algebras will be given in Chapter 6.

5.1.4 Further research

Further research in the commutative case, in theory of quasi-algebras concerning quasi-BCK algebras, quasi-RM algebras is done in [101]. Further research on quasi-BCI(P) (quasi-BCI(PR)) algebras, quasi-BCK(P) (quasi-BCK(PR)) algebras, etc. is done in [102]. Further research on quasi-algebras with quasi-negation is done in [103]. We have introduced in [103] the notion of quasi-implicative-Boole algebra.

This section will be connected further to implicative-groups and their generalizations.

5.2 Quasi-m unital magmas vs. regular-m unital magmas

By analogy with a theory of quasi-pM algebras vs. a theory of regular pM algebras started here (see [101] for the commutative case), we introduce here a new theory, the theory of quasi-m unital magmas, generalizations of unital magmas, vs. a theory of regular-m unital magmas.

Let $\mathcal{A} = (A, \odot, 1)$ be an algebra of type $(2, 0)$. Let us introduce the following properties: for all $x, y \in A$,
(pU) $1 \odot x = x = x \odot 1$, i.e. 1 is the unit element of $\odot$,
(qm-pU) $1 \odot (x \odot y) = x \odot y = (x \odot y) \odot 1$,
(U1) $1 \odot 1 = 1$.

Then we have, obviously:

Proposition 5.2.1
(i) (pU) $\Longrightarrow$ (qm-pU),
(ii) (pU) $\Longrightarrow$ (U1).

Besides the set A, let us define the following subset of A:

$$V_{pU} \stackrel{df.}{=} \{x \in A \mid x \stackrel{(pU)}{=} 1 \odot x = x \odot 1\} \subseteq A.$$

Note that, if property (pU) holds, then $A \subseteq V_{pU}$, hence $V_{pU} = A$.

5.2.1 Regular-m algebras (structures)

Definitions 5.2.2
(1m) The algebra $(A, \odot, 1)$ of type $(2, 0)$ (the structure $(A, \leq, \odot, 1)$) is called *regular-m*, if it satisfies the property (pU).

(1'm) Any algebra (structure) $\mathcal{A}' = (A, \sigma)$ whose signature σ contains $\odot$, 1 is also called *regular-m*, if it satisfies the property (pU).

(1"m) Any algebra (structure) $\mathcal{A}'' = (A, \tau)$ which is term equivalent to a regular-m algebra (structure) $\mathcal{A}' = (A, \sigma)$ is also called *regular-m.*

(2m) The operation $\odot$ from a regular-m algebra (structure) is called *regular-m operation.*

(3m) The binary relation $\leq$ of a regular-m structure is called *binary regular-m relation.*

Note that, by above Definitions 5.2.2 (1m), the most general regular-m algebra is the unital magma, recalled in Chapter 3. Other regular-m algebras (structures) were recalled also in Chapter 3.

Roughly speaking, the *regular-m* algebras are those classical algebras verifying the property (pU). All the properties verified by a regular-m algebra, including the property (pU), are called *regular-m properties.* A *theory of regular-m algebras* establishes connections existing between the property (pU) and the other regular-m properties (if a regular-m property is proved by (pU) or not), as for example the Propositions 10.1.26, 10.1.33. The regular-m properties which depend on (pU) can be modified when we generalize (pU) to (qm-pU), i.e. when we develop a theory of quasi-m algebras - see further in this section.

The ordered regular-m algebras are represented by Hasse diagrams, while all the others are represented by Hasse-type diagrams.

Remark 5.2.3 In any regular-m algebra $\mathcal{A}$ we have that: $V_{pU} = A$ and this is the basic, definable property of regular-m algebras (structures).

• **Bounded regular-m structures**

Let $\mathcal{A} = (A, \leq, \odot, 1)$ be an ordered regular-m structure and $\leq$ be the order of $\mathcal{A}$ (i.e. (pU), (Re), (An), (Tr) hold).

Definitions 5.2.4

(i) An element $l \in A$ is called *last regular-m element* or *the greatest regular-m element* of $\mathcal{A}$, if $x \leq l$, for every $x \in A$.

(ii) An element $f \in A$ is called *first regular-m element* or *the smallest regular-m element* of $\mathcal{A}$, if $f \leq x$, for every $x \in A$.

Hence, the notions of *first regular-m element* and *last regular-m element* are *dual to each other.* Note that both the first regular-m element and the last regular-m element of $\mathcal{A}$ are unique, by (An).

The special element $1 \in A$ can be the last regular-m element. In this case, we have the property:
(L) (Last) $x \leq 1$, for all $x \in A$.

The first regular-m element will be denoted by 0; hence we have the property:
(F) (First) $0 \leq x$, for all $x \in A$.

Definition 5.2.5 An ordered regular-m structure with first regular element 0 and last regular element 1 is called *bounded* and is denoted by $(A, \leq, \odot, 0, 1)$.

Remark 5.2.6 If the regular-m structure is not ordered, then one can extend the above notions.

• **Regular-m algebras (structures) with (involutive) regular-m negations**

Consider the following property:
(pIv^L) $\quad x \odot x^{\sim} = 0 = x^{-} \odot x.$

As we know, in the left-pseudo-MV algebras $(A, \odot, ^{-}, ^{\sim}, 0, 1)$, there exists two negations, $^{-}$ and $^{\sim}$, included in the signature, that are connected with $\odot$ and 0 by (pIv^L). But there are cases when the two negations are not included in the signature. So, we introduce the following new notion.

Definitions 5.2.7 (The dual case is omitted)

(i) A *regular-m algebra (structure) with regular-m negations* is a regular-m algebra $\mathcal{A} = (A, \odot, 1)$ (a regular-m structure $\mathcal{A} = (A, \leq, \odot, 1)$) (i.e. (pU) holds) and $^{-}$, $^{\sim}$ are unary operations on A, called *regular-m negations*, which are connected with $\odot$, 1 by (pIv^L), where 0 is the notation for the element $1^{-} = 1^{\sim}$.

(i') A *regular-m algebra (structure) with regular-m negations* is an algebra $\mathcal{A} = (A, \odot, ^{-}, ^{\sim}, 1)$ (a structure $\mathcal{A} = (A, \leq, \odot, ^{-}, ^{\sim}, 1)$) such that $(A, \odot, 1)$ $((A, \leq, \odot, 1))$

is a regular-m algebra (structure, respectively) (i.e. (pU) holds) and $^-$, $^\sim$ are unary operations on A, called *regular-m negations*, which are connected with $\odot$, 1 by (pIv^L), where 0 is the notation for the element $1^- = 1^\sim$.

(ii) If the regular-m negations verify (pDN^L) ($(x^-)^\sim = x = (x^\sim)^-$, for all $x \in A$), we say that $\mathcal{A}$ is a regular-m algebra (structure) with *involutive* regular-m negations.

Remark 5.2.8 If $^- = ^\sim$, and we denote this by $^{-1}$, then (pIv^L) becomes:
(pIv) $x \cdot x^{-1} = 1 = x^{-1} \cdot 1$, with $1 = 1^{-1}$
and (pDN^L) becomes:
(DN) $(x^{-1})^{-1} = x$.

Hence, a *regular-m algebra with involutive regular-m negation* verifies (pU), (pIv) and (DN).

Note that the pseudo-MV algebra is an example of regular-m algebra with involutive negations included in the signature (i.e. we are in the "world" of unital magmas with additional operations). Other examples are presented in Chapter 10.

5.2.2 Quasi-m algebras (structures) vs. regular-m algebras (structures). The equivalence (QM)

Let $\mathcal{A} = (A, \bigodot, 1)$ be an algebra of type (2,0) or let $\mathcal{A} = (A, \preceq, \bigodot, 1)$ be a structure. Let us consider the following properties: for all $x, y \in A$,
(qm-pU) $1 \bigodot (x \bigodot y) = x \bigodot y = (x \bigodot y) \bigodot 1$, for all $x, y \in A$,
(U1) $1 \bigodot 1 = 1$.

Definitions 5.2.9

(qm1) The algebra $(A, \bigodot, 1)$ (the structure $(A, \preceq, \bigodot, 1)$) is called *quasi-m algebra (structure)*, if it satisfies the properties (qm-pU) and (U1).

(qm1') Any algebra (structure) $\mathcal{A}' = (A, \sigma)$ whose signature σ contains $\bigodot$, 1 is also called *quasi-m algebra (structure)*, if it satisfies the properties (qm-pU) and (U1).

(qm1") Any algebra (structure) $\mathcal{A}'' = (A, \tau)$ which is term equivalent to a quasi-m algebra (structure) $\mathcal{A}' = (A, \sigma)$, is also called *quasi-m algebra (structure)*.

(qm2) The operation $\bigodot$ from a quasi-m algebra (structure) is called *quasi-m operation*.

(qm3) The binary relation $\preceq$ of a quasi-m structure is called *binary quasi-relation*.

Roughly speaking, the *quasi-m algebras (structures)* are those classical algebras (structures) verifying the properties (qm-pU) and (U1). All the properties verified by a quasi-m algebra (structure), including the property (qm-pU), are called *quasi-m properties*. A *theory of quasi-m algebras (structures)* establishes the connections existing between the property (qm-pU) and the other quasi-m properties (if a quasi-m property is proved by (qm-pU) or not), as for example the Proposition 5.2.16.

Besides the set A, let us define the following subset of A:

$$V_{pU} \overset{df.}{=} \{x \in A \mid x \overset{(pU)}{=} 1 \bigodot x = x \bigodot 1\} \subseteq A.$$

Remarks 5.2.10
(i) Note that:
- if (qm-pU) holds, then $V_{pU} \neq A$, i.e. $V_{pU} \subset A$;
- if (pU) holds, then $V_{pU} = A$.

(ii) Since (pU) implies (qm-pU), it follows that any quasi-m algebra will be a generalization of the corresponding regular-m algebra (structure). For example, in the commutative case, the quasi-MV algebra (in fact, the quasi-m MV algebra) introduced in [136] is a generalization of the MV algebra - the MV algebras being called in this context *regular-m algebras.*

(iii) The above remarks show the central role played by the property (pU); we obtain the general rule that, roughly speaking:

(QM) **quasi-m algebra** + (pU) $\Longleftrightarrow$ **regular-m algebra**.

In the view of the above Remarks 5.2.10, we introduce the following definitions:

Definitions 5.2.11
1. For every quasi-m algebra (structure) $\mathcal{A}$, the subset V_{pU} of A will be called the *regular-m* set of $\mathcal{A}$ and will be denoted by $Rm(A)$:

$$Rm(A) \stackrel{def.}{=} V_{pU} \subseteq A.$$

The elements of $Rm(A)$ are called the *regular-m elements* of A.

2. The quasi-m algebra (structure) $\mathcal{A}$ is called *qm-proper*, if $Rm(A) \neq A$ (i.e. (pU) $\not\Longleftrightarrow$ (qm-pU)); otherwise, $\mathcal{A}$ is a regular-m algebra (structure, respectively).

Definition 5.2.12 For every qm-proper quasi-m algebra (structure) $\mathcal{A} = (A, \bigodot, 1)$ ($\mathcal{A} = (A, \preceq, \bigodot, 1)$), any subset $S \subseteq A$ closed under $\bigodot$ and containing 1 is called a *quasi-m subalgebra (substructure)* of $\mathcal{A}$.

We then have the following important result:

Theorem 5.2.13 *Let $\mathcal{A} = (A, \bigodot, 1)$ ($\mathcal{A} = (A, \preceq, \bigodot, 1)$) be a qm-proper quasi-m algebra (structure). Then,*

$$\mathcal{R}m(\mathcal{A}) = (Rm(A), \odot, 1) \quad (\mathcal{R}m(\mathcal{A}) = (Rm(A), \leq, \odot, 1))$$

is a regular-m algebra (structure), where $\odot = \bigodot|_{Rm(A)}$, $\quad \leq = \preceq|_{Rm(A)}$.

Proof: First, we prove that the regular-m set $Rm(A)$ is closed under $\bigodot$ and that $1 \in Rm(A)$. Indeed, if $x, y \in Rm(A) \subset A$, then $x \bigodot y \stackrel{(qm-pU)}{=} 1 \bigodot (x \bigodot y)$, hence $x \bigodot y \in Rm(A)$; since $1 \bigodot 1 = 1$ (by (U1)), it follows that $1 \in Rm(A)$. Consequently, $Rm(A)$ is a quasi-m subalgebra of $\mathcal{A}$. Hence, $\mathcal{R}m(\mathcal{A}) = (Rm(A), \bigodot, 1)$ is a quasi-m algebra.

Moreover, (qm-pU) coincides with (pU) on $Rm(A)$, i.e. (qm-pU)$|_{Rm(A)} \Longleftrightarrow$ (pU), by the definition of $Rm(A)$. Consequently, $\mathcal{R}m(\mathcal{A}) = (Rm(A), \odot, 1)$ is a regular-m algebra. Similarly for structures. □

Conventions: In order to simplify the writing,
- the *quasi-m operation* $\bigodot$ of a quasi-m algebra (structure) $\mathcal{A}$ and the corresponding (its restriction to $Rm(A)$) *regular-m operation* $\odot$ of the regular-m algebra $\mathcal{R}m(\mathcal{A})$ will be denoted the same in the sequel, namely by $\odot$,
- the *binary quasi-m relation* $\preceq$ of a quasi-m structure $\mathcal{A}$ and the corresponding (its restriction to $Rm(A)$) *binary regular-m relation* $\leq$ of the regular-m structure $\mathcal{R}m(\mathcal{A})$ will be denoted the same in the sequel, namely by $\leq$.

- **The quasi-m unital magmas**

Following the above Definitions 5.2.9 (qm1), we introduce the most general quasi-m algebra, namely:

Definition 5.2.14 Let $\mathcal{A} = (A, \odot, 1)$ be an algebra of type $(2, 0)$. $\mathcal{A}$ is called *quasi-m unital magma*, or *qm-unital magma* for short, if the properties (qm-pU) and (U1) are satisfied.

Denote by **qm-unital magma** the class of all qm-unital magmas.

We immediately obtain the following Corollary of Theorem 5.2.13:

Corollary 5.2.15 *Let $\mathcal{A} = (A, \odot, 1)$ be a qm-unital magma. Then, $\mathcal{R}m(\mathcal{A}) = (Rm(A), \odot, 1)$ is a unital magma.*

Hence, we have the following equivalence, by (QM):

qm-unital magma + (pU) $\Longleftrightarrow$ **unital magma**.

- **General properties**

Consider the following properties:
(IdP) $\quad 1 \odot x = x \odot 1$,
(Ass2) $\quad x \odot (1 \odot y) = x \odot y$ and $(x \odot 1) \odot y = x \odot y$,
(qm-PI1) $\quad x \odot y = (1 \odot x) \odot y$,
(qm-PI2) $\quad x \odot y = x \odot (y \odot 1)$,
(qm-PI) $\quad x \odot y = (1 \odot x) \odot (1 \odot y), \;\; x \odot y = (x \odot 1) \odot (y \odot 1)$.

Proposition 5.2.16 *Let $(A, \odot, 1)$ be an algebra of type $(2, 0)$ or a structure $(A, \leq, \odot, 1)$. Then, we have:*
(qm-1) (qm-pU) + (Ass) $\Longrightarrow$ (IdP),
(qm-2) (qm-pU) + (Ass) $\Longrightarrow$ (Ass2),
(qm-3) (Ass2) + (IdP) $\Longrightarrow$ (qm-PI1),
(qm-4) (Ass2) + (IdP) $\Longrightarrow$ (qm-PI2),
(qm-5) (qm-pU) + (Ass) + (qm-pI1) + (qm-PI2) $\Longrightarrow$ (qm-PI).

Proof. (qm-1): $1 \odot x \overset{(qm-pU)}{=} (1 \odot x) \odot 1 \overset{(Ass)}{=} 1 \odot (x \odot 1) \overset{(qm-pU)}{=} x \odot 1$. Thus, (IdP) holds.

(qm-2): $x \odot y \overset{(qm-pU)}{=} (x \odot y) \odot 1 \overset{(Ass)}{=} x \odot (y \odot 1) \overset{(qm-pU)}{=} x \odot [1 \odot (y \odot 1)] \overset{(Ass)}{=} x \odot [(1 \odot y) \odot 1] \overset{(qm-pU)}{=} x \odot (1 \odot y)$ and
$x \odot y \overset{(qm-pU)}{=} 1 \odot (x \odot y) \overset{(Ass)}{=} (1 \odot x) \odot y \overset{(qm-pU)}{=} [(1 \odot x) \odot 1] \odot y \overset{(Ass)}{=} [1 \odot (x \odot 1)] \odot y \overset{(qm-pU)}{=} (x \odot 1) \odot y$. Thus, (Ass2) holds.

(qm-3): $x \odot y \overset{(Ass2)}{=} (x \odot 1) \odot y \overset{(IdP)}{=} (1 \odot x) \odot y$. Thus, (qm-PI1) holds.

(qm-4): $x \odot y \overset{(Ass2)}{=} x \odot (1 \odot y) \overset{(IdP)}{=} x \odot (y \odot 1)$. Thus, (qm-PI2) holds.

(qm-5): $x \odot y \overset{(qm-pI1)}{=} (1 \odot x) \odot y \overset{(qm-pU)}{=} [(1 \odot x) \odot 1] \odot y \overset{(Ass)}{=} (1 \odot x) \odot (1 \odot y)$ and
$x \odot y \overset{(qm-pI2)}{=} x \odot (y \odot 1) \overset{(qm-pU)}{=} x \odot [1 \odot (y \odot 1)] \overset{(Ass)}{=} (x \odot 1) \odot (y \odot 1)$. Thus, (qm-PI) holds. □

An important result is the following.

Proposition 5.2.17 *Let $\mathcal{A}$ be a quasi-m algebra (structure). If the property (Ass) holds, then the following properties hold: (IdP), (Ass2), (qm-PI1), (qm-PI2), (qm-PI).*

Proof. By Proposition 5.2.16. □

- **Quasi-m order**

Let $\mathcal{A} = (A, \leq, \odot, 1)$ be a qm-proper quasi-m structure (i.e. (qm-pU), which differs from (pU), and (U1) hold) containing a binary relation $\leq$ defined on A; then its substructure $\mathcal{R}m(\mathcal{A})$ is a regular-m structure (i.e. (pU) holds), by Theorem 5.2.13, containing also $\leq$.

Consider the following properties of $\leq$:

(Re)	(Reflexivity) $x \leq x$,
(An)	(Antisymmetry) $x \leq y$ and $y \leq x \Longrightarrow x = y$,
(qm-An)	(Quasi-m Antisymmetry, or qm-Antisymmetry) $(x \leq y$ and $y \leq x) \Longrightarrow (1 \odot x = 1 \odot y$ and $x \odot 1 = y \odot 1)$,
(Tr)	(Transitivity) $x \leq y$ and $y \leq z \Longrightarrow x \leq z$.

Definitions 5.2.18 We shall say that $\mathcal{A}$ is:
- *reflexive*, if property (Re) is satisfied,
- *quasi-m antisymmetric*, if property (qm-An) is satisfied,
- *transitive*, if property (Tr) is satisfied,
- *quasi-m pre-ordered*, and $\leq$ is a *quasi-m pre-order*, if it is reflexive and transitive,
- *quasi-m ordered*, and $\leq$ is a *quasi-m order* (or a *qm-order* for short), if it is reflexive, quasi-m antisymmetric and transitive.

A quasi-m ordered quasi-m structure will be simply called: "a quasi-m ordered structure".

Remarks 5.2.19 (i) (pU) + (qm-An) $\Longrightarrow$ (An).
Consequently, in the presence of (pU), any qm-order becomes a regular-m order.

(ii) The above remark (i) shows again the central role played by the property (pU): we obtain, by (QM), the general rule that, roughly speaking,
quasi-m ordered structure + (pU) $\Longleftrightarrow$ ordered regular-m structure.

Note that the quasi-MV algebras are in fact *quasi-m ordered structures* and that the MV algebras are in fact *ordered regular-m structures.*

We now present an important remark:

Remark 5.2.20 Let $\mathcal{A} = (A, \leq, \odot, 1)$ be a quasi-m ordered structure. Since (qm-pU) coincides with (pU) on $Rm(A)$, by Theorem 5.2.13, it follows that (qm-An) coincides with (An) on $Rm(A)$. Consequently,
- the qm-order relation $\leq$ on A becomes an order relation on $Rm(A)$;
- $\mathcal{R}m(\mathcal{A}) = (Rm(A), \leq, \odot, 1)$ is an ordered regular-m structure.

- **The duality principle**

Let $\mathcal{A} = (A, \leq, \odot, 1)$ be a quasi-m ordered structure and $\leq$ be the quasi-m order of $\mathcal{A}$. The relation $\geq$, defined on A as follows: for every $x, y \in A$, $x \geq y \stackrel{def.}{\Longleftrightarrow} y \leq x$, is also a qm-order, called the *inverse qm-order relation* or the *dual qm-order relation* of the qm-order relation $\leq$.

The duality principle for quasi-m ordered structures is the following:
"every statement (definition of a notion, proposition, theorem, etc.) concerning the quasi-m ordered structure $\mathcal{A} = (A, \leq, \odot, 1)$ remains valid if we replace everywhere inside it the qm-order relation $\leq$ with the inverse qm-order relation, $\geq$".

The structure $(A, \geq, \odot, 1)$ obtained in this way is also a quasi-m ordered structure, called the *dual* of $\mathcal{A}$. The statement obtained in this way (definition of a notion, proposition, theorem, etc.) is the *dual* statement of the first statement (definition of a notion, proposition, theorem, etc.). We say also that the two quasi-m ordered structures or statements are *dual to each other* or simple *dual.* More precisely, we shall say that $(A, \leq, \odot, 1)$ is a *quasi-m ordered left-structure* while $(A, \geq, \odot, 1)$ is a *quasi-m ordered right-structure.*

- **Elements of the same m-height. The m-clouds**

Definition 5.2.21 Let $\mathcal{A} = (A, \odot, 1)$ be a quasi-m algebra or $\mathcal{A} = (A, \leq, \odot, 1)$ be a quasi-m ordered left-structure.

We say that $x, y \in A$ have *the same m-height*, or are *m-parallel*, and we denote this by $x \parallel_m y$, if $1 \odot x = 1 \odot y$ and $x \odot 1 = y \odot 1$.

Note that if $x \leq y$ and $y \leq x$ then $x \parallel_m y$, by (qm-An).
Note that if (pU) holds, then $x \parallel_m y \Leftrightarrow x = y$.

Corollary 5.2.22 *If (Ass) holds, then: for all $x, y \in A$,*

$$x \parallel_m y \Longleftrightarrow 1 \odot x \; (\stackrel{(IdP)}{=} x \odot 1) = 1 \odot y \; (\stackrel{(IdP)}{=} y \odot 1).$$

Proof. $\Longrightarrow$: By Proposition 5.2.17, (IdP) holds.
$\Longleftarrow$: Obviously. □

Proposition 5.2.23 *The relation $\|_m$ is an equivalence relation of $\mathcal{A}$.*

Proof. The relation $\|_m$ is obviously reflexive and symmetrique. It is transitive also: if $x \|_m y$ and $y \|_m z$, i.e. $(1 \odot x = 1 \odot y$ and $x \odot 1 = y \odot 1)$ and $(1 \odot y = 1 \odot z$ and $y \odot 1 = z \odot 1)$, then $1 \odot x = 1 \odot z$ and $x \odot 1 = z \odot 1$, i.e. $x \|_m z$. □

Proposition 5.2.24 *If the property (Ass2) also holds, then $\|_m$ is a congruence relation of $\mathcal{A}$.*

Proof. We must prove that $\|_m$ is compatible with $\odot$. Indeed, if $x \|_m y$ and $a \|_m b$, i.e. $(1 \odot x = 1 \odot y$ and $x \odot 1 = y \odot 1)$ and $(1 \odot a = 1 \odot b$ and $a \odot 1 = b \odot 1)$, then:
$1 \odot (x \odot a) \overset{(qm-pU)}{=} x \odot a \overset{(Ass2)}{=} x \odot (1 \odot a) = x \odot (1 \odot b) \overset{(Ass2)}{=} x \odot b \overset{(Ass2)}{=} (x \odot 1) \odot b =$
$(y \odot 1) \odot b \overset{(Ass2)}{=} y \odot b \overset{(qm-pU)}{=} 1 \odot (y \odot b)$ and
$(x \odot a) \odot 1 \overset{(qm-pU)}{=} x \odot a \overset{(Ass2)}{=} \ldots \overset{(Ass2)}{=} y \odot b \overset{(qm-pU)}{=} (y \odot b) \odot 1$,
hence $(x \odot a) \|_m (y \odot b)$. □

Proposition 5.2.25 *If the property (Ass2) holds and if $b \|_m a$, then, in the table of $\odot$, we have:*
(i) the row of b coincides with the row of a;
(ii) the column of b coincides with the column of a.

Proof: Suppose $b \|_m a$, i.e. $1 \odot b = 1 \odot a$ and $b \odot 1 = a \odot 1$. Then,
- for any $y \in A$, $b \odot y \overset{(Ass2)}{=} (b \odot 1) \odot y = (a \odot 1) \odot y \overset{(Ass2)}{=} a \odot y$, i.e. the row of b coincides with the row of a;
- for any $x \in A$, $x \odot b = \overset{(Ass2)}{=} x \odot (1 \odot b) = x \odot (1 \odot a) \overset{(Ass2)}{=} x \odot a$, i.e. the column of b coincides with the column of a. □

For each $x \in A$, we denote its equivalence class by $| x |_m$:

$$| x |_m \overset{notation}{=} \{y \in A \mid y \|_m x\}$$

and we denote by $A/\|_m$ the quotient set (i.e. the set of all equivalence classes):

$$A/\|_m \overset{notation}{=} \{| x |_m \mid x \in A\}.$$

We shall call *m-clouds* the equivalence classes determined by $\|_m$. Thus, we have, for each $x \in A$:

$$C_m(x) = | x |_m .$$

We have, then:

Lemma 5.2.26 *(See Lemma 5.1.31)*
If the property (Ass) holds, then every m-cloud of $\mathcal{A}$ contains exactly one regular-m element.

Proof. Let C_m be a m-cloud, i.e. there exists $c \in A$ such that $C_m = \mid c \mid_m$ $(= C_m(c))$.

• If $c \in Rm(A)$, then $c = 1 \odot c$. Let $a \in C_m$ (i.e. $a \parallel_m c$) such that $a \neq c$. We prove that $a \notin Rm(A)$; indeed, if, by absurdum hypothesis, $a \in Rm(A)$, then $a = 1 \odot a$; then $a = 1 \odot a = 1 \odot c = c$: contradiction.

• If $c \notin Rm(A)$, we put $b = 1 \odot c$. Then $b = 1 \odot c \overset{(qm-pU)}{=} 1 \odot (1 \odot c) = 1 \odot b$, hence $b \in Rm(A)$.

We prove now that $b \in C_m$; indeed, $1 \odot b = 1 \odot (1 \odot c) \overset{(qm-pU)}{=} 1 \odot c$ and $b \odot 1 = (1 \odot c) \odot 1 \overset{(Ass)}{=} 1 \odot (c \odot 1) \overset{(qm-pU)}{=} c \odot 1$, hence $b \parallel_m c$, i.e. $b \in C_m$.
We prove now that any other element a of C_m, such that $a \neq b$, is not a regular-m element; indeed, if, by absurdum hypothesis, $a \in Rm(A)$, i.e. $a = 1 \odot a$, then, since $a \parallel_m c$, it follows that: $a = 1 \odot a = 1 \odot c = b$: contradiction. □

We define the product of m-clouds as follows: for all $x, y \in A$,

$$\mid x \mid_m \odot \mid y \mid_m \overset{def.}{=} \mid x \odot y \mid_m .$$

Then, we have the following expected result.

Proposition 5.2.27 *The quotient algebra* $(A/\parallel_m, \odot, \mid 1 \mid_m = C_m(1))$ *is a regular-m algebra, isomorphic to* $(Rm(A), \odot, 1)$.

Proof: Routine, by Lemma 5.2.26. □

Remark 5.2.28 Given a finite regular-m algebra $(X, \odot, 1)$, we can obtain, **in general**, an infinity of finite quasi-m algebras: $(A_1, \odot, 1)$, $(A_2, \odot, 1)$, ... such that $Rm(A_1) = Rm(A_2) = ... = X$, by adding one or more elements m-parallel with some (all) elements (called regular-m elements) of X.
We can quickly draw the table of $\odot$ for such a finite quasi-m algebra $(A, \odot, 1)$ with $Rm(A) = X$ by using Proposition 5.2.25.

• **Quasi-m Hasse diagrams and quasi-m Hasse-type diagrams**

If the regular-m structure has an order relation, then it is represented by a Hasse diagram.

If the quasi-m structure has a quasi-m order, then it is represented by a *quasi-m Hasse diagram*:
- a regular-m element is represented by a bullet •,
- an element m-parallel with a regular-m element is represented by a bigcirc ○, connected by a horizontal line with the regular-m element, etc.

If the binary regular-m relation from a regular-m structure is not an order relation, then the structure is represented by a Hasse-type diagram.

If the quasi-m structure has a binary quasi-m relation that is not a quasi-m order, then it is represented by a *quasi-m Hasse-type diagram*:
- a regular-m element is represented by a small circle ∘,

- an element m-parallel with a regular-m element is represented by a bigcirc $\bigcirc$, connected by a horizontal line with the regular-m element, etc.

• **Bounded quasi-m structures**

Let $\mathcal{A} = (A, \leq, \odot, 1)$ be a quasi-m ordered left-structure and $\leq$ be the quasi-order of $\mathcal{A}$ (i.e. (qm-pU), (U1), (Re), (qm-An), (Tr) hold). Then, by Theorem 5.2.13, $\mathcal{R}m(\mathcal{A}) = (Rm(A), \leq, \odot, 1)$ is a regular-m ordered left-structure and $\leq$ is a regular-m order of $Rm(A)$ (i.e. (pU), (Re), (An), (Tr) hold).

Definitions 5.2.29

(i) An element $l \in A$ is called *last quasi-m element* or *the greatest quasi-m element* of $\mathcal{A}$, if $1 \odot x \leq 1 \odot l)$, for every $x \in A$.

(ii) An element $f \in A$ is called *first quasi-m element* or *the smallest quasi-m element*) of $\mathcal{A}$, $1 \odot f \leq 1 \odot x)$, for every $x \in A$.

Hence, *the notions of first quasi-m element and last quasi-m element are dual to each other.*

Note that in a quasi-m ordered left-structure there can be more first quasi-m elements and more last quasi-m elements, i.e. the first quasi-m element and the last quasi-m element can be not unique (when they exist). Anyway, they form m-clouds: the m-cloud of first quasi-m elements and the m-cloud of last quasi-m elements.

By Lemma 5.2.26, if the propery (Ass) holds, then every m-cloud in a quasi-m ordered structure $\mathcal{A}$ contains exactly one regular-m element.

The special element $1 \in A$ is a regular-m element (i.e. $1 \in R(A)$) and it can be the last quasi-m element; in this case, we have the property: for every $x \in A$,
(L) (Last) $x \leq 1$.

Note that there can be other last quasi-m elements too, but not as regular-m elements, by Lemma 5.2.26, if the property (Ass) holds; in this case, all the last quasi-m elements of $\mathcal{A}$ form the m-cloud of 1, denoted by $C_m(1) = \mid 1 \mid$.

In the m-cloud of the first quasi-m elements, only one is a regular-m element, by Lemma 5.2.26, if the property (Ass) holds; the first quasi-m element which is a regular-m element will be denoted by 0. Consequently,
(1) $0 \in R(A)$ and, consequently, $1 \odot 0 = 0$, by (pU);
(2) we have the property: for every $x \in A$,
(F) (First) $\quad 0 \leq x$;
(3) all the first quasi-m elements of $\mathcal{A}$ form the m-cloud of 0, denoted by $C_m(0)$.

Definition 5.2.30 A quasi-m ordered left-structure with first quasi-m element 0 and last quasi-m element 1 is called *bounded* and is denoted by $\mathcal{A} = (A, \leq, \odot, 0, 1)$.

Let us introduce the new property (10-0):
(10-0) $1 \odot 0 = 0$ and $0 \odot 1 = 0$.

Remarks 5.2.31 Let $\mathcal{A} = (A, \leq, \odot, 0, 1)$ be a bounded quasi-m left-structure. Then:
(1) $\mathcal{R}m(\mathcal{A}) = (Rm(A), \leq, \odot, 0, 1)$ is a bounded regular-m left-structure;
(2) the property (10-0) holds.

• Quasi-m algebras (structures) with (involutive) quasi-m negations

Definitions 5.2.32 (See Definition 5.2.7 in the regular case)

(i) Let $\mathcal{A} = (A, \odot, 1)$ be a quasi-m algebra (or $\mathcal{A} = (A, \leq, \odot, 1)$ be a quasi-m structure) such that the associated regular-m algebra (structure) $\mathcal{R}m(\mathcal{A})$ contains the involutive regular-m negations $^-$ and $^\sim$ (i.e. (pIv^L) and (pDN^L) holds). We then extend the regular-m negations $^- : Rm(A) \longrightarrow Rm(A)$ and $^\sim : Rm(A) \longrightarrow Rm(A)$ to the corresponding so called *quasi-m negations* $^- : A \longrightarrow A$ and $^\sim : A \longrightarrow A$ such that the following properties be verified:

(qm-NegP^L) $\quad (x \odot 1)^- = x^- \odot 1 = 1 \odot x^- = (1 \odot x)^-$,
$(x \odot 1)^\sim = x^\sim \odot 1 = 1 \odot x^\sim = (1 \odot x)^\sim$ and

(pIv^L) $\quad x \odot x^\sim = 0 = x^- \odot x$, with $0 = 1^- = 1^\sim$.

If there exist such quasi-m negations, then we shall say that $\mathcal{A}$ is a *quasi-m algebra (structure) with the quasi-m negations* $^-$, $^\sim$.

(i') A *quasi-m algebra (structure) with quasi-m negations* is an algebra $\mathcal{A} = (A, \odot, ^-, ^\sim, 1)$ (a structure $\mathcal{A} = (A, \leq, \odot, ^-, ^\sim, 1)$) such that $(A, \odot, 1)$ ($(A, \leq, \odot, 1)$ respectively) is a quasi-m algebra (quasi-m structure) (i.e. (qm-pU) and (U1) hold) and $^-$, $^\sim$ are unary operations on A, called *quasi-m negations*, which are connected with $\odot$, 1 by (qm-NegP^L) and (pIv^L).

(ii) If the quasi-m negations $^-$, $^\sim$ verify (pDN^L), we shall say that $\mathcal{A}$ is a quasi-m algebra (structure) with *involutive* quasi-m negations.

Remark 5.2.33 (See the analogous Remark 5.1.42 for quasi-pseudo-algebras)

If the two quasi-m negations $^-$ and $^\sim$ exist and coincide and if we denote the unique quasi-m negation by $^-$, then it verifies the properties (i.e. (qm-NegP^L) and (pIv^L) become):

(qm-NegP) $\quad (x \cdot 1)^- = x^- \cdot 1 = 1 \cdot x^- = (1 \cdot x)^-$ and

(pIv) $\quad x \cdot x^- = 1 = x^- \cdot x$, with $1 = 1^-$.

If such an involutive quasi-m negation exists, then we shall say that $\mathcal{A}$ is a *quasi-m algebra (structure) with an involutive quasi-m negation.*

Remark 5.2.34 Given a regular-m algebra (structure) $\mathcal{A} = (A, \odot, 1)$ ($\mathcal{A} = (A, \leq, \odot, 1)$, respectively) with involutive regular-m negations, there can be more involutive quasi-m negations $^-$ and $^\sim$ verifying (qm-NegP^L), i.e. there can be more quasi-m algebras (structures) with involutive quasi-m negations associated to the given $\mathcal{A}$ (see examples further in Chapter 10).

Note that the quasi-MV algebra [136] is an example of commutative quasi-m algebra with involutive quasi-m negation (see [103]). Most probably, the quasi-pseudo-MV algebra [28], [29], [30] is an example of quasi-m algebra with involutive quasi-m negations.

Consider now the following properties:

$(\overline{pU^L})$ $\quad 1 \odot x^- = x^- = x^- \odot 1, \quad 1 \odot x^\sim = x^\sim = x^\sim \odot 1;$

(qm-$\overline{pU^L}$) $\quad 1 \odot (x \odot y)^- = (x \odot y)^- = (x \odot y)^- \odot 1;$
$1 \odot (x \odot y)^\sim = (x \odot y)^\sim = (x \odot y)^\sim \odot 1,$

(qm-$\overline{pU(1 \odot x, x \odot 1)^L}$) $\quad 1 \odot (1 \odot x)^- = (1 \odot x)^- = (1 \odot x)^- \odot 1,$
$1 \odot (1 \odot x)^\sim = (1 \odot x)^\sim = (1 \odot x)^\sim \odot 1,$
$1 \odot (x \odot 1)^- = (x \odot 1)^- = (x \odot 1)^- \odot 1,$
$1 \odot (x \odot 1)^\sim = (x \odot 1)^\sim = (x \odot 1)^\sim \odot 1.$

Proposition 5.2.35 *We have:*
(j) $\quad (\overline{pU^L}) \Longrightarrow$ *(qm-$\overline{pU^L}$)* $\Longrightarrow$ *(qm-$\overline{pU(1 \odot x, x \odot 1)^L}$).*

(jj) If (qm-NegP^L) holds, then (qm-$\overline{pU(1 \odot x, x \odot 1)^L}$) holds too.

Proof. (j): Obviously.

(jj): $1 \odot (1 \odot x)^- = 1 \odot (1 \odot x^-) \overset{(qm-pU)}{=} 1 \odot x^- = (1 \odot x)^-$. Similarly the rest.
□

Remark 5.2.36 If $^- = ^\sim$, then the above properties become:

$(\overline{pU})$ $\quad 1 \cdot x^- = x^- = x^- \cdot 1,$

(qm-$\overline{pU}$) $\quad 1 \cdot (x \cdot y)^- = (x \cdot y)^- = (x \cdot y)^- \cdot 1,$

(qm-$\overline{pU(1 \cdot x, x \cdot 1)}$) $\quad 1 \cdot (1 \cdot x)^- = (1 \cdot x)^- = (1 \cdot x)^- \cdot 1,$
$1 \cdot (x \cdot 1)^- = (x \cdot 1)^- = (x \cdot 1)^- \cdot 1.$

Consequently, we obtain, by above Proposition 5.2.35:

Proposition 5.2.37 *We have:*
(j) $\quad (\overline{pU}) \Longrightarrow$ *(qm-$\overline{pU}$)* $\Longrightarrow$ *(qm-$\overline{pU(1 \odot x, x \odot 1)}$).*

(jj) If (qm-NegP) holds, then (qm-$\overline{pU(1 \odot x, x \odot 1)}$) holds too.

5.2.3 Qm-monoids vs. monoids

Recall that a *monoid* is an algebra $\mathcal{G} = (G, \cdot, 1)$ of type $(2, 0)$ verifying: for all $x, y, z \in G$,
(pU) $1 \cdot x = x = x \cdot 1$,
(Ass) $x \cdot (y \cdot z) = (x \cdot y) \cdot z$, i.e. the operation $\cdot$ is associative.

Let us introduce the following new notion.

Definition 5.2.38 A *quasi-m monoid*, or a *qm-monoid* for short, is an algebra $\mathcal{G} = (G, \cdot, 1)$ of type $(2, 0)$ verifying: for all $x, y, z \in G$,
(qm-pU) $\quad 1 \cdot (x \cdot y) = x \cdot y = (x \cdot y) \cdot 1,$
(U1) $\quad 1 \cdot 1 = 1,$
(Ass) $\quad x \cdot (y \cdot z) = (x \cdot y) \cdot z.$

Denote by **qm-monoid** the class of all qm-monoids.

Note that a qm-monoid is a qm-unital magma verifying (Ass), i.e. we have the equivalence:

(qmUMM) **qm-unital magma** + (Ass) $\Longleftrightarrow$ **qm-monoid.**

A qm-monoid is *commutative*, if the binary operation $\cdot$ is commutative.

By above Theorem 5.2.13, we have immediately the following corollary:

Theorem 5.2.39 *Let $\mathcal{G} = (G, \cdot, 1)$ be a qm-monoid. Then,*

$$\mathcal{R}m(\mathcal{G}) = (Rm(G), \cdot, 1)$$

is a monoid.

Proof. By Theorem 5.2.13, $\mathcal{R}m(\mathcal{G}) = (Rm(G), \cdot, 1)$ is a regular-m algebra; it verifies (Ass), hence is a monoid. □

Following the general definition, a qm-monoid is *qm-proper*, if it is not a monoid. A qm-monoid is *proper*, if is not a qm-group.

Since a monoid may be: a proper monoid, or a group, we then introduce the following definitions:

Definition 5.2.40 Let $\mathcal{A} = (A, \cdot, 1)$ be a proper qm-proper qm-monoid. We shall say that $\mathcal{A}$ is:
- *pure*, if $\mathcal{R}m(\mathcal{A})$ is a proper monoid,
- of *group-type*, if $\mathcal{R}m(\mathcal{A})$ is a group.

Examples of monoids and qm-monoids are presented in Chapter 6 (Exemple 16').

Proposition 5.2.41 *Let $\mathcal{A} = (A, \cdot, 1)$ be a qm-monoid. Then, the following properties hold: (IdP), (Ass2), (qm-PI1), (qm-PI2), (qm-PI).*

Proof. By Proposition 5.2.17. □

5.2.4 Further research

Further research to do concerns the definitions of qm-po-ms, qm-po-ms(pR), etc. and their connections with the corresponding quasi-pseudo-algebras (of logic).

This section will be connected further to groups and their generalizations.

5.3 Connections

Consider the following properties:
(pDN^L) $(x^{-^L})^{\sim^L} = x = (x^{\sim^L})^{-^L}$,
(pN1^{Lb}) $0 = 1^- = 1^\sim,\ 1 = 0^- = 0^\sim$,
(pDNeg3^{Lb}) $(x \to^L y^-)^\sim = (y \rightsquigarrow^L x^\sim)^-$,

(pM^L) $1 \to^L x = x = 1 \rightsquigarrow^L x$,
(pU) $x \odot 1 = x = 1 \odot x$;
(IdEq^L) $x \to^L y = 1 \Longleftrightarrow x \rightsquigarrow^L y = 1$,
(IdEqP^{Lb}) $x \odot y^\sim = 0 \Longleftrightarrow y^- \odot x = 0$, with $0 = 1^- = 1^\sim$;
(pEq^L) $x \leq y \Longleftrightarrow x \to^L y = 1$ $(\overset{(IdEq^L)}{\Leftrightarrow}$ $x \rightsquigarrow^L y = 1)$,
(pEqP^{Lb}) $x \leq y \Longleftrightarrow x \odot y^\sim = 0$ $(\overset{(IdEqP^{Lb})}{\Longleftrightarrow}$ $y^- \odot x = 0)$;
(pDNeg1^{Lb}) $x \to^L y = y^- \rightsquigarrow^L x^-$, $x \rightsquigarrow^L y = y^\sim \to^L x^\sim$,
(pDNP1) $(x \odot y^\sim)^- = ((x^-)^- \odot y^-)^\sim$, $(y^- \odot x)^\sim = (y^\sim \odot (x^\sim)^\sim)^-$;
(pDNeg2^{Lb}) $x^\sim \to^L y = y^- \rightsquigarrow^L x$,
(pDNP2) $(x^- \odot y^-)^\sim = (x^\sim \odot y^\sim)^-$;
(pEx^L) $x \to^L (y \rightsquigarrow^L z) = y \rightsquigarrow^L (x \to^L z)$,
(Ass) $x \odot (y \odot z) = (x \odot y) \odot z$;
(pRe^L) $x \to^L x = 1$ $(\Leftrightarrow x \rightsquigarrow^L x = 1)$,
(pIv^L) $x \odot x^\sim = 0 = x^- \odot x$, where $0 = 1^- = 1^\sim$;
(q-pM^L) $1 \rightsquigarrow^L (x \to^L y) = x \to^L y$, $1 \to^L (x \rightsquigarrow^L y) = x \rightsquigarrow^L y$,
(qm-pU) $1 \odot (x \odot y) = x \odot y = (x \odot y) \odot 1$;
(q-pNeg^L) $x \to^L 0 = 1 \to^L x^- = (1 \rightsquigarrow^L x)^-$,
$x \rightsquigarrow^L 0 = 1 \rightsquigarrow^L x^\sim = (1 \to^L x)^\sim$,
(qm-NegP^L) $(x \odot 1)^- = x^- \odot 1 = 1 \odot x^- = (1 \odot x)^-$,
$(x \odot 1)^\sim = x^\sim \odot 1 = 1 \odot x^\sim = (1 \odot x)^\sim$;
(pM1^L) $1 \to^L 1 = 1$ $(\Leftrightarrow 1 \rightsquigarrow^L 1 = 1)$,
(U1) $1 \odot 1 = 1$;
(IdR^L) $1 \to^L x = 1 \rightsquigarrow^L x$,
(IdP) $1 \odot x = x \odot 1$.

We establish now a general result.

Theorem 5.3.1 *(See Theorem 12.4.1)*

(1) Let $\mathcal{A} = (A, \to^L, \rightsquigarrow^L, ^-, ^\sim, 0, 1)$ *be an algebra of type* $(2, 2, 1, 1, 0, 0)$ *verifying (pN1^{Lb}), (pDN^L) and (pDNeg3^{Lb}).*

Define $\Phi(\mathcal{A}) \overset{def.}{=} (A, \odot, ^-, ^\sim, 0, 1)$ *by:*

$x \odot y \overset{def.}{=} (x \to^L y^-)^\sim \overset{(pDNeg3^{Lb})}{=} (y \rightsquigarrow^L x^\sim)^-$.

Then, $\Phi(\mathcal{A})$ *is an algebra of type* $(2, 1, 1, 0, 0)$ *verifying (pN1^{Lb}), (pDN^L).*

(1') Conversely, let $\mathcal{A} = (A, \odot, ^-, ^\sim, 0, 1)$ *be an algebra of type* $(2, 1, 1, 0, 0)$ *verifying (pN1^{Lb}), (pDN^L).*

Define $\Psi(\mathcal{A}) \overset{def.}{=} (A, \to^L, \rightsquigarrow^L, ^-, ^\sim, 0, 1)$ *by:*

$x \to^L y \overset{def.}{=} (x \odot y^\sim)^-$, $x \rightsquigarrow^L y \overset{def.}{=} (y^- \odot x)^\sim$.

Then, $\Psi(\mathcal{A})$ *is an algebra of type* $(2, 2, 1, 1, 0, 0)$ *verifying (pN1^{Lb}), (pDN^L) and (pDNeg3^{Lb}).*

(2) The maps Φ *and* Ψ *are mutually inverse.*

(3) The following equivalences hold:

(pM^L) $\Longleftrightarrow$ *(pU),*
(IdEq^L) $\Longleftrightarrow$ *(IdEqP^{Lb}),*
(pEq^L) $\Longleftrightarrow$ *(pEqP^{Lb}),*

(pDNeg1^{Lb}) $\Longleftrightarrow$ (pDNP1),
(pDNeg2^{Lb}) $\Longleftrightarrow$ (pDNP2),
(pEx^L) + (pDNeg1^{Lb}) $\Longleftrightarrow$ (Ass) + (pDNP1),
(pRe^L) $\Longleftrightarrow$ (pIv^L);
(q-pM^L) $\Longleftrightarrow$ (qm-pU),
(q-pNeg^L) $\Longleftrightarrow$ (qm-NegP^L),
(pM1^L) + (q-pNeg^L) $\Longleftrightarrow$ (U1) + (qm-NegP^L),
(IdR^L) + (q-pNeg^L) $\Longleftrightarrow$ (IdP) + (qm-NegP^L).

Proof. (1): Obviously.

(1'): We must prove that (pDNeg3^{Lb}) holds. Indeed,
$(x \to^L y^-)^\sim = ((x \odot (y^-)^\sim)^-)^\sim \overset{(pDN^L)}{=} x \odot y$ and
$(y \rightsquigarrow^L x^\sim)^- = (((x^\sim)^- \odot y)^\sim)^- \overset{(pDN^L)}{=} x \odot y$.

(2): Let $(A, \to^L, \rightsquigarrow^L, ^-, ^\sim, 0, 1) \xrightarrow{\Phi} (A, \odot, ^-, ^\sim, 0, 1) \xrightarrow{\Psi} (A, \Rightarrow^L, \approx>^L, ^-, ^\sim, 0, 1)$. Then, for all $x, y \in A$,
$x \Rightarrow^L y = (x \odot y^\sim)^- = ((x \to^L (y^\sim)^-)^\sim)^- \overset{(pDN^L)}{=} x \to^L y$ and
$x \approx>^L y = (y^- \odot x)^\sim = ((x \rightsquigarrow^L (y^-)^\sim)^-)^\sim \overset{(pDN^L)}{=} x \rightsquigarrow^L y$.

Conversely, let $(A, \odot, ^-, ^\sim, 0, 1) \xrightarrow{\Psi} (A, \to^L, \rightsquigarrow^L, ^-, ^\sim, 0, 1) \xrightarrow{\Phi} (A, \bigodot, ^-, ^\sim, 0, 1)$. Then, for all $x, y \in A$,
$x \bigodot y = (x \to^L y^-)^\sim = ((x \odot (y^-)^\sim)^-)^\sim \overset{(pDN^L)}{=} x \odot y$.

(3): (pM^L) $\Longleftrightarrow$ (pU):
$\Longrightarrow$: $1 \odot x = (1 \to^L x^-)^\sim \overset{(pM^L)}{=} (x^-)^\sim \overset{(pDN^L)}{=} x$ and
$x \odot 1 = (1 \rightsquigarrow^L x^\sim)^- \overset{(pM^L)}{=} (x^\sim)^- \overset{(pDN^L)}{=} x$. Thus, (pU) holds.
$\Longleftarrow$: $1 \to^L x = (1 \odot x^\sim)^- \overset{(pU)}{=} (x^\sim)^- \overset{(pDN^L)}{=} x$ and
$1 \rightsquigarrow^L x = (x^- \odot 1)^\sim \overset{(pU)}{=} (x^-)^\sim \overset{(pDN^L)}{=} x$. Thus, (pM^L) holds.

(IdEq^L) $\Longleftrightarrow$ (IdEqP^{Lb}) (See the proof of Corollary 2.11.6):
$x \odot y^\sim = 0 \Leftrightarrow (x \to^L (y^\sim)^-)^\sim = 0 \overset{(pDN^L)}{\Leftrightarrow} (x \to^L y)^\sim = 0 \overset{(pN1^{Lb})}{\Leftrightarrow} x \to^L y = 1$ and
$y^- \odot x = 0 \Leftrightarrow (x \rightsquigarrow^L (y^-)^\sim)^- = 0 \overset{(pDN^L)}{\Leftrightarrow} (x \rightsquigarrow^L y)^- = 0 \overset{(pN1^{Lb})}{\Leftrightarrow} x \rightsquigarrow^L y = 1$.

(pEq^L) $\Longleftrightarrow$ (pEqP^{Lb}): By above proof,
$\Longrightarrow$: $x \leq y \Leftrightarrow x \to^L y = 1 \Leftrightarrow x \odot y^\sim = 0$, hence $x \leq y \Longleftrightarrow x \odot y^\sim = 0$.
$\Longleftarrow$: $x \leq y \Leftrightarrow x \odot y^\sim = 0 \Leftrightarrow x \to^L y = 1$, hence $x \leq y \Longleftrightarrow x \to^L y = 1$.

(pDNeg1^{Lb}) $\Longleftrightarrow$ (pDNP1) (See Corollary 2.11.6):
$x \to^L y = (x \odot y^\sim)^-$ and $y^- \rightsquigarrow^L x^- = ((x^-)^- \odot y^-)^\sim$; also
$x \rightsquigarrow^L y = (y^- \odot x)^\sim$ and $y^\sim \to^L x^\sim = (y^\sim \odot (x^\sim)^\sim)^-$.

(pDNeg2^{Lb}) $\Longleftrightarrow$ (pDNP2) (See Corollary 2.11.6):
$y^- \rightsquigarrow^L x = (x^- \odot y^-)^\sim$ and $x^\sim \to^L y = (x^\sim \odot y^\sim)^-$.

(pEx^L) + (pDNeg1^{Lb}) $\Longleftrightarrow$ (Ass) + (pDNP1):
$\Longrightarrow$: $x \odot (y \odot z) = x \odot (y \to^L z^-)^\sim = (x \to^L ((y \to^L z^-)^\sim)^-)^\sim \overset{(pDN^L)}{=}$
$(x \to^L (y \to^L z^-))^\sim \overset{(pDNeg1^{Lb})}{=} (x \to^L ((z^-)^- \rightsquigarrow^L y^-))^\sim \overset{(pEx^L)}{=}$
$((z^-)^- \rightsquigarrow^L (x \to^L y^-))^\sim$ and

$(x \odot y) \odot z = (x \to^L y^-)^\sim \odot z = ((x \to^L y^-)^\sim \to^L z^-)^\sim \overset{(pDNeg1^{Lb})}{=}$
$((z^-)^- \rightsquigarrow^L ((x \to^L y^-)^\sim)^-)^\sim \overset{(pDN^L)}{=} ((z^-)^- \rightsquigarrow^L (x \to^L y^-))^\sim$.
$\Longleftarrow$: $x \to^L (y \rightsquigarrow^L z) = x \to^L (z^- \odot y)^\sim = (x \odot ((z^- \odot y)^\sim)^\sim)^- \overset{(pDNP1)}{=}$
$((x^-)^- \odot ((z^- \odot y)^\sim)^-)^\sim \overset{(pDN^L)}{=} ((x^-)^- \odot (z^- \odot y))^\sim$ and
$y \rightsquigarrow^L (x \to^L z) = y \rightsquigarrow^L (x \odot z^\sim)^- = (((x \odot z^\sim)^-)^- \odot y)^\sim \overset{(pDNP1)}{=}$
$(((x^-)^- \odot z^-)^\sim)^- \odot y)^\sim \overset{(pDN^L)}{=} (((x^-)^- \odot z^-) \odot y)^\sim \overset{(Ass)}{=} ((x^-)^- \odot (z^- \odot y))^\sim$.

(pRe^L) $\Longleftrightarrow$ (pIv^L):
$x \odot x^\sim = 0 \Leftrightarrow (x \to^L (x^\sim)^-)^\sim = 0 \overset{(pDN^L)}{\Leftrightarrow} (x \to^L x)^\sim = 0 \overset{(pN1^{Lb})}{\Leftrightarrow} x \to^L x = 1$ and
$x^- \odot x = 0 \Leftrightarrow (x \rightsquigarrow^L (x^-)^\sim)^- = 0 \overset{(pDN^L)}{\Leftrightarrow} (x \rightsquigarrow^L x)^- = 0 \overset{(pN1^{Lb})}{\Leftrightarrow} x \rightsquigarrow^L x = 1$.

(q-pM^L) $\Longleftrightarrow$ (qm-pU) (See Proposition 5.1.18):
$\Longrightarrow$: $1 \odot (x \odot y) = 1 \odot (x \to^L y^-)^\sim = (1 \to^L [(x \to^L y^-)^\sim]^-)^\sim$
$\overset{(pDN^L)}{=} (1 \to^L (x \to^L y^-))^\sim \overset{(q-pM_d^L)}{=} (x \to^L y^-)^\sim = x \odot y$ and
$(x \odot y) \odot 1 = (y \rightsquigarrow^L x^\sim)^- \odot 1 = (1 \rightsquigarrow^L [(y \rightsquigarrow^L x^\sim)^-]^\sim)^-$
$\overset{(pDN^L)}{=} (1 \rightsquigarrow^L (y \rightsquigarrow^L x^\sim))^- \overset{(q-pM_d^L)}{=} (y \rightsquigarrow^L x^\sim)^- = x \odot y$. Thus, (qm-pU) holds.
$\Longleftarrow$: $1 \to^L (x \rightsquigarrow^L y) = 1 \to^L (y^- \odot x)^\sim = (1 \odot ((y^- \odot x)^\sim)^\sim)^- \overset{(qm-NegP^L)}{=}$
$1 \odot (((y^- \odot x)^\sim)^\sim)^- \overset{(pDN^L)}{=} 1 \odot (y^- \odot x)^\sim \overset{(qm-NegP^L)}{=} (1 \odot (y^- \odot x))^\sim \overset{(qm-pU)}{=}$
$(y^- \odot x)^\sim = x \rightsquigarrow^L y$ and
$1 \rightsquigarrow^L (x \to^L y) = 1 \rightsquigarrow^L (x \odot y^\sim)^- = (((x \odot y^\sim)^-)^- \odot 1)^\sim \overset{(qm-NegP^L)}{=}$
$(((x \odot y^\sim)^-)^-)^\sim \odot 1 \overset{(pDN^L)}{=} (x \odot y^\sim)^- \odot 1 \overset{(qm-NegP^L)}{=} ((x \odot y^\sim) \odot 1)^- \overset{(qm-pU)}{=}$
$(x \odot y^\sim)^- = x \to^L y$. Thus, (q-pM^L) holds.

(q-pNeg^L) $\Longleftrightarrow$ (qm-NegP^L):
$\Longrightarrow$: $(x \odot 1)^- = ((x \to^L 1^-)^\sim)^- \overset{(pDN^L)}{=} x \to^L 1^- \overset{(pN1^{Lb})}{=} x \to^L 0$,
$x^- \odot 1 = (1 \rightsquigarrow^L (x^-)^\sim)^- \overset{(pDN^L)}{=} (1 \rightsquigarrow^L x)^-$,
$1 \odot x^- = (x^- \rightsquigarrow^L 1^\sim)^- \overset{(pN1^{Lb})}{=} (x^- \rightsquigarrow^L 0)^- \overset{(q-pNeg^L)}{=} ((1 \to^L x^-)^\sim)^- \overset{(pDN^L)}{=}$
$1 \to^L x^-$,
$(1 \odot x)^- = ((1 \to^L x^-)^\sim)^- \overset{(pDN^L)}{=} 1 \to^L x^-$.
Also, $(x \odot 1)^\sim = ((1 \rightsquigarrow^L x^\sim)^-)^\sim \overset{(pDN^L)}{=} 1 \rightsquigarrow^L x^\sim$,
$x^\sim \odot 1 = (x^\sim \to^L 1^-)^\sim \overset{(pN1^{Lb})}{=} (x^\sim \to^L 0)^\sim \overset{(q-pNeg^L)}{=} ((1 \rightsquigarrow^L x^\sim)^-)^\sim \overset{(pDN^L)}{=}$
$1 \rightsquigarrow^L x^\sim$,
$1 \odot x^\sim = (1 \to^L (x^\sim)^-)^\sim \overset{(pDN^L)}{=} (1 \to^L x)^\sim$,
$(1 \odot x)^\sim = ((x \rightsquigarrow^L 1^\sim)^-)^\sim \overset{(pDN^L)}{=} x \rightsquigarrow^L 1^\sim \overset{(pN1^{Lb})}{=} x \rightsquigarrow^L 0$.
Then, by (q-pNeg^L), (qm-NegP^L) holds.
$\Longleftarrow$: $x \to^L 0 = (x \odot 0^\sim)^- \overset{(pN1^{Lb})}{=} (x \odot 1)^-$, $1 \to^L x^- = (1 \odot (x^-)^\sim)^- \overset{(pDN^L)}{=}$
$(1 \odot x)^-$, $(1 \rightsquigarrow^L x)^- = ((x^- \odot 1)^\sim)^- \overset{(pDN^L)}{=} x^- \odot 1$.

Also, $x \rightsquigarrow^L 0 = (0^- \odot x)^\sim \overset{(pN1^{Lb})}{=} (1 \odot x)^\sim$, $1 \rightsquigarrow^L x^\sim = ((x^\sim)^- \odot 1)^\sim \overset{(pDN^L)}{=} (x \odot 1)^\sim$, $(1 \to^L x)^\sim = ((1 \odot x^\sim)^-)^\sim \overset{(pDN^L)}{=} 1 \odot x^\sim$.
Then, by (qm-NegP^L), (q-pNeg^L) holds.

(pM1^L) + (q-pNeg^L) $\Longleftrightarrow$ (U1) + (qm-NegP^L):

$\Longrightarrow$: $1 \odot 1 = (1 \rightsquigarrow^L 1^\sim)^- \overset{(pN1^{Lb})}{=} (1 \rightsquigarrow^L 0)^- \overset{(q-pNeg^L)}{=} 1 \to^L 0^- = 1 \to^L 1 = 1$.

$\Longleftarrow$: $1 \to^L 1 = (1 \odot 1^\sim)^- \overset{(pN1^{Lb})}{=} (1 \odot 0)^- \overset{(qm-NegP^L)}{=} 1 \odot 0^- \overset{(pN1^{Lb})}{=} 1 \odot 1 = 1$.

(IdR^L) + (q-pNeg^L) $\Longleftrightarrow$ (IdP) + (qm-NegP^L):

$\Longrightarrow$: $1 \odot x = (1 \to^L x^-)^\sim \overset{(q-pNeg^L)}{=} ((1 \rightsquigarrow^x)^-)^\sim \overset{(pDN^L)}{=} 1 \rightsquigarrow^L x$ and
$x \odot 1 = (1 \rightsquigarrow^L x^\sim)^- \overset{(q-pNeg^L)}{=} ((1 \to^L x)^\sim)^- \overset{(pDN^L)}{=} 1 \to^L x$; then, by (IdR^L), (IdP) holds.

$\Longleftarrow$: $1 \to^L x = (1 \odot x^\sim)^- \overset{(qm-NegP^L)}{=} ((x \odot 1)^\sim)^- \overset{(pDN^L)}{=} x \odot 1$ and
$1 \rightsquigarrow^L x = (x^- \odot 1)^\sim \overset{(qm-NegP^L)}{=} ((1 \odot x)^-)^\sim \overset{(pDN^L)}{=} 1 \odot x$; then, by (IdP), (IdR^L) holds. □

Corollary 5.3.2

(1) Let $\mathcal{A} = (A, \leq, \to^L, \rightsquigarrow^L, ^-, ^\sim, 0, 1)$ be a bounded regular pseudo-structure with regular negations verifying (pN1^{Lb}), (pDN^L) and (pDNeg3^{Lb}).

Define $\Phi(\mathcal{A}) \overset{def.}{=} (A, \leq, \odot, ^-, ^\sim, 0, 1)$ by:
$x \odot y \overset{def.}{=} (x \to^L y^-)^\sim \overset{(pDNeg3^{Lb})}{=} (y \rightsquigarrow^L x^\sim)^-$.

Then, $\Phi(\mathcal{A})$ is a bounded regular-m structure with regular-m negations verifying (IdEqP^{Lb}), (pEqP^{Lb}), (pN1^{Lb}), (pDN^L).

(1') Conversely, let $\mathcal{A} = (A, \leq, \odot, ^-, ^\sim, 0, 1)$ be a bounded regular-m structure with regular-m negations verifying (IdEqP^{Lb}), (pEqP^{Lb}), (pN1^{Lb}), (pDN^L).

Define $\Psi(\mathcal{A}) \overset{def.}{=} (A, \leq, \to^L, \rightsquigarrow^L, ^-, ^\sim, 0, 1)$ by:
$x \to^L y \overset{def.}{=} (x \odot y^\sim)^-$, $\quad x \rightsquigarrow^L y \overset{def.}{=} (y^- \odot x)^\sim$.

Then, $\Psi(\mathcal{A})$ is a bounded regular pseudo-structure with regular negations verifying (pN1^{Lb}), (pDN^L), (pDNeg3^{Lb}).

(2) The maps Φ and Ψ are mutually inverse.

Proof. (1): By hypothesis, $\mathcal{A}$ verifies (IdEq^L), (pEq^L), (pM^L), (F), (L), (pN1^{Lb}), (pDN^L). Since $\leq$ is an order, then (pRe^L) holds. Then, by Theorem 5.3.1 (3), $\Phi(\mathcal{A})$ verifies (IdEqP^{Lb}), (pEqP^{Lb}), (pU), (F), (L), (pN1^{Lb}), (pDN^L) and (pIv^L), respectively. Hence, $\Phi(\mathcal{A})$ is a bounded regular-m structure with regular-m negations verifying (IdEqP^{Lb}), (pEqP^{Lb}), (pN1^{Lb}), (pDN^L).

(1'): By Theorem 5.3.1 (1'), $\Psi(\mathcal{A})$ verifies (pN1^{Lb}), (pDN^L) and (pDNeg3^{Lb}). By hypothesis, $\mathcal{A}$ verifies (pU), (F), (L), (pIv^L), (IdEqP^{Lb}), (pEqP^{Lb}). Then, by Theorem 5.3.1 (3), $\Psi(\mathcal{A})$ verifies (pM^L), (F), (L), (pRe^L), (IdEq^L), (pEq^L), respectively. We then obtain:
$x \to^L 0 = (x \odot 0^\sim)^- \overset{(pN1^{Lb})}{=} (x \odot 1)^- \overset{(pU)}{=} x^-$ and
$x \rightsquigarrow^L 0 = (0^- \odot x)^\sim \overset{(pN1^{Lb})}{=} (1 \odot x)^\sim \overset{(pU)}{=} x^\sim$,

hence (pNeg^L) holds too. Hence, $\Psi(\mathcal{A})$ is a bounded regular pseudo-structure with regular negations verifying (pN1^{Lb}), (pDN^L), (pDNeg3^{Lb}).

(2): By Theorem 5.3.1 (2). □

Corollary 5.3.3

(1) Let $\mathcal{A} = (A, \leq, \to^L, \rightsquigarrow^L, ^-, ^\sim, 0, 1)$ *be a bounded quasi-pseudo-structure with quasi-negations verifying (pN1^{Lb}), (pDN^L), (pDNeg3^{Lb}) and (pEx^L).*

Define $\Phi(\mathcal{A}) \overset{def.}{=} (A, \leq, \odot, ^-, ^\sim, 0, 1)$ *by:*

$x \odot y \overset{def.}{=} (x \to^L y^-)^\sim \overset{(pDNeg3^{Lb})}{=} (y \rightsquigarrow^L x^\sim)^-$.

Then, $\Phi(\mathcal{A})$ *is a bounded quasi-m structure with quasi-m negations verifying (pN1^{Lb}), (pDN^L) and also (IdEqP^{Lb}), (pEqP^{Lb}), (Ass).*

(1') Conversely, let $\mathcal{A} = (A, \leq, \odot, ^-, ^\sim, 0, 1)$ *be a bounded quasi-m structure with quasi-m negations verifying (pN1^{Lb}), (pDN^L) and (IdEqP^{Lb}), (pEqP^{Lb}), (Ass).*

Define $\Psi(\mathcal{A}) \overset{def.}{=} (A, \leq, \to^L, \rightsquigarrow^L, ^-, ^\sim, 0, 1)$ *by:*

$x \to^L y \overset{def.}{=} (x \odot y^\sim)^-, \quad x \rightsquigarrow^L y \overset{def.}{=} (y^- \odot x)^\sim$.

Then, $\Psi(\mathcal{A})$ *is a bounded quasi-pseudo-structure with quasi-negations verifying (pN1^{Lb}), (pDN^L), (pDNeg3^{Lb}) and (pEx^L).*

(2) The maps Φ *and* Ψ *are mutually inverse.*

(3) $R(A) = Rm(A)$.

(4) $C(x) = C_m(x)$, *for all* $x \in A$, *and contains exactly one regular (regular-m) element.*

Proof. (1): By hypothesis, $\mathcal{A}$ satisfies (IdEq^L), (pEq^L), (q-pM^L), (pM1^L), (q-Neg^L), (F), (L) hold and also (pN1^{Lb}), (pDN^L), (pDNeg3^{Lb}) and (pEx^L). Then, by Theorem 5.3.1 (3), $\Phi(\mathcal{A})$ verifies (IdEqP^{Lb}), (pEqP^{Lb}), (qm-pU), (U1), (qm-NegP^L), (F), (L) and also (pN1^{Lb}), (pDN^L) and (Ass), respectively. Hence, $\Phi(\mathcal{A})$ is a bounded quasi-m structure with quasi-m negations verifying the specified properties.

(1'): By hypothesis, $\mathcal{A}$ satisfies (qm-pU), (U1), (qm-NegP^L), (F), (L) and also (pN1^{Lb}), (pDN^L) and (IdEqP^{Lb}), (pEqP^{Lb}), (Ass). Then, by Theorem 5.3.1 (3), $\Psi(\mathcal{A})$ verifies (q-pM^L), (pM1^L), (q-Neg^L), (F), (L) and also (pN1^{Lb}), (pDN^L) and (IdEq^L), (pEq^L), (pEx^L), respectively. Hence, $\Psi(\mathcal{A})$ is a bounded quasi-pseudo-structure with quasi-negations verifying the specified properties.

(2): Obviously, by Theorem 5.3.1 (2).

(3): By Theorem 5.3.1 (3), (pM^L) $\Longleftrightarrow$ (pU), hence $R(A) = Rm(A)$, by definitions.

(4): $\Longrightarrow$: Since (pEx^L) holds, then (IdR^L) holds, by Proposition 5.1.19; hence, by Corollary 5.1.27,

$x \parallel y \iff 1 \to^L x \ (\overset{(IdR^L)}{=} 1 \rightsquigarrow^L x) = 1 \to^L y \ (\overset{(IdR^L)}{=} 1 \rightsquigarrow^L y)$.

But, by Theorem 5.3.1 (3), (IdR^L) + (q-pNeg^L) $\Longleftrightarrow$ (IdP) + (qm-NegP^L), hence (IdP) holds; hence, by Corollary 5.2.22,

$x \parallel_m y \iff 1 \odot x \ (\overset{(IdP)}{=} x \odot 1) = 1 \odot y \ (\overset{(IdP)}{=} y \odot 1)$.

But, $1 \to^L x = x \odot 1$ and $1 \rightsquigarrow^L x = 1 \odot x$, by (q-pNeg^L). Thus, $x \parallel y \Longrightarrow x \parallel_m y$.

$\Longleftarrow$: Since (Ass) holds, then (IdP) holds, by Proposition 5.2.17; by Corollary 5.2.22, $x \parallel_m y \iff 1 \odot x \ (\overset{(IdP)}{=} \ x \odot 1) = 1 \odot y \ (\overset{(IdP)}{=} \ y \odot 1)$.
But, by Theorem 5.3.1 (3), (IdR^L) + (q-pNeg^L) $\iff$ (IdP) + (qm-NegP^L), hence (IdR^L) holds; hence, by Corollary 5.1.27,
$x \parallel y \iff 1 \to^L x \ (\overset{(IdR^L)}{=} \ 1 \rightsquigarrow^L x) = 1 \to^L y \ (\overset{(IdR^L)}{=} \ 1 \rightsquigarrow^L y)$.
But, $x \odot 1 = 1 \to^L x$ and $1 \odot x = 1 \rightsquigarrow^L x$, by (qm-NegP^L). Thus, $x \parallel_m y \Longrightarrow x \parallel y$.
Thus, $C(x) = C_m(x)$, for all $x \in A$.
Note now that, by Lemma 5.1.31, since (pEx^L) holds, then every cloud contains exactly one regular element, and, by Lemma 5.2.26, since (Ass) holds, then every cloud-m contains exactly one regular-m element. But the clouds coincide with the m-clouds and the regular elements coincide with the regular-m elements, by above (3). □

The two "worlds", of (quasi-) pM algebras and of (quasi-m) unital magmas (with additional operation(s)), can be represented in two different ways, as the next Figures show. Examples of connections between the two "worlds" are presented in Chapter 12.

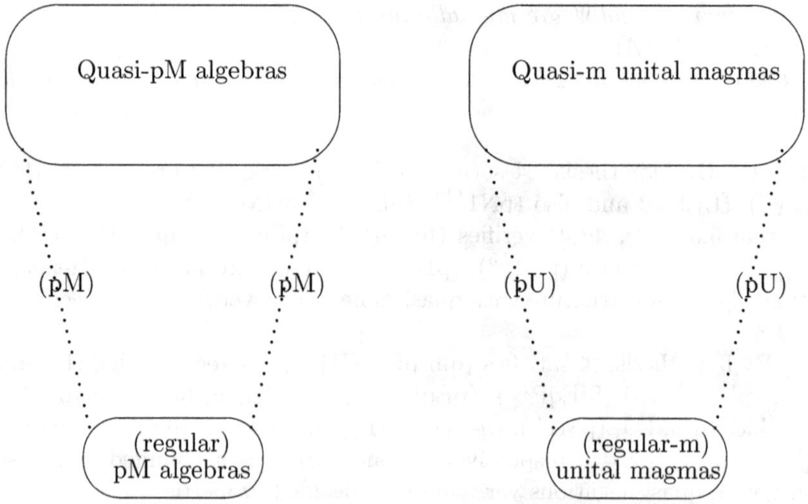

Figure 5.2: The two "worlds" - the first representation

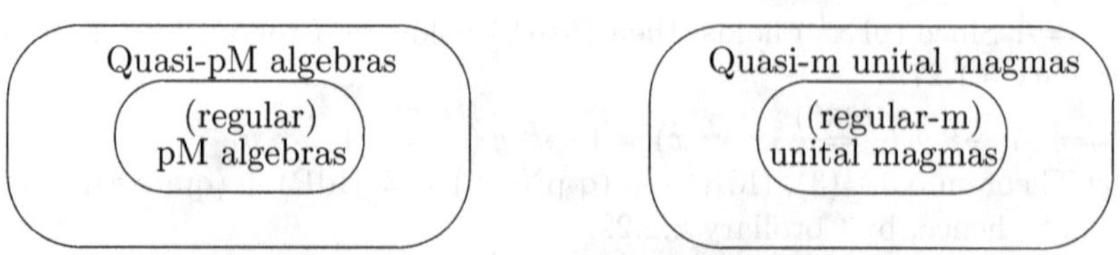

Figure 5.3: The two "worlds" - the second representation

Chapter 6

Examples of the two kinds of algebras

In this chapter, we present some examples of (quasi-) pseudo-M algebras and of (quasi-m) unital magmas. We also present some programs in the PASCAL programming language to check some properties.

The chapter has 4 sections.

6.1 Examples of (quasi-) pM algebras

In this section, we present examples of left-pseudo-M algebras (Examples 0 - 15) and examples of quasi-BCI(P) algebras (Example 16).

• Examples 0: Proper pM, pML, pME, pMEL algebras

Consider the set $A = \{a, b, 1\}$ organized as a structure in the following four ways:
$\mathcal{A}_1 = (A, \leq_1, \rightarrow_1, \rightsquigarrow_1, 1)$, $\mathcal{A}_2 = (A, \leq_2, \rightarrow_2, \rightsquigarrow_2, 1)$,
$\mathcal{A}_3 = (A, \leq_3, \rightarrow_3, \rightsquigarrow_3, 1)$, $\mathcal{A}_4 = (A, \leq_4, \rightarrow_4, \rightsquigarrow_4, 1)$,
where the tables of pseudo-implications and the Hasse-type diagrams (see Remark 2.1.21) are the following, respectively:

$\mathcal{A}_1$

$\rightarrow_1$	a	b	1
a	a	a	a
b	a	a	a
1	**a**	**b**	**1**

$\rightsquigarrow_1$	a	b	1
a	b	b	b
b	b	b	b
1	**a**	**b**	**1**

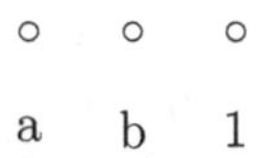

where (pEq^L) $x \leq_1 y \Longleftrightarrow x \rightarrow_1 y = 1 \overset{(IdEq^L)}{\Longleftrightarrow} x \rightsquigarrow_1 y = 1$.

$\mathcal{A}_2$

$\rightarrow_2$	a	b	1
a	a	a	1
b	a	a	1
1	**a**	**b**	**1**

$\rightsquigarrow_2$	a	b	1
a	b	b	1
b	b	b	1
1	**a**	**b**	**1**

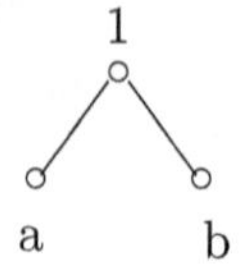

where (pEq^L) $x \leq_2 y \Longleftrightarrow x \rightarrow_2 y = 1 \overset{(IdEq^L)}{\Longleftrightarrow} x \rightsquigarrow_2 y = 1$.

$\mathcal{A}_3$

$\rightarrow_3$	a	b	1
a	a	a	a
b	a	a	a
1	**a**	**b**	**1**

$\rightsquigarrow_3$	a	b	1
a	a	a	a
b	a	a	a
1	**a**	**b**	**1**

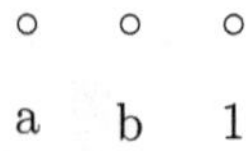

where (pEq^L) $x \leq_3 y \Longleftrightarrow x \rightarrow_3 y = 1 \overset{(IdEq^L)}{\Longleftrightarrow} x \rightsquigarrow_3 y = 1$.

$\mathcal{A}_4$

$\rightarrow_4$	a	b	1
a	a	a	1
b	a	a	1
1	**a**	**b**	**1**

$\rightsquigarrow_4$	a	b	1
a	a	a	1
b	a	a	1
1	**a**	**b**	**1**

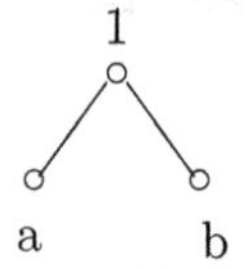

where (pEq^L) $x \leq_4 y \Longleftrightarrow x \rightarrow_4 y = 1 \overset{(IdEq^L)}{\Longleftrightarrow} x \rightsquigarrow_4 y = 1$.

Then, $\mathcal{A}_1$ is a proper left-pM algebra, $\mathcal{A}_2$ is a proper left-pML algebra, $\mathcal{A}_3$ is a proper left-pME algebra, $\mathcal{A}_4$ is a proper left-pMEL algebra.

• **Example 1: Bounded BCK(P) lattice**

Consider the bounded Ore lattice $(L_2 = \{0, 1\}, \leq, 0, 1)$, with greatest element 1 and smallest element 0, represented by the Hasse diagram given in Figure 6.1.

Figure 6.1: The bounded lattice $(\mathcal{L}_2, \leq, 0, 1)$

The equivalent $(x \leq y \Leftrightarrow x \wedge y = x \Leftrightarrow x \vee y = y)$ bounded Dedekind lattice $(L_2, \wedge = \min, \vee = \max, 0, 1)$ has the following tables for $\wedge$ and $\vee$:

$\mathcal{L}_2$

$\wedge = \min$	**0**	**1**
0	**0**	**0**
1	**0**	**1**

$\vee = \max$	**0**	**1**
0	**0**	**1**
1	**1**	**1**

Define on L_2 the following implication $\rightarrow^L$:

$\mathcal{L}_2$

$\rightarrow^L$	**0**	**1**
0	**1**	**1**
1	**0**	**1**

$\mathcal{L}_2$

$\odot (= \wedge)$	**0**	**1**
0	**0**	**0**
1	**0**	**1**

Then, $\mathcal{L}_2 = (L_2, \leq, \to^L, 0, 1)$ is a bounded BCK lattice, with product $\odot$ (with (P)), where: for all $x, y \in L_2$,
(P) (product) $\exists\, x \odot y \overset{notation}{=} \min\{z \mid x \leq y \to^L z\}$.
We thus obtain the above table of the product $\odot$.
Hence, $\mathcal{L}_2 = (L_2, \leq, \to^L, 0, 1)$ **is a bounded BCK(P) lattice** (see Example 1' in a next section). We shall see later (Example 14 in this section, Example 14' in a next section) that $\mathcal{L}_2$ is just (term-equivalent with) the standard implicative-Boolean algebra $\mathcal{L}_2$.

• **Example 2: BCI(P) algebra**
By using Theorem 2.3.16, we obtain from the above bounded BCK(P) lattice $(L_2, \leq, \to^L, 0, 1)$ a BCI(P) algebra. Namely, consider an element $b \notin L_2$ and consider the poset $(M' = L_2 \cup \{b\}, \leq)$, represented in Figure 6.2.

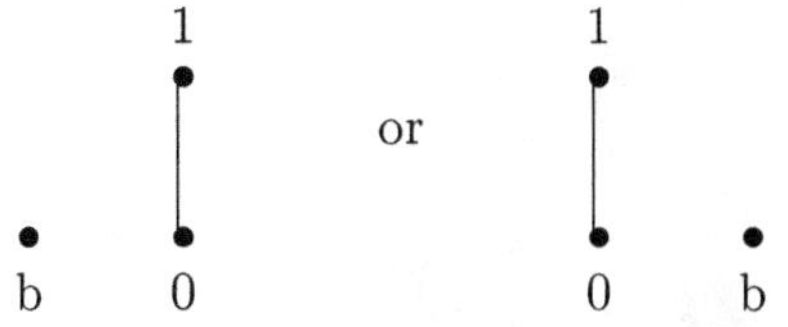

Figure 6.2: The poset $(M' = L_2 \cup \{b\}, \leq)$

Define on M' the implication $\to'$ as follows (by Theorem 2.3.16 again):

$\to'$	**0**	b	**1**
0	**1**	b	**1**
b	b	1	b
1	**0**	b	**1**

$\odot'$	**0**	b	**1**
0	**0**	b	**0**
b	b	0	b
1	**0**	b	**1**

.

Then, $\mathcal{M}' = (M', \leq, \to', 1)$ **is a BCI algebra, with product** $\odot'$ **(with (P))**, where: for all $x, y \in L_2$,
(P) (product) $\exists\, x \odot' y \overset{notation}{=} \min\{z \mid x \leq y \to' z\}$.
We thus obtain the corresponding table of the product $\odot'$ presented above.
Note that $M'^{-} = \{0, 1\}$, while $M'^{+} = \{1\}$.

• **Example 3: Left-BCK algebra (without (P))**
Consider the poset $(A = \{c, d, 1\}, \leq, 1)$ with greatest element 1, whose Hasse diagram is given in Figure 6.3.

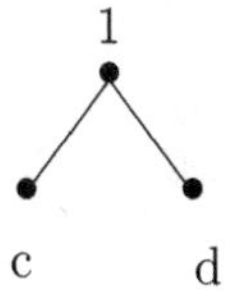

Figure 6.3: The poset $(A = \{c, d, 1\}, \leq, 1)$ with 1

Consider the implication $\to$ given in the following table:

$\mathcal{A}$

$\rightarrow$	c	d	1
c	1	d	1
d	c	1	1
1	c	d	1

.

Then $\mathcal{A} = (A, \leq, \rightarrow, 1)$ is a **left-BCK algebra**, without product (without (P)), since, for example: $c \odot d = \min\{z \mid c \leq d \rightarrow z\} = \min\{c, d, 1\}$ does not exist.

• **Example 4: Left-BCI algebra (without (P))**

By using Theorem 2.3.16, we obtain from the above left-BCK algebra $(A, \leq, \rightarrow, 1)$ a left-BCI algebra. Namely, we consider an element $m \notin A$ and the poset $(A' = \{c, d, 1\} \cup \{m\}, \leq)$, represented by the Hasse diagram from Figure 6.4.

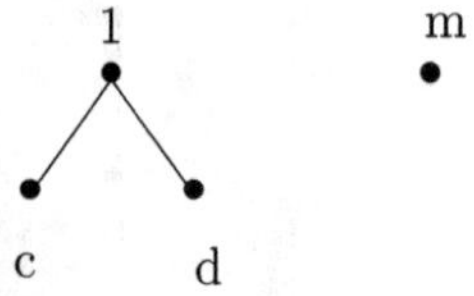

Figure 6.4: The poset $(A' = A \cup \{m\}, \leq)$

Consider the implication $\rightarrow'$ defined on A' as follows (by Theorem 2.3.16 again):

$\mathcal{A}'$

$\rightarrow'$	c	d	m	1
c	1	d	m	1
d	c	1	m	1
m	m	m	1	m
1	c	d	m	1

.

Then, $\mathcal{A}' = (A', \leq, \rightarrow', 1)$ **is a left-BCI algebra**, without product (without (P)), since: $c \odot' d = \min\{z \mid c \leq d \rightarrow' z\} = \min\{c, d, 1\}$ does not exist.

Note that $A'^{-} = \{c, d, 1\}$, while $A'^{+} = \{1\}$.

• **Example 5: Implicative structure which is not a BCI lattice**

Consider the lattice $(\mathbf{Z} = \{\ldots, -3, -2, -1, 0, 1, 2, 3, \ldots\}, \leq)$, whose Hasse diagram is presented in next Figure 6.5.

Consider also the implication $\rightarrow$ defined on $\mathbf{Z}$ by the table presented below and the structure $(\mathbf{Z}, \wedge = \min, \vee = \max, \rightarrow, \mathbf{1} = 0)$ with product $\odot$, where the product $\odot$ is defined as follows: for all $x, y \in \mathbf{Z}$,
(P) (product) $\exists\, x \odot y \overset{not.}{=} \min\{z \mid x \leq y \rightarrow^{L} z\}$,

and its table is presented below also.

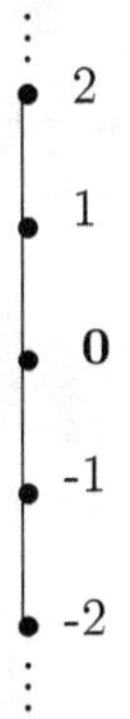

Figure 6.5: The lattice $(\mathbf{Z}, \leq)$

$\rightarrow$	...	-3	-2	-1	**0**	1	2	3	...
⋮	...	⋮	⋮	⋮	⋮	⋮	⋮	⋮	...
-3	...	**0**	1	2	3	4	5	6	...
-2	...	-1	**0**	1	2	3	4	5	...
-1	...	-2	-1	**0**	1	2	3	4	...
0	...	-3	-2	-1	**0**	1	2	3	...
1	...	-4	-3	-2	-1	**0**	1	2	...
2	...	-5	-4	-3	-2	-1	**0**	1	...
3	...	-6	-5	-4	-3	-2	-1	**0**	...
⋮	...	⋮	⋮	⋮	⋮	⋮	⋮	⋮	...

$\odot$	...	-3	-2	-1	**0**	1	2	3	...
⋮	...	⋮	⋮	⋮	⋮	⋮	⋮	⋮	...
-3	...	-6	-5	-4	-3	-2	-1	**0**	...
-2	...	-5	-4	-3	-2	-1	**0**	1	...
-1	...	-4	-3	-2	-1	**0**	1	2	...
0	...	-3	-2	-1	**0**	1	2	3	...
1	...	-2	-1	**0**	1	2	3	4	...
2	...	-1	**0**	1	2	3	4	5	...
3	...	**0**	1	2	3	4	5	6	...
⋮	...	⋮	⋮	⋮	⋮	⋮	⋮	⋮	...

Remark that the implication $\rightarrow$ verifies the properties (pBB^L), (pM^L), but does not verify the property (Eq$^{\leq}$) ($-3 \leq 2$ and $-3 \rightarrow 2 = 5 \neq 0$), hence $\rightarrow$ is not a residuum, conform with Definition 1.1.3, and hence **the above defined structure is not a BCI algebra, more precisely is not a BCI lattice**. We shall see (Example 1 of Chapter 9) that this structure is an l-implicative-group.

Note that $\mathbf{Z}^- = \{\ldots, -3, -2, -1, 0\}$ and $\mathbf{Z}^+ = \{0, 1, 2, 3, \ldots\}$.

- **Example 6: Infinite BCK(P) lattice**

Consider the lattice ($\mathbf{Z}^- = \{\ldots, -3, -2, -1, 0\}, \leq, \mathbf{1} = 0$) with greatest element 0, whose Hasse diagram is presented in Figure 6.6.

Figure 6.6: The lattice $(\mathbf{Z}^-, \leq, 0)$

Consider also the implication $\to^L$ defined on $\mathbf{Z}^-$, with the table presented below.

Then, $(\mathbf{Z}^-, \wedge = \min, \vee = \max, \to^L, \mathbf{1} = 0)$ **is a left-BCK lattice with product, i.e. is a BCK(P) lattice**, where the product $\odot$ is defined as follows: for all $x, y \in \mathbf{Z}^-$,

(P) (product) $\exists\, x \odot y \overset{notation}{=} \min\{z \mid x \leq y \to^L z\}$,

and its table is presented below also.

$\to^L$	...	-3	-2	-1	0
$\vdots$	...	$\vdots$	$\vdots$	$\vdots$	$\vdots$
-3	...	0	0	0	0
-2	...	-1	0	0	0
-1	...	-2	-1	0	0
0	...	-3	-2	-1	0

$\odot$	...	-3	-2	-1	0
$\vdots$	...	$\vdots$	$\vdots$	$\vdots$	$\vdots$
-3	...	-6	-5	-4	-3
-2	...	-5	-4	-3	-2
-1	...	-4	-3	-2	-1
0	...	-3	-2	-1	0

- **Example 7: Infinite bounded BCK(P) lattice**

Starting from the previous BCK(P) lattice $(\mathbf{Z}^-, \wedge = \min, \vee = \max, \to^L, \mathbf{1} = 0)$, let us consider the symbol $-\infty \notin \mathbf{Z}^-$ and define $Z^-_{-\infty} \overset{def.}{=} \{-\infty\} \cup \mathbf{Z}^-$. Extend the lattice order relation $\leq$ from $\mathbf{Z}^-$ to $Z^-_{-\infty}$ as presented in Figure 6.7.

Extend the operation $\to^L$ on $\mathbf{Z}^-$ to $\to^L_2$ on $Z^-_{-\infty}$ as follows:

$$x \to^L_2 y = \begin{cases} x \to^L y, & \text{if } x, y \in \mathbf{Z}^-, \\ -\infty, & \text{if } x \in \mathbf{Z}^-, y = -\infty, \\ 0, & \text{if } x = -\infty. \end{cases}$$

Hence, $\to^L_2$ has the table presented below.

Then, $(Z^-_{-\infty}, \min, \max, \to^L_2, -\infty, 0)$ **is a bounded BCK lattice with product $\odot_2$, i.e. is a bounded BCK(P) lattice**, where:

(P) (product) $\exists\, x \odot_2 y \overset{notation}{=} \min\{z \mid x \leq y \to^L_2 z\}$,

hence the resulting table of $\odot_2$ is presented below also.

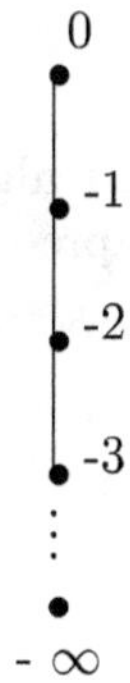

Figure 6.7: The bounded lattice $(\{-\infty\} \cup \mathbf{Z}^-, \leq, -\infty, 0)$

$\to_2^L$	$-\infty$	$\dots$	-3	-2	-1	0
$-\infty$	0	$\dots$	0	0	0	0
$\vdots$	$\vdots$	$\dots$	$\vdots$	$\vdots$	$\vdots$	$\vdots$
-3	$-\infty$	$\dots$	0	0	0	0
-2	$-\infty$	$\dots$	-1	0	0	0
-1	$-\infty$	$\dots$	-2	-1	0	0
0	$-\infty$	$\dots$	-3	-2	-1	0

$\odot_2$	$-\infty$	$\dots$	-3	-2	-1	0
$-\infty$	$-\infty$	$\dots$	$-\infty$	$-\infty$	$-\infty$	$-\infty$
$\vdots$	$\vdots$	$\dots$	$\vdots$	$\vdots$	$\vdots$	$\vdots$
-3	$-\infty$	$\dots$	-6	-5	-4	-3
-2	$-\infty$	$\dots$	-5	-4	-3	-2
-1	$-\infty$	$\dots$	-4	-3	-2	-1
0	$-\infty$	$\dots$	-3	-2	-1	0

Note that $(Z^-_{-\infty}, \min, \max, \to_2^L, -\infty, 0)$ is in fact just a linearly-ordered Hájek(P) algebra (with the product $\odot_2$) verifying (P1^L), (P2^L) (see [31]; see [94], pag. 182; see Example 1 of Chapter 9).

• **Example 8: Infinite BCI(P) algebra**

Starting from the above bounded BCK(P) lattice $(Z^-_{-\infty}, \min, \max, \to_2^L, -\infty, 0)$, we obtain a BCI(P) by Theorem 2.3.16. Namely, consider the element $m \notin Z^-_{-\infty}$ and the poset $(Z^{-'}_{-\infty} = Z^-_{-\infty} \cup \{m\}, \leq)$ represented by the Hasse diagram from Figure 6.8.

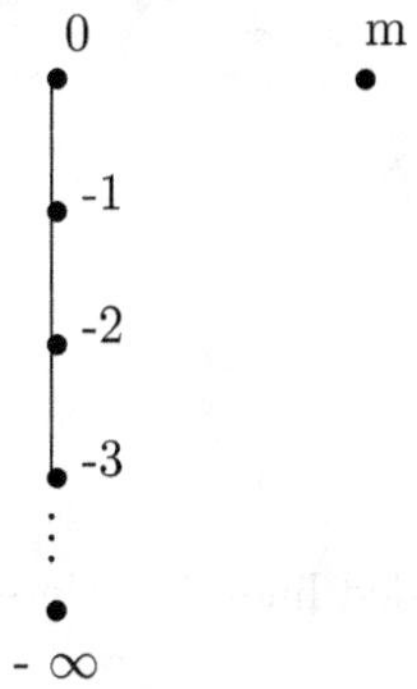

Figure 6.8: The poset $(Z^{-'}_{-\infty} = \{-\infty\} \cup \mathbf{Z}^- \cup \{m\}, \leq)$

Define on $Z^{-'}_{-\infty}$ the new operation $\to_2^{L'}$, by Theorem 2.3.16 again; the table of

$\rightarrow_2^{L'}$ is given below.

Then, $(Z_{-\infty}^{-'}, \leq, \rightarrow_2^{L'}, 1 = 0)$ **is left-BCI algebra with product** $\odot_2'$, **i.e. is a BCI(P) algebra**, the table of $\odot_2'$ being presented below also.

$\rightarrow_2^{L'}$	$-\infty$	...	-3	-2	-1	m	0
$-\infty$	0	...	0	0	0	m	0
⋮	⋮	...	⋮	⋮	⋮	⋮	⋮
-3	$-\infty$	...	0	0	0	m	0
-2	$-\infty$	...	-1	0	0	m	0
-1	$-\infty$	...	-2	-1	0	m	0
m	m	...	m	m	m	0	m
0	$-\infty$	...	-3	-2	-1	m	0

$\odot_2'$	$-\infty$	...	-3	-2	-1	m	0
$-\infty$	$-\infty$	...	$-\infty$	$-\infty$	$-\infty$	m	$-\infty$
⋮	⋮	...	⋮	⋮	⋮	⋮	⋮
-3	$-\infty$	...	-6	-5	-4	m	-3
-2	$-\infty$	...	-5	-4	-3	m	-2
-1	$-\infty$	...	-4	-3	-2	m	-1
m	m	...	m	m	m	$-\infty$	m
0	$-\infty$	...	-3	-2	-1	m	0

.

Note that $(Z_{-\infty}^{-'})^- = \{-\infty, \ldots, -3, -2, -1, 0\}$, while $(Z_{-\infty}^{-'})^+ = \{0\}$.

• **Example 9: Bounded pseudo-BCK(pP) lattice**

Consider the bounded, linearly-ordered Ore lattice $(L_4 = \{0, a, b, 1\}, \leq, 0, 1)$, whose Hasse diagram is presented in Figure 6.9.

Figure 6.9: The bounded linearly-ordered Ore lattice $(L_4, \leq, 0, 1)$

The equivalent bounded, linearly-ordered Dedekind lattice $(L_4, \wedge = \min, \vee = \max, 0, 1)$ has the following tables of $\wedge$ and $\vee$:

$\mathcal{L}_4$

$\wedge = \min$	0	a	b	1
0	0	0	0	0
a	0	a	a	a
b	0	a	b	b
1	0	a	b	1

$\vee = \max$	0	a	b	1
0	0	a	b	1
a	a	a	b	1
b	0	b	b	1
1	1	1	1	1

.

Consider also defined on L_4 the following implications $\to^L$ and $\rightsquigarrow^L$:

$\mathcal{L}_4$

$\to^L$	0	a	b	1
0	1	1	1	1
a	a	1	1	1
b	a	a	1	1
1	0	a	b	1

$\rightsquigarrow^L$	0	a	b	1
0	1	1	1	1
a	b	1	1	1
b	0	a	1	1
1	0	a	b	1

$\odot$	0	a	b	1
0	0	0	0	0
a	0	0	0	a
b	0	a	b	b
1	0	a	b	1

.

Then, $\mathcal{L}_4 = (L_4, \wedge = \min, \vee = \max, \to^L, \rightsquigarrow^L, 0, 1)$ is a bounded pBCK lattice with pseudo-product (with (pP)), where for all $x, y \in L_4$:
(pP) $\exists\, x \odot y \overset{notation}{=} \min\{z \mid x \leq y \to^L z\} = \min\{z \mid y \leq x \rightsquigarrow^L z\}$,
and the resulting table of $\odot$ is presented above also.

Hence, $\mathcal{L}_4 = (L_4, \wedge = \min, \vee = \max, \to^L, \rightsquigarrow^L, 0, 1)$ **is a bounded pBCK(pP) lattice**, namely one verifying (pprel^L):
(pprel^L) $(x \to^L y) \vee (y \to^L x) = 1 = (x \rightsquigarrow^L y) \vee (y \rightsquigarrow^L x)$,
and not verifying (pdiv^L):
(pdiv^L) $x \wedge y = (x \to^L y) \odot x = x \odot (x \rightsquigarrow^L y)$,
since, for example, $a = b \wedge a \neq (b \to^L a) \odot b = 0$.

Consequently, $\mathcal{L}_4$ is a pseudo-$\alpha\beta$ algebra, namely one verifying the property (pWNM) (pseudo-Weak Nilpotent Minimum) (see [94], pag. 491):
(pWNM) $(x \odot y)^- \vee [(x \wedge y) \to^L (x \odot y)] = 1 = (x \odot y)^\sim \vee [(x \wedge y) \rightsquigarrow^L (x \odot y)]$,
where $x^- = x \to^L 0$ and $x^\sim = x \rightsquigarrow^L 0$.

• **Example 10: Pseudo-BCI(pP) algebra**

By Theorem 2.3.16, we shall obtain from the above bounded pBCK(pP) lattice $(L_4, \min, \max, \to^L, \rightsquigarrow^L, 0, 1)$ a pBCI algebra (with (pP)). Namely, consider an element $c \notin L_4$ and the poset $(L'_4 = L_4 \cup \{c\}, \leq)$ represented in Figure 6.10.

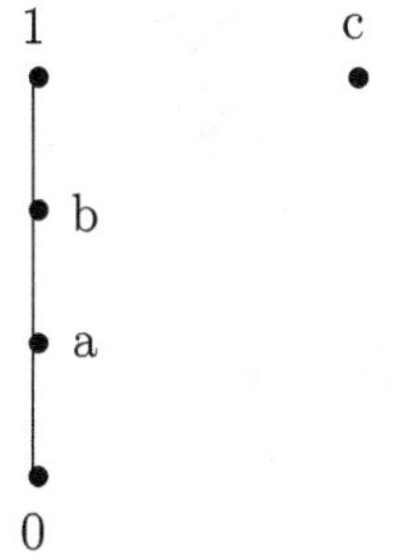

Figure 6.10: The poset $(L'_4 = L_4 \cup \{c\}, \leq)$

By Theorem 2.3.16 again, consider the new implications $\to'$ and $\rightsquigarrow'$, defined by the following tables:

$\mathcal{L}'_4$

$\to'$	0	a	b	c	1
0	1	1	1	c	1
a	a	1	1	c	1
b	a	a	1	c	1
c	c	c	c	1	c
1	0	a	b	c	1

$\rightsquigarrow'$	0	a	b	c	1
0	1	1	1	c	1
a	b	1	1	c	1
b	0	a	1	c	1
c	c	c	c	1	c
1	0	a	b	c	1

$\odot'$	0	a	b	c	1
0	0	0	0	c	0
a	0	0	0	c	a
b	0	a	b	c	b
c	c	c	c	0	c
1	0	a	b	c	1

.

Then, $\mathcal{L}'_4 = (L'_4 = \{0, a, b, c, 1\}, \leq, \to', \rightsquigarrow', 1)$ **is a pBCI(pP) algebra**, with the above pseudo-product $\odot'$ defined by: for all $x, y \in L'_4$,
(pP) (pseudo-product) $\exists\, x \odot' y \overset{notation}{=} \min\{z \mid x \leq y \to' z\} = \min\{z \mid y \leq x \rightsquigarrow' z\}$.

Note that $L_4^{'-} = \{0, a, b, 1\}$, while $L_4^{'+} = \{1\}$.

• **Example 11: Left-pBCK algebra (without (pP))**

Consider the ordinal product $\mathcal{A}_6 = \mathcal{L}_4 \bigodot \mathcal{A}$, of the bounded pBCK(pP) lattice $\mathcal{L}_4$ from Example 9 and the left-BCK algebra without (P) $\mathcal{A}$ from Example 3, whose Hasse diagram is presented in Figure 6.11 and whose operations $\to, \rightsquigarrow$ defined on $A_6 = \{0, a, b, c, d, 1\}$ are given by the next tables.

$\mathcal{A}_6$

$\to$	0	a	b	c	d	1
0	1	1	1	c	d	1
a	a	1	1	c	d	1
b	a	a	1	c	d	1
c	0	a	b	1	d	1
d	0	a	b	c	1	1
1	0	a	b	c	d	1

$\rightsquigarrow$	0	a	b	c	d	1
0	1	1	1	c	d	1
a	b	1	1	c	d	1
b	0	a	1	c	d	1
c	0	a	b	1	d	1
d	0	a	b	c	1	1
1	0	a	b	c	d	1

.

Then, $\mathcal{A}_6 = (A_6, \leq, \to, \rightsquigarrow, 1)$ is a **left-pBCK algebra**, without pseudo-product (without (pP)), since, for example, $c \odot d \overset{notation}{=} \min\{z \mid c \leq d \to z\} = \min\{c, d, 1\}$ does not exist.

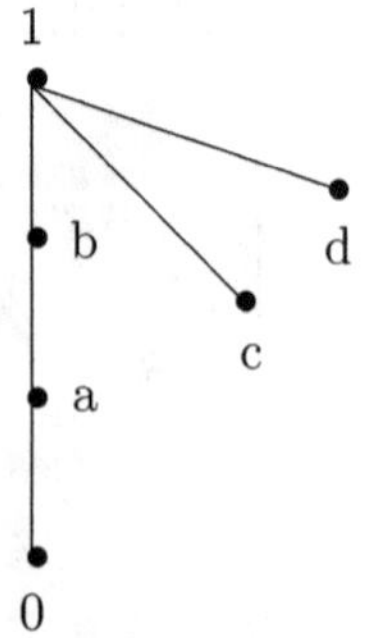

Figure 6.11: The poset $(A_6, \leq, 1)$ with greatest element 1

• **Example 12: Left-pBCI algebra (without (pP))**

By Theorem 2.3.16, we shall obtain a left-pBCI algebra from the above left-pBCK algebra $\mathcal{A}_6 = (A_6, \leq, \to, \rightsquigarrow, 1)$. Namely, consider an element $m \notin A_6 = \{0, a, b, c, d, 1\}$ and the poset $(A'_6 = A_6 \cup \{m\}, \leq)$, presented in Figure 6.12.

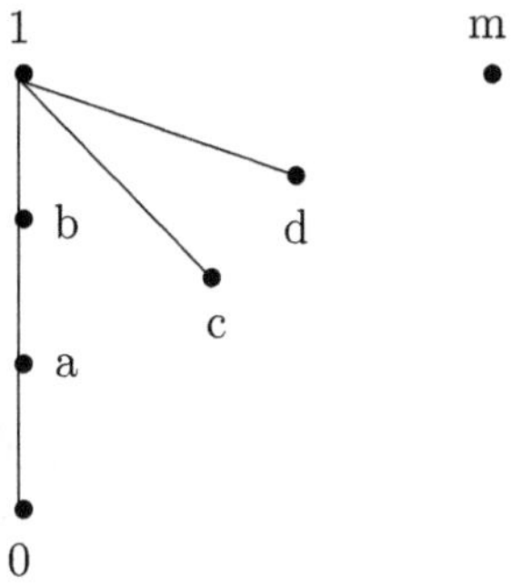

Figure 6.12: The poset $(A'_6, \leq)$

By Theorem 2.3.16 again, we define on A'_6 the following implications $\to'$, $\rightsquigarrow'$:

$\mathcal{A}'_6$

$\to'$	0	a	b	c	d	m	1
0	1	1	1	c	d	m	1
a	a	1	1	c	d	m	1
b	a	a	1	c	d	m	1
c	0	a	b	1	d	m	1
d	0	a	b	c	1	m	1
m	m	m	m	m	m	1	m
1	0	a	b	c	d	m	1

$\rightsquigarrow'$	0	a	b	c	d	m	1
0	1	1	1	c	d	m	1
a	b	1	1	c	d	m	1
b	0	a	1	c	d	m	1
c	0	a	b	1	d	m	1
d	0	a	b	c	1	m	1
m	m	m	m	m	m	1	m
1	0	a	b	c	d	m	1

.

Then, $\mathcal{A}'_6 = (A'_6, \leq, \to', \rightsquigarrow', 1)$ **is a left-pBCI algebra**, without pseudo-product (without (pP)), since, for example, $c \odot d \overset{notation}{=} \min\{c, d, 1\}$ does not exist.

Note that $A'^{-}_6 = \{0, a, b, c, d, 1\}$, while $A'^{+}_6 = \{1\}$.

• Example 13: The linearly-ordered left-Wajsberg algebra $\mathcal{L}_4$

Consider the bounded, linearly-ordered poset $(L_4 = \{0, 1, 2, 3\}, \leq, 0, 3)$. It can be organized as a left-Wajsberg algebra $\mathcal{L}_4 = (L_4, \to, ^-, 3)$, with

$$x \to y = \min\{3, y - x + 3\}, \quad x^- = x \to 0 = 3 - x,$$

whose tables are as follow:

$\mathcal{L}_4$

$\to$	**0**	1	**2**	**3**
0	**3**	**3**	**3**	**3**
1	**2**	**3**	**3**	**3**
2	1	**2**	**3**	**3**
3	**0**	1	**2**	**3**

x	x^-
0	**3**
1	2
2	1
3	**0**

.

Note that (see property (P)):

$\exists\, x \odot y \overset{notation}{=} \min\{z \mid x \leq y \to z\} = (x \to y^-)^- = \max\{0, x + y - 3\}$, hence we have the following table of $\odot$:

	$\odot$	0	1	2	3
	0	0	0	0	0
$\mathcal{L}_4$	1	0	0	0	1
	2	0	0	1	2
	3	0	1	2	3

Remark that the property (DN) ($(x^-)^- = x$, for all $x \in L_4$) is verified, hence there exist also the dual operations $\to^R$ and $\oplus$ (see Remarks 2.11.7, 2.11.8, 2.11.9).

• **Example 14: The standard left-implicative-Boolean algebra $\mathcal{L}_2$**

The standard left-implicative-Boolean algebra $\mathcal{L}_2 = (L_2 = \{0, 1\}, \to^L, ^-, 1)$ is represented by the Hasse diagram given in Figure 6.1 and has the tables of $\to^L$ and $^-$ recalled below (see Example 1 and Examples 1', 14' in a next section).

	$\to^L$	0	1
$\mathcal{L}_2$	0	1	1
	1	0	1

x	$x^- = x \to 0$
0	1
1	0

Remark that the property (DN) ($(x^-)^- = x$, for all $x \in L_2$) is verified, hence there exists also the dual operation $\to^R$ (see Remarks 2.11.7, 2.11.8, 2.11.9).

• **Example 15: The non-linearly-ordered left-implicative-Boolean algebra $\mathcal{L}_{2\times 2}$**

The non-linearly-ordered left-implicative-Boolean algebra

$$\mathcal{L}_{2\times 2} = (L_{2\times 2} = \{0, a, b, 1\}, \to^L, ^-, 1)$$

is represented by the Hasse diagram given in Figure 6.15 and has the tables of $\to^L$ and $^-$ recalled below.

	$\to^L$	0	a	b	1
	0	1	1	1	1
$\mathcal{L}_{2\times 2}$	a	b	1	b	1
	b	a	a	1	1
	1	0	a	b	1

x	$x^- = x \to 0$
0	1
a	b
b	a
1	0

Remark that the property (DN) ($(x^-)^- = x$, for all $x \in L_{2\times 2}$) is verified, hence there exist also the dual operation $\to^R$ (see Remarks 2.11.7, 2.11.8, 2.11.9).

Remark 6.1.1 One can find more examples of algebras of logic in the book [94], for example.

• **Example 16: Quasi-BCI(P) algebras**

the Consider the BCI(P) algebra $\mathcal{A}_3 = (\{a, b, 1\}, \to', 1)$ from Example 2, where the ordered regular set $(A_3, \leq)$ is recalled in the following Hasse diagram in the left side of the Figure 6.13, and the tables of $\rightarrow'$ and $\odot'$ are recalled below:

$\mathcal{A}_3$

$\rightarrow'$	a	b	1
a	1	a	a
b	a	1	1
1	a	b	1

$\odot'$	a	b	1
a	b	a	a
b	a	b	b
1	a	b	1

.

Consider now the quasi-set with 4 elements: $A_4 = \{a, b, c, 1\}$; then A_4 is quasi-ordered in three different ways: $(A_4^1, \leq)$, $(A_4^2, \leq)$, $(A_4^3, \leq)$, as shown in the quasi-Hasse diagram presented in the right side of Figure 6.13.

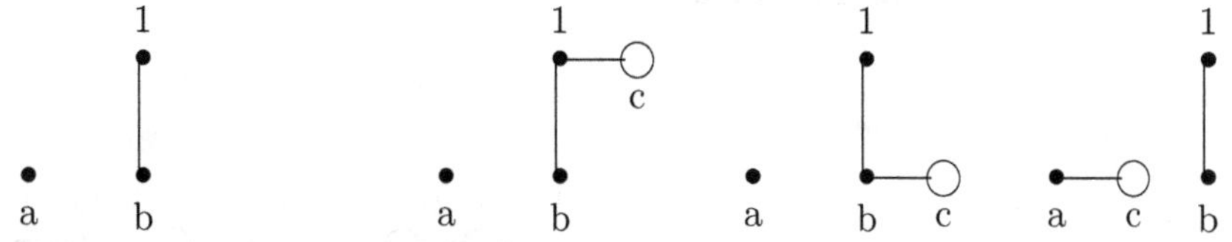

A_3 BCI $\quad$ A_4^1 q-BCI $\quad$ A_4^2 q-BCI $\quad$ A_4^3 q-BCI

Figure 6.13: The regular poset A_3 and the associated quasi-ordered sets A_4^1, A_4^2, A_4^3

We then obtain the following tables of ($\rightarrow_1$, $\odot_1$), ($\rightarrow_2$, $\odot_2$), ($\rightarrow_3$, $\odot_3$) of the quasi-BCI(P) algebras $\mathcal{A}_4^1$, $\mathcal{A}_4^2$, $\mathcal{A}_4^3$, respectively:

$\mathcal{A}_4^1$

$\rightarrow_1$	a	b	c	1
a	1	a	a	a
b	a	1	1	1
c	a	b	1	1
1	a	b	1	1

$\odot_1$	a	b	c	1
a	b	a	a	a
b	a	b	b	b
c	a	b	1	1
1	a	b	1	1

$\mathcal{A}_4^2$

$\rightarrow_2$	a	b	c	1
a	1	a	a	a
b	a	1	1	1
c	a	1	1	1
1	a	b	b	1

$\odot_2$	a	b	c	1
a	b	a	a	a
b	a	b	b	b
c	a	b	b	b
1	a	b	b	1

$\mathcal{A}_4^3$

$\rightarrow_3$	a	b	c	1
a	1	a	1	a
b	a	1	a	1
c	1	a	1	a
1	a	b	a	1

$\odot_3$	a	b	c	1
a	b	a	b	a
b	a	b	a	b
c	b	a	b	a
1	a	b	a	1

.

Hence, $\mathcal{A}_4^1 = (A_4^1, \rightarrow_1, 1)$, $\mathcal{A}_4^2 = (A_4^2, \rightarrow_2, 1)$ and $\mathcal{A}_4^3 = (A_4^3, \rightarrow_3, 1)$ are q-BCI(P) algebras such that:
$\mathcal{R}(\mathcal{A}_4^1) = \mathcal{A}_3$, $\mathcal{R}(\mathcal{A}_4^2) = \mathcal{A}_3$ and $\mathcal{R}(\mathcal{A}_4^3) = \mathcal{A}_3$.

Note that, following Propositions 5.1.30 and 5.2.25, in the tables of ($\rightarrow$, $\odot$) of:

- $\mathcal{A}_4^1$, the line (column) of c is identique with the line (column) of 1, since $c \parallel 1$ and $c \parallel_m 1$,
- $\mathcal{A}_4^2$, the line (column) of c is identique with the line (column) of b, since $c \parallel b$ and $c \parallel_m b$,
- $\mathcal{A}_4^3$, the line (column) of c is identique with the line (column) of a, since $c \parallel a$ and $c \parallel_m a$.

Consider now the quasi-set with 5 elements: $A_5 = \{a, b, c, d, 1\}$, quasi-ordered in the following four different ways: $(A_5^1, \leq)$, $(A_5^2, \leq)$, $(A_5^3, \leq)$, $(A_5^4, \leq)$, presented in the quasi-Hasse diagram from Figure 6.14.

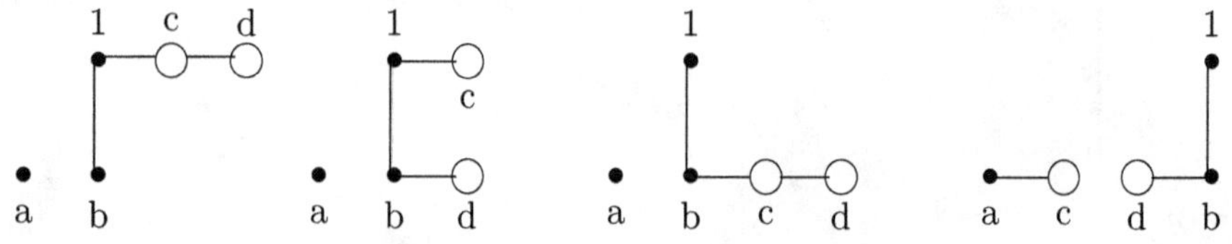

A_5^1 q-BCI A_5^2 q-BCI A_5^3 q-BCI A_5^4 q-BCI

Figure 6.14: Four quasi-ordered sets with 5 elements: A_5^1 - A_5^4

We then obtain the following tables of ($\rightarrow_1$, $\odot_1$), ($\rightarrow_2$, $\odot_2$), ($\rightarrow_3$, $\odot_3$), ($\rightarrow_4$, $\odot_4$) of the quasi-BCI(P) algebras $\mathcal{A}_5^1$, $\mathcal{A}_5^2$, $\mathcal{A}_5^3$, $\mathcal{A}_5^4$, respectively:

$\mathcal{A}_5^1$

$\rightarrow_1$	a	b	c	d	1
a	1	a	a	a	a
b	a	1	1	1	1
c	a	b	1	1	1
d	a	b	1	1	1
1	a	b	1	1	1

$\odot_1$	a	b	c	d	1
a	b	a	a	a	a
b	a	b	b	b	b
c	a	b	1	1	1
d	a	b	1	1	1
1	a	b	1	1	1

$\mathcal{A}_5^2$

$\rightarrow_2$	a	b	c	d	1
a	1	a	a	a	a
b	a	1	1	1	1
c	a	b	1	b	1
d	a	1	1	1	1
1	a	b	1	b	1

$\odot_2$	a	b	c	d	1
a	b	a	a	a	a
b	a	b	b	b	b
c	a	b	1	b	1
d	a	b	b	b	b
1	a	b	1	b	1

$\mathcal{A}_5^3$

$\rightarrow_3$	a	b	c	d	1
a	1	a	a	a	a
b	a	1	1	1	1
c	a	1	1	1	1
d	a	1	1	1	1
1	a	b	b	b	1

$\odot_3$	a	b	c	d	1
a	b	a	a	a	a
b	a	b	b	b	b
c	a	b	b	b	b
d	a	b	b	b	b
1	a	b	b	b	1

$\mathcal{A}_5^4$

$\rightarrow_4$	**a**	**b**	c	d	**1**
a	**1**	**a**	1	a	**a**
b	**a**	**1**	a	1	**1**
c	1	a	1	a	a
d	a	1	a	1	1
1	**a**	**b**	a	b	**1**

$\odot_4$	**a**	**b**	c	d	**1**
a	**b**	**a**	b	a	**a**
b	**a**	**b**	a	b	**b**
c	b	a	b	a	a
d	a	b	a	b	b
1	**a**	**b**	a	b	**1**

.

Hence, $\mathcal{A}_5^1 = (A_5^1, \rightarrow_1, 1)$ - $\mathcal{A}_5^4 = (A_5^4, \rightarrow_4, 1)$ are q-BCI(P) algebras such that: $\mathcal{R}(\mathcal{A}_5^1) = \mathcal{A}_3$, $\mathcal{R}(\mathcal{A}_5^2) = \mathcal{A}_3$, $\mathcal{R}(\mathcal{A}_5^3) = \mathcal{A}_3$ and $\mathcal{R}(\mathcal{A}_5^4) = \mathcal{A}_3$.

Note that, in the tables of $(\rightarrow, \odot)$ of:
- $\mathcal{A}_5^1$, the lines (columns) of c, d are identique with the line (column) of 1,
- $\mathcal{A}_5^2$, the line (column) of c is identique with the line (column) of 1 and the line (column) of d is identique w ith the line (column) of b,
- $\mathcal{A}_5^3$, the lines (columns) of c, d are identique with the line (column) of b,
- $\mathcal{A}_5^4$, the line (column) of c is identique with the line (column) of a and the line (column) of d is identique with the line (column) of b.

6.2 The program VERPEX1.PAS to check (pEx1)

We have written a program (VERPEX1.PAS) in PASCAL programming language to check if the property (pEx1) is verified by an algebra $(A, \rightarrow, \rightsquigarrow, 1)$, where A is a finite set containing maximum 10 elements, the last element being the element 1.

- **The program VERPEX1.PAS**

```
PROGRAM Verification_of_condition_pEx1;
{We associate variable x with i, variable y with j }
{We verify condition (pEx1): x-(y 1)=y (x-1) }
{We use sir= STRING[1],}
{ hence you must introduce the elements (which have only one character)}
{ of the matrix of operations without spaces between them}
{REMARK. If you want to use elements of 2 characters, then put }
{ sir=STRING[2] and allow 2 positions for each character (right alingnment).}
{We write on the screen only the messages if the respective condition is}
{ or is NOT verified. The rest of informations are written on a file with
internal name: fl and external name given by you when running the program.}
USES CRT;
TYPE
   int_n=1..10;
   bit=0..1;
   sir=STRING[1];
   vector= ARRAY[int_n] OF sir;
   matrice= ARRAY[int_n,int_n] OF sir;
VAR
   i, j, n, iXZ, iYZ: int_n;
```

```
   XZ, YZ, Left_side, Right_side:sir;
   VB: bit;
   S:vector;
   arrow,c_arrow: matrice;
   fl:TEXT;
   spec_fisier_lista:STRING;
FUNCTION Index(n:int_n; VAR S:vector; litera:sir) : BYTE;
VAR
   ind:int_n;
BEGIN
   ind:=1;
   WHILE NOT((ind > n) XOR (litera = S[ind])) DO
      ind:=ind+1;
   IF ind > n THEN
      BEGIN
         Index:=0; WRITELN('EROARE!!!!!!!!!!')
      END
   ELSE
      Index:=ind
END;
PROCEDURE Reading (VAR A:matrice; n:int_n);
VAR
   i,j:int_n;
BEGIN
   WRITELN('Give the matrix line by line;', ' push ENTER at the end of the
line:');
   FOR i:=1 TO n DO
      WRITE(S[i]);
   WRITELN;
   WRITELN('————————————');
   FOR i:=1 TO n DO
      BEGIN
         FOR j:=1 TO n DO
            READ(A[i,j]);
         READLN
      END;
END;
PROCEDURE Writing_fl (VAR A: matrice; n:int_n);
VAR
   i,j:int_n;
BEGIN
   WRITE(fl,' ');
   FOR i:=1 TO n DO
      WRITE (fl,' ', S[i]);
   WRITELN(fl);
   WRITELN(fl,'————————————');
```

```
   FOR i:=1 TO n DO
      BEGIN
         WRITE(fl,S[i],' ');
         FOR j:=1 TO n DO
            WRITE(fl,A[i,j],' ');
         WRITELN (fl)
      END
   READLN
END;
BEGIN
   CLRSCR;
   WRITE('Give a name to the external file list, for example A.DOC: ');
   READLN(spec_fisier_lista); ASSIGN(fl,spec_fisier_lista); REWRITE(fl);
   WRITELN('Give n <= 15:');
   READLN(n);
   WRITELN('Give the string of elements');
   FOR i:=1 TO n DO
      BEGIN
         WRITE('S(',i,')='); READLN(S[i])
      END;
   Writeln(fl,'The string of elements is:');
   FOR i:=1 TO n DO
      WRITE(fl, ' ', S[i]);
   WRITELN(fl); WRITELN;
{=== Read-Write the r_implication operation as a matrix ======}
   WRITELN('Give the matrix of r_implication');
   Citire(arrow,n);
   WRITELN(fl, 'The matrix of r_implication is:');
   Scriere_fl(arrow,n);
{========================End r_implication========}
{=== Read-Write the c_implication operation as a matrix ======}
   WRITELN('Give the matrix of c_implication');
   Citire(c_arrow,n);
   WRITELN(fl, 'The matrix of c_implication is:');
   Scriere_fl(c_arrow,n);
{========================End c_implication========}
   VB:=0;
{=== begin-condition (pEx1): x-(y 1)=y (x-1) ============}
   FOR i:=1 TO n DO
      FOR j:=1 TO n DO
         BEGIN
            YZ:=c_arrow[j,n]; iYZ:=Index(n,S,YZ);
            XZ:=arrow[i,n]; iXZ:=Index(n,S,XZ);
            Left_side:=arrow[i,iYZ]; Right_side:=c_arrow[j,iXZ];
            IF Left_side <> Right_side THEN
               BEGIN
```

```
            WRITELN(fl,'For ', S[i], ',', S[j], ':',
            ' Left-side= ', Left_side,' Right-side= ', Right_side);
            VB:=1
        END
    END;
{===end condition (pEx1)==============================}
  IF VB = 1 THEN
    BEGIN
      WRITELN(fl,' Condition (pEx1) is NOT verified');
      WRITELN(' Condition (pEx1) is NOT verified');
    END
  ELSE
    BEGIN
      WRITELN(fl,' Condition (pEx1) is verified');
      WRITELN(' Condition (pEx1) is verified');
    END;
{=====================================================}
  WRITELN(' Good Bye');
  CLOSE(fl);
  READLN
END.
```

6.3 Examples of (quasi-m) unital magmas

In this section, we present examples of left-unital magmas (Examples 0' - 15') and examples of commutative qm-monoids (Example 16').

There is a connection between the example number n' from this section and the example number n from the first section.

• **Example 0': Proper unital magma**

Recall that a unital magma is an algebra $(A, \cdot, 1)$ of type $(2, 0)$ verifying (pU) (unit element): $1 \cdot x = x = x \cdot 1$.

A unital magma is *proper*, if it does not verify the property (Ass).

Consider now the set $A = \{a, b, 1\}$ organized as an algebra as follows: $\mathcal{A} = (A, \cdot, 1)$, where the table of $\cdot$ is given below:

$\mathcal{A}$

$\cdot$	a	b	1
a	a	b	**a**
b	a	a	**b**
1	**a**	**b**	**1**

.

$\mathcal{A}$ verifies obviously the property (pU) and it does not verify the property (Ass) for $x = b$, $y = a$, $z = b$:
$a = b \cdot b = b \cdot (a \cdot b) = x \cdot (y \cdot z) \neq (x \cdot y) \cdot z = (b \cdot a) \cdot b = a \cdot b = b$.

• **Example 1': Bounded l-cim(R)**

Consider the bounded Ore lattice $(L_2 = \{0,1\}, \leq, 0, 1)$, with greatest element 1 and smallest element 0, represented by the Hasse diagram given in Figure 6.1.

The equivalent $(x \leq y \Leftrightarrow x \wedge y = x \Leftrightarrow x \vee y = y)$ bounded Dedekind lattice $(L_2, \wedge = \min, \vee = \max, 0, 1)$ was presented before (see Example 1).

Define on L_2 the following product $\odot$:

$\mathcal{L}_2$	$\odot(=\wedge)$	**0**	**1**
	0	0	0
	1	0	1

Then, $\mathcal{L}_2 = (L_2, \leq, \odot, 0, 1)$ is a bounded l-cim, with residuum (with (R)), where: for all $y, z \in L_2$,
(R) (residuum) $\exists\ y \to^L z \overset{notation}{=} \max\{x \mid x \odot y \leq z\}$.
We then obtain the following table of the residuum $\to^L$:

$\mathcal{L}_2$	$\to^L$	**0**	**1**
	0	1	1
	1	0	1

Hence, $\mathcal{L}_2 = (L_2, \leq, \odot, 0, 1)$ **is a bounded l-cim(R)**, equivalent by (iE4) with the bounded BCK(P) lattice from Example 1 in the first section. We shall see later that $\mathcal{L}_2$ is just (term-equivalent with) the standard Boolean algebra $\mathcal{L}_2$ (see Examples 14 and 14').

• **Example 2': po-cm(R)** (which is not a po-group)
By using Theorem 3.3.9, we obtain from the above bounded l-cim(R)

$$\mathcal{L}_2 = (L_2, \leq, \odot, 0, 1)$$

a po-cm(R). Namely, consider an element $b \notin L_2$ and consider the poset $(M' = L_2 \cup \{b\}, \leq)$, represented in Figure 6.2.

Define on M' the product $\odot'$ as follows (by Theorem 3.3.9 again):

$\mathcal{M}'$	$\odot'$	**0**	b	**1**
	0	**0**	b	**0**
	b	b	0	b
	1	**0**	b	**1**

$\to'$	**0**	b	**1**
0	1	b	1
b	b	1	b
1	**0**	b	1

x	$x^- = x \to' 1$	$(x^-)^-$
0	1	1
b	b	b
1	1	1

Then, $\mathcal{M}' = (M', \leq, \odot', 1)$ **is a po-cm, with residuum (with (R)), i.e. is a po-cm(R)**, where: for all $y, z \in L_2$,
(R) (residuum) $\exists\ y \to' z \overset{notation}{=} \max\{x \mid x \odot' y \leq z\}$.

We thus obtain the table of the implication $\to'$ presented above, implication that is indeed a residuum, conform to Definition 1.1.3, hence (Eq^L) holds. Consequently, by (E1), $(M', \leq, \odot', 1)$ is term equivalent with $(M', \leq, \to', 1)$ from Example 2 in the first section.

Note that the property (DN) ($(x^-)^- = x$, for all $x \in M'$) is not verified, since $(0^-)^- = 1 \neq 0$; hence, M' is not a po-group.

Note also that $M^{'-} = \{0,1\}$, while $M^{'+} = \{1\}$.

• **Example 3': Bounded left-l-cim (without (R))**

Consider the bounded non-linearly ordered Ore lattice $\mathcal{L}_{2\times2} = (L_{2\times2} = \{0,a,b,1\}, \leq, 0, 1)$, with greatest element 1 and smallest element 0, represented by the Hasse diagram given in Figure 6.15.

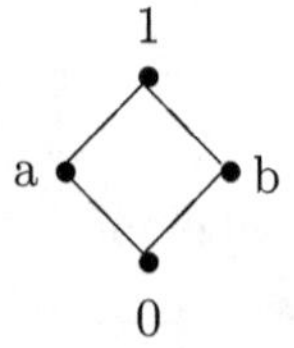

Figure 6.15: The bounded lattice $(L_{2\times2}, \leq, 0, 1)$

The equivalent ($x \leq y \Leftrightarrow x \wedge y = x \Leftrightarrow x \vee y = y$) bounded Dedekind lattice $(L_{2\times2}, \wedge, \vee, 0, 1)$ has the tables of $\wedge$ and $\vee$ given below:

$\mathcal{L}_{2\times2}$

$\wedge$	0	a	b	1
0	0	0	0	0
a	0	a	0	a
b	0	0	b	b
1	0	a	b	1

$\vee$	0	a	b	1
0	0	a	b	1
a	a	a	1	1
b	b	1	b	1
1	1	1	1	1

.

Consider also the commutative monoid $(L_{2\times2}, \odot, 1)$ with the table of $\odot$ given below:

$\mathcal{L}_{2\times2}$

$\odot$	0	a	b	1
0	0	0	0	0
a	0	a	0	a
b	0	0	0	b
1	0	a	b	1

.

Note that $\odot \neq \wedge$ and that $\leq$ is compatible with $\odot$.

Consequently, $(L_{2\times2}, \wedge, \vee, \odot, 0, 1)$ **is a bounded left-l-cim.**

Note that there is no residuum (without (R)), because, for example:

$b \to 0 \overset{notation}{=} \max\{x \mid b \odot x \leq 0\} = \max\{0, a, b\}$ does not exist.

• **Example 4': Left-po-cm (without (R))**

By Theorem 3.3.9, we shall obtain from the above bounded left-l-cim

$$(L_{2\times2}, \wedge, \vee, \odot, 0, 1)$$

a left-po-cm. Namely, consider an element $c \notin L_{2\times2}$ and the poset $(M' = L_{2\times2} \cup \{c\}, \leq)$, represented by the Hasse diagram given in Figure 6.16.

Define on $M' = \{0, a, b, c, 1\}$ the operation $\odot'$, by Theorem 3.3.9, as follows:

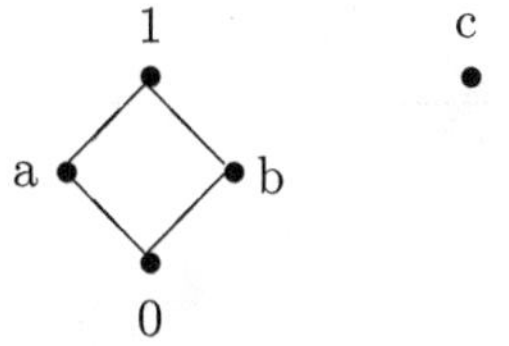

Figure 6.16: The poset $M' = (L_{2\times 2} \cup \{c\}, \leq)$

$\mathcal{M}'$

$\odot'$	0	a	b	c	1
0	0	0	0	c	0
a	0	a	0	c	a
b	0	0	0	c	b
c	c	c	c	0	c
1	0	a	b	c	1

.

Then, by Theorem 3.3.9 again, $\mathcal{M}' = (M', \leq, \odot', 1)$ **is a left-po-cm**; $\mathcal{M}'$ is without a residuum, because again, for example:
$b \to 0 \overset{notation}{=} \max\{x \mid b \odot' x \leq 0\} = \max\{0, a, b\}$ does not exist.

Note that $M'^{-} = \{0, a, b, 1\}$, while $M'^{+} = \{1\}$.

- **Example 5': Infinite l-cm(R)**

Consider the lattice $(\mathbf{Z} = \{\ldots, -3, -2, -1, \mathbf{0}, 1, 2, 3, \ldots\}, \leq)$, whose Hasse diagram is presented in Figure 6.5. Consider also the monoid $(\mathbf{Z}, \odot = +, \mathbf{1} = 0)$, where the table of $\odot = +$ is presented below.

Then, $(\mathbf{Z}, \wedge = \min, \vee = \max, \odot = +, \mathbf{1} = 0)$ **is a left-l-cm with residuum, i.e. is a l-cm(R)** (see Example 5 and (I2)), where the implication $\to$ is defined as follows: for all $y, z \in \mathbf{Z}$,
(R) (residuum) $\exists\, y \to z \overset{notation}{=} \max\{x \mid x \odot y \leq z\}$,
and the resulting table is presented below also.

$\odot = +$	...	-3	-2	-1	0	1	2	3	...
⋮	...	⋮	⋮	⋮	⋮	⋮	⋮	⋮	...
-3	...	-6	-5	-4	-3	-2	-1	0	...
-2	...	-5	-4	-3	-2	-1	0	1	...
-1	...	-4	-3	-2	-1	0	1	2	...
0	...	-3	-2	-1	0	1	2	3	...
1	...	-2	-1	0	1	2	3	4	...
2	...	-1	0	1	2	3	4	5	...
3	...	0	1	2	3	4	5	6	...
⋮	...	⋮	⋮	⋮	⋮	⋮	⋮	⋮	...

$\rightarrow$	...	-3	-2	-1	**0**	1	2	3	...
⋮	...	⋮	⋮	⋮	⋮	⋮	⋮	⋮	...
-3	...	**0**	1	2	3	4	5	6	...
-2	...	-1	**0**	1	2	3	4	5	...
-1	...	-2	-1	**0**	1	2	3	4	...
0	...	-3	-2	-1	**0**	1	2	3	...
1	...	-4	-3	-2	-1	**0**	1	2	...
2	...	-5	-4	-3	-2	-1	**0**	1	...
3	...	-6	-5	-4	-3	-2	-1	**0**	...
⋮	...	⋮	⋮	⋮	⋮	⋮	⋮	⋮	...

Remark that the implication $\rightarrow$ is not quite a residuum, conform with Definition 1.1.3, because (Eq$^{\leq}$) is not verified: for example, $1 \leq 3$ but $1 \rightarrow 3 = 2 \neq 0$.

Remark also that this l-cm(R) is in fact an l-group, where $-x = x \rightarrow 0$ (see Example 1 in Chapter 9).

Note that $\mathbf{Z}^- = \{\dots, -3, -2, -1, 0\}$ and $\mathbf{Z}^+ = \{0, 1, 2, 3, \dots\}$.

• **Example 6': Infinite l-cim(R)**

Consider the lattice $(\mathbf{Z}^- = \{\dots, -3, -2, -1, 0\}, \leq, \mathbf{1} = 0)$ with greatest element 0, whose Hasse diagram is presented in Figure 6.6. Consider also the monoid $(\mathbf{Z}^-, \odot = +, 0)$, where the table of $\odot$ is presented below.

Then, $(\mathbf{Z}^-, \min, \max, \odot, 0)$ **is a left-l-cim with residuum, i.e. is a l-cim(R)**, where the residuum $\rightarrow^L$ is defined as follows: for all $y, z \in \mathbf{Z}^-$,

(R) (residuum) $\exists\, y \rightarrow^L z \overset{notation}{=} \max\{x \mid x \odot y \leq z\}$,

and its table is presented below also.

$\odot = +$	...	-3	-2	-1	0
⋮	...	⋮	⋮	⋮	⋮
-3	...	-6	-5	-4	-3
-2	...	-5	-4	-3	-2
-1	...	-4	-3	-2	-1
0	...	-3	-2	-1	0

$\rightarrow^L$	...	-3	-2	-1	0
⋮	...	⋮	⋮	⋮	⋮
-3	...	0	0	0	0
-2	...	-1	0	0	0
-1	...	-2	-1	0	0
0	...	-3	-2	-1	0

Note that the l-cim(R) $(\mathbf{Z}^-, \min, \max, \odot, \mathbf{1} = 0)$ is equivalent, by (iE2), with the BCK(P) lattice $(\mathbf{Z}^-, \min, \max, \rightarrow^L, \mathbf{1} = 0)$ from Example 6.

• **Example 7': Infinite bounded l-cim(R)**

Starting from the previous l-cim(R) $(\mathbf{Z}^-, \min, \max, \odot, 0)$, let us consider the symbol $-\infty \notin \mathbf{Z}^-$ and define $Z^-_{-\infty} \overset{def.}{=} \{-\infty\} \cup \mathbf{Z}^-$. Extend the lattice order relation $\leq$ from $\mathbf{Z}^-$ to $Z^-_{-\infty}$ as presented in Figure 6.7.

Extend the operation $\odot$ on $\mathbf{Z}^-$ to $\odot_2$ on $Z^-_{-\infty}$ as follows:

$$x \odot_2 y = \begin{cases} x \odot y = x + y, & \text{if } x, y \in \mathbf{Z}^-, \\ -\infty, & \text{if } otherwise. \end{cases}$$

Hence, $\odot_2$ has the table presented below.

Then, $(Z^-_{-\infty}, \min, \max, \odot_2, -\infty, 0)$ **is a bounded left-l-cim with residuum** $\rightarrow^L_2$**, i.e. is a bounded l-cim(R)**, where:
(R) $\exists\ y \rightarrow^L_2 z \overset{notation}{=} \max\{x \mid x \odot_2 y \leq z\}$,
hence the resulting table of $\rightarrow^L_2$ is presented below.

$\odot_2$	$-\infty$	...	-3	-2	-1	0
$-\infty$	$-\infty$	...	$-\infty$	$-\infty$	$-\infty$	$-\infty$
⋮	⋮	...	⋮	⋮	⋮	⋮
-3	$-\infty$	...	-6	-5	-4	-3
-2	$-\infty$	...	-5	-4	-3	-2
-1	$-\infty$	...	-4	-3	-2	-1
0	$-\infty$	...	-3	-2	-1	0

$\rightarrow^L_2$	$-\infty$	...	-3	-2	-1	0
$-\infty$	0	...	0	0	0	0
⋮	⋮	...	⋮	⋮	⋮	⋮
-3	$-\infty$	...	0	0	0	0
-2	$-\infty$	...	-1	0	0	0
-1	$-\infty$	...	-2	-1	0	0
0	$-\infty$	...	-3	-2	-1	0

Note that this bounded l-cim(R) is equivalent by (iE4) to the bounded BCK(P) lattice from Example 7.

Note that $(Z^-_{-\infty}, \min, \max, \odot_2, \rightarrow^L_2, -\infty, 0)$ is in fact just a linearly-ordered left-product algebra (see [31]; see [94], pag. 182; see Example 1 of Chapter 9).

- **Example 8': Infinite po-cm(R)** (which is not a po-group)

Starting from the above bounded l-cim(R) $(Z^-_{-\infty}, \min, \max, \odot_2, -\infty, 0)$, we obtain a po-cm(R) by Theorem 3.3.9. Namely, consider the element $m \notin Z^-_{-\infty}$ and the poset $(Z^{-'}_{-\infty} = Z^-_{-\infty} \cup \{m\}, \leq)$ represented by the Hasse diagram from Figure 6.8. Define on $Z^{-'}_{-\infty}$ the new operation $\odot'_2$, by Theorem 3.3.9 again; the table of $\odot'_2$ is given below.

Then, $(Z^{-'}_{-\infty}, \leq, \odot'_2, \mathbf{1} = 0)$ **is left-po-cm with residuum** $\rightarrow^{L'}_2$**, i.e. is a po-cm(R)**, the table of $\rightarrow^{L'}_2$ being presented below also.

$\odot'_2$	$-\infty$	...	-3	-2	-1	m	0
$-\infty$	$-\infty$	...	$-\infty$	$-\infty$	$-\infty$	m	$-\infty$
⋮	⋮	...	⋮	⋮	⋮	⋮	⋮
-3	$-\infty$	...	-6	-5	-4	m	-3
-2	$-\infty$	...	-5	-4	-3	m	-2
-1	$-\infty$	...	-4	-3	-2	m	-1
m	m	...	m	m	m	$-\infty$	m
0	$-\infty$	...	-3	-2	-1	m	0

$\rightarrow^{L'}_2$	$-\infty$	...	-3	-2	-1	m	0
$-\infty$	0	...	0	0	0	m	0
⋮	⋮	...	⋮	⋮	⋮	⋮	⋮
-3	$-\infty$	...	0	0	0	m	0
-2	$-\infty$	...	-1	0	0	m	0
-1	$-\infty$	...	-2	-1	0	m	0
m	m	...	m	m	m	0	m
0	$-\infty$	...	-3	-2	-1	m	0

Remark that the implication $\to_2^{L'}$ is indeed a residuum, conform with Definition 1.1.3, hence (Eq$\leq$) holds. Consequently, by (E1), this po-cm(R) is equivalent to the BCI(P) algebra from Example 8.

Note that the property (DN) ($-(-x) = x$, for all $x \in \mathbf{Z}^{-'}_{-\infty}$, where $-x = x \to_2^{L'} 0$) is not verified, since, for example, $-(-2) = 0 \neq -2$; hence, $\mathbf{Z}^{-'}_{-\infty}$ is not a po-group. Note that $(Z^{-'}_{-\infty})^- = \{-\infty, \ldots, -3, -2, -1, 0\}$, while $(Z^{-'}_{-\infty})^+ = \{0\}$.

- **Example 9': Bounded l-im(pR)**

Consider the bounded linearly-ordered Ore lattice $(L_4 = \{0, a, b, 1\}, \leq, 0, 1)$, whose Hasse diagram is presented in Figure 6.9.

Consider also defined on L_4 the following product $\odot$:

$\mathcal{L}_4$

$\odot$	0	a	b	1
0	0	0	0	0
a	0	0	0	a
b	0	a	b	b
1	0	a	b	1

$\to^L$	0	a	b	1
0	1	1	1	1
a	a	1	1	1
b	a	a	1	1
1	0	a	b	1

$\rightsquigarrow^L$	0	a	b	1
0	1	1	1	1
a	b	1	1	1
b	0	a	1	1
1	0	a	b	1

.

Then, $\mathcal{L}_4 = (L_4, \wedge = \min, \vee = \max, \odot, 0, 1)$ **is a bounded l-im with pseudo-residuum (with (pR)), i.e. a bounded l-im(pR)**, where for all $y, z \in L_4$:

(pR) $\exists\, y \to^L z \overset{notation}{=} \max\{x \mid x \odot y \leq z\}$,

$\exists\, y \rightsquigarrow^L z \overset{notation}{=} \max\{x \mid y \odot x \leq z\}$

and the resulting tables of $\to^L$ and $\rightsquigarrow^L$ are presented above also.

Note that $\mathcal{L}_4 = (L_4, \wedge = \min, \vee = \max, \odot, 0, 1)$ is a bounded l-im(pR) verifying (pprel^L):

(pprel^L) $(x \to^L y) \vee (y \to^L x) = 1 = (x \rightsquigarrow^L y) \vee (y \rightsquigarrow^L x)$,

and not verifying (pdiv^L):

(pdiv^L) $x \wedge y = (x \to^L y) \odot x = x \odot (x \rightsquigarrow^L y)$,

since, for example, $a = b \wedge a \neq (b \to^L a) \odot b = 0$.

Consequently, $\mathcal{L}_4$ is a pseudo-MTL algebra, namely one verifying the property (pWNM) (pseudo-Weak Nilpotent Minimum):

(pWNM) $(x \odot y)^- \vee [(x \wedge y) \to^L (x \odot y)] = 1 = (x \odot y)^\sim \vee [(x \wedge y) \rightsquigarrow^L (x \odot y)]$,

where $x^- = x \to^L 0$ and $x^\sim = x \rightsquigarrow^L 0$ (see [94], pag. 491).

Note that this bounded l-im(pR) is equivalent by (iE4) with the bounded pBCK(pP) lattice from Example 9.

- **Example 10': Left-po-m(pR)** (which is not a po-group)

By Theorem 3.3.9, we shall obtain from the above bounded l-im(pR)

$$(L_4, \wedge = \min, \vee = \max, \odot, 0, 1)$$

a left-po-m (with (pR)). Namely, consider an element $c \notin L_4$ and the poset $(L'_4 = L_4 \cup \{c\}, \leq)$ represented in Figure 6.10. By Theorem 3.3.9 again, consider the new product $\odot'$ defined by the following table:

$\mathcal{L}'_4$

$\odot'$	0	a	b	c	1
0	0	0	0	c	0
a	0	0	0	c	a
b	0	a	b	c	b
c	c	c	c	0	c
1	0	a	b	c	1

$\to'$	0	a	b	c	1
0	1	1	1	c	1
a	a	1	1	c	1
b	a	a	1	c	1
c	c	c	c	1	c
1	0	a	b	c	1

$\rightsquigarrow'$	0	a	b	c	1
0	1	1	1	c	1
a	b	1	1	c	1
b	0	a	1	c	1
c	c	c	c	1	c
1	0	a	b	c	1

.

Then, $\mathcal{L}'_4 = (L'_4 = \{0, a, b, c, 1\}, \leq, \odot', 1)$ is a left-po-m, with the above presented pseudo-implication $(\to', \rightsquigarrow')$ defined by: for all $y, z \in L'_4$,
(pR) (pseudo-residuum) $\exists\, y \to' z \overset{notation}{=} \max\{x \mid x \odot y \leq z\}$,
$\exists\, y \rightsquigarrow' z \overset{notation}{=} \max\{x \mid y \odot x \leq z\}$.

Hence, $\mathcal{L}'_4 = (L'_4 = \{0, a, b, c, 1\}, \leq, \odot', 1)$ **is a po-m(pR)**.

Remark that the pseudo-implication $(\to', \rightsquigarrow')$ is indeed a pseudo-residuum, conform with Definition 1.1.3, hence (pEq^L) holds. Consequently, this po-m(pR) is equivalent by (E1) to the pBCI(pP) algebra from Example 10.

Note that the property (pDN) $((x^-)^\sim = x = (x^\sim)^-$, for all $x \in L'_4$, where $x^- = x \to' 1$, $x^\sim = x \rightsquigarrow' 1)$ is not verified, since, for example, $(a^-)^\sim = 1^\sim = 1 \neq a$; hence, L'_4 is not a po-group.

Note also that $L_4^{'-} = \{0, a, b, 1\}$, while $L_4^{'+} = \{1\}$.

• **Example 11': Bounded left-l-im (without (pR))**

Consider the ordinal product $\mathcal{A}_7 = \mathcal{L}_4 \bigodot \mathcal{L}_{2\times 2}$, of the bounded l-im(pR) $\mathcal{L}_4$ from Example 9' and the bounded left-l-cim without (R) $\mathcal{L}_{2\times 2}$ from Example 3', organized as a lattice as in Figure 6.17 and whose operation $\odot$ defined on $A_7 = \{0, a, b, c, d, f, 1\}$ is given by the next table.

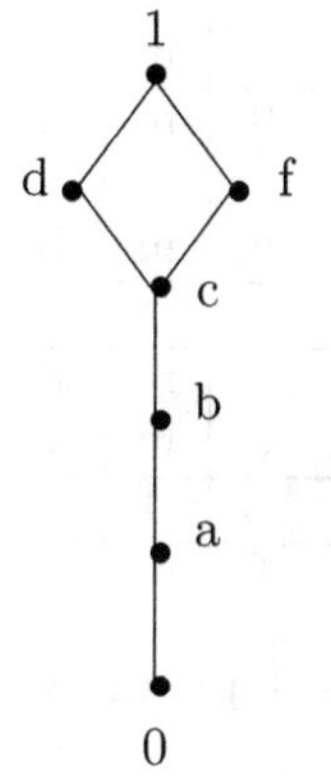

Figure 6.17: The bounded lattice $(A_7, \leq, 0, 1)$

$\mathcal{A}_7$

$\odot$	0	a	b	c	d	f	1
0	0	0	0	0	0	0	0
a	0	0	0	a	a	a	a
b	0	a	b	b	b	b	b
c	0	a	b	c	c	c	c
d	0	a	b	c	d	c	d
f	0	a	b	c	c	c	f
1	0	a	b	c	d	f	1

Then, $\mathcal{A}_7 = (A_7, \leq, \odot, 1)$ **is a bounded left-l-im**, without pseudo-residuum (without (pR)), since, for example:
$f \to c \overset{notation}{=} \max\{x \mid f \odot x \leq c\} = \max\{0, a, b, c, d, f\}$ does not exist.

• **Example 12': Left-po-m (without (pR))**

By Theorem 3.3.9, we shall obtain a left-po-m from the above bounded left-l-im $\mathcal{A}_7 = (A_7, \leq, \odot, 1)$. Namely, consider an element $m \notin A_7 = \{0, a, b, c, d, f, 1\}$ and the poset $(A'_7 = A_7 \cup \{m\}, \leq)$, presented in Figure 6.18.

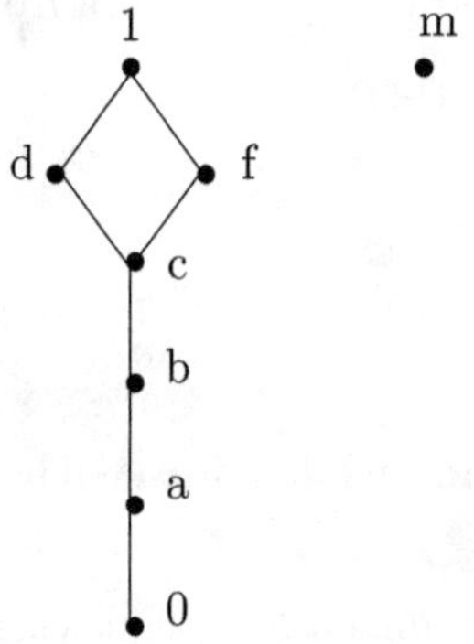

Figure 6.18: The poset $(A'_7, \leq)$

By Theorem 3.3.9 again, we define on A'_7 the following pseudo-product $\odot'$:

$\mathcal{A}'_7$

$\odot'$	0	a	b	c	d	f	m	1
0	0	0	0	0	0	0	m	0
a	0	0	0	a	a	a	m	a
b	0	a	b	b	b	b	m	b
c	0	a	b	c	c	c	m	c
d	0	a	b	c	d	c	m	d
f	0	a	b	c	c	c	m	f
m	m	m	m	m	m	m	0	m
1	0	a	b	c	d	f	m	1

Then, $\mathcal{A}'_7 = (A'_7, \leq, \odot', 1)$ **is a left-po-m**, without pseudo-residuum (without (pR)), since, for example:
$f \to c \overset{notation}{=} \max\{x \mid f \odot' x \leq c\} = \max\{0, a, b, c, d, f\}$ does not exist.

Note that $A_7^{'-} = \{0, a, b, c, d, f, 1\}$, while $A_7^{'+} = \{1\}$.

• **Example 13': The linearly-ordered left-MV algebra $\mathcal{L}_4$**
Consider the bounded, linearly-ordered poset $(L_4 = \{0, 1, 2, 3\}, \leq, 0, 3)$. It can be organized as a left-MV algebra $\mathcal{L}_4 = (L_4, \odot, ^-, 3)$, with

$$x \odot y = \max\{0, x + y - 3\}, \quad x^- = 3 - x,$$

whose tables are as follow:

$\mathcal{L}_4$

$\odot$	0	1	2	3
0	0	0	0	0
1	0	0	0	1
2	0	0	1	2
3	0	1	2	3

x	x^-
0	3
1	2
2	1
3	0

.

Note that (see condition (R)):
$\exists\, y \to z \overset{notation}{=} \max\{x \mid x \odot y \leq z\} = (x \odot y^-)^- = \min\{3, y - x + 3\}$,
hence we have the following resulting table of $\to$:

$\mathcal{L}_4$

$\to$	0	1	2	3
0	3	3	3	3
1	2	3	3	3
2	1	2	3	3
3	0	1	2	3

.

Remark that $x^- = x \to 0$, for all $x \in L_4$ and that the property (DN) ($(x^-)^- = x$, for all $x \in L_4$) is verified. Hence, there exist also the dual operations $\oplus$ and $\to^R$ (see Remarks 2.11.7, 2.11.8 and 2.11.9).

Note that the left-MV algebra $(L_4, \odot, ^-, 3)$ is term equivalent by (iE9') to the left-Wajsberg algebra $(L_4, \to, ^-, 3)$ from Example 13.

• **Example 14': The standard left-Boolean algebra $\mathcal{L}_2$**
The standard left-Boolean algebra $\mathcal{L}_2 = (L_2 = \{0, 1\}, \wedge, \vee, ^-, 0, 1)$ is represented by the Hasse diagram given in Figure 6.1 and has the tables of $\wedge, \vee$ and $^-$ recalled below (see Example 1).

$\mathcal{L}_2$

$\odot = \wedge$	0	1
0	0	0
1	0	1

$\vee$	0	1
0	0	1
1	1	1

x	x^-
0	1
1	0

.

Remark that the property (DN) ($(x^-)^- = x$, for all $x \in L_2$) is verified. Hence, there exists also the dual operation $\oplus = \vee$ (see Remarks 2.11.7, 2.11.8 and 2.11.9).

This left-Boolean algebra $(L_2, \wedge, \vee, ^-, 0, 1)$ is term equivalent by (iE11') to the left-implicative-Boolean algebra $(L_2, \to^L, ^-, 1)$ from Example 14 (see also Examples 1 and 1').

• Example 15': The non-linearly-ordered left-Boolean algebra $\mathcal{L}_{2\times 2}$
The non-linearly-ordered left-Boolean algebra

$$\mathcal{L}_{2\times 2} = (L_{2\times 2} = \{0, a, b, 1\}, \wedge, \vee, ^-, 0, 1)$$

is represented by the Hasse diagram given in Figure 6.15 and has the tables of $\wedge, \vee$ and $^-$ recalled below.

$\mathcal{L}_{2\times 2}$

$\odot = \wedge$	0	a	b	1
0	0	0	0	0
a	0	a	0	a
b	0	0	b	b
1	0	a	b	1

$\vee$	0	a	b	1
0	0	a	b	1
a	a	a	1	1
b	b	1	b	1
1	1	1	1	1

x	x^-
0	1
a	b
b	a
1	0

Remark that the property (DN) ($(x^-)^- = x$, for all $x \in L_{2\times 2}$) is verified, hence there exists also the dual operation $\oplus = \vee$ (see Remarks 2.11.7, 2.11.8 and 2.11.9).

Note that the left-Boolean algebra $(L_{2\times 2}, \wedge, \vee, ^-, 0, 1)$ is term equivalent by (iE11') to the left-implicative-Boolean algebra $(L_{2\times 2}, \rightarrow^L, ^-, 1)$ from Example 15.

• Example 16': Commutative qm-monoids
Consider the (regular-m) monoid with 2 elements $\mathcal{A}_2 = (\{a, 1\}, \odot, 1)$, whose (regular-m) set A_2 is represented in the left part of Figure 6.19 and whose table of $\odot$ is recalled below.

Consider then the quasi-m set with 3 elements $A_3 = \{a, b, 1\}$, with $Rm(A_3) = A_2$, represented in two ways: $b \parallel_m a$ (A_3^1) and $b \parallel_m 1$ (A_3^2), in the right part of Figure 6.19 also.

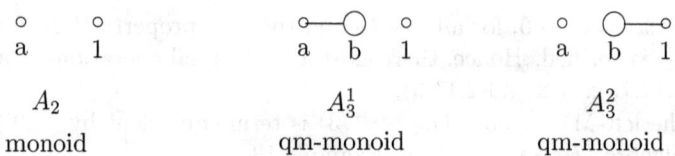

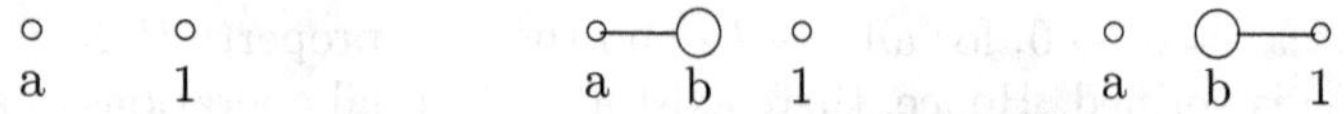

A_2 monoid $\qquad$ A_3^1 qm-monoid $\qquad$ A_3^2 qm-monoid

Figure 6.19: The monoid with 2 elements and the associated two qm-monoids with 3 elements

Then, the tables of the corresponding quasi-m operations $\odot^1$ and $\odot^2$ of $\mathcal{A}_3^1$ and $\mathcal{A}_3^2$, respectively, are the following (see Remark 5.2.28):

$\mathcal{A}_2$

$\odot$	a	1
a	a	a
1	a	1

$\mathcal{A}_3^1$

$\odot^1$	a	b	1
a	a	a	a
b	a	a	a
1	a	a	1

$\mathcal{A}_3^2$

$\odot^2$	a	b	1
a	a	a	a
b	a	1	1
1	a	1	1

Hence, $\mathcal{A}_3^1 = (A_3^1, \odot^1, 1)$ and $\mathcal{A}_3^2 = (A_3^2, \odot^2, 1)$ are two qm-monoids such that: $\mathcal{R}m(\mathcal{A}_3^1) = \mathcal{A}_2$ and $\mathcal{R}m(\mathcal{A}_3^2) = \mathcal{A}_2$.

Note that, following Proposition 5.2.25, in the table of:
- $\mathcal{A}_3^1$, the line (column) of b is identique with the line (column) of a,
- $\mathcal{A}_3^2$, the line (column) of b is identique with the line (column) of 1.

Consider further the quasi-m set with 4 elements A_4, with $Rm(A_4) = A_2$, represented in the three distinct ways as in next Figure 6.20.

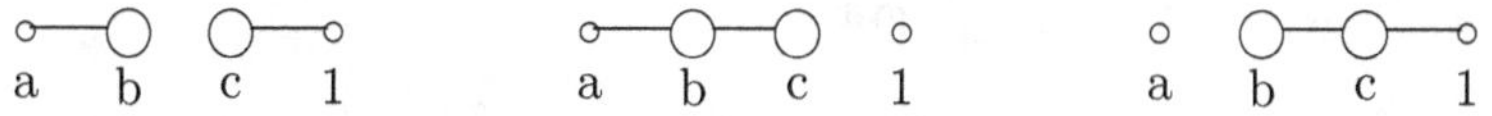

$\mathcal{A}_4^1$ qm-monoid $\qquad$ $\mathcal{A}_4^2$ qm-monoid $\qquad$ $\mathcal{A}_4^3$ qm-monoid

Figure 6.20: The associated three qm-monoids with 4 elements

Then, the tables of the corresponding quasi-m operations $\odot_1$, $\odot_2$ and $\odot_3$ of $\mathcal{A}_4^1$, $\mathcal{A}_4^2$ and $\mathcal{A}_4^3$, respectively, are the following (see Remark 5.2.28):

$\mathcal{A}_4^1$	$\odot_1$	a	b	c	1
	a	a	a	a	a
	b	a	a	a	a
	c	a	a	1	1
	1	a	a	1	1

$\mathcal{A}_4^2$	$\odot_2$	a	b	c	1
	a	a	a	a	a
	b	a	a	a	a
	c	a	a	a	a
	1	a	a	a	1

$\mathcal{A}_4^3$	$\odot_3$	a	b	c	1
	a	a	a	a	a
	b	a	1	1	1
	c	a	1	1	1
	1	a	1	1	1

Hence, $\mathcal{A}_4^1 = (A_4^1, \odot_1, 1)$, $\mathcal{A}_4^2 = (A_4^2, \odot_2, 1)$ and $\mathcal{A}_4^3 = (A_4^3, \odot_3, 1)$ are three qm-monoids such that:
$\mathcal{R}m(\mathcal{A}_4^1) = \mathcal{A}_2$, $\mathcal{R}m(\mathcal{A}_4^2) = \mathcal{A}_2$ and $\mathcal{R}m(\mathcal{A}_4^3) = \mathcal{A}_2$.

Note that in the table of:
- $\mathcal{A}_4^1$, the line (column) of b is identique with the line (column) of a and the line (column) of c is identique with the line (column) of 1,
- $\mathcal{A}_4^2$, the lines (columns) of b, c are identique with the line (column) of a,
- $\mathcal{A}_4^3$, the lines (columns) of b, c are identique with the line (column) of 1.

6.4 The PASCAL program VERASS.PAS to check the associativity

We have written a program (VERASS.PAS), in PASCAL programming language, which verifies the property (Ass) (Associativity) in an algebra $(A, \odot, 1)$, where A is a finite set with maximum 10 elements, containing as last element in the list the element 1, and the operation $\odot$ (product) is given as a matrix, line by line.

- **The program VERASS.PAS**

```
PROGRAM Verification_of_condition_associativity;
{We verify (Ass): x.(y.z)=(x.y).z }
{We associate variable x with i, variable y with j, variable z with k. }
   ⋮
USES CRT;
TYPE
   int_n=1..10;
```

```
   bit=0..1;
   sir=STRING[1];
   vector= ARRAY[int_n] OF sir;
   matrice= ARRAY[int_n,int_n] OF sir;
VAR
   i, j, k, n, iXY, iYZ: int_n;
   XY, YZ, t1, t2:sir;
   VB: bit;
   S:vector;
   product: matrice;
   fl:TEXT;
   spec_fisier_lista:STRING;
FUNCTION Index(n:int_n; VAR S:vector; litera:sir) : BYTE;
   :
END;
PROCEDURE Reading (VAR A:matrice; n:int_n);
   :
END;
PROCEDURE Writing_fl(VAR A: matrice; n:int_n);
   :
END;
BEGIN
   CLRSCR;
   WRITE('Give a name to the external file list, for example A.DOC: ');
   READLN(spec_fisier_lista); ASSIGN(fl,spec_fisier_lista); REWRITE(fl);
   WRITELN('Give n <= 15:');
   READLN(n);
   WRITELN('Give the string of elements');
   FOR i:=1 TO n DO
      BEGIN
         WRITE('S(',i,')='); READLN(S[i])
      END;
   Writeln(fl,'The string of elements is:');
   FOR i:=1 TO n DO
      WRITE(s[i]);
   WRITELN; WRITELN(fl);
{==== Read-Write the pseudo-product operation as a matrix ===}
   WRITELN('Give the matrix of pseudo-product');
   Reading(product,n);
   WRITELN(fl, 'The matrix of pseudo-product is:');
   Writing_fl(product,n);
{====================End pseudo-product==========}
   VB:=0;
{=== begin-condition (Ass): x.(y.z)=(x.y).z ==============}
```

```
  FOR i:=1 TO n DO
    FOR j:=1 TO n DO
      FOR k:=1 TO n DO
        BEGIN
          XY:=product[i,j]; iXY:=Index(n,S,XY);
          YZ:=product[j,k]; iYZ:=Index(n,S,YZ);
          t1:=product[i,iYZ]; t2:=product[iXY,k];
          IF t1 <> t2 THEN
            BEGIN
              WRITELN(fl,'For ', S[i], ',', S[j], ',', S[k],':', ' t1= ', t1,' t2=
', t2);
              VB:=1
            END
        END;
{=== end-condition (Ass) ============================}
  IF VB=1 THEN
    BEGIN
      WRITELN('Condition (Ass) is NOT satisfied');
      WRITELN(fl,'Condition (Ass) is NOT satisfied')
    END
  ELSE
    BEGIN
      WRITELN('Condition (Ass) is satisfied');
      WRITELN(fl,'Condition (Ass) is satisfied')
    END;
  WRITELN(fl,'——————————————-');
{==============================================}
  WRITELN(' Good Bye');
  CLOSE(fl);
  READLN
END.
```

Chapter 7

Groups, po-groups, l-groups

In this chapter, we recall [97], [100] the group, the po-group and the l-group, the intermediary notions of *X-po-group*, *X-l-group* and *X-group*, respectively, and prove the equivalences (EG1), (EG2), (EG3) between these notions, respectively.

The chapter has three sections.

7.1 Groups

7.1.1 The group

The most used definition of the group is the following:

Definition 7.1.1 (Definition 1 of groups)

A *group* is an algebra (in multiplicative (left) notation) $\mathcal{G} = (G, \cdot, ^{-1}, 1)$ of type $(2, 1, 0)$, verifying: for all $x, y, z \in G$,
(pU) $x \cdot 1 = x = 1 \cdot x$,
(pIv) $x \cdot x^{-1} = 1 = x^{-1} \cdot x$, i.e. each element x has an inverse element, x^{-1},
(Ass) $x \cdot (y \cdot z) = (x \cdot y) \cdot z$.

Denote by **group** the class of all groups.

The group is said to be *commutative*, or *abelian*, if $x \cdot y = y \cdot x$, for all $x, y \in G$.

Recall now the following equivalent definition of the group, very useful in this monograph for the connection with the new notion of *hub*, defined in Chapter 10.

Definition 7.1.2 (Definition 1' of groups)

A *group* is an algebra $\mathcal{G} = (G, \cdot, 1)$ of type $(2, 0)$, verifying: for all $x, y, z \in G$,
(pU) $x \cdot 1 = x = 1 \cdot x$,
(pIv') for all $x \in G$, there exists $y \in G$, such that $x \cdot y = 1 = y \cdot x$, i.e. each element x has an inverse element, y,
(Ass) $x \cdot (y \cdot z) = (x \cdot y) \cdot z$.

Note that the inverse element of x is **unique**, because of (pU) and (Ass); indeed, if there exist x^{-} and $x^{\sim}$ such that $x \cdot x^{-} = x^{-} \cdot x = 1$ and $x \cdot x^{\sim} = x^{\sim} \cdot x = 1$, then:

$x^- \overset{(pU)}{=} x^- \cdot 1 = x^- \cdot (x \cdot x^\sim) \overset{(Ass)}{=} (x^- \cdot x) \cdot x^\sim = 1 \cdot x^\sim \overset{(pU)}{=} x^\sim$.
The unique inverse of x is denoted by x^{-1}, thus there exists an unary operation on G, denoted by $^{-1}$, introduced in the signature of $\mathcal{G}$.

Remark 7.1.3 (See Remark 2.12.8)

We have called *hub* an algebra $(G, \cdot, 1)$ of type $(2, 0)$ verifying (pU) and (pIv'), introduced in Chapter 10. Then, the group defined by Definition 7.1.1 is connected with the hub by the following result:
"*The groups are categorically equivalent to the associative hubs.* "

Following this result, we obtain the equivalent Definition 7.1.2 of groups. Also, following this result, we obtain examples of groups from the examples of associative hubs.

Proposition 7.1.4 *Let* $\mathcal{G} = (G, \cdot, ^{-1}, 1)$ *be a group. Then, we have: for all* $x, y \in G$,

(pIvP)	$x \cdot y = 1 (\Leftrightarrow y \cdot x = 1) \Longleftrightarrow y = x^{-1}$,
(DN)	$(x^{-1})^{-1} = x$,
(N1)	$1^{-1} = 1$,
(pIvG)	$(x \cdot y^{-1})^{-1} = y \cdot x^{-1}, \quad (y^{-1} \cdot x)^{-1} = x^{-1} \cdot y$,
(NegP)	$(x \cdot y)^{-1} = y^{-1} \cdot x^{-1}$,
(qm-NegP)	$(1 \cdot x)^{-1} = 1 \cdot x^{-1} = x^{-1} \cdot 1 = (x \cdot 1)^{-1}$,
(IdEqP)	$y \cdot x^{-1} = 1 \Longleftrightarrow x^{-1} \cdot y = 1$,
(pEqP=)	$x = y \Longleftrightarrow y \cdot x^{-1} = 1 \overset{(IdEqP)}{\Longleftrightarrow} x^{-1} \cdot y = 1$.

We also have the equivalence:
(pIvP) $\Longleftrightarrow$ *(pEqP=).*

Proof.

(pIvP): Suppose that $x \cdot y = 1$; then, $y \overset{(pU)}{=} 1 \cdot y \overset{(pIv)}{=} (x^{-1} \cdot x) \cdot y \overset{(Ass)}{=} x^{-1} \cdot (x \cdot y) = x^{-1} \cdot 1 \overset{(pU)}{=} x^{-1}$. Conversely, suppose that $y = x^{-1}$; then, $x \cdot y = x \cdot x^{-1} \overset{(pIv)}{=} 1$. Thus, $x \cdot y = 1 \Longleftrightarrow y = x^{-1}$. Similarly, $y \cdot x = 1 \Longleftrightarrow y = x^{-1}$. So, (pIvP) holds.

(DN): By (pIv), $x^{-1} \cdot x = 1$; then, by (pIvP), $x = (x^{-1})^{-1}$; thus, (DN) holds.

(N1): $1^{-1} \overset{(pU)}{=} 1 \cdot 1^{-1} \overset{(pIv)}{=} 1$.

(pIvG): $(x \cdot y^{-1}) \cdot (y \cdot x^{-1}) \overset{(Ass)}{=} x \cdot (y^{-1} \cdot y) \cdot x^{-1} \overset{(pIv)}{=} x \cdot 1 \cdot x^{-1} \overset{(Ass)}{=} (x \cdot 1) \cdot x^{-1} \overset{pU)}{=} x \cdot x^{-1} \overset{(pIv)}{=} 1$ and
$(y \cdot x^{-1}) \cdot (x \cdot y^{-1}) \overset{(Ass)}{=} y \cdot (x^{-1} \cdot x) \cdot y^{-1} \overset{(pIv)}{=} y \cdot 1 \cdot y^{-1} \overset{(Ass)}{=} (y \cdot 1) \cdot y^{-1} \overset{(pU)}{=} y \cdot y^{-1} \overset{(pIv)}{=} 1$; then, by (pIvP), (pIvG) holds.

(NegP): $(x \cdot y)^{-1} \overset{(DN)}{=} (x \cdot (y^{-1})^{-1})^{-1} \overset{(pIvG)}{=} y^{-1} \cdot x^{-1}$.

(qm-NegP): Obviously, by (pU).

(IdEqP): Suppose $y \cdot x^{-1} = 1$; then, $y^{-1} \overset{(pU)}{=} y^{-1} \cdot 1 = y^{-1} \cdot (y \cdot x^{-1}) \overset{(Ass)}{=} (y^{-1} \cdot y) \cdot x^{-1} \overset{(pIv)}{=} 1 \cdot x^{-1} \overset{(pU)}{=} x^{-1}$; then, $x^{-1} \cdot y = y^{-1} \cdot y \overset{(pIv)}{=} 1$. Conversely, suppose $x^{-1} \cdot y = 1$; then, $y^{-1} \overset{(pU)}{=} 1 \cdot y^{-1} = (x^{-1} \cdot y) \cdot y^{-1} \overset{(Ass)}{=} x^{-1} \cdot (y \cdot y^{-1}) \overset{(pIv)}{=} x^{-1} \cdot 1 \overset{(pU)}{=} x^{-1}$; then, $y \cdot x^{-1} = y \cdot y^{-1} \overset{(pIv)}{=} 1$. Thus, (IdEqP) holds.

(pEqP=): If $x = y$, then $y \cdot x^{-1} = x \cdot x^{-1} \stackrel{(pIv)}{=} 1$. Conversely, if $y \cdot x^{-1} = 1$, then
$x \stackrel{(pU)}{=} 1 \cdot x = (y \cdot x^{-1}) \cdot x \stackrel{(Ass)}{=} y \cdot (x^{-1} \cdot x) \stackrel{(pIv)}{=} y \cdot 1 = y$. Thus, (pEqP=) holds.
(pIvP) $\Longleftrightarrow$ (pEqP=), by (DN). □

Note that, in the non-commutative case, the property (DN) is the analogous of the properties (pDN^L) and (pDN^R) (pseudo-Double Negation), where the two negations coincide: $x^{-1} = x^{-} = x^{\sim}$. Hence, the group is an involutive algebra.

Remark 7.1.5 (See Remark 5.2.8)
The group (Definition 1) is a regular-m algebra with involutive regular-m negation, by Definition 5.2.7, since (pU), (pIv), (DN) hold.

- **New operations in groups: $\to$ and $\rightsquigarrow$**

The groups have the following very interesting property.

Proposition 7.1.6 *Let $(G, \cdot, ^{-1}, 1)$ be a group. For all $x \in G$, define the mapings $f_x : G \longrightarrow G$ and ${}_xf : G \longrightarrow G$ by: for all $g \in G$,*

$$f_x(g) = g \cdot x, \quad {}_xf(g) = x \cdot g.$$

Then, f_x and ${}_xf$ are bijective.

Proof.
Injectivity: Let $g_1, g_2 \in G$, $g_1 \neq g_2$. We must prove that $f_x(g_1) \neq f_x(g_2)$ and ${}_xf(g_1) \neq {}_xf(g_2)$.
Suppose, by absurdum hypothesis, that $f_x(g_1) = f_x(g_2)$, i.e. $g_1 \cdot x = g_2 \cdot x$; then, $(g_1 \cdot x) \cdot x^{-1} = (g_2 \cdot x) \cdot x^{-1}$ and hence, by (Ass), $g_1 = g_1 \cdot 1 = g_1 \cdot (x \cdot x^{-1}) = g_2 \cdot (x \cdot x^{-1}) = g_2 \cdot 1 = g_2$: contradiction.
Similarly, suppose, by absurdum hypothesis, that ${}_xf(g_1) = {}_xf(g_2)$, i.e. $x \cdot g_1 = x \cdot g_2$; then, $x^{-1} \cdot (x \cdot g_1) = x^{-1} \cdot (x \cdot g_2)$ and hence, by (Ass), $g_1 = 1 \cdot g_1 = (x^{-1} \cdot x) \cdot g_1 = (x^{-1} \cdot x) \cdot g_2 = 1 \cdot g_2 = g_2$: contradiction.
Surjectivity of f_x: We must prove that for all $y \in G$, there exists $z \in G$, such that $y = f_x(z) = z \cdot x$. Indeed, let $y \in G$ and take $z = y \cdot x^{-1}$; then, $f_x(z) = z \cdot x = (y \cdot x^{-1}) \cdot x \stackrel{(Ass)}{=} y \cdot (x^{-1} \cdot x) \stackrel{(pIv)}{=} y \cdot 1 \stackrel{(pU)}{=} y$.
Surjectivity of ${}_xf$: We must prove that for all $y \in G$, there exists $u \in G$, such that $y = {}_xf(u) = x \cdot u$. Indeed, let $y \in G$ and take $u = x^{-1} \cdot y$; then, ${}_xf(u) = x \cdot u = x \cdot (x^{-1} \cdot y) \stackrel{(Ass)}{=} (x \cdot x^{-1}) \cdot y \stackrel{(pIv)}{=} 1 \cdot y \stackrel{(pU)}{=} y$. □

We then have immediately the following corollary.

Corollary 7.1.7 *Let $(G, \cdot, ^{-1}, 1)$ be a group. For all $x, y \in G$,*
- there exists a unique $z \in G$, $z = y \cdot x^{-1}$, such that $y = f_x(z) = z \cdot x$ and
- there exists a unique $u \in G$, $u = x^{-1} \cdot y$, such that $y = {}_xf(u) = x \cdot u$.

By above Corollary 7.1.7, we can introduce two binary operations on $(G, \cdot, ^{-1}, 1)$ as follows: for all $x, y \in G$,

$$x \to y \stackrel{def.}{=} z = y \cdot x^{-1}, \quad x \rightsquigarrow y \stackrel{def.}{=} u = x^{-1} \cdot y,$$

and they satisfy the following property: for all $x, y \in G$,

(pR=) $\exists\ x \to y = z$ such that $z \cdot x = y$ and $\exists\ x \rightsquigarrow y = u$ such that $x \cdot u = y$.

Remark 7.1.8 Recall from Chapter 3 the dual properties: for all x, y,
(pR) $\exists\ x \to^L y = \max\{z \mid z \odot x \leq y\}$ and $\exists\ x \rightsquigarrow^L y = \max\{u \mid x \odot u \leq y\}$,
(pcoR) $\exists\ x \to^R y = \min\{z \mid z \oplus x \geq y\}$ and $\exists\ x \rightsquigarrow^R y = \min\{u \mid x \oplus u \leq y\}$.

Recall also the property (p-s^L) $(x \leq y \Longrightarrow x = y)$. Then, note that:
(pR) + (p-s^L) $\Longrightarrow$ (pR=) and, dually, (pcoR) + (p-s^L) $\Longrightarrow$ (pR=).

Since the group is an involutive non-commutative algebra, then, by analogy with (4.1), we introduce the new operations $\to$ and $\rightsquigarrow$ on G (the pair $(\to, \rightsquigarrow)$ is called "pseudo-implication") defined by: for all $x, y \in G$,

$$(7.1)\quad x \to y \stackrel{def.}{=} (x \cdot y^{-1})^{-1} \stackrel{(pIvG)}{=} y \cdot x^{-1}, \quad x \rightsquigarrow y \stackrel{def.}{=} (y^{-1} \cdot x)^{-1} \stackrel{(pIvG)}{=} x^{-1} \cdot y.$$

Note that:

(i) in additive (right) notation, a group is an algebra $(G, +, -, 0)$ and, hence, (7.1) becomes:

$$x \to y \stackrel{def.}{=} -(x + (-y)), \quad x \rightsquigarrow y \stackrel{def.}{=} -(-y + x).$$

(ii) In ([56], pag. 161), the implication $\rightsquigarrow$ is denoted by $\backslash$ $(x \backslash y = x \rightsquigarrow y)$ and the implication $\to$ is replaced by its inverse, denoted by $/$ (i.e. $x/y = y \to x$).

(iii) If the group is commutative, then the two implications coincide: $\to = \rightsquigarrow$.

(iv) The group may be defined equivalently as an algebra $(G, \cdot, \to, \rightsquigarrow, ^{-1}, 1)$, where $\to$ and $\rightsquigarrow$ are expressed in terms of $\cdot$ and $^{-1}$:

$$x \to y = (x \cdot y^{-1})^{-1} = y \cdot x^{-1}, \quad x \rightsquigarrow y = (y^{-1} \cdot x)^{-1} = x^{-1} \cdot y.$$

Remarks 7.1.9 (By analogy to Remarks 2.11.7, 2.11.8)

Since the group is an involutive algebra (with an unique negation), then we can define the analogous additional operations: $\oplus$ and $\Rightarrow$, $\approx>$ in two alternative ways.

(1) The first way (see Remarks 2.11.7): for all $x, y \in G$,
$x \oplus y \stackrel{def.}{=} (y^{-1} \cdot x^{-1})^{-1} = x \cdot y$, by (2.28), and
$x \Rightarrow y \stackrel{def.}{=} (x \oplus y^-)^- = (x \cdot y^{-1})^{-1} = y \cdot x^{-1} = x \to y$ and
$x \approx> y \stackrel{def.}{=} (y^- \oplus x)^- = (y^{-1} \cdot x)^{-1} = x^{-1} \cdot y = x \rightsquigarrow y$, by (2.29).

Hence, the old operations $\to$, $\rightsquigarrow$ are expressed in term of the "new" operations $\to$, $\rightsquigarrow$:
$x \to y = (x^- \approx> y^-)^- = (x^{-1} \rightsquigarrow y^{-1})^{-1}$ and
$x \rightsquigarrow y = (x^- \Rightarrow y^-)^- = (x^{-1} \to y^{-1})^{-1}$, by (2.32).

(2) The second alternative way (see Remarks 2.11.8): for all $x, y \in G$,
$x \oplus y \stackrel{def.}{=} (x^{-1} \cdot y^{-1})^{-1} = y \cdot x$, by (2.40), and

$x \Rightarrow y \overset{def.}{=} (x \oplus y^-)^- = (y^{-1} \cdot x)^{-1} = x^{-1} \cdot y = x \rightsquigarrow y$ and
$x \approx> y \overset{def.}{=} (y^- \oplus x)^- = (x \cdot y^{-1})^{-1} = y \cdot x^{-1} = x \to y$, by (2.41).

Hence, the old operations $\to, \rightsquigarrow$ are expressed in term of the "new" operations $\to, \rightsquigarrow$:
$x \to y = (x^- \Rightarrow y^-)^- = (x^{-1} \rightsquigarrow y^{-1})^{-1}$, by (2.44),
$x \rightsquigarrow y = (x^- \approx> y^-)^- = (x^{-1} \to y^{-1})^{-1}$, by (2.45).

Note that, in both ways, the new additional operations coincide with the old operations. Hence, because of an unique negation, we have an unique addition and two unique implications. We can say that the addition $\cdot$ is selfdual and that the dual of $\to$ is $\rightsquigarrow$ and the dual of $\rightsquigarrow$ is $\to$. Moreover, the implication $\to$ can be expressed in terms of $\rightsquigarrow$ and viceversa:

(7.2) $$x \to y = (x^{-1} \rightsquigarrow y^{-1})^{-1}, \quad x \rightsquigarrow y = (x^{-1} \to y^{-1})^{-1}.$$

Consequently, now one may better understand the results from papers [54] and [76] concerning algebras $(G, \circ)$ of type (2), with two (one respectively) equations, that are termwise equivalent to groups.

Remark 7.1.10 Note that in an involutive BCK(P) algebra (or, equivalently, by (iE5), an involutive po-cim(pR)), (7.1) does not hold, i.e. $(x \odot y^-)^- \neq y \odot x^-$. Indeed, take as example the following particular case (see [94], pag. 163): the Boolean algebra $L_{2\times 2} = \{0, a, b, 1\}$, with $0 < a, b < 1$, organized as a BCK(P) algebra with the operation $\to$ and $\odot$ as in the following tables ($x^- = x \to 0$):

$\mathcal{L}_{2\times 2}$

$\to$	0	a	b	1
0	1	1	1	1
a	b	1	b	1
b	a	a	1	1
1	0	a	b	1

$\odot$	0	a	b	1
0	0	0	0	0
a	0	a	0	a
b	0	0	b	b
1	0	a	b	1

.

Then, for $x = a$ and $y = 1$, we obtain $1 = 0^- = (a \odot 0)^- = (a \odot 1^-)^- \neq 1 \odot a^- = 1 \odot b = b$.

Remarks 7.1.11 Let $(G, \cdot, ^{-1}, 1)$ be a group.

(1) We have the *special property* (see (2.24)): for all $x, y \in G$,

(7.3) $$x \cdot y = (x \to y^{-1})^{-1} = y^{-1} \to x, \quad x \cdot y = (y \rightsquigarrow x^{-1})^{-1} = x^{-1} \rightsquigarrow y.$$

Indeed, $(x \to y^{-1})^{-1} = (y^{-1} \cdot x^{-1})^{-1} \overset{(pIv)}{=} x \cdot y$ and $y^{-1} \to x = x \cdot (y^{-1})^{-1} \overset{(DN)}{=} x \cdot y$ and $(y \rightsquigarrow x^{-1})^{-1} = (y^{-1} \cdot x^{-1})^{-1} = x \cdot y$ and $x^{-1} \rightsquigarrow y = (x^{-1})^{-1} \cdot y = x \cdot y$.

(2) It follows that we also have the property: for all $x, y \in G$,
(pDNeg2) $y^{-1} \to x = x^{-1} \rightsquigarrow y (= x \cdot y)$.

Remark 7.1.12 (See Remark 7.1.10)

Recall that in an involutive pBCK(pP) algebra, the properties (2.24) and (pDNeg2^{Lb}) hold, but the analogous of (7.3) does not hold. Indeed, take the commutative case, i.e. an involutive BCK(P) algebra, and here take the Boolean algebra $\mathcal{L}_{2\times 2}$ from the previous Remark 7.1.10. We prove that $(x \to (y^-))^- \neq y^- \to x$. Indeed, for $x = a$ and $y = b$ we obtain $0 = 1^- = (a \to a)^- = (a \to (b^-))^- \neq b^- \to a = a \to a = 1$.

Proposition 7.1.13 *Let $(G, \cdot, ^{-1}, 1)$ be a group. Then, we have: for all $x, y, z \in G$,*

(Id) $x \to 1 = x \rightsquigarrow 1$,
(pBB=) $y \to z = (z \to x) \rightsquigarrow (y \to x)$, $y \rightsquigarrow z = (z \rightsquigarrow x) \to (y \rightsquigarrow x)$,
(pD=) $(y \to x) \rightsquigarrow x = y = (y \rightsquigarrow x) \to x$,
(IdEq) $x \to y = 1 \Leftrightarrow x \rightsquigarrow y = 1$,
(pEq=) $x = y \Leftrightarrow x \to y = 1 \overset{(IdEq)}{\Leftrightarrow} x \rightsquigarrow y = 1$,
(pM) $1 \to x = x = 1 \rightsquigarrow x$,
(pEx) $z \rightsquigarrow (y \to x) = y \to (z \rightsquigarrow x)$,
(pB=) $z \to x = (y \to z) \to (y \to x)$, $z \rightsquigarrow x = (y \rightsquigarrow z) \rightsquigarrow (y \rightsquigarrow x)$,
(pRe') $x \to x = 1 = x \rightsquigarrow x$,
(pGa=) $x \cdot y = z \Leftrightarrow x = y \to z \Leftrightarrow y = x \rightsquigarrow z$ *(see [56], page 161),*
(p@) $x \to y = (x \cdot z) \to (y \cdot z)$, $x \rightsquigarrow y = (z \cdot x) \rightsquigarrow (z \cdot y)$.

Proof. (Id): $x \to 1 \overset{def.}{=} 1 \cdot x^{-1} \overset{(pU)}{=} x^{-1}$ and $x \rightsquigarrow 1 \overset{def.}{=} x^{-1} \cdot 1 \overset{(pU)}{=} x^{-1}$. Thus, (Id) holds.

(pBB=): $(z \to x) \rightsquigarrow (y \to x) = (x \cdot z^{-1}) \rightsquigarrow (x \cdot y^{-1}) = (x \cdot z^{-1})^{-1} \cdot (x \cdot y^{-1}) = (z \cdot x^{-1}) \cdot (x \cdot y^{-1}) \overset{(Ass),(pIv)}{=} z \cdot 1 \cdot y^{-1} \overset{(Ass),(pU)}{=} z \cdot y^{-1} = y \to z$ and
$(z \rightsquigarrow x) \to (y \rightsquigarrow x) = (z^{-1} \cdot x) \to (y^{-1} \cdot x) = (y^{-1} \cdot x) \cdot (z^{-1} \cdot x)^{-1} = (y^{-1} \cdot x) \cdot (x^{-1} \cdot z) \overset{(Ass),(pIv)}{=} y^{-1} \cdot 1 \cdot z \overset{(Ass),(pU)}{=} y^{-1} \cdot z = y \rightsquigarrow z$.

(pD=): $(y \to x) \rightsquigarrow x = (x \cdot y^{-1}) \rightsquigarrow x = (x \cdot y^{-1})^{-1} \cdot x = (y \cdot x^{-1}) \cdot x = y$ and $(y \rightsquigarrow x) \to x = (y^{-1} \cdot x) \to x = x \cdot (y^{-1} \cdot x)^{-1} = x \cdot (x^{-1} \cdot y) = y$.

(IdEq): If $x \to y = 1$, i.e. $y \cdot x^{-1} = 1$, then $y \overset{(pU)}{=} y \cdot 1 \overset{(pIv)}{=} y \cdot (x^{-1} \cdot x) \overset{(Ass)}{=} (y \cdot x^{-1}) \cdot x = 1 \cdot x \overset{(pU)}{=} x$, and hence $x \rightsquigarrow y = x^{-1} \cdot y = x^{-1} \cdot x = 1$. Similarly, if $x \rightsquigarrow y = 1$, then $x \to y = 1$.

(pEq=): If $x = y$, then $1 \overset{(pIv)}{=} x \cdot x^{-1} = y \cdot x^{-1} = x \to y$; if $x \to y = 1$, i.e. $y \cdot x^{-1} = 1$, then $y \overset{(pU)}{=} y \cdot 1 \overset{(pIv)}{=} y \cdot (x^{-1} \cdot x) \overset{(Ass)}{=} (y \cdot x^{-1}) \cdot x = 1 \cdot x \overset{(pU)}{=} x$. Similarly, $x = y \Leftrightarrow x \rightsquigarrow y = 1$.

(pM): $1 \to x = x \cdot 1^{-1} \overset{(N1)}{=} x \cdot 1 \overset{(pU)}{=} x$ and $1 \rightsquigarrow x = 1^{-1} \cdot x \overset{(N1)}{=} 1 \cdot x \overset{(pU)}{=} x$.

(pEx): $z \rightsquigarrow (y \to x) = z \rightsquigarrow (x \cdot y^{-1}) = z^{-1} \cdot (x \cdot y^{-1})$; $y \to (z \rightsquigarrow x) = y \to (z^{-1} \cdot x) = (z^{-1} \cdot x) \cdot y^{-1} \overset{(Ass)}{=} z^{-1} \cdot (x \cdot y^{-1})$; thus (pEx) holds.

(pB=): $(y \to z) \to (y \to x) = (x \cdot y^{-1}) \cdot (z \cdot y^{-1})^{-1} = (x \cdot y^{-1}) \cdot (y \cdot z^{-1}) = x \cdot z^{-1} = z \to x$ and $(y \rightsquigarrow z) \rightsquigarrow (y \rightsquigarrow x) = (y^{-1} \cdot z)^{-1} \cdot (y^{-1} \cdot x) = (z^{-1} \cdot y) \cdot (y^{-1} \cdot x) = z^{-1} \cdot x = z \rightsquigarrow x$, by (Ass).

(pRe'): $x \to x = x \cdot x^{-1} \overset{(pIv)}{=} 1$ and $x \rightsquigarrow x = x^{-1} \cdot x \overset{(pIv)}{=} 1$.

(pGa=): $x \cdot y = z$ implies $x = z \cdot y^{-1} = y \to z$ and $y = x^{-1} \cdot z = x \rightsquigarrow z$, by (pIv); conversely, $x = y \to z$, i.e. $x = z \cdot y^{-1}$, implies $x \cdot y = z$ and, similarly, $y = x \rightsquigarrow z$ implies $x \cdot y = z$ too, by (pIv).

(p@): $(x \cdot z) \to (y \cdot z) = (y \cdot z) \cdot (x \cdot z)^{-1} = (y \cdot z) \cdot (z^{-1} \cdot x^{-1}) = y \cdot x^{-1} = x \to y$, $(z \cdot x) \rightsquigarrow (z \cdot y) = (z \cdot x)^{-1} \cdot (z \cdot y) = (x^{-1} \cdot z^{-1}) \cdot (z \cdot y) = x^{-1} \cdot y = x \rightsquigarrow y$. □

Proposition 7.1.14 *(See Proposition 2.3.25 and Lemma 2.11.2)*

Let $(G, \cdot, ^{-1}, 1)$ be a group. Then, we have: for all $x, y, z \in G$,
(pNg) $(x \to y)^{-1} = y \to x$ *and* $(x \rightsquigarrow y)^{-1} = y \rightsquigarrow x$,
(q-pNeg) $x \to 1 = 1 \to x^- = (1 \rightsquigarrow x)^- = x \rightsquigarrow 1 = 1 \rightsquigarrow x^- = (1 \to x)^-$,
(pNg1) $x \to y^{-1} = y \rightsquigarrow x^{-1}$,
(pDNeg1) $x \to y = y^{-1} \rightsquigarrow x^{-1}$ *and* $x \rightsquigarrow y = y^{-1} \to x^{-1}$,
(pDNeg2) $y^{-1} \to x = x^{-1} \rightsquigarrow y (= x \cdot y)$.

Proof.
(pNg): $(x \to y)^{-1} \overset{def.}{=} (y \cdot x^{-1})^{-1} \overset{(NegP)}{=} (x^{-1})^{-1} \cdot y^{-1} \overset{(DN)}{=} x \cdot y^{-1} \overset{def.}{=} y \to x$ and $(x \rightsquigarrow y)^{-1} \overset{def.}{=} (x^{-1} \cdot y)^{-1} \overset{(NegP)}{=} y^{-1} \cdot (x^{-1})^{-1} \overset{(DN)}{=} y^{-1} \cdot x \overset{def.}{=} y \rightsquigarrow x$.
(q-pNeg): Routine, by using (pU), (N1), (DN) and (qm-NegP).
(pNg1): $x \rightsquigarrow y^{-1} = x^{-1} \cdot y^{-1} = y \to x^{-1}$.
(pDNeg1): $y^{-1} \to x^{-1} = x^{-1} \cdot (y^{-1})^{-1} = x^{-1} \cdot y = x \rightsquigarrow y$ and $y^{-1} \rightsquigarrow x^{-1} = (y^{-1})^{-1} \cdot x^{-1} = y \cdot x^{-1} = x \to y$.
(pDNeg2): $(y^{-1}) \to x = x \cdot (y^{-1})^{-1} = x \cdot y$ and $(x^{-1}) \rightsquigarrow y = (x^{-1})^{-1} \cdot y = x \cdot y$, by (DN). □

Proposition 7.1.15 *Let $(G, \cdot, ^{-1}, 1)$ be a group. Then, we have: for all $x, y, z \in G$,*

$$x \cdot y = (x \to 1) \rightsquigarrow y = (y \rightsquigarrow 1) \to x, \tag{7.4}$$

$$(x \cdot y) \to z = x \to (y \to z), \quad (y \cdot x) \rightsquigarrow z = x \rightsquigarrow (y \rightsquigarrow z), \tag{7.5}$$

$$(z \to x) \cdot y = x \cdot (z \rightsquigarrow y), \tag{7.6}$$

$$z \to (x \cdot y) = x \cdot (z \to y), \quad z \rightsquigarrow (x \cdot y) = (z \rightsquigarrow x) \cdot y, \tag{7.7}$$

$$[(y \to x) \rightsquigarrow x] \to x = y \to x, \quad [(y \rightsquigarrow x) \to x] \rightsquigarrow x = y \rightsquigarrow x, \tag{7.8}$$

$$x^{-1} = x \to 1 = x \rightsquigarrow 1 \quad (\textit{see Proposition } 2.3.21), \tag{7.9}$$

$$x = y \Leftrightarrow y^{-1} = x^{-1}, \tag{7.10}$$

$$(x \to 1)^{-1} = x = (x \rightsquigarrow 1)^{-1}, \tag{7.11}$$

$$(y \cdot x) \to x = y^{-1} = (x \cdot y) \rightsquigarrow x, \tag{7.12}$$

Proof. (7.4): $x \cdot y = (x^{-1})^{-1} \cdot y = (1 \cdot x^{-1})^{-1} \cdot y = (x \to 1) \rightsquigarrow y$ and $x \cdot y = x \cdot (y^{-1})^{-1} = x \cdot (y^{-1} \cdot 1)^{-1} = (y \rightsquigarrow 1) \to x$.

(7.5): $x \to (y \to z) = x \to (z \cdot y^{-1}) = (z \cdot y^{-1}) \cdot x^{-1} \overset{(Ass)}{=} z \cdot (y^{-1} \cdot x^{-1}) = z \cdot (x \cdot y)^{-1} = (x \cdot y) \to z$ and
$x \rightsquigarrow (y \rightsquigarrow z) = x \rightsquigarrow (y^{-1} \cdot z) = x^{-1} \cdot (y^{-1} \cdot z) = [x^{-1} \cdot y^{-1}] \cdot z = (y \cdot x)^{-1} \cdot z = (y \cdot x) \rightsquigarrow z$.

(7.6): $(z \to x) \cdot y = (x \cdot z^{-1}) \cdot y \overset{(Ass)}{=} x \cdot (z^{-1} \cdot y) = x \cdot (z \rightsquigarrow y)$.

(7.7): $z \to (x \cdot y) = (x \cdot y)z^{-1} \overset{(Ass)}{=} x \cdot (y \cdot z^{-1}) = x \cdot (z \to y)$ and $z \rightsquigarrow (x \cdot y) = (z^{-1}) \cdot (x \cdot y) \overset{(Ass)}{=} ((z^{-1}) \cdot x) \cdot y = (z \rightsquigarrow x) \cdot y$.

(7.8): $[(y \to x) \rightsquigarrow x] \to x = [(x \cdot y^{-1}) \rightsquigarrow x] \to x = [(x \cdot y^{-1})^{-1} \cdot x] \to x = [(y \cdot x^{-1}) \cdot x] \to x = [y \cdot (x^{-1} \cdot x)] \to x = [y \cdot 1] \to x = y \to x$ and $[(y \rightsquigarrow x) \to x] \rightsquigarrow x = [(y^{-1} \cdot x) \to x] \rightsquigarrow x = [x \cdot (y^{-1} \cdot x)^{-1}] \rightsquigarrow x = [x \cdot (x^{-1} \cdot y)] \rightsquigarrow x = [(x \cdot (x^{-1})) \cdot y] \rightsquigarrow x = [1 \cdot y] \rightsquigarrow x = y \rightsquigarrow x$.

(7.9): $x \to 1 = 1 \cdot x^{-1} = x^{-1}$ and $x \rightsquigarrow 1 = x^{-1} \cdot 1 = x^{-1}$, by (pU).

(7.10): (i) If $x = y$, then, obviously, $y^{-1} = x^{-1}$; (ii) if $y^{-1} = x^{-1}$, then, by (i), $(x^{-1})^{-1} = (y^{-1})^{-1}$, i.e. $x = y$, by (DN).

(7.11): $(x \to 1)^{-1} = (1 \cdot x^{-1})^{-1} = x \cdot 1^{-1} \overset{(N1)}{=} x \cdot 1 \overset{(pU)}{=} x$ and $(x \rightsquigarrow 1)^{-1} = (x^{-1} \cdot 1)^{-1} \overset{(pU)}{=} (x^{-1})^{-1} \overset{(DN)}{=} x$.

(7.12): by (7.5). □

Proposition 7.1.16 *Let* $(G, \cdot, ^{-1}, 1)$ *be a group. Then, we have: for all* $x, y \in G$,
(α) $(x \to y) \cdot x = y = x \cdot (x \rightsquigarrow y)$,
(β) $x \to (y \cdot x) = y = x \rightsquigarrow (x \cdot y)$,

(7.13) $$y \to (x \to (y \cdot x)) = 1 = y \rightsquigarrow (x \rightsquigarrow (x \cdot y)).$$

Proof. (α): by (7.6).

(β): by (7.7).

(7.13): $y \to (x \to (y \cdot x)) = y \to y = 1$ and $y \rightsquigarrow (x \rightsquigarrow (x \cdot y)) = y \rightsquigarrow y = 1$, by (β) and (pRe'). □

Remark 7.1.17 Note that:
- ($pGa^=$) is a join of the analogous of the properties: (pPR) and (pScoR),
- (α) is the analogous of both the properties (pRR) and (pcoRR),
- (β) is the analogous of both the properties (pPP) and (pSS).

Proposition 7.1.18 *In a group* $(G, \cdot, ^{-1}, 1)$*, the following properties hold: for all* $x, y, z, x_1, x_2, \ldots, x_n \in G$,
(a) $(y \to z) \cdot (x \to y) = x \to z$, $(x \rightsquigarrow y) \cdot (y \rightsquigarrow z) = x \rightsquigarrow z$,
(b) $(x_{n-1} \to x_n) \cdot \ldots \cdot (x_2 \to x_3) \cdot (x_1 \to x_2) = x_1 \to x_n$,
$(x_1 \rightsquigarrow x_2) \cdot (x_2 \rightsquigarrow x_3) \cdot \ldots \cdot (x_{n-1} \rightsquigarrow x_n) = x_1 \rightsquigarrow x_n$.

Proof.
(a): $(x \rightsquigarrow y) \cdot (y \rightsquigarrow z) = (x^{-1} \cdot y) \cdot (y^{-1} \cdot z) = x^{-1} \cdot (y \cdot y^{-1}) \cdot z = x^{-1} \cdot 1 \cdot z = x^{-1} \cdot z = x \rightsquigarrow z$. The rest has similar proof. □

7.1.2 The X-group - an intermediary notion

Definition 7.1.19 We shall name *X-group* an algebra $(G, \cdot, \to, \rightsquigarrow, 1)$ of type $(2, 2, 2, 0)$ such that ((IdEq) follows) $(G, \cdot, 1)$ is a monoid (i.e. (Ass) and (pU) hold) and the operation $\cdot$ is connected to the pseudo-implication $(\to, \rightsquigarrow)$ by the following property: for all $x, y, z \in G$,
($pGa^=$) $x \cdot y = z \Leftrightarrow x = y \to z \Leftrightarrow y = x \rightsquigarrow z$.

Remark 7.1.20 Note that (pGa$^=$) is a particular non-commutative Galois connection, with $=$ instead of $\leq$ or $\geq$.

Denote by **X-group** the class of all X-groups.
First, we present the following general lemma:

Lemma 7.1.21 *Let $\mathcal{A} = (A, \cdot, \to, \rightsquigarrow)$ (or $\mathcal{A} = (A, \to, \rightsquigarrow, \cdot)$) be an algebra of type $(2, 2, 2)$. Then, we have the following equivalence:*

$$(pGa^=) \Leftrightarrow (\alpha) + (\beta),$$

where: for all $x, y \in A$,

$$(\alpha) \qquad (x \to y) \cdot x = y = x \cdot (x \rightsquigarrow y),$$

$$(\beta) \qquad x \to (y \cdot x) = y = x \rightsquigarrow (x \cdot y).$$

Proof. (pGa$^=$) $\Longrightarrow$ $(\alpha) + (\beta)$:
(α): $(x \to y) \cdot x = y \overset{(pGa^=)}{\Leftrightarrow} x \to y = x \to y$ and $x \cdot (x \rightsquigarrow y) = y \overset{(pGa^=)}{\Leftrightarrow} x \rightsquigarrow y = x \rightsquigarrow y$, which are true.
(β): $y = x \to (y \cdot x) \overset{(pGa^=)}{\Leftrightarrow} y \cdot x = y \cdot x$ and $y = x \rightsquigarrow (x \cdot y) \overset{(pGa^=)}{\Leftrightarrow} x \cdot y = x \cdot y$, which are true.

$(\alpha) + (\beta) \Longrightarrow$ (pGa$^=$):
Suppose $x \cdot y = z$; then $y \to z = y \to (x \cdot y) \overset{(\beta)}{=} x$ and $x \rightsquigarrow z = x \rightsquigarrow (x \cdot y) \overset{(\beta)}{=} y$.
Conversely, suppose $x = y \to z$; then $x \cdot y = (y \to z) \cdot y \overset{(\alpha)}{=} z$; suppose $y = x \rightsquigarrow z$; then $x \cdot y = x \cdot (x \rightsquigarrow z) \overset{(\alpha)}{=} z$. □

Proposition 7.1.22 *Let $(G, \cdot, \to, \rightsquigarrow, 1)$ be an X-group. Then, the properties (α) and (β) hold and (pGa$^=$) $\Leftrightarrow$ $(\alpha) + (\beta)$.*

Proof. By above Lemma 7.1.21. □

Proposition 7.1.23 *Let $(G, \cdot, \to, \rightsquigarrow, 1)$ be an X-group. Then, the following properties hold: for all $x, y, z \in G$:*
(Id) $\quad x \to 1 = x \rightsquigarrow 1,$
(pEx) $\quad x \to (y \rightsquigarrow z) = y \rightsquigarrow (x \to z),$
(pM) $\quad 1 \to x = x = 1 \rightsquigarrow x,$
(IdEq) $\quad x \to y = 1 \Leftrightarrow x \rightsquigarrow y = 1,$
(pEq=) $\quad x = y \Longleftrightarrow x \to y = 1 \; (\overset{(IdEq)}{\Leftrightarrow} x \rightsquigarrow y = 1 \,).$

Proof.
(Id): $x \to 1 \overset{(pU)}{=} (x \to 1) \cdot 1 \overset{(\alpha)}{=} (x \to 1) \cdot [x \cdot (x \rightsquigarrow 1)] \overset{(Ass)}{=}$
$[(x \to 1) \cdot x] \cdot (x \rightsquigarrow 1) \overset{(\alpha)}{=} 1 \cdot (x \rightsquigarrow 1) \overset{(pU)}{=} x \rightsquigarrow 1.$

(pEx): $x \to (y \rightsquigarrow z) = a \overset{(pGa^=)}{\Leftrightarrow} x = a \rightsquigarrow (y \rightsquigarrow z) \overset{(pGa^=)}{\Leftrightarrow} a \cdot x = y \rightsquigarrow z$
$\overset{(pGa^=)}{\Leftrightarrow} y \cdot (a \cdot x) = z \overset{(Ass)}{\Leftrightarrow} (y \cdot a) \cdot x = z \overset{(pGa^=)}{\Leftrightarrow} y \cdot a = x \to z$
$\overset{(pGa^=)}{\Leftrightarrow} a = y \rightsquigarrow (x \to z$; thus, (pEx) holds.

(pM): $1 \to x \overset{(pU)}{=} (1 \to x) \cdot 1 = x \overset{(pGa^{=})}{\Leftrightarrow} 1 \to x = 1 \to x$, that is true.

Also, $1 \rightsquigarrow x \overset{(pU)}{=} 1 \cdot (1 \rightsquigarrow x) = x \overset{(pGa^{=})}{\Leftrightarrow} 1 \rightsquigarrow x = 1 \rightsquigarrow x$, that is true. Thus, (pM) holds.

(IdEq): $x = y \overset{(pU)}{\Leftrightarrow} 1 \cdot x = y \overset{(pGa^{=})}{\Leftrightarrow} 1 = x \to y$ and $x = y \overset{(pU)}{\Leftrightarrow} x \cdot 1 = y \overset{(pGa^{=})}{\Leftrightarrow} 1 = x \rightsquigarrow y$. Thus, (IdEq) holds.

(pEq=): By above proof. □

Note that the X-group may be defined equivalently as an algebra $(G, \cdot, \to, \rightsquigarrow, ^{-1}, 1)$, where $^{-1}$ can be expressed in terms of $\to$, $\rightsquigarrow$ and 1: $x^{-1} = x \to 1 \overset{(Id)}{=} x \rightsquigarrow 1$.

7.1.3 The basic equivalence (EG1)

Here we present the basic equivalence (EG1) between groups and X-groups.

We prove that the groups are termwise equivalent to the X-groups (result which can be found in ([56], page 161) in a different form; we prefer the following form):

Theorem 7.1.24

(1) Let $\mathcal{G} = (G, \cdot, ^{-1}, 1)$ be a group.

Define $\rho(\mathcal{G}) \overset{def.}{=} (G, \cdot, \to, \rightsquigarrow, 1)$, where $x \to y \overset{def.}{=} y \cdot x^{-1}$ and $x \rightsquigarrow y \overset{def.}{=} x^{-1} \cdot y$.

Then, $\rho(\mathcal{G})$ is an X-group.

(1') Conversely, let $\mathcal{G} = (G, \cdot, \to, \rightsquigarrow, 1)$ be an X-group.

Define $\rho^*(\mathcal{G}) \overset{def.}{=} (G, \cdot, ^{-1}, 1)$, where $x^{-1} \overset{def.}{=} x \to 1 \overset{(Id)}{=} x \rightsquigarrow 1$.

Then, $\rho^*(\mathcal{G})$ is a group.

(2) The above defined mappings ρ and ρ^* are mutually inverse.

Proof.

(1): By Proposition 7.1.13.

(1'): Since $(G, \cdot, 1)$ is monoid, it follows that (Ass), (pU) hold. It remains to prove that (pIv) holds too. Indeed, $(x^{-1}) \cdot x = (x \to 1) \cdot x \overset{(\alpha)}{=} 1$ and $x \cdot (x^{-1}) = x \cdot (x \rightsquigarrow 1) \overset{(\alpha)}{=} 1$; thus, (pIv) holds.

(2): If $(G, \cdot, ^{-1}, 1) \xrightarrow{\rho} (G, \cdot, \to, \rightsquigarrow, 1) \xrightarrow{\rho^*} (G, \cdot, \sim, 1)$, then $\sim x = x \to 1 = 1 \cdot x^{-1} = x^{-1}$.

Let now $(G, \cdot, \to, \rightsquigarrow, 1) \xrightarrow{\rho^*} (G, \cdot, ^{-1}, 1) \xrightarrow{\rho} (G, \cdot, \Rightarrow, \approx>, 1)$.

First we prove that $x \Rightarrow y = x \to y$. Indeed,

$x \Rightarrow y = y \cdot x^{-1} = y \cdot (x^{-1}) = y \cdot (x \to 1) \overset{(Id)}{=} y \cdot (x \rightsquigarrow 1) \overset{(\alpha)}{=}$
$[(x \to y) \cdot x] \cdot (x \rightsquigarrow 1) \overset{(Ass)}{=} (x \to y) \cdot [x \cdot (x \rightsquigarrow 1)] \overset{(\alpha)}{=} (x \to y) \cdot 1 \overset{(pU)}{=} x \to y$.

Now we shall prove that $x \approx> y = x \rightsquigarrow y$. Indeed,

$x \approx> y = x^{-1} \cdot y = (x \to 1) \cdot y \overset{(\alpha)}{=} (x \to 1) \cdot [x \cdot (x \rightsquigarrow y)] \overset{(Ass)}{=}$
$[(x \to 1) \cdot x] \cdot (x \rightsquigarrow y) \overset{(\alpha)}{=} 1 \cdot (x \rightsquigarrow y) \overset{(pU)}{=} x \rightsquigarrow y$. □

Hence, we have the basic equivalence:

(EG1) **X-group** $\Leftrightarrow$ **group**.

7.2 Po-groups

7.2.1 The po-group

Definition 7.2.1 A *partially-ordered group*, or a *po-group* for short, is a structure $\mathcal{G} = (G, \leq, \cdot, ^{-1}, 1)$, such that $(G, \cdot, ^{-1}, 1)$ is a group, $(G, \leq)$ is a poset and $\leq$ is compatible with $\cdot$, i.e. the property (pCp) holds: for all $x, y \in G$,
(pCp) $x \leq y \Longrightarrow a \cdot x \leq a \cdot y$ and $x \cdot a \leq y \cdot a$.

Remarks 7.2.2 The presence of the order relation $\leq$ implies the presence of the *duality principle*. It follows that there are two dual po-groups, $\mathcal{G}_1$ and $\mathcal{G}_2$.

(i) If their support sets coincide ($G = G_1 = G_2$), we shall say that G is *self-dual*, i.e. the dual of $(G, \leq, \cdot, ^{-1}, 1)$ is $(G, \geq, \cdot, ^{-1}, 1)$ and vice-versa. Namely, if $(G, \leq, \cdot, ^{-1}, 1)$ is a po-group, then note that $\cdot$ is a pseudo-t-norm on the poset $(G, \leq)$ containing 1, while in the dual po-group $(G, \geq, \cdot, ^{-1}, 1)$, $\cdot$ is a pseudo-t-conorm on the poset $(G, \geq)$ containing 1, by Definition 1.1.1. It follows that we can say that $(G, \leq, \cdot, ^{-1}, 1)$ is the left-po-group, while $(G, \geq, \cdot, ^{-1}, 1)$ is the right-po-group.

(ii) If their support sets differ ($G_1 \neq G_2$), then their Unit elements 1_1, 1_2 differ and suppose that $1_1 < 1_2$ in the union set $G_1 \cup G_2$; we shall then call $\mathcal{G}_1$ as the *left-po-group* and $\mathcal{G}_2$ as the *right-po-group*.

Denote by **po-group** the class of all po-groups, namely:
- by **po-group**^{L}, the class of all left-po-groups and
- by **po-group**^{R}, the class of all right-l-groups.

Proposition 7.2.3 *Let $\mathcal{G}$ be a po-group. Then, we have: for all $x, y, a, b \in G$, if $x \leq y$ and $a \leq b$, then $x \cdot a \leq y \cdot b$ and $a \cdot x \leq b \cdot y$.*

Proposition 7.2.4 *(See Lemma 2.11.2)*

Let $\mathcal{G}$ be a po-group. Then (we recall the old results from literature): for all $x, y, a \in G$,
(ImNeg) $x \leq y \Longrightarrow y^{-1} \leq x^{-1}$,
(EqNeg) $x \leq y \Leftrightarrow y^{-1} \leq x^{-1}$,
(pEqCp) $x \leq y \Leftrightarrow a \cdot x \leq a \cdot y \Leftrightarrow x \cdot a \leq y \cdot a$.

Proof. We recall only the proof of (ImNeg): let $x \leq y$; then, by (pCp), (pIv), $1 = x \cdot (x^{-1}) \leq y \cdot (x^{-1})$; then, by (pCp) again and (pU), (Ass), (pIv), (pU), $y^{-1} = (y^{-1}) \cdot 1 \leq (y^{-1}) \cdot [y \cdot (x^{-1})] = [(y^{-1}) \cdot y] \cdot (x^{-1}) = 1 \cdot (x^{-1}) = x^{-1}$. $\Box$

Proposition 7.2.5 *Let $\mathcal{G}$ be a po-group. Then, we have: for all $x, y, a, z \in G$,*
(pGa$^{\leq}$) $x \cdot y \leq z \Leftrightarrow x \leq y \to z \Leftrightarrow y \leq x \rightsquigarrow z$ *and, dually,*
(pGa$^{\geq}$) $x \cdot y \geq z \Leftrightarrow x \geq y \to z \Leftrightarrow y \geq x \rightsquigarrow z$.

Proof.

(pGa$^{\leq}$): $x \leq y \to z \Leftrightarrow x \leq z \cdot y^{-1} \Longrightarrow x \cdot y \leq (z \cdot y^{-1}) \cdot y = z \cdot (y^{-1} \cdot y) = z \cdot 1 = z$, by (pCp), (Ass), (pIv), (pU) and $x \cdot y \leq z \Longrightarrow x = x \cdot 1 = x \cdot (y \cdot y^{-1}) = (x \cdot y) \cdot (y^{-1}) \leq z \cdot (y^{-1}) = y \to z$, by (pCp). Similarly, $y \leq x \rightsquigarrow z \Leftrightarrow y \leq x^{-1} \cdot z \Leftrightarrow x \cdot y \leq z$, by (pCp).

(pGa$^{\geq}$): dually. □

Remark 7.2.6 Note that:
- (pGa$^{\leq}$) is the analogous of the property (pPR)=(pRP),
- (pGa$^{\geq}$) is the analogous of the property (pScoR)=(pcoRS).

Theorem 7.2.7

(1) Let $\mathcal{G} = (G, \leq, \cdot, ^{-1}, 1)$ be a po-group. Define $\phi(\mathcal{G}) \stackrel{def.}{=} (G, \leq, \cdot, \to, \rightsquigarrow, 1)$ by

$$x \to y \stackrel{def.}{=} (x \cdot y^{-1})^{-1} = y \cdot x^{-1}, \quad x \rightsquigarrow y \stackrel{def.}{=} (y^{-1} \cdot x)^{-1} = x^{-1} \cdot y.$$

Then, $\phi(\mathcal{G})$ is a po-m (pGa$^{\leq}$) (verifying (pIv), where $x^{-1} = x \to 1 = x \rightsquigarrow 1$, by (7.9)).

(1') Let $\mathcal{G} = (G, \leq, \cdot, \to^L, \rightsquigarrow^L, 1)$ be a po-m(pPR), verifying (pIv), where $x^{-1} = x \to^L 1$. Define $\psi(\mathcal{G}) \stackrel{def.}{=} (G, \leq, \cdot, ^{-1}, 1)$.
Then, $\psi(\mathcal{G})$ is a po-group.

(2) The above defined mappings ϕ and ψ are mutually inverse and (pPR)= (pGa$^{\leq}$).

Proof. (1): By Proposition 7.2.5, the property (pGa$^{\leq}$) holds, hence $\phi(\mathcal{G})$ is a po-m(pPR), with (pPR)=(pGa$^{\leq}$) .

(1'): Let $\mathcal{G} = (G, \leq, \cdot, \to^L, \rightsquigarrow^L, 1)$ be a po-m(pPR), verifying (pIv) ($x \cdot x^{-1} = 1 = x^{-1} \cdot x$), where $x^{-1} = x \to^L 1$. We must prove that $\psi(\mathcal{G}) \stackrel{def.}{=} (G, \leq, \cdot, ^{-1}, 1)$ is a po-group. It remains to prove that (pCp) holds. Indeed, by Proposition 3.6.2, (pCp) holds.

(2): First, we have:

$(G, \leq, \cdot, ^{-1}, 1)$	$\stackrel{\phi}{\longrightarrow}$	$(G, \leq, \cdot, \to, \rightsquigarrow, 1)$	$\stackrel{\psi}{\longrightarrow}$	$(G, \leq, \cdot, ', 1)$
po-group		po-m(pPR) where $x \to y = y \cdot x^{-1}$ $x \rightsquigarrow y = x^{-1} \cdot y$		po-group where $x' = x \to 1$

We must prove that $x' = x^{-1}$, for all $x \in G$. Indeed, $x' = x \to 1 = 1 \cdot (x^{-1}) \stackrel{(pU)}{=} x^{-1}$.

Then, we have:

$(G, \leq, \cdot, \to^L, \rightsquigarrow^L, 1)$	$\stackrel{\psi}{\longrightarrow}$	$(G, \leq, \cdot, ^{-1}, 1)$	$\stackrel{\phi}{\longrightarrow}$	$(G, \leq, \cdot, \to, \rightsquigarrow, 1)$
po-m(pPR) verifying (pIv), where $x^{-1} = x \to^L 1$		po-group where $x^{-1} = x \to^L 1$		po-m(pGa$^{\leq}$) verifying (pIv), where $x^{-1} = x \to 1 = x \rightsquigarrow 1$

where:
(pPR) $x \cdot y \leq z \Leftrightarrow x \leq y \to^L z \Leftrightarrow y \leq x \rightsquigarrow^L z$
(pGa$^{\leq}$) $x \cdot y \leq z \Leftrightarrow x \leq y \to z \Leftrightarrow y \leq x \rightsquigarrow z$.
We must prove that:

$$x \to y = x \to^L y, \quad x \rightsquigarrow y = x \rightsquigarrow^L y.$$

We prove that $x \to y = x \to^L y$. Indeed,
$x \to y \leq x \to^L y \overset{(pPR)}{\Leftrightarrow} y \overset{(\alpha)}{=} (x \to y) \cdot x \leq y$, that is true by the reflexivity of $\leq$;
$x \to^L y \leq x \to y \overset{(pGa^{\leq})}{\Leftrightarrow} (x \to^L y) \cdot x \leq y$, that is true by (pRR), by Lemma 2.6.3.
Hence, by the antisymmetry of $\leq$, we obtain that $x \to y = x \to^L y$.

We prove now that $x \rightsquigarrow y = x \rightsquigarrow^L y$. Indeed,
$x \rightsquigarrow y \leq x \rightsquigarrow^l y \overset{(pPR)}{\Leftrightarrow} y \overset{(\alpha)}{=} x \cdot (x \rightsquigarrow y) \leq y$, that is true by the reflexivity of $\leq$;
$x \rightsquigarrow^L y \leq x \rightsquigarrow y \overset{(pGa^{\leq})}{\Leftrightarrow} x \cdot (x \rightsquigarrow^L y) \leq y$, that is true by (pRR).
Hence, we obtain that $x \rightsquigarrow y = x \rightsquigarrow^L y$.

Consequently, (pGa$^{\leq}$) = (pPR). □

Remark 7.2.8 By above Theorem 7.2.7 and its dual, we have the dual equivalences:

(EGM) **po-group**^{L} $\Leftrightarrow$ **po-m(pGa$^{\leq}$)** + (pIv) and, dually,
po-group^R $\Leftrightarrow$ **po-m(pGa$^{\geq}$)** + (pIv) .

Corollary 7.2.9 *A po-group is a po-m(pR).*

Proof. Let $(G, \leq, \cdot, ^{-1}, 1)$ be a po-group. By Theorem 7.2.7 (1), $(G, \leq, \cdot, \to, \rightsquigarrow, 1)$ is a po-m(pPR), hence $(G, \leq, \cdot, 1)$ is a po-m(pR), by Theorem 3.7.1 (1') (see (EM1)). □

Corollary 7.2.10 *Let $\mathcal{G}$ be a po-group. Then, for all $x, y \in G$, we have:*
(i) $y \leq 1 \Leftrightarrow x \leq y \to x \Leftrightarrow x \leq y \rightsquigarrow x$ and, dually,
(i') $y \geq 1 \Leftrightarrow x \geq y \to x \Leftrightarrow x \geq y \rightsquigarrow x$;
(ii) $y \leq z \Leftrightarrow 1 \leq y \to z \Leftrightarrow 1 \leq y \rightsquigarrow z$ and, dually,
(ii') $y \geq z \Leftrightarrow 1 \geq y \to z \Leftrightarrow 1 \geq y \rightsquigarrow z$.

Proof.
(i): take $z = x$ and then $z = y$ in (pGa$^{\leq}$). (i'): take $z = x$ and then $z = y$ in (pGa$^{\geq}$).
(ii): take $x = 1$ and then $y = 1$ in (pGa$^{\leq}$). (ii'): take $x = 1$ and then $y = 1$ in (pGa$^{\geq}$). □

Note that the properties (DN) and (EqNeg) make the operation $^{-1}$ be an *involution.*

Proposition 7.2.11 *Let $\mathcal{G}$ be a po-group. Then, for all $x, y, z \in G$, we have:*
(p^{*L}) $\quad x \leq y \implies z \to x \leq z \to y$ *and* $z \rightsquigarrow x \leq z \rightsquigarrow y$,
(p^{**L}) $\quad x \leq y \implies y \to z \leq x \to z$ *and* $y \rightsquigarrow z \leq x \rightsquigarrow z$.

Proof. ($p*^L$): Let $x \leq y$; then $z \to x = x \cdot z^{-1} \overset{(pCp)}{\leq} y \cdot z^{-1} = z \to y$ and $z \rightsquigarrow x = z^{-1} \cdot x \overset{(pCp)}{\leq} z^{-1} \cdot y = z \rightsquigarrow y$.

($p**^L$): Let $x \leq y$; then by (ImNeg), $y^{-1} \leq x^{-1}$; hence, $y \to z = z \cdot y^{-1} \overset{(pCp)}{\leq} z \cdot x^{-1} = x \to z$ and $y \rightsquigarrow z = y^{-1} \cdot z \overset{(pCp)}{\leq} x^{-1} \cdot z = x \rightsquigarrow z$. □

Corollary 7.2.12 *Let $\mathcal{G}$ be a po-group. Then, we have: for all $x, y \in G$, if $x \leq y$ then $x \to y \geq 1$, $x \rightsquigarrow y \geq 1$ and $y \to x \leq 1$, $y \rightsquigarrow x \leq 1$.*

Proof. Let $x \leq y$; then,
- by ($p*^L$), $x \to x \leq x \to y$, i.e., by (pRe'), $1 \leq x \to y$; similarly, $1 \leq x \rightsquigarrow y$;
- by ($p**^L$), $y \to x \leq x \to x = 1$, i.e. $y \to x \leq 1$; similarly, $y \rightsquigarrow x \leq 1$. □

7.2.2 The negative and the positive cones of po-groups

Let $\mathcal{G} = (G, \leq, \cdot, ^{-1}, 1)$ be a po-group (left-po-group or right-po-group).

Define the *negative cone* of $\mathcal{G}$ as follows: $G^{-} \overset{def.}{=} \{x \in G \mid x \leq 1\}$.

Define the *positive cone* of $\mathcal{G}$ as follows: $G^{+} \overset{def.}{=} \{x \in G \mid x \geq 1\}$.

Then, we have the following result.

Lemma 7.2.13 *(See Lemma 3.3.4)*
G^{-} and G^{+} are closed under $\cdot$.

Proof. Let $x, y \in G^{-}$, i.e. $x \leq 1$, $y \leq 1$; then, by (pCp), (pU), $x \cdot y \leq 1 \cdot y = y$; since $y \leq 1$, it follows, by transitivity of $\leq$, that $x \cdot y \leq 1$, i.e. $x \cdot y \in G^{-}$. Similarly, G^{+} is closed under $\cdot$. □

Note that G^{-} and G^{+} are not closed under $^{-1}$.

Note that a po-group $(G, \leq, \cdot, ^{-1}, 1)$ is a po-m $(G, \leq, \cdot, 1)$ verifying (pIv). By Lemma 7.2.13, we immediately obtain:

Proposition 7.2.14 *(See Proposition 3.3.5)*
Let $\mathcal{G} = (G, \leq, \cdot, ^{-1}, 1)$ be a po-group. Then
(1) $\mathcal{G}^{-} = (G^{-}, \leq, \odot = \cdot, \mathbf{1} = 1)$ is a left-po-im;
(1') $\mathcal{G}^{+} = (G^{+}\cdot, \leq, \oplus = \cdot, \mathbf{0} = 1)$ is a right-po-im.

If the partial order $\leq$ is linear (total), then $\mathcal{G}$ is a *linearly-ordered group* or a *totally-ordered group*. Note that in a linearly-ordered group $\mathcal{G}$ we have either $x \leq 1$ or $x \geq 1$ (i.e. $1 \leq x$), for any $x \in G$, hence $G = G^{-} \cup G^{+}$.

7.2.3 The X-po-group - an intermediary notion

Definition 7.2.15 We shall name *X-po-group* a structure $(G, \leq, \cdot, \to, \rightsquigarrow, 1)$ such that $(G, \cdot, 1)$ is a monoid (i.e. (Ass), (pU) hold), $(G, \leq)$ is a poset and the operation $\cdot$ and the pseudo-implication $(\to, \rightsquigarrow)$ verify the following two properties: for all $x, z \in G$,

(pGa$^\leq$) $x \cdot y \leq z \Leftrightarrow x \leq y \to z \Leftrightarrow y \leq x \rightsquigarrow z$,
(pGa$^\geq$) $x \cdot y \geq z \Leftrightarrow x \geq y \to z \Leftrightarrow y \geq x \rightsquigarrow z$.

Denote by **X-po-group** the class of all X-po-groups, namely:
- by **X-po-group**^{L}, the class of all left-X-po-groups,
- by **X-po-group**^{R}, the class of all right-X-po-groups.

Remark 7.2.16 Note that an X-po-group is in the same time a po-m(pGa$^\leq$) and a po-m(pGa$^\geq$). Hence, we can write:

$$\textbf{X-po-group} = \textbf{po-m(pGa}^{\leq}\textbf{)} + \textbf{po-m(pGa}^{\geq}\textbf{)}.$$

First, we give some general results as a general lemma:

Lemma 7.2.17 *(see Lemma 7.1.21 and Lemma 2.6.3)*
Let $\mathcal{A} = (A, \leq, \cdot, \to, \rightsquigarrow)$ (or $\mathcal{A} = (A, \leq, \to, \rightsquigarrow, \cdot)$) be a structure such that:
(a) $(A, \leq)$ is a poset;
(b) the properties (pGa$^\leq$) and (pGa$^\geq$) hold.
Then, we have:
(i) The following properties hold: for all $x, y, z \in A$,
(pGa$^=$) $\quad x \cdot y = z \Leftrightarrow x = y \to z \Leftrightarrow y = x \rightsquigarrow z$,
(α) $\quad (x \to y) \cdot x = y = x \cdot (x \rightsquigarrow y)$,
(β) $\quad x \to (y \cdot x) = y = x \rightsquigarrow (x \cdot y)$.
(pEqCp) $\quad x \leq y \Leftrightarrow z \cdot x \leq z \cdot y \Leftrightarrow x \cdot z \leq y \cdot z$,
(p^{*L}) $\quad x \leq y$ *implies* $z \to x \leq z \to y$, $z \rightsquigarrow x \leq z \rightsquigarrow y$,
(p^{**L}) $\quad x \leq y$ *implies* $y \to z \leq x \to z$, $y \rightsquigarrow z \leq x \rightsquigarrow z$.
(ii) The following equivalence holds:

$$(pGa^{\leq}) + (pGa^{\geq}) \Leftrightarrow (pGa^{=}) + (pCp) + (p*^{L}).$$

Proof. (i):
(pGa$^=$): The properties (pGa$^\leq$) and (pGa$^\geq$) imply the property (pGa$^=$) since:
$(p \leftrightarrow q$ and $r \leftrightarrow s)$ imply $(p$ and $r) \leftrightarrow (q$ and $s)$
and since the order relation $\leq$ is antisymmetrique.
(α) and (β): By Lemma 2.6.3, since (pGa$^\leq$) holds, it follows that we have:
(pPP) $y \leq x \to (y \cdot x)$, $x \leq y \rightsquigarrow (y \cdot x)$,
(pRR) $(y \to x) \cdot y \leq x$, $y \cdot (y \rightsquigarrow x) \leq x$.
By the dual of Lemma 2.6.3, since (pGa$^\geq$) holds, it follows that we have:
(pSS) $\quad y \geq x \to (y \cdot x)$, $x \geq y \rightsquigarrow (y \cdot x)$,
(pcoRR) $\quad (y \to x) \cdot y \geq x$, $y \cdot (y \rightsquigarrow x) \geq x$.
Hence, by the antisymmetry of $\leq$, we obtain:
$y = x \to (y \cdot x)$, $x = y \rightsquigarrow (y \cdot x)$,
$(y \to x) \cdot y = x$, $y \cdot (y \rightsquigarrow x) = x$,
i.e. (α) and (β) hold.
(pEqCp): Suppose $x \leq y$; by (β), $y = z \rightsquigarrow (z \cdot y)$, hence $x \leq z \rightsquigarrow (z \cdot y)$ and $x \leq z \rightsquigarrow (z \cdot y) \overset{(pGa^{\leq})}{\Leftrightarrow} z \cdot x \leq z \cdot y$, i.e. the first part of (pCp) holds. Conversely, $z \cdot x \leq z \cdot y \overset{(pGa^{\leq})}{\Leftrightarrow} x \leq z \rightsquigarrow (z \cdot y) \overset{(\beta)}{=} y$.

Suppose again $x \leq y$; by (β), $y = z \to (y \cdot z)$, hence $x \leq z \to (y \cdot z)$ and $x \leq z \to (y \cdot z) \overset{(pGa^{\leq})}{\Leftrightarrow} x \cdot z \leq y \cdot z$; thus (pCp) holds. Conversely, $x \cdot z \leq y \cdot z \overset{(pGa^{\leq})}{\Leftrightarrow} x \leq z \to (y \cdot z) \overset{(\beta)}{=} y$.

(p^{*L}): By Lemma 2.6.3, since ($pGa^{\leq}$) holds.

(p^{**L}): By Lemma 2.6.3, since ($pGa^{\leq}$) holds.

(ii): By (i), it remains to prove that (pCp) + (p^{*L}) + ($pGa^{=}$) imply ($pGa^{\leq}$) + ($pGa^{\geq}$). Indeed, (pCp) + (p^{*L}) + ($pGa^{=}$) imply ($pGa^{\leq}$):

- Suppose $x \cdot y \leq z$; then by (p^{*L}), $y \to (x \cdot y) \leq y \to z$ and $x \rightsquigarrow (x \cdot y) \leq x \rightsquigarrow z$; hence, by ($\beta$), we obtain that $x \leq y \to z$ and $y \leq x \rightsquigarrow z$.

- Conversely, suppose $x \leq y \to z$; then by (pCp), $x \cdot y \leq (y \to z) \cdot y \overset{(\alpha)}{=} z$; suppose $y \leq x \rightsquigarrow z$; then by (pCp), $x \cdot y \leq x \cdot (x \rightsquigarrow z) \overset{(\alpha)}{=} z$.

Similarly, (pCp) + (p^{*L}) + ($pGa^{=}$) imply ($pGa^{\geq}$). □

Note that if $(G, \leq, \cdot, \to, \rightsquigarrow, 1)$ is an X-po-group, then $(G, \cdot, \to, \rightsquigarrow, 1)$ is an X-group, by above Lemma.

Proposition 7.2.18 *Let $(G, \leq, \cdot, \to, \rightsquigarrow, 1)$ be an X-po-group. Then the properties ($pGa^{=}$), (α), (β), (pEqCp), (p^{*L}), (p^{**L}) and (Id), (pBB=), (pM), ($pEq^{=}$) hold.*

Proof. By above Lemma 7.2.17, the properties ($pGa^{=}$), (α), (β), (pEqCp), (p^{*L}), (p^{**L}) hold. Hence, $(G, \cdot, \to, \rightsquigarrow, 1)$ is an X-group and hence, by Proposition 7.1.23, the properties (Id), (pBB=), (pM), ($pEq^{=}$) also hold. □

7.2.4 The equivalences (EG2)

Here we present the equivalences (EG2) between the po-groups and the X-po-groups.

We prove that the po-groups are termwise equivalent to the X-po-groups.

Theorem 7.2.19 *(See Theorem 7.1.24)*

(1) Let $\mathcal{G} = (G, \leq, \cdot, ^{-1}, 1)$ be a po-group.

Define $\rho'(\mathcal{G}) \overset{def.}{=} (G, \leq, \cdot, \to, \rightsquigarrow, 1)$, with $(G, \cdot, \to, \rightsquigarrow, 1) = \rho(G, \cdot, ^{-1}, 1)$ from Theorem 7.1.24 (1).

Then, $\rho'(\mathcal{G})$ is an X-po-group.

(1') Conversely, let $\mathcal{G} = (G, \leq, \cdot, \to, \rightsquigarrow, 1)$ be an X-po-group.

Define ${\rho^*}'(\mathcal{G}) \overset{def.}{=} (G, \leq, \cdot, ^{-1}, 1)$, with $(G, \cdot, ^{-1}, 1) = \rho^*(G, \cdot, \to, \rightsquigarrow, 1)$ from Theorem 7.1.24 (1').

Then, ${\rho^*}'(\mathcal{G})$ is a po-group.

(2) The above defined mappings ρ' and ${\rho^*}'$ are mutually inverse.

Proof.

(1): follows by Theorem 7.1.24 (1) and by Proposition 7.2.5.

(1'): by Theorem 7.1.24 (1'), $(G, \cdot, ^{-1}, 1)$ is a group; (pCp) holds by above Lemma 7.2.17.

(2): follows by Theorem 7.1.24 (2). □

Hence, we have the dual equivalences (EG2) (the analogous of (iEM5)):

(EG2) **X-po-group**^{L} $\Leftrightarrow$ **po-group**^{L} and, dually,
X-po-group^R $\Leftrightarrow$ **po-group**^{R}.

Remarks 7.2.20
Note that (EG2) can be written as follows, by using Remark 7.2.16:
po-group $\Leftrightarrow$ **po-m(pGa**$^{\leq}$**)** + **po-m(pGa**$^{\geq}$**)**.
In the same time, note that, by (EGM), we have:
po-group^L $\Leftrightarrow$ **po-m(pGa**$^{\leq}$**)** + (pIv) and, dually,
po-group^R $\Leftrightarrow$ **po-m(pGa**$^{\geq}$**)** + (pIv).

7.3 *l*-groups

7.3.1 The *l*-group

Definition 7.3.1 Let $\mathcal{G} = (G, \leq, \cdot, ^{-1}, 1)$ be a po-group. If the partial order $\leq$ is a lattice order, then the po-group is called *lattice-ordered group*, or *l-group* for short, and is denoted generally by: $\mathcal{G} = (G, \vee, \wedge, \cdot, ^{-1}, 1)$, namely:
- by $\mathcal{G}^L = (G^L, \wedge, \vee, \odot, ^{-1}, 1)$, if it is a left-*l*-group,
- by $\mathcal{G}^R = (G^R, \vee, \wedge, \oplus, -, 0)$, if it is a right-*l*-group.

Denote by ***l*-group** the class of all *l*-groups, namely:
- by ***l*-group**^{L} the class of all left-*l*-groups and
- by ***l*-group**^{R} the class of all right-*l*-groups.

An introduction in l-groups is [6], see also [38].

Note that an *l*-group may be linearly-ordered or not, while a linearly-ordered group is an *l*-group.

Proposition 7.3.2 *(See Propositions 2.8.2, 2.10.4, 2.8.4 and Corollary 2.8.5)*

Let $\mathcal{G}$ be an l-group. Then the following properties hold (old results from the literature): for all $x, y, a, b \in G$,
(1) $a \cdot (x \vee y) \cdot b = (a \cdot x \cdot b) \vee (a \cdot y \cdot b)$ *and, dually,*
(1') $a \cdot (x \wedge y) \cdot b = (a \cdot x \cdot b) \wedge (a \cdot y \cdot b)$;
(2) $(x \vee y)^{-1} = (x^{-1}) \wedge (y^{-1})$ *and, dually,*
(2') $(x \wedge y)^{-1} = (x^{-1}) \vee (y^{-1})$;
(3) $x \vee y = x \cdot (x \wedge y)^{-1} \cdot y = [(x \wedge y) \to x] \cdot y = x \cdot [(x \wedge y) \rightsquigarrow y]$ *and, dually,*
(3') $x \wedge y = x \cdot (x \vee y)^{-1} \cdot y = [(x \vee y) \to x] \cdot y = x \cdot [(x \vee y) \rightsquigarrow y]$;
(4) The lattice $(G, \vee, \wedge)$ is distributive.

Theorem 7.3.3 *(See Theorem 7.2.7)*

(1) Let $\mathcal{G} = (G, \vee, \wedge, \cdot, ^{-1}, 1)$ be an l-group.
Define $\phi(\mathcal{G}) \stackrel{def.}{=} (G, \vee, \wedge, \cdot, \to, \rightsquigarrow, 1)$ *by:*

$$x \to y \stackrel{def.}{=} (x \cdot y^{-1})^{-1} = y \cdot x^{-1}, \quad x \rightsquigarrow y \stackrel{def.}{=} (y^{-1} \cdot x)^{-1} = x^{-1} \cdot y.$$

Then, $\phi(\mathcal{G})$ is an l-m (pGa$^{\leq}$) (verifying (pIv), where $x^{-1} = x \to 1 = x \rightsquigarrow 1$, by (7.9)).

(1') Let $\mathcal{G} = (G, \vee, \wedge, \cdot, \to^L, \rightsquigarrow^L, 1)$ be a l-m(pPR), verifying (pIv), where $x^{-1} = x \to^L 1$.
Define $\psi(\mathcal{G}) \stackrel{def.}{=} (G, \leq, \cdot, ^{-1}, 1)$.
Then, $\psi(\mathcal{G})$ is an l-group.

(2) The above defined mappings ϕ and ψ are mutually inverse and (pPR)= (pGa$^{\leq}$).

Corollary 7.3.4 *(See [63])*
The l-groups are term equivalent to residuated lattices verifying (pIv).

Proof. By Theorem 7.3.3, the l-groups are term equivalent to l-ms(pPR) verifying (pIv), while **l-m(pPR)=pR-L**^{L}, by Definition 3.9.2. □

Remark 7.3.5 (See Remark 7.2.8) By above Theorem 7.3.3, we have the dual equivalences:

(EGM-l) l-**group**^{L} $\Leftrightarrow$ l-**m(pGa**$^{\leq}$**)** + (pIv) and, dually,
l-**group**^{R} $\Leftrightarrow$ l-**m(pGa**$^{\geq}$**)** + (pIv).

By Corollary 7.2.9, we immediately obtain:

Corollary 7.3.6 *An l-group is a l-m(pR).*

Proposition 7.3.7 *(See Lemma 2.4.2 and its dual)*
In an l-group $(G, \vee, \wedge, \cdot, ^{-1}, 1)$, the following properties hold: for all $x, y, z \in G$,

(7.14) $$z \to (x \vee y) = (z \to x) \vee (z \to y), \quad z \rightsquigarrow (x \vee y) = (z \rightsquigarrow x) \vee (z \rightsquigarrow y)$$

and, dually,

(7.15) $$z \to (x \wedge y) = (z \to x) \wedge (z \to y), \quad z \rightsquigarrow (x \wedge y) = (z \rightsquigarrow x) \wedge (z \rightsquigarrow y);$$

(7.16) $$(x \vee y) \to z = (x \to z) \wedge (y \to z), \ (x \vee y) \rightsquigarrow z = (x \rightsquigarrow z) \wedge (y \rightsquigarrow z)$$

and, dually,

(7.17) $$(x \wedge y) \to z = (x \to z) \vee (y \to z), \quad (x \wedge y) \rightsquigarrow z = (x \rightsquigarrow z) \vee (y \rightsquigarrow z).$$

Proof.

(7.14):
$y \to (x \vee z) = (x \vee z) \cdot y^{-1} = (x \cdot y^{-1}) \vee (z \cdot y^{-1}) = (y \to x) \vee (y \to z)$,
$y \rightsquigarrow (x \vee z) = y^{-1} \cdot (x \vee z) = (y^{-1} \cdot x) \vee (y^{-1} \cdot z) = (y \rightsquigarrow x) \vee (y \rightsquigarrow z)$.

(7.15):
$y \to (x \wedge z) = (x \wedge z) \cdot y^{-1} = (x \cdot y^{-1}) \wedge (z \cdot y^{-1}) = (y \to x) \wedge (y \to z)$,
$y \rightsquigarrow (x \wedge z) = y^{-1} \cdot (x \wedge z) = (y^{-1} \cdot x) \wedge (y^{-1} \cdot z) = (y \rightsquigarrow x) \wedge (y \rightsquigarrow z)$.

(7.16):
$(x \vee z) \rightarrow y = y \cdot (x \vee z)^{-1} = y \cdot [(x^{-1}) \wedge (z^{-1})] = (y \cdot x^{-1}) \wedge (y \cdot z^{-1}) = (x \rightarrow y) \wedge (z \rightarrow y)$,
$(x \vee z) \rightsquigarrow y = (x \vee z)^{-1} \cdot y = [(x^{-1}) \wedge (z^{-1})] \cdot y = (x^{-1} \cdot y) \wedge (z^{-1} \cdot y) = (x \rightsquigarrow y) \wedge (z \rightsquigarrow y)$.
(7.17):
$(x \wedge z) \rightarrow y = y \cdot (x \wedge z)^{-1} = y \cdot [(x^{-1}) \vee (z^{-1})] = (y \cdot x^{-1}) \vee (y \cdot z^{-1}) = (x \rightarrow y) \vee (z \rightarrow y)$,
$(x \wedge z) \rightsquigarrow y = (x \wedge z)^{-1} \cdot y = [(x^{-1}) \vee (z^{-1})] \cdot y = (x^{-1} \cdot y) \vee (z^{-1} \cdot y) = (x \rightsquigarrow y) \vee (z \rightsquigarrow y)$. □

Proposition 7.3.8 *In an l-group $(G, \vee, \wedge, \cdot, ^{-1}, 1)$, the following properties also hold: for all $x, y, z \in G$,*

(7.18) $$[(x \wedge 1) \rightsquigarrow 1] \wedge 1 = 1, \quad [(x \wedge 1) \rightarrow 1] \wedge 1 = 1$$

and, dually,

(7.19) $$[(x \vee 1) \rightsquigarrow 1] \vee 1 = 1, \quad [(x \vee 1) \rightarrow 1] \vee 1 = 1.$$

(7.20) $$(x \vee y) \rightarrow (x \wedge y) = (x \rightarrow y) \wedge (y \rightarrow x) \wedge 1 \leq 1 \text{ and}$$

(7.21) $$(x \vee y) \rightsquigarrow (x \wedge y) = (x \rightsquigarrow y) \wedge (y \rightsquigarrow x) \wedge 1 \leq 1$$

and, dually,

(7.22) $$(x \wedge y) \rightarrow (x \vee y) = (x \rightarrow y) \vee (y \rightarrow x) \vee 1 \geq 1 \text{ and}$$

(7.23) $$(x \wedge y) \rightsquigarrow (x \vee y) = (x \rightsquigarrow y) \vee (y \rightsquigarrow x) \vee 1 \geq 1;$$

(7.24) $$x \rightarrow (x \wedge y) = 1 \wedge (x \rightarrow y), \quad x \rightsquigarrow (x \wedge y) = 1 \wedge (x \rightsquigarrow y),$$

(7.25) $$(x \wedge y) \rightarrow x = 1 \vee (y \rightarrow x), \quad (x \wedge y) \rightsquigarrow x = 1 \vee (y \rightsquigarrow x).$$

Proof.
(7.18):
$[(x \wedge 1) \rightsquigarrow 1] \wedge 1 = [(x \wedge 1)^{-1} \cdot 1] \wedge 1 = [(x^{-1} \vee -1) \cdot 1] \wedge 1 = [x^{-1} \vee 1] \wedge 1 \overset{absorption}{=} 1$,
$[(x \wedge 1) \rightarrow 1] \wedge 1 = [1(x \wedge 1)^{-1}] \wedge 1 = [1 \cdot (x^{-1} \vee -1)] \wedge 1 = [x^{-1} \vee 1] \wedge 1 \overset{absorption}{=} 1$.
(7.19):
$[(x \vee 1) \rightsquigarrow 1] \vee 1 = [(x \vee 1)^{-1} \cdot 1] \vee 1 = [(x^{-1} \wedge -1) \cdot 1] \vee 1 = [x^{-1} \wedge 1] \vee 1 \overset{absorption}{=} 1$,
$[(x \vee 1) \rightarrow 1] \vee 1 = [1 \cdot (x \vee 1)^{-1}] \vee 1 = [1 \cdot (x^{-1} \wedge -1)] \vee 1 = [x^{-1} \wedge 1] \vee 1 \overset{absorption}{=} 1$.
(7.20): $(x \vee y) \rightarrow (x \wedge y) = [x \rightarrow (x \wedge y)] \wedge [y \rightarrow (x \wedge y)] = [(x \rightarrow x) \wedge (x \rightarrow y)] \wedge [(y \rightarrow x) \wedge (y \rightarrow y)] = [1 \wedge (x \rightarrow y)] \wedge [(y \rightarrow x) \wedge 1] = (x \rightarrow y) \wedge (y \rightarrow x) \wedge 1 \leq 1$, by (7.15), (7.16).
(7.21): similarly.
(7.22): similarly, by (7.14), (7.17).
(7.23): similarly.
(7.24): by (7.15).
(7.25): by (7.17). □

7.3.2 The negative and the positive cones of l-groups

Lemma 7.3.9 *(See Lema 11.1.41)*
Let $\mathcal{G} = (G, \vee, \wedge, \cdot, ^{-1}, 1)$ be an l-group (left-l-group or right-l-group). Then, G^- and G^+ are closed under $\vee$ and $\wedge$.

Proof.
Let $x, y \in G^-$, i.e. $x \leq 1$ and $y \leq 1$. Then, 1 is an upper bound of $\{x, y\}$; hence, $1 \geq x \vee y$, i.e. $x \vee y \in G^-$; since $1 \geq x \vee y \geq x \wedge y$, it follows that $1 \geq x \wedge y$, i.e. $x \wedge y \in G^-$ too.
Let now $x, y \in G^+$, i.e. $x \geq 1$ and $y \geq 1$. Then, 1 is a lower bound of $\{x, y\}$; hence, $1 \leq x \wedge y$, i.e. $x \wedge y \in G^+$; since $1 \leq x \wedge y \leq x \vee y$, it follows that $1 \leq x \vee y$, i.e. $x \vee y \in G^+$ too. □

Lemma 7.3.10 *(See Lemma 7.2.13)*
Let $\mathcal{G} = (G, \vee, \wedge, \cdot, ^{-1}, 1)$ be an l-group (left-l-group or right-l-group). Then, G^- and G^+ are closed under $\cdot$.

Corollary 7.3.11 *Let $\mathcal{G}$ be an l-group. Then we have: for all $x, y \in G$,*
(i) $x, y \in G^-$ imply $x \cdot y \leq x \wedge y$,
(i') $x, y \in G^+$ imply $x \cdot y \geq x \vee y$.

Proof. By Proposition 7.3.2 (3') and (3),
$x \wedge y = [(x \vee y) \to x] \cdot y$ and $x \vee y = [(x \wedge y) \to x] \cdot y$.
(i): if $x, y \leq 1$, then $x \vee y \leq 1$, hence by Corollary 7.2.10, $x \leq (x \vee y) \to x$; then, by (pCp), we obtain that $x \cdot y \leq [(x \vee y) \to x] \cdot y = x \wedge y$.
(i'): if $x, y \geq 1$, then $x \wedge y \geq 1$, hence by Corollary 7.2.10, $x \geq (x \wedge y) \to x$; then, by (pCp), we obtain that $x \cdot y \geq [(x \wedge y) \to x] \cdot y = x \vee y$. □

We have obviously the following result.

Proposition 7.3.12 *(See Proposition 7.2.14)*
Let $\mathcal{G} = (G, \vee, \wedge, \cdot, ^{-1}, 1)$ be an l-group. Then:
(1) $\mathcal{G}^- = (G^-, \wedge, \vee, \odot = \cdot, \mathbf{1} = 1)$ is a left-l-im;
(1') $\mathcal{G}^+ = (G^+, \vee, \wedge, \oplus = \cdot, \mathbf{0} = 1)$ is a right-l-im.

7.3.3 The X-l-group - an intermediary notion

Definition 7.3.13 An *X-l-group* is an X-po-group $\mathcal{G} = (G, \leq, \cdot, \to, \rightsquigarrow, 1)$, where the partial order $\leq$ is a lattice order. An *X-l-group* is denoted by

$$(G, \vee, \wedge, \cdot, \to, \rightsquigarrow, 1).$$

Denote by **X-l-group** the class of all X-l-groups, namely:
- by **X-l-group**^{L}, the class of all left-X-l-groups and
- by **X-l-group**^{R}, the class of all right-X-l-groups.

Remark 7.3.14 (See Remark 7.2.16
Note that an X-l-group is in the same time an l-m(pGa$^{\leq}$) and an l-m(pGa$^{\geq}$). Hence, we can write:

$$\textbf{X-}l\textbf{-group} = l\textbf{-m(pGa}^{\leq}\textbf{)} + l\textbf{-m(pGa}^{\geq}\textbf{)}.$$

7.3.4 The equivalences (EG3)

Here we present the equivalences (EG3) between X-*l*-groups and *l*-groups.

We obviously obtain, by (EG2), the dual equivalences (EG3) (the analogous of (iEM6)):

(EG3) **X-*l*-group**^{L} $\Leftrightarrow$ ***l*-group**^{L} and, dually,
X-*l*-group^R $\Leftrightarrow$ ***l*-group**^{R}.

Remarks 7.3.15 (See Remarks 7.2.20)

Note that (EG3) can be written as follows, by using Remark 7.3.14:

***l*-group** $\Leftrightarrow$ ***l*-m(pGa**$^{\leq}$**)** + ***l*-m(pGa**$^{\geq}$**)**.

In the same time, note that, by (EGM-l), we have:

***l*-group** $\Leftrightarrow$ ***l*-m(pGa**$^{\leq}$**)** + (pIv).

Chapter 8

Implicative-groups, po-(l-)implicative-groups

In this chapter, we recall and study [97], [100] the implicative-group, the po-implicative-group, the l-implicative-group and the intermediary notions of X-implicative-group, X-po-implicative-group, X-l-implicative-group, respectively, and prove the equivalences (EI1), (EI2), (EI3) between these notions, respectively.

The chapter has three sections.

8.1 Implicative-groups

8.1.1 The implicative-group

Definition 8.1.1 (Definition 1 of implicative-groups)

An *implicative-group*, or an *i-group* for short, is an algebra $\mathcal{G} = (G, \to, \rightsquigarrow, 1)$ of type $(2, 2, 0)$ such that the following axioms hold: for all $x, y, z \in G$,
(pM) $1 \to x = x = 1 \rightsquigarrow x$,
(pBB=) $y \to z = (z \to x) \rightsquigarrow (y \to x), \quad y \rightsquigarrow z = (z \rightsquigarrow x) \to (y \rightsquigarrow x)$.

Denote by **i-group** the class of all i-groups.

The i-group is said to be *commutative*, or *abelian*, if $x \to y = x \rightsquigarrow y$, for all $x, y \in G$.

• **General properties**

Let $(G, \to = \to^L, \rightsquigarrow = \rightsquigarrow^L, 1)$ be a left-pseudo-algebra, as defined in Chapter 2 by Definition 2.1.1 (the dual case is omitted), where:
(IdEq) $x \to y = 1 \Longleftrightarrow x \rightsquigarrow y = 1$.
We can then define on G an *internal* (*natural*) binary relation, $\leq$, by: for all $x, y \in G$,

$$(8.1) \qquad (pdfrelR) \qquad x \leq y \overset{def.}{\Longleftrightarrow} x \to y = 1 \ (\overset{(IdEq)}{\Longleftrightarrow} x \rightsquigarrow y = 1).$$

Equivalently, let $(G, \leq, \to = \to^L, \rightsquigarrow = \rightsquigarrow^L, 1)$ be a left-pseudo-structure, as defined in Chapter 2 by Definition 2.1.1 (the dual case is omitted), where $\leq$ and $\to$, $\rightsquigarrow$, 1 are connected by:

(pEq^L) $\quad x \leq y \Longleftrightarrow x \to y = 1 \; (\overset{(IdEq)}{\Leftrightarrow} \; x \rightsquigarrow y = 1)$.

Consider the following properties:

(pB=) $\quad z \to x = (y \to z) \to (y \to x), \;\; z \rightsquigarrow x = (y \rightsquigarrow z) \rightsquigarrow (y \rightsquigarrow x)$,
(pC^L) $\quad x \to (y \rightsquigarrow z) \leq y \rightsquigarrow (x \to z), \;\; x \rightsquigarrow (y \to z) \leq y \to (x \rightsquigarrow z)$,
(pD=) $\quad (y \to x) \rightsquigarrow x = y = (y \rightsquigarrow x) \to x$,
(Id) $\quad x \to 1 = x \rightsquigarrow 1$,
(IdEq) $\quad x \to y = 1 \Leftrightarrow x \rightsquigarrow y = 1$;
(pEq=) $\quad x = y \Longleftrightarrow x \to y = 1 \; (\overset{(IdEq)}{\Leftrightarrow} \; x \rightsquigarrow y = 1)$;
(pEq#$^=$) $\quad x = y \to z \Leftrightarrow y = x \rightsquigarrow z$,
(pEx) $\quad x \to (y \rightsquigarrow z) = y \rightsquigarrow (x \to z)$;
(pRe') $\quad x \to x = 1 = x \rightsquigarrow x$;
(p-s^L) $\quad$ (p-semisimple) $x \leq y \Longrightarrow x = y$,
(pBB^L) $\quad y \to z \leq (z \to x) \rightsquigarrow (y \to x), \;\; y \rightsquigarrow z \leq (z \rightsquigarrow x) \to (y \rightsquigarrow x)$,
(pB^L) $\quad z \to x \leq (y \to z) \to (y \to x), \;\; z \rightsquigarrow x \leq (y \rightsquigarrow z) \rightsquigarrow (y \rightsquigarrow x)$,
(pD^L) $\quad y \leq (y \to x) \rightsquigarrow x, \;\; y \leq (y \rightsquigarrow x) \to x$,
(An) $\quad x \leq y$ and $y \leq x \Longrightarrow x = y$,
(Tr) $\quad x \leq y$ and $y \leq z \Longrightarrow x \leq z$,
(p*^{L}) $\quad x \leq y \Longrightarrow (z \to x \leq z \to y$ and $z \rightsquigarrow x \leq z \rightsquigarrow y)$,
(p**^{L}) $\quad x \leq y \Longrightarrow (y \to z \leq x \to z$ and $y \rightsquigarrow z \leq x \rightsquigarrow z)$.

Proposition 8.1.2 *(See Proposition 2.1.8)*
Let $(G, \to, \rightsquigarrow, 1)$ be a left-pseudo-algebra (or, equivalently, let $(G, \leq, \to, \rightsquigarrow, 1)$ be a left-pseudo-structure). Then, we have:
(0) (pD=) + (pM) $\Longrightarrow$ (p-s^L);
(1) (pM) + (pBB=) $\Longrightarrow$ (pRe'),
(1') (pEq=) + (IdEq) $\Longrightarrow$ (pRe'),
(2) (pM) + (pBB=) $\Longrightarrow$ (pD=),
(2') (pEx) + (pRe') + (pEq=) + (IdEq) $\Longrightarrow$ (pD=):
(3) (pD=) + (pM) + (pRe') $\Longrightarrow$ (IdEq),
(4) (pRe') + (pD=) $\Longrightarrow$ (IdEq),
(4') (pRe') + (pD=) $\Longrightarrow$ (pEq=),
(5) (pBB=) + (pRe') + (pD=) + (pdfrelR) + (IdEq) $\Longrightarrow$ (pC^L),
(5') (pBB=) + (pRe') + (pD=) + (pEq=) $\Longrightarrow$ (pEx),
(6) (pC^L) + (p-s^L) $\Longrightarrow$ (pEx),
(7) (pBB=) + (pEx) + (pEq=) + (IdEq) $\Longrightarrow$ (pB=),
(7') (pB=) + (pEx) + (pEq=) + (IdEq) $\Longrightarrow$ (pBB=),
(7") (pEx) + (pEq=) + (IdEq) $\Longrightarrow$ (pBB=) $\Leftrightarrow$ (pB=),
(8) (pEx) + (pRe') $\Longrightarrow$ (Id),
(9) (pEq=) + (IdEq) + (pEx) $\Longrightarrow$ (pEq#$^=$),
(10) (pD=) + (pEq=) + (IdEq) $\Longrightarrow$ (pM),
(11) (pEx) + (pD=) $\Longrightarrow$ (pBB=),

(12) (pBB=) + (pEq=) $\Longrightarrow$ (pBB^L),
(13) (pB=) + (pEq=) $\Longrightarrow$ (pB^L),
(14) (pD=) + (pEq=) $\Longrightarrow$ (pD^L),
(15) (pBB^L) + (p-s^L) $\Longrightarrow$ (pBB=),
(16) (pB^L) + (p-s^L) $\Longrightarrow$ (pB=),
(17) (pD^L) + (p-s^L) $\Longrightarrow$ (pD=),
(18) (pEq=) $\Longrightarrow$ (p-s^L),
(19) (pEq=) $\Longrightarrow$ (Tr),
(20) (pEq=) $\Longrightarrow$ (An),
(21) (pEq=) $\Longrightarrow$ (p^{*L}), (p^{**L}).

Proof. (0): If $x \leq y$, i.e. $x \to y = 1$, then:
$x \overset{(pD=)}{=} (x \to y) \rightsquigarrow y = 1 \rightsquigarrow y \overset{(pM)}{=} y$. Thus, (p-s^L) holds.

(1): $x \to x \overset{(pM)}{=} (1 \rightsquigarrow x) \to (1 \rightsquigarrow x) \overset{(pBB=)}{=} 1 \rightsquigarrow 1 \overset{(pM)}{=} 1$ and
$x \rightsquigarrow x \overset{(pM)}{=} (1 \to x) \rightsquigarrow (1 \to x) \overset{(pBB=)}{=} 1 \to 1 \overset{(pM)}{=} 1$. Thus, (pRe') holds.

(1'): Obviously, $x = x$ implies $x \to x = 1 = x \rightsquigarrow x$, by (pEq=), (IdEq). Thus, (pRe') holds.

(2): $(y \to x) \rightsquigarrow x \overset{(pM)}{=} (y \to x) \rightsquigarrow (1 \to x) \overset{(pBB=)}{=} 1 \to y \overset{(pM)}{=} y$ and
$(y \rightsquigarrow x) \to x \overset{(pM)}{=} (y \rightsquigarrow x) \to (1 \rightsquigarrow x) \overset{(pBB=)}{=} 1 \rightsquigarrow y \overset{(pM)}{=} y$. Thus, (pD=) holds.

(2'): $y \to ((y \to x) \rightsquigarrow x) \overset{(pEx)}{=} (y \to x) \rightsquigarrow (y \to x) \overset{(pRe')}{=} 1$; hence,
$y = (y \to x) \rightsquigarrow x$, by (pEq=);
$y \rightsquigarrow ((y \rightsquigarrow x) \to x) \overset{(pEx)}{=} (y \rightsquigarrow x) \to (y \rightsquigarrow x) \overset{(pRe')}{=} 1$; hence, $y \to ((y \rightsquigarrow x) \to x) = 1$, by (IdEq); hence, $y = (y \rightsquigarrow x) \to x$, by (pEq=) again. Thus, (pD=) holds.

(3): If $x \to y = 1$, then $x \overset{(pD=)}{=} (x \to y) \rightsquigarrow y = 1 \rightsquigarrow y \overset{(pM)}{=} y$, hence $x \rightsquigarrow y = y \rightsquigarrow y \overset{(pRe')}{=} 1$. Conversely, if $x \rightsquigarrow y = 1$, then $x \overset{(pD=)}{=} (x \rightsquigarrow y) \to y = 1 \to y \overset{(pM)}{=} y$, hence $x \to y = y \to y \overset{(pRe')}{=} 1$. Thus, (IdEq) holds.

(4): If $x \to y = 1$, then $x \overset{(pD=)}{=} (x \to y) \rightsquigarrow y = 1 \rightsquigarrow y \overset{(pRe')}{=} (y \to y) \rightsquigarrow y \overset{(pD=)}{=} y$, hence $x \rightsquigarrow y = y \rightsquigarrow y \overset{(pRe')}{=} 1$. Conversely, if $x \rightsquigarrow y = 1$, then $x \overset{(pD=)}{=} (x \rightsquigarrow y) \to y = 1 \to y \overset{(pRe')}{=} (y \rightsquigarrow y) \to y \overset{(pD=)}{=} y$, hence $x \to y = y \to y \overset{(pRe')}{=} 1$. Thus, (IdEq) holds.

(4'): If $x = y$, then $x \to y = y \to y \overset{(pRe')}{=} 1$. Conversely, if $x \to y = 1$, then $x \overset{(pD=)}{=} (x \to y) \rightsquigarrow y = 1 \rightsquigarrow y \overset{(pRe')}{=} (y \to y) \rightsquigarrow y \overset{(pD=)}{=} y$. Thus, (pEq=) holds.

(5): First, by (pBB=), we have: $Y \rightsquigarrow Z = (Z \rightsquigarrow X) \to (Y \rightsquigarrow X)$;
take $X = u \to x$, $Y = y$, $Z = z \to x$, to obtain:
$y \rightsquigarrow (z \to x) = ((z \to x) \rightsquigarrow (u \to x)) \to (y \rightsquigarrow (u \to x))$,
which, by (pBB=) again, becomes:
(a) $y \rightsquigarrow (z \to x) = (u \to z) \to (y \rightsquigarrow (u \to x))$.
Take now, in (a), $z = y \rightsquigarrow x$ and $u = z$, to obtain:
$1 \overset{(pRe')}{=} y \rightsquigarrow y \overset{(pD=)}{=} y \rightsquigarrow [(y \rightsquigarrow x) \to x] = (z \to (y \rightsquigarrow x)) \to (y \rightsquigarrow (z \to x))$,
hence $z \to (y \rightsquigarrow x) \leq y \rightsquigarrow (z \to x)$, by (pdfrelR).

Then, by (pBB=) again, we have: $Y \to Z = (Z \to X) \rightsquigarrow (Y \to X)$;
take $X = u \rightsquigarrow x$, $Y = y$, $Z = z \rightsquigarrow x$, to obtain:
$y \to (z \rightsquigarrow x) = ((z \rightsquigarrow x) \to (u \rightsquigarrow x)) \rightsquigarrow (y \to (u \rightsquigarrow x))$,
which, by (pBB=) again, becomes:
(b) $y \to (z \rightsquigarrow x) = (u \rightsquigarrow z) \rightsquigarrow (y \to (u \rightsquigarrow x))$.
Take now, in (b), $z = y \to x$ and $u = z$, to obtain:
$1 \overset{(pRe')}{=} y \to y \overset{(pD=)}{=} y \to [(y \to x) \rightsquigarrow x] = (z \rightsquigarrow (y \to x)) \rightsquigarrow (y \to (z \rightsquigarrow x))$,
hence, by (IdEq), $(z \rightsquigarrow (y \to x)) \to (y \to (z \rightsquigarrow x)) = 1$,
hence, by (pdfrelR), $z \rightsquigarrow (y \to x) \leq y \to (z \rightsquigarrow x)$. Thus, (p$C^L$) holds.

(5'): By (pBB=), we have: $Y \rightsquigarrow Z = (Z \rightsquigarrow X) \to (Y \rightsquigarrow X)$;
take $X = u \to x$, $Y = y$, $Z = z \to x$, to obtain:
$y \rightsquigarrow (z \to x) = ((z \to x) \rightsquigarrow (u \to x)) \to (y \rightsquigarrow (u \to x))$,
which, by (pBB=) again, becomes:
(a) $y \rightsquigarrow (z \to x) = (u \to z) \to (y \rightsquigarrow (u \to x))$.
Take now, in (a), $z = y \rightsquigarrow x$ and $u = z$, to obtain:
$1 \overset{(pRe')}{=} y \rightsquigarrow y \overset{(pD=)}{=} y \rightsquigarrow [(y \rightsquigarrow x) \to x] = (z \to (y \rightsquigarrow x)) \to (y \rightsquigarrow (z \to x))$,
hence $z \to (y \rightsquigarrow x) = y \rightsquigarrow (z \to x)$, by (pEq=).

(6): By (pC^L), we have: $x \to (y \rightsquigarrow z) \leq y \rightsquigarrow (x \to z)$.
Hence, by (p-s^L), we have: $x \to (y \rightsquigarrow z) = y \rightsquigarrow (x \to z)$. Thus, (pEx) holds.

(7): By (pBB=), we have:
$y \to z = (z \to x) \rightsquigarrow (y \to x)$ and $y \rightsquigarrow z = (z \rightsquigarrow x) \to (y \rightsquigarrow x)$,
which, by (pEq=) and (IdEq), become:
$(y \to z) \to [(z \to x) \rightsquigarrow (y \to x)] = 1$ and $(y \rightsquigarrow z) \rightsquigarrow [(z \rightsquigarrow x) \to (y \rightsquigarrow x)] = 1$,
which, by (pEx), become:
$(z \to x) \rightsquigarrow [(y \to z) \to (y \to x)] = 1$ and $(z \rightsquigarrow x) \to [(y \rightsquigarrow z) \rightsquigarrow (y \rightsquigarrow x)] = 1$,
which, by (pEq=) and (IdEq), become:
$z \to x = (y \to z) \to (y \to x)$ and $z \rightsquigarrow x = (y \rightsquigarrow z) \rightsquigarrow (y \rightsquigarrow x)$, i.e. (pB=) holds.

(7'): By (pB=), we have:
$z \to x = (y \to z) \to (y \to x)$ and $z \rightsquigarrow x = (y \rightsquigarrow z) \rightsquigarrow (y \rightsquigarrow x)$,
which, by (pEq=) and (IdEq), become:
$(z \to x) \rightsquigarrow [(y \to z) \to (y \to x)] = 1$ and $(z \rightsquigarrow x) \to [(y \rightsquigarrow z) \rightsquigarrow (y \rightsquigarrow x)] = 1$,
which, by (pEx), become:
$(y \to z) \to [(z \to x) \rightsquigarrow (y \to x)] = 1$ and $(y \rightsquigarrow z) \rightsquigarrow [(z \rightsquigarrow x) \to (y \rightsquigarrow x)] = 1$,
which, by (pEq=) and (IdEq), become:
$y \to z = (z \to x) \rightsquigarrow (y \to x)$ and $y \rightsquigarrow z = (z \rightsquigarrow x) \to (y \rightsquigarrow x)$, i.e. (pBB=) holds.

(7"): By (7) and (7').

(8): $x \to 1 \overset{(pRe')}{=} x \to (x \rightsquigarrow x) \overset{(pEx)}{=} x \rightsquigarrow (x \to x) \overset{(pRe')}{=} x \rightsquigarrow 1$. Thus, (Id) holds.

(9): $x = y \to z \overset{(pEq=)}{\Leftrightarrow} x \to (y \to z) = 1 \overset{(IdEq)}{\Leftrightarrow} x \rightsquigarrow (y \to z) = 1 \overset{(pEx)}{\Leftrightarrow} y \to (x \rightsquigarrow z) = 1 \overset{(pEq=)}{\Leftrightarrow} y = x \rightsquigarrow z$. Thus, (pEq#=) holds.

(10): $1 \to x = x \overset{(pEq=)}{\Longleftrightarrow} (1 \to x) \to x = 1 \overset{(IdEq)}{\Leftrightarrow} (1 \to x) \rightsquigarrow x = 1$, that is true by (pD=), and
$1 \rightsquigarrow x = x \overset{(pEq=)}{\Longleftrightarrow} (1 \rightsquigarrow x) \to x = 1$, that is true by (pD=) again. Thus, (pM)

holds.

(11): $(z \to x) \rightsquigarrow (y \to x) \overset{(pEx)}{=} y \to ((z \to x) \rightsquigarrow x) \overset{(pD=)}{=} y \to z$ and $(z \rightsquigarrow x) \to (y \rightsquigarrow x) \overset{(pEx)}{=} y \rightsquigarrow ((z \rightsquigarrow x) \to x) \overset{(pD=)}{=} y \rightsquigarrow z$. Thus, (pBB=) holds.

(12) - (21): Immediately. □

Proposition 8.1.3 *Let* $(G, \to, \rightsquigarrow, 1)$ *be an i-group. Then, the following properties hold: for all* $x, y, z \in G$,

(pRe') $\quad x \to x = 1 = x \rightsquigarrow x$,
(pD=) $\quad y = (y \to x) \rightsquigarrow x, \quad y = (y \rightsquigarrow x) \to x$,
(IdEq) $\quad x \to y = 1 \Longleftrightarrow x \rightsquigarrow y = 1$,
(pEq=) $\quad x = y \Longleftrightarrow x \to y = 1 \ (\overset{(IdEq)}{\Leftrightarrow} x \rightsquigarrow y = 1)$.

Proof. (pRe'): By Proposition 8.1.2 (1), (pM) + (pBB=) $\Longrightarrow$ (pRe').

(pD=): By Proposition 8.1.2, (2), (pM) + (pBB=) $\Longrightarrow$ (pD=).

(IdEq): By Proposition 8.1.2 (4), (pRe') + (pD=) $\Longrightarrow$ (IdEq).

(pEq=): By Proposition 8.1.2 (4'), (pRe') + (pD=) $\Longrightarrow$ (pEq=). □

The following theorem provides an equivalent definition of i-groups.

Theorem 8.1.4

(1) Let $\mathcal{G} = (G, \to, \rightsquigarrow, 1)$ *be an i-group. Then, the properties (IdEq), (pEq=), (pM), (pBB=) hold.*

(1') Let $\mathcal{G} = (G, \to, \rightsquigarrow, 1)$ *be an algebra of type* $(2, 2, 0)$ *verifying the properties (IdEq), (pEq=), (pM), (pBB=). Then,* $\mathcal{G}$ *is an i-group.*

Proof. (1): By Proposition 8.1.3, (IdEq) and (pEq=) hold; the other two properties hold by Definition 1 of i-groups.

(1'): Obviously, by Definition 1 of i-groups. □

It follows that we can define equivalently the i-groups as follows.

Definition 8.1.5 (Definition 1' of i-groups)

An *implicative-group*, or an *i-group* for short, is an algebra $\mathcal{G} = (G, \to, \rightsquigarrow, 1)$ of type $(2, 2, 0)$ such that the following axioms hold: (pM), (pBB=), (IdEq), (pEq=).

Proposition 8.1.6 *Let* $\mathcal{G} = (G, \to, \rightsquigarrow, 1)$ *be an i-group. Then, the following properties hold: (p-s*^{L}*), (pEq=), (pC*^{L}*), (pEx), (pB=), (Id), (pEq#*$^=$*), (pBB*^{L}*), (pB*^{L}*), (pD*^{L}*), (An), (Tr), (p*^{*L}*), (p*^{**L}*).*

Proof. By Proposition 8.1.2. □

We then have the following important remarks.

Remarks 8.1.7

(1) The i-group, as defined by Definition 1', is a special left-pseudo-algebra, namely a left-pseudo-algebra (of logic) $(G, \to = \to^L, \rightsquigarrow = \rightsquigarrow^L, 1)$ for which the binary relation $\leq$ defined by (8.1) becomes the equality relation $=$ (by (pEq^L) and (pEq=)), and so the property (p-s^L) holds.

(2) The i-group can be defined, equivalently, as a special left-pseudo-structure: $(G, \leq, \rightarrow=\rightarrow^L, \rightsquigarrow=\rightsquigarrow^L, 1)$, where (pEq^L) becomes (pEq=), by Proposition 8.1.6; therefore, as left-pseudo-structure, the i-group $(G, =, \rightarrow, \rightsquigarrow, 1)$ is denoted also by $(G, \rightarrow, \rightsquigarrow, 1)$.

By above Proposition 8.1.6, we immediately obtain the following important result (the dual one is omitted).

Proposition 8.1.8 *Any i-group is a left-pBCI algebra.*

- **Four other equivalent definitions of the i-group**

The following theorem provides an equivalent definition of the i-group.

Theorem 8.1.9 *(See Theorem 2.3.7)*
Let $\mathcal{G} = (G, \rightarrow, \rightsquigarrow, 1)$ be an algebra of type $(2, 2, 0)$. The following are equivalent:
(a) $\mathcal{G}$ is an i-group (i.e. (pBB=) and (pM) hold).
(b) (pBB=), (pD=), (IdEq) and (pEq=) hold.

Proof. (a) $\Longrightarrow$ (b): By Proposition 8.1.3, (pD=) and (IdEq), (pEq=) hold.
(b) $\Longrightarrow$ (a): By Proposition 8.1.2 (10), (pD=) + (pEq=) imply (pM); thus, (pM) holds. □

Hence, we have the following equivalent definition of the i-group:

Definition 8.1.10 (Definition 2 of i-groups) (See Definition 1 of pBCI algebras)
An *implicative-group*, or an *i-group* for short, is an algebra $\mathcal{G} = (G, \rightarrow, \rightsquigarrow, 1)$ of type $(2, 2, 0)$, such that the following axioms hold: (pBB=), (pD=), (IdEq), (pEq=).

To introduce another two equivalent definition of i-groups, we need the following results.

Theorem 8.1.11
(1) Let $(G, \rightarrow, \rightsquigarrow, 1)$ be an i-group. Then, the properties (pEx), (IdEq), (pEq=) hold.
(1') Let $\mathcal{G} = (G, \rightarrow, \rightsquigarrow, 1)$ be an algebra of type $(2, 2, 0)$ verifying the properties (pEx), (IdEq), (pEq=). Then, $\mathcal{G}$ is an i-group.

Proof. (1): By Propositions 8.1.3 and 8.1.6.
(1'): We use Definition 1 of i-groups, hence we must prove that (pM) and (pBB=) hold. First, by Proposition 8.1.2 (1'), (pEq=) + (IdEq) imply (pRe'). Then, by Proposition 8.1.2 (2'), (pEx) + (pRe') + (pEq=) $\Longrightarrow$ (pD=). By Proposition 8.1.2 (10), (pD=) + (pEq=) + (IdEq) $\Longrightarrow$ (pM), thus (pM) holds. Finally, by Proposition 8.1.2 (11), (pEx) + (pD=) $\Longrightarrow$ (pBB=). Thus, (pBB=) holds. □

By above Theorem 8.1.11, we obtain the following equivalent definition of i-groups.

Definition 8.1.12 (Definition 3 of i-groups) (See Definition 3 of pBCI (pBCK algebras))

An *implicative-group*, or an *i-group* for short, is an algebra $\mathcal{G} = (G, \rightarrow, \rightsquigarrow, 1)$ of type $(2, 2, 0)$ verifying the following axioms: (pEx), (IdEq), (pEq=).

Note that we shall sometimes use in the sequel Definition 3 of i-groups, in order to better see the analogies with pBCI algebras (pBCK algebras), but also the analogies with groups: note that (pM), (pEx), (pRe') corresponds to (pU), (Ass), (pIv) respectively (see Theorem 12.4.1).

Theorem 8.1.13

(1) Let $(G, \rightarrow, \rightsquigarrow, 1)$ be an i-group. Then, the properties (pEx), (pM), (IdEq), (pEq=) hold.

(1') Let $\mathcal{G} = (G, \rightarrow, \rightsquigarrow, 1)$ be an algebra of type $(2, 2, 0)$ verifying the properties (pEx), (pM), (IdEq), (pEq=). Then, $\mathcal{G}$ is an i-group.

Proof. (1): We use Definition 3 of i-groups, i.e. (pEx), (IdEq), (pEq=) hold. We must prove that (pM) holds. Indeed, by Proposition 8.1.2 (1'), (pEq=) + (IdEq) imply (pRe'); by (2'), (pEx) + (pRe') + (pEq=) imply (pD=) and, by (10), (pD=) + (IdEq) + (pEq=) imply (pM).

(1'): Obviously. □

Definition 8.1.14 (Definition 4 of i-groups) An *implicative-group*, or an *i-group* for short, is algebra $\mathcal{G} = (G, \rightarrow, \rightsquigarrow, 1)$ of type $(2, 2, 0)$ verifying the following axioms: (pEx), (pM), (IdEq), (pEq=).

Theorem 8.1.15

(1) Let $(G, \rightarrow, \rightsquigarrow, 1)$ be an i-group. Then, the properties (pEx), (pM), (pRe'), (IdEq), (pEq=) hold.

(1') Let $\mathcal{G} = (G, \rightarrow, \rightsquigarrow, 1)$ be an algebra of type $(2, 2, 0)$ verifying the properties (pEx), (pM), (pRe'), (IdEq), (pEq=). Then, $\mathcal{G}$ is an i-group.

Proof. (1): Suppose (pEx), (IdEq), (pEq=) hold, by Definition 3. We must prove that (pM) and (pRe') hold. Indeed, by Proposition 8.1.2 (1'), (pEq=) + (IdEq) imply (pRe'). By Proposition 8.1.2 (2'), (pEx) + (pRe') + (pEq=) + (IdEq) $\Longrightarrow$ (pD=); by Proposition 8.1.2 (10), (pD=) + (pEq=) + (IdEq) $\Longrightarrow$ (pM).

(1'): Obviously. □

By above Theorem 8.1.15, we obtain the following equivalent definition of i-groups.

Definition 8.1.16 (Definition 5)

An *implicative-group*, or an *i-group* for short, is an algebra $\mathcal{G} = (G, \rightarrow, \rightsquigarrow, 1)$ of type $(2, 2, 0)$ verifying the following axioms: (pEx), (pM), (pRe'), (IdEq), (pEq=).

- **One involutive negation in i-groups:** $^{-1}$

Let $\mathcal{G} = (G, \to, \rightsquigarrow, 1)$ be an i-group. Define two *negations* on G by: for all $x \in G$,

$$x^- \stackrel{def.}{=} x \to 1, \quad x^\sim \stackrel{def.}{=} x \rightsquigarrow 1.$$

But (Id) holds, by Proposition 8.1.6, hence $^- = ^\sim$, i.e. there exists in fact only one negation (see the similar one negation in pBCI algebras), defined as follows: for all $x \in G$,

$$x^{-1} \stackrel{def.}{=} x \to 1 \stackrel{(Id)}{=} x \rightsquigarrow 1. \tag{8.2}$$

Proposition 8.1.17 *(See Lemma 2.11.2 and Proposition 2.11.3)*
Let $\mathcal{G}$ be an i-group. Then, we have: for all $x, y \in G$,
(N1) $1^{-1} = 1$,
(DN) $(x^{-1})^{-1} = x$,
(pNg) $(x \to y)^{-1} = y \to x, \quad (x \rightsquigarrow y)^{-1} = y \rightsquigarrow x$,
(q-pNeg) $x \to 1 = 1 \to x^- = (1 \rightsquigarrow x)^- = x \rightsquigarrow 1 = 1 \rightsquigarrow x^- = (1 \to x)^-$,
(pNg1) $x \to y^{-1} = y \rightsquigarrow x^{-1}$,
(pDNeg1) $y \to z = z^{-1} \rightsquigarrow y^{-1}$ *and* $y \rightsquigarrow z = z^{-1} \to y^{-1}$,
(pDNeg2) $y^{-1} \to x = x^{-1} \rightsquigarrow y$,
(pDNeg3) $(x \to y^{-1})^{-1} = (y \rightsquigarrow x^{-1})^{-1}$.

Proof. (N1): Take $x = 1$ in (pM).
(DN): Take $x = 1$ in (pD=): $y = (y \to 1) \rightsquigarrow 1 = (y^{-1})^{-1}$.
(pNg): $(x \to y)^{-1} = (x \to y) \rightsquigarrow 1 \stackrel{(pRe')}{=} (x \to y) \rightsquigarrow (y \to y) \stackrel{(pBB=)}{=} y \to x$ and $(x \rightsquigarrow y)^{-1} = (x \rightsquigarrow y) \to 1 = (x \rightsquigarrow y) \to (y \rightsquigarrow y) = y \rightsquigarrow x$.
(q-pNeg): Routine, by (pNg), (pM), (Id).
(pNg1): $y \rightsquigarrow x^{-1} \stackrel{def.}{=} y \rightsquigarrow (x \to 1) \stackrel{(pEx)}{=} x \to (y \rightsquigarrow 1) = x \to y^{-1}$.
(pDNeg1): Take $x = 1$ in (pBB=).
(pDNeg2): $x^{-1} \rightsquigarrow y \stackrel{(DN)}{=} x^{-1} \rightsquigarrow (y^{-1})^{-1} \stackrel{(pNg1)}{=} y^{-1} \to (x^{-1})^{-1} = y^{-1} \to x$.
(pDNeg3): By (pNg1). □

Remark 8.1.18 The negation $^{-1}$ is *involutive*, by (DN).

Note that we can define equivalently the i-group as follows.

Definition 8.1.19 (Definition 3' of i-groups) (See Definition 3 of i-groups)
An *implicative-group*, or an *i-group* for short, is an algebra $(G, \to, \rightsquigarrow, ^{-1}, 1)$ of type $(2, 2, 1, 0)$ verifying (IdEq), (pEq=), (pM) (not necessary), (pEx), (DN) (not necessary) and: (pNeg) $\quad x^{-1} = x \to 1 \ (\stackrel{(Id)}{=} x \rightsquigarrow 1)$.

Remark 8.1.20 (See Remark 7.1.3)
We have called *implicative-hub* an algebra $(G, \to, \rightsquigarrow, 1)$ of type $(2, 2, 0)$ verifying (IdEq), (pEq=), (pM), introduced in Chapter 11. Then, the implicative-group defined by Definition 8.1.19 (Definition 3') is connected the the implicative-hub by the following result:

"The implicative-groups are categorically equivalent to the implicative-hubs verifying (pEx), (DN) and (pNeg)."

Following this result, we obtain examples of implicative-groups from the examples of implicative-hubs verifying (pEx), (DN) and (pNeg).

Remark 8.1.21 (See Remark 5.1.6)

Note that the *equality*: $x^{-1} = x \to 1 \ (\overset{(Id)}{=} x \rightsquigarrow 1)$
can be used:
- either as the *definition* of the negation $^{-1}$ in the implicative-hub $(A, \to, \rightsquigarrow, 1)$,
- or as the *connection* (pNeg) between the negation $^{-1}$ and $\to, \rightsquigarrow,\ 1$ in the implicative-group (Definition 3').

Remark 8.1.22 (See Remark 5.1.8)

The i-group (Definition 3') is a regular pseudo-algebra with involutive regular negation, by Definition 5.1.7, since (pM), (pNeg), (DN) hold.

Proposition 8.1.23 *Let $(G, \to, \rightsquigarrow, 1)$ be an i-group. Then, we have: for all $x, y \in G$,*

$$(8.3) \qquad x \to y = (x^{-1} \rightsquigarrow y^{-1})^{-1}, \quad x \rightsquigarrow y = (x^{-1} \to y^{-1})^{-1} \quad (\textit{see } (7.2)),$$

$$(8.4) \qquad (x \to y^{-1})^{-1} = y^{-1} \to x, \quad (y \rightsquigarrow x^{-1})^{-1} = x^{-1} \rightsquigarrow y.$$

Proof. (8.3): $x \to y \overset{(pNg)}{=} (y \to x)^{-1} \overset{(pDNeg1)}{=} (x^{-1} \rightsquigarrow y^{-1})^{-1}$
and $x \rightsquigarrow y = (y \rightsquigarrow x)^{-1} = (x^{-1} \to y^{-1})^{-1}$.

(8.4): by (pNg). □

- **New operation in i-groups:** $\cdot$

The i-groups have a very interesting property.

Proposition 8.1.24 *Let $(G, \to, \rightsquigarrow, 1)$ be an i-group and $x^{-1} \overset{def.}{=} x \to 1 = x \rightsquigarrow 1$, for all $x \in G$. For all $x \in G$, define the mapings $h_x : G \longrightarrow G$ and ${}_xh : G \longrightarrow G$ by: for all $g \in G$,*

$$h_x(g) = x \to g, \quad {}_xh(g) = x \rightsquigarrow g.$$

Then, h_x and ${}_xh$ are bijective.

Proof.

Injectivity: Let $g_1,\ g_2 \in G$, $g_1 \neq g_2$. We must prove that $h_x(g_1) \neq h_x(g_2)$ and ${}_xh(g_1) \neq {}_xh(g_2)$.
Suppose, by absurdum hypothesis, that $h_x(g_1) = h_x(g_2)$, i.e. $x \to g_1 = x \rightsquigarrow g_2$; then, $g_1 \to g_2 \overset{(pB=)}{=} (x \to g_1) \to (x \to g_2) \overset{(pRe')}{=} 1$, hence, by (pEq=), $g_1 = g_2$: contradiction.
Similarly, suppose, by absurdum hypothesis, that ${}_xh(g_1) = {}_xh(g_2)$, i.e. $x \rightsquigarrow g_1 =$

$x \rightsquigarrow g_2$; then, $g_1 \rightsquigarrow g_2 \overset{(pB=)}{=} (x \rightsquigarrow g_1) \rightsquigarrow (x \rightsquigarrow g_2) \overset{(pRe')}{=} 1$, hence $g_1 = g_2$: contradiction.

Surjectivity of h_x: We must prove that for all $y \in G$, there exists $z \in G$, such that $y = h_x(z) = x \to z$. Indeed, let $y \in G$ and take $z = y^{-1} \rightsquigarrow x$; then, $h_x(z) = x \to z = x \to (y^{-1} \rightsquigarrow x) \overset{(pEx)}{=} y^{-1} \rightsquigarrow (x \to x) \overset{(pRe')}{=} y^{-1} \rightsquigarrow 1 = (y^{-1})^{-1} \overset{(DN)}{=} y$.

Surjectivity of ${}_xh$: We must prove that for all $y \in G$, there exists $u \in G$, such that $y = {}_xh(u) = x \rightsquigarrow u$. Indeed, let $y \in G$ and take $u = y^{-1} \to x$; then, ${}_xh(u) = x \rightsquigarrow u = x \rightsquigarrow (y^{-1} \to x) \overset{(pEx)}{=} y^{-1} \to (x \rightsquigarrow x) \overset{(pRe')}{=} y^{-1} \to 1 = (y^{-1})^{-1} \overset{(DN)}{=} y$. □

We then have immediately the following corollary.

Corollary 8.1.25 *Let $(G, \to, \rightsquigarrow, 1)$ be an i-group. For all $x, y \in G$,*
- there exists a unique $z \in G$, $z = y^{-1} \rightsquigarrow x$, such that $y = h_x(z) = x \to z$ and
- there exists a unique $u \in G$, $u = y^{-1} \to x$, such that $y = {}_xh(u) = x \rightsquigarrow u$.

By above Corollary 8.1.25, we can introduce a binary operation $\cdot$ on $(G, \to, \rightsquigarrow, 1)$ as follows: for all $x, y \in G$,

$$y \cdot x \overset{def.}{=} z = y^{-1} \rightsquigarrow x, \quad x \cdot y \overset{def.}{=} u = y^{-1} \to x,$$

i.e.

(8.5) $$x \cdot y \overset{def.}{=} y^{-1} \to x \overset{(pDNeg2)}{=} x^{-1} \rightsquigarrow y.$$

Note that $\cdot$ satisfies the following property: for all $x, y \in G$,

(pP=) $\exists\, y \cdot x = z$ such that $x \to z = y$ and $x \cdot y = u$ such that $x \rightsquigarrow u = y$.

Remark 8.1.26 Recall from Chapter 2 the dual properties: for all x, y,
(pP) $\exists\, y \odot x = \min\{z \mid y \le x \to^L z\}$ and $x \odot y = \min\{u \mid y \le x \rightsquigarrow^L u\}$,
(pS) $\exists\, y \oplus x = \max\{z \mid y \ge x \to^R z\}$ and $x \oplus y = \max\{u \mid y \ge x \rightsquigarrow^R u\}$.

Recall also the property (p-s^L) ($x \le y \Longrightarrow x = y$). Then, note that:
(pP) + (p-s^L) $\Longrightarrow$ (pP=) and, dually, (pS) + (p-s^R) $\Longrightarrow$ (pS=).

Since the i-group is, by (DN), an involutive algebra (where there is only one involutive negation), we then define, by analogy to (2.24), the new operation $\cdot$ on G alternatively by: for all $x, y \in G$,

(8.6) $$x \cdot y \overset{def.}{=} (x \to y^{-1})^{-1} \overset{(pDNeg3)}{=} (y \rightsquigarrow x^{-1})^{-1}.$$

Indeed, it is the same operation as defined by (8.5), because, by (8.4), $(x \to y^{-1})^{-1} = y^{-1} \to x$ and $(y \rightsquigarrow x^{-1})^{-1} = x^{-1} \rightsquigarrow y$.

Remark 8.1.27 Note that the i-group may be defined equivalently as an algebra $(G, \to, \rightsquigarrow, \cdot, ^{-1}, 1)$, where $^{-1}$, $\cdot$ are expressed in terms of $\to$, $\rightsquigarrow$, 1 by (8.2), (8.6).

Remark 8.1.28 Let $(G, \to, \rightsquigarrow, 1)$ be an i-group. We have the *special* properties (see (2.25)): for all $x, y \in G$,

$$x \to y = (x \cdot y^{-1})^{-1}, \quad x \rightsquigarrow y = (y^{-1} \cdot x)^{-1}, \tag{8.7}$$

$$x \to y = y \cdot x^{-1}, \quad x \rightsquigarrow y = x^{-1} \cdot y. \tag{8.8}$$

Indeed, $(x \cdot y^{-1})^{-1} \overset{(8.6)}{=} ((x \to (y^{-1})^{-1})^{-1})^{-1} \overset{(DN)}{=} x \to y$ and $(y^{-1} \cdot x)^{-1} = ((x \rightsquigarrow (y^{-1})^{-1})^{-1})^{-1} = x \rightsquigarrow y$. Thus, (8.7) holds.

Also, $y \cdot x^{-1} \overset{(8.6)}{=} (y \to (x^{-1})^{-1})^{-1} \overset{(DN)}{=} (y \to x)^{-1} \overset{(pNg)}{=} x \to y$ and $x^{-1} \cdot y \overset{(8.6)}{=} (y \rightsquigarrow (x^{-1})^{-1})^{-1} \overset{(DN)}{=} (y \rightsquigarrow x)^{-1} \overset{(pNg)}{=} x \rightsquigarrow y$. Thus, (8.8) holds.

Theorem 8.1.29 *Let $(G, \to, \rightsquigarrow, 1)$ be an i-group and consider the operation $\cdot$ defined by (8.6). Then, the property (pGa$^=$) holds, where for all $x, y, z \in G$,*
(pGa$^=$) $\quad x = y \to z \Leftrightarrow y = x \rightsquigarrow z \Leftrightarrow x \cdot y = z.$

Proof. The equivalence $x = y \to z \Leftrightarrow y = x \rightsquigarrow z$ is just the property (pEq#$^=$), that holds by Proposition 8.1.6. We shall prove that $x \cdot y = z \Leftrightarrow x = y \to z$. Indeed, $x \cdot y = z \overset{(8.5)}{\Leftrightarrow} x^{-1} \rightsquigarrow y = z \overset{(pEq\#^=)}{\Leftrightarrow} z \to y = x^{-1} \Leftrightarrow y \to z = x$: indeed, if $z \to y = x^{-1}$, then $(z \to y)^{-1} = (x^{-1})^{-1}$, i.e. $y \to z = x$, by (pNg), (DN), and if $y \to z = x$, then $(y \to z)^{-1} = x^{-1}$, i.e. $z \to y = x^{-1}$. □

Remark 8.1.30 Note that the property (pGa$^=$) is a particular non-commutative Galois connection, where we have $=$ instead of $\leq$ or $\geq$. Therefore we called it (pGa$^=$). Consequently, note that an i-group is an involutive algebra (since the negation $^{-1}$ is involutive) that verifies the property (pP) (see Theorem 2.11.4).

We then immediately obtain the following important result.

Corollary 8.1.31 *(See Proposition 8.1.8)*
Any i-group is a pBCI(pP) algebra.

The following result will be useful in Chapter 9 to prove the equivalence between groups and i-groups.

Proposition 8.1.32 *Let $(G, \to, \rightsquigarrow, 1)$ be an i-group.*
Then, the following properties hold: for all $x, y, z \in G$,
(Ass) $\quad x \cdot (y \cdot z) = (x \cdot y) \cdot z,$
(pU) $\quad x \cdot 1 = x = 1 \cdot x,$
(pIv) $\quad x \cdot x^{-1} = 1 = x^{-1} \cdot x,$
(α) $\quad (x \to y) \cdot x = y = x \cdot (x \rightsquigarrow y),$
(β) $\quad x \to (y \cdot x) = y = x \rightsquigarrow (x \cdot y).$

Proof. (Ass): $x \cdot (y \cdot z) \overset{(8.5)}{=} x \cdot (z^{-1} \to y) = x^{-1} \rightsquigarrow (z^{-1} \to y)$ and $(x \cdot y) \cdot z = (x^{-1} \rightsquigarrow y) \cdot z = z^{-1} \to (x^{-1} \rightsquigarrow y) \overset{(pEx)}{=} x^{-1} \rightsquigarrow (z^{-1} \to y)$; thus, (Ass) holds.

(pU): $x \cdot 1 \overset{(8.5)}{=} 1^{-1} \to x \overset{(N1)}{=} 1 \to x \overset{(pM)}{=} x$
and $1 \cdot x = 1^{-1} \rightsquigarrow x = 1 \rightsquigarrow x = x$.

(pIv): $x \cdot x^{-1} \overset{(8.5)}{=} x^{-1} \rightsquigarrow x^{-1} \overset{(pRe')}{=} 1$ and $x^{-1} \cdot x = x^{-1} \to x^{-1} = 1$.

(α): $(x \to y) \cdot x \overset{(8.5)}{=} (x \to y)^{-1} \rightsquigarrow x \overset{(pNg)}{=} (y \to x) \rightsquigarrow x \overset{(pD=)}{=} y$
and $x \cdot (x \rightsquigarrow y) \overset{(8.5)}{=} (x \rightsquigarrow y)^{-1} \to x \overset{(pNg)}{=} (y \rightsquigarrow x) \to x = y$. Thus, ($\alpha$) holds.

(β): $x \to (y \cdot x) \overset{(8.8)}{=} (y \cdot x) \cdot x^{-1} \overset{(Ass)}{=} y \cdot (x \cdot x^{-1}) \overset{(pIv)}{=} y \cdot 1 \overset{(pU)}{=} y$
and $x \rightsquigarrow (x \cdot y) = x^{-1} \cdot (x \cdot y) = (x^{-1} \cdot x) \cdot y = 1 \cdot y = y$. Thus, ($\beta$) holds. □

Corollary 8.1.33 *Let* $(G, \to, \rightsquigarrow, 1)$ *be an i-group. Then,* $(G, \cdot, ^{-1}, 1)$ *is a group.*

Proof. By Proposition 8.1.32. □

8.1.2 The X-implicative-group - an intermediary notion

Definition 8.1.34 (Definition 1)
An *X-implicative-group*, or an *X-i-group* for short, is an algebra

$$(G, \to, \rightsquigarrow, \cdot, 1)$$

of type $(2, 2, 2, 0)$ such that (pEx), (pM) hold and the pseudo-implication $(\to, \rightsquigarrow)$ and the operation $\cdot$ verify the following property: for all $x, y, z \in G$,
(pGa$^=$) $\quad x = y \to z \Leftrightarrow y = x \rightsquigarrow z \Leftrightarrow x \cdot y = z$.

Denote by **X-i-group** the class of all X-i-groups.

Proposition 8.1.35 *Let* $(G, \to, \rightsquigarrow, \cdot, 1)$ *be an X-i-group. Then,*

$$(pGa^=) \Leftrightarrow (\alpha) + (\beta)$$

and the properties (α) *and* (β) *hold.*

Proof. By Lemma 7.1.21, $(pGa^=) \Leftrightarrow (\alpha) + (\beta)$. Since $(pGa^=)$ holds, then (α) and (β) hold too. □

Proposition 8.1.36 *Let* $(G, \to, \rightsquigarrow, \cdot, 1)$ *be an X-i-group. Then we have, for all* $x, y, z \in G$,
(pD=) $\quad y = (y \to x) \rightsquigarrow x, \quad y = (y \rightsquigarrow x) \to x,$
(pBB=) $\quad y \to z = (z \to x) \rightsquigarrow (y \to x), \quad y \rightsquigarrow z = (z \rightsquigarrow x) \to (y \rightsquigarrow x),$
(pU) $\quad x \cdot 1 = x = 1 \cdot x,$
(Ass) $\quad x \cdot (y \cdot z) = (x \cdot y) \cdot z,$
(IdEq) $\quad x \to y = 1 \Longleftrightarrow x \rightsquigarrow y = 1,$
(pEq=) $\quad x = y \Longleftrightarrow x \to y = 1 \overset{(IdEq)}{\Longleftrightarrow} x \rightsquigarrow y = 1.$

Proof. (pD=): $y = (y \to x) \rightsquigarrow x \overset{(pGa^=)}{\Leftrightarrow} y \to x = y \to x$, which is true and $y = (y \rightsquigarrow x) \to x \overset{(pGa^=)}{\Leftrightarrow} y \rightsquigarrow x = y \rightsquigarrow x$, which is also true; thus, (pD=) holds.

(pBB=): $(z \to x) \rightsquigarrow (y \to x) \overset{(pEx)}{=} y \to ((z \to x) \rightsquigarrow x) \overset{(pD=)}{=} y \to z$ and $(z \rightsquigarrow x) \to (y \rightsquigarrow x) \overset{(pEx)}{=} y \rightsquigarrow ((z \rightsquigarrow x) \to x) \overset{(pD=)}{=} y \rightsquigarrow z$; thus, (pBB=) holds.

(pU): $x \cdot 1 = x \overset{(pGa^=)}{\Leftrightarrow} x = 1 \to x$ and $1 \cdot x = x \overset{(pGa^=)}{\Leftrightarrow} x = 1 \rightsquigarrow x$, which are true by (pM).

(Ass): $x \cdot (y \cdot z) = a \overset{(pGa^=)}{\Leftrightarrow} y \cdot z = x \rightsquigarrow a \overset{(pGa^=)}{\Leftrightarrow} y = z \to (x \rightsquigarrow a) \overset{(pEx)}{=} x \rightsquigarrow (z \to a) \overset{(pGa^=)}{\Leftrightarrow} x \cdot y = z \to a \overset{(pGa^=)}{\Leftrightarrow} (x \cdot y) \cdot z = a$.

(IdEq): $1 = x \to y \overset{(pGa^=)}{\Leftrightarrow} 1 \cdot x = y \overset{(pU)}{\Leftrightarrow} x = y$ and $1 = x \rightsquigarrow y \overset{(pGa^=)}{\Leftrightarrow} x \cdot 1 = y \overset{(pU)}{\Leftrightarrow} x = y$; thus, (IdEq) holds.

(pEq=): The same as above. □

Remark 8.1.37 If $(G, \to, \rightsquigarrow, \cdot, 1)$ is an X-i-group, then $(G, \cdot, \to, \rightsquigarrow, 1)$ is an X-group, since (Ass), (pU) hold.

8.1.3 The basic equivalence (EI1)

Here we present the equivalence (EI1) between the i-groups and the X-i-groups.

We prove that the i-groups are termwise equivalent to the X-i-groups:

Theorem 8.1.38

(1) Let $\mathcal{G} = (G, \to, \rightsquigarrow, 1)$ be an i-group.

Define $\pi(\mathcal{G}) \overset{def.}{=} (G, \to, \rightsquigarrow, \cdot, 1)$ *by:* $x \cdot y \overset{def.}{=} (x \to y^{-1})^{-1} \overset{(pNg1)}{=} (y \rightsquigarrow x^{-1})^{-1}$, *where* $x^{-1} \overset{def.}{=} x \to 1 \overset{(Id)}{=} x \rightsquigarrow 1$.

Then, $\pi(\mathcal{G})$ is an X-i-group.

(1') Conversely, let $\mathcal{G} = (G, \to, \rightsquigarrow, \cdot, 1)$ be an X-i-group.

Define $\pi^*(\mathcal{G}) \overset{def.}{=} (G, \to, \rightsquigarrow, 1)$.

Then, $\pi^*(\mathcal{G})$ is an i-group.

(2) The above defined mappings are mutually inverse.

Proof.

(1): By Definition 1, (pM) holds, by Proposition 8.1.6, (pEx) holds and by Theorem 8.1.29, (pGa$^=$) holds.

(1'): By Proposition 8.1.36, (pBB=) holds.

(2) If $(G, \to, \rightsquigarrow, 1) \xrightarrow{\pi} (G, \to, \rightsquigarrow, \cdot, 1) \xrightarrow{\pi^*} (G, \to, \rightsquigarrow, 1)$,
then there is nothing to prove.

If $(G, \to, \rightsquigarrow, \cdot, 1) \xrightarrow{\pi^*} (G, \to, \rightsquigarrow, 1) \xrightarrow{\pi} (G, \to, \rightsquigarrow, \oplus, 1)$,
then we have to prove that $x \oplus y = x \cdot y$, for all $x, y \in G$. Indeed,
$x \oplus y = (x \to y^{-1})^{-1} = [x \to (y \to 1)] \rightsquigarrow 1 = x \cdot y \overset{(pGa^=)}{\Leftrightarrow}$
$x = y \to ([x \to (y \to 1)] \rightsquigarrow 1)$, which is true, since
$y \to ([x \to (y \to 1)] \rightsquigarrow 1) = [x \to (y \to 1)] \rightsquigarrow (y \to 1) = x$, by (pEx) and (pD=). □

Hence, we have the basic equivalence:

(EI1) **i-group** $\Leftrightarrow$ **X-i-group.**

8.2 Po-implicative-groups

8.2.1 The po-implicative-group

Definition 8.2.1 A *partially-ordered implicative-group*, or a *po-implicative-group* or even a *po-i-group* for short, is a structure

$$\mathcal{G} = (G, \leq, \to, \rightsquigarrow, 1),$$

such that $(G, \to, \rightsquigarrow, 1)$ is an i-group, $(G, \leq)$ is a poset and $\leq$ is compatible with $\to, \rightsquigarrow$, i.e. we have: for all $x, y, z \in G$,
(p*^{L}) $x \leq y$ implies $z \to x \leq z \to y$ and $z \rightsquigarrow x \leq z \rightsquigarrow y$.

Note that in a po-i-group we can define a negation $^{-1}$ by (8.2) and a new operation $\cdot$ by (8.5).

Remark 8.2.2 Note that a po-i-group may be defined equivalently as a structure

$$\mathcal{G} = (G, \leq, \to, \rightsquigarrow, \cdot, ^{-1}, 1),$$

by Remark 8.1.27.

The presence of the order relation implies the presence of the *Duality Principle.* It follows that there are two dual po-i-groups. If their support sets coincide ($G = G_1 = G_2$), we say that they are *self-dual*, i.e. $(G, \leq, \to, \rightsquigarrow, 1)$ is in the same time left-po-i-group and right-po-i-group; if their support sets differ ($G_1 \neq G_2$), then their unit elements differ and say that $1_1 \leq 1_2$ in the union set $G_1 \cup G_2$; we then call $\mathcal{G}_1$ as the *left-po-i-group* and $\mathcal{G}_2$ as the *right-po-i-group.*

Denote by **po-i-group** the class of all po-i-groups, namely:
- by **po-i-group**^{L} the class of all left-po-i-groups and
- by **po-i-group**^{R} the class of all right-po-i-groups.

Remark 8.2.3 (See Corollary 7.2.9)

While a left-po-group is a left-po-m, a left-po-i-group is not a left-pBCI algebra, because the property (pEq^L) is not verified (See Example 5 from Chapter 6). The same, in the dual case.

Remark 8.2.4

If $(G^L, \leq, \to^L, \rightsquigarrow^L, 1)$ is a left-po-i-group, then $(\to^L, \rightsquigarrow^L)$ is not a pseudo-residuum on the poset $(G^L, \leq)$ containing 1, by Definition 1.1.3, because the property (pEq^L) is not verified ($\leq$ is not the natural, internal order relation).

Dually, if $(G^R, \geq, \to^R, \rightsquigarrow^R, 1)$ is a right-po-i-group, then $(\to^R, \rightsquigarrow^R)$ is not a pseudo-coresiduum on the poset $(G^R, \geq)$ containing 1, by Definition 1.1.3, because the property (pEq^R) is not verified.

If the partial order relation $\leq$ is linear (total), then $\mathcal{G}$ is a linearly-ordered po-i-group.

Proposition 8.2.5 *Let* $(G, \leq, \to, \rightsquigarrow, 1)$ *be a po-i-group. Then, the following properties hold: for all* $x, y, a \in G$,

(ImNeg) $x \leq y$ *implies* $y^{-1} \leq x^{-1}$,
(pCp) $x \leq y$ *implies* $a \cdot x \leq a \cdot y$ *and* $x \cdot a \leq y \cdot a$,
(p*Eq) $x \leq y \Leftrightarrow z \to x \leq z \to y \Leftrightarrow z \rightsquigarrow x \leq z \rightsquigarrow y$,
(pGa$^=$) $x \cdot y = z \Leftrightarrow x = y \to z \Leftrightarrow y = x \rightsquigarrow z$,
(pEq#$^\leq$) $x \leq y \to z \Leftrightarrow y \leq x \rightsquigarrow z$,
(pEq#$^\geq$) $x \geq y \to z \Leftrightarrow y \geq x \rightsquigarrow z$.

Proof. (ImNeg): Let $x \leq y$; then, by (p$*^L$), (pRe'), we have $1 = x \to x \leq x \to y$; then, by (p$*^L$) again and by (pEx), (pRe'), we obtain $y^{-1} = y \rightsquigarrow 1 \leq y \rightsquigarrow (x \to y) = x \to (y \rightsquigarrow y) = x \to 1 = x^{-1}$.

(pCp): Let $x \leq y$; then $y^{-1} \leq x^{-1}$, by (ImNeg); by (p$*^L$), $a \to (y^{-1}) \leq a \to (x^{-1})$, hence by (ImNeg) again, we obtain $a \cdot x = (a \to x^{-1})^{-1} \leq (a \to y^{-1})^{-1} = a \cdot y$; by (p$*^L$), $a \rightsquigarrow (y^{-1}) \leq a \rightsquigarrow (x^{-1})$, hence by (ImNeg) again, we obtain $x \cdot a = (a \rightsquigarrow x^{-1})^{-1} \leq (a \rightsquigarrow y^{-1})^{-1} = y \cdot a$.

(p*Eq): By (p$*^L$), it is sufficient to prove that $z \to x \leq z \to y$ implies $x \leq y$ and that $z \rightsquigarrow x \leq z \rightsquigarrow y$ implies $x \leq y$. Indeed, $z \to x \leq z \to y$ implies, by above (pCp), that $x \overset{(\alpha)}{=} (z \to x) \cdot z \leq (z \to y) \cdot z \overset{(\alpha)}{=} y$ and $z \rightsquigarrow x \leq z \rightsquigarrow y$ implies, by above (pCp), that $x \overset{(\alpha)}{=} z \cdot (z \rightsquigarrow x) \leq z \cdot (z \rightsquigarrow y) \overset{(\alpha)}{=} y$.

(pGa$^=$): follows by Theorem 8.1.29.

(pEq#$^\leq$): If $x \leq y \to z$, then, by (ImNeg), $z \to y \overset{(pNg)}{=} (y \to z)^{-1} \leq x^{-1}$; then, by (pCp), $y \overset{(\alpha)}{=} (z \to y) \cdot z \leq x^{-1} \cdot z \overset{(8.5)}{=} ((x^{-1})^{-1}) \rightsquigarrow z \overset{(DN)}{=} x \rightsquigarrow z$.
Similarly, if $y \leq x \rightsquigarrow z$, then, by (ImNeg), $z \rightsquigarrow x \overset{(pNg)}{=} (x \rightsquigarrow z)^{-1} \leq y^{-1}$; then, by (pCp), $x \overset{(\alpha)}{=} z \cdot (z \rightsquigarrow x) \leq z \cdot (y^{-1}) \overset{(8.5)}{=} ((y^{-1})^{-1}) \to z \overset{(DN)}{=} y \to z$.

(pEq#$^\geq$): dually. □

Theorem 8.2.6 *Let* $(G, \leq, \to, \rightsquigarrow, 1)$ *be a po-i-group. Then the properties (pGa$^\leq$) and (pGa$^\geq$) hold, where for all* $x, y, z \in G$,
(pGa$^\leq$) $x \leq y \to z \Leftrightarrow y \leq x \rightsquigarrow z \Leftrightarrow x \cdot y \leq z$,
(pGa$^\geq$) $x \geq y \to z \Leftrightarrow y \geq x \rightsquigarrow z \Leftrightarrow x \cdot y \geq z$
and $\cdot$ *is the operation defined by (8.6), via (8.2).*

Proof. (pGa$^\leq$): The equivalence $x \leq y \to z \Leftrightarrow y \leq x \rightsquigarrow z$ is just (pEq#$^\leq$) from above Proposition 8.2.5. We prove now that $x \cdot y \leq z \Leftrightarrow x \leq y \to z$. Indeed, if $x \cdot y \leq z$, then, by (pCp), $x \overset{(pU)}{=} x \cdot 1 \overset{(pIv)}{=} x \cdot (y \cdot (y^{-1})) \overset{(Ass)}{=} (x \cdot y) \cdot (y^{-1}) \leq z \cdot (y^{-1}) \overset{(8.5)}{=} ((y^{-1})^{-1}) \to z \overset{(DN)}{=} y \to z$.
Conversely, if $x \leq y \to z$, then, by (pCp), $x \cdot y \leq (y \to z) \cdot y \overset{(\alpha)}{=} z$.

(pGa$^\geq$): dually. □

Remark 8.2.7 Note that:
- (pGa$^\leq$) is the analogous of the (pRP)=(pPR) from algebras of logic,
- (pGa$^\geq$) is the analogous of the (pcoRS)=(pScoR) from algebras of logic,
and that both are non-commutative Galois connections.

Corollary 8.2.8 *Let* $(G, \leq, \rightarrow, \rightsquigarrow, 1)$ *be a po-i-group. Then,* $(G, \leq, \cdot, ^{-1}, 1)$ *is a po-group.*

Proof. By Corollary 8.1.33 and Proposition 8.2.5. □

8.2.2 The negative and the positive cones of po-i-groups

Let $\mathcal{G} = (G, \leq, \rightarrow, \rightsquigarrow, 1)$ be a po-i-group (left-po-i-group or right-po-i-group).

Define the *negative cone* of $\mathcal{G}$ as follows: $G^- \stackrel{def.}{=} \{x \in G \mid x \leq 1\}$.

Define the *positive cone* of $\mathcal{G}$ as follows: $G^+ \stackrel{def.}{=} \{x \in G \mid x \geq 1\}$.

Then, we have:

Lemma 8.2.9 *Let* $\mathcal{G} = (G, \leq, \rightarrow, \rightsquigarrow, 1)$ *be a po-i-group.*

(1) For all $x \in G$, *we have:*
- if $x \in G^-$, *then* $x^{-1} \in G^+$ *and*
- if $x \in G^+$, *then* $x^{-1} \in G^-$.

(2) G^-, G^+ *are not closed under* $\rightarrow$, $\rightsquigarrow$.

Proof. (1): Let $x \in G^-$, i.e. $x \leq 1$; then, $x \rightarrow x \leq x \rightarrow 1$ and $x \rightsquigarrow x \leq x \rightsquigarrow 1$, by (p$*^L$); hence, $1 \leq x \rightarrow 1$ and $1 \leq x \rightsquigarrow 1$, by (pRe'), hence $x^{-1} = x \rightarrow 1 = x \rightsquigarrow 1 \in G^+$. Similarly, if $x \in G^+$, then $x^{-1} = x \rightarrow 1 = x \rightsquigarrow 1 \in G^-$.

(2): Let $x, y \in G^-$, i.e. $x \leq 1$ and $y \leq 1$; then, $x \rightarrow y \stackrel{(p*^L)}{\leq} x \rightarrow 1$ and $x \rightsquigarrow y \stackrel{(p*^L)}{\leq} x \rightsquigarrow 1$; but, $x \rightarrow 1$, $x \rightsquigarrow 1 \in G^+$, by above (1); hence, G^- is not closed under $\rightarrow, \rightsquigarrow$. Similarly, G^+ is not closed under $\rightarrow$, $\rightsquigarrow$. □

8.2.3 The X-po-implicative-group - an intermediary notion

Definition 8.2.10 We shall name *X-po-implicative-group*, or *X-po-i-group* for short, a structure

$$(G, \leq, \rightarrow, \rightsquigarrow, \cdot, 1)$$

such that the following properties (pM) and (pEx) hold: for all $x, y, z \in G$,
(pM) $1 \rightarrow x = x = 1 \rightsquigarrow x$,
(pEx) $x \rightarrow (y \rightsquigarrow z) = y \rightsquigarrow (x \rightarrow z)$,
$(G, \leq)$ is a poset and the pseudo-implication $(\rightarrow, \rightsquigarrow)$ and $\cdot$ verify the two properties: for all $x, y, z \in G$,
(pGa$^{\leq}$) $x \leq y \rightarrow z \Leftrightarrow y \leq x \rightsquigarrow z \Leftrightarrow x \cdot y \leq z$,
(pGa$^{\geq}$) $x \geq y \rightarrow z \Leftrightarrow y \geq x \rightsquigarrow z \Leftrightarrow x \cdot y \geq z$.

Denote by **X-po-i-group** the class of all X-po-i-groups, namely:
- by **X-po-i-group**^{L} the class of all left-X-po-i-groups and
- by **X-po-i-group**^{R} the class of all right-X-po-i-groups.

Corollary 8.2.11 *Let* $(G, \leq, \rightarrow, \rightsquigarrow, \cdot, 1)$ *be an X-po-i-group. Then,* $(G, \rightarrow, \rightsquigarrow, \cdot, 1)$ *is an X-i-group.*

Proof. Obviously, by Lemma 7.2.17. □

Proposition 8.2.12 *Let $(G, \leq, \rightarrow, \rightsquigarrow, \cdot, 1)$ be an X-po-i-group. Then, the properties (pGa=), (α), (β), (pEqCp), (p^{*L}), (p^{**L}) hold.*

Proof. By Lemma 7.2.17. □

Proposition 8.2.13 *Let $(G, \leq, \rightarrow, \rightsquigarrow, \cdot, 1)$ be an X-po-i-group. Then, the properties (Ass), (pU), (pD=), (pBB=), (Id), (IdEq), (pEq=) hold.*

Proof. By Corollary 8.2.11, $(G, \rightarrow, \rightsquigarrow, \cdot, 1)$ is an X-i-group, hence, by Proposition 8.1.36, (Ass), (pU), (pD=), (pBB=), (Id), (IdEq) and (pEq=) hold. □

8.2.4 The equivalences (EI2)

Here we present the equivalences (EI2) between the po-i-groups and the X-po-i-groups.

We prove that the po-i-groups are termwise equivalent to the X-po-i-groups.

Theorem 8.2.14 *(See Theorem 8.1.38)*

(1) Let $\mathcal{G} = (G, \leq, \rightarrow, \rightsquigarrow, 1)$ be a po-i-group.

Define $\pi'(\mathcal{G}) \stackrel{def.}{=} (G, \leq, \rightarrow, \rightsquigarrow, \cdot, 1)$, where $(G, \rightarrow, \rightsquigarrow, \cdot, 1) = \pi(G, \rightarrow, \rightsquigarrow, 1)$ from Theorem 8.1.38 (1).

Then, $\pi'(\mathcal{G})$ is an X-po-i-group.

(1') Conversely, let $\mathcal{G} = (G, \leq, \rightarrow, \rightsquigarrow, \cdot, 1)$ be an X-po-i-group.

Define $\pi^{*'}(\mathcal{G}) \stackrel{def.}{=} (G, \leq, \rightarrow, \rightsquigarrow, 1)$.

Then, $\pi^{*'}(\mathcal{G})$ is a po-i-group.

(2) The above defined mappings π' and $\pi^{*'}$ are mutually inverse.

Proof. (1): By Theorems 8.1.38 (1) and 8.2.6.

(1'): By Corollary 8.2.11, $(G, \rightarrow, \rightsquigarrow, \cdot, 1)$ is an X-i-group and then, by Theorem 8.1.38 (1'), $(G, \rightarrow, \rightsquigarrow, 1)$ is an i-group. It remains to prove that (p^{*L}) holds, which follows by above Proposition 8.2.12.

(2) follows by Theorem 8.1.38 (2). □

Hence, we have the equivalences (EI2) (the analogous of (iEL5)):

(EI2) **po-i-group**^{L} $\Leftrightarrow$ **X-po-i-group**^{L} and, dually,
po-i-group^R $\Leftrightarrow$ **X-po-i-group**^{R}.

8.3 l-implicative-groups

8.3.1 The l-implicative-group

Definition 8.3.1 Let $\mathcal{G} = (G, \leq, \rightarrow, \rightsquigarrow, 1)$ be a po-i-group. If the partial order relation $\leq$ is a lattice order relation, with the lattice operations $\wedge$ and $\vee$ ($x \leq$

$y \Leftrightarrow x \wedge y = x \Leftrightarrow x \vee y = y$), then $\mathcal{G}$ is a *lattice-ordered implicative-group*, or an *l-implicative-group* or even an *l-i-group* for short, denoted by:

$$\mathcal{G} = (G, \vee, \wedge, \rightarrow, \rightsquigarrow, 1).$$

Denote by **l-i-group** the class of all l-i-groups, namely:
- by **l-i-group^L**, the class of all left-l-i-groups and
- by **l-i-group^R**, the class of all right-l-i-groups.

Note that an l-i-group may be linearly-ordered or not, while a linearly-ordered i-group is an l-i-group.

Remark 8.3.2 Note that an l-i-group may be defined equivalently as an algebra

$$\mathcal{G} = (G, \vee, \wedge, \rightarrow, \rightsquigarrow, \cdot, ^{-1}, 1),$$

by Remark 8.2.2.

8.3.2 The negative and the positive cones of l-i-groups

Lemma 8.3.3 *Let $\mathcal{G} = (G, \vee, \wedge, \rightarrow, \rightsquigarrow, 1)$ be an l-i-group (left-l-i-group or right-l-i-group). Then, G^- and G^+ are closed under $\vee$ and $\wedge$.*

Proof. See the proof of Lema 7.3.9. □

Lemma 8.3.4 *(See Lemma 8.2.9)*

Let $\mathcal{G} = (G, \vee, \wedge, \rightarrow, \rightsquigarrow, 1)$ be an l-i-group (left-l-i-group or right-l-i-group).

(1) Define, for all $x, y \in G^-$:

(8.9) $$x \rightarrow^L y \overset{def.}{=} (x \rightarrow y) \wedge 1, \quad x \rightsquigarrow^L y \overset{def.}{=} (x \rightsquigarrow y) \wedge 1.$$

Then, we have:
(i) If $x \in G^-$, then $x \rightarrow^L 1 = 1$, $x \rightsquigarrow^L 1 = 1 \in G^-$;
(ii) G^- is closed under $\rightarrow^L$, $\rightsquigarrow^L$ and we have, for all $x, y, z \in G^-$:

(8.10) $$x \leq y \rightarrow z \Leftrightarrow x \leq y \rightarrow^L z, \quad y \leq x \rightsquigarrow z \Leftrightarrow y \leq x \rightsquigarrow^L z,$$

(8.11) $$x \leq y \rightarrow^L z \Leftrightarrow y \leq x \rightsquigarrow^L z.$$

(1') Dually, define, for all $x, y \in G^+$:

(8.12) $$x \rightarrow^R y \overset{def.}{=} (x \rightarrow y) \vee 1, \quad x \rightsquigarrow^R y \overset{def.}{=} (x \rightsquigarrow y) \vee 1.$$

Then, we have:
(i') If $x \in G^+$, then $x \rightarrow^R 1 = 1$, $x \rightsquigarrow^R 1 = 1 \in G^+$;
(ii') G^+ is closed under $\rightarrow^R$, $\rightsquigarrow^R$ and we have, for all $x, y, z \in G^+$:

(8.13) $$x \geq y \rightarrow z \Leftrightarrow x \geq y \rightarrow^R z, \quad y \geq x \rightsquigarrow z \Leftrightarrow y \geq x \rightsquigarrow^R z,$$

(8.14) $$x \geq y \rightarrow^R z \Leftrightarrow y \geq x \rightsquigarrow^R z.$$

Proof. (1) (i): Let $x \in G^-$, i.e. $x \leq 1$; then, $1 = x \to x \leq x \to 1$ and $1 = x \rightsquigarrow x \leq x \rightsquigarrow 1$, by (p*^{L}), (pRe'); hence, $x \to^L 1 \overset{(8.9)}{=} (x \to 1) \wedge 1 = 1$ and $x \rightsquigarrow^L 1 = (x \rightsquigarrow 1) \wedge 1 = 1$.

(ii): Let $x, y \in G^-$, i.e. $x \leq 1$ and $y \leq 1$. Then, $x \to y \leq x \to 1$ and $x \rightsquigarrow y \leq x \rightsquigarrow 1$, by (p*^{L}). Then, $x \to^L y \overset{(8.9)}{=} (x \to y) \wedge 1 \leq (x \to 1) \wedge 1 = 1$ and $x \rightsquigarrow^L y = (x \rightsquigarrow y) \wedge 1 \leq (x \rightsquigarrow 1) \wedge 1 = 1$; hence $x \to^L y \in G^-$ and $x \rightsquigarrow^L y \in G^-$.

(8.10): If $x \leq y \to z$ then, since $x \leq 1$, we obtain $x \leq (y \to z) \wedge 1 = y \to^L z$, i.e. $x \leq y \to^L z$; conversely, if $x \leq y \to^L z$ then, since $y \to^L z = (y \to z) \wedge 1 \leq y \to z$, we obtain $x \leq y \to z$; thus, $x \leq y \to z \Leftrightarrow x \leq y \to^L z$. The second part has a similar proof. Thus, (8.10) holds.

(8.11): By Theorem 8.2.6, the property (pGa$^{\leq}$) holds; then apply (8.10).

(1'): dually. □

8.3.3 The X-l-implicative-group - an intermediary notion

Definition 8.3.5 Let $\mathcal{G} = (G, \leq, \to, \rightsquigarrow, \cdot, 1)$ be an X-po-i-group. If the partial order $\leq$ is a lattice order, then the X-po-i-group $\mathcal{G}$ is called an *X-l-i-group* and is denoted by:

$$\mathcal{G} = (G, \vee, \wedge, \to, \rightsquigarrow, \cdot, 1).$$

Denote by **X-l-i-group** the class of all X-l-i-groups, namely:
- by **X-l-i-group^L**, the class of all left-X-l-i-groups and
- by **X-l-i-group^R**, the class of all right-X-l-i-groups.

8.3.4 The equivalences (EI3)

Here we present the equivalences (EI3) between the l-i-groups and the X-l-i-groups.

We obtain obviously, by (EI2), the equivalences (EI3) (the analogous of (iEL6)):

(EI3) **l-i-group^L** $\Leftrightarrow$ **X-l-i-group^L** and, dually,
l-i-group^R $\Leftrightarrow$ **X-l-i-group^R**,

i.e. the l-i-groups are term-equivalent to the X-l-i-groups.

Chapter 9

Connections between groups and implicative-groups

In this chapter, we present [97], [100] the equivalences (EIG1), (EIG2), (EIG3) between the intermediary notions, in Section 1, and the equivalences (Eq1), (Eq2), (Eq3) between the corresponding four notions, in Section 2.

We also present some examples, in Section 3:
- an example of non-commutative group (implicative-group),
- some examples of l-groups (l-implicative-groups).

The chapter has 3 sections.

9.1 Connections between the two intermediary notions

9.1.1 The basic equivalence (EIG1)

Theorem 9.1.1 *(See Theorem 4.1.2 for the idea)*

(1) Let $\mathcal{G} = (G, \to, \rightsquigarrow, \cdot, 1)$ be an X-i-group.
Define $\Phi'(\mathcal{G}) \stackrel{def.}{=} (G, \cdot, \to, \rightsquigarrow, 1)$.

Then, $\Phi'(\mathcal{G})$ is an X-group.

(1') Let $\mathcal{G} = (G, \cdot, \to, \rightsquigarrow, 1)$ be an X-group.
Define $\Psi'(\mathcal{G}) \stackrel{def.}{=} (G, \to, \rightsquigarrow, \cdot, 1)$.

Then, $\Psi'(\mathcal{G})$ is an X-i-group.

(2) The maps Φ' and Ψ' are mutually inverse.

Proof. (1): By Proposition 8.1.36.

(1'): By Proposition 7.1.23.

(2): Obviously. □

Hence, we have the equivalence (see (iELM1)):

(EIG1) **X-i-group** $\Leftrightarrow$ **X-group**,

i.e. the X-implicative-groups and the X-groups are termwise equivalent.

9.1.2 The equivalences (EIG2)

Theorem 9.1.2 *(See Theorem 4.1.2 for the idea)*
(1) Let $\mathcal{G} = (G, \leq, \rightarrow, \rightsquigarrow, \cdot, 1)$ *be an X-po-i-group.*
Define $\Phi'(\mathcal{G}) \stackrel{def.}{=} (G, \leq, \cdot, \rightarrow, \rightsquigarrow, 1)$.
Then, $\Phi'(\mathcal{G})$ *is an X-po-group.*
(1') Let $\mathcal{G} = (G, \leq, \cdot, \rightarrow, \rightsquigarrow, 1)$ *be an X-po-group.*
Define $\Psi'(\mathcal{G}) \stackrel{def.}{=} (G, \leq, \rightarrow, \rightsquigarrow, \cdot, 1)$.
Then, $\Psi'(\mathcal{G})$ *is an X-po-i-group.*
(2) The maps Φ' *and* Ψ' *are mutually inverse.*

Proof. (1): By Proposition 8.2.13.
(1'): By Proposition 7.2.18.
(2): Obviously. □

Hence, we have the dual equivalences (EIG2) (the analogous of (iELM5)):

(EIG2) **X-po-i-group**^{L} $\Leftrightarrow$ **X-po-group**^{L} and, dually,
X-po-i-group^R $\Leftrightarrow$ **X-po-group**^{R},

i.e. the X-po-implicative-groups and the X-po-groups are termwise equivalent.

9.1.3 The equivalences (EIG3)

Here we present the equivalences (EIG3) between the X-l-i-groups and the X-l-groups. The announced result follows immediately by Theorem 9.1.2.
Hence we have the dual equivalences (EIG3) (the analogous of (iELM6)):

(EIG3) **X-l-i-group**^{L} $\Leftrightarrow$ **X-l-group**^{L} and, dually,
X-l-i-group^R $\Leftrightarrow$ **X-l-group**^{R},

i.e. the X-l-implicative-groups are termwise equivalent to the X-l-groups.

9.2 Connections between the four notions

9.2.1 The basic equivalences (Eq1), at group level

By Theorems 7.1.24, 8.1.38 and 9.1.1, we obtain:

Corollary 9.2.1 *The groups, the X-groups, the i-groups and the X-i-groups are all termwise equivalent.*

Hence, by (EG1), (EI1) and (EIG1), we obtain the following basic equivalences:

(Eq1) **i-group** $\Leftrightarrow$ **X-i-group** $\Leftrightarrow$ **X-group** $\Leftrightarrow$ **group**,

that are illustrated in Figure 9.1 (we use Definition 3 of i-groups).

i-group	$\Leftrightarrow$	**X-i-group**	$\Leftrightarrow$	**X-group**	$\Leftrightarrow$	**group**
$(G, \rightarrow, \rightsquigarrow, 1)$ (pEx), (IdEq), (pEq=)		$(G, \rightarrow, \rightsquigarrow, \cdot, 1)$ (pM), (pEx), ($\text{pGa}^=$)		$(G, \cdot, \rightarrow, \rightsquigarrow, 1)$ (pU), (Ass), ($\text{pGa}^=$)		$(G, \cdot, ^{-1}, 1)$ (pU), (Ass), (pIv)
$x^{-1} = x \rightarrow 1,$ $= x \rightsquigarrow 1$ $x \cdot y$ $= (x \rightarrow y^{-1})^{-1}$ $= (y \rightsquigarrow x^{-1})^{-1}$		$x^{-1} = x \rightarrow 1$ $= x \rightsquigarrow 1,$		$x^{-1} = x \rightarrow 1$ $= x \rightsquigarrow 1$		$x \rightarrow y =$ $(x \cdot y^{-1})^{-1}$ $= y \cdot x^{-1},$ $x \rightsquigarrow y =$ $(y^{-1} \cdot x)^{-1}$ $= x^{-1} \cdot y$
(I.1)		(I.2)		(II.2)		(II.1)

Figure 9.1: The equivalences (Eq1)

Remark 9.2.2 (See Remark 7.1.9) For all $x \in G$, we have:
$1 = x \rightarrow x = x \rightsquigarrow x$, $x^{-1} = x \rightarrow 1 = x \rightsquigarrow 1$ and $(x^{-1})^{-1} = x$
(there is only one **involutive** negation).

Hence, we have that groups are termwise equivalent to i-groups, i.e. we have the following theorem, for which we present a direct proof.

Theorem 9.2.3
(1) Let $\mathcal{G} = (G, \cdot, ^{-1}, 1)$ be a group.
Define $\Psi(\mathcal{G}) \stackrel{def.}{=} (G, \rightarrow, \rightsquigarrow, 1)$ by: for all $x, y \in G$,

$$x \rightarrow y \stackrel{def.}{=} ((x \cdot y^{-1})^{-1} \stackrel{(pIvG)}{=})\ y \cdot x^{-1},$$

$$x \rightsquigarrow y \stackrel{def.}{=} ((y^{-1} \cdot x)^{-1} \stackrel{(pIvG)}{=})\ x^{-1} \cdot y.$$

Then $\Psi(\mathcal{G})$ is an i-group.

(1') Conversely, let $\mathcal{G} = (G, \to, \rightsquigarrow, 1)$ be an i-group.
Define $\Phi(\mathcal{G}) \overset{def.}{=} (G, \cdot, ^{-1}, 1)$ by: for all $x, y \in G$,

$$x^{-1} \overset{def.}{=} x \to 1 \overset{(Id)}{=} x \rightsquigarrow 1,$$

$$x \cdot y \overset{def.}{=} ((x \to y^{-1})^{-1} \overset{(8.4)}{=}) \, y^{-1} \to x \overset{(pDNeg2)}{=} ((y \rightsquigarrow x^{-1})^{-1} \overset{(8.4)}{=}) \, x^{-1} \rightsquigarrow y.$$

Then $\Phi(\mathcal{G})$ is a group.
(2) The maps Φ and Ψ are mutually inverse.

Proof. (1): By Proposition 7.1.13.
(1'): By Corollary 8.1.33.
(2): Let $(G, \cdot, ^{-1}, 1) \xrightarrow{\Psi} (G, \to, \rightsquigarrow, 1) \xrightarrow{\Phi} (G, \odot, ^*, 1)$. Then, for all $x, y \in G$, $x^* = x \to 1 = (x \cdot 1^{-1})^{-1} = (x \cdot 1)^{-1} = x^{-1}$ and
$x \odot y = (x \to y^*)^* = (x \to y^{-1})^{-1} = ((x \cdot ((y^{-1})^{-1}))^{-1})^{-1} = x \cdot y$, by (DN).
Let now $(G, \to, \rightsquigarrow, 1) \xrightarrow{\Phi} (G, \cdot, ^{-1}, 1) \xrightarrow{\Psi} (G, \Rightarrow, \approx>, 1)$. Then, for all $x, y \in G$,
$x \Rightarrow y = (x \cdot y^{-1})^{-1} = ((x \to ((y^{-1})^{-1}))^{-1})^{-1} = x \to y$ and
$x \approx> y = (y^{-1} \cdot x)^{-1} = ((x \rightsquigarrow ((y^{-1})^{-1}))^{-1})^{-1} = x \rightsquigarrow y.$ □

Corollary 9.2.4 *The i-group is commutative iff the termwise equivalent group is commutative, i.e. $x \to y = x \rightsquigarrow y$ for all x, y if and only if $x \cdot y = y \cdot x$ for all x, y.*

Proof. $x \to y = x \rightsquigarrow y$ for all x, y implies $x^{-1} \to y = x^{-1} \rightsquigarrow y \Leftrightarrow y \cdot (x^{-1})^{-1} = (x^{-1})^{-1} \cdot y$, i.e. $y \cdot x = x \cdot y$, by (DN). Conversely, $x \cdot y = y \cdot x$ for all x, y implies $x^{-1} \cdot y = y \cdot x^{-1}$, i.e. $x \rightsquigarrow y = x \to y$. □

9.2.2 The equivalences (Eq2), at po-group level

By Theorems 7.2.19, 8.2.14 and 9.1.2, we obtain:

Corollary 9.2.5 *The po-groups, the X-po-groups, the po-i-groups and the X-po-i-groups are all termwise equivalent.*

Hence, by (EG2), (EI2), (EIG2), we obtain the dual equivalences:

(Eq2) **po-i-group**^{L} $\Leftrightarrow$ **X-po-i-group**^{L} $\Leftrightarrow$ **X-po-group**^{L} $\Leftrightarrow$ **po-group**^{L} and
po-i-group^R $\Leftrightarrow$ **X-po-i-group**^{R} $\Leftrightarrow$ **X-po-group**^{R} $\Leftrightarrow$ **po-group**^{R},

that are illustrated in Figure 9.2 (see Figure 9.1 and Figure 4.2, where note that (pBB^L) + (pM^L) + (pEq^L) $\Longrightarrow$ (p*^{L})).

Hence, we have that po-groups are termwise equivalent to po-implicative-groups, i.e. we have the following theorem, for which we present a direct proof.

Theorem 9.2.6 *(See Theorem 9.2.3)*
(1) Let $\mathcal{G} = (G, \leq, \cdot, ^{-1}, 1)$ be a po-group.
Define $\Psi'(\mathcal{G}) \overset{def.}{=} (G, \leq, \to, \rightsquigarrow, 1)$, where $(G, \to, \rightsquigarrow, 1) = \Psi(G, \cdot, ^{-1}, 1)$, with Ψ from Theorem 9.2.3(1).

po-i-group^L	$\Leftrightarrow$	**X-po-i-group^L**	$\Leftrightarrow$	**X-po-g.^{L}**	$\Leftrightarrow$	**po-group^L**
$(G, \leq, \rightarrow, \rightsquigarrow, 1)$		$(G, \leq, \rightarrow, \rightsquigarrow, \cdot, 1)$		$(G, \leq, \cdot, \rightarrow, \rightsquigarrow, 1)$		$(G, \leq, \cdot, ^{-1}, 1)$
poset		poset		poset		poset
(pEx),		(pM),		(pU),		(pU),
(IdEq),		(pEx),		(Ass),		(Ass),
(pEq=),						(pIv),
(p*^{L})		(pGa$^\leq$), (pGa$^\geq$)		(pGa$^\leq$), (pGa$^\geq$)		**(pCp)**
$x^{-1} = x \rightarrow 1 = x \rightsquigarrow 1$, $x \cdot y = (x \rightarrow y^{-1})^{-1} = (y \rightsquigarrow x^{-1})^{-1}$		$x^{-1} = x \rightarrow 1 = x \rightsquigarrow 1$		$x^{-1} = x \rightarrow 1 = x \rightsquigarrow 1$		$x \rightarrow y = (x \cdot y^{-1})^{-1}$, $x \rightsquigarrow y = (y^{-1} \cdot x)^{-1}$
(I.1)		(I.2)		(II.2)		(II.1)

Figure 9.2: The equivalences (Eq2) (see the analogous (iE5))

Then $\Psi'(\mathcal{G})$ is a po-i-group.

(1') Conversely, let $\mathcal{G} = (G, \leq, \rightarrow, \rightsquigarrow, 1)$ be a po-i-group. Define $\Phi'(\mathcal{G}) \stackrel{def.}{=} (G, \leq, \cdot, ^{-1}, 1)$, where $(G, \cdot, ^{-1}, 1) = \Phi(G, \rightarrow, \rightsquigarrow, 1)$, with Φ from Theorem 9.2.3(1').

Then $\Phi'(\mathcal{G})$ is a po-group.

(2) The maps Φ' and Ψ' are mutually inverse.

Proof. To prove (1), by Theorem 9.2.3 (1), it remains to prove (p*^{L}), which holds by Proposition 7.2.11.

(1') follows by Corollary 8.2.8.

(2) follows by Theorem 9.2.3 (2). □

9.2.3 The equivalences (Eq3), at l-group level

By the analagous in lattice-ordered case of Theorems 7.2.19, 8.2.14 and 9.1.2, we obtain:

Corollary 9.2.7 *The l-groups, the X-l-groups, the l-i-groups and the X-l-i-groups are all termwise equivalent.*

Hence (by (EG3), (EI3), (EIG3)), we have the dual equivalences:

(Eq3) **l-i-group^L** $\Leftrightarrow$ **X-l-i-group^L** $\Leftrightarrow$ **X-l-group^L** $\Leftrightarrow$ **l-group^L** and, dually,
l-i-group^R $\Leftrightarrow$ **X-l-i-group^R** $\Leftrightarrow$ **X-l-group^R** $\Leftrightarrow$ **l-group^R**,

that are represented in Figure 9.3 (see Figures 9.1, 9.2).

l-i-group^L	$\Leftrightarrow$	**X-l-i-group^L**	$\Leftrightarrow$	**X-l-group^L**	$\Leftrightarrow$	**l-group^L**
$(G, \wedge, \vee,$ $\rightarrow, \rightsquigarrow, 1)$ (pEx), (IdEq), (pEq=), **(p*^{L})**		$(G, \wedge, \vee,$ $\rightarrow, \rightsquigarrow, \cdot, 1)$ (pM), (pEx), (pGa$^\leq$), (pGa$^\geq$)		$(G, \wedge, \vee,$ $\cdot, \rightarrow, \rightsquigarrow, 1)$ (pU), (Ass), (pGa$^\leq$), (pGa$^\geq$)		$(G, \wedge, \vee,$ $\cdot, ^{-1}, 1)$ (pU), (Ass), (pIv), **(pCp)**
$x^{-1} = x \rightarrow 1$ $= x \rightsquigarrow 1,$ $x \cdot y$ $= (x \rightarrow y^{-1})^{-1}$ $= (y \rightsquigarrow x^{-1})^{-1}$		$x^{-1} = x \rightarrow 1$ $= x \rightsquigarrow 1$		$x^{-1} = x \rightarrow 1$ $= x \rightsquigarrow 1$		$x \rightarrow y =$ $(x \cdot y^{-1})^{-1},$ $x \rightsquigarrow y =$ $(y^{-1} \cdot x)^{-1}$
(I.1)		(I.2)		(II.2)		(II.1)

Figure 9.3: The equivalences (Eq3) (see the analogous (iE6))

9.3 Examples

In this section, we present some exemples.

9.3.1 Example of non-commutative group (implicative-group)

Consider the particular poset (where $\leq$ is $=$) $(G_6 = \{a, b, c, d, f, 1\}, \leq)$ represented by the Hasse diagram from Figure 9.4.

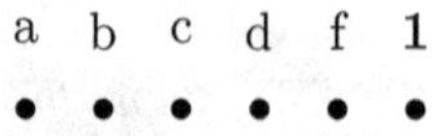

Figure 9.4: The poset $(G_6, \leq)$

Define on G_6 the following product $\cdot$ and negation (inverse) $^{-1}$ by the following tables:

$\mathcal{G}_{6gr}$

$\cdot$	a	b	c	d	f	**1**
a	**1**	c	b	f	d	a
b	d	**1**	f	a	c	b
c	f	a	d	**1**	b	c
d	b	f	**1**	c	a	d
f	c	d	a	b	**1**	f
1	a	b	c	d	f	**1**

x	x^{-1}
a	a
b	b
c	d
d	c
f	f
1	**1**

.

Then, $\mathcal{G}_{6gr} = (G_6, \cdot, ^{-1}, 1)$ is a (non-commutative) group.

The term-equivalent (by (Eq1)) (non-commutative) implicative-group is $\mathcal{G}_{6igr} = (G_6, \rightarrow, \rightsquigarrow, 1)$, where the implications $\rightarrow$ and $\rightsquigarrow$ have the following tables [51]:

$\rightarrow$	a	b	c	d	f	**1**
a	**1**	d	f	b	c	a
b	c	**1**	a	f	d	b
c	f	a	**1**	c	b	d
d	b	f	d	**1**	a	c
f	d	c	b	a	**1**	f
1	a	b	c	d	f	**1**

$\rightsquigarrow$	a	b	c	d	f	**1**
a	**1**	c	b	f	d	a
b	d	**1**	f	a	c	b
c	b	f	**1**	c	a	d
d	f	a	d	**1**	b	c
f	c	d	a	b	**1**	f
1	a	b	c	d	f	**1**

x	x^{-1}
a	a
b	b
c	d
d	c
f	f
1	**1**

.

9.3.2 Examples of l-groups (l-implicative-groups)

• **Example 1.** Consider the lattice

$$(\mathbf{Z} = \{\ldots, -3, -2, -1, \mathbf{0}, 1, 2, 3, \ldots\}, \wedge = \min, \vee = \max),$$

whose Hasse diagram is presented in Figure 6.5. Then,

$$\mathcal{Z}_g = (\mathbf{Z}, \wedge = \min, \vee = \max, +, -, \mathbf{0})$$

is a self-dual, linearly-ordered, commutative l-group, where the tables of + and − are given below.

+	...	-3	-2	-1	**0**	1	2	3	...
⋮	...	⋮	⋮	⋮	⋮	⋮	⋮	⋮	...
-3	...	-6	-5	-4	-3	-2	-1	**0**	...
-2	...	-5	-4	-3	-2	-1	**0**	1	...
-1	...	-4	-3	-2	-1	**0**	1	2	...
0	...	-3	-2	-1	**0**	1	2	3	...
1	...	-2	-1	**0**	1	2	3	4	...
2	...	-1	**0**	1	2	3	4	5	...
3	...	**0**	1	2	3	4	5	6	...
⋮	...	⋮	⋮	⋮	⋮	⋮	⋮	⋮	...

x	$-x\ (= x \rightarrow 0)$
⋮	⋮
-3	3
-2	2
-1	1
0	**0**
1	-1
2	-2
3	-3
⋮	⋮

Note that the l-group $\mathcal{Z}_g$ is a particular case of a l-cm(R), by Corollary 7.3.6, namely is a particular case of the infinite l-cm(R) $(\mathbf{Z}, \min, \max, +, \mathbf{0})$, presented in Example 5', Chapter 6, where the residuum $\rightarrow$ is defined as follows: for all $y, z \in \mathbf{Z}$,

(R) (residuum) $\exists\, y \to z \overset{notation}{=} \max\{x \mid x \odot y \leq z\}$,
and the resulting table is presented again below.

$\to$	...	-3	-2	-1	**0**	1	2	3	...
⋮	...	⋮	⋮	⋮	⋮	⋮	⋮	⋮	...
-3	...	**0**	1	2	3	4	5	6	...
-2	...	-1	**0**	1	2	3	4	5	...
-1	...	-2	-1	**0**	1	2	3	4	...
0	...	-3	-2	-1	**0**	1	2	3	...
1	...	-4	-3	-2	-1	**0**	1	2	...
2	...	-5	-4	-3	-2	-1	**0**	1	...
3	...	-6	-5	-4	-3	-2	-1	**0**	...
⋮	...	⋮	⋮	⋮	⋮	⋮	⋮	⋮	...

Note that the residuum $\to$ is just the implication: $x \to y \overset{(7.1)}{=} y + (-x)$.
Then,

$$\mathcal{Z}_{ig} = (\mathbf{Z}, \min, \max, \to, 0)$$

is the term equivalent commutative l-implicative-group (by (Eq3)) (see Example 5, Chapter 6).

• **Example 2.** $\mathcal{R}_g = (\mathbf{R}, \wedge = \min, \vee = \max, +, -, 0)$ is a self-dual, linearly-ordered, commutative l-group, where for all $x, y \in \mathbf{R}$,

$$x \to y \overset{(7.1)}{=} y + (-x).$$

Then, $\mathcal{R}_{ig} = (\mathbf{R}, \wedge = \min, \vee = \max, \to, 0)$ is the term-equivalent commutative l-implicative-group (see (Eq3)).

• **Example 3.** $\mathcal{D}_g = (D = (0, +\infty) = \{x \in \mathbf{R} \mid x > 0\}, \wedge, \vee, \cdot, ^{-1}, 1)$ is a self-dual, linearly-ordered, commutative l-group, where for all $x, y \in D$,

$$x \to y \overset{(7.1)}{=} y \cdot \frac{1}{x} = \frac{y}{x}.$$

Then, $\mathcal{D}_{ig} = (D = (0, +\infty) = \{x \in \mathbf{R} \mid x > 0\}, \wedge, \vee, \to, 1)$ is the term-equivalent commutative l-implicative-group (see (Eq3)).

• **Example 4.** Consider the structure $\mathcal{G}_g^L = (G^L = (-\infty, 0) \times \mathbf{R}, \leq, +, -, 0_{G^L})$ [96], defined as follows: for all $(-a, b)$, $(-c, d) \in G^L$,

$$(-a, b) + (-c, d) \overset{def.}{=} (-ac, bc + ad), \;\; 0_{G^L} = (-1, 0), \quad -(-a, b) \overset{def.}{=} (-\frac{1}{a}, -\frac{b}{a^2})$$

and with the lexicographic order $\leq$.
Then, $\mathcal{G}_g^L$ is a linearly-ordered, commutative l-group, where:

$$(-a, b) \to (-c, d) \overset{(7.1)}{=} (-\frac{c}{a}, \frac{ad - bc}{a^2}).$$

Then, $\mathcal{G}_{ig}^L = (G^L = (-\infty, 0) \times \mathbf{R}, \leq, \rightarrow, 0_{G^L})$ is the term-equivalent commutative l-implicative-group (see (Eq3)).

• **Example 4'.** Dually, consider the structure [52]:

$$\mathcal{G}_g^R = (G^R = (0, \infty) \times \mathbf{R}, \leq, +, -, 0_{G^R}),$$

defined as follows: for all (a, b), $(c, d) \in G^R$,

$$(a, b) + (c, d) \overset{def.}{=} (ac, bc + ad), \ \ 0_{G^R} = (1, 0), \quad -(a, b) \overset{def.}{=} (\frac{1}{a}, -\frac{b}{a^2})$$

and with the lexicographic order $\leq$.
Then, $\mathcal{G}_g^R$ is a linearly-ordered, commutative l-group, where:

$$(a, b) \rightarrow (c, d) \overset{(7.1)}{=} (\frac{c}{a}, \frac{ad - bc}{a^2}).$$

Then, $\mathcal{G}_{ig}^R = (G^R = (0, \infty) \times \mathbf{R}, \leq, \rightarrow, 0_{G^R})$ is the term-equivalent commutative l-implicative-group (see (Eq3)).

Since $(-1, 0) < (1, 0)$ in the set $G^L \cup G^R$, it follows that $\mathcal{G}_g^L$ is the left-l-group and $\mathcal{G}_g^R$ is the right-l-group.

• **Example 5.** Let $G^L = (-\infty, 0) \times \mathbf{R}$ and define a binary operation "+" on G^L by: for all $(-a, b)$, $(-c, d) \in G^L$,

$$(-a, b) + (-c, d) \overset{def.}{=} (-ac, bc + d).$$

Then, the operation "+" is associative, non-commutative and $0_{G^L} = (-1, 0)$ is the unit element.

Define a negation "-" by:

$$-(-a, b) \overset{def.}{=} (-\frac{1}{a}, -\frac{b}{a}).$$

Hence, $-(-a, b)$ is the inverse of $(-a, b)$, since $(-a, b) + (-\frac{1}{a}, -\frac{b}{a}) = (-1, \frac{b}{a} - \frac{b}{a}) = (-1, 0)$.

Consider the lexicographic order $\leq$ ($(a, b) < (c, d)$ iff $a < c$ or ($a = c$ and $b < d$)) as the order relation; it makes G^L a lattice.

Then, the structure $\mathcal{G}_g^L = (G^L, \leq, +, -, 0_{G^L})$ is a linearly-ordered, non-commutative l-group [96], where:

$$(-a, b) \rightarrow (-c, d) \overset{(7.1)}{=} (-\frac{c}{a}, \frac{d - b}{a}), \quad (-a, b) \rightsquigarrow (-c, d) \overset{(7.1)}{=} (-\frac{c}{a}, \frac{ad - bc}{a}).$$

Then, $\mathcal{G}_{ig}^L = (G^L, \leq, \rightarrow, \rightsquigarrow, 0_{G^L})$ is the term-equivalent non-commutative l-implicative-group (see (Eq3)).

• **Example 5'.** Dually, let $G^R = (0, \infty) \times \mathbf{R}$ and define a binary operation "+" on G^R by:

$$(a, b) + (c, d) \stackrel{def.}{=} (ac, bc + d).$$

Then, the operation "+" is associative, non-commutative and $0_{G^R} = (1, 0)$ is the unit element.

Define a negation "-" by:

$$-(a, b) \stackrel{def.}{=} (\frac{1}{a}, -\frac{b}{a}).$$

Consider the lexicographic order $\leq$ as the order relation; it makes G^R a lattice.

Then, the structure $\mathcal{G}_g^R = (G^R, \leq, +, -, 0_{G^R})$ is a linearly-ordered, non-commutative l-group [52], where:

$$(a, b) \to (c, d) \stackrel{(7.1)}{=} (\frac{c}{a}, \frac{d - b}{a}), \quad (a, b) \rightsquigarrow (c, d) \stackrel{(7.1)}{=} (\frac{c}{a}, \frac{ad - bc}{a}).$$

Then, $\mathcal{G}_{ig}^R = (G^R, \leq, \to, \rightsquigarrow, 0_{G^R})$ is the term-equivalent non-commutative l-implicative-group (see (Eq3)).

Since $(-1, 0) < (1, 0)$ in the set $G_L \cup G_R$, it follows that $\mathcal{G}_g^L$ is the left-l-group and $\mathcal{G}_g^R$ is the right-l-group.

Chapter 10

Generalizations of groups

In this chapter, we introduce and study the new notions of hub, moon and goop and we show their connections with the two definitions of the groups. We also introduce and study the new notions of qm-hub, qm-moon, qm-group and qm-goop and we show the connections between all these generalizations of the groups. Final connections of some generalizations of the groups involved in this monograph are presented.

The chapter has 3 sections.

10.1 Hubs, moons. Groups, goops

In this section, we introduce and study the new notions of *hub*[1], of involutive hub, of associative (involutive) hub and of sharp (involutive) hub - all in the "world" of unital magmas. In parallel, we also introduce and study the new notions of *moon*[2], of involutive moon, of associative (involutive) moon and of sharp (involutive) moon - all in the "world" of unital magmas with additional unary operation. We finally introduce the new notion of *goop*[3]. The connections between all these generalizations of the groups are presented.

10.1.1 The (involutive) hub, moon

• In the "world" of unital magmas

[1]We have introduced the notion of hub in the summer of 2016, first under the name of *loop*. Then, Paul Flondor has remarked, in a private discussion on June, 18 2017, that the new notion (called now *hub*) and the *loop* are incomparable, both being particular cases of unital magmas and generalizations of groups. Finally, we decided the name *hub* on November, 1st 2017, with the help of George Georgescu and Denisa Diaconescu.

[2]We have introduced the notion of *moon* in the night of November 30/31, 2017, in order to complete the hierarchies of algebras connected to the two definitions of the groups (presented in Chapter 7).

[3]We have introduced the *goop* on November 22, 2017. The name *goop* was suggested by Toma Albu.

Definitions 10.1.1

1. A *hub* is an algebra $\mathcal{A} = (A, \cdot, 1)$ of type $(2, 0)$ verifying:

(pU) $x \cdot 1 = x = 1 \cdot x$, for all $x \in A$,

(pIv') for all $x \in A$, there exists $y \in A$, such that $x \cdot y = 1 = y \cdot x$.

2. An *involutive hub* is an algebra $\mathcal{A} = (A, \cdot, 1)$ of type $(2, 0)$ verifying (pU) and:

(pIv") for all $x \in A$, there exists a unique $y = y_x \in A$, such that $x \cdot y = 1 = y \cdot x$.

Denote by **(involutive) hub** the class of all (involutive, respectively) hubs.

A hub is *commutative*, if $x \cdot y = y \cdot x$, for all x, y.

Note that a hub is a unital magma verifying (pIv'), i.e. we have the equivalence:

(UMH) **unital magma** + (pIv') = **hub.**

Obviously, an involutive hub is a hub, because (pIv") implies (pIv'). Hence, a hub verifying (pIv") is an involutive hub.

Proposition 10.1.2 *(See Remark 5.2.8)*

Let $\mathcal{A} = (A, \cdot, 1)$ be an involutive hub. Then, there exists an involutive regular-m negation $^{-1}$ (i.e. an unary operation defined on A veryfying (pIv) and (DN)).

Proof. By (pIv"), for all $x \in A$, there exists a unique $y = y_x \in A$ such that $x \cdot y = 1 = y \cdot x$; it follows that there exists a unary operation $^{-1} : A \longrightarrow A$, defined as follows: for all $x \in A$, $x^{-1} \overset{def.}{=} y$ and we have:

$$(pIv) \qquad x \cdot x^{-1} = 1 = x^{-1} \cdot x.$$

Hence, for each $x \in A$, there exists a unique $y \in A$, denoted $y = x^{-1}$, such that $x \cdot y = 1 = y \cdot x$, and for this $y \in A$, there exists also a unique $z \in A$, denoted $z = y^{-1}$, such that $y \cdot z = 1 = z \cdot y$; then, $z = x$, since, otherwise, for $y \in A$, there exists two different elements, x and z, such that $y \cdot x = 1 = x \cdot y$ and $y \cdot z = 1 = z \cdot y$: contradiction with (pIv"). Thus, $z = y^{-1} = (x^{-1})^{-1} = x$. Hence, (DN) holds. □

Remark 10.1.3 The involutive hub $(A, \cdot, 1)$, with $^{-1}$ the involutive regular-m negation, is a regular-m algebra with involutive regular-m negation, by Remark 5.2.8.

Examples 10.1.4

(1) The algebra $\mathcal{H}_1 = (\{a, b, 1\}, \cdot, 1)$, with

·	a	b	1
a	a	1	a
b	1	1	b
1	a	b	1

, is a **commutative hub that is not involutive**, since, for $x = b$, there exists $y = a$ s.t. $x \cdot y = y \cdot x = 1$ and there exists also $z = b$, s.t. $x \cdot z = z \cdot x = 1$. (2) The algebra $\mathcal{H}_2 = (\{a, b, 1\}, \cdot, 1)$, with

·	a	b	1
a	1	a	a
b	1	1	b
1	a	b	1

, is a **non-commutative involutive hub**, with $(a, b, 1)^{-1} = (a, b, 1)$.

• In the "world" of unital magmas with additional unary operation

Definitions 10.1.5

1. A *moon* is an algebra $\mathcal{A} = (A, \cdot, ^{-1}, 1)$ of type $(2, 1, 0)$ verifying: for all $x \in A$,

(pU) $x \cdot 1 = x = 1 \cdot x$,

(pIv) $x \cdot x^{-1} = 1 = x^{-1} \cdot x$.

2. A moon is *involutive*, if verifies: for all $x \in A$,

(DN) $(x^{-1})^{-1} = x$.

Denote by **(involutive) moon** the class of all (involutive, respectively) moons.

A moon is *commutative*, if $x \cdot y = y \cdot x$, for all x, y.

Note that a moon is a unital magma with an additional unary operation, $^{-1}$.

Note also that, if $(A, \cdot, ^{-1}, 1)$ is a moon, then $(A, \cdot, 1)$ is a hub, since (pIv) implies (pIv').

Remark 10.1.6 The involutive moon $(A, \cdot, ^{-1}, 1)$ is a regular-m algebra with involutive regular-m negation, by Remark 5.2.8.

The connection between the involutive hubs and the involutive moons is the following.

Proposition 10.1.7 *Let $\mathcal{A} = (A, \cdot, 1)$ be an involutive hub with $^{-1}$ the involutive regular-m negation. Define $F(\mathcal{A}) = (A, \cdot, ^{-1}, 1)$. Then, $F(\mathcal{A})$ is an involutive moon.*

Proof. Obviously, by Proposition 10.1.2. □

The converse of Proposition 10.1.7 does not hold, as the following example (2) shows.

Examples 10.1.8

(1) The algebra $\mathcal{M}_1 = (\{a, b, 1\}, \cdot, ^{-1}, 1)$, with

$\cdot$	a	b	1
a	a	1	a
b	1	1	b
1	a	b	**1**

and $(a, b, 1)^{-1} = (b, b, 1)$, is a **commutative moon that is not involutive**, since $(a^{-1})^{-1} = b^{-1} = b \neq a$. Note that $(\{a, b, 1\}, \cdot, 1)$ is the hub $\mathcal{H}_1$ that is not involutive from above Examples 10.1.4.

(2) The algebra $\mathcal{M}_2 = (\{a, b, 1\}, \cdot, ^{-1}, 1)$, with

$\cdot$	a	b	1
a	a	1	a
b	1	1	b
1	a	b	**1**

and $(a, b, 1)^{-1} = (b, a, 1)$, is a **commutative involutive moon**. Note that $(\{a, b, 1\}, \cdot, 1)$ is the hub $\mathcal{H}_1$ that is not involutive from above Examples 10.1.4.

(3) The algebra $\mathcal{M}_3 = (\{a, b, 1\}, \cdot, ^{-1}, 1)$, with

$\cdot$	a	b	1
a	1	a	a
b	1	1	b
1	a	b	1

and $(a, b, 1)^{-1} = (a, b, 1)$, is a **non-commutative involutive moon**. Note that $(\{a, b, 1\}, \cdot, 1)$ is the involutive hub $\mathcal{H}_2$ from above Examples 10.1.4.

10.1.2 The associative (involutive) hub, moon vs. the group

We shall analyse here in some details the two definitions of the groups (presented in Chapter 7) and their connections with the hubs and the moons, in order to better understand the generalizations of the groups from this chapter.

- **In the "world" of unital magmas**

Lemma 10.1.9 *Let $(A, \cdot, 1)$ be a unital magma. If (Ass) holds, then:*

$$(pIv') \iff (pIv'').$$

Proof. (pIv") $\Longrightarrow$ (pIv'): Obviously.

(pIv') $\Longrightarrow$ (pIv"): Let $x \in A$; by (pIv'), there exists $y \in A$ s.t. $x \cdot y = y \cdot x = 1$. Suppose by absurdum hypothesis that y is not unique, i.e. there exists also $z \in A$ s.t. $x \cdot z = z \cdot x = 1$; then, $z \overset{(pU)}{=} z \cdot 1 = z \cdot (x \cdot y) \overset{(Ass)}{=} (z \cdot x) \cdot y = 1 \cdot y \overset{(pU)}{=} y$: contradiction. Hence, (pIv") holds. □

Definitions 10.1.10

1. An *associative hub* is a hub verifying (Ass).
2. An *associative involutive hub* is an involutive hub verifying (Ass).

Denote by **associative hub** and **associative involutive hub** the corresponding classes of algebras.

Note that the *associative involutive hubs* coincide with the *involutive associative hubs* (algebras $(A, \cdot, 1)$ verifying (pU), (Ass), (pIv")).

Remark 10.1.11 Note that the *associative hub* is just the *group* (Definition 1') (see also Remark 7.1.3).

Hence, we have:
(HG) **hub** + (Ass) = **associative hub** = **group** (Definition 1').

We shall say that a hub is *proper*, if it is not a group (Definition 1'), i.e. if (Ass) is not verified.

By above Lemma 10.1.9, we obtain the following equivalence:
(H1) **group** = **associative hub** $\equiv$ **associative involutive hub**,
where $\equiv$ is the notation for the equivalence when the signatures coincides.

Example 10.1.12 The algebra $\mathcal{G} = (\{a, b, 1\}, \cdot, 1)$, with

$\cdot$	a	b	1
a	b	1	a
b	1	a	b
1	a	b	1

, is a *commutative associative (involutive) hub* (with $(a, b, 1)^{-1} = (b, a, 1)$), **namely is the only group with 3 elements (Definition 1').**

- **In the "world" of unital magmas with additional unary operation**

Definition 10.1.13 An *associative moon* is a moon verifying (Ass).

Note that the *associative involutive moons* coincide with the *involutive associative moons.*

Denote by **associative (involutive) moon** the class of all associative (involutive, respectively) moons.

Remark 10.1.14 Note that the *associative moon* is just the *group* (Definition 1).

Hence, we have:

(MG) **moon** + (Ass) = **associative moon** = **group** (Definition 1).

We shall say that a moon is *proper*, if it is not a group (Definition 1), i.e. if (Ass) is not verified.

Proposition 10.1.15 *Let $\mathcal{A} = (A, \cdot, ^{-1}, 1)$ be an associative moon. Then, (DN) holds.*

Proof. Obviously, since $\mathcal{A}$ is a group. □

By above Proposition, any associative moon is involutive, hence we obtain the equivalence:
(M1) **group** = **associative moon** $\equiv$ **associative involutive moon.**

The connectionn between the associative involutive hubs and the associative involutive moons is the following:

Theorem 10.1.16

(1) Let $\mathcal{A} = (A, \cdot, 1)$ be an associative involutive hub, with $^{-1}$ the involutive regular-m negation. Define $F(\mathcal{A}) = (A, \cdot, ^{-1}, 1)$. Then, $F(\mathcal{A})$ is an associative involutive moon.

(1') Conversely, let $\mathcal{A} = (A, \cdot, ^{-1}, 1)$ be an associative involutive moon. Define $G(\mathcal{A}) = (A, \cdot, 1)$. Then, $G(\mathcal{A})$ is an associative involutive hub and its involutive regular-m negation is $^{-1}$.

(2) The mapps F and G are mutually inverse.

Proof. (1): By Proposition 10.1.7.

(1'): We must prove that (pIv") holds. Indeed, let $x \in A$; put $y = x^{-1}$, hence we have $x \cdot y = 1 = y \cdot x$, by (pIv). Suppose, by absurdum hypothesis that there exists also $z \in A$ s.t. $z \neq y$ and $x \cdot z = 1 = z \cdot x$. Then, $z \overset{(pU)}{=} 1 \cdot z = (y \cdot x) \cdot z \overset{(Ass)}{=} y \cdot (x \cdot z) = y \cdot 1 \overset{(pU)}{=} y$: contradiction. Thus, (pIv") holds. The rest follows obviously.

(2): Immediately. □

By Theorem 10.1.16, we obtain the equivalence:
(HM1) **associative involutive hub** $\Longleftrightarrow$ **associative involutive moon.**

By equivalences (H1), (HM1) and (M1), we have the equivalences:

associative hub $\equiv$ **associative involutive hub**
$\Longleftrightarrow$ **associative involutive moon** $\equiv$ **associative moon,**
hence we obtain the equivalence:

group (Def. 1')= **associative hub** $\Longleftrightarrow$ **associative moon** =**group** (Def. 1),

which is the base for the two definitions of the groups.

Example 10.1.17 The algebra $\mathcal{G} = (\{a, b, 1\}, \cdot, ^{-1}, 1)$, with

$\cdot$	a	b	1
a	b	1	a
b	1	a	b
1	a	b	**1**

and $(a, b, 1)^{-1} = (b, a, 1)$, is a *commutative associative (involutive) moon, namely is the only group with 3 elements (Definition 1).*

Remark 10.1.18 The Definition 1' of groups, as associative hubs, is very useful when looking for examples obtained by programs (because the negation is missing) (see the examples in Chapters 18, 19).

Note that we have the hierarchy No. 1, hence the hierarchy No. 1', from Figures 10.1, 10.2, respectively.

10.1.3 The sharp (involutive) hub, moon vs. the goop

- **In the "world" of unital magmas**

Definitions 10.1.19

1. A hub $(A, \cdot, 1)$ is *sharp*, if it verifies the additional property:
(pIv"') for all $x \in A$, there exists a unique $y \in A$, such that $x \cdot y = 1$
(i.e. there exists only one element 1 on each line of the table (matrix) of $\cdot$).
2. An involutive hub $(A, \cdot, 1)$ is *sharp*, if it verifies (pIv"').

Note that the *sharp involutive hubs* coincide with the *involutive sharp hubs* (algebras $(A, \cdot, 1)$ verifying (pU), (pIv'), (pIv"'), (pIv")).

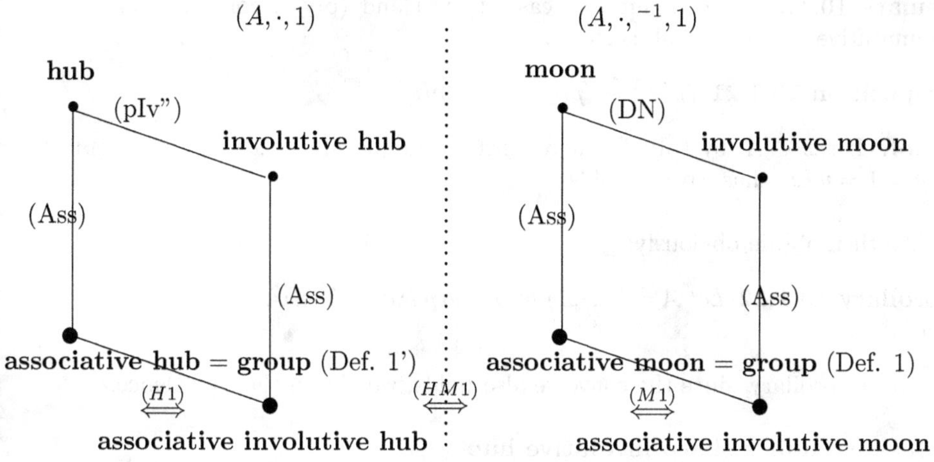

Figure 10.1: The hierarchy No. 1

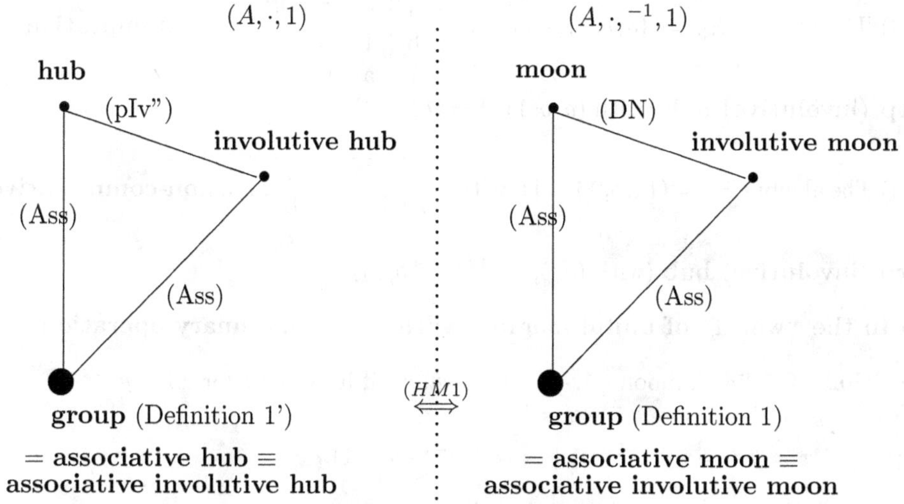

Figure 10.2: The hierarchy No. 1'

Denote by **sharp hub** and **sharp involutive hub** the corresponding classes of algebras.

Remark 10.1.20 In commutative case, (pIv") and (pIv"') coincide, hence any commutative involutive hub is sharp.

Proposition 10.1.21 *(pIv') + (pIv"')* $\Longrightarrow$ *(pIv").*

Proof. Let $x \in A$; by (pIv"'), there exists a unique y, s.t. $x \cdot y = 1$; by (pIv'), $x \cdot y = 1 = y \cdot x$; thus, (pIv") holds. □

We then obtain obviously:

Corollary 10.1.22 *Let* $\mathcal{A} = (A, \cdot, 1)$ *be a sharp hub. Then,* $\mathcal{A}$ *is a sharp involutive hub.*

By above corollary, since the converse also holds, we obtain the equvalence:

(H2) **sharp hub** $\equiv$ **sharp involutive hub.**

Examples 10.1.23

(1) The **commutative hub** $\mathcal{H}_1$ from Examples 10.1.4 **is not sharp**, since for the element $x = b$ (the second line of the matrix), there are two elements, a and b, s.t. $x \cdot a = 1 = x \cdot b$.

(2) The **non-commutative involutive hub** $\mathcal{H}_2$ from Examples 10.1.4 **is not sharp**, since for $x = b$, there are two elements, a and b, s.t. $x \cdot a = 1 = x \cdot b$.

(3) The algebra $\mathcal{H}_3 = (\{a, b, 1\}, \cdot, 1)$, with

·	a	b	1
a	a	1	a
b	1	a	b
1	a	b	1

, is a **commutative sharp (involutive) hub** (with $(a, b, 1)^{-1} = (b, a, 1)$).

(4) The algebra $\mathcal{H}_4 = (\{a, b, 1\}, \cdot, 1)$, with

·	a	b	1
a	1	a	a
b	b	1	b
1	a	b	1

, is a **non-commutative sharp (involutive) hub** (with $(a, b, 1)^{-1} = (a, b, 1)$).

• **In the "world" of unital magmas with additional unary operation**

Definition 10.1.24 A moon $(A, \cdot, ^{-1}, 1)$ is *sharp*, if it verifies: for all $x, y \in A$,
(IdEqP) $y \cdot x^{-1} = 1 \Longleftrightarrow x^{-1} \cdot y = 1$,
(pEqP=) $x = y \Longleftrightarrow y \cdot x^{-1} = 1$ $(\overset{(IdEqP)}{\Longleftrightarrow}$ $x^{-1} \cdot y = 1)$.

Denote by **sharp moon** the class of all sharp moons.

Note that the *sharp involutive moons* coincide with the *involutive sharp moons*. Denote by **sharp involutive moon** the class of all sharp involutive moons.

The goop

Definition 10.1.25 A *goop* is an algebra $(A, \cdot, ^{-1}, 1)$ of type $(2, 1, 0)$ verifying (pU), (IdEqP) and (pEqP=).

Denote by **goop** the class of all goops.
Note that we have:

goop $\supseteq$ sharp moon $\supseteq$ sharp involutive moon.

Consider the following properties: for all x, y,
(N1) $1^{-1} = 1$,
(Ass1) $x \cdot (1 \cdot y) = (x \cdot 1) \cdot y$,
(Ass2) $x \cdot (1 \cdot y) = x \cdot y$ and $(x \cdot 1) \cdot y = x \cdot y$.

Proposition 10.1.26 *Let $(A, \cdot, ^{-1}, 1)$ be a goop. Then,*
(i) (pEqP=) $\Longrightarrow$ (pIv) $\Longrightarrow$ (pIv'),
(ii) (pIv) + (pEqP=) $\Longrightarrow$ (DN),
(iii) (pEqP=) + (DN) $\Longrightarrow$ (pIv"'),
(iv) (pU) + (pIv) $\Longrightarrow$ (N1),
(v) (pU) $\Longrightarrow$ (Ass 2),
(vi) (Ass2) $\Longrightarrow$ (Ass1),
hence (pIv), (pIv'), (DN), (pIv"'), (N1), (Ass2) and (Ass1) hold.

Proof. (i) By (pEqP=), $x = x \Longleftrightarrow x \cdot x^{-1} = 1 \overset{(IdEqP)}{\Longleftrightarrow} x^{-1} \cdot x = 1$; thus, (pIv) holds. (pIv') follows immediately by (pIv).

(ii): By (pIv), $(x^{-1})^{-1} \cdot x^{-1} = 1 = x^{-1} \cdot (x^{-1})^{-1}$; then, by (pEqP=), $x = (x^{-1})^{-1}$. Thus, (DN) holds.

(iii): Let $x \in A$; then, there exists $y = x^{-1}$, such that $x \cdot y = x \cdot x^{-1} = 1$, by (pEqP=) (for $y = x$); suppose, by absurdum hypothesis, that there exists also $z \in A$, s.t. $z \neq y$ and $x \cdot z = 1$; then, $1 = x \cdot z \overset{(DN)}{=} (x^{-1})^{-1} \cdot z = y^{-1} \cdot z$; hence, by (pEqP=), we obtain that $z = y$: contradiction. Thus, (pIv"') holds.

(iv): $1^{-1} \overset{(pU)}{=} 1 \cdot 1^{-1} \overset{(pIv)}{=} 1$. Thus, (N1) holds.

(v): Obviously.

(vi): Obviously. □

Remark 10.1.27 Since (pIv) and (DN) hold, then it follows that the goop is a regular-m algebra with involutive regular-m negation, by Remark 5.2.8.

Theorem 10.1.28 *The goops, the sharp moons and the sharp involutive moons are equivalent.*

Proof. Let $\mathcal{A} = (A, \cdot, ^{-1}, 1)$ be a goop. By Proposition 10.1.26 (i), (ii), the properties (pIv) and (DN) are verified also. Hence, $\mathcal{A}$ is a sharp involutive moon. The rest is obvious. □

By Theorem 10.1.28, we have the equivalences:

(M2) **goop** $\equiv$ **sharp moon** $\equiv$ **sharp involutive moon.**

The connection between the sharp involutive hubs and the sharp involutive moons is the following:

Theorem 10.1.29
(1) Let $\mathcal{A} = (A, \cdot, 1)$ *be a sharp involutive hub with* $^{-1}$ *the involutive regular-m negation. Define* $F(\mathcal{A}) = (A, \cdot, ^{-1}, 1)$*. Then,* $F(\mathcal{A})$ *is a sharp involutive moon.*
(1') Conversely, let $\mathcal{A} = (A, \cdot, ^{-1}, 1)$ *be a sharp involutive moon. Define* $G(\mathcal{A}) = (A, \cdot, 1)$*. Then,* $G(\mathcal{A})$ *is a sharp involutive hub and its involutive regular-m negation is* $^{-1}$*.*
(2) The mapps F *and* G *are mutually inverse.*

Proof. (1): By Proposition 10.1.7, $F(\mathcal{A})$ is an involutive moon. Then, (IdEqP) and (pEqP=) follow rather obviously.
(1'): Let $\mathcal{A} = (A, \cdot, ^{-1}, 1)$ be a sharp involutive moon, hence a goop. Then, by Proposition 10.1.26, (pIv') and (pIv"') hold, hence $G(\mathcal{A})$ is a sharp (involutive) hub.
(2): Obviously. $\Box$

By Theorem 10.1.29, we obtain the equivalence:

(HM2) **sharp involutive hub** $\Longleftrightarrow$ **sharp involutive moon.**

By (H2), (HM2) and (M2), we obtain the equivalences:

sharp hub $\equiv$ **sharp involutive hub**
$\Longleftrightarrow$ **sharp involutive moon** $\equiv$ **sharp moon** $\equiv$ **goop.**

Remark 10.1.30 Just as the group (Definition 1) is defined equivalently as an associative hub (Definition 1'), the same we could define the goop equivalently, by above equivalences, as a sharp involutive hub (or even as a sharp hub). In fact, we find examples of goops by finding examples of sharp involutive hubs.

Note that we have the hierarchy No. 2, hence the hierarchy No. 2', from Figures 10.3, 10.4, respectively.

Examples 10.1.31
(1) The **commutative moon** $\mathcal{M}_1$ from Exemples 10.1.8 **is not sharp**, since for $x = b$ and $y = a$, we have $x \neq y$ and $y \cdot x^{-1} = a \cdot b^{-1} = a \cdot b = 1$; thus (EqP=) does not hold.
(2) The **commutative involutive moon** $\mathcal{M}_2$ from Exemples 10.1.8 **is not sharp**, since for $x = a$ and $y = b$, we have $x \neq y$ and $y \cdot x^{-1} = b \cdot a^{-1} = b \cdot b = 1$.
(3) The **non-commutative involutive moon** $\mathcal{M}_3$ from Exemples 10.1.8 **is not sharp**, since for $x = a$ and $y = b$, we have $x \neq y$ and $y \cdot x^{-1} = b \cdot a^{-1} = b \cdot a = 1$.

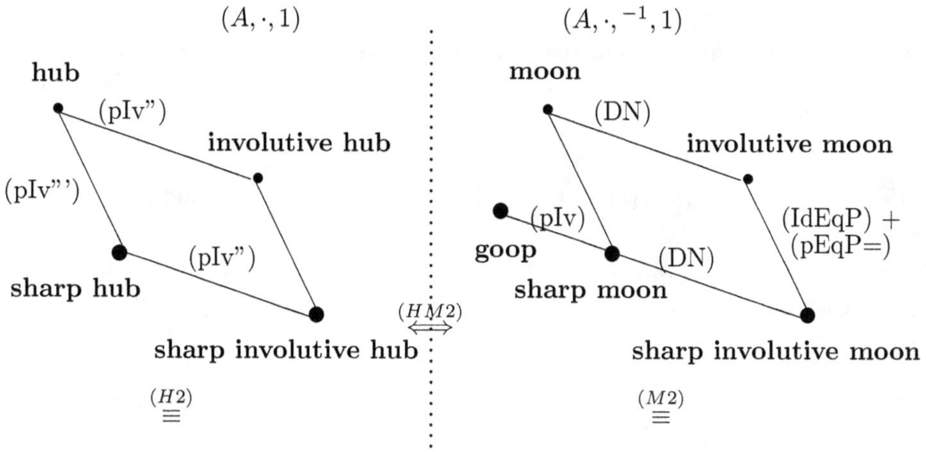

Figure 10.3: The hierarchy No. 2

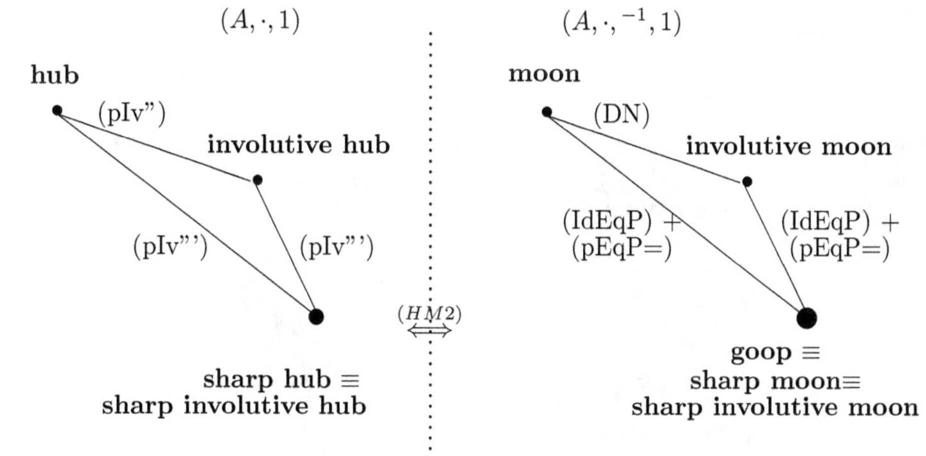

Figure 10.4: The hierarchy No. 2'

(4) The algebra $\mathcal{M}_4 = (\{a, b, 1\}, \cdot, ^{-1}, 1)$, with

$\cdot$	a	b	1
a	a	1	a
b	1	a	b
1	a	b	1

and $(a, b, 1)^{-1} = (b, a, 1)$, is a **commutative sharp (involutive) moon, hence a commutative goop**, s.t. $(\{a, b, 1\}, \cdot, 1)$ is the commutative sharp (involutive) hub $\mathcal{H}_3$ from Exemples 10.1.23.

(5) The algebra $\mathcal{M}_5 = (\{a, b, 1\}, \cdot, ^{-1}, 1)$, with

$\cdot$	a	b	1
a	1	a	a
b	b	1	b
1	a	b	1

and $(a, b, 1)^{-1} = (a, b, 1)$, is a **non-commutative sharp (involutive) moon, hence a non-commutative goop**, s.t. $(\{a, b, 1\}, \cdot, 1)$ is the non-commutative sharp (involutive) hub $\mathcal{H}_4$ from Exemples 10.1.23.

Note that any group (Definition 1') is a sharp involutive hub and a sharp involutive hub verifying (Ass) is a group. Hence, we have the equivalence:

sharp involutive hub + (Ass) $\equiv$ **group** (Definition 1').

Note that any group (Definition 1) is a goop, by Proposition 7.1.4, and a goop verifying (Ass) is a group. Hence, we have the equivalence:

(GoGr) **goop** + (Ass) $\equiv$ **group** (Definition 1).

By hierarchies No. 1' and No. 2', we finally have the resuming hierarchy No. 3 from Figure 10.5.

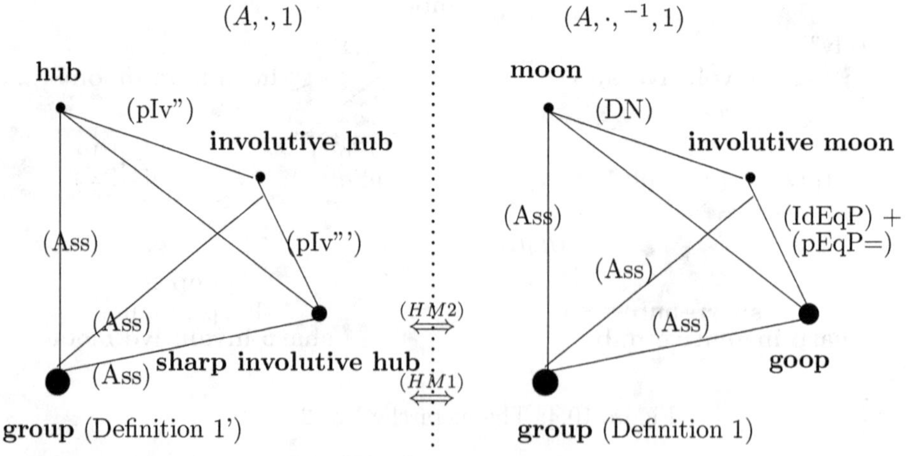

Figure 10.5: The hierarchy No. 3

Remark 10.1.32 We could name the sharp involutive hub and the goop as the "non-associative groups".

We shall introduce now new operations in goops: $\to$ and $\rightsquigarrow$.

Let $\mathcal{A} = (A, \cdot, ^{-1}, 1)$ be a goop. We shall extend to $\mathcal{A}$ the new operations $\to$ and $\rightsquigarrow$ defined on a group by (7.1). Thus, we define the new operations $\to$ and $\rightsquigarrow$ on A as follows: for all $x, y \in A$,

$$x \to y \overset{def.}{=} y \cdot x^{-1}, \quad x \rightsquigarrow y \overset{def.}{=} x^{-1} \cdot y. \tag{10.1}$$

Consider the following properties: for all $x, y \in A$,

(Id) $x \to 1 = x \rightsquigarrow 1$,
(pM) $1 \to x = x = 1 \rightsquigarrow x$,
(pEx1) $x \to (y \rightsquigarrow 1) = y \rightsquigarrow (x \to 1)$,
(pNeg) $x^{-1} = x \to 1 \overset{(Id)}{=} x \rightsquigarrow 1$,
(IdEq) $x \to y = 1 \Longleftrightarrow x \rightsquigarrow y = 1$,
(pEq=) $x = y \Longleftrightarrow x \to y = 1 \overset{(IdEq)}{\Longleftrightarrow} x \rightsquigarrow y = 1$.

The following result is useful further, in Chapter 12.

Proposition 10.1.33 *Let $(A, \cdot, ^{-1}, 1)$ be a goop. Then,*
(1) (pU) $\Longrightarrow$ (Id),
(2) (pU) + (N1) $\Longrightarrow$ (pM),
(3) (pU) $\Longrightarrow$ (pEx1),
(4) (pU) + (Id) $\Longrightarrow$ (pNeg),
(5) (IdEqP) $\Longrightarrow$ (IdEq),
(6) (pEqP=) $\Longrightarrow$ (pEq=),
hence (Id), (pM), (pEx1), (pNeg), (IdEq) and (pEq=) hold.

Proof. (1): $x \to 1 \overset{(10.1)}{=} 1 \cdot x^{-1} \overset{(pU)}{=} x^{-1} \overset{(pU)}{=} x^{-1} \cdot 1 \overset{(10.1)}{=} x \rightsquigarrow 1$. Thus, (Id) holds.

(2): $1 \to x \overset{(10.1)}{=} x \cdot 1^{-1} \overset{(N1)}{=} x \cdot 1 \overset{(pU)}{=} x$ and $1 \rightsquigarrow x \overset{(10.1)}{=} 1^{-1} \cdot x \overset{(N1)}{=} 1 \cdot x \overset{(pU)}{=} x$. Thus, (pM) holds.

(3): $x \to (y \rightsquigarrow 1) \overset{(10.1)}{=} x \to (y^{-1} \cdot 1) \overset{(pU)}{=} x \to y^{-1} \overset{(10.1)}{=} y^{-1} \cdot x^{-1}$ and $y \rightsquigarrow (x \to 1) \overset{(10.1)}{=} y \rightsquigarrow (1 \cdot x^{-1}) \overset{(pU)}{=} y \rightsquigarrow x^{-1} \overset{(10.1)}{=} y^{-1} \cdot x^{-1}$; hence $x \to (y \rightsquigarrow 1) = y \rightsquigarrow (x \to 1)$, i.e. (pEx1) holds.

(4): $x \to 1 \overset{(10.1)}{=} 1 \cdot x^{-1} \overset{(pU)}{=} x^{-1}$ and $x \rightsquigarrow 1 \overset{(10.1)}{=} x^{-1} \cdot 1 \overset{(pU)}{=} x^{-1}$. Thus, (pNeg) holds.

(5): By (IdEqP), $y \cdot x^{-1} = 1 \Longleftrightarrow x^{-1} \cdot y = 1$; hence, by (10.1), $x \to y = 1 \Longleftrightarrow x \rightsquigarrow y = 1$, i.e. (IdEq) holds.

(6): By (pEqP=), $x = y \Longleftrightarrow y \cdot x^{-1} = 1 \overset{(IdEqP)}{\Longleftrightarrow} x^{-1} \cdot y = 1$; hence, by (10.1), we obtain: $x = y \Longleftrightarrow x \to y = 1 \overset{(IdEq)}{\Longleftrightarrow} x \rightsquigarrow y = 1$, i.e. (pEq=) holds. $\square$

10.2 (Super) qm-hubs, qm-moons. Qm-groups, qm-goops

In this section, we introduce and study the new notions of *qm-hubs* and *super qm-hubs* as related generalizations of the hubs; we also introduce and study the involutive qm-hubs, the associative (involutive) qm-hubs and the sharp (involutive) qm-hubs- all in the "world" of qm-unital magmas. In parallel, we also introduce and study the new notions of *qm-moons* and *super qm-moons* as related generalizations of the moons; we introduce and study the involutive qm-moons, the associative (involutive) qm-moons and the qm-groups, the sharp (involutive) qm-moons and the qm-goops - all in the "world" of qm-unital magmas with additional unary operation. Connections between these generalizations of the groups are presented.

10.2.1 The (involutive) (super) qm-hub, qm-moon

- **In the "world" of qm-unital magmas**

There are two generalizations of the hubs as quasi-m algebras: the quasi-m hubs and the super quasi-m hubs. The idea came since there are two related generalizations of the implicative-hubs as quasi-pseudo-algebras (introduced in Chapter 11), fact which was obvious by the connection existing between the implicative-hubs and the pseudo-*aRM** algebras revealed in Chapter 13.

Definitions 10.2.1

1. A *quasi-m hub*, or a *qm-hub* for short, is an algebra $\mathcal{A} = (A, \cdot, 1)$ of type $(2, 0)$ verifying the properties: for all $x, y \in A$,
(qm-pU) $1 \cdot (x \cdot y) = x \cdot y = (x \cdot y) \cdot 1$,
(U1) $1 \cdot 1 = 1$,
(pIv') for all $x \in A$, there exists $y \in A$, such that $x \cdot y = y \cdot x = 1$.

2. We say that the qm-hub is *super*, if the following additional property is satisfied: for all $x, y \in A$,
(Ass2) $x \cdot (1 \cdot y) = x \cdot y$ and $(x \cdot 1) \cdot y = x \cdot y$.

Denote by **(super) qm-hub** the class of all (super) qm-hubs.

A (super) qm-hub is *commutative*, if $\cdot$ is commutative.

Note that (Ass2) implies (Ass1).

Note that a (super) qm-hub is, by definition, a quasi-m algebra, as this notion was defined in Chapter 5.

Note also that a qm-hub is a qm-unital magma verifying (pIv'). Hence, we have:

(qmUMH) **qm-unital magma** + (pIv') = **qm-hub**.

We analyse now the connection between the (super) qm-hubs and the hubs.

By Proposition 5.2.1, any hub is a (super) qm-hub. Conversely, any (super) qm-hub verifying (pU) is a hub. Hence, we have the following equivalences, by

(QM):

qm-hub + (pU) ≡ **hub**, **super qm-hub** + (pU) ≡ **hub**.

Following the general definition, a (super) qm-hub is *qm-proper*, if it is not a hub (i.e. if (pU) is not verified).

A qm-hub is *proper*, if (Ass) is not verified and if is not super. A super qm-hub is *proper*, if (Ass) is not verified.

We have the following result.

Theorem 10.2.2 *Let $\mathcal{A} = (A, \cdot, 1)$ be a proper qm-proper (super) qm-hub and consider the regular-m set $Rm(A) = \{x \in A \mid x = 1 \cdot x = x \cdot 1\}$. Then,*

$$\mathcal{R}m(\mathcal{A}) = (Rm(A), \cdot, 1)$$

is a unital magma.

Proof. By Theorem 5.2.13. □

We introduce now the following definitions.

Definition 10.2.3 Let $\mathcal{A} = (A, \cdot, 1)$ be a proper qm-proper (super) qm-hub. We shall say that $\mathcal{A}$ is:
- *pure*, if $\mathcal{R}m(\mathcal{A})$ is a proper hub,
- of *group type*, if $\mathcal{R}m(\mathcal{A})$ is a group (as associative hub),
- of *monoid type*, if $\mathcal{R}m(\mathcal{A})$ is a proper monoid,
- of *unital magma type*, if $\mathcal{R}m(\mathcal{A})$ is a proper unital magma.

The involutive qm-hub

Let $\mathcal{A} = (A, \cdot, 1)$ be a pure qm-hub. Then, $\mathcal{R}m(\mathcal{A}) = (Rm(A), \cdot, 1)$ is a hub. If $\mathcal{R}m(\mathcal{A})$ is an involutive hub, with $^{-1}$ the involutive regular-m negation, we then extend the regular-m negation $^{-1} : Rm(A) \longrightarrow Rm(A)$ to a quasi-m negation $^{-} : A \longrightarrow A$, such that the following properties be verified (see Remark 5.2.33): for all $x \in A$,
(qm-NegP) $(x \cdot 1)^{-} = x^{-} \cdot 1 = 1 \cdot x^{-} = (1 \cdot x)^{-}$,
(pIv) $x \cdot x^{-} = x^{-} \cdot x = 1$.

If the quasi-m negation $^{-}$ exists and if it verifies (DN) ($(x^{-})^{-} = x$, for all $x \in A$), then $\mathcal{A}$ is called *involutive* (qm-hub), with $^{-}$ its involutive quasi-m negation.

Remark 10.2.4 The involutive qm-hub $(A, \cdot, 1)$, with $^{-}$ its involutive quasi-m negation, is a quasi-m algebra with an involutive quasi-m negation, by Remark 5.2.33.

Denote by **involutive qm-hub** the class of all involutive qm-hubs.

Note that we have the following equivalence, by (QM):

involutive qm-hub + (pU) ≡ **involutive hub**.

Examples 10.2.5

(1) The commutative proper qm-proper super qm-hub $\mathcal{QMH}^1 = (A_4 = \{a, b, c, 1\}, \odot_{11}, 1)$,

where

$\odot_{11}$	a	b	c	1
a	a	1	a	a
b	1	1	b	b
c	a	b	1	1
1	**a**	**b**	1	**1**

. Note that $Rm(A_4) = \{a, b, 1\}$,

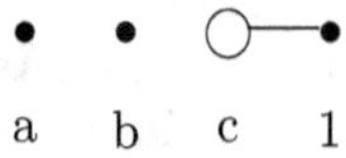

and

$\odot_{11}$	a	b	1
a	a	1	a
a	1	1	b
1	a	b	**1**

, hence $(\{a, b, 1\}, \odot_{11}, 1)$ is the commutative hub $\mathcal{H}_1$ that is not involutive from Examples 10.1.4. Hence, **$\mathcal{QMH}^1$ is a super pure qm-hub that is not involutive.**

(2) The non-commutative proper qm-proper super qm-hub $\mathcal{QMH}^2 = (A_4 = \{a, b, c, 1\}, \odot_{35}, 1)$,

where

$\odot_{35}$	a	b	c	1
a	1	a	a	a
b	1	1	1	b
c	1	1	1	b
1	**a**	**b**	b	**1**

. Note that $Rm(A_4) = \{a, b, 1\}$,

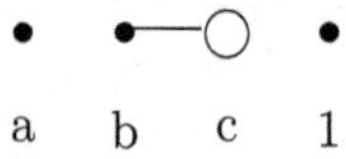

and

$\odot_{35}$	a	b	1
a	1	a	a
a	1	1	b
1	a	b	**1**

, hence $(\{a, b, 1\}, \odot_{35}, 1)$ is the non-commutative involutive hub $\mathcal{H}_2$ from Examples 10.1.4, with $(a, b, 1)^{-1} = (a, b, 1)$.

The quasi-m negation $^-$, if exists, must verify (qm-NegP) and (pIv). By (qm-NegP),
$b = b^- = (c \odot 1)^- = (1 \odot c)^- = b^- = b$; hence, $1 \odot c^- = c^- \odot 1 = b$; hence, $c^- = b, c$:
- for $c^- = b$, we obtain $(a, b, c, 1)^{-_1} = (\mathbf{a}, \mathbf{b}, b, \mathbf{1})$, which verifies (qm-NegP) and (pIv); hence is a quasi-m negation, a non-involutive one;
- for $c^- = c$, we obtain $(a, b, c, 1)^{-_2} = (\mathbf{a}, \mathbf{b}, c, \mathbf{1})$, which verifies (qm-NegP) and (pIv); hence is a quasi-m negation, an involutive one; thus, **the super pure qm-hub $\mathcal{QMH}^2$ is involutive**, with $^{-_2}$ its involutive quasi-m negation.

(2') The non-commutative proper qm-proper super qm-hub $\mathcal{QMH}^{2'} = (A_4 = \{a, b, c, 1\}, \odot_{44}, 1)$,

where

$\odot_{44}$	a	b	c	1
a	1	a	1	a
b	1	1	1	b
c	1	1	1	b
1	**a**	**b**	a	**1**

. Note that $Rm(A_4) = \{a, b, 1\}$,

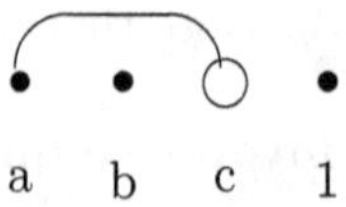

and

$\odot_{44}$	a	b	1
a	1	a	a
a	1	1	b
1	a	b	**1**

, hence $(\{a, b, 1\}, \odot_{44}, 1)$ is again the non-commutative involutive hub $\mathcal{H}_2$ from Examples 10.1.4, with $(a, b, 1)^{-1} = (a, b, 1)$.

The quasi-m negation $^-$, if exists, must verify (qm-NegP) and (pIv). But, by (qm-NegP), $b = b^- = (c \odot 1)^- \neq (1 \odot c)^- = a^- = a$; hence, a quasi-negation does not exist. Thus, **the super pure qm-hub $\mathcal{QMH}^{2'}$ is not involutive.**

(3) The commutative proper qm-proper qm-hub $\mathcal{QMH}^3 = (A_4 = \{a, b, c, 1\}, \odot_{16}, 1)$, where

$\odot_{16}$	a	b	c	1
a	a	a	1	a
b	a	a	1	b
c	1	1	a	b
1	**a**	**b**	b	**1**

. Note that $Rm(A_4) = \{a, b, 1\}$,

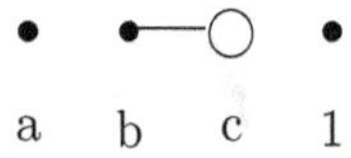

and

$\odot_{16}$	a	b	1
a	a	a	a
a	a	a	b
1	a	b	**1**

, hence $(\{a, b, 1\}, \odot_{16}, 1)$ is a monoid. Hence, **the qm-hub $\mathcal{WMH}^3$ is of monoid type.**

Note that the qm-hub $\mathcal{QMH}^3$ is **not super**, since (Ass2) is not verified: $a = a \odot_{16} b = a \odot_{16} (1 \odot_{16} c) \neq a \odot_{16} c = 1$.

(4) The non-commutative proper qm-proper qm-hub $\mathcal{QMH}^4 = (A_4 = \{a, b, c, 1\}, \odot, 1)$, where

$\odot$	a	b	c	1
a	a	1	1	a
b	1	1	a	b
c	a	a	1	1
1	**a**	**b**	b	**1**

. Note that $Rm(A_4) = \{a, b, 1\}$,

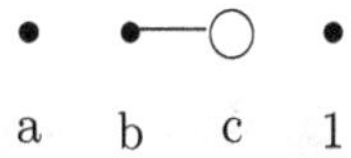

and

$\odot$	a	b	1
a	a	1	a
a	1	1	b
1	a	b	**1**

, hence $(\{a, b, 1\}, \odot, 1)$ is again the commutative hub $\mathcal{H}_1$ that is not involutive from Examples 10.1.4.

Note that **the pure qm-hub $\mathcal{QMH}^4$ is not super**, since (Ass2) is not verified: $1 = b \odot b = b \odot (1 \odot c) \neq b \odot c = a$.

• In the "world" of qm-unital magmas with additional operation

Definitions 10.2.6

1. A *quasi-m moon*, or a *qm-moon* for short, is an algebra $\mathcal{A} = (A, \cdot, ^-, 1)$ of type $(2, 1, 0)$ verifying the properties: for all $x, y \in A$,
(qm-pU) $\quad 1 \cdot (x \cdot y) = x \cdot y = (x \cdot y) \cdot 1$,
(U1) $\quad 1 \cdot 1 = 1$,
(pIv) $\quad x \cdot x^- = 1 = x^- \cdot x$.

2. We say that the qm-moon is *super*, if the following additional property is satisfied: for all $x, y \in A$,
(Ass2) $x \cdot (1 \cdot y) = x \cdot y$ and $(x \cdot 1) \cdot y = x \cdot y$.

Denote by **(super) qm-moon** the class of all (super) qm-moons.

A (super) qm-moon is *commutative*, if $\cdot$ is commutative.

Note that a (super) qm-moon is by definition a quasi-m algebra, as this notion was defined in Chapter 5.

Note that if $\mathcal{A} = (A, \cdot, ^-, 1)$ is a qm-moon, then $(A, \cdot, 1)$ is a qm-hub, since (pIv) implies (pIv').

We analyse now the connection between the (super) qm-moons and the moons.

By Proposition 5.2.1, any moon is a (super) qm-moon. Conversely, any (super) qm-moon verifying (pU) is a moon. Hence, we have the following equivalences, by (QM):

qm-moon + (pU) $\equiv$ **moon**, **super qm-moon** + (pU) $\equiv$ **moon**.

Following the general definition, a (super) qm-moon is *qm-proper*, if it is not a moon (i.e. if (pU) is not verified). A qm-moon is *proper*, if (Ass) is not verified and if is not super. A super qm-moon is *proper*, if (Ass) is not verified.

We have the following result.

Theorem 10.2.7 *Let $\mathcal{A} = (A, \cdot, ^-, 1)$ be a proper qm-proper (super) qm-moon and consider the regular-m set $Rm(A) = \{x \in A \mid x = 1 \cdot x = x \cdot 1\}$. Then,*

$$\mathcal{R}m(\mathcal{A}) = (Rm(A), \cdot, ^-, 1)$$

is a unital magma with additional operation.

Proof. By Theorem 5.2.13. □

We introduce now the following definitions.

Definition 10.2.8 Let $\mathcal{A} = (A, \cdot, ^-, 1)$ be a proper qm-proper (super) qm-moon. We shall say that $\mathcal{A}$ is:
- *pure*, if $\mathcal{R}m(\mathcal{A})$ is a proper moon,
- of *group type*, if $\mathcal{R}m(\mathcal{A})$ is a group (as associative moon).

The involutive qm-moon

Definition 10.2.9 A qm-moon $\mathcal{A} = (A, \cdot, ^-, 1)$ is *involutive*, if the following properties are satisfied: for all $x \in A$,
(qm-NegP) $(x \cdot 1)^- = x^- \cdot 1 = 1 \cdot x^- = (1 \cdot x)^-$,
(DN) $(x^-)^- = x$.

Denote by *involutive moon* the class of all involutive moons.

Remark 10.2.10 The involutive qm-moon $(A, \cdot, ^-, 1)$ is a quasi-m algebra with an involutive quasi-m negation, by Remark 5.2.33.

Proposition 10.2.11 *Let $\mathcal{A} = (A, \cdot, 1)$ be an involutive qm-hub, with $^-$ its involutive quasi-m negation. Define $F(\mathcal{A}) = (A, \cdot, ^-, 1)$. Then, $F(\mathcal{A})$ is an involutive qm-moon.*

Proof. Obviously. □

The converse of Proposition 10.2.11 does not hold, as the following example (1) shows.

Examples 10.2.12

(1) The algebra $\mathcal{QMM}^1 = (\{a, b, c, 1\}, \odot_{11}, ^-, 1)$, with

$\odot_{11}$	a	b	c	1
a	a	1	a	a
b	1	1	b	b
c	a	b	1	1
1	a	b	1	1

and $(a, b, c, 1)^- = (b, a, c, 1)$, is an **involutive super commutative qm-moon**, such that $(\{a, b, c, 1\}, \odot_{11}, 1)$ is the super qm-hub $\mathcal{QMH}^1$ that is not involutive from Examples 10.2.5 (1).

(2) The algebra $\mathcal{QMM}^2 = (\{a, b, c, 1\}, \odot_{35}, ^-, 1)$, with

$\odot_{35}$	a	b	c	1
a	1	a	a	a
b	1	1	1	b
c	1	1	1	b
1	a	b	b	1

and $(a, b, c, 1)^- = (a, b, c, 1)$, is an **involutive super non-commutative qm-moon** (by Proposition 10.2.11), such that $(\{a, b, c, 1\}, \odot_{35}, 1)$ is the involutive super qm-hub $\mathcal{QMH}^2$ from Examples 10.2.5 (2).

(3) The algebra $\mathcal{QMM}^3 = (\{a, b, c, 1\}, \odot_{11}, ^-, 1)$, with

$\odot_{11}$	a	b	c	1
a	a	1	a	a
b	1	1	b	b
c	a	b	1	1
1	a	b	1	1

and $(a, b, c, 1)^- = (b, b, 1, 1)$, is a **super commutative qm-moon that is not involutive**, such that $(\{a, b, c, 1\}, \odot_{11}, 1)$ is the super qm-hub $\mathcal{QMH}^1$ that is not involutive from Examples 10.2.5 (1).

Proposition 10.2.13 *Let $(A, \cdot, ^-, 1)$ be an involutive qm-moon. Then, we have: for all $x, y \in A$,*
(N1) $1^- = 1$,
(IdP) $1 \cdot x = x \cdot 1$,
(qm-$\overline{pU}$) $\quad 1 \cdot (x \cdot y)^- = (x \cdot y)^- = (x \cdot y)^- \cdot 1$,
(qm-$\overline{pU}(1 \cdot x, x \cdot 1)$) $\quad 1 \cdot (1 \cdot x)^- = (1 \cdot x)^- = (1 \cdot x)^- \cdot 1$,
$\quad 1 \cdot (x \cdot 1)^- = (x \cdot 1)^- = (x \cdot 1)^- \cdot 1$.

Proof. (N1): $1^- \overset{(U1)}{=} (1 \cdot 1)^- \overset{(qm-NegP)}{=} 1^- \cdot 1 \overset{(pIv)}{=} 1$.

(IdP): By (qm-NegP), $(1 \cdot x)^- = (x \cdot 1)^-$; then, $((1 \cdot x)^-)^- = ((x \cdot 1)^-)^-$, hence $1 \cdot x = x \cdot 1$, by (DN).

(qm-$\overline{pU}$): $(x \cdot y)^- \overset{(qm-pU)}{=} (1 \cdot (x \cdot y))^- \overset{(qm-NegP)}{=} 1 \cdot (x \cdot y)^-$ and $(x \cdot y)^- \overset{(qm-pU)}{=} ((x \cdot y) \cdot 1)^- \overset{(qm-NegP)}{=} (x \cdot y)^- \cdot 1$. Thus, (qm-$\overline{pU}$) holds.

(qm-$\overline{pU}(1 \cdot x, x \cdot 1)$): Obviously, by (qm-$\overline{pU}$) (see Proposition 5.2.37). □

10.2.2 The associative (involutive) qm-hub, qm-moon vs. the qm-group

We introduce now and study the associative qm-hub and the associative involutive qm-hub. We also introduce and study the associative qm-moon and the associative involutive qm-moon. Finally, we introduce the qm-group and show their connections.

- **In the "world" of qm-unital magmas**

Definition 10.2.14 An *associative qm-hub* is a qm-hub verifying (Ass).

Denote by **associative qm-hub** the class of all associative qm-hubs.

Note that any associative qm-hub is a qm-monoid and that a qm-monoid verifying (pIv') is an associative qm-hub. Hence, we obviously have:

(qmMAH) **qm-monoid** + (pIv') = **associative qm-hub.**

Proposition 10.2.15 *Let* $\mathcal{G} = (G, \cdot, 1)$ *be an associative qm-hub. Then: for all* $x, y \in G$,
(IdP) $1 \cdot x = x \cdot 1$,
(Ass2) $x \cdot y = x \cdot (1 \cdot y)$ *and* $x \cdot y = (x \cdot 1) \cdot y$,
(qm-PI) $x \cdot y = (1 \cdot x) \cdot (1 \cdot y)$, $x \cdot y = (x \cdot 1) \cdot (y \cdot 1)$.

Proof. By Proposition 5.2.41, since $\mathcal{G}$ is a qm-monoid. □

Corollary 10.2.16 *Any associative qm-hub is super.*

The connections between these classes are shown in Figure 10.6.

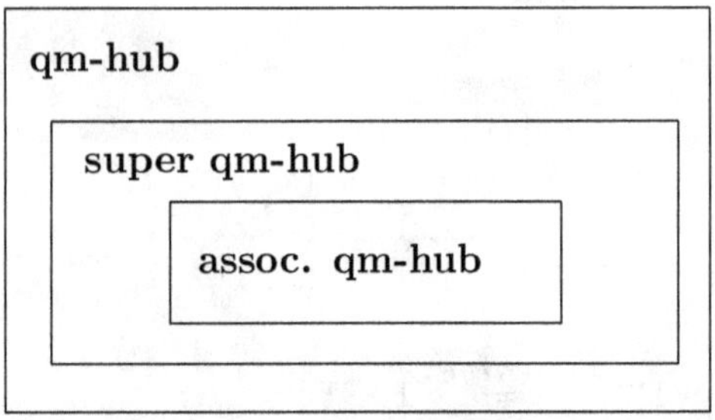

Figure 10.6: Connections

We shall establish now the connection between the associative qm-hubs and the groups.

By Proposition 5.2.1, any group (i.e. associative hub, by Definition 1' and Remark 7.1.3) is an associative qm-hub. Conversely, any associative qm-hub verifying (pU) is a group. Hence, we have the following equivalences, by (QM):

associative qm-hub + (pU) $\equiv$ **associative hub** = **group** (Definition 1'),

associative involutive qm-hub + (pU) $\equiv$ **associative involutive hub** $\equiv$
$\equiv$ **associative hub** = **group** (Definition 1').

Note that, indeed,

associative qm-hub + (pU) $\stackrel{(qmMAH)}{\Longleftrightarrow}$ **qm-monoid** + (pIv') + (pU) $\Longleftrightarrow$ **monoid** + (pIv') $\stackrel{(MG)}{\Longleftrightarrow}$ **group.**

We say that an associative qm-hub is *qm-proper*, if is not a group (associative hub) (i.e. if (pU) is not verified).

Theorem 10.2.17 *Let $\mathcal{G} = (G, \cdot, 1)$ be a qm-proper associative qm-hub and consider the regular-m set $Rm(G) = \{x \in G \mid x = 1 \cdot x = x \cdot 1\}$. Then,*

$$\mathcal{R}m(\mathcal{G}) = (Rm(G), \cdot, 1)$$

is a group (as associative hub, by Definition 1').

Proof. By Theorem 5.2.13, $\mathcal{R}m(\mathcal{G})$ is regular-m algebra. It verifies (Ass), obviously. Suppose, by absurdum hypothesis, that it does not verify (pIv') (i.e. it is only a monoid); hence, there is a row a in the table of $\cdot$ of $\mathcal{R}m(\mathcal{G})$ not containing the element 1. On the other hand, $\mathcal{G}$ verifies (Ass2), by Proposition 10.2.15; then, by Proposition 5.2.25, any row b in the table of $\cdot$ of $\mathcal{G}$, such that $b \parallel_m a$, will coincide with the row a, hence row b would not contain the element 1: contradiction, because $\mathcal{G}$ verifies (pIv'). Thus, $\mathcal{R}m(\mathcal{G})$ verifies also (pIv'), hence it is a group. □

Lemma 10.2.18 *Let $\mathcal{G} = (G, \cdot, 1)$ be a qm-proper associative qm-hub. For any $x \in G$, if $x \notin Rm(G)$, then*

$$r_x \stackrel{notation}{=} 1 \cdot x \stackrel{(IdP)}{=} x \cdot 1 \in Rm(G).$$

Proof. $1 \cdot r_x = 1 \cdot (1 \cdot x) \stackrel{(qm-pU)}{=} 1 \cdot x = r_x$ and
$r_x \cdot 1 = (x \cdot 1) \cdot 1 = x \cdot 1 = r_x$; thus, $r_x \in Rm(G)$. □

We shall introduce now new operations in associative qm-hubs: $\to$ and $\rightsquigarrow$.

Let $\mathcal{G} = (G, \cdot, 1)$ be an associative qm-hub. Hence, $\mathcal{R}m(\mathcal{G}) = (Rm(G), \cdot, 1)$ is a group (Definition 1'). We shall extend to $\mathcal{G}$ the new operations $\to$ and $\rightsquigarrow$ defined on the group $\mathcal{R}m(\mathcal{G})$ by (7.1). Thus, we define the new operations $\to$ and $\rightsquigarrow$ on G

by using Lemma 10.2.18, as follows: for all $x, y \in G$,

$$x \to y \overset{def.}{=} \begin{cases} x \to y, & \text{if} \quad x, y \in Rm(G), \\ r_x \to y, & \text{if} \quad x \notin Rm(G),\ r_x = 1 \cdot x \in Rm(G), \quad y \in Rm(G), \\ x \to r_y, & \text{if} \quad x \in Rm(G), \quad y \notin Rm(G),\ r_y = 1 \cdot y \in Rm(G), \\ r_x \to r_y, & \text{if} \quad x, y \notin Rm(G), r_x = 1 \cdot x,\ r_y = 1 \cdot y,\ r_x, r_y \in Rm(G), \end{cases}$$

$$x \rightsquigarrow y \overset{def.}{=} \begin{cases} x \rightsquigarrow y, & \text{if} \quad x, y \in Rm(G), \\ r_x \rightsquigarrow y, & \text{if} \quad x \notin Rm(G),\ r_x = 1 \cdot x \in Rm(G), \quad y \in Rm(G), \\ x \rightsquigarrow r_y, & \text{if} \quad x \in Rm(G), \quad y \notin Rm(G),\ r_y = 1 \cdot y \in Rm(G), \\ r_x \rightsquigarrow r_y, & \text{if} \quad x, y \notin Rm(G), r_x = 1 \cdot x,\ r_y = 1 \cdot y,\ r_x, r_y \in Rm(G). \end{cases}$$

The following result is useful further in Chapter 12.

Proposition 10.2.19 *Let $\mathcal{G} = (G, \cdot, 1)$ be an associative qm-hub and consider the new operations $\to$ and $\rightsquigarrow$ above defined on G. Then, (pRe'), (pBB=), (IdEq) hold.*

Proof. By Theorem 10.2.17, $\mathcal{R}m(\mathcal{G})$ is a group (i.e. associative hub); hence, (pRe'), (pBB=), (IdEq) hold on $\mathcal{R}m(\mathcal{G})$, by Proposition 7.1.13.

(pRe'): If $x \in Rm(G)$, then $x \to x = 1 = x \rightsquigarrow x$.
If $x \notin Rm(G)$, then $r_x = 1 \cdot x \in Rm(G)$, by Lemma 10.2.18; hence,
$x \to x \overset{def.}{=} r_x \to r_x = 1 = r_x \rightsquigarrow r_x \overset{def.}{=} x \rightsquigarrow x$. Thus, (pRe') holds.

(pBB=): Similarly.

(IdEq): Similarly. □

The associative involutive qm-hub

Given a q-proper associative qm-hub $\mathcal{G} = (G, \cdot, 1)$, its regular algebra $(Rm(G), \cdot, 1)$ is a group (= associative hub), with an involutive regular-m negation $^{-1}$, verifying (pIv):
$x \cdot x^{-1} = x^{-1} \cdot x = 1$, for all $x \in G$.

If the extended quasi-negation $^{-}$ on G exists (and we shall prove in Proposition 10.2.20 that it always exists) and if is *involutive*, then we have an *involutive associative qm-hub*, with $^{-}$ its involutive quasi-m negation.

Note that the *associative involutive qm-hubs* coincide with the *involutive associative qm-hubs* (algebras $(G, \cdot, 1)$ verifying (qm-pU), (U1), (pIv'), (Ass), (qm-NegP), (pIv), (DN)). Denote by **associative involutive qm-hub** the class of all associative involutive qm-hubs.

Proposition 10.2.20 *Any qm-proper associative qm-hub admits a quasi-m negation.*

Proof. Let $\mathcal{G} = (G, \cdot, 1)$ be a qm-proper associative qm-hub.

First note that, by Proposition 10.2.15, (IdP) and (qm-PI) hold.

Its regular algebra $(Rm(G), \cdot, 1)$ is a group, with $^{-1}$ its involutive regular-m negation verifying (pIv). We must prove that $^{-1}$ can be extended to $^{-} : G \longrightarrow G$ verifying: for all $x \in G$,

(qm-NegP) $(x \cdot 1)^- = x^- \cdot 1 = 1 \cdot x^- = (1 \cdot x)^-$ and
(pIv) $x \cdot x^- = x^- \cdot x = 1$.

Note that (qm-NegP) and (pIv) are verified obviously by all $x \in Rm(G)$. It remains to prove that (qm-NegP) and (pIv) are verified by all $x \notin Rm(G)$.

Let $x \notin Rm(G)$ and let $r_x \overset{def.}{=} 1 \cdot x \overset{(IdP)}{=} x \cdot 1$; then, $r_x \in Rm(G)$, by Lemma 10.2.18.

We shall prove first that:

$$(1 \cdot x)^- = (x \cdot 1)^-. \tag{10.2}$$

Indeed, $(1 \cdot x)^- = r_x^- = r_x^{-1}$ and $(x \cdot 1)^- = r_x^- = r_x^{-1}$; thus, (10.2) holds. Hence, by $1 \cdot x^- = x^- \cdot 1 = r_x^{-1}$, we obtain x^-.

It remains to prove that (pIv) is verified. Indeed,
$x \cdot x^- \overset{(qm-PI)}{=} (1 \cdot x) \cdot (1 \cdot x^-) = r_x \cdot r_x^{-1} = 1$ and $x^- \cdot x \overset{(qm-PI)}{=} (1 \cdot x^-) \cdot (1 \cdot x) = r_x^{-1} \cdot r_x = 1$.

Thus, the quasi-m negation exists. □

Examples 10.2.21 (See Remarks 10.2.22)

(1) The **commutative associative qm-hub** $\mathcal{A}ff\mathcal{QMH}^1 = (A_4 = \{a, b, c, 1\}, \odot_3, 1)$,

where

$\odot_3$	a	b	c	1
a	b	1	b	a
b	1	a	1	b
c	b	1	b	a
1	**a**	**b**	a	**1**

. Note that $Rm(A_4) = \{a, b, 1\}$,

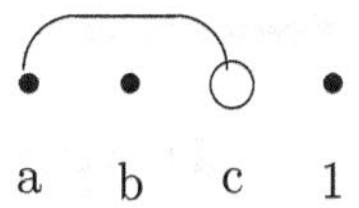

and

$\odot_3$	a	b	1
a	b	1	a
a	1	a	b
1	a	b	**1**

, hence $(\{a, b, 1\}, \odot_3, 1)$ is the commutative group (associative hub) with 3 elements, with the involutive regular-m negation $(a, b, 1)^{-1} = (b, a, 1)$.

By (qm-NegP), $b = a^- = (1 \odot_3 c)^- = 1 \odot_3 c^-$, hence $c^- = b$; thus, the quasi-m negation exists, it is $(a, b, c, 1)^- = (\mathbf{b}, \mathbf{a}, b, \mathbf{1})$, but it is not involutive. Thus, $\mathcal{A}ff\mathcal{QMH}^1$ is **not involutive**.

(2) The **commutative associative qm-hub** $\mathcal{A}ff\mathcal{QMH}^2 = (A_4 = \{a, b, c, 1\}, \odot_2, 1)$,

where

$\odot_2$	a	b	c	1
a	b	1	a	a
b	1	a	b	b
c	a	b	1	1
1	**a**	**b**	1	**1**

. Note that $Rm(A_4) = \{a, b, 1\}$,

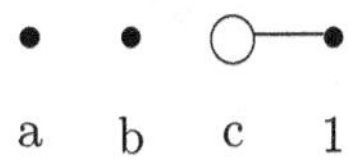

and

$\odot_2$	a	b	1
a	b	1	a
a	1	a	b
1	a	b	**1**

, hence $(\{a, b, 1\}, \odot_2, 1)$ is the commutative group (associative hub) with 3 elements, with the involutive regular-m negation $(a, b, 1)^{-1} = (b, a, 1)$.

By (qm-NegP), $1 = 1^- = (1 \odot_2 c)^- = 1 \odot_2 c^-$, hence $c^- = c$ or $c^- = 1$; thus, the quasi-m negation $(a, b, c, 1)^- = (\mathbf{b}, \mathbf{a}, c, \mathbf{1})$ is an involutive one. Thus, $\mathcal{A}\int\int\mathcal{QMH}^2$ **is involutive.**

Remarks 10.2.22

(i) There exist *associative qm-hubs* for which an involutive quasi-m negation does not exist, i.e. for which there is no an associated *involutive associative qm-hub.*

(ii) There exist *associative qm-hubs* for which there exist more than one involutive quasi-m negation, i.e. for which there exist more than one associated *involutive associative qm-hubs.*

• In the "world" of qm-unital magmas with additional operation

Definition 10.2.23 An *associative qm-moon* is a qm-moon verifying (Ass).

Denote by **associative qm-moon** the class of all associative qm-moons.

Note that the *associative involutive qm-moons* coincide with the *involutive associative qm-moons.* Denote by **associative involutive qm-moon** the class of all associative involutive qm-moons.

We have the equivalences:

associative qm-moon + (pU) $\equiv$ **associative moon** = **group** (Definition 1),

associative involutive qm-moon + (pU) $\equiv$ **associative involutive moon** $\equiv$ $\equiv$ **associative moon** = **group** (Definition 1).

The connection between the involutive associative qm-hubs and the involutive associative qm-moons is the following:

Theorem 10.2.24

(1) Let $\mathcal{A} = (A, \cdot, 1)$ be an associative involutive qm-hub, with the involutive quasi-m negation denoted by $^-$. Define $F(\mathcal{A}) = (A, \cdot, ^-, 1)$. Then, $F(\mathcal{A})$ is an associative involutive qm-moon.

(1') Conversely, let $\mathcal{A} = (A, \cdot, ^-, 1)$ be an associative involutive qm-moon. Define $G(\mathcal{A}) = (A, \cdot, 1)$. Then, $G(\mathcal{A})$ is an associative involutive qm-hub and $^-$ is an involutive quasi-m negation.

(2) The mapps F and G are mutually inverse.

Proof. Obviously, by definitions. □

By Theorem 10.2.24, we have the equivalence:

(qmHM1) **involutive assoc. qm-hub** $\Longleftrightarrow$ **involutive assoc. qm-moon.**

Theorem 10.2.25 *Let $\mathcal{G} = (G, \cdot, ^-, 1)$ be an associative qm-moon. Then,*

$$\mathcal{R}m(\mathcal{G}) = (Rm(G), \cdot, ^{-1}, 1)$$

is a group (Definition 1), where $^- \mid_{Rm(G)} = ^{-1}$.

Proof. By Theorem 5.2.13. □

Consider now the following properties:
(qm-pIvP) $(u\cdot v)\cdot(x\cdot y)=1(\Leftrightarrow(x\cdot y)\cdot(u\cdot v)=1)\Longleftrightarrow x\cdot y=(u\cdot v)^-$,
(pIvG) $(x\cdot y^-)^-=y\cdot x^-,\quad (y^-\cdot x)^-=x^-\cdot y$,
(IdEqP) $y\cdot x^-=1\Longleftrightarrow x^-\cdot y=1$,
(qm-pEqP=) $1\cdot x(\overset{(IdP)}{=}x\cdot 1)=1\cdot y(\overset{(IdP)}{=}y\cdot 1)\Longleftrightarrow y\cdot x^-=1\ (\overset{(IdEqP)}{\Longleftrightarrow}x^-\cdot y=1)$.

Proposition 10.2.26 *(See Proposition 10.2.13)*
Let $(G,\cdot,^-,1)$ be an involutive associative qm-moon. Then,
(qm-n1) (qm-pU) + (pIv) + (Ass) + (qm-$\overline{pU}$) $\Longrightarrow$ (qm-pIvP),
(qm-n2) (qm-pIvP) + (Ass) + (pIv) + (IdP) + (U1) $\Longrightarrow$ (pIvG),
(qm-n3) (qm-PI) + (pIv) + (Ass) + (IdP) $\Longrightarrow$ (IdEqP) + (qm-pEqP=),
i.e. (qm-pIvP), (pIvG), (IdEqP), (qm-pEqP=) hold.

Proof. (qm-n1): Suppose that $(u\cdot v)\cdot(x\cdot y)=1$; then, $x\cdot y\overset{(qm-pU)}{=}1\cdot(x\cdot y)\overset{(pIv)}{=}[(u\cdot v)^-\cdot(u\cdot v)]\cdot(x\cdot y)\overset{(Ass)}{=}(u\cdot v)^-\cdot[(u\cdot v)\cdot(x\cdot y)]=(u\cdot v)^-\cdot 1\overset{(qm-\overline{pU})}{=}(u\cdot v)^-$.
Conversely, suppose that $x\cdot y=(u\cdot v)^-$; then, $(u\cdot v)\cdot(x\cdot y)=(u\cdot v)\cdot(u\cdot v)^-\overset{(pIv)}{=}1$.
Thus, $(u\cdot v)\cdot(x\cdot y)=1\Longleftrightarrow x\cdot y=(u\cdot v)^-$.
Similarly, $(x\cdot y)\cdot(u\cdot v)=1\Longleftrightarrow x\cdot y=(u\cdot v)^-$. So, (qm-pIvP) holds.

(qm-n2): To prove that $(x\cdot y^-)^-=y\cdot x^-$, we have to prove, by (qm-pIvP), that:
(a) $(x\cdot y^-)\cdot(y\cdot x^-)=1$
Indeed, $(x\cdot y^-)\cdot(y\cdot x^-)\overset{(Ass)}{=}x\cdot(y^-\cdot y)\cdot x^-\overset{(pIv)}{=}x\cdot 1\cdot x^-\overset{(Ass)}{=}(x\cdot 1)\cdot x^-\overset{(IdP)}{=}(1\cdot x)\cdot x^-\overset{(Ass)}{=}1\cdot(x\cdot x^-)\overset{(pIv)}{=}1\cdot 1\overset{(U1)}{=}1$; thus (a) holds.
Similarly we prove that $(y^-\cdot x)^-=x^-\cdot y$. Thus, (pIvG) holds.

(qm-n3): Suppose that $1\cdot x=1\cdot y$; then,
$y\cdot x^-\overset{(qm-PI)}{=}(1\cdot y)\cdot(1\cdot x^-)=(1\cdot x)\cdot(1\cdot x^-)\overset{(qm-PI)}{=}x\cdot x^-\overset{(pIv)}{=}1$.

Conversely, suppose that $y\cdot x^-=1$; then,
$1\cdot x=(y\cdot x^-)\cdot x\overset{(Ass)}{=}y\cdot(x^-\cdot x)\overset{(pIv)}{=}y\cdot 1\overset{(IdP)}{=}1\cdot y$. Thus, $1\cdot x=1\cdot y\Longleftrightarrow y\cdot x^-=1$.

Similarly, $1\cdot x=1\cdot y\Longleftrightarrow x^-\cdot y=1$. Thus, (IdEqP) and (qm-pEqP=) hold. □

We shall introduce now new operations in involutive associative qm-moons: $\to$ and $\rightsquigarrow$.

Let $\mathcal{G}=(G,\cdot,^-,1)$ be an involutive associative qm-moon. Hence, $\mathcal{Rm}(\mathcal{G})=(Rm(G),\cdot,^{-1},1)$ is a group. We extend the new operations $\to$ and $\rightsquigarrow$ defined on $Rm(G)$ (the pair $(\to,\rightsquigarrow)$ is called "pseudo-implication") by (7.1) to G as follows: for all $x,y\in G$,

$$x\to y\overset{def.}{=}(x\cdot y^-)^-\overset{(pIvG)}{=}y\cdot x^-,\quad x\rightsquigarrow y\overset{def.}{=}(y^-\cdot x)^-\overset{(pIvG)}{=}x^-\cdot y. \tag{10.3}$$

Then, we have the following Proposition, that generalizes Proposition 7.1.13; it will be useful in Chapter 12; note that all the properties from Proposition 7.1.13

which are proved by using (pU), must now be proved by using (qm-pU) and (U1) and, hence, reformulated, eventually.

Proposition 10.2.27 *Let $(G, \cdot, ^-, 1)$ be an involutive associative qm-moon and let $\to$, $\rightsquigarrow$ be the new operations defined on G by (10.3). Then, we have: for all $x, y, z \in G$,*
(Id) $x \to 1 = x \rightsquigarrow 1$,
(pBB=) $y \to z = (z \to x) \rightsquigarrow (y \to x), \quad y \rightsquigarrow z = (z \rightsquigarrow x) \to (y \rightsquigarrow x)$,
(pEx) $x \to (y \rightsquigarrow z) = y \rightsquigarrow (x \to z)$,
(qm-pD=) $(y \to x) \rightsquigarrow x = y \cdot 1 \overset{(IdP)}{=} 1 \cdot y = (y \rightsquigarrow x) \to x$,
(IdEq) $x \to y = 1 \Leftrightarrow x \rightsquigarrow y = 1$,
(qm-pEq=) $1 \cdot x (\overset{(IdP)}{=} x \cdot 1) = y \cdot 1 (\overset{(IdP)}{=} 1 \cdot y) \Leftrightarrow x \to y = 1 \overset{(IdEq)}{\Leftrightarrow} x \rightsquigarrow y = 1$,
(q-pM) $1 \to (x \rightsquigarrow y) = x \rightsquigarrow y, \quad 1 \rightsquigarrow (x \to y) = x \to y$,
(pRe') $x \to x = 1 = x \rightsquigarrow x$,
(pM1) $1 \to 1 = 1 = 1 \rightsquigarrow 1$,
(q-pNeg) $x \to 1 = 1 \to x^- = (1 \rightsquigarrow x)^- = x \rightsquigarrow 1 = 1 \rightsquigarrow x^- = (1 \to x)^-$.

Proof. (Id): $x \to 1 \overset{def.}{=} 1 \cdot x^- \overset{(qm-NegP)}{=} x^- \cdot 1 \overset{def.}{=} x \rightsquigarrow 1$.

(pBB=): $(z \to x) \rightsquigarrow (y \to x) \overset{def.}{=} (x \cdot z^-) \rightsquigarrow (x \cdot y^-)$
$\overset{def.}{=} (x \cdot z^-)^- \cdot (x \cdot y^-) \overset{(pIvG)}{=} (z \cdot x^-) \cdot (x \cdot y^-)$
$\overset{(Ass)}{=} z \cdot (x^- \cdot x) \cdot y^- \overset{(pIv)}{=} z \cdot 1 \cdot y^-$
$\overset{(Ass)}{=} (z \cdot 1) \cdot y^- \overset{(IdP)}{=} (1 \cdot z) \cdot y^-$
$\overset{(Ass)}{=} 1 \cdot (z \cdot y^-) \overset{(qm-pU)}{=} z \cdot y^- \overset{def.}{=} y \to z$.

Also, $(z \rightsquigarrow x) \to (y \rightsquigarrow x) \overset{def.}{=} (z^- \cdot x) \to (y^- \cdot x) \overset{def.}{=} (y^- \cdot x) \cdot (z^- \cdot x)^- \overset{(pIvG)}{=}$ $(y^- \cdot x) \cdot (x^- \cdot z) \overset{(Ass)}{=} y^- \cdot (x \cdot x^-) \cdot z \overset{(pIv)}{=} y^- \cdot 1 \cdot z \overset{(Ass)}{=} y^- \cdot (1 \cdot z) \overset{(IdP)}{=} y^- \cdot (z \cdot 1) \overset{(Ass)}{=}$ $(y^- \cdot z) \cdot 1 \overset{(qm-pU)}{=} y^- \cdot z \overset{def.}{=} y \rightsquigarrow z$. Thus, (pBB=) holds.

(pEx): $x \to (y \rightsquigarrow z) = x \to (y^- \cdot z) = (y^- \cdot z) \cdot x^- \overset{(Ass)}{=} y^- \cdot (z \cdot x^-) = y \rightsquigarrow$ $(z \cdot x^-) = y \rightsquigarrow (x \to z)$.

(qm-pD=): $(y \to x) \rightsquigarrow x \overset{def.}{=} (x \cdot y^-) \rightsquigarrow x \overset{def.}{=} (x \cdot y^-)^- \cdot x \overset{(pIvG)}{=} (y \cdot x^-) \cdot x \overset{(Ass)}{=}$ $y \cdot (x^- \cdot x) \overset{(pIv)}{=} y \cdot 1$ and
$(y \rightsquigarrow x) \to x \overset{def.}{=} (y^- \cdot x) \to x \overset{def.}{=} x \cdot (y^- \cdot x)^- \overset{(pIvG)}{=} x \cdot (x^- \cdot y) \overset{(Ass)}{=} (x \cdot x^-) \cdot y \overset{(pIv)}{=} 1 \cdot y$.
Thus, by (IdP), (qm-pD=) holds.

(IdEq): If $x \to y = 1$, i.e. $y \cdot x^- = 1$, then $1 \cdot x = (y \cdot x^-) \cdot x \overset{(Ass)}{=} y \cdot (x^- \cdot x) \overset{(pIv)}{=} y \cdot 1$; then, $x \rightsquigarrow y \overset{def.}{=} x^- \cdot y \overset{(qm-pU)}{=} (x^- \cdot y) \cdot 1 \overset{(Ass)}{=} x^- \cdot (y \cdot 1) = x^- \cdot (1 \cdot x) \overset{(IdP)}{=}$ $x^- \cdot (x \cdot 1) \overset{(Ass)}{=} (x^- \cdot x) \cdot 1 \overset{(pIv)}{=} 1 \cdot 1 \overset{(U1)}{=} 1$.
Conversely, if $x \rightsquigarrow y = 1$, i.e. $x^- \cdot y = 1$, then $x \cdot 1 = x \cdot (x^- \cdot y) \overset{(Ass)}{=} (x \cdot x^-) \cdot y \overset{(pIv)}{=} 1 \cdot y$; then, $x \to y \overset{def.}{=} y \cdot x^- \overset{(qm-pU)}{=} 1 \cdot (y \cdot x^-) \overset{(Ass)}{=} (1 \cdot y) \cdot x^- = (x \cdot 1) \cdot x^- \overset{(IdP)}{=}$ $(1 \cdot x) \cdot x^- \overset{(Ass)}{=} 1 \cdot (x \cdot x^-) \overset{(pIv)}{=} 1 \cdot 1 \overset{(U1)}{=} 1$. Thus, (IdEq) holds.

(qm-pEq=): If $1 \cdot x = y \cdot 1$, then $x \rightsquigarrow y \overset{def.}{=} x^- \cdot y \overset{(qm-pU)}{=} (x^- \cdot y) \cdot 1 \overset{(Ass)}{=}$

$x^- \cdot (y \cdot 1) = x^- \cdot (1 \cdot x) \overset{(IdP)}{=} x^- \cdot (x \cdot 1) \overset{(Ass)}{=} (x^- \cdot x) \cdot 1 \overset{(pIv)}{=} 1 \cdot 1 \overset{(U1)}{=} 1$.
Conversely, if $x \rightsquigarrow y = 1$, i.e. $x^- \cdot y = 1$, then $x \cdot 1 = x \cdot (x^- \cdot y) \overset{(Ass)}{=} (x \cdot x^-) \cdot y \overset{(pIv)}{=} 1 \cdot y$; then, $1 \cdot x \overset{(IdP)}{=} x \cdot 1 = 1 \cdot y \overset{(IdP)}{=} y \cdot 1$. Thus, (qm-pEq=) holds.

(q-pM): $1 \to (x \rightsquigarrow y) \overset{def.}{=} 1 \to (x^- \cdot y) \overset{def.}{=} (x^- \cdot y) \cdot 1^- \overset{(N1)}{=} (x^- \cdot y) \cdot 1 \overset{(qm-pU)}{=} x^- \cdot y \overset{def.}{=} x \rightsquigarrow y$ and
$1 \rightsquigarrow (x \to y) \overset{def.}{=} 1 \rightsquigarrow (y \cdot x^-) \overset{def.}{=} 1^- \cdot (y \cdot x^-) \overset{(N1)}{=} 1 \cdot (y \cdot x^-) \overset{(qm-pU)}{=} y \cdot x^- \overset{def.}{=} x \to y$.
Thus, (q-pM) holds.

(pRe'): $x \to x \overset{def.}{=} x \cdot x^- \overset{(pIv)}{=} 1$ and $x \rightsquigarrow x \overset{def.}{=} x^- \cdot x \overset{(pIv)}{=} 1$.

(pM1): By (pRe').

(q-pNeg): First, we prove (a): $x \to 1 = 1 \to x^- = (1 \to x)^- = 1 \cdot x^-$.
Indeed, $x \to 1 \overset{def.}{=} 1 \cdot x^-$; $1 \to x^- \overset{def.}{=} x^- \cdot 1^- \overset{(N1)}{=} x^- \cdot 1$;
$(1 \to x)^- \overset{def.}{=} (x \cdot 1^-)^- \overset{(N1)}{=} (x \cdot 1)^- \overset{(qm-NegP)}{=} 1 \cdot x^-$
(or $(1 \to x)^- \overset{def.}{=} ((1 \cdot x^-)^-)^- \overset{(DN)}{=} 1 \cdot x^-$); thus, (a) holds.

Second, we prove (b): $x \rightsquigarrow 1 = 1 \rightsquigarrow x^- = (1 \rightsquigarrow x)^- = x^- \cdot 1$.
Indeed, $x \rightsquigarrow 1 \overset{def.}{=} x^- \cdot 1$; $1 \rightsquigarrow x^- \overset{def.}{=} 1^- \cdot x^- \overset{(N1)}{=} 1 \cdot x^-$;
$(1 \rightsquigarrow x)^- \overset{def.}{=} (1^- \cdot x)^- \overset{(N1)}{=} (1 \cdot x)^- \overset{(qm-NegP)}{=} x^- \cdot 1$
(or $(1 \rightsquigarrow x)^- \overset{def.}{=} ((x^- \cdot 1)^-)^- \overset{(DN)}{=} x^- \cdot 1$); thus, (b) holds.

Now, by (a), (b) and (IdP) or (qm-NegP), (q-pNeg) holds. □

The qm-group

Definition 10.2.28 A *quasi-m group*, or a *qm-group* for short, is an algebra $\mathcal{G} = (G, \cdot, ^-, 1)$ of type $(2, 1, 0)$ verifying the following properties: for all $x, y, z \in G$,
(qm-pU) $1 \cdot (x \cdot y) = x \cdot y = (x \cdot y) \cdot 1$,
(U1) $1 \cdot 1 = 1$,
(pIv) $x \cdot x^- = 1 = x^- \cdot x$, where $1 = 1^-$;
(Ass) $x \cdot (y \cdot z) = (x \cdot y) \cdot z$,
$\overline{(\text{qm-}pU(1 \cdot x, x \cdot 1))}$ $1 \cdot (1 \cdot x)^- = (1 \cdot x)^- = (1 \cdot x)^- \cdot 1$,
$1 \cdot (x \cdot 1)^- = (x \cdot 1)^- = (x \cdot 1)^- \cdot 1$,
(DN) $(x^-)^- = x$.

Denote by **qm-group** the class of all qm-groups.

Proposition 10.2.29 *Let $\mathcal{G} = (G, \cdot, ^-, 1)$ be a qm-group. Then,*
(qm-ng1) (qm-pU) + (Ass) $\Longrightarrow$ (IdP),
(qm-ng2) (IdP) + (Ass) + (pIv) + (U1) + $\overline{(\text{qm-}pU(1 \cdot x, x \cdot 1))}$ + (qm-pU) $\Longrightarrow$ (qm-NegP),
i.e. (IdP) and (qm-NegP) hold.

Proof. (qm-ng1): By Proposition 5.2.16 (qm-1), (qm-pU) + (Ass) $\Longrightarrow$ (IdP), thus (IdP) holds.

(qm-ng2): $(x\cdot 1)\cdot(x^-\cdot 1) \overset{(IdP)}{=} (1\cdot x)\cdot(x^-\cdot 1) \overset{(Ass)}{=} 1\cdot(x\cdot x^-)\cdot 1 \overset{(pIv)}{=} 1\cdot 1\cdot 1 \overset{(U1)}{=} 1$.
$(x\cdot 1)^- \overset{(qm-\overline{pU(1\cdot x, x\cdot 1)})}{=} (x\cdot 1)^-\cdot 1 = (x\cdot 1)^-\cdot[(x\cdot 1)\cdot(x^-\cdot 1)] \overset{(Ass)}{=}$
$[(x\cdot 1)^-\cdot(x\cdot 1)]\cdot(x^-\cdot 1) \overset{(pIv)}{=} 1\cdot(x^-\cdot 1) \overset{(qm-pU)}{=} x^-\cdot 1$.
Then, by (IdP), (qm-NegP) holds. □

Hence we have the following connections:

qm-group + (pU) $\equiv$ **group** (Definition 1),

qm-group $\subseteq$ **associative qm-moon** $\supseteq$ **involutive associative qm-moon**.

Theorem 10.2.30 *The qm-groups are equivalent to the involutive associative qm-moons.*

Proof. Let $\mathcal{G} = (G, \cdot, ^-, 1)$ be a qm-group, i.e. (qm-pU), (U1), (pIv), (Ass), (qm-$\overline{pU(1\cdot x, x\cdot 1)}$), (DN) hold. We must prove that $\mathcal{G}$ is an involutive associative qm-moon, i.e. (qm-pU), (U1), (Ass), (qm-NegP), (pIv), (DN) hold. Thus, all we have to prove is that (qm-NegP) hold. Indeed, (qm-NegP) follows by Proposition 10.2.29.

Conversely, let $\mathcal{G} = (G, \cdot, ^-, 1)$ be an associative involutive qm-moon. We must prove that $\mathcal{G}$ is a qm-group. Thus, all we have to prove is that (qm-$\overline{pU(1\cdot x, x\cdot 1)}$) holds. Indeed, by Proposition 5.2.37 (jj), (qm-NegP) $\Longrightarrow$ (qm-$\overline{pU(1\cdot x, x\cdot 1)}$) (or by Proposition 10.2.13). □

By Theorems 10.2.24 and 10.2.30, we obtain:

Theorem 10.2.31 *The qm-groups are categorically equivalent to the involutive associative qm-hubs.*

Remark 10.2.32 By above Theorem 10.2.31, we could define, equivalently, the qm-group as an involutive associative qm-hub. In fact, we find examples of qm-groups by finding examples of involutive associative qm-hubs.

Note that we have the hierarchy No. qm1 from Figure 10.7.

Examples 10.2.33 (See Examples 10.2.21)
(1) The algebra $\mathcal{A}\!\int\!\!\int\mathcal{QMM}^1 = (A_4 = \{a, b, c, 1\}, \odot_3, ^-, 1)$, where

$\odot_3$	a	b	c	1
a	b	1	b	a
b	1	a	1	b
c	b	1	b	a
1	**a**	**b**	a	**1**

and $(a, b, c, 1)^- = (b, a, b, 1)$, is an **associative commutative qm-moon that is not involutive**, because (DN) is not verified. Note that the reduct $(A_4 = \{a, b, c, 1\}, \odot_3, 1)$ is $\mathcal{A}\!\int\!\!\int\mathcal{QMH}^1$ from Examples 10.2.21.

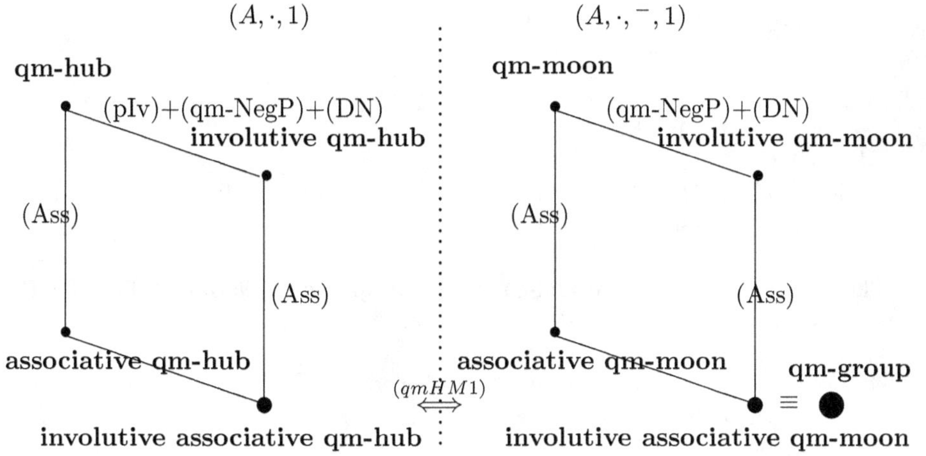

Figure 10.7: The hierarchy No. qm1

(2) The algebra $\mathcal{A}\!\int\!\!\int\mathcal{QMM}^2 = (A_4 = \{a,b,c,1\}, \odot_2, ^-, 1)$, where

$\odot_2$	a	b	c	1
a	b	1	a	a
b	1	a	b	b
c	a	b	1	1
1	**a**	**b**	1	**1**

and $(a,b,c,1)^- = (b,a,c,1)$ is an **involutive associative commutative qm-moon**, hence a **commutative qm-group**: $\mathcal{QMG}_4^2$. Note that the reduct $(A_4 = \{a,b,c,1\}, \odot_2, 1)$ is $\mathcal{A}\!\int\!\!\int\mathcal{QMH}^2$ from Examples 10.2.21.

Example 10.2.34 (Non-commutative qm-group)
Starting from the non-commutative group (= associative hub)

$$\mathcal{G}_{6g} = (G_6 = \{a,b,c,d,f,1\}, \odot, 1)$$

presented in Chapter 9, with $(a,b,c,d,f,1)^{-1} = (a,b,d,c,f,1)$, we can obtain an infinity of associative qm-hubs with minimum 7 elements (by adding quasi-m elements parallel-m to the regular-m elements $a,b,c,d,f,1$), which contain (as their regular-m part) the group $\mathcal{G}_{6g}$. For example, if we add one quasi-m element w, parallel-m to the regular-m element f, we obtain the particular poset (where $\leq$ is $=$) $(G_7 = \{a,b,c,d,f,w,1\}, \leq)$ represented by the quasi-Hasse diagram in Figure 10.8.

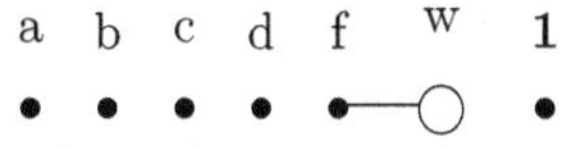

Figure 10.8: $(G_7, \leq)$

Consider the algebra $\mathcal{G}_7 = (G_7, \odot, 1)$, where the product $\odot$ has the following table, by Proposition 5.2.25 ($w \parallel_m f$):

$\mathcal{G}_7$

$\odot$	a	b	c	d	f	w	1
a	**1**	c	b	f	d	d	a
b	d	**1**	f	a	c	c	b
c	f	a	d	**1**	b	b	c
d	b	f	**1**	c	a	a	d
f	c	d	a	b	**1**	**1**	f
w	c	d	a	b	**1**	**1**	f
1	a	b	c	d	f	f	**1**

Then, $\mathcal{G}_7$ is an associative non-commutative qm-hub. We are looking for the quasi-m negations. By (qm-NegP), we must have:
$f = f^- = (w \odot 1)^- = (1 \odot w)^- = f^- = f$,
hence $1 \odot w^- = w^- \odot 1 = f$, hence $w^- = f, w$:
- if $w^- = f$, then the quasi-m negation is $(a, b, c, d, f, w, 1)^{-1} = (a, b, d, c, f, f, 1)$ and is not involutive;
- if $w^- = w$, then the quasi-m negation is $(a, b, c, d, f, w, 1)^{-2} = (a, b, d, c, f, w, 1)$ and is involutive; hence, there exists one **involutive associative non-commutative qm-hub**, hence there exists one **non-commutative qm-group** $\mathcal{QMG}_7 = (G_7, \odot, ^{-2}, 1)$, associated to $\mathcal{G}_7$.

Remarks 10.2.35 As we have said (see Remarks 10.2.22):
(i) There exist *associative qm-hubs* for which an involutive quasi-m negation does not exist, i.e. for which there is no an associated *involutive associative qm-hub* (hence no associated *qm-group*).
(ii) There exist *associative qm-hubs* for which there exist more than one involutive quasi-m negation, i.e. for which there exist more than one associated *involutive associative qm-hubs* (i.e. associated *qm-groups*).

10.2.3 The sharp (involutive) qm-hub, qm-moon vs. the qm-goop

In this subsection, we introduce and study the sharp (super) qm-hub/qm-moon and the sharp involutive (super) qm-hub/qm-moon. We also introduce and study the new notion of *qm-goop*.

- **In the "world" of qm-unital magmas**

Definition 10.2.36 A qm-hub $(A, \cdot, 1)$ is *sharp*, if there exists a quasi-m negation $^-$ verifying (qm-NegP), (pIv) and verifying also: for all $x, y \in A$,
(IdEqP) $\quad y \cdot x^- = 1 \Longleftrightarrow x^- \cdot y = 1$,
(qm-pEqP=) $(1 \cdot x = 1 \cdot y \text{ and } x \cdot 1 = y \cdot 1) \Longleftrightarrow y \cdot x^- = 1 \ (\overset{(IdEqP)}{\Longleftrightarrow} x^- \cdot y = 1)$,
(Ass1) $\quad x \cdot (1 \cdot y) = (x \cdot 1) \cdot y$.

Denote by **sharp qm-hub** the class of all sharp qm-hubs.

Note that the *sharp involutive qm-hubs* coincide with the *involutive sharp qm-hubs* (i.e. sharp qm-hubs verifyng (DN)). Denote by **sharp involutive qm-hub** the class of all sharp involutive qm-hubs.

• **In the "world" of qm-unital magmas with additional operation**

Definition 10.2.37 A *sharp qm-moon* is an algebra $(A, \cdot, ^-, 1)$ of type $(2, 1, 0)$ verifying: for all $x, y \in A$,
(qm-pU) $1 \cdot (x \cdot y) = x \cdot y = (x \cdot y) \cdot 1$,
(U1) $1 \cdot 1 = 1$,
(IdEqP) $y \cdot x^- = 1 \Longleftrightarrow x^- \cdot y = 1$,
(qm-pEqP=) $(1 \cdot x = 1 \cdot y$ and $x \cdot 1 = y \cdot 1) \Longleftrightarrow y \cdot x^- = 1$ $(\overset{(IdEqP)}{\Longleftrightarrow}$ $x^- \cdot y = 1)$,
(Ass1) $x \cdot (1 \cdot y) = (x \cdot 1) \cdot y$.

Proposition 10.2.38 *Let $\mathcal{A} = (A, \cdot, ^-, 1)$ be a sharp qm-moon. Then, $\mathcal{A}$ is a qm-moon.*

Proof. We must prove that (pIv) holds. Indeed, let $x \in A$; then, $1 \cdot x = 1 \cdot x$ and $x \cdot 1 = x \cdot 1$; hence, by (qm-pEqP=), $x \cdot x^- = 1$ $(\Leftrightarrow x^- \cdot x = 1)$; hence, (pIv) holds. □

Then, we obviously have that:

Corollary 10.2.39 *Any qm-moon verifying (IdEqP), (qm-pEqP=), (Ass1) is a sharp qm-moon.*

Definition 10.2.40 An involutive qm-moon is *sharp*, if (IdEqP), (qm-pEqP=), (Ass1) hold.

Note that the sharp involutive qm-moons coincide with the involutive sharp qm-moons (i.e. sharp qm-moons verifyng (qm-NegP), (DN)).

Denote by **sharp qm-moon** and by **sharp involutive qm-moon** the corresponding classes of algebras.

The connection between the involutive sharp qm-hubs and the involutive sharp qm-moons is the following:

Theorem 10.2.41

(1) Let $\mathcal{A} = (A, \cdot, 1)$ be a sharp involutive qm-hub, with the involutive quasi-m negation denoted by $^-$. Define $F(\mathcal{A}) = (A, \cdot, ^-, 1)$. Then, $F(\mathcal{A})$ is a sharp involutive qm-moon.

(1') Conversely, let $\mathcal{A} = (A, \cdot, ^-, 1)$ be a sharp involutive qm-moon. Define $G(\mathcal{A}) = (A, \cdot, 1)$. Then, $G(\mathcal{A})$ is a sharp involutive qm-hub and $^-$ is its involutive quasi-m negation.

(2) The mapps F and G are mutually inverse.

Proof. Obviously, by definitions. □

By Theorem 10.2.41, we have the equivalence:

(qmHM2) **sharp involutive qm-hub** $\Longleftrightarrow$ **sharp involutive qm-moon.**

We analyse now the connections between the (involutive) sharp qm-moons and the (involutive) moons. We have:

sharp qm-moon + (pU) $\equiv$ **goop**,

sharp involutive qm-moon + (pU) $\equiv$ **sharp involutive moon** $\equiv$ **goop**.

Note that there are examples of qm-hubs/qm-moons that are:
1 - neither involutive nor sharp,
2 - sharp, but not involutive,
3 - involutive, but not sharp,
4 - sharp and involutive,
as the following examples show.

Examples 10.2.42 The following examples of qm-hubs/qm-moons are pure.

(1) Consider the commutative proper qm-proper qm-hub (not super) with 4 elements:

$\mathcal{A}_{925} = (A_4, \odot_{925}, 1)$ with

$\odot_{925}$	a	b	c	**1**
a	b	1	a	a
b	1	b	1	b
c	a	1	a	b
1	a	b	b	**1**

,

where $\mathcal{R}m(\mathcal{A}_{925}) = (\{a, b, 1\}, \cdot, 1)$, with

$\cdot$	a	b	1	x^{-1}
a	b	1	a	**b**
b	1	b	b	**a**
1	a	b	1	**1**

, is a commutative sharp proper involutive hub.

By (qm-NegP), $a = b^- = (1 \odot_{925} c)^- = c^- \odot_{925} 1$, hence $c^- = a$; then, $(a, b, c, 1)^- = (\mathbf{b}, \mathbf{a}, a, \mathbf{1})$, hence the qm-negation exists, but (DN) is not satisfied. Note that the qm-negation does not verify (pIv): $c \odot_{925} c^- = c \odot_{925} a = a \neq 1$; hence, $(A_4, \odot_{925}, ^-, 1)$ is not a qm-moon.

If we take $(a, b, c, 1)^\sim = (\mathbf{b}, \mathbf{a}, b, \mathbf{1})$, then (pIv) is verified, hence $(A_4, \odot_{925}, ^\sim, 1)$ is a qm-moon. But $^\sim$ does not verify (qm-NegP): $a = b^\sim = (1 \odot_{925} c)^\sim \neq c^\sim \odot_{925} 1 = b \odot_{925} 1 = b$. Thus, the qm-moon $(A_4, \odot_{925}, ^\sim, 1)$ is not involutive.

Note also that $1 \odot_{925} b = 1 \odot_{925} c = b$, but $c \odot_{925} b^\sim = c \odot_{925} a = a \neq 1$, hence (qm-pEqP=) is not satisfied. The property (Ass1) is also not satisfied: for b, c, $1 = b \odot_{925} c = (b \odot_{925} 1) \odot_{925} c \neq b \odot_{925} (1 \odot_{925} c) = b \odot_{925} b = b$. Thus, the qm-moon $(A_4, \odot_{925}, ^\sim, 1)$ is not sharp.

Hence, the (not super) qm-moon $(A_4, \odot_{925}, ^\sim, 1)$ is **neither involutive, nor sharp**. Also, the (not super) qm-hub $(A_4, \odot_{925}, 1)$ is **neither involutive, nor sharp**.

(1') Consider the commutative proper qm-proper qm-hub (not super) with 4 elements:

$\mathcal{A}_{924} = (A_4, \odot_{924}, 1)$ with

$\odot_{924}$	a	b	c	**1**
a	b	1	a	a
b	1	b	1	b
c	a	1	a	a
1	a	b	a	**1**

,

where $\mathcal{R}m(\mathcal{A}_{924}) = (\{a, b, 1\}, \cdot, 1)$, with

$\cdot$	a	b	1	x^{-1}
a	b	1	a	**b**
b	1	b	b	**a**
1	a	b	1	**1**

, is a commutative sharp proper involutive hub.

By (qm-NegP), $b = a^- = (1 \odot_{924} c)^- = c^- \odot_{924} 1$, hence $c^- = b$; then, $(a, b, c, 1)^- = (\mathbf{b}, \mathbf{a}, b, \mathbf{1})$, hence the qm-negation exists, but (DN) is not satisfied. Note that the qm-negation verifies (pIv), hence $(A_4, \odot_{925}, ^-, 1)$ is a qm-moon, a non-involutive one.

Note that $1 \odot_{924} a = 1 \odot_{924} c = a$, and $a \odot_{924} c^- = a \odot_{924} b = 1$ and $c \odot_{924} a^- = c \odot_{924} b = 1$, hence (qm-pEqP=) is satisfied. But, (Ass1) is not satisfied: for a, c, $a = a \odot_{924} c = (a \odot_{924} 1) \odot_{924} c \neq a \odot_{924} (1 \odot_{924} c) = a \odot_{924} a = b$. Thus, the qm-moon $(A_4, \odot_{924}, ^-, 1)$ is not sharp.

Hence, the (not super) qm-moon $(A_4, \odot_{924}, ^-, 1)$ is **neither involutive, nor sharp**. Also, the (not super) qm-hub $(A_4, \odot_{924}, 1)$ is **neither involutive, nor sharp**.

(2) Consider the commutative proper qm-proper super qm-hub with 4 elements:

$\mathcal{A}_7 = (A_4, \odot_7, 1)$

$\odot_7$	a	b	c	**1**
a	a	1	a	a
b	1	a	1	b
c	a	1	a	a
1	a	b	a	**1**

,

where $\mathcal{R}m(\mathcal{A}_7) = (\{a, b, 1\}, \cdot, 1)$, with

$\cdot$	a	b	1	x^{-1}
a	a	1	a	**b**
b	1	a	b	**a**
1	a	b	1	**1**

, is a commutative sharp proper involutive hub.

By (qm-NegP), $b = a^- = (1 \odot_7 c)^- = c^- \odot_7 1$, hence $c^- = b$; then, $(a, b, c, 1)^- = (\mathbf{b}, \mathbf{a}, b, \mathbf{1})$, hence the qm-negation exists, but (DN) is not satisfied. Note that the qm-negation verifies (pIv), hence $(A_4, \odot_{925}, ^-, 1)$ is a super qm-moon that is not involutive.

Note that $1 \odot_7 c = 1 \odot_7 1 = 1$, and $c \odot_7 1^- = c \odot_7 1 = 1$ and $1 \odot_7 c^- = 1 \odot_7 c = 1$, hence (qm-pEqP=) is satisfied. (Ass1) is also satisfied, because (Ass2) implies (Ass1). Thus, $(A_4, \odot_7, ^-, 1)$ is sharp.

Hence, the super qm-moon $(A_4, \odot_7, ^-, 1)$ is **sharp, but not involutive**. Also, the super qm-hub $(A_4, \odot_7, 1)$ is **sharp, but not involutive**.

(3) Consider the non-commutative proper qm-proper super qm-hub with 4 elements:

$\mathcal{A}_{26} = (A_4, \odot_{26}, 1)$

$\odot_{26}$	a	b	c	**1**
a	1	a	a	a
b	1	1	b	b
c	a	b	1	1
1	a	b	1	**1**

,

where $\mathcal{R}m(\mathcal{A}_{26}) = (\{a, b, 1\}, \cdot, 1)$, with

$\cdot$	a	b	1	x^{-1}
a	1	a	a	**a**
b	1	1	b	**b**
1	a	b	1	**1**

, is a non-commutative involutive, non-sharp hub.

By (qm-NegP), $1 = 1^- = (1 \odot_{26} c)^- = c^- \odot_{26} 1 = 1 \odot_{26} c^- = (c \odot_{26} 1)^- = 1^- = 1$, hence $c^- = c, 1$; we take $c^- = c$; hence, the qm-negation exists and $(a, b, c, 1)^- = (\mathbf{a}, \mathbf{b}, c, \mathbf{1})$ verifies (DN). Note that the qm-negation verifies (pIv), hence $(A_4, \odot_{26}, ^-, 1)$ is a super qm-moon that is involutive.

Note that $b \odot_{26} a = 1$, i.e. $b \odot_{26} a^- = 1$, or $b^- \odot_{26} a = 1$, but $b = 1 \odot_{26} b \neq 1 \odot_{26} a = a$, hence (qm-pEqP=) is not satisfied. Note that (Ass1) is satisfied, because (Ass2) implies (Ass1). Thus, $(A_4, \odot_{26}, ^-, 1)$ is not sharp.

Hence, the super qm-moon $(A_4, \odot_{26}, ^-, 1)$ is **involutive, but not sharp**. Also, the super qm-hub $(A_4, \odot_{26}, 1)$ is **involutive, but not sharp**, $^-$ being its involutive quasi-m negation.

(4) Consider the commutative proper qm-proper super qm-hub with 4 elements:

$\mathcal{A}_{17} = (A_4, \odot_{17}, 1)$

$\odot_{17}$	a	b	c	**1**
a	b	1	a	a
b	1	b	b	b
c	a	b	1	1
1	a	b	1	**1**

,

where $\mathcal{R}m(\mathcal{A}_{17}) = (\{a, b, 1\}, \cdot, 1)$, with

$\cdot$	a	b	1	x^{-1}
a	b	1	a	**b**
b	1	b	b	**a**
1	a	b	1	**1**

, is a commutative sharp proper involutive hub.

By (qm-NegP), $1 = 1^- = (1 \odot_{17} c)^- = c^- \odot_{17} 1$, hence $c^- = c$; then, $(a, b, c, 1)^- = (\mathbf{b}, \mathbf{a}, c, \mathbf{1})$, hence the qm-negation exists and (DN) is satisfied. Note that the qm-negation verifies (pIv), hence $(A_4, \odot_{17}, ^-, 1)$ is a super qm-moon that is involutive.

Note that $1 \odot_{17} c = 1 \odot_{17} 1 = 1$, and $c \odot_{17} 1^- = c \odot_{17} 1 = 1$ and $1 \odot_{17} c^- = 1 \odot_{17} c = 1$, hence (qm-pEqP=) is satisfied. (Ass1) is also satisfied, because (Ass2) implies (Ass1). Thus, $(A_4, \odot_{17}, ^-, 1)$ is sharp.

Thus, the super qm-moon $(A_4, \odot_{17}, ^-, 1)$ is **sharp and involutive**. Also, the super qm-hub $(A_4, \odot_{17}, 1)$ is **sharp and involutive**, with $^-$ its involutive quasi-m negation.

Open problem 10.2.43 *Find examples of sharp involutive qm-hubs/qm-moons that are not super.*

We introduce now in sharp involutive qm-moons the new operations: $\to$, $\rightsquigarrow$.

Let $\mathcal{A} = (A, \cdot, ^-, 1)$ be a sharp involutive qm-moon. Then, $(Rm(A), \cdot, ^{-1}, 1)$ is a sharp involutive moon, hence a goop. We extend the new operations $\to$ and $\rightsquigarrow$ defined on $Rm(A)$ to A as follows: for all $x, y \in A$,

$$x \to y \stackrel{def.}{=} y \cdot x^-, \quad x \rightsquigarrow y \stackrel{def.}{=} x^- \cdot y. \tag{10.4}$$

The following result is useful in Chapter 12.

Proposition 10.2.44 *Let $\mathcal{A} = (A, \cdot, ^-, 1)$ be a sharp involutive qm-moon and consider the new operations $\to$ and $\rightsquigarrow$ defined by (10.4). Then,*

(1) the following properties are verified: for all $x, y \in A$,

(q-pM) $1 \to (x \rightsquigarrow y) = x \rightsquigarrow y$ *and* $1 \rightsquigarrow (x \to y) = x \to y$,

(pM1) $1 \to 1 = 1 = 1 \rightsquigarrow 1$,

(IdEq) $x \to y = 1 \Longleftrightarrow x \rightsquigarrow y = 1$,

(q-pEq=) $(1 \to x = 1 \to y$ *and* $1 \rightsquigarrow x = 1 \rightsquigarrow y) \Longleftrightarrow x \to y = 1 \stackrel{(IdEq)}{\Longleftrightarrow} x \rightsquigarrow y = 1$,

(q-pNeg) $x \to 1 = 1 \to x^- = (1 \rightsquigarrow x)^- = x \rightsquigarrow 1 = 1 \rightsquigarrow x^- = (1 \to x)^-$,

(DN) $(x^-)^- = x$,

(Id) $x \to 1 = x \rightsquigarrow 1$,

(pEx1) $x \to (y \rightsquigarrow 1) = y \rightsquigarrow (x \to 1)$,

(pDNeg2) $y^- \to x = x^- \rightsquigarrow y$,

(i.e. $(A, \to, \rightsquigarrow, ^-, 1)$ is a strong involutive q-implicative-moon with (pDNeg2));

(2) if, moreover, $\mathcal{A}$ is super, then we also have:

(pEx2) $y^- \to (x^- \rightsquigarrow 1) = y^- \to x$ *and* $x^- \rightsquigarrow (y^- \to 1) = x^- \rightsquigarrow y$

(i.e. $(A, \to, \rightsquigarrow, ^-, 1)$ is super too).

Proof. (1): We have (qm-pU), (U1), (pIv), (DN), (qm-NegP), (IdEqP), (qm-pEqP=), (Ass1) verified, by hypothesis. Hence, (N1) holds too, by Proposition 10.2.13.

(q-pM): $1 \to (x \rightsquigarrow y) \stackrel{def.}{=} 1 \to (x^- \cdot y) \stackrel{def.}{=} (x^- \cdot y) \cdot 1^- \stackrel{(N1)}{=} (x^- \cdot y) \cdot 1 \stackrel{(qm-pU)}{=} x^- \cdot y \stackrel{def.}{=} x \rightsquigarrow y$ and
$1 \rightsquigarrow (x \to y) \stackrel{def.}{=} 1 \rightsquigarrow (y \cdot x^-) \stackrel{def.}{=} 1^- \cdot (y \cdot x^-) \stackrel{(N1)}{=} 1 \cdot (y \cdot x^-) \stackrel{(qm-pU)}{=} y \cdot x^- \stackrel{def.}{=} x \to y$.

(pM1): $1 \to 1 \stackrel{def.}{=} 1 \cdot 1^- \stackrel{(N1)}{=} 1 \cdot 1 \stackrel{(U1)}{=} 1$ and $1 \rightsquigarrow 1 \stackrel{def.}{=} 1^- \cdot 1 \stackrel{(N1)}{=} 1 \cdot 1 \stackrel{(U1)}{=} 1$.

(IdEq): $x \to y = 1 \stackrel{def.}{\Leftrightarrow} y \cdot x^- = 1 \stackrel{(IdEqP)}{\Leftrightarrow} x^- \cdot y = 1 \stackrel{def.}{\Leftrightarrow} x \rightsquigarrow y = 1$.

(q-pEq=): Suppose that $(1 \to x = 1 \to y$ and $1 \rightsquigarrow x = 1 \rightsquigarrow y)$, i.e., by definition, $(x \cdot 1^- = y \cdot 1^-$ and $1^- \cdot x = 1^- \cdot y)$; then, by (N1), $(x \cdot 1 = y \cdot 1$ and $1 \cdot x = 1 \cdot y)$; then, by (q-pEq=P), $y \cdot x^- = 1$, i.e. $x \to y = 1$.

Suppose that $x \to y = 1$, i.e. $y \cdot x^- = 1$, by definition; then, by (q-pEq=P), $(1 \cdot x = 1 \cdot y$ and $x \cdot 1 = y \cdot 1)$; then, by (N1), $(1^- \cdot x = 1^- \cdot y$ and $x \cdot 1^- = y \cdot 1^-)$; then, by definition, $(1 \rightsquigarrow x = 1 \rightsquigarrow y$ and $1 \to x = 1 \to y)$. Thus, (q-pEq=) holds.

(q-pNeg): $x \to 1 \stackrel{def.}{=} 1 \cdot x^-$; $1 \to x^- \stackrel{def.}{=} x^- \cdot 1^- \stackrel{(N1)}{=} x^- \cdot 1$; $(1 \to x)^- \stackrel{def.}{=} (x \cdot 1^-)^- \stackrel{(N1)}{=} (x \cdot 1)^-$; $x \rightsquigarrow 1 \stackrel{def.}{=} x^- \cdot 1$; $1 \rightsquigarrow x^- \stackrel{def.}{=} 1^- \cdot x^- \stackrel{(N1)}{=} 1 \cdot x^-$; $(1 \rightsquigarrow x)^- \stackrel{def.}{=} (1^- \cdot x)^- \stackrel{(N1)}{=} (1 \cdot x)^-$. Then, apply (qm-NegP).

(DN): obviously, because $^-$ is the same.

(Id): $x \to 1 \stackrel{def.}{=} 1 \cdot x^- \stackrel{(qm-NegP)}{=} x^- \cdot 1 \stackrel{def.}{=} x \rightsquigarrow 1$.

(pEx1): $x \to (y \rightsquigarrow 1) \stackrel{def.}{=} x \to (y^- \cdot 1) \stackrel{def.}{=} (y^- \cdot 1) \cdot x^- \stackrel{(Ass1)}{=} y^- \cdot (1 \cdot x^-) \stackrel{def.}{=} y \rightsquigarrow (1 \cdot x^-) \stackrel{def.}{=} y \rightsquigarrow (x \to 1)$.

(pDNeg2): $y^- \to x \stackrel{def.}{=} x \cdot (y^-)^- \stackrel{(DN)}{=} x \cdot y$ and $x^- \rightsquigarrow y \stackrel{def.}{=} (x^-)^- \cdot y \stackrel{(DN)}{=} x \cdot y$.

(2): Suppose now that $\mathcal{A}$ is super, i.e. (Ass2) holds. Then,
$y^- \to (x^- \rightsquigarrow 1) \stackrel{def.}{=} y^- \to ((x^-)^- \cdot 1) \stackrel{(DN)}{=} y^- \to (x \cdot 1) \stackrel{def.}{=} (x \cdot 1) \cdot (y^-)^- \stackrel{(DN)}{=} (x \cdot 1) \cdot y \stackrel{(Ass2)}{=} x \cdot y \stackrel{(DN)}{=} x \cdot (y^-)^- \stackrel{def.}{=} y^- \to x$ and
$x^- \rightsquigarrow (y^- \to 1) \stackrel{def.}{=} x^- \rightsquigarrow (1 \cdot (y^-)^-) \stackrel{(DN)}{=} x^- \rightsquigarrow (1 \cdot y) \stackrel{def.}{=} (x^-)^- \cdot (1 \cdot y) \stackrel{(DN)}{=} x \cdot (1 \cdot y) \stackrel{(Ass2)}{=} x \cdot y \stackrel{(DN)}{=} (x^-)^- \cdot y \stackrel{def.}{=} x^- \rightsquigarrow y$. Thus, (pEx2) holds. □

The (super) qm-goop

Definitions 10.2.45

1. A *qm-goop* is an algebra $(A, \cdot, ^-, 1)$ of type $(2, 1, 0)$ such that the following properties hold: (qm-pU), (U1), (IdEqP), (qm-pEqP=), (qm-$\overline{pU(1 \cdot x, x \cdot 1)}$), (IdP), (DN), (Ass1).
2. A qm-goop is *super*, if (Ass2) holds.

Denote by **(super) qm-goop** the class of all qm-goops (super qm-goops, respectively).

We have:

(super) qm-goop + (pU) ≡ **goop**.

Proposition 10.2.46 *Let $(A, \cdot, ^-, 1)$ be a qm-goop. Then,*
(qm-ngo1) (U1) + (qm-pEqP=) + (qm-$\overline{pU(1 \cdot x, x \cdot 1)}$) $\Longrightarrow$ (N1),
(qm-ngo2) (IdP) + (qm-pU) + (qm-pEqP=) + (DN) + (IdEqP) + (qm-$\overline{pU(1 \cdot x, x \cdot 1)}$) $\Longrightarrow$ (qm-NegP),
i.e. (N1) and (qm-NegP) hold.

Proof. (qm-ngo1): Since $1 \cdot 1 = 1 \cdot 1$, then, by (qm-pEqP=), we obtain $1 \cdot 1^- = 1$; then, by (U1), $1 \cdot (1 \cdot 1)^- = 1$; then, by (qm-$\overline{pU(1 \cdot x, x \cdot 1)}$), $(1 \cdot 1)^- = 1$, i.e. $1^- = 1$, by (U1) again. Thus, (N1) holds.

(qm-ngo2): $1 \cdot x \stackrel{(IdP)}{=} x \cdot 1 \stackrel{(qm-pU)}{=} 1 \cdot (x \cdot 1)$; then, by (qm-pEqP=), $(x \cdot 1) \cdot x^- = 1$; then, by (DN), $((x \cdot 1)^-)^- \cdot x^- = 1$, hence $x^- \cdot ((x \cdot 1)^-)^- = 1$, by (IdEqP); then, by (qm-pEqP=) again, $1 \cdot (x \cdot 1)^- = 1 \cdot x^-$, hence $(x \cdot 1)^- = 1 \cdot x^-$, by (qm-$\overline{pU(1 \cdot x, x \cdot 1)}$); then apply (IdP). Thus, (qm-NegP) holds. □

Theorem 10.2.47 *The qm-goops are equivalent to the sharp involutive qm-moons.*

Proof. Let $\mathcal{A} = (A, \cdot, ^-, 1)$ be a qm-goop, i.e. (qm-pU), (U1), (IdEqP), (qm-pEqP=), (qm-$\overline{pU(1 \cdot x, x \cdot 1)}$), (IdP), (DN), (Ass1) hold. We must prove that $\mathcal{A}$ is a sharp involutive qm-moon, i.e. (qm-pU), (U1), (qm-NegP), (pIv), (DN), (IdEqP), (qm-pEqP=), (Ass1) hold. Hence, it remains to prove that (qm-NegP), (pIv) hold.

Indeed, (qm-NegP) follows by Proposition 10.2.46. (pIv) follows by (IdEqP), (qm-pEqP=) for $x = y$.

Conversely, let $\mathcal{A} = (A, \cdot, ^-, 1)$ be a sharp involutive qm-moon. We must prove that $\mathcal{A}$ is a qm-goop. Hence, it remains to prove that (qm-$pU(1 \cdot x, x \cdot 1)$), (IdP) hold. Indeed, apply Proposition 10.2.13. □

Hence, we have the hierarchy No. qm2 from Figure 10.9.

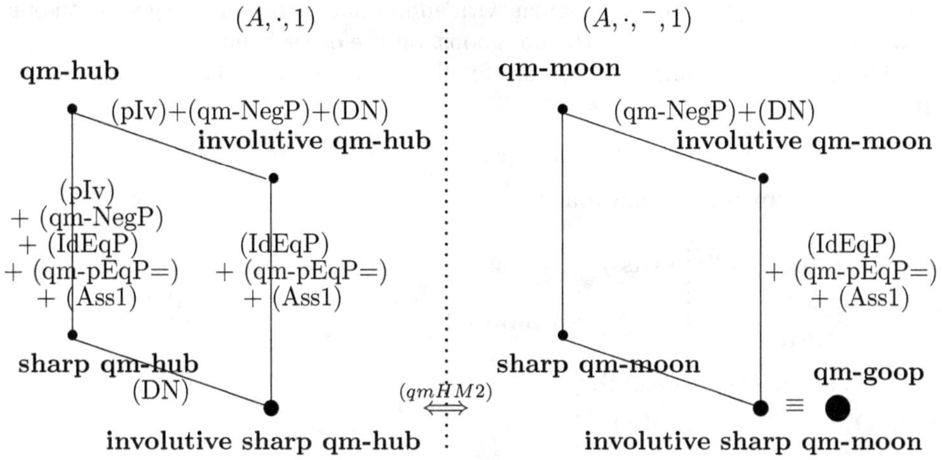

Figure 10.9: The hierarchy No. qm2

Example 10.2.48 The commutative super qm-moon $(A_4, \odot_{17}, ^-, 1)$ from Examples 10.2.42 (4) is a **super qm-goop**.

Theorem 10.2.49

(1) Let $\mathcal{G} = (G, \cdot, ^-, 1)$ be a qm-group. Then, $\mathcal{G}$ is a qm-goop verifying (Ass).

(1') Conversely, let $\mathcal{A} = (A, \cdot, ^-, 1)$ be a qm-goop verifying (Ass). Then, $\mathcal{A}$ is a qm-group.

Proof. (1): Let $\mathcal{G} = (G, \cdot, ^-, 1)$ be a qm-group, i.e. (qm-pU), (U1), (pIv), (Ass), (qm-$pU(1 \cdot x, x \cdot 1)$), (DN) hold. To prove that $\mathcal{G}$ is a qm-goop, it remains to prove that (IdEqP), (qm-pEqP=), (IdP) and (Ass1) hold. By Proposition 10.2.29, (IdP) holds. By Theorem 10.2.30, the qm-groups are equivalent to the associative involutive qm-moons; then, by Proposition 10.2.26, (IdEqP) and (qm-pEqP=) hold. Finally, (Ass) implies (Ass1).

(1'): Let $\mathcal{A} = (A, \cdot, ^-, 1)$ be a qm-goop verifying (Ass). To prove that $\mathcal{A}$ is a qm-group, it remains to prove that (pIv) holds. Indeed, (pIv) follows by (IdEqP), (qm-pEqP=) for $x = y$. □

By above theorem, we obtain the equivalences:

(qmGoGr) **(super) qm-goop** + (Ass) ≡ **qm-group**.

10.3 Final connections

By above considerations, there exist mainly the following generalizations of the groups (Definition 1') (the loops, recalled in Chapter 3, are omitted), in the "world" of qm-unital (unital) magmas: the monoids, the hubs, the unital magmas, on one hand, and the associative qm-hubs, the qm-monoids, the (super) qm-hubs, the qm-unital magmas, on the other hand.

We also have the following generalizations of the groups (Definition 1), in the "world" of qm-unital (unital) magmas with additional operation: the goops, on one hand, and the qm-groups and the qm-goops, on the other hand.

These generalizations of groups are finally connected as in the following Figure 10.10.

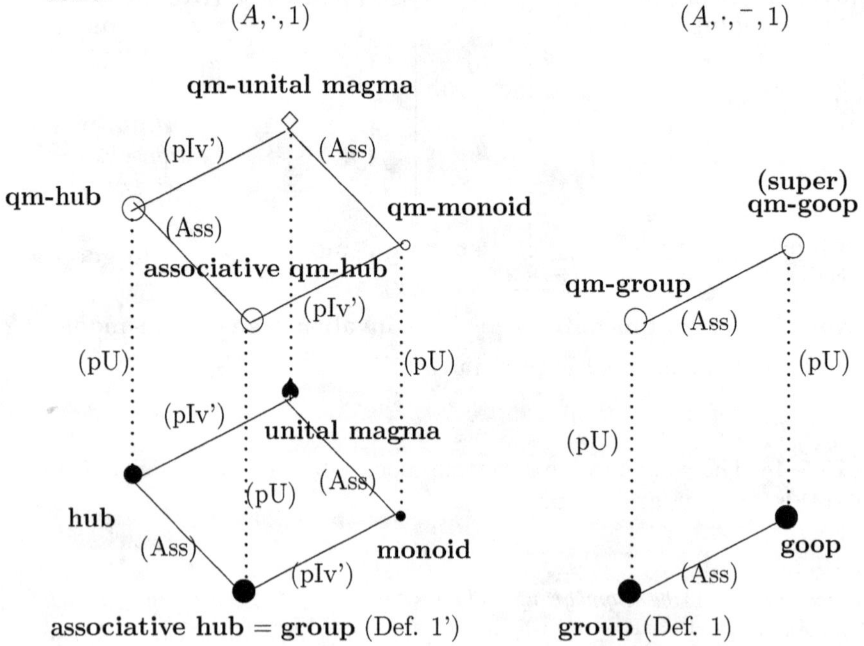

Figure 10.10: Generalizations of groups

Chapter 11

Generalizations of implicative-groups

In this chapter, we introduce and study the analogous new notions of those from Chapter 10, namely:
- the implicative-hub, the implicative-moon and the implicative-goop, and we show their connections with the definitions 5, 5' of the implicative-group (in Section 1),
- the q-implicative-hub, the q-implicative-moon, the q-implicative-group and the q-implicative-goop, and we show the connections between all these generalizations of the implicative-goups (in Section 3).

We also introduce the (strong) quasi-pseudo-*aRM** algebras (in Section 2). Final connections of some generalizations of the implicative-groups are also presented.

The chapter has 4 sections.

11.1 I-hubs, i-moons. I-groups, i-goops

In this section, we introduce and study the new notions of *implicative-hub*, of involutive implicative-hub, of exchanged (involutive) i-hub and of strong (involutive) implicative-hub - all in the "world" of pseudo-M algebras. In parallel, we also introduce and study the new notions of *implicative-moon*, of involutive implicative-moon, of exchanged (involutive) i-moon and of strong (involutive) implicative-moon - all in the "world" of pseudo-M algebras with additional unary operation. We finaly introduce and study the new notion of *implicative-goop*. The connections between all these generalizations of the implicative-groups are presented.

11.1.1 The (involutive) i-hub, i-moon

- **In the "world" of pM algebras**

Definition 11.1.1 An *implicative-hub*, or an *i-hub* for short, is an algebra $\mathcal{A} = (A, \to, \rightsquigarrow, 1)$ of type $(2,2,0)$ verifying: for all $x, y \in A$,
(pM) $1 \to x = x = 1 \rightsquigarrow x$,
(IdEq) $x \to y = 1 \Leftrightarrow x \rightsquigarrow y = 1$,
(pEq=) $x = y \Longleftrightarrow x \to y = 1 \; (\overset{(IdEq)}{\Longleftrightarrow} x \rightsquigarrow y = 1)$.

Denote by **i-hub** the class of all i-hubs.

An i-hub is *commutative*, if $\to = \rightsquigarrow$. In this case, the property (IdEq) is superfluous, the property (pM) becomes (M) ($1 \to x = x$, for all x) and the property (pEq=) becomes (Eq=) ($x = y \Leftrightarrow x \to y = 1$, for all x, y).

General properties

Let $(A, \to = \to^L, \rightsquigarrow = \rightsquigarrow^L, 1)$ be a left-pseudo-algebra (i.e. verifying (IdEq)), as defined in Chapter 2 by Definition 2.1.1 (the dual case is omitted).

We can then define on A an *internal* (or *natural*) binary relation $\leq$ by: for all $x, y \in A$,

$$x \leq y \overset{def.}{\Longleftrightarrow} x \to y = 1 \overset{(IdEq)}{\Longleftrightarrow} x \rightsquigarrow y = 1. \tag{11.1}$$

Equivalently, let $(A, \leq, \to = \to^L, \rightsquigarrow = \rightsquigarrow^L, 1)$ be a left-pseudo-structure, as defined in Chapter 2 by Definition 2.1.1 (the dual case is omitted), where $\leq$ and $\to$, $\rightsquigarrow$, 1 are connected by:
(pEq^L) $x \leq y \Longleftrightarrow x \to y = 1 \overset{(IdEq)}{\Leftrightarrow} x \rightsquigarrow y = 1$.

Consider the following properties:
(Id) $x \to 1 = x \rightsquigarrow 1$,
(pEx) $x \to (y \rightsquigarrow z) = y \rightsquigarrow (x \to z)$,
(pRe') $x \to x = 1 = x \rightsquigarrow x$,
(p-s^L) (p-semisimple) $x \leq y \Longrightarrow x = y$ (in fact, $x \leq y \Longleftrightarrow x = y$),
(pBB=) $y \to z = (z \to x) \rightsquigarrow (y \to x)$, $y \rightsquigarrow z = (z \rightsquigarrow x) \to (y \rightsquigarrow x)$,
(pB=) $z \to x = (y \to z) \to (y \to x)$, $z \rightsquigarrow x = (y \rightsquigarrow z) \rightsquigarrow (y \rightsquigarrow x)$,
(pD=) $(y \to x) \rightsquigarrow x = y = (y \rightsquigarrow x) \to x$,
(An) $(x \leq y$ and $y \leq x) \Longrightarrow x = y$,
(Tr) $(x \leq y$ and $y \leq z) \Longrightarrow x \leq z$,
(p*^{L}) $x \leq y \Longrightarrow (z \to x \leq z \to y$ and $z \rightsquigarrow x \leq z \rightsquigarrow y)$,
(p*^{L}') $x \to y = 1 \Longrightarrow ((z \to x) \to (z \to y) = 1$ and $(z \rightsquigarrow x) \to (z \rightsquigarrow y) = 1)$,
(p**^{L}) $x \leq y \Longrightarrow (y \to z \leq x \to z$ and $y \rightsquigarrow z \leq x \rightsquigarrow z)$,
(p**^{L}') $x \to y = 1 \Longrightarrow ((y \to z) \to (x \to z) = 1$ and $(y \rightsquigarrow z) \to (x \rightsquigarrow z) = 1)$.

Proposition 11.1.2 *Let $\mathcal{A} = (A, \to, \rightsquigarrow, 1)$ be an i-hub. Then, the properties (pRe'), (p-s^L), (Tr), (An), (p*^{L}), (p**^{L}) hold.*

Proof. By Proposition 8.1.2. □

Remarks 11.1.3

(1) The i-hub is a special left-pseudo-algebra, namely a left-pseudo-algebra (of logic) $(A, \to = \to^L, \rightsquigarrow = \rightsquigarrow^L, 1)$ for which the binary relation $\leq$ defined by (11.1) becomes the equality relation $=$ (by (pEq=) and (pEq^L)), and so the property (p-s^L) holds.

(2) The i-hub can be defined, equivalently, as a special left-pseudo-structure: $(A, \leq, \to = \to^L, \rightsquigarrow = \rightsquigarrow^L, 1)$, where (pEq^L) becomes (pEq=); therefore, as left-pseudo-structure, the i-hub $(A, =, \to, \rightsquigarrow, 1)$ is denoted also by $(A, \to, \rightsquigarrow, 1)$.

By above Proposition and above Remarks, we immediately obtain the following important result (the dual one is omitted).

Proposition 11.1.4

(1) Any i-hub is a left-pM algebra verifying (p-s^L).

*(2) Any i-hub is a left-p*aRM** algebra verifying (p-s^L).*

The i-hubs have a special property:

Proposition 11.1.5

(1) Let $\mathcal{A} = (A, \to, \rightsquigarrow, 1)$ be an i-hub. Then, $(A, \to, 1)$ and $(A, \rightsquigarrow, 1)$ are commutative i-hubs.

(1') Conversely, let $(A, \to, 1)$ and $(A, \rightsquigarrow, 1)$ be any two commutative i-hubs verifying (IdEq). Then, $\mathcal{A} = (A, \to, \rightsquigarrow, 1)$ is an i-hub.

Proof. Obviously. □

Note that any i-group (Definition 1') is an i-hub and that any i-hub verifying (pBB=) is an i-group. Hence, we have:

i-hub + (pBB=) = **i-group**.

Moreover, the following important result holds for i-hubs.

Theorem 11.1.6 *(See Theorem 2.1.9)*

Let $\mathcal{A} = (A, \to, \rightsquigarrow, 1)$ be an i-hub. Then,

$$(pEx) \iff (pBB =) \iff (pB =).$$

Proof.

(pEx) $\Longrightarrow$ (pBB=) $\Leftrightarrow$ (pB=): By Proposition 8.1.2 (7"), (pEx) + (pEq=) $\Longrightarrow$ (pBB=) $\Leftrightarrow$ (pB=); thus, the implication holds.

(pBB=) $\Longrightarrow$ (pEx): By Proposition 8.1.2 (2), (pM) + (pBB=) $\Longrightarrow$ (pD=), and by Proposition 8.1.2 (5'), (pBB=) + (pRe') + (pD=) + (pEq=) $\Longrightarrow$ (pEx); thus, the implication holds and this completes the proof of the theorem. □

Consequently, we obtain the following equivalences:

(IHIG) **i-group** = **i-hub** + (pBB=) $\equiv$ **i-hub** + (pEx) $\equiv$ **i-hub** + (pB=),
where $\equiv$ is the notation for the equivalence when the signatures coincide.

We shall say that an i-hub is *proper*, if it is not an i-group, i.e. if (pEx) is not verified.

The involutive i-hub

Let $\mathcal{A} = (A, \to, \rightsquigarrow, 1)$ be an i-hub. We can define two negations: $^-$ and $^\sim$, as follows: for all $x \in A$, $x^- \stackrel{def.}{=} x \to 1$, $x^\sim \stackrel{def.}{=} x \rightsquigarrow 1$. Then, in general, the two negations are different. But, there are i-hubs $\mathcal{A}$ such that, for all $x \in A$, $x^- = x^\sim$, i.e. (Id) holds ($x \to 1 = x \rightsquigarrow 1$, for all $x \in A$); one can define then a unique negation $^{-1}$: for all $x \in A$, $x^{-1} \stackrel{def.}{=} x^- \stackrel{(Id)}{=} x^\sim$. Hence, the negation $^{-1}$ verifies: for all $x \in A$,

(pNeg) $\quad x^{-1} = x \to 1 \stackrel{(Id)}{=} x \rightsquigarrow 1$.

If this unique negation is *involutive* (i.e. verifies (DN): $(x^{-1})^{-1} = x$, for all $x \in A$), then we say that $\mathcal{A}$ is *involutive* and that $^{-1}$ is its involutive regular negation.

Remark 11.1.7 The involutive i-hub, with its involutive regular negation, is a regular left-pseudo-algebra with involutive regular negation, by Remark 5.1.8.

The involutive i-hubs also have a special property:

Proposition 11.1.8

(1) Let $\mathcal{A} = (A, \to, \rightsquigarrow, 1)$ be an involutive i-hub and $^{-1}$ be its involutive regular negation. Then, $(A, \to, 1)$ and $(A, \rightsquigarrow, 1)$ are involutive commutative i-hubs and $^{-1}$ is their involutive regular negation.

(1') Conversely, let $(A, \to, 1)$, with $x^- = x \to 1$, for all $x \in A$, and $(A, \rightsquigarrow, 1)$, with $x^\sim = x \rightsquigarrow 1$, for all $x \in A$, be two involutive commutative i-hubs. If (IdEq) holds and $x^- = x^\sim$, for all $x \in A$, then, $\mathcal{A} = (A, \to, \rightsquigarrow, 1)$ is an involutive i-hub and $x^{-1} \stackrel{def.}{=} x^- = x^\sim$ is its involutive regular negation.

Proof. Obviously. □

Examples 11.1.9

(1) $i\mathcal{H}_1 = (\{a, b, 1\}, \to, 1)$, with

$\to$	a	b	1
a	1	a	a
b	a	1	a
1	a	b	**1**

, is a **commutative i-hub that is not involutive**, since, $(a, b, 1)^{-1} = (a, a, 1)$ does not satisfy (DN).

(2) $i\mathcal{H}_2 = (\{a, b, 1\}, \to, \rightsquigarrow, 1)$, with

$\to$	a	b	1
a	1	a	a
b	b	1	b
1	a	b	**1**

,

$\rightsquigarrow$	a	b	1
a	1	a	b
b	b	1	b
1	a	b	**1**

, is a **non-commutative i-hub that is not involutive**, because (Id) does not hold.

(3) $i\mathcal{H}_3 = (\{a, b, 1\}, \rightarrow, 1)$, with

$\rightarrow$	a	b	1
a	1	a	b
b	a	1	a
1	a	b	1

, is a **commutative involutive i-hub**, with $(a, b, 1)^{-1} = (b, a, 1)$.

• **In the "world" of pM algebras with additional unary operation**

Definitions 11.1.10

1. An *implicative-moon*, or an *i-moon* for short, is an algebra $\mathcal{A} = (A, \rightarrow, \rightsquigarrow, ^{-1}, 1)$ of type $(2, 2, 1, 0)$ verifying: for all $x, y \in A$,
(pM) $1 \rightarrow x = x = 1 \rightsquigarrow x$,
(IdEq) $x \rightarrow y = 1 \Leftrightarrow x \rightsquigarrow y = 1$,
(pEq=) $x = y \Longleftrightarrow x \rightarrow y = 1 \ (\overset{(IdEq)}{\Longleftrightarrow} x \rightsquigarrow y = 1)$,
(Id) $x \rightarrow 1 = x \rightsquigarrow 1$,
(pNeg) $x^{-1} = x \rightarrow 1 \ (\overset{(Id)}{=} x \rightsquigarrow 1)$.

2. An i-moon is *involutive*, if (DN) ($(x^{-1})^{-1} = x$) holds.

Denote by **(involutive) i-moon** the class of all (involutive, respectively) i-moons.

An i-moon is *commutative*, if $\rightarrow = \rightsquigarrow$.

Note that an i-moon is a pM algebra with an additional unary operation, $^{-1}$.

Note also that, if $\mathcal{A} = (A, \rightarrow, \rightsquigarrow, ^{-1}, 1)$ is an i-moon, then $(A, \rightarrow, \rightsquigarrow, 1)$ is an i-hub. The converse is not true in general (see $i\mathcal{H}_2$ from Examples 11.1.9). But, in commutative case, since (Id) always holds, we have:

commutative i-hub $\Longleftrightarrow$ commutative i-moon.

Corollary 11.1.11 *Let $\mathcal{A} = (A, \rightarrow, \rightsquigarrow, ^{-1}, 1)$ be an i-moon. Then,*

$$(pEx) \Longleftrightarrow (pBB=) \Longleftrightarrow (pB=).$$

Proof. By Theorem 11.1.6. □

Remarks 11.1.12

(i) Note that (see Remark 5.1.6) the identity: $x^{-1} = x \rightarrow 1 \ (\overset{(Id)}{=} x \rightsquigarrow 1)$ can be used:
- either as the *definition* of the negation $^{-1}$, in the involutive implicative-hub $(A, \rightarrow, \rightsquigarrow, 1)$,
- or as the *connection* (pNeg) between the negation $^{-1}$ and $\rightarrow, \rightsquigarrow$, 1, in the (involutive) implicative-moon $(A, \rightarrow, \rightsquigarrow, ^{-1}, 1)$.

(ii) The involutive i-moon $(A, \rightarrow, \rightsquigarrow, ^{-1}, 1)$ is a regular left-pseudo-algebra with involutive regular negation, by Remark 5.1.8.

The following theorem establishes the categorically equivalence between the involutive i-hubs and the involutive i-moons.

Theorem 11.1.13

(1) Let $\mathcal{A} = (A, \rightarrow, \rightsquigarrow, 1)$ be an involutive i-hub and $^{-1}$ be its involutive regular negation. Define $F(\mathcal{A}) = (A, \rightarrow, \rightsquigarrow, ^{-1}, 1)$.

Then, $F(\mathcal{A})$ is an involutive i-moon.

(1') Conversely, let $(A, \rightarrow, \rightsquigarrow, ^{-1}, 1)$ be an involutive i-moon. Define $G(\mathcal{A}) = (A, \rightarrow, \rightsquigarrow, 1)$.

Then, $G(\mathcal{A})$ is an involutive i-hub and $^{-1}$ is its involutive regular negation.

(2) The mapps F and G are mutually inverse.

Proof. Obviously. □

Hence, we have the equivalence:

(IHM0) **involutive i-hub** $\Longleftrightarrow$ **involutive i-moon.**

Remark 11.1.14 In the commutative case, (Id) is always verified, so the i-hubs and the i-moons are also categorically equivalent.

Examples 11.1.15

(1) $i\mathcal{M}_1 = (\{a, b, 1\}, \rightarrow, ^{-1}, 1)$, with

→	a	b	1
a	1	a	a
b	a	1	a
1	a	b	1

and $(a, b, 1)^{-1} = (a, a, 1)$, is a **commutative i-moon that is not involutive**, since (DN) is not verified. Note that $(\{a, b, 1\}, \rightarrow, 1)$ is the i-hub $i\mathcal{H}_1$ from Examples 11.1.9.

(2) $i\mathcal{M}_2 = (\{a, b, 1\}, \rightarrow, ^{-1}, 1)$, with

→	a	b	1
a	1	a	b
b	a	1	a
1	a	b	1

and $(a, b, 1)^{-1} = (b, a, 1)$, is a **commutative involutive i-moon.** Note that $(\{a, b, 1\}, \rightarrow, 1)$ is the i-hub $i\mathcal{H}_3$ from Examples 11.1.9.

(3) $i\mathcal{M}_3 = (\{a, b, 1\}, \rightarrow, \rightsquigarrow, ^{-1}, 1)$, with

→	a	b	1
a	1	a	a
b	a	1	a
1	a	b	1

,

⇝	a	b	1
a	1	a	a
b	b	1	a
1	a	b	1

and $(a, b, 1)^{-1} = (a, a, 1)$, is a **non-commutative i-moon that is not involutive,** because (DN) does not hold.

(4) $i\mathcal{M}_4 = (\{a, b, 1\}, \rightarrow, \rightsquigarrow, ^{-1}, 1)$, with

→	a	b	1
a	1	a	b
b	a	1	a
1	a	b	1

,

⇝	a	b	1
a	1	b	b
b	a	1	a
1	a	b	1

and $(a, b, 1)^{-1} = (b, a, 1)$, is an **involutive non-commutative i-moon.**

11.1.2 The exchanged (involutive) i-hubs, i-moons vs. the i-group

We shall analyse here in some details the two definitions 5 and 5' of the i-groups (presented in Chapter 8) and their connections with the i-hubs and the i-moons, in order to better understand the generalizations of the i-groups from this chapter.

- **In the "world" of pM algebras**

Definition 11.1.16 An *exchanged i-hub* is an i-hub verifying (pEx).

Denote by **exchanged i-hub** the class of all exchanged i-hubs.

Note that the exchanged involutive i-hubs coincide with the involutive exchanged i-hubs. Denote by **exchanged involutive i-hub** the class of all exchanged involutive i-hubs.

Remark 11.1.17 Note that the *exchanged i-hub* is just the *i-group* (Definition 4).

Hence, we have:

(IHG) **i-hub** + (pEx) = **exchanged i-hub** = **i-group** (Definition 4).

We shall say that an i-hub is *proper*, if it is not an i-group (Def. 4), i.e. if (pEx) is not verified.

Proposition 11.1.18 *Let* $\mathcal{A} = (A, \to, \rightsquigarrow, 1)$ *be an exchanged i-hub. Then, (Id) and (DN) hold.*

Proof. By Proposition 8.1.6, (Id) holds. By Proposition 8.1.17, (DN) holds. □

By above Proposition, any exchanged i-hub is involutive, hence we have the equivalence:

(IH1) **i-group** (Def. 5) = **exchanged i-hub** ≡ **exchanged involutive i-hub.**

Example 11.1.19 The algebra $\mathcal{G} = (\{a, b, 1\}, \to, 1)$, with

$\to$	a	b	1
a	1	a	b
b	b	1	a
1	a	b	**1**

, is a **commutative exchanged (involutive) i-hub** (with $(a, b, 1)^{-1} = (b, a, 1)$), **namely is the only i-group with 3 elements (Definition 5).**

- **In the "world" of pM algebras with additional unary operation**

Definition 11.1.20 An *exchanged i-moon* is an i-moon verifying (pEx).

Denote by **exchanged i-moon** the class of all exchanged i-moons.

Note that the exchanged involutive i-moons coincide with the involutive exchanged i-moons. Denote by **exchanged involutive i-moon** the class of all exchanged involutive i-moons.

Remark 11.1.21 Note that the *exchanged i-moon* is just the *i-group* (Definition 3').

Hence, we have:

(IMG) **i-moon** + (pEx) = **exchanged i-moon** = **i-group** (Definition 3').

We shall say that an i-moon is *proper*, if it is not an i-group (Definition 3'), i.e. if (pEx) is not verified.

Proposition 11.1.22 *Let* $\mathcal{A} = (A, \rightarrow, \rightsquigarrow, ^{-1}, 1)$ *be an exchanged i-moon. Then, (DN) holds.*

Proof. Note that $(A, \rightarrow, \rightsquigarrow, 1)$ is an i-group; hence, by Proposition 8.1.17, (DN) holds. □

By above Proposition, any exchanged i-moon is involutive, hence we have the equivalence:

(IM1) **i-group** = **exchanged i-moon** $\equiv$ **exchanged involutive i-moon.**

The connectionn between the exchanged involutive i-hubs and the exchanged involutive i-moons is the following:

Theorem 11.1.23

(1) Let $\mathcal{A} = (A, \rightarrow, \rightsquigarrow, 1)$ *be an exchanged involutive i-hub, with* $^{-1}$ *the involutive regular negation. Define* $F(\mathcal{A}) = (A, \rightarrow, \rightsquigarrow, ^{-1}, 1)$*. Then,* $F(\mathcal{A})$ *is an exchanged involutive i-moon.*

(1') Conversely, let $\mathcal{A} = (A, \rightarrow, \rightsquigarrow, ^{-1}, 1)$ *be an exchanged involutive i-moon. Define* $G(\mathcal{A}) = (A, \rightarrow, \rightsquigarrow, 1)$*. Then,* $G(\mathcal{A})$ *is an exchanged involutive i-hub and its involutive regular negation is* $^{-1}$*.*

(2) The mapps F and G are mutually inverse.

Proof. By Theorem 11.1.13. □

By Theorem 11.1.23, we have the equivalence:

(IHM1) **exchanged involutive i-hub** $\Longleftrightarrow$ **exchanged involutive i-moon.**

By equivalences (IH1), (IHM1) and (IM1), we have the equivalences:

exchanged i-hub $\equiv$ **exchanged involutive i-hub**
$\Longleftrightarrow$ **exchanged involutive i-moon** $\equiv$ **exchanged i-moon,**
hence we obtain the equivalence:

i-group (Df. 5)=**exchanged i-hub** $\Longleftrightarrow$ **exchanged i-moon**=**i-group** (Df. 5'),

which is the base for the two definitions 5 and 5' of the i-groups.

Example 11.1.24 The algebra $\mathcal{G} = (\{a, b, 1\}, \rightarrow, ^{-1}, 1)$, with

$\rightarrow$	a	b	1
a	1	a	b
b	b	1	a
1	a	b	1

and $(a, b, 1)^{-1} = (b, a, 1)$, is a **commutative exchanged (involutive) i-moon, namely is the only i-group with 3 elements (Definition 3').**

Remark 11.1.25 The Definition 4 of i-groups, as exchanged i-hubs, is very useful when looking for examples obtained by programs (because the negation is missing) (see the examples in Chapters 18, 21).

Note that we have the hierarchy No. i-1, hence the hierarchy No. i-1', from Figures 11.1, 11.2, respectively.

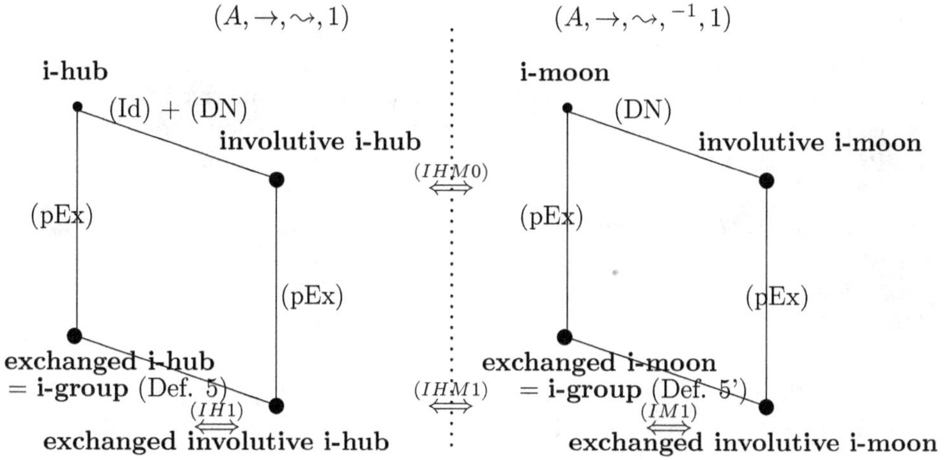

Figure 11.1: The hierarchy No. i-1

11.1.3 The strong (involutive) i-hub, i-moon vs. the i-goop

- **In the "world" of pM algebras**

Consider the following property (pEx1), weaker than the pseudo-exchange property (pEx) verified by i-groups:
(pEx1) $\quad x \rightarrow (y \rightsquigarrow 1) = y \rightsquigarrow (x \rightarrow 1)$, for all $x, y \in A$.

Definition 11.1.26 (Definition 1 of strong i-hubs)
An i-hub $\mathcal{A} = (A, \rightarrow, \rightsquigarrow, 1)$ is *strong*, if the properties (Id) and (pEx1) are verified.

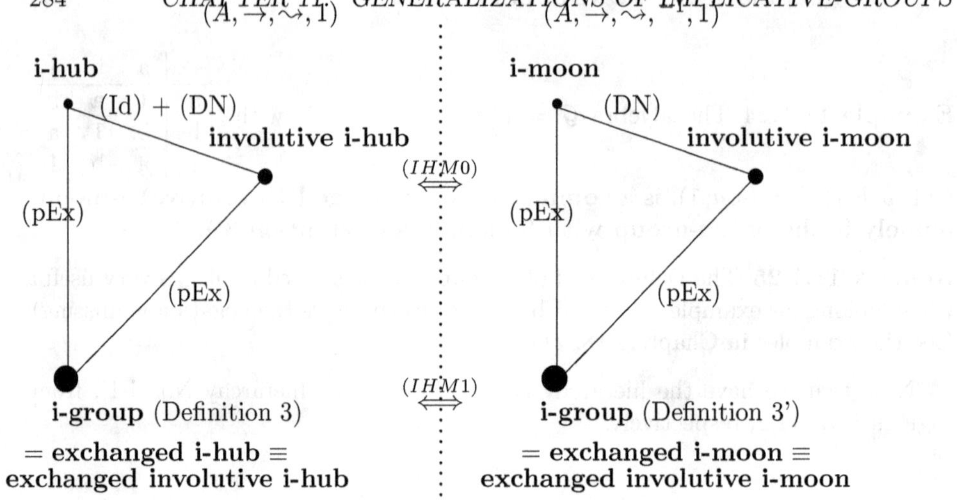

Figure 11.2: The hierarchy No. i-1'

Denote by **strong i-hub** the class of all strong i-hubs.

Note that the strong involutive i-hubs coincide with the involutive strong i-hubs. Denote by **strong involutive i-hub** the class of all strong involutive i-hubs.

Note also that in the commutative case, the property (Id) always holds.

By Proposition 11.1.4, we obtain immediately that:

Proposition 11.1.27 *Any strong i-hub is a strong left-p*aRM** algebra verifying (p-s^L).*

One involutive negation: $^{-1}$

Let $\mathcal{A} = (A, \to, \rightsquigarrow 1)$ be a strong i-hub. Define a negation $^{-1} : A \longrightarrow A$ as follows: for all $x \in A$, $x^{-1} \overset{def.}{=} x \to 1 \overset{(Id)}{=} x \rightsquigarrow 1$.

Proposition 11.1.28 *Let $\mathcal{A} = (A, \to, \rightsquigarrow, 1)$ be a strong i-hub. Then, we have: for all $x, y \in A$,*

(N1) $1^{-1} = 1$,

(pD1=) $x = (x \to 1) \rightsquigarrow 1, \quad x = (x \rightsquigarrow 1) \to 1$,

(DN) $(x^{-1})^{-1} = x$,

(q-pNeg) $x \to 1 = 1 \to x^- = (1 \rightsquigarrow x)^- \overset{(Id)}{=} x \rightsquigarrow 1 = 1 \rightsquigarrow x^- = (1 \to x)^-$,

(pEx2) $y^{-1} \to (x^{-1} \rightsquigarrow 1) = y^{-1} \to x$, $x^{-1} \rightsquigarrow (y^{-1} \to 1) = x^{-1} \rightsquigarrow y$,

(pNg1) $x \to y^{-1} = y \rightsquigarrow x^{-1}$,

(pDNeg2) $y^{-1} \to x = x^{-1} \rightsquigarrow y$.

Proof.

(N1): $1^{-1} = 1 \to 1 \overset{(pM)}{=} 1$.

(pD1=): $x \to [(x \to 1) \rightsquigarrow 1] \overset{(pEx1)}{=} (x \to 1) \rightsquigarrow (x \to 1) \overset{(pRe')}{=} 1$, hence $x = (x \to 1) \rightsquigarrow 1)$, by (pEq=) and

$x \rightsquigarrow [(x \rightsquigarrow 1) \to 1] \stackrel{(pEx1)}{=} (x \rightsquigarrow 1) \to (x \rightsquigarrow 1) \stackrel{(pRe')}{=} 1$, hence $x = (x \rightsquigarrow 1) \to 1)$, by (pEq=).

(DN): $(x^{-1})^{-1} \stackrel{def.}{=} (x \to 1) \rightsquigarrow 1 \stackrel{(pD1=)}{=} x$.

(q-pNeg): $x \to 1 = x^{-1}$, $1 \to x^{-1} = x^{-1}$, by (pM), and $(1 \to x)^{-1} = x^{-1}$, by (pM); $x \rightsquigarrow 1 = x^{-1}$, $1 \rightsquigarrow x^{-1} = x^{-1}$, by (pM), and $(1 \rightsquigarrow x)^{-1} = x^{-1}$, by (pM).

(pEx2): $x^{-1} \rightsquigarrow 1 = (x^{-1})^{-1} = x$, by (DN), and $y^{-1} \to 1 = (y^{-1})^{-1} = y$, by (DN).

(pNg1): $x \to y^{-1} \stackrel{def.}{=} x \to (y \rightsquigarrow 1) \stackrel{(pEx1)}{=} y \rightsquigarrow (x \to 1) \stackrel{def.}{=} y \rightsquigarrow x^{-1}$.

(pDNeg2): $y^{-1} \to x \stackrel{(DN)}{=} y^{-1} \to (x^{-1})^{-1} \stackrel{(pNg1)}{=} x^{-1} \rightsquigarrow (y^{-1})^{-1} = x^{-1} \rightsquigarrow y$. □

By above Proposition, we immediately obtain that:

Corollary 11.1.29 *Any strong i-hub is involutive.*

By above corollary, we obtain the equivalence:

(IH2) **strong i-hub ≡ strong involutive i-hub.**

Examples 11.1.30

(1) Consider the **involutive commutative i-hub** with 4 elements:
$\mathcal{A}_4 = (A_4 = \{a, b, c, 1\}, \to = \to_{577}, 1)$ with

$\to_{577}$	a	b	c	1
a	1	a	b	c
b	a	1	b	b
c	a	a	1	a
1	a	b	c	1

with

x	$x^{-1} = x \to_{577} 1$
a	c
b	b
c	a
1	1

It is **not strong**, because (pEx1) is not verified for $x = a$, $y = b$:
$a = a \to b = a \to (b \to 1) \neq b \to (a \to 1) = b \to c = b$.

(2) Consider the **strong commutative (involutive) i-hub** with 3 elements:

$\mathcal{A}_3^1 = (A_3 = \{a, b, 1\}, \to, 1)$ with

$\to$	a	b	1
a	1	a	a
b	a	1	b
1	a	b	1

(3) Consider the following **(non-commutative) i-hub**:

$\mathcal{A}_3^2 = (A_3 = \{a, b, 1\}, \to, \rightsquigarrow, 1)$ with

$\to$	a	b	1
a	1	a	a
b	b	1	b
1	a	b	1

$\rightsquigarrow$	a	b	1
a	1	b	a
b	a	1	b
1	a	b	1

It is **involutive**, with $(a, b, 1)^{-1} = (a, b, 1)$, and it is **strong**.

Proposition 11.1.31 *Let $\mathcal{A} = (A, \to, \rightsquigarrow, 1)$ be an i-hub. The following are equivalent:*
(a) $\mathcal{A}$ is strong,
(b) the properties (Id), (DN), (pDNeg2) hold, where $x^{-1} \stackrel{def.}{=} x \to 1 \stackrel{(Id)}{=} x \rightsquigarrow 1$, i.e. $\mathcal{A}$ is involutive with (pDNeg2).

Proof. (a) $\Longrightarrow$ (b): by Proposition 11.1.28.

(b) $\Longrightarrow$ (a): we must prove that (pEx1) holds; indeed,

$x \to (y \rightsquigarrow 1) \stackrel{(DN)}{=} (x^{-1})^{-1} \to y^{-1} \stackrel{(pDNeg2)}{=} (y^{-1})^{-1} \rightsquigarrow x^{-1} \stackrel{(DN)}{=} y \rightsquigarrow (x \to 1)$. □

By Proposition 11.1.31, we obviously can give another equivalent definition of strong i-hubs, as follows.

Definition 11.1.32 (Definition 2 of strong i-hubs)

An i-hub $\mathcal{A} = (A, \to, \rightsquigarrow, 1)$ is *strong*, if the properties (Id), (DN) and (pDNeg2) hold, where $x^{-1} \stackrel{def.}{=} x \to 1 \stackrel{(Id)}{=} x \rightsquigarrow 1$.

We obviously have the following result.

Corollary 11.1.33 *Let $\mathcal{A} = (A, \to, \rightsquigarrow, 1)$ be an involutive i-hub. Then,*

$$(pEx1) \Longleftrightarrow (pDNeg2).$$

- **In the "world" of pM algebras with additional unary operation**

Definition 11.1.34 (Definition 1 of strong i-moons)

An i-moon is *strong*, if the property (pEx1) is verified.

Denote by **strong i-moon** the class of all strong i-moons.

Note that strong involutive i-moons coincide with the involutive strong i-moons. Denote by **strong involutive i-moon** the class of all strong involutive i-moons.

The implicative-goop

Definition 11.1.35 An *implicative-goop*, or an *i-goop* for short, is a strong i-moon.

Denote by **i-goop** the class of all i-goops.

Remark 11.1.36 Since (pNeg), (Id) and (DN) hold, it follows that the i-goop is a regular pseudo-algebra with involutive regular negation, by Remark 5.1.8.

Corollary 11.1.37 *Any strong i-moon (= i-goop) is involutive.*

Proof. Let $\mathcal{A} = (A, \to, \rightsquigarrow, ^{-1}, 1)$ be a strong i-moon. Then, $(A, \to, \rightsquigarrow, 1)$ is a strong i-hub. Then, by Proposition 11.1.28, (DN) holds., hence $\mathcal{A}$ is involutive. □

By above Corollary, we obtain the equivalence:

(IM2) **i-goop** = **strong i-moon** $\equiv$ **strong involutive i-moon**.

By (IHM0), we obtain the equivalence:

(IHM2) **strong involutive i-hub** $\Longleftrightarrow$ **strong involutive i-moon**.

Theorem 11.1.38 *Let $\mathcal{A} = (A, \to, \rightsquigarrow, ^{-1}, 1)$ be a strong involutive i-moon. Then,*

$$(pEx1) \Longleftrightarrow (pDNeg2).$$

Proof. $\Longrightarrow$: First we prove:

$$(pNeg) + (pEx1) \Longrightarrow (pNg1).$$

Indeed, $x \to y^{-1} \overset{(pNeg)}{=} x \to (y \rightsquigarrow 1) \overset{(pEx1)}{=} y \rightsquigarrow (x \to 1) = y \rightsquigarrow x^{-1}$.

Now we prove:

$$(pNg1) + (DN) \Longrightarrow (pDNeg2).$$

Indeed, $y^{-1} \to x \overset{(DN)}{=} y^{-1} \to (x^{-1})^{-1} \overset{(pNg1)}{=} x^{-1} \rightsquigarrow (y^{-1})^{-1} \overset{(DN)}{=} x^{-1} \rightsquigarrow y$.

$\Longleftarrow$: First we prove:

$$(DN) + (pDNeg2) \Longrightarrow (pNg1).$$

Indeed, $x \to y^{-1} \overset{(DN)}{=} (x^{-1})^{-1} \to y^{-1} \overset{(pDNeg2)}{=} (y^{-1})^{-1} \rightsquigarrow x^{-1} \overset{(DN)}{=} y \rightsquigarrow x^{-1}$.

Now we prove:

$$(pNg1) \Longrightarrow (pEx1).$$

Indeed, $x \to (y \rightsquigarrow 1) = x \to y^{-1} \overset{(pNg1)}{=} y \rightsquigarrow x^{-1} = y \rightsquigarrow (x \to 1)$. □

By (IH2), (IHM2) and (IM2), we obtain the equivalences:

strong i-hub $\equiv$ strong involutive i-hub
$\Longleftrightarrow$ strong involutive i-moon $\equiv$ strong i-moon $=$ i-goop.

Remark 11.1.39 Just as the i-group (Definition 3') is defined equivalently as an exchanged i-hub (Definition 4), the same we could define the i-goop, equivalently, by above equivalences, as a strong involutive i-hub (or even as a strong i-hub). In fact, we find examples of i-goops by finding examples of strong (involutive) i-hubs.

Note that we have the hierarchy No. i-2, hence the hierarchy No. i-2', from Figures 11.3, 11.4, respectively.

Examples 11.1.40 (See Examples 11.1.30)

(1) $\mathcal{A}_4 = (A_4 = \{a, b, c, 1\}, \to = \to_{577}, ^{-1}, 1)$, with

$\to_{577}$	a	b	c	1
a	1	a	b	c
b	a	1	b	b
c	a	a	1	a
1	a	b	c	1

and $(a, b, c, 1)^{-1} = (c, b, a, 1)$,

is an **involutive commutative i-moon that is not strong.**

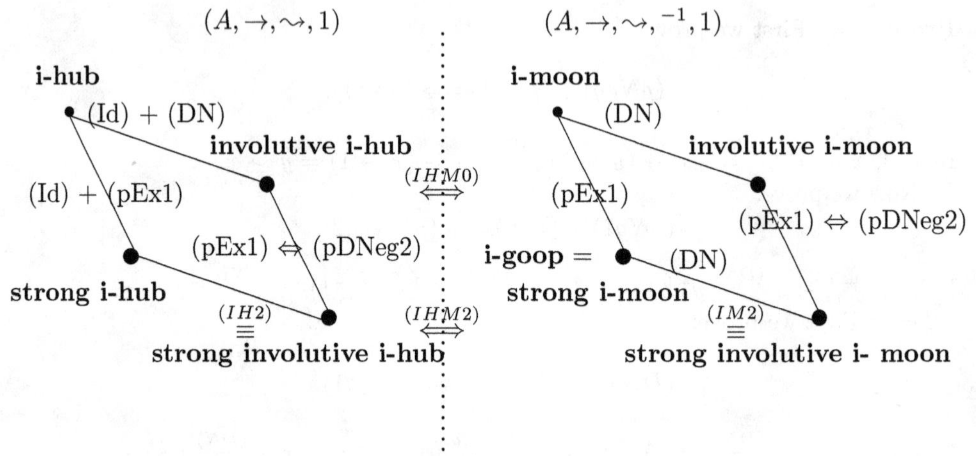

Figure 11.3: The hierarchy No. i-2

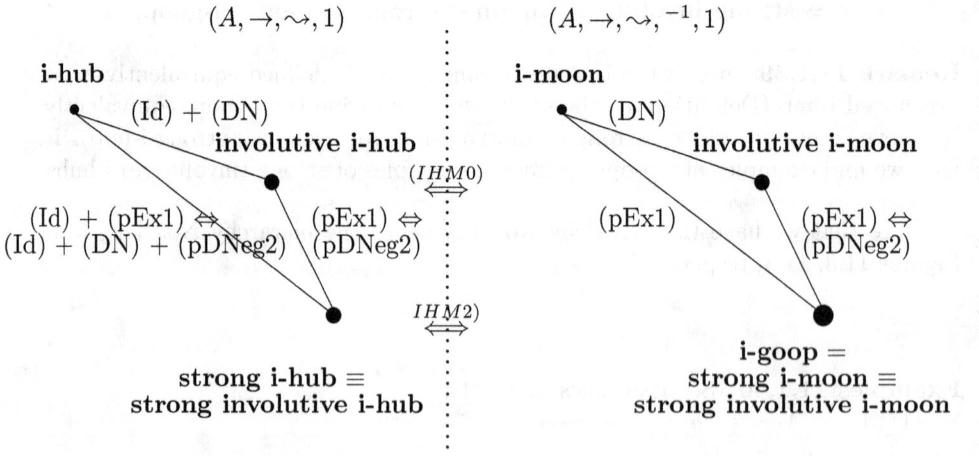

Figure 11.4: The hierarchy No. i-2'

(2) $\mathcal{A}_3^1 = (A_3 = \{a, b, 1\}, \rightarrow, ^{-1}, 1)$, with

$\rightarrow$	a	b	1
a	1	a	a
b	a	1	b
1	a	b	**1**

and $(a, b, 1)^{-1} = (a, b, 1)$, is a **commutative i-goop**.

(3) $\mathcal{A}_3^2 = (A_3 = \{a, b, 1\}, \rightarrow, \rightsquigarrow, ^{-1}, 1)$, with

$\rightarrow$	a	b	1
a	1	a	a
b	b	1	b
1	a	b	**1**

$\rightsquigarrow$	a	b	1
a	1	b	a
b	a	1	b
1	a	b	**1**

and $(a, b, 1)^{-1} = (a, b, 1)$,

is an **(non-commutative) i-goop**.

Proposition 11.1.41 *Let $\mathcal{G} = (G, \rightarrow, \rightsquigarrow, 1)$ be an i-group. Then, $\mathcal{G}$ is a strong i-hub.*

Proof. By Proposition 8.1.6, (Id) and (pEx) hold, hence (pEx1) holds. □

Note that by above Proposition 11.1.41, any i-group (Definition 3) is a strong i-hub, hence is a strong involutive i-hub and, conversely, any strong involutive i-hub verifying (pEx) is an i-group. Hence, we have the equivalence:

strong involutive i-hub + (pEx) $\equiv$ **i-group** (Definition 3).

Note also that any i-group (Definition 3') is an i-goop, by Proposition 8.1.6, and an i-goop verifying (pEx) is an i-group. Hence, we have the equivalence:

(IGoGr) **i-goop** + (pEx) $\equiv$ **i-group** (Definition 3').

By hierarchies No. i-1' and 1-2', we finally obtain the resuming hierarchy No. i-3 from Figure 11.5.

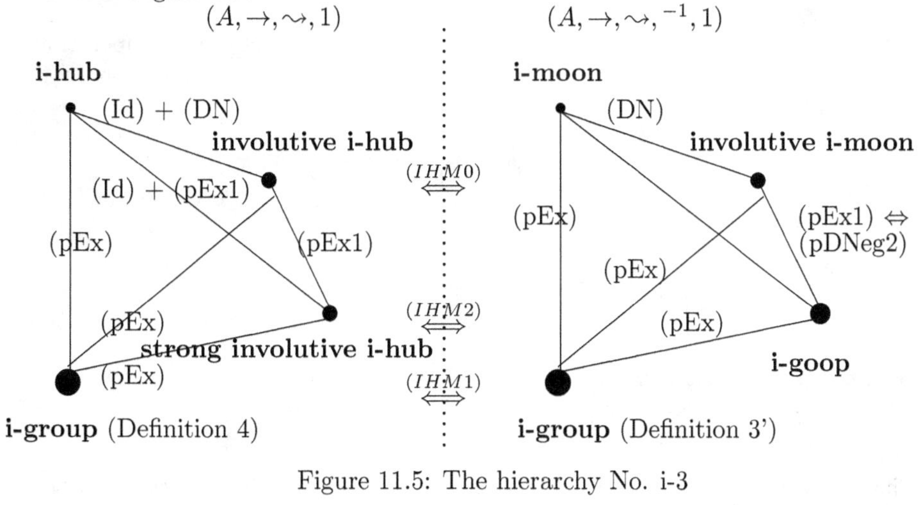

Figure 11.5: The hierarchy No. i-3

Remark 11.1.42 (See Remark 10.1.32)
We could name the strong involutive i-hub and the i-goop as the "non-exchanged i-groups".

Consider now the following new property: for all y, z,
(pBB1=) $\quad y \to z = (z \to 1) \rightsquigarrow (y \to 1), \quad y \rightsquigarrow z = (z \rightsquigarrow 1) \to (y \rightsquigarrow 1)$.

Then, we have:

Proposition 11.1.43 *(See Proposition 2.1.18)*
Let $\mathcal{A} = (A, \to, \rightsquigarrow, 1)$ be an algebra of type $(2,2,0)$ verifying (IdEq) (i.e. a left-pseudo-algebra). If (pD1=) holds, then we have:

$$(pEx1) \iff (pBB1=).$$

Proof. (pEx1) $\Longrightarrow$ (pBB1=):
$(z \to 1) \rightsquigarrow (y \to 1) \overset{(pEx1)}{=} y \to ((z \to 1) \rightsquigarrow 1) \overset{(pD1=)}{=} y \to z$ and
$(z \rightsquigarrow 1) \to (y \rightsquigarrow 1) \overset{(pEx1)}{=} y \rightsquigarrow ((z \rightsquigarrow 1) \to 1) \overset{(pD1=)}{=} y \rightsquigarrow z$.
(pBB1=) $\Longrightarrow$ (pEx1):
$x \to (y \rightsquigarrow 1) \overset{(pBB1=)}{=} ((y \rightsquigarrow 1) \to 1) \rightsquigarrow (x \to 1) \overset{(pD1=)}{=} y \rightsquigarrow (x \to 1)$. □

Corollary 11.1.44 *(See Corollary 2.1.19)*
Let $\mathcal{A} = (A, \to, \rightsquigarrow, ^{-1}, 1)$ be an i-goop. Then, we have:

$$(pEx1) \iff (pBB1=).$$

Proof. By Proposition 11.1.28, (pD1=) holds. Then, apply Proposition 11.1.43. □

- **New operation in i-goops:** $\cdot$

Let $\mathcal{A} = (A, \to, \rightsquigarrow, ^{-1}, 1)$ be an i-goop. We shall extend to $\mathcal{A}$ the new operation $\cdot$ defined on an i-group by (8.5). Thus, we define the new operation $\cdot$ on A as follows: for all $x, y \in A$,

$$y \cdot x \overset{def.}{=} y^{-1} \rightsquigarrow x, \quad x \cdot y \overset{def.}{=} y^{-1} \to x,$$

i.e.

$$x \cdot y \overset{def.}{=} y^{-1} \to x \overset{(pDNeg2)}{=} x^{-1} \rightsquigarrow y. \tag{11.2}$$

Consider the following properties: for all $x, y \in A$,
(pU) $\quad x \cdot 1 = x = 1 \cdot x$,
(IdEqP) $\quad y \cdot x^{-1} = 1 \iff x^{-1} \cdot y = 1$,
(pEqP=) $\quad x = y \iff y \cdot x^{-1} = 1 \overset{(IdEqP)}{\iff} x^{-1} \cdot y = 1$.
The following result will be usefull to prove the term-equivalence between goops and i-goops, in Chapter 12.

Proposition 11.1.45 *(See proposition 8.1.32 for i-groups)*
Let $(A, \rightarrow, \rightsquigarrow, ^{-1}, 1)$ be an i-goop and let $\cdot$ be the new operation defined on A by (11.2). Then,
(i) (N1) + (pM) $\Longrightarrow$ (pU),
(ii) (IdEq) $\Longrightarrow$ (IdEqP),
(iii) (pEq=) $\Longrightarrow$ (pEqP=),
hence (pU), (IdEqP), (pEqP=) hold.

Proof.
(i): $x \cdot 1 \stackrel{(11.2)}{=} 1^{-1} \rightarrow x \stackrel{(N1)}{=} 1 \rightarrow x \stackrel{(pM)}{=} x$
and $1 \cdot x = 1^{-1} \rightsquigarrow x = 1 \rightsquigarrow x = x$. Thus, (pU) holds.
(ii) By (IdEq), $x \rightarrow y = 1 \Longleftrightarrow x \rightsquigarrow y = 1$; but,
$y \cdot x^{-1} = 1 \stackrel{(11.2)}{\Longleftrightarrow} (x^{-1})^{-1} \rightarrow y = 1 \stackrel{(DN)}{\Longleftrightarrow} x \rightarrow y = 1$ and
$x^{-1} \cdot y = 1 \stackrel{(11.2)}{\Longleftrightarrow} (x^{-1})^{-1} \rightsquigarrow y = 1 \stackrel{(DN)}{\Longleftrightarrow} x \rightsquigarrow y = 1$. So, (IdEqP) holds.
(iii): By (pEq=), we have: $x = y \Longleftrightarrow x \rightarrow y = 1 \stackrel{(IdEq)}{\Longleftrightarrow} x \rightsquigarrow y = 1$; by above (ii), (IdEq) implies (IdEqP); so, (pEqP=) holds. □

11.2 (Strong) quasi-p*aRM** algebras

In this section, we introduce and study the quasi-p*aRM** algebras and the strong quasi-p*aRM** algebras. We also establish the connections existing between the quasi-pBCI algebras (presented in Chapter 5) and the (strong) quasi-p*aRM** algebras.

11.2.1 The quasi-p*aRM** algebra

We introduce the following definitions (the dual case is omitted):

Definitions 11.2.1 (See Definitions 2.1.7)
Let $\mathcal{A}^L = (A^L, \leq, \rightarrow^L, \rightsquigarrow^L, 1)$ be a left-pseudo-structure, where $\leq$ is a binary relation, $\rightarrow^L$ and $\rightsquigarrow^L$ are binary operations on A^L and $1 \in A^L$, all connected by (pEq^L). We say that $\mathcal{A}^L$ is:
- a *left-quasi-pseudo-aRM algebra*, or a *left-q-paRM algebra* for short, if the properties (IdEq^L), (pEq^L), (q-pAn^L), (Re), (q-pM^L) hold,
- a *left-quasi-pseudo-oRM algebra*, or a *left-q-poRM algebra* for short, if it is a left-q-paRM algebra verifying (Tr) ("o" comes from "ordered"),
- a *left-quasi-pseudo-*aRM algebra*, or a *left-q-p*aRM algebra* for short, if it is a left-q-paRM algebra verifying (p*^{L}),
- a *left-quasi-pseudo-aRM** algebra*, or a *left-q-paRM** algebra* for short, if it is a left-q-paRM algebra verifying (p**^{L}),
- a *left-quasi-pseudo-*aRM** algebra*, or a *left-q-p*aRM** algebra* for short, if it is a left-q-paRM algebra verifying (p*^{L}), (p**^{L}).

We are more interested in this monograph by the left-q-p*aRM** algebras. Note that a left-q-p*aRM** algebra $\mathcal{A}^L$ is a left-quasi-pseudo-structure, by above definition, since (Re) implies (pM1).

We can define equivalently the left-quasi-p*aRM** algebra as a left-quasi-pseudo-algebra (the dual case is omitted):

Definition 11.2.2 (Definition 1' of left-q-p*aRM** algebras)
A *left-quasi-pseudo-*aRM** algebra*, or a *left-q-p*aRM** algebra* for short, is an algebra $\mathcal{A}^L = (A^L, \to^L, \leadsto^L, 1)$ of type 2, 2, 0) verifying: for all $x, y, z \in A^L$,

(q-pAn^{L}') (quasi-Antisymmetry) $(x \to^L y = 1$ and $y \to^L x = 1) \Longrightarrow$ $(1 \to^L x = 1 \to^L y$ and $1 \leadsto^L x = 1 \leadsto^L y)$

(pRe^L) $x \to^L x = 1 (\Leftrightarrow x \leadsto^L x = 1)$ (Reflexivity),

(q-pM^L) $1 \to^L (x \leadsto^L y) = x \leadsto^L y), \ 1 \leadsto^L (x \to^L y) = x \to^L y$,

(p*^{L}') $x \to^L y = 1 \Longrightarrow$ $((z \to^L x) \to^L (z \to^L y) = 1$ and $(z \leadsto^L x) \leadsto^L (z \leadsto^L y) = 1)$,

(p**^{L}') $x \to^L y = 1 \Longrightarrow$ $((y \to^L z) \to^L (x \to^L z) = 1$ and $(y \leadsto^L z) \leadsto^L (x \leadsto^L z) = 1)$,

(IdEq^L) $x \to^L y = 1 \Longleftrightarrow x \leadsto^L y = 1$,

where $x \leq y \overset{def.}{\Longleftrightarrow} x \to^L y = 1 \overset{(IdEq^L)}{\Longleftrightarrow} x \leadsto^L y = 1$.

A left-q-p*aRM** algebra is *commutative*, if $\to^L = \leadsto^L$ (dually, if $\to^R = \leadsto^R$). A commutative left-q-p*aRM** algebra is a left-q-*aRM** algebra.

Denote by **q-p*aRM****^{L} the class of all left-q-p*aRM** algebras. Dually, denote by **q-p*aRM****^{R} the class of all right-q-p*aRM** algebras.

The connection between the q-p*aRM** and the p*aRM** algebras

Any left-p*aRM** algebra is a left-q-p*aRM** algebra and, conversely, every left-q-p*aRM** algebra verifying (pM^L) is a left-p*aRM** algebra. Hence, we can write, by (Q), the special equivalences:

q-p*aRM**^{L} + (pM^L) $\Longleftrightarrow$ **p*aRM****^{L} and, dually,
q-p*aRM**^{R} + (pM^R) $\Longleftrightarrow$ **p*aRM****^{R}.

Following the general definition from Chapter 5, a left-q-p*aRM** algebra is:
- *q-proper*, if it is not a left-p*aRM** algebra, i.e. if the property (pM^L) is not verified;
- *proper*, if the properties (pEx^L), (pB^{L}') and (L) are not verified.

Theorem 11.2.3 *Let $\mathcal{A}^L = (A^L, \leq, \to^L, \leadsto^L, 1)$ be a q-proper left-q-p*aRM** algebra and consider the regular set $R(A^L) = \{x \in A^L \mid x = 1 \to^L x = 1 \leadsto^L x\}$. Then,*

$$\mathcal{R}(\mathcal{A}^L) = (R(A^L), \leq, \to^L, \leadsto^L, 1)$$

*is a left-p*aRM** algebra.*

Proof. By Theorem 5.1.13. □

A proper q-proper left-q-p*aRM** algebra is:
- *pure*, if its regular part is a proper left-p*aRM** algebra,
- of *implicative-hub type*, if its regular part is a proper implicative-hub,

- of *implicative-group type*, if its regular part is an implicative-group,
- of *implicative-Boole type*, if its regular part is the implicative-Boolean algebra with 2 elements.

Let $\mathcal{A}^L = (A^L, \leq, \rightarrow^L, \rightsquigarrow^L, 1)$ be a left-q-p*aRM** algebra. Consider the following properties: for all $x, y \in A^L$,
(Id^L) $\quad x \rightarrow^L 1 = x \rightsquigarrow^L 1$,
(pEx1^L) $\quad x \rightarrow^L (y \rightsquigarrow^L 1) = y \rightsquigarrow^L (x \rightarrow^L 1)$.

11.2.2 The strong quasi-p*aRM** algebra

Definition 11.2.4 A left-q-p*aRM** algebra $\mathcal{A}^L = (A^L, \leq, \rightarrow^L, \rightsquigarrow^L, 1)$ is *strong*, if the properties (Id^L) and (pEx1^L) are verified.

Denote by **strong q-p*aRM**^{**L}** the class of all strong left-q-p*aRM** algebras. Dually, denote by **strong q-p*aRM**^{**R}** the class of all strong right-q-p*aRM** algebras.

Note that, if $\mathcal{A}^L = (A^L, \leq, \rightarrow^L, \rightsquigarrow^L, 1)$ is a strong left-q-p*aRM** algebra, then $\mathcal{R}(\mathcal{A}^L) = (R(A^L), \leq, \rightarrow^L, \rightsquigarrow^L, 1)$ is a strong left-p*aRM** algebra.

11.2.3 Connections between quasi-pBCI algebras and (strong) quasi-p*aRM** algebras

Proposition 11.2.5
*(1) Any left-q-pBCI algebra is a (strong) left-q-p*aRM** algebra.*
*(2) Conversely, any (strong) left-q-p*aRM** algebra verifying (pEx^L) and (pB^L) is a left-q-pBCI algebra.*

Proof. (1): We must prove that (p*^{L}) and (p**^{L}) hold; this follows by Proposition 5.1.49. Note that any left-q-pBCI algebra verifies (pEx^L), hence verifies (pEx1^L).
(2): We must prove that (pBB^L) and (pD^L) hold. Indeed,
- by Proposition 2.1.8 (6), (pEx^L) + (pEq^L) + (pB^L) imply (pBB^L) and
- by Proposition 2.1.8 (7), (Re) + (pEx^L) + (pEq^L) imply (pD^L). □

By above Proposition and by previous results, we obtain the hierarchies from the following Figure 11.6.

11.3 (Super) q-i-hubs, q-i-moons. Q-i-groups, q-i-goops

In this section, we introduce and study the new notions of *qm-i-hubs* and *super qm-i-hubs* as related generalizations of the i-hubs; we also introduce and study the involutive q-i-hubs, the exchanged (involutive) q-i-hubs and the strong (involutive) q-i-hubs - all in the "world" of quasi-pseudo-M algebras. In parallel, we also introduce and study the new notions of *q-i-moons* and *super q-i-moons* as related

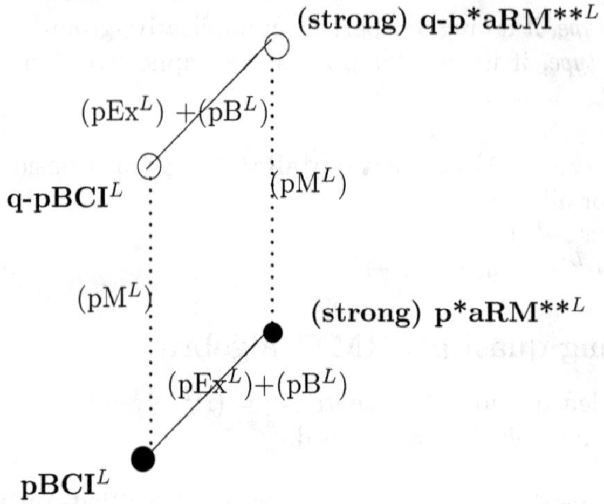

Figure 11.6: Connections between left-(quasi-)pBCI algebras and (strong) left-(quasi-)p*aRM** algebras

generalizations of the i-moons; we introduce and study the involutive q-i-moons, the exchanged (involutive) q-i-moons and the q-i-groups, the strong (involutive) q-i-moons and the q-i-goops - all in the "world" of quasi-pseudo-M algebras with additional unary operation. Connections between these generalizations of the i-groups are presented.

11.3.1 The (involutive) (super) q-i-hub, q-i-moon

- **In the "world" of q-pM algebras**

We introduce and study the *q-i-hubs* and the *super q-i-hubs* as two generalizations of the implicative-hubs as quasi-pseudo-algebras, fact which was obvious to us by the connection existing between the implicative-hubs and the pseudo-*aRM** algebras revealed in Chapter 13. The notions of *(super) quasi-implicative-hub* are the analogous, in the "world" of quasi-pM algebras (of logic), of the notions of *(super) quasi-m hub*, in the "worl"' of qm-unital magmas.

Definitions 11.3.1

1. A *quasi-implicative-hub*, or a *q-implicative-hub* or even a *q-i-hub* for short, is an algebra $\mathcal{A} = (A, \rightarrow, \rightsquigarrow, 1)$ of type $(2, 2, 0)$ verifying: for all $x, y \in A$,
(q-pM) $1 \rightarrow (x \rightsquigarrow y) = x \rightsquigarrow y, \quad 1 \rightsquigarrow (x \rightarrow y) = x \rightarrow y$,
(pM1) $1 \rightarrow 1 = 1 = 1 \rightsquigarrow 1$,
(IdEq) $x \rightarrow y = 1 \Leftrightarrow x \rightsquigarrow y = 1$,
(q-pEq=) $(1 \rightarrow x = 1 \rightarrow y$ and $1 \rightsquigarrow x = 1 \rightsquigarrow y) \Longleftrightarrow x \rightarrow y = 1\ (\overset{(IdEq)}{\Longleftrightarrow}\ x \rightsquigarrow y = 1)$.

2. We say that the q-i-hub is *super*, or is a *q-i-hub$^{**}_{*}$*, if the following additional properties are verified: for all $x, y, z \in A$,

(p*') $x \to y = 1$ implies $((z \to x) \to (z \to y) = 1$ and $(z \rightsquigarrow x) \to (z \rightsquigarrow y) = 1)$,
(p**') $x \to y = 1$ implies $((y \to z) \to (x \to z) = 1$ and $(y \rightsquigarrow z) \to (x \rightsquigarrow z) = 1)$.

Denote by **(super) q-i-hub** the class of all (super) q-i-hubs.
A (super) q-i-hub is *commutative*, if $\to = \rightsquigarrow$.

General properties

Let $(A, \to = \to^L, \rightsquigarrow = \rightsquigarrow^L, 1)$ be a left-pseudo-algebra, as defined in Chapter 2 by Definition 2.1.1 (the dual case is omitted), where:
(IdEq) $x \to y = 1 \Longleftrightarrow x \rightsquigarrow y = 1$.
We can then define on A an *internal* (or *natural*) binary relation $\leq$ by:

$$x \leq y \overset{def.}{\Longleftrightarrow} x \to y = 1 \, (\overset{(IdEq)}{\Longleftrightarrow} x \rightsquigarrow y = 1). \tag{11.3}$$

Equivalently, let $(A, \leq, \to = \to^L, \rightsquigarrow = \rightsquigarrow^L, 1)$ be a left-pseudo-structure, as defined in Chapter 2 by Definition 2.1.1 (the dual case is omitted), where $\leq$ and $\to$, $\rightsquigarrow$, 1 are connected by:
(pEq^L) $x \leq y \Longleftrightarrow x \to y = 1 \, (\overset{(IdEq)}{\Leftrightarrow} x \rightsquigarrow y = 1)$.

Consider the following properties: for all $x, y, z \in A$,
(p*^{L}) $x \leq y \Longrightarrow (z \to x \leq z \to y, \; z \rightsquigarrow x \leq z \rightsquigarrow y)$,
(p**^{L}) $x \leq y \Longrightarrow (y \to z \leq x \to z, \; y \rightsquigarrow z \leq x \rightsquigarrow z)$,
(pRe') $x \to x = 1 = x \rightsquigarrow x$,
(Tr) (Transitivity) $x \leq y$ and $y \leq z \Longrightarrow x \leq z$,
(q-pR1) $(1 \rightsquigarrow x) \to x = 1, \; (1 \to x) \rightsquigarrow x = 1$,
(pC^L) $x \to (y \rightsquigarrow z) \leq y \rightsquigarrow (x \to z), \; x \rightsquigarrow (y \to z) \leq y \to (x \rightsquigarrow z)$,
(pEx) $x \to (y \rightsquigarrow 1) = y \rightsquigarrow (x \to 1)$,
(pEx1) $x \to (y \rightsquigarrow z) = y \rightsquigarrow (x \to z)$;
(pBB^L) $y \to z \leq (z \to x) \rightsquigarrow (y \to x), \; y \rightsquigarrow z \leq (z \rightsquigarrow x) \to (y \rightsquigarrow x)$,
(pBB=) $y \to z = (z \to x) \rightsquigarrow (y \to x), \quad y \rightsquigarrow z = (z \rightsquigarrow x) \to (y \rightsquigarrow x)$,
(pB^L) $z \to x \leq (y \to z) \to (y \to x), \; z \rightsquigarrow x \leq (y \rightsquigarrow z) \rightsquigarrow (y \rightsquigarrow x)$,
(pB=) $z \to x = (y \to z) \to (y \to x), \; z \rightsquigarrow x = (y \rightsquigarrow z) \rightsquigarrow (y \rightsquigarrow x)$;
(pD^L) $y \leq (y \to x) \rightsquigarrow x, \; y \leq (y \rightsquigarrow x) \to x$,
(q-pD=) $1 \to y = (y \to x) \rightsquigarrow x, \; 1 \rightsquigarrow y = (y \rightsquigarrow x) \to x$,
(q-pD1=) $1 \to x = (x \to 1) \rightsquigarrow 1, \; 1 \rightsquigarrow x = (x \rightsquigarrow 1) \to 1$;
(IdR) $1 \to x = 1 \rightsquigarrow x$,
(q-pII) $x \to y = (1 \to x) \to (1 \to y), \quad x \rightsquigarrow y = (1 \rightsquigarrow x) \rightsquigarrow (1 \rightsquigarrow y)$,
(q-pI1) $(1 \to x) \rightsquigarrow 1 = (1 \rightsquigarrow x) \to 1$,
(Id) $x \to 1 = x \rightsquigarrow 1$,
(q-pAn^L) (quasi-pseudo-Antisymmetry)
$(x \leq y$ (and $y \leq x)) \Longrightarrow (1 \to x = 1 \to y$ and $1 \rightsquigarrow x = 1 \rightsquigarrow y)$
(q-p-s^L) (quasi-p-semisimple) $x \leq y \Longleftrightarrow (1 \to x = 1 \to y$ and $1 \rightsquigarrow x = 1 \rightsquigarrow y)$.
We then have the following result (the dual case is omitted).

Proposition 11.3.2 *Let $(A, \to, \rightsquigarrow, 1)$ be a left-pseudo-algebra or, equivalently, let $(A, \leq, \to, \rightsquigarrow, 1)$ be a left-pseudo-structure. Then,*
(q1=) (pBB=) + (q-pM) $\Longrightarrow$ (q-pAn^L),
(q2=) (pBB=) $\Longrightarrow$ (p^{**L}),
(q3=) (pBB=) + (q-pM) $\Longrightarrow$ (Tr),
(q4=) (q-p-s^L) $\Longrightarrow$ (q-pAn^L),
(q5=) (pBB=) + (pRe') + (q-pM) + (IdEq) $\Longrightarrow$ (q-pR1),
(q6=) (pBB=) + (pRe') + (q-pM) + (q-pR1) $\Longrightarrow$ (pD^L),
(q7=) (pD^L) + (q-pAn^L) + (q-pM) $\Longrightarrow$ (q-pD=),
(q8=) (pBB=) + (pD^L) $\Longrightarrow$ (pC^L),
(q9=) (q-pEq=) $\Longleftrightarrow$ (q-p-s^L),
(q10=) (pBB=) + (pD^L) + (pRe') + (q-pM) $\Longrightarrow$ (pB^L),
(q11=) (pB^L) + (q-pAn^L) + (q-pM) $\Longrightarrow$ (pB=),
(q12=) (pB=) $\Longrightarrow$ (p^{*L}),
(q12=') (pB=) $\Longrightarrow$ (q-pII),
(q13=) (pEx) + (q-pM) $\Longrightarrow$ (IdR),
(q14=) (pEx) + (pRe') + (pB=) + (q-pM) + (IdR) + (q-pAn^L) $\Longrightarrow$ (q-pI1),
(q15=) (pB=) +(pRe') + (IdR) + (q-pI1) $\Longrightarrow$ (Id),
(q16=) (pB=) + (pRe') + (q-pAn^L) $\Longrightarrow$ (q-pEq=),
(q17=) (pBB^L) + (q-pAn^L) + (q-pM) + (Id) $\Longrightarrow$ (pBB=),
(q18=) (q-pEq=) $\Longrightarrow$ (pRe'),
(q19=) (q-pEq=) $\Longrightarrow$ (Tr),
(q20=) (q-pEq=) $\Longrightarrow$ (q-pAn^L),
(q21=) (pBB=) + (pEx) + (q-pM) + (q-pEq=) $\Longrightarrow$ (pB=),
(q21=') (pB=) + (pEx) + (q-pM) + (q-pEq=) $\Longrightarrow$ (pBB=),
(q21=") (pEx) + (q-pM) + (q-pEq=) $\Longrightarrow$ (pBB=) $\Leftrightarrow$ (pB=),
(q22=) (pBB=) + (q-pEq=) + (q-pM) + (Id) $\Longrightarrow$ (pBB^L),
(q23=) (pB=) + (q-pEq=) + (q-pM) $\Longrightarrow$ (pB^L);
(q24=) (pEx) + (pRe') + (q-pEq=) + (q-pM) $\Longrightarrow$ (q-pD=),
(q24=') (pEx1) + (pRe') + (q-pEq=) + (q-pM) $\Longrightarrow$ (q-pD1=);
(q25=) (pEx) + (q-pD=) + (q-pM) + (IdR) $\Longrightarrow$ (pBB=).

Proof. (q1=): Suppose that $x \leq y$, i.e. $x \to y = 1$ ($\Leftrightarrow x \rightsquigarrow y = 1$); then:
$1 \to x \overset{(pBB=)}{=} (x \to y) \rightsquigarrow (1 \to y) = 1 \rightsquigarrow (1 \to y) \overset{(q-pM)}{=} 1 \to y$ and
$1 \rightsquigarrow x \overset{(pBB=)}{=} (x \rightsquigarrow y) \to (1 \rightsquigarrow y) = 1 \to (1 \rightsquigarrow y) \overset{(q-pM)}{=} 1 \rightsquigarrow y$. Thus, (q-pAn^L) holds.

(q2=): Suppose that $x \leq y$, hence $x \to y = 1 = x \rightsquigarrow y$, by (pEq^L); then:
$1 = x \to y \overset{(pBB=)}{=} (y \to z) \rightsquigarrow (x \to z)$, hence $y \to z \leq x \to z$ and
$1 = x \rightsquigarrow y \overset{(pBB=)}{=} (y \rightsquigarrow z) \to (x \rightsquigarrow z)$, hence $y \rightsquigarrow z \leq x \rightsquigarrow z$. Thus, (p^{**L}) holds.

(q3=): Suppose that $x \leq y$ and $y \leq z$, hence $x \to y = 1 = x \rightsquigarrow y$ and $y \to z = 1 = y \rightsquigarrow z$, by (pEq^L);
then, $1 = x \to y \overset{(pBB=)}{=} (y \to z) \rightsquigarrow (x \to z) = 1 \rightsquigarrow (x \to z) \overset{(q-pM)}{=} x \to z$, hence $x \leq z$. Thus, (Tr) holds.

(q4=): Obviously.

(q5=): First, we prove that $(1 \rightsquigarrow x) \to x = 1$. Indeed,
- if x is a regular element, i.e. $1 \rightsquigarrow x \overset{(pM)}{=} x$, then $(1 \rightsquigarrow x) \to x = x \to x \overset{(pRe')}{=} 1$;
- if x is not a regular element, i.e. $1 \rightsquigarrow x = y \neq x$, with y a regular element, then we have

$$1 \rightsquigarrow y \overset{(pM)}{=} y = 1 \rightsquigarrow x; \tag{11.4}$$

then, $1 \rightsquigarrow y \overset{(pBB=)}{=} (y \rightsquigarrow x) \to (1 \rightsquigarrow x) \overset{(11.4)}{=} (y \rightsquigarrow x) \to (1 \rightsquigarrow y)$, hence
$1 \overset{(pRe')}{=} (1 \rightsquigarrow y) \rightsquigarrow (1 \rightsquigarrow y) = [(y \rightsquigarrow x) \to (1 \rightsquigarrow y)] \rightsquigarrow (1 \rightsquigarrow y)$
$\overset{(q-pM)}{=} [(y \rightsquigarrow x) \to (1 \rightsquigarrow y)] \rightsquigarrow [1 \to (1 \rightsquigarrow y)] \overset{(pBB=)}{=} 1 \to (y \rightsquigarrow x) \overset{(q-pM)}{=} y \rightsquigarrow x$;
thus, $y \rightsquigarrow x = 1$; then, by (IdEq), $y \to x = 1$; hence, $(1 \rightsquigarrow x) \to x \overset{(11.4)}{=} y \to x = 1$.

Now we prove that $(1 \to x) \rightsquigarrow x = 1$. Indeed,
- if x is a regular element, i.e. $1 \to x \overset{(pM)}{=} x$, then $(1 \to x) \rightsquigarrow x = x \rightsquigarrow x \overset{(pRe')}{=} 1$;
- if x is not a regular element, i.e. $1 \to x = y \neq x$, with y a regular element, then we have

$$1 \to y \overset{(pM)}{=} y = 1 \to x; \tag{11.5}$$

then, $1 \to y \overset{(pBB=)}{=} (y \to x) \rightsquigarrow (1 \to x) \overset{(11.5)}{=} (y \to x) \rightsquigarrow (1 \to y)$, hence
$1 \overset{(pRe')}{=} (1 \to y) \to (1 \to y) = [(y \to x) \rightsquigarrow (1 \to y)] \to (1 \to y)$
$\overset{(q-pM)}{=} [(y \to x) \rightsquigarrow (1 \to y)] \to [1 \rightsquigarrow (1 \to y)] \overset{(pBB=)}{=} 1 \rightsquigarrow (y \to x) \overset{(q-pM)}{=} y \to x$;
thus, $y \to x = 1$; then, by (IdEq), $y \rightsquigarrow x = 1$; hence, $(1 \to x) \rightsquigarrow x \overset{(11.5)}{=} y \rightsquigarrow x = 1$.
Thus, (q-pR1) holds.

(q6=): First, we prove that $y \leq (y \to x) \rightsquigarrow x$. Indeed,
$y \to [(y \to x) \rightsquigarrow x] \overset{(pBB=)}{=} ([(y \to x) \rightsquigarrow x] \to x) \rightsquigarrow (y \to x)$
$\overset{(q-pM)}{=} ([(y \to x) \rightsquigarrow x] \to x) \rightsquigarrow [1 \rightsquigarrow (y \to x)]$
$\overset{(pBB=)}{=} ([(y \to x) \rightsquigarrow x] \to x) \rightsquigarrow ([(y \to x) \rightsquigarrow x] \to (1 \rightsquigarrow x))$
$\overset{(pBB=)}{=} ([(y \to x) \rightsquigarrow x] \to x) \rightsquigarrow ([(1 \rightsquigarrow x) \to x] \rightsquigarrow ([(y \to x) \rightsquigarrow x] \to x))$
$\overset{(q-pR1)}{=} ([(y \to x) \rightsquigarrow x] \to x) \rightsquigarrow (1 \rightsquigarrow ([(y \to x) \rightsquigarrow x] \to x))$
$\overset{(q-pM)}{=} ([(y \to x) \rightsquigarrow x] \to x) \rightsquigarrow ([(y \to x) \rightsquigarrow x] \to x) \overset{(pRe')}{=} 1$;
thus, $y \leq (y \to x) \rightsquigarrow x$.

Now we prove that $y \leq (y \rightsquigarrow x) \to x$. Indeed,
$y \rightsquigarrow [(y \rightsquigarrow x) \to x] \overset{(pBB=)}{=} ([(y \rightsquigarrow x) \to x] \rightsquigarrow x) \to (y \rightsquigarrow x)$
$\overset{(q-pM)}{=} ([(y \rightsquigarrow x) \to x] \rightsquigarrow x) \to [1 \to (y \rightsquigarrow x)]$
$\overset{(pBB=)}{=} ([(y \rightsquigarrow x) \to x] \rightsquigarrow x) \to ([(y \rightsquigarrow x) \to x] \rightsquigarrow (1 \to x))$
$\overset{(pBB=)}{=} ([(y \rightsquigarrow x) \to x] \rightsquigarrow x) \to ([(1 \to x) \rightsquigarrow x] \to ([(y \rightsquigarrow x) \to x] \rightsquigarrow x))$
$\overset{(q-pR1)}{=} ([(y \rightsquigarrow x) \to x] \rightsquigarrow x) \to (1 \to ([(y \rightsquigarrow x) \to x] \rightsquigarrow x))$
$\overset{(q-pM)}{=} ([(y \rightsquigarrow x) \to x] \rightsquigarrow x) \to ([(y \rightsquigarrow x) \to x] \rightsquigarrow x) \overset{(pRe')}{=} 1$;
thus, $y \leq (y \rightsquigarrow x) \to x$. Thus, (pD^L) holds.

(q7=): By (pD^L), $y \leq (y \to x) \rightsquigarrow x$ and $y \leq (y \rightsquigarrow x) \to x$; hence, by (q-pAn^L) and (q-pM), we obtain:
$1 \to y = 1 \to [(y \to x) \rightsquigarrow x] = (y \to x) \rightsquigarrow x$ and
$1 \rightsquigarrow y = 1 \rightsquigarrow [(y \rightsquigarrow x) \to x] = (y \rightsquigarrow x) \to x$. Thus, (q-pD=) holds.

(q8=): First, by (pBB=), we have: $Y \rightsquigarrow Z = (Z \rightsquigarrow X) \to (Y \rightsquigarrow X)$.
Take $X = u \to x$, $Y = y$, $Z = z \to x$; we obtain:
$y \rightsquigarrow (z \to x) = ((z \to x) \rightsquigarrow (u \to x)) \to (y \rightsquigarrow (u \to x))$
$\overset{(pBB=)}{=} (u \to z) \to (y \rightsquigarrow (u \to x))$.
Take now $z = y \rightsquigarrow x$ and $u = z$; we obtain:
$1 \overset{(pD^{\leq})}{=} y \rightsquigarrow [(y \rightsquigarrow x) \to x] = (z \to (y \rightsquigarrow x)) \to (y \rightsquigarrow (z \to x))$,
hence $z \to (y \rightsquigarrow x) \leq y \rightsquigarrow (z \to x)$.

Then, by (pBB=) again, we have: $Y \to Z = (Z \to X) \rightsquigarrow (Y \to X)$.
Take $X = u \rightsquigarrow x$, $Y = y$, $Z = z \rightsquigarrow x$; we obtain:
$y \to (z \rightsquigarrow x) = ((z \rightsquigarrow x) \to (u \rightsquigarrow x)) \rightsquigarrow (y \to (u \rightsquigarrow x))$
$\overset{(pBB=)}{=} (u \rightsquigarrow z) \rightsquigarrow (y \to (u \rightsquigarrow x))$.
Take now $z = y \to x$ and $u = z$; we obtain:
$1 \overset{(pD^{\leq})}{=} y \to [(y \to x) \rightsquigarrow x] = (z \rightsquigarrow (y \to x)) \rightsquigarrow (y \to (z \rightsquigarrow x))$,
hence $z \rightsquigarrow (y \to x) \leq y \to (z \rightsquigarrow x)$. Thus, (pC^L) holds.

(q9=): Obviously.

(q10=): We prove first that $z \to x \leq (y \to z) \to (y \to x)$.

By (pBB=), we have: $X \rightsquigarrow Y = (Y \rightsquigarrow z) \to (X \rightsquigarrow z)$.
Take $X = u$ and $Y = (u \rightsquigarrow v) \to v$; we obtain:
$1 \overset{(pD^{\leq})}{=} u \rightsquigarrow [(u \rightsquigarrow v) \to v] = ([(u \rightsquigarrow v) \to v] \rightsquigarrow z) \to (u \rightsquigarrow z)$.
After renaming variables, we obtain:
(a1) $([(x \rightsquigarrow y) \to y] \rightsquigarrow z) \to (x \rightsquigarrow z) = 1$.

Next, by (pBB=) again, we have: $X \to Y = (Y \to z) \rightsquigarrow (X \to z)$.
Take $X = u \to v$ and $Y = (v \to w) \rightsquigarrow (u \to w)$; we obtain:
$1 \overset{(pBB=),(pRe')}{=} (u \to v) \to [(v \to w) \rightsquigarrow (u \to w)]$
$= ([(v \to w) \rightsquigarrow (u \to w)] \to z) \rightsquigarrow ((u \to v) \to z)$.
After renaming variables, we obtain:
(b1) $([(x \to y) \rightsquigarrow (u \to y)] \to z) \rightsquigarrow ((u \to x) \to z) = 1$.

Take $z = u \to y$ in (b1) to obtain:
(c1) $([(x \to y) \rightsquigarrow (u \to y)] \to (u \to y)) \rightsquigarrow ((u \to x) \to (u \to y)) = 1$.

Now, in (a1), take $x = v \to w$, $y = t \to w$, $z = (t \to v) \to (t \to w)$ to obtain:
$([((v \to w) \rightsquigarrow (t \to w)) \to (t \to w)] \rightsquigarrow ((t \to v) \to (t \to w))) \to$
$\to ((v \to w) \rightsquigarrow ((t \to v) \to (t \to w))) = 1$,
hence, by (c1):
$1 \to ((v \to w) \rightsquigarrow ((t \to v) \to (t \to w))) = 1$,
hence, by (q-pM):
$(v \to w) \rightsquigarrow ((t \to v) \to (t \to w)) = 1$, i.e. $v \to w \leq (t \to v) \to (t \to w)$.
After renaming variables, we obtain: $z \to x \leq (y \to z) \to (y \to x)$.

Similarly, we prove now that $z \rightsquigarrow x \leq (y \rightsquigarrow z) \rightsquigarrow (y \rightsquigarrow x)$.

By (pBB=), we have: $X \to Y = (Y \to z) \rightsquigarrow (X \to z)$.

Take $X = u$ and $Y = (u \to v) \rightsquigarrow v$; we obtain:
$1 \overset{(pD^{\leq})}{=} u \to [(u \to v) \rightsquigarrow v] = ([(u \to v) \rightsquigarrow v] \to z) \rightsquigarrow (u \to z)$.
After renaming variables, we obtain:
(a2) $([(x \to y) \rightsquigarrow y] \to z) \rightsquigarrow (x \to z) = 1$.

Next, by (pBB=) again, we have: $X \rightsquigarrow Y = (Y \rightsquigarrow z) \to (X \rightsquigarrow z)$.
Take $X = u \rightsquigarrow v$ and $Y = (v \rightsquigarrow w) \to (u \rightsquigarrow w)$; we obtain:
$1 \overset{(pBB=),(pRe')}{=} (u \rightsquigarrow v) \rightsquigarrow [(v \rightsquigarrow w) \to (u \rightsquigarrow w)]$
$= ([(v \rightsquigarrow w) \to (u \rightsquigarrow w)] \rightsquigarrow z) \to ((u \rightsquigarrow v) \rightsquigarrow z)$.
After renaming variables, we obtain:
(b2) $([(x \rightsquigarrow y) \to (u \rightsquigarrow y)] \rightsquigarrow z) \to ((u \rightsquigarrow x) \rightsquigarrow z) = 1$.

Take $z = u \rightsquigarrow y$ in (b2) to obtain:
(c2) $([(x \rightsquigarrow y) \to (u \rightsquigarrow y)] \rightsquigarrow (u \rightsquigarrow y)) \to ((u \rightsquigarrow x) \rightsquigarrow (u \rightsquigarrow y)) = 1$.

Now, in (a2), take $x = v \rightsquigarrow w$, $y = t \rightsquigarrow w$, $z = (t \rightsquigarrow v) \rightsquigarrow (t \rightsquigarrow w)$ to obtain:
$([((v \rightsquigarrow w) \to (t \rightsquigarrow w)) \rightsquigarrow (t \rightsquigarrow w)] \to ((t \rightsquigarrow v) \rightsquigarrow (t \rightsquigarrow w))) \rightsquigarrow$
$\rightsquigarrow ((v \rightsquigarrow w) \to ((t \rightsquigarrow v) \rightsquigarrow (t \rightsquigarrow w))) = 1$,
hence, by (c2):
$1 \rightsquigarrow ((v \rightsquigarrow w) \to ((t \rightsquigarrow v) \rightsquigarrow (t \rightsquigarrow w))) = 1$,
hence, by (q-pM):
$(v \rightsquigarrow w) \to ((t \rightsquigarrow v) \rightsquigarrow (t \rightsquigarrow w)) = 1$, i.e. $v \rightsquigarrow w \leq (t \rightsquigarrow v) \rightsquigarrow (t \rightsquigarrow w)$.
After renaming variables, we get: $z \rightsquigarrow x \leq (y \rightsquigarrow z) \rightsquigarrow (y \rightsquigarrow x)$. Thus, (pB^L) holds.

(q11=): By (pB^L), we have: $z \to x \leq (y \to z) \to (y \to x)$. Then, by (q-pAn^L), we have: $1 \rightsquigarrow (z \to x) = 1 \rightsquigarrow ((y \to z) \to (y \to x))$, hence, by (q-pM), we obtain: $z \to x = (y \to z) \to (y \to x)$.

Similarly, by (pB^L) again, we have: $z \rightsquigarrow x \leq (y \rightsquigarrow z) \rightsquigarrow (y \rightsquigarrow x)$. Then, by (q-pAn^L), we have: $1 \to (z \rightsquigarrow x) = 1 \to ((y \rightsquigarrow z) \rightsquigarrow (y \rightsquigarrow x))$, hence, by (q-pM), we obtain: $z \rightsquigarrow x = (y \rightsquigarrow z) \rightsquigarrow (y \rightsquigarrow x)$. Thus, (pB=) holds.

(q12=): Suppose that $x \leq y$, hence $x \to y = 1 = x \rightsquigarrow y$, by (11.3); then,
$1 = x \to y \overset{(pB=)}{=} (z \to x) \to (z \to y)$, hence $z \to x \leq z \to y$ and
$1 = x \rightsquigarrow y \overset{(pB=)}{=} (z \rightsquigarrow x) \rightsquigarrow (z \rightsquigarrow y)$, hence $z \rightsquigarrow x \leq z \rightsquigarrow y$. Thus, (p^{*L}) holds.

(q12='): $x \to y \overset{(pB=)}{=} (1 \to x) \to (1 \to y)$ and $x \rightsquigarrow y \overset{(pB=)}{=} (1 \rightsquigarrow x) \rightsquigarrow (1 \rightsquigarrow y)$. Thus, (q-pII) holds.

(q13=): By Proposition 5.1.19, or directly: $1 \to x \overset{(q-pM)}{=} 1 \rightsquigarrow (1 \to x) \overset{(pEx)}{=} 1 \to (1 \rightsquigarrow x) \overset{(q-pM)}{=} 1 \rightsquigarrow x$. Thus, (IdR) holds.

(q14=): First we prove that $(1 \to x) \rightsquigarrow 1 \leq (1 \rightsquigarrow x) \to 1$. Indeed, by (11.3) or by (pEq^L):
$[(1 \to x) \rightsquigarrow 1] \rightsquigarrow [(1 \rightsquigarrow x) \to 1] \overset{(pEx)}{=} (1 \rightsquigarrow x) \to ([(1 \to x) \rightsquigarrow 1] \rightsquigarrow 1)$
$\overset{(pRe')}{=} (1 \rightsquigarrow x) \to ([(1 \to x) \rightsquigarrow 1] \rightsquigarrow [(1 \to x) \rightsquigarrow (1 \to x)])$
$\overset{(pB=)}{=} (1 \rightsquigarrow x) \to [1 \rightsquigarrow (1 \to x)] \overset{(q-pM)}{=} (1 \rightsquigarrow x) \to (1 \to x)$
$\overset{(IdR)}{=} (1 \rightsquigarrow x) \to (1 \rightsquigarrow x) \overset{(pRe')}{=} 1$.

Now, by (q-pAn^L), we obtain:

$(1 \to x) \rightsquigarrow 1 \overset{(q-pM)}{=} 1 \to [(1 \to x) \rightsquigarrow 1] = 1 \to [(1 \rightsquigarrow x) \to 1]$ and

$$1 \rightsquigarrow [(1 \to x) \rightsquigarrow 1] = 1 \rightsquigarrow [(1 \rightsquigarrow x) \to 1] \overset{(q-pM)}{=} (1 \rightsquigarrow x) \to 1 .$$

Hence, $(1 \to x) \rightsquigarrow 1 = 1 \to [(1 \rightsquigarrow x) \to 1] = 1 \to [1 \rightsquigarrow [(1 \rightsquigarrow x) \to 1]]$
$\overset{(q-pM)}{=} 1 \rightsquigarrow [(1 \rightsquigarrow x) \to 1] \overset{(q-pM)}{=} (1 \rightsquigarrow x) \to 1$. Thus, (q-pI1) holds.

(q15=): $x \to 1 \overset{(pB=)}{=} (1 \to x) \to (1 \to 1) \overset{(pRe')}{=} (1 \to x) \to 1 \overset{(IdR)}{=} (1 \rightsquigarrow x) \to 1$
$\overset{(q-pI1)}{=} (1 \to x) \rightsquigarrow 1 \overset{(IdR)}{=} (1 \rightsquigarrow x) \rightsquigarrow 1 \overset{(pRe')}{=} (1 \rightsquigarrow x) \rightsquigarrow (1 \rightsquigarrow 1) \overset{(pB=)}{=} x \rightsquigarrow 1$.
Thus, (Id) holds.

(q16=): Suppose $1 \to x = 1 \to y$; then, $x \to y \overset{(pB=)}{=} (1 \to x) \to (1 \to y) = (1 \to x) \to (1 \to x) \overset{(pRe')}{=} 1$. Conversely, suppose $x \to y = 1$, i.e. $x \le y$, by (11.3) or by (pEq^L) ; then, by (q-pAn^L), $1 \to x = 1 \to y$ and $1 \rightsquigarrow x = 1 \rightsquigarrow y$.

Suppose $1 \rightsquigarrow x = 1 \rightsquigarrow y$; then, $x \rightsquigarrow y \overset{(pB=)}{=} (1 \rightsquigarrow x) \rightsquigarrow (1 \rightsquigarrow y) = (1 \rightsquigarrow x) \rightsquigarrow (1 \rightsquigarrow x) \overset{(pRe')}{=} 1$. Conversely, suppose $x \rightsquigarrow y = 1$, i.e. $x \le y$, by (11.3) or by (pEq^L); then, by (q-pAn^L), $1 \rightsquigarrow x = 1 \rightsquigarrow y$ and $1 \to x = 1 \to y$. Thus, (q-pEq=) holds.

(q17=): By (pBB^L), we have: $y \to z \le (z \to x) \rightsquigarrow (y \to x)$. Then, by (q-$pAn^L$), we have: $1 \rightsquigarrow (y \to z) = 1 \rightsquigarrow ((z \to x) \rightsquigarrow (y \to x)) \overset{(Id)}{=} 1 \to ((z \to x) \rightsquigarrow (y \to x))$, hence, by (q-pM), we obtain: $y \to z = (z \to x) \rightsquigarrow (y \to x)$.

Similarly, by (pBB^L) again, we have: $y \rightsquigarrow z \le (z \rightsquigarrow x) \to (y \rightsquigarrow x)$. Then, by (q-$pAn^L$), we have: $1 \to (y \rightsquigarrow z) = 1 \to ((z \rightsquigarrow x) \to (y \rightsquigarrow x)) \overset{(Id)}{=} 1 \rightsquigarrow ((z \rightsquigarrow x) \to (y \rightsquigarrow x))$, hence, by (q-pM), we obtain: $y \rightsquigarrow z = (z \rightsquigarrow x) \to (y \rightsquigarrow x)$. Thus, (pBB=) holds.

(q18=): Obviously, for $x = y$.

(q19=): Immediately.

(q20=): Obviously.

(q21=): First, by (q13=), (pEx) + (q-pM) imply (IdR).
Then, by (pBB=), we have:
(a) $y \to z = (z \to x) \rightsquigarrow (y \to x)$ and (b) $y \rightsquigarrow z = (z \rightsquigarrow x) \to (y \rightsquigarrow x)$.

Now, by (q-pM), (IdR) and (a), we obtain:
$1 \rightsquigarrow (y \to z) = y \to z = (z \to x) \rightsquigarrow (y \to x) = 1 \to [(z \to x) \rightsquigarrow (y \to x)] = 1 \rightsquigarrow [(z \to x) \rightsquigarrow (y \to x)]$ and
$1 \to (y \to z) = 1 \rightsquigarrow (y \to z) = y \to z = (z \to x) \rightsquigarrow (y \to x) = 1 \to [(z \to x) \rightsquigarrow (y \to x)]$.
Then, by (q-pEq=), $(y \to z) \to [(z \to x) \rightsquigarrow (y \to x)] = 1$;
then, by (pEx), $(z \to x) \rightsquigarrow [(y \to z) \to (y \to x)] = 1$;
then, by (q-pEq=), $1 \rightsquigarrow (z \to x) = 1 \rightsquigarrow [(y \to z) \to (y \to x)]$;
hence, by (q-pM), $z \to x = (y \to z) \to (y \to x)$.

Then, by (q-pM), (IdR) and (b), we obtain:
$1 \to (y \rightsquigarrow z) = y \rightsquigarrow z = (z \rightsquigarrow x) \to (y \rightsquigarrow x) = 1 \rightsquigarrow [(z \rightsquigarrow x) \to (y \rightsquigarrow x)] = 1 \to [(z \rightsquigarrow x) \to (y \rightsquigarrow x)]$ and
$1 \rightsquigarrow (y \rightsquigarrow z) = 1 \to (y \rightsquigarrow z) = y \rightsquigarrow z = (z \rightsquigarrow x) \to (y \rightsquigarrow x) = 1 \rightsquigarrow [(z \rightsquigarrow x) \to (y \rightsquigarrow x)]$.
Then, by (q-pEq=), $(y \rightsquigarrow z) \rightsquigarrow [(z \rightsquigarrow x) \to (y \rightsquigarrow x)] = 1$;

then, by (pEx), $(z \rightsquigarrow x) \to [(y \rightsquigarrow z) \rightsquigarrow (y \rightsquigarrow x)] = 1$;
then, by (q-pEq=), $1 \to (z \rightsquigarrow x) = 1 \to [(y \rightsquigarrow z) \rightsquigarrow (y \rightsquigarrow x)]$;
then, by (q-pM), $z \rightsquigarrow x = (y \rightsquigarrow z) \rightsquigarrow (y \rightsquigarrow x)$. So, (pB=) holds.

(q21='): First, by (q13=), (pEx) + (q-pM) imply (IdR).
Then, by (pB=), we have:
$z \to x = (y \to z) \to (y \to x)$ and $z \rightsquigarrow x = (y \rightsquigarrow z) \rightsquigarrow (y \rightsquigarrow x)$; then, by (q-pM),
$1 \rightsquigarrow (z \to x) = z \to x = (y \to z) \to (y \to x) = 1 \rightsquigarrow [(y \to z) \to (y \to x)]$ and
$1 \to (z \rightsquigarrow x) = z \rightsquigarrow x = (y \rightsquigarrow z) \rightsquigarrow (y \rightsquigarrow x) = 1 \to [(y \rightsquigarrow z) \rightsquigarrow (y \rightsquigarrow x)]$;
then, by (q-pEq=):
$(z \to x) \rightsquigarrow [(y \to z) \to (y \to x)] = 1$ and $(z \rightsquigarrow x) \to [(y \rightsquigarrow z) \rightsquigarrow (y \rightsquigarrow x)] = 1$;
then, by (pEx):
$(y \to z) \to [(z \to x) \rightsquigarrow (y \to x)] = 1$ and $(y \rightsquigarrow z) \rightsquigarrow [(z \rightsquigarrow x) \to (y \rightsquigarrow x)] = 1$;
then, by (q-pEq=), we have:
$1 \rightsquigarrow (y \to z) = 1 \rightsquigarrow [(z \to x) \rightsquigarrow (y \to x)] \overset{(IdR)}{=} 1 \to [(z \to x) \rightsquigarrow (y \to x)]$ and
$1 \to (y \rightsquigarrow z) = 1 \to [(z \rightsquigarrow x) \to (y \rightsquigarrow x)] \overset{(IdR)}{=} 1 \rightsquigarrow [(z \rightsquigarrow x) \to (y \rightsquigarrow x)]$;
then, by (q-pM):
$y \to z = (z \to x) \rightsquigarrow (y \to x)$ and $y \rightsquigarrow z = (z \rightsquigarrow x) \to (y \rightsquigarrow x)$. So, (pBB=) holds.

(q21=''): By (q21=) and (q21=').

(q22): By (pBB=), $y \to z = (z \to x) \rightsquigarrow (y \to x)$. Then,
$1 \rightsquigarrow (y \to z) \overset{(q-pM)}{=} y \to z = (z \to x) \rightsquigarrow (y \to x) \overset{(q-pM)}{=}$
$1 \to ((z \to x) \rightsquigarrow (y \to x)) \overset{(Id)}{=} 1 \rightsquigarrow ((z \to x) \rightsquigarrow (y \to x))$. Now, by (q-pEq=),
$(y \to z) \to ((z \to x) \rightsquigarrow (y \to x)) = 1$, i.e. $y \to z \leq (z \to x) \rightsquigarrow (y \to x)$.

By (pBB=) again, $y \rightsquigarrow z = (z \rightsquigarrow x) \to (y \rightsquigarrow x)$. Then,
$1 \to (y \rightsquigarrow z) \overset{(q-pM)}{=} y \rightsquigarrow z = (z \rightsquigarrow x) \to (y \rightsquigarrow x) \overset{(q-pM)}{=}$
$1 \rightsquigarrow ((z \rightsquigarrow x) \to (y \rightsquigarrow x)) \overset{(Id)}{=} 1 \to ((z \rightsquigarrow x) \to (y \rightsquigarrow x))$.
Now, by (q-pEq=), $(y \rightsquigarrow z) \rightsquigarrow ((z \rightsquigarrow x) \to (y \rightsquigarrow x)) = 1$, i.e.
$y \rightsquigarrow z \leq (z \rightsquigarrow x) \to (y \rightsquigarrow x)$. Thus, (pBB^L) holds.

(q23): By (pB=), $z \to x = (y \to z) \to (y \to x)$. Then,
$1 \rightsquigarrow (z \to x) \overset{(q-pM)}{=} z \to x = (y \to z) \to (y \to x) \overset{(q-pM)}{=}$
$1 \rightsquigarrow ((y \to z) \to (y \to x))$. Now, by (q-pEq=),
$(z \to x) \to ((y \to z) \to (y \to x)) = 1$, i.e. $z \to x \leq (y \to z) \to (y \to x)$.

By (pB=), $z \rightsquigarrow x = (y \rightsquigarrow z) \rightsquigarrow (y \rightsquigarrow x)$. Then,
$1 \to (z \rightsquigarrow x) \overset{(q-pM)}{=} z \rightsquigarrow x = (y \rightsquigarrow z) \rightsquigarrow (y \rightsquigarrow x) \overset{(q-pM)}{=}$
$1 \to ((y \rightsquigarrow z) \rightsquigarrow (y \rightsquigarrow x))$. Now, by (q-pEq=),
$(z \rightsquigarrow x) \to ((y \rightsquigarrow z) \rightsquigarrow (y \rightsquigarrow x)) = 1$, i.e. $z \rightsquigarrow x \leq (y \rightsquigarrow z) \rightsquigarrow (y \rightsquigarrow x)$. Thus, (pB^L) holds.

(q24=): $y \to ((y \to x) \rightsquigarrow x) \overset{(pEx)}{=} (y \to x) \rightsquigarrow (y \to x) \overset{(pRe')}{=} 1$; then, by (q-pEq=), $1 \to y = 1 \to ((y \to x) \rightsquigarrow x) \overset{(q-pM)}{=} (y \to x) \rightsquigarrow x$.
$y \rightsquigarrow ((y \rightsquigarrow x) \to x) \overset{(pEx)}{=} (y \rightsquigarrow x) \to (y \rightsquigarrow x) \overset{(pRe')}{=} 1$; then, by (q-pEq=), $1 \rightsquigarrow y = 1 \rightsquigarrow ((y \rightsquigarrow x) \to x) \overset{(q-pM)}{=} (y \rightsquigarrow x) \to x$. Thus, (q-pD=) holds.

(q24='): $y \to ((y \to 1) \rightsquigarrow 1) \overset{(pEx1)}{=} (y \to 1) \rightsquigarrow (y \to 1) \overset{(pRe')}{=} 1$; then, by

(q-pEq=), $1 \to y = 1 \to ((y \to 1) \rightsquigarrow 1) \overset{(q-pM)}{=} (y \to 1) \rightsquigarrow 1$.
$y \rightsquigarrow ((y \rightsquigarrow 1) \to 1) \overset{(pEx1)}{=} (y \rightsquigarrow 1) \to (y \rightsquigarrow 1) \overset{(pRe')}{=} 1$; then, by (q-pEq=), $1 \rightsquigarrow y = 1 \rightsquigarrow ((y \rightsquigarrow 1) \to 1) \overset{(q-pM)}{=} (y \rightsquigarrow 1) \to 1$. Thus, (q-pD1=) holds.

(q25=): $(z \to x) \rightsquigarrow (y \to x) \overset{(pEx)}{=} y \to ((z \to x) \rightsquigarrow x) \overset{(q-pD=)}{=} y \to (1 \to z) \overset{(IdR)}{=} y \to (1 \rightsquigarrow z) \overset{(pEx)}{=} 1 \rightsquigarrow (y \to z) \overset{(q-pM)}{=} y \to z$ and $(z \rightsquigarrow x) \to (y \rightsquigarrow x) \overset{(pEx)}{=} y \rightsquigarrow ((z \rightsquigarrow x) \to x) \overset{(q-pD=)}{=} y \rightsquigarrow (1 \rightsquigarrow z) \overset{(IdR)}{=} y \rightsquigarrow (1 \to z) \overset{(pEx)}{=} 1 \to (y \rightsquigarrow z) \overset{(q-pM)}{=} y \rightsquigarrow z$. Thus, (pBB=) holds. □

Consider now the following properties: for all $x, y \in A$,

(pD=)	$y = (y \to x) \rightsquigarrow x, \quad y = (y \rightsquigarrow x) \to x,$
(pM)	$1 \to x = x = 1 \rightsquigarrow x,$
(p-s^L) (p-semisimple)	$x \le y \Longrightarrow x = y$ (in fact, $x \le y \Longleftrightarrow x = y$),
(DN)	$(x^{-1})^{-1} = x.$

Proposition 11.3.3
(1) (q-pD=) + (pM) $\Longrightarrow$ (pD=),
(2) (q-p-s^L) + (pM) $\Longrightarrow$ (p-s^L),
(3) (q-pAn^L) + (pM) $\Longrightarrow$ (An),
(4) (q-pR1) + (pM) $\Longrightarrow$ (pRe'),
(5) (IdR) + (pM) $\Longrightarrow$ $x = x$,
(6) (q-pI1) + (pM) $\Longrightarrow$ (Id);
(7) (pD=) $\Longrightarrow$ (DN),
(8) (pBB=) + (pRe') + (DN) $\Longrightarrow$ (pM),
(9) (pBB=) + (pRe') + (pD=) $\Longrightarrow$ (pM).

Proof. (1) - (6): Obviously.

(7): Take $x = 1$ in (pD=); we obtain $y = (y^{-1})^{-1}$.

(8): $1 \to x \overset{(pBB=)}{=} (x \to 1) \rightsquigarrow (1 \to 1) \overset{(pRe')}{=} (x \to 1) \rightsquigarrow 1 \overset{(DN)}{=} x$ and $1 \rightsquigarrow x \overset{(pBB=)}{=} (x \rightsquigarrow 1) \to (1 \rightsquigarrow 1) \overset{(pRe')}{=} (x \rightsquigarrow 1) \to 1 \overset{(DN)}{=} x$.

(9): Obviously, by above (7) and (8). □

Proposition 11.3.4 *Let $\mathcal{A} = (A, \to, \rightsquigarrow, 1)$ be a (super) q-i-hub. Then, (pRe'), (Tr), (q-pAn^L) and (q-p-s^L) hold.*

Proof. Define a binary relation $\le$ on A by (11.3). By Proposition 11.3.2 (q18=), (q19=), (q20=), (q9=), the properties (pRe'), (Tr), (q-pAn^L) and (q-p-s^L) are verified. □

We then have the following important remarks (the dual case is omitted).

Remarks 11.3.5

(1) The (super) q-i-hub, as defined by Definition 11.3.1, is a special left-quasi-pseudo-algebra as defined in Chapter 5, namely a left-quasi-pseudo-algebra (of

logic) for which $x \leq y$, defined by (11.3), becomes $1 \to x$ $(= 1 \rightsquigarrow x) = 1 \to y$ $(= 1 \rightsquigarrow y)$, and so the property (q-p-s^L) holds.

(2) The (super) q-i-hub can be defined, equivalently, as a special left-quasi-pseudo-structure: $(A, \leq, \to, \rightsquigarrow, 1)$, where (pEq^L) becomes (q-pEq=); therefore, as left-quasi-pseudo-structure, the q-i-hub is denoted also by
$(A, \to, \rightsquigarrow, 1)$.

Consequently,
- if $\mathcal{A}$ is only a q-i-hub, then $(A, \leq, \to^L = \to, \rightsquigarrow^L = \rightsquigarrow, 1)$ is only a left-quasi-poRM algebra ("o" comes from "ordered"),
- if $\mathcal{A}$ is a super q-i-hub, then $(A, \leq, \to^L = \to, \rightsquigarrow^L = \rightsquigarrow, 1)$ is left-quasi-p*aRM** algebra,

as these quasi-pseudo-algebras were defined in the previous section.

So, we have the following result (the dual case is omitted).

Proposition 11.3.6

(1) Any q-i-hub is a left-q-poRM algebra verifying (q-p-s^L).

*(2) Any super q-i-hub is a left-q-p*aRM** algebra verifying (q-p-s^L).*

Examples 11.3.7 Consider the set $A_4 = \{a, b, c, 1\}$ and the following tables of $\to$ (the commutative case) verifying (q-M), (M1) and (q-Eq=), hence (Re), (Tr) and (q-An) (these are q-i-hubs):

$\to_1$	a	b	c	1
a	1	a	a	a
b	a	1	a	b
c	a	a	1	1
1	a	b	1	1

$\to_2$	a	b	c	1
a	1	1	a	a
b	1	1	a	c
c	a	a	1	c
1	a	a	c	1

$\to_3$	a	b	c	1
a	1	a	a	b
b	a	1	a	a
c	a	b	1	1
1	a	b	1	1

.

Then, note that $\to_1$ does not verify (*'), (**'); $\to_2$ verifies (*'), but does not verify (**'); $\to_3$ verifies (**'), but does not verify (*'). Hence, $(A_4, \to_1, 1)$, $(A_4, \to_2, 1)$, $(A_4, \to_3, 1)$ are quasi-i-hubs, but not super ones, following the above definitions, i.e. $(A_4, \leq, \to_1, 1)$, $(A_4, \leq, \to_2, 1)$, $(A_4, \leq, \to_3, 1)$ are only left-quasi-oRM algebras and not left-quasi-*aRM** algebras.

Examples of super q-i-hubs are given later in this section.

We analyse now the connection between the (super) q-i-hubs and the i-hubs.

Any i-hub is a (super) q-i-hub. Conversely, any (super) q-i-hub verifying (pM) is an i- hub. Hence, we have the following equivalence, by (Q):

(super) q-i-hub + (pM) $\equiv$ **i-hub.**

Following the general definition, a (super) q-i-hub is *q-proper*, if it is not an i-hub (i.e. if (pM) is not verified).

A q-i-hub is *proper*, if (pEx) is not verified and if is not super. A super q-i-hub is *proper*, if (pEx) is not verified.

We have the following result.

Theorem 11.3.8 *Let $\mathcal{A} = (A, \to, \leadsto, 1)$ be a proper q-proper (super) q-i-hub and consider the regular set $R(A) = \{x \in A \mid x = 1 \to x = 1 \leadsto x\}$. Then,*

$$\mathcal{R}(\mathcal{A}) = (R(A), \to, \leadsto, 1)$$

is an i-hub.

Proof. By Theorem 5.1.13. □

Since an implicative-hub may be a proper implicative-hub or an implicative-group, we then introduce the following definitions:

Definition 11.3.9 Let $\mathcal{A} = (A, \to, \leadsto, 1)$ be a proper q-proper (super) q-i-hub. We shall say that $\mathcal{A}$ is:
- *pure*, if $\mathcal{R}(\mathcal{A})$ is a proper implicative-hub,
- of *implicative-group type*, if $\mathcal{R}(\mathcal{A})$ is an implicative-group.

We have not found examples of proper q-proper (super) q-i-hubs of implicative-group type.

Proposition 11.3.10 *Let $\mathcal{A} = (A, \to, \leadsto, 1)$ be a (super) q-i-hub. Then, the following equivalences hold:*

$$(pEx) \iff (pBB =) \iff (pB =).$$

Proof. By Proposition 11.3.2 (q21="),
(pEx) + (q-pM) + (q-pEq=) imply (pBB=) ⇔ (pB=).
Conversely,
by Proposition 11.3.2 (q5=), (pBB=) + (pRe') + (q-pM) + (IdEq) imply (q-pR1),
by Proposition 11.3.2 (q6=), (pBB=) + (pRe') + (q-pM) + (q-pR1) imply (pD^L),
by Proposition 11.3.2 (q8=), (pBB=) + (pD^L) imply (pC^L),
by Proposition 5.1.25 (q-2), (pC^L) + (q-pM) + (q-pAn^L) imply (pEx^L). □

The involutive (super) q-i-hubs

Given a pure q-i-hub $\mathcal{A} = (A, \to, \leadsto, 1)$, its regular algebra $(R(A), \to, \leadsto, 1)$ is an i-hub. If (Id) holds, then a regular negation $^{-1} : R(A) \longrightarrow R(A)$ exists: $x^{-1} \overset{def.}{=} x \to 1 \overset{(Id)}{=} x \leadsto 1$.

Definition 11.3.11 A *quasi-negation* $^-$ on A is an extension $^- : A \longrightarrow A$ of the regular negation $^{-1}$ defined on $R(A)$ (i.e. $^- |_{R(A)} = {}^{-1}$), verifying: for all $x \in A$,
(q-pNeg) $\quad x \to 1 = 1 \to x^- = (1 \leadsto x)^- = x \leadsto 1 = 1 \leadsto x^- = (1 \to x)^-$.

If the quasi-negation exists, it can be *involutive* or not, i.e. verify or not (DN).

If the quasi-negation $^-$ exists and is involutive (i.e. verifies (DN): $(x^-)^- = x$, for all x), then we say that $\mathcal{A}$ is an *involutive* q-i-hub, with $^-$ its involutive quasi-negation.

Remark 11.3.12 The involutive q-i-hub $(A, \to, \leadsto, 1)$, with $^-$ its quasi-negation, is a left-quasi-pseudo-algebra with an involutive quasi-negation, by Remark 5.1.42.

Denote by **involutive q-i-hub** the class of all involutive q-i-hubs.

Note that an involutive q-i-hub may satisfy or may not satisfy the property (pDNeg2) ($y^- \to x = x^- \rightsquigarrow y$, for all x, y).

Note also that we have:

involutive q-i-hub + (pM) $\equiv$ **involutive i-hub.**

Examples 11.3.13 Consider the following tables:

$\to_1$	a	b	c	1
a	1	a	a	a
b	a	1	a	a
c	a	a	1	1
1	a	b	1	1

$\to_{1174}$	a	b	c	1
a	1	1	a	a
b	1	1	a	c
c	a	a	1	c
1	a	a	c	1

$\to_4$	a	b	c	1
a	1	a	a	a
b	a	1	a	b
c	a	a	1	1
1	a	b	1	1

.

(1) ($A_4 = \{a, b, c, 1\}, \to_1, 1$) is a **proper q-proper commutative q-i-hub, non-involutive**, because $(a, b, 1)^{-1} = (a, a, 1)$ does not verify (DN), even if the quasi-negation $^-$ exists (if (q-Neg) is verified) and $(a, b, c, 1)^- = (a, a, c, 1)$.

(2) ($A_4 = \{a, b, c, 1\}, \to = \to_{1174}, 1$) is a **proper q-proper commutative q-i-hub, non-involutive**, because $(a, c, 1)^{-1} = (a, c, 1)$ verifies (DN), but the quasi-negation does not exist ((q-Neg) is not verified for $x = b$: $c = b \to 1 = 1 \to b^- \neq (1 \to b)^- = a^- = a$).

(3) ($A_4 = \{a, b, c, 1\}, \to_4, 1$) is a **proper q-proper commutative involutive q-i-hub**, with $(a, b, 1)^{-1} = (a, b, 1)$ and $(a, b, c, 1)^- = (\mathbf{a}, \mathbf{b}, c, \mathbf{1})$.

Examples 11.3.14 Consider now the following tables:

$\to_1$	a	b	c	1
a	1	a	a	a
b	a	1	a	a
c	a	b	1	1
1	a	b	1	1

$\to_{12}$	a	b	c	1
a	1	a	a	a
b	b	1	b	b
c	a	b	1	1
1	a	b	1	1

$\to_{72}$	a	b	c	1
a	1	b	1	b
b	a	1	a	a
c	1	b	1	b
1	a	b	a	1

.

(1) ($A_4 = \{a, b, c, 1\}, \to_1, 1$) is a **proper q-proper commutative super q-i-hub, non-involutive**, because $(a, b, 1)^{-1} = (a, a, 1)$ does not verify (DN), but the quasi-negation $^-$ exists ((q-Neg) holds) and $(a, b, c, 1)^- = (a, a, c, 1)$.

(2) ($A_4 = \{a, b, c, 1\}, \to_{12}, 1$) is a **proper q-proper commutative involutive super q-i-hub**, with $(a, b, 1)^{-1} = (a, b, 1)$ and $(a, b, c, 1)^- = (\mathbf{a}, \mathbf{b}, c, \mathbf{1})$.

(3) ($A_4 = \{a, b, c, 1\}, \to_{72}, 1$) is a **proper q-proper commutative super q-i-hub, non-involutive**, because $(a, b, 1)^{-1} = (b, a, 1)$ verifies (DN), the quasi-negation $^-$ exists, but $(a, b, c, 1)^- = (b, a, b, 1)$ does not verify (DN).

• In the "world" of q-pM algebras with additional unary operation

Definitions 11.3.15

1. A *quasi-implicative-moon*, or *q-implicative-moon* or even *q-i-moon* for short, is an algebra $(A, \to, \rightsquigarrow, ^-, 1)$ of type $(2, 2, 1, 0)$ verifying (q-pM), (pM1), (IdEq), (q-pEq=), (Id), (q-pNeg).
2. A q-i-moon is *super*, if (p*') and (p**') hold.
3. A q-i-moon is *involutive*, if (DN) holds.

Denote by **(involutive, super) q-i-moon** the class of all (involutive, super, respectively) q-i-moons.

A (super) q-i-moon is commutative, if $\rightarrow=\rightsquigarrow$.

Note that if $(A, \rightarrow, \rightsquigarrow, ^-, 1)$ is a (super) q-i-moon, then $(A, \rightarrow, \rightsquigarrow, 1)$ is a (super, respectively) q-i-hub.

Note also that a (super) q-i-moon is a left-quasi-pseudo-algebra with quasi-negation, as this notion was defined in Chapter 5.

We analyse now the connection between the (super) q-i-moons and the i-moons. We obviously have:

(super) q-i-moon + (pM) $\equiv$ **i-moon**.

Following the general definition, a (super) q-i-moon is *q-proper*, if it is not an i-moon (i.e. if (pM) is not verified).

A q-i-moon is *proper*, if (pEx) is not verified and if is not super. A super q-i-moon is *proper*, if (pEx) is not verified.

We have the following result.

Theorem 11.3.16 *Let $\mathcal{A} = (A, \rightarrow, \rightsquigarrow, ^-, 1)$ be a proper q-proper (super) q-i-moon and consider the regular set $R(A) = \{x \in A \mid x = 1 \rightarrow x = 1 \rightsquigarrow x\}$. Then,*

$$\mathcal{R}(\mathcal{A}) = (R(A), \rightarrow, \rightsquigarrow, ^{-1}, 1)$$

is an i-moon.

Proof. By Theorem 5.1.13. □

Since an i-moon may be a proper i-moon or an i-group (= exchanged i-moon (Definition 5')), we then introduce the following definitions:

Definition 11.3.17 Let $\mathcal{A} = (A, \rightarrow, \rightsquigarrow, ^-, 1)$ be a proper q-proper (super) q-i-moon. We shall say that $\mathcal{A}$ is:
- *pure*, if $\mathcal{R}(\mathcal{A})$ is a proper i-moon,
- of *implicative-group type*, if $\mathcal{R}(\mathcal{A})$ is an i-group (Definition 5').

Proposition 11.3.18 *Let $(A, \rightarrow, \rightsquigarrow, ^-, 1)$ be a q-i-moon. Then, (N1) ($1^- = 1$) holds.*

Proof. $1^- \overset{(pM1)}{=} (1 \rightarrow 1)^- \overset{(q-pNeg)}{=} 1 \rightarrow 1 \overset{(pM1)}{=} 1.$ □

The following theorem establishes the categorically equivalence between the involutive q-i-hubs and the involutive q-i-moons.

Theorem 11.3.19

(1) Let $\mathcal{A} = (A, \rightarrow, \rightsquigarrow, 1)$ be an involutive i-hub and $^-$ be its involutive quasi-negation. Define $F(\mathcal{A}) = (A, \rightarrow, \rightsquigarrow, ^-, 1)$.

Then, $F(\mathcal{A})$ is an involutive q-i-moon.

(1') Conversely, let $(A, \rightarrow, \rightsquigarrow, ^-, 1)$ be an involutive q-i-moon. Define $G(\mathcal{A}) = (A, \rightarrow, \rightsquigarrow, 1)$.

Then, $G(\mathcal{A})$ is an involutive q-i-hub and $^-$ is its involutive quasi-negation.

(2) The mapps F and G are mutually inverse.

Proof. Obviously, by definitions. □

Hence, we have the equivalence:

(qIHM0) **involutive q-i-hub** $\Longleftrightarrow$ **involutive q-i-moon.**

Examples 11.3.20 (See Examples 11.3.13)

Consider the following tables:

$\rightarrow_1$	a	b	c	1
a	1	a	a	a
b	a	1	a	a
c	a	a	1	1
1	a	b	1	1

$\rightarrow_4$	a	b	c	1
a	1	a	a	a
b	a	1	a	b
c	a	a	1	1
1	a	b	1	1

(1) $(A_4 = \{a,b,c,1\}, \rightarrow_1, ^-, 1)$ is a **proper q-proper commutative q-i-moon, non-involutive**, with $(a,b,c,1)^- = (a,a,c,1)$.

(2) $(A_4 = \{a,b,c,1\}, \rightarrow_4, 1)$ is a **proper q-proper commutative involutive q-i-moon**, with $(a,b,c,1)^- = (\mathbf{a},\mathbf{b},c,\mathbf{1})$.

Consider the following properties:

(q-$\overline{pM}$) $\quad 1 \rightarrow (x \rightsquigarrow y)^- = (x \rightsquigarrow y)^-, \quad 1 \rightsquigarrow (x \rightarrow y)^- = (x \rightarrow y)^-$,

(q-$pM(1 \rightarrow x, 1 \rightsquigarrow x)$) $\quad 1 \rightarrow (1 \rightsquigarrow x)^- = (1 \rightsquigarrow x)^-, 1 \rightsquigarrow (1 \rightarrow x)^- = (1 \rightarrow x)^-$,

(q-pNeg) $\quad x \rightarrow 1 = 1 \rightarrow x^- = (1 \rightsquigarrow x)^- = x \rightsquigarrow 1 = 1 \rightsquigarrow x^- = (1 \rightarrow x)^-$,

(IdR) $\quad 1 \rightarrow x = 1 \rightsquigarrow x$.

Proposition 11.3.21 *(See Proposition 8.1.17)*

Let $\mathcal{G} = (G, \rightarrow, \rightsquigarrow, ^-, 1)$ be an involutive q-i-moon. Then,

(q-n1) (q-pM) + (q-pNeg) $\Longrightarrow$ (q-$\overline{pM}$),

(q-n2) (q-$\overline{pM}$) $\Longrightarrow$ (q-$pM(1 \rightarrow x, 1 \rightsquigarrow x)$),

(q-n3) (q-pNeg) + (DN) $\Longrightarrow$ (IdR),

i.e. the properties (q-$\overline{pM}$), (q-$pM(1 \rightarrow x, 1 \rightsquigarrow x)$), (IdR) hold.

Proof. (q-n1): $(x \rightsquigarrow y)^- \stackrel{(q-pM)}{=} (1 \rightarrow (x \rightsquigarrow y))^- \stackrel{(q-pNeg)}{=} 1 \rightarrow (x \rightsquigarrow y)^-$ and $(x \rightarrow y)^- \stackrel{(q-pM)}{=} (1 \rightsquigarrow (x \rightarrow y))^- \stackrel{(q-pNeg)}{=} 1 \rightsquigarrow (x \rightarrow y)^-$; thus, (q-$\overline{pM}$) holds.

(q-n2): Obviously.

(qn-3): By (q-pNeg), $(1 \rightarrow x)^- = (1 \rightsquigarrow x)^-$; then, $((1 \rightarrow x)^-)^- = ((1 \rightsquigarrow x)^-)^-$, i.e. $1 \rightarrow x = 1 \rightsquigarrow x$, by (DN); thus, (IdR) holds. □

Consider now the following properties:

(pNg) $\quad (x \rightarrow y)^- = y \rightarrow x, \quad (x \rightsquigarrow y)^- = y \rightsquigarrow x$,

(pNg1) $\quad x \rightarrow y^- = y \rightsquigarrow x^-$,

(pDNeg1) $\quad x \rightarrow y = y^- \rightsquigarrow x^-, \quad x \rightsquigarrow y = y^- \rightarrow x^-$,

(pDNeg2) $\quad y^- \rightarrow x = x^- \rightsquigarrow y$,

(pDNeg3) $\quad (x \rightarrow y^-)^- = (y \rightsquigarrow x^-)^-$,

(Id) $\quad x \rightarrow 1 = x \rightsquigarrow 1$.

Proposition 11.3.22 *Let $\mathcal{G} = (G, \to, \rightsquigarrow, ^-, 1)$ be an algebra of type $(2,2,1,0)$. Then,*
(q-ng0) (pBB=) + (pM1) + (DN) + (pDNeg2) + (N1) + (IdR) + (q-$\overline{\text{pM}}$)
$\Longrightarrow$ (pNg),
(q-ng0') (pNg) + (N1) + (pNg1) + (IdR) $\Longrightarrow$ (q-pNeg),
(q-ng1) (pDNeg2) + (DN) $\Longrightarrow$ (pDNeg1),
(q-ng1') (pDNeg2) + (DN) + (IdR) $\Longrightarrow$ (Id),
(q-ng2) (pDNeg1) + (DN) $\Longrightarrow$ (pNg1),
(q-ng3) (pNg1) $\Longrightarrow$ (pDNeg3),
(q-ng4) (pNg1) + (N1) + (q-pII) + (pM1) + (pDNeg1) + (IdR) + (q-$\overline{\text{pM}(1 \to x, 1 \rightsquigarrow x)}$)
$\Longrightarrow$ (q-pNeg),
(q-ng4') (pNg1) + (N1) + (q-pD1=) + (pDNeg1) + (IdR) + (q-$\overline{\text{pM}(1 \to x, 1 \rightsquigarrow x)}$)
$\Longrightarrow$ (q-pNeg).

Proof. (q-ng0): $y \to x \overset{(pBB=)}{=} (x \to y) \rightsquigarrow (y \to y) \overset{(pM1)}{=} (x \to y) \rightsquigarrow 1 \overset{(DN)}{=}$ $((x \to y)^-)^- \rightsquigarrow 1 \overset{(pDNeg2)}{=} 1^- \to (x \to y)^- \overset{(N1)}{=} 1 \to (x \to y)^- \overset{(IdR)}{=}$ $1 \rightsquigarrow (x \to y)^- \overset{(q-\overline{pM})}{=} (x \to y)^-$ and

$y \rightsquigarrow x \overset{(pBB=)}{=} (x \rightsquigarrow y) \to (y \rightsquigarrow y) \overset{(pM1)}{=} (x \rightsquigarrow y) \to 1 \overset{(DN)}{=}$ $((x \rightsquigarrow y)^-)^- \to 1 \overset{(pDNeg2)}{=} 1^- \rightsquigarrow (x \rightsquigarrow y)^- \overset{(N1)}{=} 1 \rightsquigarrow (x \rightsquigarrow y)^- \overset{(IdR)}{=}$ $1 \to (x \rightsquigarrow y)^- \overset{(q-\overline{pM})}{=} (x \rightsquigarrow y)^-$. Thus, (pNg) holds.

(q-ng0'): $1 \to x^- \overset{(pNg1)}{=} x \rightsquigarrow 1^- \overset{(N1)}{=} x \rightsquigarrow 1$ and
$x \rightsquigarrow 1 \overset{(pNg)}{=} (1 \rightsquigarrow x)^-$. Then apply (IdR) to obtain (q-pNeg).

(q-ng1): Take $y = y^-$ in (pDNeg2), to obtain: $y \to x \overset{(DN)}{=} (y^-)^- \to x = x^- \rightsquigarrow y^-$; also, take $x = x^-$ in (pDNeg2), to obtain: $x \rightsquigarrow y \overset{(DN)}{=} (x^-)^- \rightsquigarrow y = y^- \to x^-$; thus, (pDNeg1) holds.

(q-ng1'): $x \to 1 \overset{(DN)}{=} (x^-)^- \to 1 \overset{(pDNeg2)}{=} 1^- \rightsquigarrow x^- \overset{(N1)}{=} 1 \rightsquigarrow x^- \overset{(IdR)}{=} 1 \to x^- \overset{(N1)}{=} 1^- \to x^- \overset{(pDNeg2)}{=} (x^-)^- \rightsquigarrow 1 \overset{(DN)}{=} x \rightsquigarrow 1$; thus, (Id) holds.

(q-ng2): $x \to y^- \overset{(pDNeg1)}{=} (y^-)^- \rightsquigarrow x^- \overset{(DN)}{=} y \rightsquigarrow x^-$; thus, (pNg1) holds.

(q-ng3): Obviously.

(q-ng4): $1 \to x^- \overset{(pNg1)}{=} x \rightsquigarrow 1^- \overset{(N1)}{=} x \rightsquigarrow 1$ and
$x \rightsquigarrow 1 \overset{(q-pII)}{=} (1 \rightsquigarrow x) \rightsquigarrow (1 \rightsquigarrow 1) \overset{(pM1)}{=} (1 \rightsquigarrow x) \rightsquigarrow 1 \overset{(pDNeg1)}{=} 1^- \to (1 \rightsquigarrow x)^- \overset{(N1)}{=}$ $1 \to (1 \rightsquigarrow x)^- \overset{(q-\overline{pM(1\to x, 1\rightsquigarrow x)})}{=} (1 \rightsquigarrow x)^-$. Then, apply (IdR) to obtain (q-pNeg).

(q-ng4'): $1 \to x^- \overset{(pNg1)}{=} x \rightsquigarrow 1^- \overset{(N1)}{=} x \rightsquigarrow 1$ and
$(1 \rightsquigarrow x)^- \overset{(q-\overline{pM(1\to x, 1\rightsquigarrow x)})}{=} 1 \to (1 \rightsquigarrow x)^- \overset{(N1)}{=} 1^- \to (1 \rightsquigarrow x)^- \overset{(pDNeg1)}{=}$ $(1 \rightsquigarrow x) \rightsquigarrow 1 \overset{(pDNeg1)}{=} (x^- \to 1^-) \rightsquigarrow 1 \overset{(N1)}{=} (x^- \to 1) \rightsquigarrow 1 \overset{(q-pD1=)}{=} 1 \to x^-$. Then, apply (IdR) to obtain (q-pNeg). □

11.3.2 The exchanged (involutive) q-i-hub, q-i-moon vs. the q-i-group

We introduce now and study the exchanged q-i-hub (four definitions) (introduced under the name of *residoid* in [105]) and the exchanged involutive q-i-hub. We also introduce and study the exchanged q-i-moon and the exchanged involutive q-i-moon. Finally, we introduce the q-i-group and show their connections.

- **In the "world" of q-pM algebras**

Definition 11.3.23 (Definition 1 of exchanged q-i-hubs)
An *exchanged q-implicative-hub*, or *exchanged q-i-hub* for short, is an algebra $\mathcal{G} = (G, \rightarrow, \rightsquigarrow, 1)$ of type $(2, 2, 0)$ verifying: for all $x, y, z \in G$,
(IdEq) $x \rightarrow y = 1 \Leftrightarrow x \rightsquigarrow y = 1$,
(pBB=) $y \rightarrow z = (z \rightarrow x) \rightsquigarrow (y \rightarrow x)$, $y \rightsquigarrow z = (z \rightsquigarrow x) \rightarrow (y \rightsquigarrow x)$,
(pRe') $x \rightarrow x = 1 = x \rightsquigarrow x$.

Denote by **exchanged q-i-hub** the class of all exchanged q-i-hubs.
An exchanged q-i-hub is *commutative* if $\rightarrow = \rightsquigarrow$.

Proposition 11.3.24 *Let $\mathcal{G}$ be an exchanged q-i-hub. Then, the following properties hold: for all $x, y \in G$,*
(pM1) $1 \rightarrow 1 = 1 = 1 \rightsquigarrow 1$,
(q-pM) $1 \rightarrow (x \rightsquigarrow y) = x \rightsquigarrow y$, $1 \rightsquigarrow (x \rightarrow y) = x \rightarrow y$,
(q-pN) $1 \rightarrow (x \rightsquigarrow y) = 1 \Longrightarrow x \rightsquigarrow y = 1$, $1 \rightsquigarrow (x \rightarrow y) = 1 \Longrightarrow x \rightarrow y = 1$.

Proof. (pM1): Take $x = 1$ in (pRe').
(q-pM): $1 \rightarrow (x \rightsquigarrow y) \overset{(pRe')}{=} (y \rightsquigarrow y) \rightarrow (x \rightsquigarrow y) \overset{(pBB=)}{=} x \rightsquigarrow y$ and $1 \rightsquigarrow (x \rightarrow y) \overset{(pRe')}{=} (y \rightarrow y) \rightsquigarrow (x \rightarrow y) \overset{(pBB=)}{=} x \rightarrow y$.
(q-pN): By (q-pM). □

Remark 11.3.25 The properties (q-pM) and (pM1) say that an exchanged q-i-hub is a *left-quasi-pseudo-algebra*, as this notion was defined in Chapter 5. Hence, Propositions 5.1.19 and 5.1.25 hold.

Theorem 11.3.26
(1) Let $\mathcal{G} = (G, \rightarrow, \rightsquigarrow, 1)$ be an algebra of type $(2, 2, 0)$ verifying (IdEq), (q-pM), (pM1) and (pBB=). Then, $\mathcal{G}$ is an exchanged q-i-hub.
(1') An exchanged q-i-hub verifies the properties (IdEq), (q-pM), (pM1), (pBB=).

Proof. (1): We must prove that (pRe') holds.
First, we prove that the following properties hold:
(q-pRe($1 \rightarrow x$)) $(1 \rightarrow x) \rightsquigarrow (1 \rightarrow x) = 1 (\Leftrightarrow (1 \rightarrow x) \rightarrow (1 \rightarrow x) = 1)$,
(q-pRe($1 \rightsquigarrow x$)) $(1 \rightsquigarrow x) \rightarrow (1 \rightsquigarrow x) = 1 (\Leftrightarrow (1 \rightsquigarrow x) \rightsquigarrow (1 \rightsquigarrow x) = 1)$.
Indeed, $(1 \rightarrow x) \rightsquigarrow (1 \rightarrow x) \overset{(pBB=)}{=} 1 \rightarrow 1 \overset{(pM1)}{=} 1$; then, by (IdEq), we also have that $(1 \rightarrow x) \rightarrow (1 \rightarrow x) = 1$; thus, (q-pRe($1 \rightarrow x$)) holds.

Similarly, $(1 \rightsquigarrow x) \to (1 \rightsquigarrow x) \overset{(pBB=)}{=} 1 \rightsquigarrow 1 \overset{(pM1)}{=} 1$; then, by (IdEq), we also have that $(1 \rightsquigarrow x) \rightsquigarrow (1 \rightsquigarrow x) = 1$; thus, (q-pRe($1 \rightsquigarrow x$)) holds.

Now, we prove hat (pRe') holds. Indeed,
$x \to x \overset{(pBB=)}{=} (x \to x) \rightsquigarrow (x \to x) \overset{(q-pM)}{=}$
$(1 \rightsquigarrow (x \to x)) \rightsquigarrow (1 \rightsquigarrow (x \to x)) \overset{(q-pRe(1\rightsquigarrow x))}{=} 1$ or, similarly,
$x \rightsquigarrow x \overset{(pBB=)}{=} (x \rightsquigarrow x) \to (x \rightsquigarrow x) \overset{(q-pM)}{=}$
$(1 \to (x \rightsquigarrow x)) \to (1 \to (x \rightsquigarrow x)) \overset{(q-pRe(1\to x))}{=} 1$. Then apply (IdEq).

(1'): By definition and by Proposition 11.3.24. □

Hence, we have the following equivalent definition of exchanged q-i-hubs.

Definition 11.3.27 (Definition 2 of exchanged q-i-hubs)
An *exchanged q-i-hub* is an algebra $(G, \to, \rightsquigarrow, 1)$ of type $(2, 2, 0)$ verifying the properties: (IdEq), (q-pM), (pM1), (pBB=).

We shall prove now the following result about the exchanged q-i-hubs (the dual case is obvious).

Proposition 11.3.28 *Let $\mathcal{G} = (G, \to, \rightsquigarrow, 1)$ be an exchanged q-i-hub. Then:*
*(i) the following properties hold: (q-pAn^L), (p**^{L}), (Tr), (q-pR1), (pD^L), (q-pD=), (pC^L), (pEx), (pB^L), (pBB^L), (pB=), (p*^{L}), (IdR), (q-pII), (q-pI1), (Id), (q-pEq=), (pRe'), (q-pD=), (q-p-s^L).*
(ii) The property (q-pEq=) has the particular form:

$$1 \to x (= 1 \rightsquigarrow x) = 1 \to y (= 1 \rightsquigarrow y) \Longleftrightarrow x \to y = 1 \; (\overset{(IdEq)}{\Longleftrightarrow} x \rightsquigarrow y = 1).$$

(iii) The following equivalences hold:

$$(pEx) \Longleftrightarrow (pBB =) \Longleftrightarrow (pB =).$$

Proof. (i): By Proposition 11.3.2 and by Proposition 5.1.25 (q-2) ((pC^L) + (q-pM) + (q-pAn) imply (pEx)).
(ii): By (IdR).
(iii) By Proposition 11.3.2 (see Proposition 11.3.10). □

We then have the following important remarks (the dual case is omitted).

Remarks 11.3.29
(1) The exchanged q-i-hub, as defined by Definition 1, is a special left-quasi-pseudo-algebra as defined in Chapter 5, namely a left-quasi-pseudo-algebra (of logic) for which $x \leq y$, defined by (11.3), becomes $1 \to x$ $(= 1 \rightsquigarrow x) = 1 \to y$ $(= 1 \rightsquigarrow y)$, and so the property (q-p-s^L) holds.
(2) The exchanged q-i-hub can be defined, equivalently, as a special left-quasi-pseudo-structure: $(G, \leq, \to, \rightsquigarrow, 1)$, where (pEq^L) becomes (q-pEq=), by Proposition 11.3.28; therefore, as left-quasi-pseudo-structure, the exchanged q-i-hub is denoted also by $(G, \to, \rightsquigarrow, 1)$.

By Proposition 11.3.28, since (p^{*L}) and (p^{**L}) hold, it follows that:

Corollary 11.3.30 *Any exchanged q-i-hub is super.*

Hence, we have the connections shown in Figure 11.7.

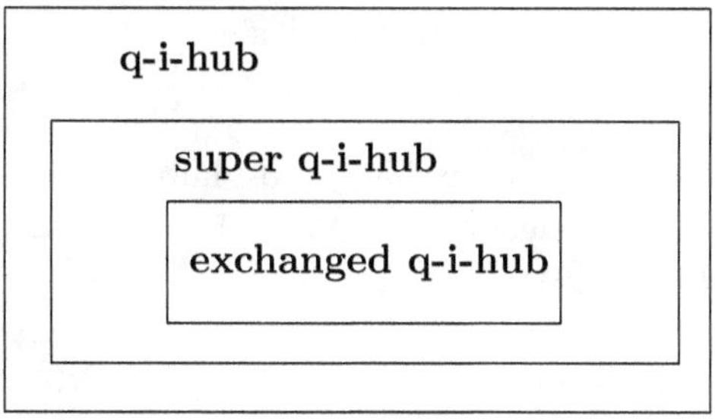

Figure 11.7: Connections

Two other equivalent definitions of exchanged q-i-hubs

There are many equivalent definitions of exchanged q-i-hubs. The following theorems provides two of them.

Theorem 11.3.31
(1) Let $\mathcal{G} = (G, \to, \rightsquigarrow, 1)$ be an algebra of type $(2, 2, 0)$ verifying (pBB=), (q-pD=), (q-pM), (IdEq), (q-pEq=). Then, $\mathcal{G}$ is a exchanged q-i-hub.
(1') An exchanged q-i-hub verifies the properties (pBB=), (q-pD=), (q-pM), (IdEq), (q-pEq=).

Proof. (1): We must prove that (pRe') holds; this follows by (q-pEq=): $1 \to x = 1 \to x$, hence $x \to x = 1 (\overset{(IdEq)}{\Longleftrightarrow} x \rightsquigarrow x = 1)$.
(1'): By Proposition 11.3.28. □

Hence, we have the following equivalent definition of exchanged q-i-hubs.

Definition 11.3.32 (Definition 3 of exchanged q-i-hubs)
An *exchanged q-i-hub* is an algebra $(G, \to, \rightsquigarrow, 1)$ of type $(2, 2, 0)$ verifying the properties: (pBB=), (q-pD=), (q-pM), (IdEq), (q-pEq=).

Theorem 11.3.33
(1) Let $\mathcal{G} = (G, \to, \rightsquigarrow, 1)$ be an algebra of type $(2, 2, 0)$ verifying (pEx), (q-pM), (pM1), (IdEq), (q-pEq=). Then, $\mathcal{G}$ is an exchanged q-i-hub.
(1') An exchanged q-i-hub verifies the properties (pEx), (q-pM), (pM1), (IdEq), (q-pEq=).

Proof. (1): We must prove that (pRe') and (pBB=) hold. Indeed, by Proposition 11.3.2 (q18=), (q-pEq=) implies (pRe'); thus, (pRe') holds.

By Proposition 11.3.2 (q13=), (pEx) + (q-pM) imply (IdR); by Proposition 11.3.2 (q24=), (pEx) + (pRe') + (q-pEq=) + (q-pM) imply (q-pD=);
by Proposition 11.3.2 (q25), (pEx) + (q-pD=) + (q-pM) + (IdR) imply (pBB=); thus, (pBB=) holds.

(1'): By Proposition 11.3.28. □

Hence, we have the following important equivalent definition of exchanged q-i-hubs.

Definition 11.3.34 (Definition 4 of exchanged q-i-hubs)

An *exchanged q-i-hub* is an algebra $(G, \rightarrow, \rightsquigarrow, 1)$ of type $(2, 2, 0)$ verifying the properties: (q-pM), (pM1), (IdEq), (q-pEq=), (pEx).

Remark 11.3.35 Since an algebra $(G, \rightarrow, \rightsquigarrow, 1)$ of type $(2, 2, 0)$ verifying the properties (q-pM), (pM1), (IdEq), (q-pEq=) is just the q-i-hub, it follows, by Definition 4, that an exchanged q-i-hubs is a q-i-hub verifying the *pseudo-Exchange property*, (pEx). This explains the name "exchanged q-i-hub".

Remark 11.3.36 By Defintion 4 of exchanged q-i-hubs and Proposition 11.3.28, any exchanged q-i-hub is a (super) q-i-hub and any (super) q-i-hub verifying (pEx) is a exchanged q-i-hub. Hence, we have the equivalence:

(super) q-i-hub + (pEx) = **exchanged q-i-hub.**

Consequently, we obtain the following equivalences:

(qIHIG) **exchanged q-i-hub** = **q-i-hub** + (pEx) ≡ **q-i-hub** + (pBB=)
≡ **q-i-hub** + (pB=).

We obtain immediately, by Proposition 11.3.28, the following important result (the dual case is omitted).

Proposition 11.3.37 *Any exchanged q-i-hub is a left-q-pBCI algebra verifying (q-p-s^L).*

The connection between the exchanged q-i-hubs and the i-groups

Any i-group is an exchanged q-i-hub (Definition 1), since (IdEq), (pBB=), (pRe') hold. Conversely, we have obviously that any exchanged q-i-hub verifying (pM) is an i-group. Hence, we have the equivalence, by (Q):

exchanged q-i-hub + (pM) ≡ **exchanged i-hub** = **i-group** (Def. 5).

We shall say that an exchanged q-i-hub is *q-proper*, if it is not an i-group (i.e. if (pM) is not verified).

Theorem 11.3.38 *Let $\mathcal{G} = (G, \to, \rightsquigarrow, 1)$ be a q-proper exchanged q-i-hub and consider the regular set $R(G) = \{x \in G \mid x = 1 \to x = 1 \rightsquigarrow x\}$. Then,*

$$\mathcal{R}(\mathcal{G}) = (R(G), \to, \rightsquigarrow, 1)$$

is an i-group (as exchanged i-hub, by Definition 5).

Proof. By Theorem 5.1.13, $\mathcal{R}(\mathcal{G})$ is a regular pseudo-algebra; it verifies (pM) and (pBB=), hence it is an implicative-group. □

Lemma 11.3.39 *Let $\mathcal{G} = (G, \to, \rightsquigarrow, 1)$ be a q-proper exchanged q-i-hub. For any $x \in G$, if $x \notin R(G)$, then*

$$r_x \overset{notation}{=} 1 \to x \overset{(IdR)}{=} 1 \rightsquigarrow x \in R(G).$$

Proof. $1 \to r_x = 1 \to (1 \rightsquigarrow x) \overset{(q-pM)}{=} 1 \rightsquigarrow x = r_x$ and $1 \rightsquigarrow r_x = 1 \rightsquigarrow (1 \to x) \overset{(q-pM)}{=} 1 \to x = r_x$; thus, $r_x \in R(G)$. □

- **New operation in exchanged q-i-hubs:** $\cdot$

Let $\mathcal{G} = (G, \to, \rightsquigarrow, 1)$ be an exchanged q-i-hub. Hence, $(R(G), \to, \rightsquigarrow, 1)$ is an i-group (= exchanged i-hub), by Theorem 11.3.38. We shall extend to $\mathcal{G}$ the new operation $\cdot$ defined on $\mathcal{R}(\mathcal{G})$ by (8.5). Thus, we define the new operation $\cdot$ on G, by using Lemma 11.3.39, as follows: for all $x, y \in G$,

$$x \cdot y \overset{def.}{=} \begin{cases} x \cdot y, & \text{if} \quad x, y \in R(G), \\ r_x \cdot y, & \text{if} \quad x \notin R(G),\ r_x = 1 \to x \in R(G), \quad y \in R(G), \\ x \cdot r_y, & \text{if} \quad x \in R(G), \quad y \notin R(G),\ r_y = 1 \to y \in R(G), \\ r_x \cdot r_y, & \text{if} \quad x, y \notin R(G), r_x = 1 \to x,\ r_y = 1 \to y,\ r_x, r_y \in R(G). \end{cases}$$

The following result is useful further in Chapter 12.

Proposition 11.3.40 *Let $\mathcal{G} = (G, \to, \rightsquigarrow, 1)$ be an exchanged q-i-hub and consider the new operation $\cdot$ above defined on G. Then, the properties (qm-pU), (U1), (Ass), (pIv') hold.*

Proof. By Theorem 11.3.38, $(R(G), \to, \rightsquigarrow, 1)$ is an i-group (hence $(R(G), \cdot, 1)$ is a group). By Proposition 8.1.32, (pU), (pIv) and (Ass) hold in $R(G)$; then, by Proposition 5.2.1, (pU) implies (qm-pU) and (U1), and since (pIv) holds, it follows immediately that (pIv') holds. □

Corollary 11.3.41 *Let $\mathcal{G} = (G, \to, \rightsquigarrow, 1)$ be an exchanged q-i-hub and consider the operation $\cdot$ above defined. Then, $(G, \cdot, 1)$ is an associative qm-hub.*

The involutive exchanged q-i-hub

As we have seen (see Remark 11.3.35), an *exchanged q-i-hub* is just a q-i-hub verifying (pEx).

Given a q-proper exchanged q-i-hub $\mathcal{G} = (G, \rightarrow, \rightsquigarrow, 1)$, its regular algebra $(R(G), \rightarrow, \rightsquigarrow, 1)$ is an i-group (= exchanged i-hub), with an involutive regular negation:
$x^{-1} \overset{def.}{=} x \rightarrow 1 \overset{(Id)}{=} x \rightsquigarrow 1$.

If the extended quasi-negation $^-$ on G exists (and we shall prove in Proposition 11.3.42 that it always exists) and if is *involutive*, then we have an *involutive exchanged q-i-hub*, with $^-$ its involutive quasi-negation.

Note that the *exchanged involutive q-i-hubs* coincide with the *involutive exchanged q-i-hubs.* Denote by **exchanged involutive q-i-hub** the class of all exchanged involutive q-i-hubs.

Proposition 11.3.42 *Any q-proper exchanged q-i-hub admits a quasi-negation.*

Proof. Let $\mathcal{G} = (G, \rightarrow, \rightsquigarrow, 1)$ be a q-proper exchanged q-i-hub.

First note that, by Proposition 11.3.28, the properties (Id), (IdR), (pEx), (q-pEq=), (pRe'), (q-pD=) hold.

Its regular algebra $(R(G), \rightarrow, \rightsquigarrow, 1)$ is an i-group, with $^{-1}$ its involutive regular negation defined by: for all $x \in G$,
$x^{-1} \overset{def.}{=} x \rightarrow 1 \overset{(Id)}{=} x \rightsquigarrow 1$.

We must prove that $^{-1}$ can be extended to $^- : G \longrightarrow G$ verifying: for all $x \in G$,
(q-pNeg) $\quad x \rightarrow 1 = 1 \rightarrow x^- = (1 \rightsquigarrow x)^- \;=\; x \rightsquigarrow 1 = 1 \rightsquigarrow x^- = (1 \rightarrow x)^-$.

Note that (q-pNeg) is verified obviously by all $x \in R(G)$. It remains to prove that (q-pNeg) is verified by all $x \notin R(G)$.

Let $x \notin R(G)$ and let $r_x \overset{def.}{=} 1 \rightarrow x \overset{(IdR)}{=} 1 \rightsquigarrow x$; then, $r_x \in R(G)$, by Lemma 11.3.39. We shall prove first that:

$$x \rightarrow 1 = (1 \rightsquigarrow x)^-. \tag{11.6}$$

Indeed, $(1 \rightsquigarrow x)^- = r_x^- = r_x^{-1} = r_x \rightarrow 1 = (1 \rightsquigarrow x) \rightarrow 1$. Then,
$(x \rightarrow 1) \rightsquigarrow [(1 \rightsquigarrow x) \rightarrow 1] \overset{(pEx)}{=} (1 \rightsquigarrow x) \rightarrow [(x \rightarrow 1) \rightsquigarrow 1] \overset{(q-pD=)}{=}$
$(1 \rightsquigarrow x) \rightarrow (1 \rightarrow x) \overset{(IdR)}{=} (1 \rightarrow x) \rightarrow (1 \rightarrow x) \overset{(pRe')}{=} 1$.

Hence, by (q-pEq=), $1 \rightsquigarrow (x \rightarrow 1) = 1 \rightsquigarrow [(1 \rightsquigarrow x) \rightarrow 1]$, i.e.
$x \rightarrow 1 = (1 \rightsquigarrow x) \rightarrow 1$, by (q-pM); thus, (11.6) holds.

Now, by (IdR) and (Id),
$Q \overset{notation}{=} x \rightarrow 1 = (1 \rightsquigarrow x)^- \;=\; x \rightsquigarrow 1 = 1 \rightsquigarrow x^- \;=\; (1 \rightarrow x)^-$; hence, the quasi-negation $^-$ exists. It is (they are) determined by: $1 \rightarrow x^- = 1 \rightsquigarrow x^- = Q$. □

Examples 11.3.43

(1) The **commutative exchanged q-i-hub** $\mathcal{EXQIH}^1 = (A_4 = \{a, b, c, 1\}, \rightarrow_2, 1)$,

where

$\rightarrow_2$	a	b	c	1
a	1	a	a	b
b	b	1	1	a
c	b	1	1	a
1	**a**	**b**	b	**1**

. Note that $R(A_4) = \{a, b, 1\}$,

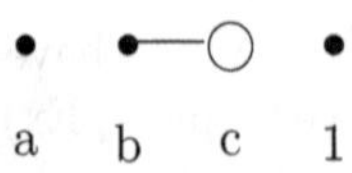

and

$\rightarrow_2$	a	b	1
a	1	a	b
a	b	1	a
1	a	b	1

, hence $(\{a, b, 1\}, \rightarrow_2, 1)$ is the commutative i-group (exchanged i-hub) with 3 elements, with the involutive regular negation $(a, b, 1)^{-1} = (b, a, 1)$.

By (q-Neg), $a = c \rightarrow_2 1 = 1 \rightarrow_2 c^- = (1 \rightarrow_2 c)^- = b^- = a$, hence $c^- = a$; thus, the unique quasi-negation $(a, b, c, 1)^- = (\mathbf{b}, \mathbf{a}, a, \mathbf{1})$ is not involutive. Thus, $\mathcal{EXQIH}^2$ is **not involutive**.

(2) The **commutative exchanged q-i-hub** $\mathcal{EXQIH}^2 = (A_4 = \{a, b, c, 1\}, \rightarrow_4, 1)$,

where

$\rightarrow_4$	a	b	c	1
a	1	a	b	b
b	b	1	a	a
c	a	b	1	1
1	**a**	**b**	1	**1**

. Note that $R(A_4) = \{a, b, 1\}$,

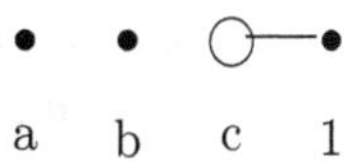

and

$\rightarrow_4$	a	b	1
a	1	a	b
a	b	1	a
1	a	b	1

, hence $(\{a, b, 1\}, \rightarrow_4, 1)$ is the commutative i-group (exchanged i-hub) with 3 elements, with $(a, b, 1)^{-1} = (b, a, 1)$.

By (q-Neg), $1 = c \rightarrow_4 1 = 1 \rightarrow_4 c^- = (1 \rightarrow_4 c)^- = 1^- = 1$, hence $c^- = c, 1$; thus, there are two quasi-negations:
- $(a, b, c, 1)^{-_1} = (\mathbf{b}, \mathbf{a}, c, \mathbf{1})$ is involutive;
- $(a, b, c, 1)^{-_2} = (\mathbf{b}, \mathbf{a}, 1, \mathbf{1})$ is not involutive.

Thus, $\mathcal{EXQIH}^2$ is **involutive**, with $^{-_1}$ its involutive quasi-negation.

Remarks 11.3.44 (See Remarks 10.2.22)

(i) There exist *exchanged q-i-hubs* for which an involutive quasi-negation does not exist, i.e. for which there is no an associated *involutive exchanged q-i-hub.*

(ii) There exist *exchanged q-i-hubs* for which there exist more than one involutive quasi-negations, i.e. for which there exist more than one associated *involutive exchanged q-i-hubs.*

- **In the "world" of q-pM algebras with additional unary operation**

Definition 11.3.45 An *exchanged q-i-moon* is a q-i-moon verifying (pEx).

Denote by **exchanged q-i-moon** the class of all exchanged q-i-moons.

By Proposition 11.3.42, we immediately obtain that:

Corollary 11.3.46 *Any q-proper exchanged q-i-hub has at least one associated exchanged q-i-moon.*

Note that the *exchanged involutive q-i-moons* coincide with the *involutive exchanged q-i-moons.* Denote by **involutive exchanged q-i-moon** the class of all involutive exchanged q-i-moons.

Note that the (involutive) exchanged q-i-moon is also a quasi-pseudo-algebra and that, by (Q), we have:

(involutive) exchanged q-i-moon + (pM) $\equiv$ **exchanged i-moon** = **i-group**.

The connection between the involutive exchanged q-i-hubs and the involutive exchanged q-i-moons is the following.

Theorem 11.3.47
(1) Let $\mathcal{A} = (A, \to, \leadsto, 1)$ be an involutive exchanged q-i-hub, with $^-$ the involutive quasi-negation. Define $F(\mathcal{A}) = (A, \to, \leadsto, ^-, 1)$. Then, $F(\mathcal{A})$ is an involutive exchanged q-i-moon.
(1') Conversely, let $\mathcal{A} = (A, \to, \leadsto, ^-, 1)$ be an involutive exchanged q-i-moon. Define $G(\mathcal{A}) = (A, \to, \leadsto, 1)$. Then, $G(\mathcal{A})$ is an involutive exchanged q-i-hub and its involutive quasi-negation is $^-$.
(2) The mapps F and G are mutually inverse.

Proof. By Theorem 11.3.19. □

By above Theorem 11.3.47, we have the equivalence (see (qIHM0)):

(qIHM1) **involutive exchanged q-i-hub** $\Longleftrightarrow$ **involutive exchanged q-i-moon**.

Theorem 11.3.48 *Let $\mathcal{G} = (G, \to, \leadsto, ^-, 1)$ be an involutive exchanged q-i-moon. Then,*

$$\mathcal{R}(\mathcal{G}) = (R(G), \to, \leadsto, ^{-1}, 1)$$

is an i-group (= exchanged i-moon, by Definition 5'), where $^- \mid_{R(G)} = {}^{-1}$ and $x^{-1} = x \to 1 \overset{(Id)}{=} x \leadsto 1$.

Proof. By Theorem 5.1.13. □

Consider now the following properties:
(pNg) $(x \to y)^- = y \to x, \quad (x \leadsto y)^- = y \leadsto x,$
(pNg1) $x \to y^- = y \leadsto x^-,$
(pDNeg1) $x \to y = y^- \leadsto x^-, \quad x \leadsto y = y^- \to x^-,$
(pDNeg2) $y^- \to x = x^- \leadsto y,$
(pDNeg3) $(x \to y^-)^- = (y \leadsto x^-)^-.$

Proposition 11.3.49 *(See Propositions 8.1.17, 11.3.21)*
Let $\mathcal{G} = (G, \to, \leadsto, ^-, 1)$ be an involutive exchanged q-i-moon. Then,
(q-n4) (pBB=) + (pRe') + (q-pNeg) + (q-pM) $\Longrightarrow$ (pNg),
(q-n5) (q-pM) + (pEx) + (q-pNeg) $\Longrightarrow$ (pNg1),
(q-n6) (DN) + (pNg1) $\Longrightarrow$ (pDNeg1),
(q-n7) (DN) + (pNg1) $\Longrightarrow$ (pDNeg2),
(q-n8) (pNg1) $\Longrightarrow$ (pDNeg3),
i.e. the properties (pNg), (pNg1), (pDNeg1), (pDNeg2), (pDNeg3) hold.

Proof. (q-n4): $y \to x \overset{(pBB=)}{=} (x \to y) \rightsquigarrow (y \to y) \overset{(pRe')}{=} (x \to y) \rightsquigarrow 1 \overset{(q-pNeg)}{=}$ $(1 \rightsquigarrow (x \to y))^- \overset{(q-pM)}{=} (x \to y)^-$ and $y \rightsquigarrow x \overset{(pBB=)}{=} (x \rightsquigarrow y) \to (y \rightsquigarrow y) \overset{(pRe')}{=}$ $(x \rightsquigarrow y) \to 1 \overset{(q-pNeg)}{=} (1 \to (x \rightsquigarrow y))^- \overset{(q-pM)}{=} (x \rightsquigarrow y)^-$; thus, (pNg) holds.

(q-n5): $x \to y^- \overset{(q-pM)}{=} 1 \rightsquigarrow (x \to y^-) \overset{(pEx)}{=} x \to (1 \rightsquigarrow y^-) \overset{(q-pNeg)}{=}$ $x \to (y \rightsquigarrow 1)$ and $y \rightsquigarrow x^- \overset{(q-pM)}{=} 1 \to (y \rightsquigarrow x^-) \overset{(pEx)}{=} y \rightsquigarrow (1 \to x^-) \overset{(q-pNeg)}{=}$ $y \rightsquigarrow (x \to 1) \overset{(pEx)}{=} x \to (y \rightsquigarrow 1)$; thus, (pNg1) holds.

(q-n6): $x \to y \overset{(DN)}{=} x \to (y^-)^- \overset{(pNg1)}{=} y^- \rightsquigarrow x^-$ and

$x \rightsquigarrow y \overset{(DN)}{=} x \rightsquigarrow (y^-)^- \overset{(pNg1)}{=} y^- \to x^-$; thus, (pDNeg1) holds.

(q-n7): $y^- \to x \overset{(DN)}{=} y^- \to (x^-)^- \overset{(pNg1)}{=} x^- \rightsquigarrow (y^-)^- \overset{(DN)}{=} x^- \rightsquigarrow y$; thus, (pDNeg2) holds.

(q-n8): Obviously. □

Proposition 11.3.50 *(See Proposition 8.1.23)*
Let $(G, \to, \rightsquigarrow, ^-, 1)$ be an involutive exchanged q-i-moon. Then, we have: for all $x, y \in G$,

(11.7) $$(x \to y^-)^- = y^- \to x, \quad (y \rightsquigarrow x^-)^- = x^- \rightsquigarrow y.$$

Proof. By (pNg). □

- **New operation in involutive exchanged q-i-moons:** $\cdot$

Since the involutive exchanged q-i-moon is, by (DN), an involutive algebra, we then extend (8.6), by defining the new operation $\cdot$ on G as follows: for all $x, y \in G$,

(11.8) $$x \cdot y \overset{def.}{=} (x \to y^-)^- \overset{(pDNeg3)}{=} (y \rightsquigarrow x^-)^-.$$

Then, we have:

Lemma 11.3.51

(11.9) $$x \cdot y = y^- \to x \overset{(pDNeg2)}{=} x^- \rightsquigarrow y.$$

Proof. $x \cdot y \overset{def.}{=} (x \to y^-)^- \overset{(11.7)}{=} y^- \to x$. □

The following result will be used in Chapter 12.

Proposition 11.3.52 *(See Proposition 8.1.32)*
Let $(G, \to, \rightsquigarrow, ^-, 1)$ be an involutive exchanged q-i-moon. Then, the following properties hold: for all $x, y, z \in G$,

(Ass) $x \cdot (y \cdot z) = (x \cdot y) \cdot z$,
(qm-pU) $(x \cdot y) \cdot 1 = x \cdot y = 1 \cdot (x \cdot y)$,
(pIv) $x \cdot x^- = 1 = x^- \cdot x$,
(U1) $1 \cdot 1 = 1$.

Proof. (Ass): $x \cdot (y \cdot z) \stackrel{(11.9)}{=} x \cdot (z^- \to y) \stackrel{(11.9)}{=} x^- \rightsquigarrow (z^- \to y)$
and $(x \cdot y) \cdot z = (x^- \rightsquigarrow y) \cdot z = z^- \to (x^- \rightsquigarrow y) \stackrel{(pEx)}{=} x^- \rightsquigarrow (z^- \to y)$; thus, (Ass) holds.

(qm-pU): $(x \cdot y) \cdot 1 \stackrel{(11.9)}{=} (x^- \rightsquigarrow y) \cdot 1 \stackrel{(11.9)}{=} 1^- \to (x^- \rightsquigarrow y) \stackrel{(N1)}{=}$
$1 \to (x^- \rightsquigarrow y) \stackrel{(q-pM)}{=} x^- \rightsquigarrow y \stackrel{(11.9)}{=} x \cdot y$
and $1 \cdot (x \cdot y) \stackrel{(11.9)}{=} 1 \cdot (y^- \to x) = 1^- \rightsquigarrow (y^- \to x) = 1 \rightsquigarrow (y^- \to x) \stackrel{(q-pM)}{=}$
$y^- \to x = x \cdot y$.

(pIv): $x \cdot x^- \stackrel{(11.9)}{=} x^- \rightsquigarrow x^- \stackrel{(pRe')}{=} 1$ and $x^- \cdot x = x^- \to x^- = 1$.

(U1): $1 \cdot 1 \stackrel{(11.9)}{=} 1^- \to 1 \stackrel{(N1)}{=} 1 \to 1 \stackrel{(pM1)}{=} 1$. □

- **The q-implicative-group**

We introduce now the new notion of *q-implicative-group*, the analogous, in the "world" of algebras of logic (of q-pM algebras, more precisely), of the notion of *qm-group* from the "world" of qm-unital magmas.

Definition 11.3.53 A *quasi-implicative-group*, or *q-implicative-group* or even *q-i-group* for short, is an algebra $\mathcal{G} = (G, \to, \rightsquigarrow, ^-, 1)$ of type $(2, 2, 1, 0)$ verifying the following properties: for all $x, y, z \in G$,
(q-pM) $1 \to (x \rightsquigarrow y) = x \rightsquigarrow y$, $1 \rightsquigarrow (x \to y) = x \to y$,
(pM1) $1 \to 1 = 1 = 1 \rightsquigarrow 1$,
(IdEq) $x \to y = 1 \Longleftrightarrow x \rightsquigarrow y = 1$,
(q-pEq=) $(1 \to x = 1 \to y$ and $1 \rightsquigarrow x = 1 \rightsquigarrow y) \Longleftrightarrow x \to y = 1$ $(\stackrel{(IdEq)}{\Longleftrightarrow} x \rightsquigarrow y = 1)$,
(pEx) $x \to (y \rightsquigarrow z) = y \rightsquigarrow (x \to z)$,
(q-$pM(1 \to x, 1 \rightsquigarrow x)$) $1 \to (1 \rightsquigarrow x)^- = (1 \rightsquigarrow x)^-$, $1 \rightsquigarrow (1 \to x)^- = (1 \to x)^-$,
(pDNeg2) $y^- \to x = x^- \rightsquigarrow y$,
(N1) $1^- = 1$,
(DN) $(x^-)^- = x$.

Proposition 11.3.54 *Let $\mathcal{G} = (G, \to, \rightsquigarrow, ^-, 1)$ be a q-i-group. Then, (pDNeg1), (pNg1), (pDNeg3), (q-pNeg) hold.*

Proof. Since (q-pM), (pM1), (IdEq), (q-pEq=), (pEx) hold, then $(G, \to, \rightsquigarrow, 1)$ is an exchanged q-i-hub (Definition 4). Thus, (pBB=), (pRe') and all the properties from Proposition 11.3.28 hold, including (IdR), (q-pII). Then apply Proposition 11.3.22. □

Hence, we have:

q-i-group + (pM) ≡ **i-group** (Definition 5'),
q-i-group ⊆ **exchanged q-i-moon** ⊇ **involutive exchanged q-i-moon**.

Theorem 11.3.55 *The q-i-groups are equivalent to the involutive exchanged q-i-moons.*

Proof. Let $\mathcal{G} = (G, \rightarrow, \rightsquigarrow, ^-, 1)$ be a q-i-group. Since (q-pM), (pM1), (IdEq), (q-pEq=), (pEx) hold, then $(G, \rightarrow, \rightsquigarrow, 1)$ is an exchanged q-i-hub (Definition 4); then, by Proposition 11.3.28, (Id) holds; by Proposition 11.3.54, the property (q-pNeg) holds; then, $(G, \rightarrow, \rightsquigarrow, 1)$ is an involutive exchanged q-i-hub, since (DN) also holds. Hence, $\mathcal{G}$ is an involutive echanged q-i-moon, by Theorem 11.3.47.

Conversely, let $\mathcal{G} = (G, \rightarrow, \rightsquigarrow, ^-, 1)$ be an involutive exchanged q-i-moon, i.e. the properties (q-pM), (pM1), (IdEq), (q-pEq=), (Id), (q-pNeg) and also (pEx) and (DN) hold. By Propositions 11.3.18, 11.3.21, 11.3.49, the properties (N1), (q-$pM(1 \rightarrow x, 1 \rightsquigarrow x)$), (pDNeg2) also hold. Then, $\mathcal{G}$ is a q-i-group. □

By Theorems 11.3.55 and 11.3.47, we obtain:

Theorem 11.3.56 *The q-i-groups are categorically equivalent to the involutive exchanged q-i-hubs.*

Remark 11.3.57 By this Theorem 11.3.56, we could define, equivalently, the q-i-group as an involutive exchanged q-i-hub. In fact, we find examples of q-i-groups by finding examples of involutive exchanged q-i-hubs.

Note that we have the hierarchy No. q1 from Figure 11.8.

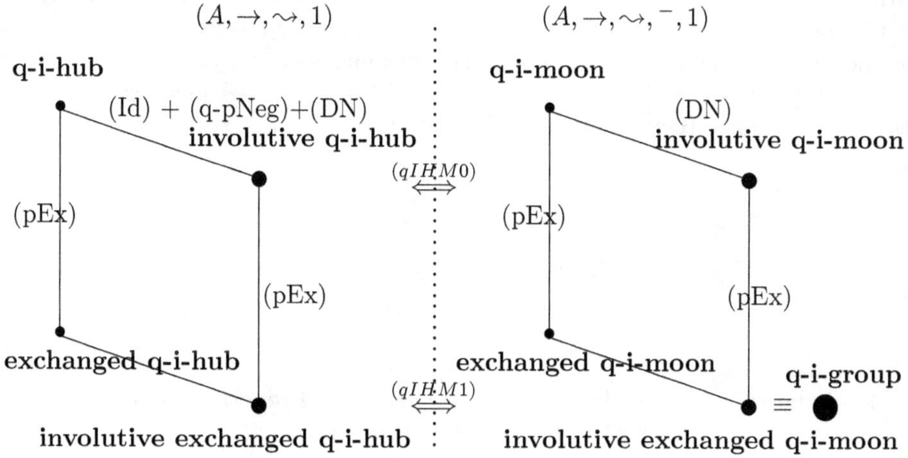

Figure 11.8: The hierarchy No. q1

Examples 11.3.58 (See Examples 11.3.43)

(1) $\mathcal{EXQIM}^1 = (A_4 = \{a, b, c, 1\}, \rightarrow_2, ^-, 1)$, where

$\rightarrow_2$	a	b	c	1
a	1	a	a	b
b	b	1	1	a
c	b	1	1	a
1	**a**	**b**	b	**1**

and $(a, b, c, 1)^- = (\mathbf{b}, \mathbf{a}, a, \mathbf{1})$, is a commutative exchanged q-i-moon that is **not involutive**, associated to $\mathcal{EXQIH}^1$ from Examples 11.3.43.

(2) $\mathcal{EXQIM}^2 = (A_4 = \{a, b, c, 1\}, \to_4, ^{-1}, 1)$, where

$\to_4$	a	b	c	1
a	1	a	b	b
b	b	1	a	a
c	a	b	1	1
1	**a**	**b**	1	**1**

and $(a, b, c, 1)^{-1} = (\mathbf{b}, \mathbf{a}, c, \mathbf{1})$, is an **involutive commutative exchanged q-i-moon**, hence is a **commutative q-i-group**, associated to $\mathcal{EXQIH}^2$ from Examples 11.3.43.

(3) $\mathcal{EXQIM}^3 = (A_4 = \{a, b, c, 1\}, \to_4, ^{-2}, 1)$, where

$\to_4$	a	b	c	1
a	1	a	b	b
b	b	1	a	a
c	a	b	1	1
1	**a**	**b**	1	**1**

and $(a, b, c, 1)^{-2} = (\mathbf{b}, \mathbf{a}, \mathbf{1}, \mathbf{1})$, is a commutative exchanged q-i-moon that is **not involutive**, associated also to $\mathcal{EXQIH}^2$ from Examples 11.3.43.

Example 11.3.59 ((Non-commutative) exchanged q-i-hub and q-i-group)

Starting from the (non-commutative) implicative-group (= exchanged implicative-hub)

$$\mathcal{G}_{6ig} = (G_6 = \{a, b, c, d, f, 1\}, \to, \rightsquigarrow, 1)$$

with $(a, b, c, d, f, 1)^{-1} = (a, b, d, c, f, 1)$ presented in Chapter 9, we can obtain an infinity of exchanged q-implicative-hubs with minimum 7 elements, which contain (as their regular part) the i-group $\mathcal{G}_{6ig}$. For example, consider the particular poset (where $\leq$ is $=$) $(G_7 = \{a, b, c, d, f, w, 1\}, \leq)$ with the quasi-element $w \parallel f$ represented by the quasi-Hasse diagram in Figure 11.9.

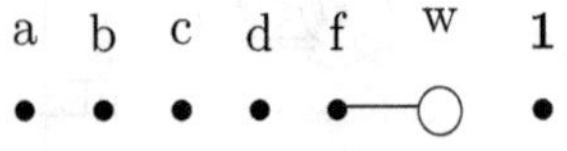

Figure 11.9: $(G_7, \leq)$

Consider the algebra $\mathcal{G}_7 = (G_7, \to, \rightsquigarrow, 1)$, where the implications $\to$ and $\rightsquigarrow$ have the following tables (note that, since $w \parallel f$, then by Proposition 5.1.30, the line (column) of w coincides with the line (column) of f):

$\to$	a	b	c	d	f	w	1
a	**1**	d	f	b	c	c	a
b	c	**1**	a	f	d	d	b
c	f	a	**1**	c	b	b	d
d	b	f	d	**1**	a	a	c
f	d	c	b	a	**1**	1	f
w	d	c	b	a	1	**1**	f
1	a	b	c	d	f	f	**1**

$\rightsquigarrow$	a	b	c	d	f	w	1
a	**1**	c	b	f	d	d	a
b	d	**1**	f	a	c	c	b
c	b	f	**1**	c	a	a	d
d	f	a	d	**1**	b	b	c
f	c	d	a	b	**1**	1	f
w	c	d	a	b	1	**1**	f
1	a	b	c	d	f	f	**1**

Then, $\mathcal{G}_7 = (G_7 = \{a, b, c, d, f, w, 1\}, \to, \rightsquigarrow, 1)$ is an *exchanged q-i-hub*, which contains as its regular part the implicative-group $\mathcal{G}_{6ig}$: $\mathcal{R}(\mathcal{G}_7) = \mathcal{G}_{6ig}$, with $R(G_7) = G_6$.

Let us determine now the *involutive exchanged q-i-hub(s)*, hence the *q-i-group(s)* associated to the *exchanged q-i-hub* $\mathcal{G}_7$. For this, we extend the *regular negation* $^{-1}: R(G_7) = G_6 \longrightarrow R(G_7) = G_6$ to a *quasi-negation* $^- : G_7 \longrightarrow G_7$ which must verify: for all $x \in G_7$,
(q-pNeg) $x \to 1 = 1 \to x^- = (1 \rightsquigarrow x)^- = x \rightsquigarrow 1 = 1 \rightsquigarrow x^- = (1 \to x)^-$.
Since $f = w \to 1 = (1 \rightsquigarrow w)^- = f^- = f^{-1} = f$ and $f = w \rightsquigarrow 1 = (1 \to w)^- = f^- = f$, then $1 \to w^- = 1 \rightsquigarrow w^- = f$. Hence, we obtain that: $w^- = f, w$. Thus, there are two quasi-negations:

x	x^{-1}		x	x^{-2}	
a	a		a	a	
b	b		b	b	
c	d	non-involutive	c	d	involutive.
d	c		d	c	
f	f		f	f	
w	f		w	w	
1	1		1	1	

Consequently, we have **one** *involutive exchanged q-i-hub*, hence **one** *q-i-group*:

$$\mathcal{QIG}_7 = (G_7, \to, \rightsquigarrow, ^{-2}, 1)$$

associated to the *exchanged q-i-hub* $\mathcal{G}_7$.

Note that this q-i-group corresponds, by Theorem 12.2.3, to the qm-group $\mathcal{QMG}_7$ presented in Chapter 10.

Remarks 11.3.60 As we have said (see Remarks 11.3.44):

(i) There exist *exchanged q-i-hubs* for which an involutive quasi-negation does not exist, i.e. for which there is no an associated *involutive exchanged q-i-hub* (hence associated *q-i-group*).

(ii) There exist *exchanged q-i-hubs* for which there exist more than one involutive quasi-negations, i.e. for which there exist more than one associated *involutive exchanged q-i-hubs* (hence one associated *q-i-group*).

11.3.3 The strong (involutive) q-i-hub, q-i-moon vs. the q-i-goop

In this section, we introduce and study the strong (super) q-i-hub/q-i-moon and the strong involutive (super) q-i-hub/q-i-moon. We also introduce and study the new notion of *q-i-goop*.

- **In the "world" of q-pM algebras**

Definition 11.3.61 A (super) q-i-hub is *strong*, if the properties (Id) and (pEx1) hold, where:
(pEx1) $x \to (y \rightsquigarrow 1) = y \rightsquigarrow (x \to 1)$, for all x, y.

Denote by **strong q-i-hub** the class of all strong q-i-hubs.

Note that the *strong involutive q-i-hubs* coincide with the *involutive strong q-i-hubs.* Denote by **involutive strong q-i-hub** the class of all involutive strong q-i-hubs.

We have, obviously:

strong (involutive) q-i-hub + (pM) $\equiv$ **strong (involutive) i-hub**.

Following By Proposition 11.3.6, it folllows that any strong super q-i-hub is a strong left-q-p*aRM** algebra (or a right-one).

- **In the "world" of q-pM algebras with additional unary operation**

Definition 11.3.62 A (super) q-i-moon is *strong*, if the propery (pEx1) holds.

Denote by **strong q-i-moon** the class of all strong q-i-moons.

Note that the *strong involutive q-i-moons* coincide with the *involutive strong q-i-moons.* Denote by **involutive strong q-i-moon** the class of all involutive strong q-i-moons.

We have, obviously:

strong (involutive) q-i-moon + (pM) $\equiv$ **strong (involutive) i-moon.**

The connection between the involutive strong q-i-hubs and the involutive strong q-i-moons is the following:

Theorem 11.3.63

(1) Let $\mathcal{A} = (A, \to, \rightsquigarrow, 1)$ *be an involutive strong q-i-hub, with the involutive quasi-negation denoted by* $^-$. *Define* $F(\mathcal{A}) = (A, \to, \rightsquigarrow, ^-, 1)$. *Then,* $F(\mathcal{A})$ *is an involutive strong q-i-moon.*

(1') Conversely, let $\mathcal{A} = (A, \to, \rightsquigarrow, ^-, 1)$ *be an involutive strong q-i-moon. Define* $G(\mathcal{A}) = (A, \to, \rightsquigarrow, 1)$. *Then,* $G(\mathcal{A})$ *is an involutive strong q-i-hub and* $^-$ *is its involutive quasi-negation.*

(2) The mapps F and G are mutually inverse.

Proof. Obviously, by definitions. □

By Theorem 11.3.63, we have the equivalence:

(qIHM2) **involutive strong q-i-hub** $\Longleftrightarrow$ **involutive strong q-i-moon.**

Consider the following new property: for all $x, y \in A$,

(pEx2) $y^- \to (x^- \rightsquigarrow 1) = y^- \to x$ and $x^- \rightsquigarrow (y^- \to 1) = x^- \rightsquigarrow y$.

Theorem 11.3.64 *Let* $\mathcal{A} = (A, \to, \rightsquigarrow, ^-, 1)$ *be an involutive q-i-moon. Then,*

(0) (p') $\Longrightarrow$ (pEx2);*

(1) if (pDNeg2) holds, then (pEx2) $\Longrightarrow$ (p');*

*(2) if (pEx1) holds (i.e. $\mathcal{A}$ is strong), then (pEx2) $\Longrightarrow$ (p**').*

Proof. (0): We prove that (p*') $\Longrightarrow$ (pEx2):
By (q-pEq=), $(1 \to X = 1 \to Y$ and $1 \rightsquigarrow X = 1 \rightsquigarrow Y) \Longleftrightarrow X \to Y = 1$.

We shall prove first: (a) $(x^- \rightsquigarrow 1) \to x = 1$.
Indeed, put $X = x^- \rightsquigarrow 1$ and $Y = x$. Then:
$1 \to X = 1 \to (x^- \rightsquigarrow 1) \overset{(q-pM)}{=} x^- \rightsquigarrow 1 \overset{(Id)}{=} x^- \to 1 \overset{(q-pNeg)}{=} 1 \to (x^-)^- \overset{(DN)}{=} 1 \to x$, $1 \to Y = 1 \to x$ and
$1 \rightsquigarrow X = 1 \rightsquigarrow (x^- \rightsquigarrow 1) \overset{(Id)}{=} 1 \rightsquigarrow (x^- \to 1) \overset{(q-pM)}{=} x^- \to 1 \overset{(Id)}{=} x^- \rightsquigarrow 1 \overset{(q-pNeg)}{=} 1 \rightsquigarrow (x^-)^- \overset{(DN)}{=} 1 \rightsquigarrow x$, $1 \rightsquigarrow Y = 1 \rightsquigarrow x$;
thus, $1 \to X = 1 \to Y$ and $1 \rightsquigarrow X = 1 \rightsquigarrow Y$, hence, by (q-pEq=), $X \to Y = 1$, i.e. (a) holds.
By (p*'), $(y^- \to X) \to (y^- \to Y) = 1$, i.e. $(y^- \to (x^- \rightsquigarrow 1)) \to (y^- \to x) = 1$.
Then, by (q-pEq=) again, we obtain that $1 \rightsquigarrow [y^- \to (x^- \rightsquigarrow 1)] = 1 \rightsquigarrow [y^- \to x]$, i.e. $y^- \to (x^- \rightsquigarrow 1) = y^- \to x$, by (q-pM). Thus, the first part of (pEx2) holds.

Now, we shall prove similarly: (b) $(x^- \to 1) \to x = 1$.
Indeed, put $X = x^- \to 1$ and $Y = x$. Then:
$1 \to X = 1 \to (x^- \to 1) \overset{(Id)}{=} 1 \to (x^- \rightsquigarrow 1) \overset{(q-pM)}{=} x^- \rightsquigarrow 1 \overset{(Id)}{=} x^- \to 1 \overset{(q-pNeg)}{=} 1 \to (x^-)^- \overset{(DN)}{=} 1 \to x$, $1 \to Y = 1 \to x$ and
$1 \rightsquigarrow X = 1 \rightsquigarrow (x^- \to 1) \overset{(q-pM)}{=} x^- \to 1 \overset{(Id)}{=} x^- \rightsquigarrow 1 \overset{(q-pNeg)}{=} 1 \rightsquigarrow (x^-)^- \overset{(DN)}{=} 1 \rightsquigarrow x$, $1 \rightsquigarrow Y = 1 \rightsquigarrow x$;
thus, $1 \to X = 1 \to Y$ and $1 \rightsquigarrow X = 1 \rightsquigarrow Y$, hence, by (q-pEq=), $X \to Y = 1$, i.e. (b) holds.
By (p*'), $(y^- \rightsquigarrow X) \to (y^- \rightsquigarrow Y) = 1$, i.e. $(y^- \rightsquigarrow (x^- \to 1)) \to (y^- \rightsquigarrow x) = 1$.
Then, by (q-pEq=) again, we obtain that $1 \to [y^- \rightsquigarrow (x^- \to 1)] = 1 \to [y^- \rightsquigarrow x]$, i.e. $y^- \rightsquigarrow (x^- \to 1) = y^- \rightsquigarrow x$, by (q-pM). Thus, the second part of (pEx2) holds too. Thus, (pEx2) holds.

(1): Suppose that (pDNeg2) holds. We prove that (pEx2) $\Longrightarrow$ (p*'):

Indeed, suppose $x \to y = 1$; then,
$1 \to x = 1 \to y$ and $1 \rightsquigarrow x = 1 \rightsquigarrow y$, by (q-pEq=);
$1^- \to x = 1^- \to y$ and $1^- \rightsquigarrow x = 1^- \rightsquigarrow y$, by (N1);
$x^- \rightsquigarrow 1 = y^- \rightsquigarrow 1$ and $x^- \to 1 = y^- \to 1$, by (pDNeg2);
$z \to (x^- \rightsquigarrow 1) = z \to (y^- \rightsquigarrow 1)$ and $z \rightsquigarrow (x^- \to 1) = z \rightsquigarrow (y^- \to 1)$;
$z \to x = z \to y$ and $z \rightsquigarrow x = z \rightsquigarrow y$, by (pEx2);
$(z \to x) \to (z \to y) = 1$ and $(z \rightsquigarrow x) \to (z \rightsquigarrow y) = 1$, by (pRe'). Thus, (p*') holds.

(2): Suppose now that $\mathcal{A}$ is strong and that $x \to y = 1$; then,
$1 \to x = 1 \to y$ and $1 \rightsquigarrow x = 1 \rightsquigarrow y$, by (q-pEq=);
$(1 \to x)^- = (1 \to y)^-$ and $(1 \rightsquigarrow x)^- = (1 \rightsquigarrow y)^-$;
$x \to 1 = y \to 1$ and $x \rightsquigarrow 1 = y \rightsquigarrow 1$, by (q-pNeg);
$z^- \rightsquigarrow (x \to 1) = z^- \rightsquigarrow (y \to 1)$ and $z^- \to (x \rightsquigarrow 1) = z^- \to (y \rightsquigarrow 1)$;
$x \to (z^- \rightsquigarrow 1) = y \to (z^- \rightsquigarrow 1)$ and $x \rightsquigarrow (z^- \to 1) = y \rightsquigarrow (z^- \to 1)$, by (pEx1);
$x \to z = y \to z$ and $x \rightsquigarrow z = y \rightsquigarrow z$, by (pEx2);
$(y \to z) \to (x \to z) = 1$ and $(y \rightsquigarrow z) \to (x \rightsquigarrow z) = 1$, by (pRe'). Thus, (p**') holds. □

By above theorem, we immediately obtain:

Corollary 11.3.65 *Let $(A, \to, \rightsquigarrow, ^-, 1)$ be a strong involutive q-i-hub with (pDNeg2). Then,*

$$(p*') + (p**') \Longleftrightarrow (pEx2).$$

Corollary 11.3.66 *Let $\mathcal{A} = (A, \to, \rightsquigarrow, ^-, 1)$ be an involutive super q-i-hub with (pDNeg2). Then, $\mathcal{A}$ is strong.*

Proof. By above Theorem 11.3.64, (pEx2) holds, i.e.
$y^- \to (x^- \rightsquigarrow 1) = y^- \to x$ and $x^- \rightsquigarrow (y^- \to 1) = x^- \rightsquigarrow y$, for all $x, y \in A$;
then, $y^- \to (x^- \rightsquigarrow 1) = x^- \rightsquigarrow (y^- \to 1)$, by (pDNeg2);
then, $y \to (x \rightsquigarrow 1) = x \rightsquigarrow (y \to 1)$, by (DN), i.e. (pEx1) holds. □

- **New operation in strong involutive q-i-moon with (pDNeg2):** $\cdot$

Let $(A, \to, \rightsquigarrow, ^-, 1)$ be a strong involutive q-i-moon with (pDNeg2). Then, $(R(A), \to, \rightsquigarrow, ^{-1}, 1)$ is a strong i-moon. We extend the new operation $\cdot$ defined on $R(A)$ by (11.2) to A as follows: for all $x, y \in A$,

$$x \cdot y \overset{def.}{=} y^- \to x \overset{(pDNeg2)}{=} x^- \rightsquigarrow y. \tag{11.10}$$

Then, we have the following result, useful in Chapter 12.

Proposition 11.3.67 *Let $\mathcal{A} = (A, \to, \rightsquigarrow, ^-, 1)$ be a strong involutive q-i-moon with (pDNeg2) and the new operation $\cdot$ defined by (11.10). Then,*

(1) the following properties hold: (qm-pU), (U1), (pIv'), (pIv), (qm-NegP), (DN), (IdEqP), (qm-pEqP=), (Ass1) (i.e. $(A, \cdot, ^-, 1)$ is a sharp involutive qm-moon);

(2) if, additionally, $\mathcal{A}$ is super, then (Ass2) holds too (i.e. $(A, \cdot, ^-, 1)$ is super too).

Proof. (1): The following properties hold: (q-pM), (pM1), (IdEq), (q-pEq=), (q-pNeg), (DN), (Id), (pEx1), (pDNeg2), by hypothesis. Hence, (N1) and (pRe') also hold.

(qm-pU): $1 \cdot (x \cdot y) \overset{def.}{=} 1 \cdot (y^- \to x) \overset{def.}{=} 1^- \rightsquigarrow (y^- \to x) \overset{(N1)}{=}$
$1 \rightsquigarrow (y^- \to x) \overset{(q-pM)}{=} y^- \to x \overset{def.}{=} x \cdot y$ and
$(x \cdot y) \cdot 1 \overset{def.}{=} (x^- \rightsquigarrow y) \cdot 1 \overset{def.}{=} 1^- \to (x^- \rightsquigarrow y) \overset{(N1)}{=} 1 \to (x^- \rightsquigarrow y) \overset{(q-pM)}{=}$
$x^- \rightsquigarrow y \overset{def.}{=} x \cdot y$.

(U1): $1 \cdot 1 \overset{def.}{=} 1^- \to 1 \overset{(N1)}{=} 1 \to 1 \overset{(pM1)}{=} 1$.

(pIv'): Let $x \in A$; there exists $y = x^- \in A$, such that:
$x \cdot y = x \cdot x^- \overset{def.}{=} (x^-)^- \to x \overset{(DN)}{=} x \to x \overset{(pRe')}{=} 1$ and
$y \cdot x = x^- \cdot x \overset{def.}{=} (x^-)^- \rightsquigarrow x \overset{(DN)}{=} x \rightsquigarrow x \overset{(pRe')}{=} 1$.

(pIv): The same proof as above.

(qm-NegP): We must prove that $(1 \cdot x)^- = x^- \cdot 1 = 1 \cdot x^- = (x \cdot 1)^-$. Indeed,
$(1 \cdot x)^- \overset{def.}{=} (1^- \rightsquigarrow x)^- \overset{(N1)}{=} (1 \rightsquigarrow x)^-$; $x^- \cdot 1 \overset{def.}{=} (x^-)^- \rightsquigarrow 1 \overset{(DN)}{=} x \rightsquigarrow 1$;

$1 \cdot x^- \overset{def.}{=} (x^-)^- \to 1 \overset{(DN)}{=} x \to 1$; $(x \cdot 1)^- \overset{def.}{=} (1^- \to x)^- \overset{(N1)}{=} (1 \to x)^-$:
but $x \to 1 = 1 \to x^- = (1 \to x)^- = x \rightsquigarrow 1 = 1 \rightsquigarrow x^- = (1 \rightsquigarrow x)^-$, by (q-pNeg); hence, (qm-NegP) holds.

(DN): is (DN), because the negation $^-$ is the same.

(IdEqP): $y \cdot x^- = 1 \overset{def.}{\Leftrightarrow} (x^-)^- \to y = 1 \overset{(DN)}{\Leftrightarrow} x \to y = 1 \overset{(IdEq)}{\Leftrightarrow}$
$x \rightsquigarrow y = 1 \overset{(DN)}{\Leftrightarrow} (x^-)^- \rightsquigarrow y = 1 \overset{def.}{\Leftrightarrow} x^- \cdot y = 1$.

(qm-pEqP=): Suppose that $(1 \cdot x = 1 \cdot y$ and $x \cdot 1 = y \cdot 1)$, i.e., by definition, $(1^- \rightsquigarrow x = 1^- \rightsquigarrow y$ and $1^- \to x = 1^- \to y)$; then, by (N1),
$(1 \rightsquigarrow x = 1 \rightsquigarrow y$ and $1 \to x = 1 \to y)$; then, by (q-pEq=),
$x \to y = 1 \overset{(DN)}{\Leftrightarrow} (x^-)^- \to y = 1 \overset{def.}{\Leftrightarrow} y \cdot x^- = 1$; thus, $y \cdot x^- = 1$.

Suppose now that $y \cdot x^- = 1$, i.e. by definition, $(x^-)^- \to y = 1$; hence, $x \to y = 1$, by (DN);
then, by (q-pEq=), $(1 \to x = 1 \to y$ and $1 \rightsquigarrow x = 1 \rightsquigarrow y)$; then, by (N1),
$(1^- \to x = 1^- \to y$ and $1^- \rightsquigarrow x = 1^- \rightsquigarrow y)$, i.e., by definition,
$(x \cdot 1 = y \cdot 1$ and $1 \cdot x = 1 \cdot y)$. Thus, (qm-pEqP=) holds.

(Ass1): $x \cdot (1 \cdot y) \overset{def.}{=} x \cdot (y^- \to 1) \overset{def.}{=} x^- \rightsquigarrow (y^- \to 1) \overset{(pEx1)}{=}$
$y^- \to (x^- \rightsquigarrow 1) \overset{def.}{=} y^- \to (x \cdot 1) \overset{def.}{=} (x \cdot 1) \cdot y$.

(2): Suppose now that $\mathcal{A}$ is super, i.e. (p*') and (p**') hold. Then, by Theorem 11.3.64 (1), (pEx2) holds. Then,
$(x \cdot 1) \cdot y \overset{def.}{=} (x^- \rightsquigarrow 1) \cdot y \overset{def.}{=} y^- \to (x^- \rightsquigarrow 1) \overset{(pEx2)}{=} y^- \to x \overset{def.}{=} x \cdot y$ and
$x \cdot (1 \cdot y) \overset{def.}{=} x \cdot (y^- \to 1) \overset{def.}{=} x^- \rightsquigarrow (y^- \to 1) \overset{(pEx2)}{=} x^- \rightsquigarrow y \overset{def.}{=} x \cdot y$. Thus, (Ass2) holds too. □

- **The q-implicative-goop**

We introduce now the new notion of *q-implicative-goop*, the analogous, in the "world" of q-pM algebras, of the notion of *qm-goop* from the "world" of qm-unital magmas.

Definition 11.3.68 A *quasi-implicative-goop*, or *q-implicative-goop* or even *q-i-goop* for short, is an algebra $\mathcal{A} = (A, \to, \rightsquigarrow, ^-, 1)$ of type $(2, 2, 1, 0)$ verifying the following properties: for all $x, y, z \in A$,
(q-pM) $1 \to (x \rightsquigarrow y) = x \rightsquigarrow y$, $1 \rightsquigarrow (x \to y) = x \to y$,
(pM1) $1 \to 1 = 1 = 1 \rightsquigarrow 1$,
(IdEq) $x \to y = 1 \Longleftrightarrow x \rightsquigarrow y = 1$,
(q-pEq=) $(1 \to x = 1 \to y$ and $1 \rightsquigarrow x = 1 \rightsquigarrow y) \Longleftrightarrow x \to y = 1$ $(\overset{(IdEq)}{\Longleftrightarrow} x \rightsquigarrow y = 1)$,
(IdR) $1 \to x = 1 \rightsquigarrow x$,
(pEx1) $x \to (y \rightsquigarrow 1) = y \rightsquigarrow (x \to 1)$,
(q-$pM(1 \to x, 1 \rightsquigarrow x)$) $1 \to (1 \rightsquigarrow x)^- = (1 \rightsquigarrow x)^-$, $1 \rightsquigarrow (1 \to x)^- = (1 \to x)^-$,
(pDNeg2) $y^- \to x = x^- \rightsquigarrow y$,
(N1) $1^- = 1$,
(DN) $(x^-)^- = x$.

Denote by **q-i-goop** the class of all q-i-goops.

Proposition 11.3.69 *Let $\mathcal{A} = (A, \to, \rightsquigarrow, ^-, 1)$ be a q-i-goop. Then, (Id), (pDNeg1), (pNg1), (pDNeg3), (q-pNeg) hold.*

Proof. (Id): by Proposition 11.3.22 (q-ng1'), (pDNeg2) + (DN) + (IdR) $\Longrightarrow$ (Id).
(pDNeg1): by Proposition 11.3.22 (q-ng1), (pDNeg2) + (DN) $\Longrightarrow$ (pDNeg1).
(pNg1): by Proposition 11.3.22 (q-ng2), (pDNeg1) + (DN) $\Longrightarrow$ (pNg1).
(pDNeg3): by Proposition 11.3.22 (q-ng3), (pNg1) $\Longrightarrow$ (pDNeg3).
(q-pNeg): by Proposition 11.3.2 (q18=), (q-pEq=) $\Longrightarrow$ (pRe'); by Proposition 11.3.2 (q24='), (pEx1) + (pRe') + (q-pEq=) + (q-pM) $\Longrightarrow$ (q-pD1=); by Proposition 11.3.22 (q-ng4'), (pNg1) + (N1) + (q-pD1=) + (pDNeg1) + (IdR) + (q-$pM(1 \to x, 1 \rightsquigarrow x)$) $\Longrightarrow$ (q-pNeg). □

Theorem 11.3.70 *The q-i-goops are equivalent to the involutive strong q-i-moons verifying (pDNeg2).*

Proof. Let $\mathcal{A} = (A, \to, \rightsquigarrow, ^-, 1)$ be a q-i-goop. To prove that $\mathcal{A}$ is an involutive strong q-i-moon verifying (pDNeg2), it remains to prove that (Id) and (q-pNeg) hold. Indeed, this follows by above Proposition 11.3.69.

Conversely, let $\mathcal{A} = (A, \to, \rightsquigarrow, ^-, 1)$ be an involutive strong q-i-moon verifying (pDNeg2). To prove that $\mathcal{A}$ is a q-i-goop, it remains to prove that (N1), (q-$pM(1 \to x, 1 \rightsquigarrow x)$) and (IdR) hold. Indeed, by Proposition 11.3.18, (N1) holds. Then, by Proposition 11.3.21, (q-$pM(1 \to x, 1 \rightsquigarrow x)$) and (IdR) hold. □

By Theorems 11.3.70 and 11.3.63, we obtain:

Theorem 11.3.71 *The q-i-goops are categorically equivalent to the involutive strong q-i-hubs verifying (pDNeg2).*

Remark 11.3.72 By this Theorem 11.3.71, we could define, equivalently, the q-i-goops as an involutive strong q-i-hub verifying (pDNeg2). In fact, we find examples of q-i-goops by finding examples of involutive strong q-i-hubs verifying (pDNeg2).

Note that there are examples of q-i-hubs/q-i-moons that are:
1 - neither involutive nor strong,
2 - strong, but not involutive,
3 - involutive, but not strong,
4 - strong and involutive,
as the following examples show.

Examples 11.3.73

(1) **Commutative super q-i-hub that is neither strong nor involutive**

Consider the following proper q-proper commutative super q-i-hub $\mathcal{A}_4^1 = (A_4 = \{a, b, c, 1\}, \to_1, ^-, 1)$:

$\mathcal{A}_4^1$

$\to_1$	a	b	c	1
a	1	a	a	a
b	a	1	a	a
c	a	b	1	1
1	a	b	1	1

, with $\mathcal{R}(\mathcal{A}_4^1) = (\{a, b, 1\}, \to_1, 1)$, an implicative-hub (that is not involutive), with

$\to_1$	a	b	1
a	1	a	a
b	a	1	a
1	a	b	1

and $(a,b,1)^{-1} = (a,a,1)$.

Note that $\mathcal{A}_4^1$ is not strong, because the property (Ex1) is not satisfied: for $x = a$, $y = b$, $1 = a \to a = a \to (b \to 1) \neq b \to (a \to 1) = b \to a = a$.

(2) **Strong commutative super q-i-hub that is not involutive**

Consider the following strong proper q-proper commutative super q-i-hub $\mathcal{A}_4^2 = (A_4 = \{a,b,c,1\}, \to_2, 1)$:

$\mathcal{A}_4^5$

$\to_5$	a	b	c	1
a	1	a	a	b
b	a	1	1	a
c	a	1	1	a
1	a	b	b	1

, with $\mathcal{R}(\mathcal{A}_4^5) = (A_3 = \{a,b,1\}, \to_5, 1)$, a strong commutative implicative-hub, with

$\to_5$	a	b	1
a	1	a	b
b	a	1	a
1	a	b	1

and $(a,b,1)^{-1} = (b,a,1)$ ($x^{-1} = x \to_5 1$, for all $x \in A_3$).

If a quasi-negation $^-$ exists, it must verify (q-Neg) ($x \to_5 1 = 1 \to_5 x^- = (1 \to_5 x)^-$) and must extend $^{-1}$. We obtain:
$a = c \to_5 1 = 1 \to_5 c^- = (1 \to_5 c)^- = b^-$, hence $c^- = a$. Thus, $(a,b,c,1)^- = (b,a,a,1)$, hence (DN) is not satisfied; hence, $\mathcal{A}_4^5$ is not involutive.

(3) **Involutive commutative super q-i-hub that is not strong**

Consider the following involutive proper q-proper commutative super q-i-hub $\mathcal{A}_4^{12} = (A_4 = \{a,b,c,1\}, \to_{12}, ^-, 1)$:

$\mathcal{A}_4^{12}$

$\to_{12}$	a	b	c	1
a	1	a	a	a
b	b	1	b	b
c	a	b	1	1
1	a	b	1	1

, with $\mathcal{R}(\mathcal{A}_4^{12}) = (\{a,b,1\}, \to_4, 1)$, an involutive implicative-hub (that is not strong), with

$\to_4$	a	b	1
a	1	a	a
b	b	1	b
1	a	b	1

and $(a,b,1)^{-1} = (a,b,1)$. Then, the involutive quasi-negation $^-$ is the following: $(a,b,c,1)^- = (a,b,c,1)$. Note that it does not verify (q-DNeg2): for $x = a$, $y = b$, $b = b \to a = b^- \to a \neq a^- \to b = a \to b = a$.

Note that $\mathcal{A}_4^{12}$ is not strong, because the property (Ex1) is not satisfied: for $x = a$, $y = b$, $a = a \to b = a \to (b \to 1) \neq b \to (a \to 1) = b \to a = b$.

(4) **Strong commutative super q-i-hub that is involutive**

Consider the following strong proper q-proper commutative super q-i-hub $\mathcal{A}_4^7 = (A_4 = \{a,b,c,1\}, \to_7, ^-, 1)$:

$\mathcal{A}_4^7$

$\rightarrow_7$	a	b	c	1
a	1	a	b	b
b	a	1	a	a
c	a	b	1	1
1	a	b	1	1

, with $\mathcal{R}(\mathcal{A}_4^7) = (\{a, b, 1\}, \rightarrow_7, 1)$, a strong implicative-hub, with

$\rightarrow_7$	a	b	1
a	1	a	b
b	a	1	a
1	a	b	1

and $(a, b, 1)^{-1} = (b, a, 1)$.

If a quasi-negation $^-$ exists, it must verify (q-Neg) and must extend $^{-1}$. We obtain: $1 = c \rightarrow_7 1 = 1 \rightarrow_7 c^- = (1 \rightarrow_7 c)^- = 1^-$, hence $c^- = c, 1$; hence, for $c^- = c$, we obtain $(a, b, c, 1)^- = (b, a, c, 1)$; then, (DN) and (q-DNeg2) are satisfied; hence, $\mathcal{A}_4^7$ is involutive.

Open problem 11.3.74 It remains an open problem to prove that any involutive strong q-i-hub/q-i-moon verifies (pDNeg2) or to find an example of involutive strong q-i-hub/q-i-moon not verifying (pDNeg2).

Hence, we have the hierarchy No. q2 from Figure 11.10.

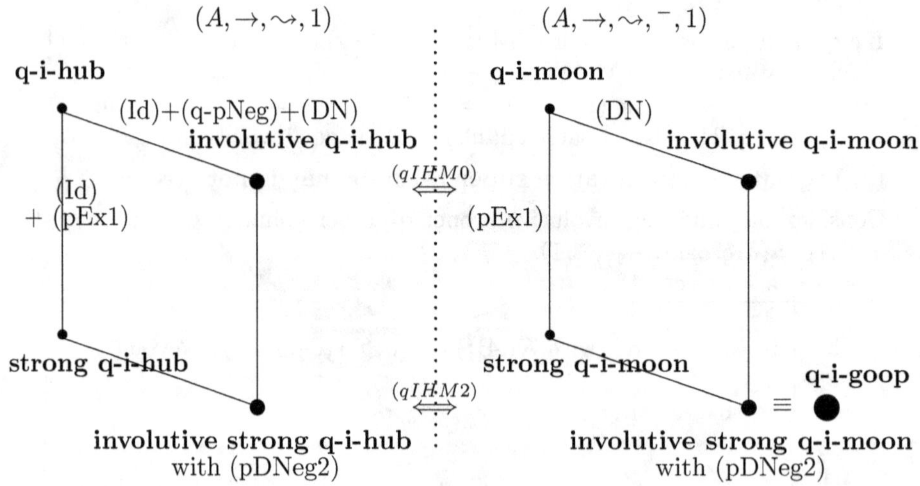

Figure 11.10: The hierarchy No. q2

11.4 Final connections

By above considerations, it follows that the implicative-groups have the following generalizations: the (strong) implicative-hubs, on one hand, and the (involutive) exchanged q-i-hubs and the (strong involutive) (super) q-i-hubs, on the other hand. These generalizations are connected as in the following Figure 11.11.

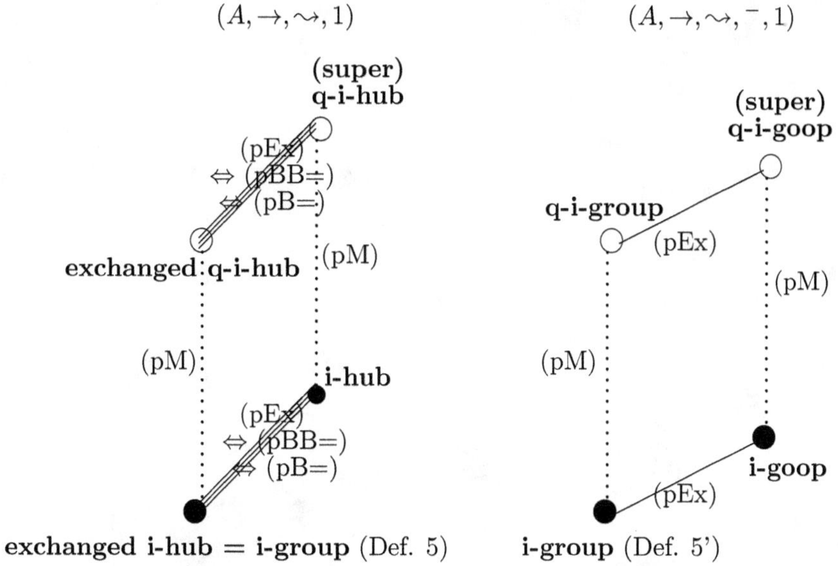

Figure 11.11: Generalizations of implicative-groups

Note that there exist also other generalizations of the implicative-groups, less visible: the pBCI algebras and the (strong) p*aRM** algebras, on one hand, and the q-pBCI algebras and the (strong) q-p*aRM** algebras, on the other hand. Their connections are presented in Chapter 13.

Chapter 12

Connections between generalizations of groups, implicative-groups

In this chapter, we establish the connections between:
- the goops and the implicative-goops,
- the qm-groups and the q-implicative-groups,
- the (super) qm-goops and the (super) q-implicative-goops.

We prove final connections and we present resuming connections.

The chapter has 5 sections.

12.1 Implicative-goops vs. goops

The following theorem says that the implicative-goops are termwise equivalent to the goops. Thus, it generalizes Theorem 9.2.3 saying that the implicative-groups are termwise equivalent to the groups.

Theorem 12.1.1 *(See Theorem 9.2.3)*

(1) Let $\mathcal{A} = (A, \rightarrow, \rightsquigarrow, ^{-1}, 1)$ *be an i-goop.*

Define $\Phi(\mathcal{A}) \overset{def.}{=} (A, \cdot, ^{-1}, 1)$ *by: for all* $x, y \in G$,

$$x \cdot y \overset{def.}{=} y^{-1} \rightarrow x \overset{(pDNeg2)}{=} x^{-1} \rightsquigarrow y.$$

Then, $\Phi(\mathcal{A})$ *is a goop.*

(1') Conversely, let $\mathcal{A} = (A, \cdot, ^{-1}, 1)$ *be a goop.*

Define $\Psi(\mathcal{A}) \overset{def.}{=} (A, \rightarrow, \rightsquigarrow, ^{-1}, 1)$ *by: for all* $x, y \in A$,

$$x \rightarrow y \overset{def.}{=} y \cdot x^{-1}, \quad x \rightsquigarrow y \overset{def.}{=} x^{-1} \cdot y.$$

Then, $\Psi(\mathcal{A})$ *is an i-goop.*

(2) The maps Φ *and* Ψ *are mutually inverse.*

Proof.

(1): By Proposition 11.1.45.

(1'): By Proposition 10.1.33.

(2): Let $(A, \cdot, ^{-1}, 1) \xrightarrow{\Psi} (A, \to, \rightsquigarrow, ^{-1}, 1) \xrightarrow{\Phi} (A, \circ, ^{-1}, 1)$.
Then, for all $x, y \in A$,
$x \circ y = y^{-1} \to x = x \cdot (y^{-1})^{-1} \stackrel{(DN)}{=} x \cdot y$.

Conversely, let $(A, \to, \rightsquigarrow, ^{-1}, 1) \xrightarrow{\Phi} (A, \cdot, ^{-1}, 1) \xrightarrow{\Psi} (A, \Rightarrow, \approx>, ^{-1}, 1)$.
Then, for all $x, y \in A$,
$x \Rightarrow y = y \cdot x^{-1} = (x^{-1})^{-1} \to y \stackrel{(DN)}{=} x \to y$ and
$x \approx> y = x^{-1} \cdot y = (x^{-1})^{-1} \rightsquigarrow y \stackrel{(DN)}{=} x \rightsquigarrow y$. □

By above theorem, we have the equivalence:

(Eq4) **i-goop** $\Longleftrightarrow$ **goop.**

Corollary 12.1.2 *The implicative-goop is commutative if and only if the termwise equivalent goop is commutative*
(i.e. $x \to y = x \rightsquigarrow y$*, for all* x, y*, if and only if* $x \cdot y = y \cdot x$*, for all* x, y*).*

Proof. $x \to y = x \rightsquigarrow y$, for all x, y, implies $x^{-1} \to y = x^{-1} \rightsquigarrow y \Leftrightarrow y \cdot (x^{-1})^{-1} = (x^{-1})^{-1} \cdot y$, i.e. $y \cdot x = x \cdot y$, by (DN). Conversely, $x \cdot y = y \cdot x$, for all x, y, implies $x^{-1} \cdot y = y \cdot x^{-1}$, i.e. $x \rightsquigarrow y = x \to y$. □

12.2 Q-implicative-groups vs. qm-groups

Note that the formulas from Theorem 9.2.3, concerning the connections between implicative-groups and groups, are preserved (extended) in the more general cases of associative qm-hubs and exchanged q-implicative-hubs and of q-implicative-groups and qm-groups.

First, we have the following important result.

Theorem 12.2.1

(1) Let $\mathcal{G} = (G, \to, \rightsquigarrow, 1)$ *be an exchanged q-i-hub.*
Define $\Phi(\mathcal{G}) \stackrel{def.}{=} (G, \cdot, 1)$ *by: for all* $x, y \in G$,

$$x \cdot y \stackrel{def.}{=} \begin{cases} x \cdot y, & \text{if} \quad x, y \in R(G), \\ r_x \cdot y, & \text{if} \quad x \notin R(G),\ r_x = 1 \to x \in R(G), \quad y \in R(G), \\ x \cdot r_y, & \text{if} \quad x \in R(G), \quad y \notin R(G),\ r_y = 1 \to y \in R(G), \\ r_x \cdot r_y, & \text{if} \quad x, y \notin R(G), r_x = 1 \to x,\ r_y = 1 \to y,\ r_x, r_y \in R(G). \end{cases}$$

Then, $\Phi(\mathcal{G})$ *is an associative qm-hub and* $Rm(G) = R(G)$.

(1') Conversely, let $\mathcal{G} = (G, \cdot, 1)$ be an associative qm-hub.
Define $\Psi(\mathcal{G}) \overset{def.}{=} (G, \to, \rightsquigarrow, 1)$ *by: for all* $x, y \in G$,

$$x \to y \overset{def.}{=} \begin{cases} x \to y, & \text{if} \quad x, y \in Rm(G), \\ r_x \to y, & \text{if} \quad x \notin Rm(G),\ r_x = 1 \cdot x \in Rm(G), \quad y \in Rm(G), \\ x \to r_y, & \text{if} \quad x \in Rm(G), \quad y \notin Rm(G),\ r_y = 1 \cdot y \in Rm(G), \\ r_x \to r_y, & \text{if} \quad x, y \notin Rm(G), r_x = 1 \cdot x,\ r_y = 1 \cdot y,\ r_x, r_y \in Rm(G), \end{cases}$$

$$x \rightsquigarrow y \overset{def.}{=} \begin{cases} x \rightsquigarrow y, & \text{if} \quad x, y \in Rm(G), \\ r_x \rightsquigarrow y, & \text{if} \quad x \notin Rm(G),\ r_x = 1 \cdot x \in Rm(G), \quad y \in Rm(G), \\ x \rightsquigarrow r_y, & \text{if} \quad x \in Rm(G), \quad y \notin Rm(G),\ r_y = 1 \cdot y \in Rm(G), \\ r_x \rightsquigarrow r_y, & \text{if} \quad x, y \notin Rm(G), r_x = 1 \cdot x,\ r_y = 1 \cdot y,\ r_x, r_y \in Rm(G). \end{cases}$$

Then, $\Psi(\mathcal{G})$ *is an exchanged q-i-hub and* $R(G) = Rm(G)$.
(2) The maps Φ *and* Ψ *are mutually inverse.*

Proof. (1): By Proposition 11.3.40, $\Phi(\mathcal{G})$ is an associative qm-hub. The same proof as above for $Rm(G) = R(G)$.

(1'): By Proposition 10.2.19, $\Psi(\mathcal{G})$ is an exchanged q-i-hub.
We prove now that $R(G) = Rm(G)$:
Let $x \in R(G)$, i.e. $x = 1 \to x = 1 \rightsquigarrow x$; then, $1 \cdot x \overset{(8.5)}{=} 1^{-1} \rightsquigarrow x = 1 \rightsquigarrow x = x$ and $x \cdot 1 \overset{(8.5)}{=} 1^{-1} \to x = 1 \to x = x$; thus, $x \in Rm(G)$.
Let $x \in Rm(G)$, i.e. $x = 1 \cdot x = x \cdot 1$; then, $1 \to x \overset{(7.1)}{=} x \cdot 1^{-1} = x \cdot 1 = x$ and $1 \rightsquigarrow x \overset{(7.1)}{=} 1^{-1} \cdot x = 1 \cdot x = x$; thus, $x \in R(G)$.
(2): Obviously, by (q-pII) and (qm-PI). □

By above theorem, we have the equivalence:

(Eq5) **exchanged q-i-hub $\Longleftrightarrow$ associative qm-hub.**

Second, we have the following result.

Theorem 12.2.2
(1) Let $\mathcal{G} = (G, \to, \rightsquigarrow, ^-, 1)$ *be an involutive exchanged q-i-moon.*
Define $\Phi(\mathcal{G}) \overset{def.}{=} (G, \cdot, ^-, 1)$ *by: for all* $x, y \in G$,

$$x \cdot y \overset{def.}{=} y^- \to x \overset{(pDNeg2)}{=} x^- \rightsquigarrow y.$$

Then, $\Phi(\mathcal{G})$ *is an involutive associative qm-moon.*
(1') Conversely, let $\mathcal{G} = (G, \cdot, ^-, 1)$ *be an involutive associative qm-moon.*
Define $\Psi(\mathcal{G}) \overset{def.}{=} (G, \to, \rightsquigarrow, ^-, 1)$ *by: for all* $x, y \in G$,

$$x \to y \overset{def.}{=} (x \cdot y^-)^- = y \cdot x^-, \quad x \rightsquigarrow y \overset{def.}{=} (y^- \cdot x)^- = x^- \cdot y.$$

Then, $\Psi(\mathcal{G})$ *is an involutive exchanged q-i-moon.*
(2) The maps Φ *and* Ψ *are mutually inverse.*

Proof.
(1): By Proposition 11.3.52.
(1'): By Proposition 10.2.27.
(2): As in Theorem 12.1.1. □

By above theorem, we have the equivalence:

(Eq6) **involutive exchanged q-i-moon** $\Longleftrightarrow$ **involutive associative qm-moon**.

Finally, we have:

Theorem 12.2.3
(1) Let $\mathcal{G} = (G, \to, \rightsquigarrow, ^-, 1)$ *be a q-i-group.*
Define $\Phi(\mathcal{G}) \stackrel{def.}{=} (G, \cdot, ^-, 1)$ *by: for all* $x, y \in G$,

$$x \cdot y \stackrel{def.}{=} y^- \to x \stackrel{(pDNeg2)}{=} x^- \rightsquigarrow y.$$

Then, $\Phi(\mathcal{G})$ *is a qm-group.*
(1') Conversely, let $\mathcal{G} = (G, \cdot, ^-, 1)$ *be a qm-group.*
Define $\Psi(\mathcal{G}) \stackrel{def.}{=} (G, \to, \rightsquigarrow, ^-, 1)$ *by: for all* $x, y \in G$,

$$x \to y \stackrel{def.}{=} (x \cdot y^-)^- = y \cdot x^-, \quad x \rightsquigarrow y \stackrel{def.}{=} (y^- \cdot x)^- = x^- \cdot y.$$

Then, $\Psi(\mathcal{G})$ *is a q-i-group.*
(2) The maps Φ *and* Ψ *are mutually inverse.*

Proof. By Theorem 12.2.2 and Theorems 10.2.30 and 11.3.55. □

By above theorem, we have the equivalence:

(Eq6') **q-i-group** $\Longleftrightarrow$ **qm-group**.

12.3 (Super) q-i-goops vs. (super) qm-goops

First, we have the following two results.

Theorem 12.3.1
(1) Let $\mathcal{A} = (A, \to, \rightsquigarrow, ^-, 1)$ *be a strong involutive q-i-moon with (pDNeg2).*
Define $\Phi(\mathcal{A}) \stackrel{def.}{=} (A, \cdot, ^-, 1)$ *by: for all* $x, y \in G$,

$$x \cdot y \stackrel{def.}{=} y^- \to x \stackrel{(pDNeg2)}{=} x^- \rightsquigarrow y.$$

Then, $\Phi(\mathcal{A})$ *is a sharp involutive qm-moon.*

(1') Conversely, let $\mathcal{A} = (A, \cdot, ^-, 1)$ be a sharp involutive qm-moon. Define $\Psi(\mathcal{A}) \stackrel{def.}{=} (A, \rightarrow, \rightsquigarrow, ^-, 1)$ by: for all $x, y \in A$,

$$x \rightarrow y \stackrel{def.}{=} y \cdot x^-, \quad x \rightsquigarrow y \stackrel{def.}{=} x^- \cdot y.$$

Then, $\Psi(\mathcal{A})$ is a strong involutive q-i-moon with (pDNeg2).

(2) The maps Φ and Ψ are mutually inverse.

Proof.

(1): By Proposition 11.3.67 (1).

(1'): By Proposition 10.2.44 (1).

(2): As in Theorem 12.1.1. □

By above theorem, we have the equivalence:

(Eq7) **strong inv. q-i-moon** with (pDNeg2) $\Longleftrightarrow$ **sharp inv. qm-moon.**

Theorem 12.3.2

(1) Let $\mathcal{A} = (A, \rightarrow, \rightsquigarrow, ^-, 1)$ be a strong involutive super q-i-moon with (pDNeg2). Define $\Phi(\mathcal{A}) \stackrel{def.}{=} (A, \cdot, ^-, 1)$ by: for all $x, y \in G$,

$$x \cdot y \stackrel{def.}{=} y^- \rightarrow x \stackrel{(pDNeg2)}{=} x^- \rightsquigarrow y.$$

Then, $\Phi(\mathcal{A})$ is a sharp involutive super qm-moon.

(1') Conversely, let $\mathcal{A} = (A, \cdot, ^-, 1)$ be a sharp involutive super qm-moon. Define $\Psi(\mathcal{A}) \stackrel{def.}{=} (A, \rightarrow, \rightsquigarrow, ^-, 1)$ by: for all $x, y \in A$,

$$x \rightarrow y \stackrel{def.}{=} y \cdot x^-, \quad x \rightsquigarrow y \stackrel{def.}{=} x^- \cdot y.$$

Then, $\Psi(\mathcal{A})$ is a strong involutive super q-i-moon with (pDNeg2).

(2) The maps Φ and Ψ are mutually inverse.

Proof.

(1): By Proposition 11.3.67 (2).

(1'): By Proposition 10.2.44 (2).

(2): As in Theorem 12.1.1. □

By above theorem, we have the equivalence:

(Eq8) **strong involutive super q-i-moon** with (pDNeg2) $\Longleftrightarrow$ **sharp involutive super qm-moon.**

Then, we have the following two results.

Theorem 12.3.3

(1) Let $\mathcal{A} = (A, \to, \rightsquigarrow, ^-, 1)$ *be a q-i-goop with (pDNeg2). Define* $\Phi(\mathcal{A}) \stackrel{def.}{=} (A, \cdot, ^-, 1)$ *by: for all* $x, y \in G$,

$$x \cdot y \stackrel{def.}{=} y^- \to x \stackrel{(pDNeg2)}{=} x^- \rightsquigarrow y.$$

Then, $\Phi(\mathcal{A})$ *is a qm-goop.*

(1') Conversely, let $\mathcal{A} = (A, \cdot, ^-, 1)$ *be a qm-goop. Define* $\Psi(\mathcal{A}) \stackrel{def.}{=} (A, \to, \rightsquigarrow, ^-, 1)$ *by: for all* $x, y \in A$,

$$x \to y \stackrel{def.}{=} y \cdot x^-, \quad x \rightsquigarrow y \stackrel{def.}{=} x^- \cdot y.$$

Then, $\Psi(\mathcal{A})$ *is a q-i-goop with (pDNeg2).*

(2) The maps Φ *and* Ψ *are mutually inverse.*

Proof. By Theorem 12.3.1 and Theorems 10.2.47 and 11.3.70. □

By above theorem, we have the equivalence:

(Eq7') **q-i-goop** with (pDNeg2) $\Longleftrightarrow$ **qm-goop**.

Theorem 12.3.4

(1) Let $\mathcal{A} = (A, \to, \rightsquigarrow, ^-, 1)$ *be a super q-i-goop with (pDNeg2). Define* $\Phi(\mathcal{A}) \stackrel{def.}{=} (A, \cdot, ^-, 1)$ *by: for all* $x, y \in G$,

$$x \cdot y \stackrel{def.}{=} y^- \to x \stackrel{(pDNeg2)}{=} x^- \rightsquigarrow y.$$

Then, $\Phi(\mathcal{A})$ *is a super qm-goop.*

(1') Conversely, let $\mathcal{A} = (A, \cdot, ^-, 1)$ *be a super qm-goop. Define* $\Psi(\mathcal{A}) \stackrel{def.}{=} (A, \to, \rightsquigarrow, ^-, 1)$ *by: for all* $x, y \in A$,

$$x \to y \stackrel{def.}{=} y \cdot x^-, \quad x \rightsquigarrow y \stackrel{def.}{=} x^- \cdot y.$$

Then, $\Psi(\mathcal{A})$ *is a super q-i-goop with (pDNeg2).*

(2) The maps Φ *and* Ψ *are mutually inverse.*

Proof. By Theorem 12.3.2 and Theorems 10.2.47 and 11.3.70. □

By above theorem, we have the equivalence:

(Eq8') **super q-i-goop** with (pDNeg2) $\Longleftrightarrow$ **super qm-goop**.

12.4 Final connections

Consider the following sets:
$R(A) = \{x \in A \mid 1 \to x = x = 1 \rightsquigarrow x\}$,
$Rm(A) = \{x \in A \mid x \cdot 1 = x = 1 \cdot x\}$.

Consider the following properties: for all x, y, z,
(N1) $1^- = 1$,
(DN) $(x^-)^- = x$,
(pDNeg2) $y^- \to x = x^- \rightsquigarrow y$,

(IdR) $1 \to x = 1 \rightsquigarrow x$,
(IdP) $x \cdot 1 = 1 \cdot x$;

(pM) $1 \to x = x = 1 \rightsquigarrow x$,
(pU) $x \cdot 1 = x = 1 \cdot x$;
(IdEq) $x \to y = 1 \Longleftrightarrow x \rightsquigarrow y = 1$,
(IdEqP) $y \cdot x^- = 1 \Longleftrightarrow x^- \cdot y = 1$;
(pEq=) $x = y \Longleftrightarrow x \to y = 1$ ($\overset{(IdEq)}{\Longleftrightarrow}$ $x \rightsquigarrow y = 1$),
(pEqP=) $x = y \Longleftrightarrow y \cdot x^- = 1$ ($\overset{(IdEqP)}{\Longleftrightarrow}$ $x^- \cdot y = 1$);
(pRe') $x \to x = 1 = x \rightsquigarrow x$,
(pIv) $x \cdot x^- = 1 = x^- \cdot x$;
(pEx) $x \to (y \rightsquigarrow z) = y \rightsquigarrow (x \to z)$,
(Ass) $x \cdot (y \cdot z) = (x \cdot y) \cdot z$;
(q-pM) $1 \to (x \rightsquigarrow y) = x \rightsquigarrow y$, $1 \rightsquigarrow (x \to y) = x \to y$,
(qm-pU) $1 \cdot (x \cdot y) = x \cdot y = (x \cdot y) \cdot 1$;
(pM1) $1 \to 1 = 1 = 1 \rightsquigarrow 1$,
(U1) $1 \cdot 1 = 1$;
(pEx1) $x \to (y \rightsquigarrow 1) = y \rightsquigarrow (x \to 1)$,
(Ass1) $x \cdot (1 \cdot y) = (x \cdot 1) \cdot y$;
(pEx2) $y^- \to (x^- \rightsquigarrow 1) = y^- \to x$, $\quad x^- \rightsquigarrow (y^- \to 1) = x^- \rightsquigarrow y$,
(Ass2) $x \cdot (1 \cdot y) = x \cdot y$, $\quad (x \cdot 1) \cdot y = x \cdot y$;
(q-pNeg) $x \to 1 = 1 \to x^- = (1 \rightsquigarrow x)^- = x \rightsquigarrow 1 = 1 \rightsquigarrow x^- = (1 \to x)^-$,
(qm-NegP) $(x \cdot 1)^- = x^- \cdot 1 = 1 \cdot x^- = (1 \cdot x)^-$;

(q-$\overline{pM(1 \to x, 1 \rightsquigarrow x)}$) $1 \to (1 \rightsquigarrow x)^- = (1 \rightsquigarrow x)^-$, $1 \rightsquigarrow (1 \to x)^- = (1 \to x)^-$,

(qm-$\overline{pU(1 \cdot x, x \cdot 1)}$) $1 \cdot (1 \cdot x)^- = (1 \cdot x)^- = (1 \cdot x)^- \cdot 1$,
$1 \cdot (x \cdot 1)^- = (x \cdot 1)^- = (x \cdot 1)^- \cdot 1$;
(q-pEq=) ($1 \to x = 1 \to y$ and $1 \rightsquigarrow x = 1 \rightsquigarrow y$) $\Longleftrightarrow x \to y = 1$ ($\overset{(IdEq)}{\Longleftrightarrow}$ $x \rightsquigarrow y = 1$),
(qm-pEqP=) ($x \cdot 1 = y \cdot 1$ and $1 \cdot x = 1 \cdot y$) $\Longleftrightarrow y \cdot x^- = 1$ ($\overset{(IdEqP)}{\Longleftrightarrow}$ $x^- \cdot y = 1$);
(q-pM(·)) $1 \to (x \cdot y) = x \cdot y = 1 \rightsquigarrow (x \cdot y)$,
(qm-pU($\to$,$\rightsquigarrow$)) $1 \cdot (x \to y) = x \to y = (x \to y) \cdot 1$,
$1 \cdot (x \rightsquigarrow y) = x \rightsquigarrow y = (x \rightsquigarrow y) \cdot 1$.

Then, we have the following final result:

Theorem 12.4.1 *(See Theorem 5.3.1)*
(1) Let $\mathcal{A} = (A, \to, \rightsquigarrow, ^-, 1)$ *be an algebra of type* $(2, 2, 1, 0)$ *verifying (N1), (DN) and (pDNeg2). Define* $\Phi(\mathcal{A}) \stackrel{def.}{=} (A, \cdot, ^-, 1)$ *by: for all* $x, y \in G$,

$$x \cdot y \stackrel{def.}{=} y^- \to x \stackrel{(pDNeg2)}{=} x^- \rightsquigarrow y.$$

Then, $\Phi(\mathcal{A})$ *is an algebra of type* $(2, 1, 0)$ *verifying (N1), (DN).*
(1') Conversely, let $\mathcal{A} = (A, \cdot, ^-, 1)$ *be an algebra of type* $(2, 1, 0)$ *verifying (N1) and (DN). Define* $\Psi(\mathcal{A}) \stackrel{def.}{=} (A, \to, \rightsquigarrow, ^-, 1)$ *by: for all* $x, y \in A$,

$$x \to y \stackrel{def.}{=} y \cdot x^-, \quad x \rightsquigarrow y \stackrel{def.}{=} x^- \cdot y.$$

Then, $\Psi(\mathcal{A})$ *is an algebra of type* $(2, 2, 1, 0)$ *verifying (N1), (DN) and (pDNeg2).*
(2) The maps Φ *and* Ψ *are mutually inverse.*
(3) $R(A) = Rm(A)$.
(4) The following equivalences hold:
(IdR) $\Longleftrightarrow$ *(IdP), (pM)* $\Longleftrightarrow$ *(pU), (IdEq)* $\Longleftrightarrow$ *(IdEqP), (pEq=)* $\Longleftrightarrow$ *(pEqP=),*
(pEx) $\Longleftrightarrow$ *(Ass), (pRe')* $\Longleftrightarrow$ *(pIv);*
(q-pM) $\Longleftrightarrow$ *(qm-pU), (pM1)* $\Longleftrightarrow$ *(U1),*
(pEx1) $\Longleftrightarrow$ *(Ass1), (pEx2)* $\Longleftrightarrow$ *(Ass2), (q-pNeg)* $\Longleftrightarrow$ *(qm-NegP),*
(q-pM$(1 \to x, 1 \rightsquigarrow x)$*)* $\Longleftrightarrow$ *(qm-pU*$(1 \cdot x, x \cdot 1)$*),*
(q-pEq=) $\Longleftrightarrow$ *(qm-pEqP=).*
(5) If (IdR) ($\Longleftrightarrow$ *(IdP)) holds, then the following equivalences hold:*
(q-pM) $\Longleftrightarrow$ *(q-pM(*$\cdot$*))* $\Longleftrightarrow$ *(qm-pU(*$\to, \rightsquigarrow$*))* $\Longleftrightarrow$ *(qm-pU).*

Proof.
(1): Obviously.
(1'): All we have to prove is: $y^- \to x = x \cdot (y^-)^- \stackrel{(DN)}{=} x \cdot y$ and $x^- \rightsquigarrow y = (x^-)^- \cdot y \stackrel{(DN)}{=} x \cdot y$; thus, (pDNeg2) holds.
(2): Let $(A, \to, \rightsquigarrow, ^-, 1) \xrightarrow{\Phi} (A, \cdot, ^-, 1) \xrightarrow{\Psi} (A, \Rightarrow, \approx>, ^-, 1)$.
Then, for all $x, y \in A$,
$x \Rightarrow y = y \cdot x^- = (x^-)^- \to y \stackrel{(DN)}{=} x \to y$ and
$x \approx> y = x^- \cdot y = (x^-)^- \rightsquigarrow y \stackrel{(DN)}{=} x \rightsquigarrow y$.
Conversely, let $(A, \cdot, ^-, 1) \xrightarrow{\Psi} (A, \to, \rightsquigarrow, ^-, 1) \xrightarrow{\Phi} (A, \circ, ^-, 1)$.
Then, for all $x, y \in A$,
$x \circ y = y^- \to x = x \cdot (y^-)^- \stackrel{(DN)}{=} x \cdot y$.
(3): Let $x \in R(A)$, i.e. $x = 1 \to x = 1 \rightsquigarrow x$; then, $1 \cdot x = 1^- \rightsquigarrow x \stackrel{(N1)}{=} 1 \rightsquigarrow x = x$ and $x \cdot 1 = 1^- \to x \stackrel{(N1)}{=} 1 \to x = x$; thus, $x \in Rm(A)$.
Let $x \in Rm(A)$, i.e. $x = 1 \cdot x = x \cdot 1$; then, $1 \to x = x \cdot 1^- = x \cdot 1 = x$ and $1 \rightsquigarrow x = 1^{-1} \cdot x \stackrel{(N1)}{=} 1 \cdot x = x$; thus, $x \in R(A)$.

(4): (IdR) $\Longleftrightarrow$ (IdP):

$\Longrightarrow$: $x \cdot 1 = 1^- \to x \overset{(N1)}{=} 1 \to x \overset{(IdR)}{=} 1 \rightsquigarrow x \overset{(N1)}{=} 1^- \rightsquigarrow x = 1 \cdot x$.

$\Longleftarrow$: $1 \to x = x \cdot 1^- \overset{(N1)}{=} x \cdot 1 \overset{(IdP)}{=} 1 \cdot x \overset{(N1)}{=} 1^- \cdot x = 1 \rightsquigarrow x$.

(pM) $\Longleftrightarrow$ (pU):

$\Longrightarrow$: $x \cdot 1 = 1^- \to x \overset{(N1)}{=} 1 \to x \overset{(pM)}{=} x$ and $1 \cdot x = 1^- \rightsquigarrow x \overset{(N1)}{=} 1 \rightsquigarrow x \overset{(pM)}{=} x$.

$\Longleftarrow$: $1 \to x = x \cdot 1^- \overset{(N1)}{=} x \cdot 1 \overset{(pU)}{=} x$ and $1 \rightsquigarrow x = 1^- \cdot x \overset{(N1)}{=} 1 \cdot x \overset{(pU)}{=} x$.

(IdEq) $\Longleftrightarrow$ (IdEqP):

$\Longrightarrow$: $y \cdot x^- = 1 \Leftrightarrow (x^-)^- \to y = 1 \overset{(DN)}{\Leftrightarrow} x \to y = 1 \overset{(IdEq)}{\Leftrightarrow}$
$x \rightsquigarrow y = 1 \overset{(DN)}{\Leftrightarrow} (x^-)^- \rightsquigarrow y = 1 \Leftrightarrow x^- \cdot y = 1$.

$\Longleftarrow$: $x \to y = 1 \overset{def.}{\Leftrightarrow} y \cdot x^- = 1 \overset{(IdEqP)}{\Leftrightarrow} x^- \cdot y = 1 \overset{def.}{\Leftrightarrow} x \rightsquigarrow y = 1$.

(pEq=) $\Longleftrightarrow$ (pEqP=):

$\Longrightarrow$: $x = y \overset{(pEq=)}{\Leftrightarrow} x \to y = 1 \Leftrightarrow y \cdot x^- = 1$, hence $x = y \Leftrightarrow y \cdot x^- = 1$.

$\Longleftarrow$: $x = y \overset{(pEqP=)}{\Leftrightarrow} y \cdot x^- = 1 \Leftrightarrow (x^-)^- \to y = 1 \overset{(DN)}{\Leftrightarrow} x \to y = 1$, hence $x = y \Leftrightarrow x \to y = 1$.

(pEx) $\Longleftrightarrow$ (Ass):

$\Longrightarrow$: $x \cdot (y \cdot z) = x \cdot (z^- \to y) = x^- \rightsquigarrow (z^- \to y)$
and $(x \cdot y) \cdot z = (x^- \rightsquigarrow y) \cdot z = z^- \to (x^- \rightsquigarrow y) \overset{(pEx)}{=} x^- \rightsquigarrow (z^- \to y)$.

$\Longleftarrow$: $x \to (y \rightsquigarrow z) = x \to (y^- \cdot z) = (y^- \cdot z) \cdot x^- \overset{(Ass)}{=} y^- \cdot (z \cdot x^-) = y \rightsquigarrow (z \cdot x^-) = y \rightsquigarrow (x \to z)$.

(pRe') $\Longleftrightarrow$ (pIv):

$\Longrightarrow$: $x \cdot x^- = 1 \Leftrightarrow (x^-)^- \to x = 1 \overset{(DN)}{\Leftrightarrow} x \to x = 1$ and $x^- \cdot x = 1 \Leftrightarrow (x^-)^- \rightsquigarrow x = 1 \overset{(DN)}{\Leftrightarrow} x \rightsquigarrow x = 1$.

$\Longleftarrow$: $x \to x = 1 \Leftrightarrow x \cdot x^- = 1$ and $x \rightsquigarrow x = 1 \Leftrightarrow x^- \cdot x = 1$.

(q-pM) $\Longleftrightarrow$ (qm-pU):

$\Longrightarrow$: $1 \cdot (x \cdot y) = 1 \cdot (y^- \to x) = 1^- \rightsquigarrow (y^- \to x) \overset{(N1)}{=}$
$1 \rightsquigarrow (y^- \to x) \overset{(q-pM)}{=} y^- \to x = x \cdot y$ and
$(x \cdot y) \cdot 1 = (x^- \rightsquigarrow y) \cdot 1 = 1^- \to (x^- \rightsquigarrow y) \overset{(N1)}{=} 1 \to (x^- \rightsquigarrow y) \overset{(q-pM)}{=}$
$x^- \rightsquigarrow y = x \cdot y$.

$\Longleftarrow$: $1 \to (x \rightsquigarrow y) = 1 \to (x^- \cdot y) = (x^- \cdot y) \cdot 1^- \overset{(N1)}{=} (x^- \cdot y) \cdot 1 \overset{(qm-pU)}{=} x^- \cdot y = x \rightsquigarrow y$ and
$1 \rightsquigarrow (x \to y) = 1 \rightsquigarrow (y \cdot x^-) = 1^- \cdot (y \cdot x^-) \overset{(N1)}{=} 1 \cdot (y \cdot x^-) \overset{(qm-pU)}{=} y \cdot x^- = x \to y$.

(pM1) $\Longleftrightarrow$ (U1):

$\Longrightarrow$: $1 \cdot 1 = 1^- \to 1 \overset{(N1)}{=} 1 \to 1 \overset{(pM1)}{=} 1$.

$\Longleftarrow$: $1 \to 1 = 1 \cdot 1^- \overset{(N1)}{=} 1 \cdot 1 \overset{(U1)}{=} 1$ and $1 \rightsquigarrow 1 = 1^- \cdot 1 \overset{(N1)}{=} 1 \cdot 1 \overset{(U1)}{=} 1$.

(pEx1) $\Longleftrightarrow$ (Ass1):

$\Longrightarrow$: $x \cdot (1 \cdot y) = x \cdot (y^- \to 1) = x^- \rightsquigarrow (y^- \to 1) \overset{(pEx1)}{=}$
$y^- \to (x^- \rightsquigarrow 1) = y^- \to (x \cdot 1) = (x \cdot 1) \cdot y$.

$\Longleftarrow$: $x \to (y \rightsquigarrow 1) = x \to (y^- \cdot 1) = (y^- \cdot 1) \cdot x^- \overset{(Ass1)}{=} y^- \cdot (1 \cdot x^-) = y \rightsquigarrow (1 \cdot x^-) = y \rightsquigarrow (x \to 1)$.

(pEx2) $\Longleftrightarrow$ (Ass2):
$\Longrightarrow$: $(x \cdot 1) \cdot y = (x^- \rightsquigarrow 1) \cdot y = y^- \to (x^- \rightsquigarrow 1) \overset{(pEx2)}{=} y^- \to x = x \cdot y$ and
$x \cdot (1 \cdot y) = x \cdot (y^- \to 1) = x^- \rightsquigarrow (y^- \to 1) \overset{(pEx2)}{=} x^- \rightsquigarrow y = x \cdot y$.
$\Longleftarrow$: $y^- \to (x^- \rightsquigarrow 1) = y^- \to ((x^-)^- \cdot 1) \overset{(DN)}{=} y^- \to (x \cdot 1) = (x \cdot 1) \cdot (y^-)^- \overset{(DN)}{=}$ $(x \cdot 1) \cdot y \overset{(Ass2)}{=} x \cdot y \overset{(DN)}{=} x \cdot (y^-)^- = y^- \to x$ and
$x^- \rightsquigarrow (y^- \to 1) = x^- \rightsquigarrow (1 \cdot (y^-)^-) \overset{(DN)}{=} x^- \rightsquigarrow (1 \cdot y) = (x^-)^- \cdot (1 \cdot y) \overset{(DN)}{=}$ $x \cdot (1 \cdot y) \overset{(Ass2)}{=} x \cdot y \overset{(DN)}{=} (x^-)^- \cdot y = x^- \rightsquigarrow y$.

(q-pNeg) $\Longleftrightarrow$ (qm-NegP):
$\Longrightarrow$: $(1 \cdot x)^- = (1^- \rightsquigarrow x)^- \overset{(N1)}{=} (1 \rightsquigarrow x)^-$; $x^- \cdot 1 = (x^-)^- \rightsquigarrow 1 \overset{(DN)}{=} x \rightsquigarrow 1$;
$1 \cdot x^- = (x^-)^- \to 1 \overset{(DN)}{=} x \to 1$; $(x \cdot 1)^- = (1^- \to x)^- \overset{(N1)}{=} (1 \to x)^-$;
but $x \to 1 = 1 \to x^- = (1 \to x)^- = x \rightsquigarrow 1 = 1 \rightsquigarrow x^- = (1 \rightsquigarrow x)^-$, by (q-pNeg.
$\Longleftarrow$: $x \to 1 = 1 \cdot x^-$; $1 \to x^- = x^- \cdot 1^- \overset{(N1)}{=} x^- \cdot 1$; $(1 \to x)^- = (x \cdot 1^-)^- \overset{(N1)}{=} (x \cdot 1)^-$;
$x \rightsquigarrow 1 = x^- \cdot 1$; $1 \rightsquigarrow x^- = 1^- \cdot x^- \overset{(N1)}{=} 1 \cdot x^-$; $(1 \rightsquigarrow x)^- = (1^- \cdot x)^- \overset{(N1)}{=} (1 \cdot x)^-$;
then, apply (qm-NegP).

(q-$\overline{pM(1 \to x, 1 \rightsquigarrow x)}$) $\Longleftrightarrow$ (qm-$\overline{pU(1 \cdot x, x \cdot 1)}$):
$\Longrightarrow$: $1 \cdot (1 \cdot x)^- = 1^- \rightsquigarrow (1 \cdot x)^- = 1 \rightsquigarrow (1 \cdot x)^- = 1 \rightsquigarrow (1^- \rightsquigarrow x)^- = 1 \rightsquigarrow (1 \rightsquigarrow x)^- = (1 \rightsquigarrow x)^- = (1 \cdot x)^-$ and similarly the rest.
$\Longleftarrow$: $1 \to (1 \rightsquigarrow x)^- = (1 \rightsquigarrow x)^- \cdot 1^- = (1 \rightsquigarrow x)^- \cdot 1 = (1^- \cdot x)^- \cdot 1 = (1 \cdot x)^- \cdot 1 = (1 \cdot x)^- = (1 \rightsquigarrow x)^-$
$1 \rightsquigarrow (1 \to x)^- = 1^- \cdot (1 \to x)^- = 1 \cdot (1 \to x)^- = 1 \cdot (x \cdot 1^-)^- = 1 \cdot (x \cdot 1)^- = (x \cdot 1)^- = (1 \to x)^-$.

(q-pEq=) $\Longleftrightarrow$ (qm-pEqP=):
$\Longrightarrow$: Suppose that ($1 \cdot x = 1 \cdot y$ and $x \cdot 1 = y \cdot 1$), i.e., by definition,
($1^- \rightsquigarrow x = 1^- \rightsquigarrow y$ and $1^- \to x = 1^- \to y$); then, by (N1),
($1 \rightsquigarrow x = 1 \rightsquigarrow y$ and $1 \to x = 1 \to y$); then, by (q-pEq=),
$x \to y = 1 \overset{(DN)}{\Leftrightarrow} (x^-)^- \to y = 1 \overset{def.}{\Leftrightarrow} y \cdot x^- = 1$; thus, $y \cdot x^- = 1$.
Suppose now that $y \cdot x^- = 1$, i.e. by definition, $(x^-)^- \to y = 1$; hence, $x \to y = 1$, by (DN); then, by (q-pEq=), ($1 \to x = 1 \to y$ and $1 \rightsquigarrow x = 1 \rightsquigarrow y$); then, by (N1), ($1^- \to x = 1^- \to y$ and $1^- \rightsquigarrow x = 1^- \rightsquigarrow y$), i.e., by definition,
($x \cdot 1 = y \cdot 1$ and $1 \cdot x = 1 \cdot y$).
$\Longleftarrow$: Suppose that ($1 \to x = 1 \to y$ and $1 \rightsquigarrow x = 1 \rightsquigarrow y$), i.e., by definition, ($x \cdot 1^- = y \cdot 1^-$ and $1^- \cdot x = 1^- \cdot y$); then, by (N1),
($x \cdot 1 = y \cdot 1$ and $1 \cdot x = 1 \cdot y$); then, by (q-pEq=P), $y \cdot x^- = 1$, i.e. $x \to y = 1$.
Suppose that $x \to y = 1$, i.e. $y \cdot x^- = 1$, by definition; then, by (q-pEq=P),
($1 \cdot x = 1 \cdot y$ and $x \cdot 1 = y \cdot 1$); then, by (N1),
($1^- \cdot x = 1^- \cdot y$ and $x \cdot 1^- = y \cdot 1^-$); then, by definition,
($1 \rightsquigarrow x = 1 \rightsquigarrow y$ and $1 \to x = 1 \to y$).

(5): (q-pM) $\Longrightarrow$ (q-pM($\cdot$)):
$1 \to (x \cdot y) = 1 \to (y \rightsquigarrow x^-)^- = 1 \to (x^- \rightsquigarrow y) \overset{(q-pM)}{=} x^- \rightsquigarrow y = x \cdot y$ and
$1 \rightsquigarrow (x \cdot y) = 1 \rightsquigarrow (x \to y^-)^- = 1 \rightsquigarrow (y^- \to x) \overset{(q-pM)}{=} y^- \to x = x \cdot y$.

(q-pM($\cdot$)) $\Longrightarrow$ (q-pM):

$1 \to (x \rightsquigarrow y) = 1 \to (x^- \cdot y) \overset{(q-pM(\cdot))}{=} x^- \cdot y = x \rightsquigarrow y$ and
$1 \rightsquigarrow (x \to y) = 1 \rightsquigarrow (y \cdot x^-) = y \cdot x^- = x \to y$.

(qm-pU) $\Longrightarrow$ (qm-pU($\to,\rightsquigarrow$)):
$1 \cdot (x \to y) = 1 \cdot (x \cdot y^{-1}) \overset{(qm-pU)}{=} x \cdot y^{-1} = x \to y$ and
$(x \to y) \cdot 1 = (x \cdot y^{-1}) \cdot 1 = x \cdot y^{-1} = x \to y$.
$1 \cdot (x \rightsquigarrow y) = 1 \cdot (x^{-1} \cdot y) = x^{-1} \cdot y = x \rightsquigarrow y$ and
$(x \rightsquigarrow y) \cdot 1 = (x^{-1} \cdot y) \cdot 1 = x^{-1} \cdot y = x \rightsquigarrow y$.

(qm-pU($\to,\rightsquigarrow$)) $\Longrightarrow$ (qm-pU):
$1 \cdot (x \cdot y) = 1 \cdot (x \to y^{-1})^{-1} = 1 \cdot (y^{-1} \to x) = y^{-1} \to x = x \cdot y$ and
$(x \cdot y) \cdot 1 = (y^{-1} \to x) \cdot 1 = y^{-1} \to x = x \cdot y$.

(qm-pU) $\Longrightarrow$ (q-pM($\cdot$)):
$1 \to (x \cdot y) = (x \cdot y) \cdot 1^{-1} = (x \cdot y) \cdot 1 \overset{(qm-pU)}{=} x \cdot y$ and
$1 \rightsquigarrow (x \cdot y) = 1^{-1} \cdot (x \cdot y) = 1 \cdot (x \cdot y) = x \cdot y$.

(q-pM) $\Longrightarrow$ (qm-pU($\to,\rightsquigarrow$)): since (IdR) holds ($1 \to x = 1 \rightsquigarrow x$),
$1 \cdot (x \to y) = 1^{-1} \rightsquigarrow (x \to y) = 1 \rightsquigarrow (x \to y) \overset{(q-pM)}{=} x \to y$ and
$(x \to y) \cdot 1 = 1^{-1} \to (x \to y) = 1 \to (x \to y) \overset{(IdR)}{=} 1 \rightsquigarrow (x \to y) = x \to y$.
$1 \cdot (x \rightsquigarrow y) = 1^{-1} \rightsquigarrow (x \rightsquigarrow y) 1 \rightsquigarrow (x \rightsquigarrow y) \overset{(IdR)}{=} 1 \to (x \rightsquigarrow y) = x \rightsquigarrow y$ and
$(x \rightsquigarrow y) \cdot 1 = 1^{-1} \to (x \rightsquigarrow y) = 1 \to (x \rightsquigarrow y) \overset{(q-pM)}{=} x \rightsquigarrow y$. □

12.5 Resuming connections

Resuming, by Figures 10.10 and 11.11, we have the connections from Figure 12.1:

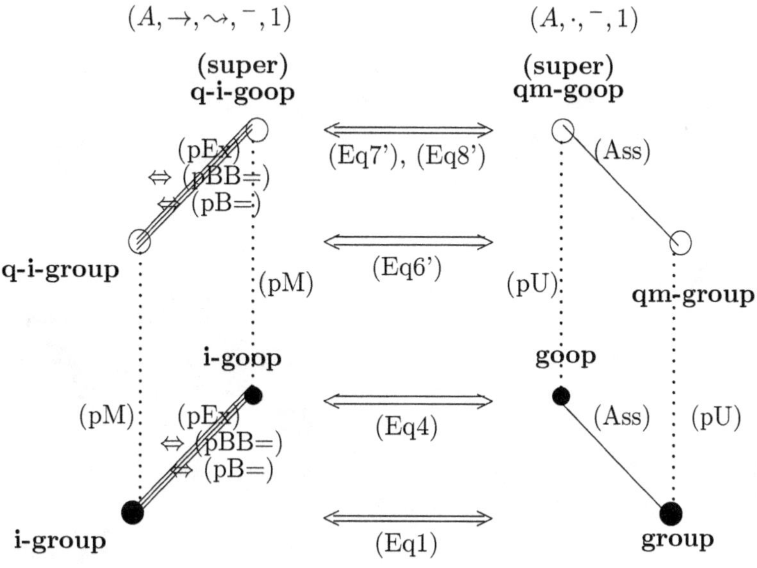

Figure 12.1: Connections between generalizations

Chapter 13

(Quasi-) p-semisimple (quasi-) pseudo-algebras

In this chapter, Section 1, we present some p-semisimple pseudo-algebras, as for example the p-semisimple pBCI, pBCI(pP), pBCI(pRP) algebras (which are the implicative-groups) and the p-semisimple (strong) p*aRM** algebras (which are the (strong) i-hubs). In Section 2, we present some quasi-p-semisimple quasi-pseudo-algebras, as the quasi-p-semisimple q-pBCI (which are the exchanged q-i-hubs) and the quasi-p-semisimple q-p*aRM** algebras (which are the super q-i-hubs). In Section 3, we present resuming connections.

13.1 p-semisimple pseudo-algebras

We have introduced in Chapter 2 the definition of *p-semisimple* pseudo-algebra (pseudo-structure) - Definition 2.1.3.

13.1.1 p-semisimple pseudo-BCI algebras (lattices)

G. Dymek [50] (see [138] for the commutative case) made the connection between the pseudo-BCI algebras (defined as reversed right-structures $(A, \leq, \star, \circ, 0)$) and the groups (defined additively), by introducing, as in the commutative case [138], the subclass of p-semisimple pseudo-BCI algebras and by proving that these are equivalent with the groups.

Definition 13.1.1 (Definition 1 of p-semisimple pBCI algebra (lattice))
(i) Let $\mathcal{A}^L = (A^L, \leq, \to^L, \rightsquigarrow^L, 1)$ be a left-pBCI algebra (lattice).
We say that $\mathcal{A}^L$ is *p-semisimple*, if $x \leq 1$ implies $x = 1$, for any $x \in A^L$ [50].
(i') Dually, let $\mathcal{A}^R = (A^R, \geq, \to^R, \rightsquigarrow^R, 0)$ be a right-pBCI algebra (lattice).
We say that $\mathcal{A}^R$ is *p-semisimple*, if $x \geq 0$ implies $x = 0$, for any $x \in A^R$.

Proposition 13.1.2 *(See [50])*

Let $\mathcal{A}^L$ be a left-pBCI algebra (lattice). Then, the following are equivalent:
(i) $\mathcal{A}^L$ is p-semisimple;
(ii) for all $x, y \in A^L$, if $x \leq y$, then $x = y$;
(iii) for all $x, y \in A^L$, $(x \to^L 1) \rightsquigarrow^L y = (y \rightsquigarrow^L 1) \to^L x$.

Note that, by above Proposition 13.1.2, we can define equivalently a p-semisimple left-pBCI algebra (lattice) as follows.

Definition 13.1.3 (Definition 2 of p-semisimple pBCI algebra (lattice))
(i) A left-pBCI algebra (lattice) $\mathcal{A}^L = (A^L, \leq, \to^L, \rightsquigarrow^L, 1)$ is *p-semisimple*, if $x \leq y$ implies $x = y$, for any $x, y \in A^L$, i.e. if (p-s^L) holds.
(i') Dually, a right-pBCI algebra (lattice) $\mathcal{A}^R = (A^R, \geq, \to^R, \rightsquigarrow^R, 0)$ is *p-semisimple*, if $x \geq y$ implies $x = y$, for any $x, y \in A^R$, i.e. if (p-s^R) holds.

Remark 13.1.4 Note that we could write:
(p-s^L) $x \leq y \Longleftrightarrow x = y$ and, dually,
(p-s^R) $x \geq y \Longleftrightarrow x = y$.

Then, we have the following important result (the dual one is omitted).

Theorem 13.1.5 *(See [50])*
(1) Let $\mathcal{A}^L = (A^L, \leq, \to^L, \rightsquigarrow^L, 1)$ be a p-semisimple left-pBCI algebra.
Define $\Phi_{p-s}(\mathcal{A}^L) \stackrel{def.}{=} (A^L, \cdot, ^{-1}, 1)$ by: for any $x, y \in A^L$,
$x^{-1} \stackrel{def.}{=} x \to^L 1 \stackrel{(Id^L)}{=} x \rightsquigarrow^L 1, \quad x \cdot y \stackrel{def.}{=} x^{-1} \rightsquigarrow^L y \stackrel{(iii)}{=} y^{-1} \to^L x.$
Then, $\Phi_{p-s}(\mathcal{A}^L)$ is a group.
(1') Conversely, let $\mathcal{A} = (A, \cdot, ^{-1}, 1)$ be a group.
Define $\Psi_{p-s}(\mathcal{A}) \stackrel{def.}{=} (A, \leq, \to^L, \rightsquigarrow^L, 1)$ by: for any $x, y \in A$,
$x \to^L y \stackrel{def.}{=} y \cdot x^{-1}, \quad x \rightsquigarrow^L y \stackrel{def.}{=} x^{-1} \cdot y$ *and*
$x \leq y \stackrel{def.}{\Leftrightarrow} x \to^L y = 1 \stackrel{(IdEq^L)}{\Longleftrightarrow} x \rightsquigarrow^L y = 1.$
Then, $\Psi_{p-s}(\mathcal{A})$ is a p-semisimple left-pBCI algebra.
(2) The above defined mappings Φ_{p-s} and Ψ_{p-s} are mutually inverse.

By Theorems 13.1.5 and 9.2.3, we obtain immediately the following Theorem (the dual case is omitted).

Theorem 13.1.6
(1) Let $\mathcal{A}^L = (A^L, \leq, \to^L, \rightsquigarrow^L, 1)$ be a p-semisimple left-pBCI algebra.
Define $\alpha_{ig}(\mathcal{A}^L) \stackrel{def.}{=} (A^L, \to^L, \rightsquigarrow^L, 1)$.
Then, $\alpha_{ig}(\mathcal{A}^L)$ is an implicative-group.
(1') Conversely, let $\mathcal{A} = (A, \to, \rightsquigarrow, 1)$ be an implicative-group.
Define $\beta_{ig}(\mathcal{A}) \stackrel{def.}{=} (A, \leq, \to, \rightsquigarrow, 1)$ by: for any $x, y \in A$,
$x \leq y \stackrel{def.}{\Leftrightarrow} x \to y = 1 \stackrel{(IdEq)}{\Longleftrightarrow} x \rightsquigarrow y = 1.$
Then, $\beta_{ig}(\mathcal{A})$ is a p-semisimple left-pBCI algebra.
(2) The above defined mappings α_{ig} and β_{ig} are mutually inverse.

Note that, by above theorem, the p-semisimple left-pBCI algebras coincide with the implicative-groups and, dually, the p-semisimple right-pBCI algebras coincide with the implicative-groups.

Hence, we have the following equivalences:
(pBCIIG) $\mathbf{pBCI}^L$ + (p-s^L) $\Longleftrightarrow$ **implicative-group** and, dually,
$\mathbf{pBCI}^R$ + (p-s^R) $\Longleftrightarrow$ **implicative-group**.

Remarks 13.1.7 Recall the following two equivalent definitions of left-pBCI algebras (the right-case is omitted).

- Definition 1 of pBCI algebras (see Definition 2.3.1):
A *left-pBCI algebra* is a structure $\mathcal{A}^L = (A^L, \leq, \to^L, \rightsquigarrow^L, 1)$, where $\leq$ is a binary relation on A^L, $\to^L$ and $\rightsquigarrow^L$ are binary operations on A^L and 1 is an element of A^L verifying: for all $x, y, z \in A^L$,

(pBB^L) $y \to^L z \leq (z \to^L x) \rightsquigarrow^L (y \to^L x),\ y \rightsquigarrow^L z \leq (z \rightsquigarrow^L x) \to^L (y \rightsquigarrow^L x)$,
(pD^L) $y \leq (y \to^L x) \rightsquigarrow^L x,\ y \leq (y \rightsquigarrow^L x) \to^L x$,
(Re) $x \leq x$ (Reflexivity),
(An) $x \leq y$ and $y \leq x \Longrightarrow x = y$ (Antisymmetry),
(IdEq^L) $x \to^L y = 1 \Longleftrightarrow x \rightsquigarrow^L y = 1$,
(pEq^L) $x \leq y \Longleftrightarrow x \to^L y = 1 \overset{(IdEq^L)}{\Longleftrightarrow} x \rightsquigarrow^L y = 1$.

- Definition 2 of pBCI algebras (see Definition 2.3.8):
A *left-pBCI algebra* is a structure $\mathcal{A}^L = (A^L, \leq, \to^L, \rightsquigarrow^L, 1)$ s.t. $(A^L, \leq)$ is a poset (with maximal element 1) and the properties (pBB^L), (pM^L), (IdEq^L), (pEq^L) hold.

Recall also that we can define a negation (that is not involutive) on a left-pBCI algebra $\mathcal{A}^L$ as follows: for all $x \in A^L$, $x^{-1} \overset{def.}{=} x \to^L 1 \overset{(Id^L)}{=} x \rightsquigarrow^L 1$.

(1) Following the Definition 1 of pBCI algebras (in left-case and in the right-case), a p-semisimple pBCI algebra is defined equivalently as an algebra $(A, \to, \rightsquigarrow, 1)$ verifying: for all $x, y, z \in A$,

(pBB=) $y \to z = (z \to x) \rightsquigarrow (y \to x),\ y \rightsquigarrow z = (z \rightsquigarrow x) \to (y \rightsquigarrow x)$,
(pD=) $y = (y \to x) \rightsquigarrow x,\ y = (y \rightsquigarrow x) \to x$,
(IdEq) $x \to y = 1 \Longleftrightarrow x \rightsquigarrow y = 1$,
(pEq=) $x = y \Longleftrightarrow x \to y = 1 \overset{(IdEq)}{\Longleftrightarrow} x \rightsquigarrow y = 1$.

Note that such an algebra is just the implicative-group, as it is defined in Chapter 8, Definition 3.

(2) Following the Definition 2 of pBCI algebras (in left-case and in the right-case), a p-semisimple pBCI algebra is defined equivalently as an algebra $(A, \to, \rightsquigarrow, 1)$ verifying (pBB=), (pM) ($1 \to x = x = 1 \rightsquigarrow x$), (IdEq), (pEq=).

Note that such an algebra is just the implicative-group, as it is defined in Chapter 8, Definition 4.

(3) Following another equivalent definition of pBCI algebras, we obtain the corresponding equivalent definition of p-semisimple pBCI algebras, i.e. of implicative-groups.

(4) Remark that the negation $^{-1}$ defined on p-semisimple pBCI algebras (= i-groups) is *involutive*: $(x^{-1})^{-1} \stackrel{def.}{=} (x \to 1) \rightsquigarrow 1 \stackrel{(pD=)}{=} x$, for all x.

Concerning the p-semisimple left-pBCI lattices, note that $\beta_{ig}(\mathcal{A})$ is only a p-semisimple left-pBCI algebra, not a lattice, hence we have only the following result in this case:

Theorem 13.1.8

(1) Let $\mathcal{A}^L = (A^L, \vee, \wedge, \to^L, \rightsquigarrow^L, 1)$ be a p-semisimple left-pBCI lattice. Define $\alpha_{ig}(\mathcal{A}^L) \stackrel{def.}{=} (A^L, \to^L, \rightsquigarrow^L, 1)$.

Then, $\alpha_{ig}(\mathcal{A}^L)$ is an implicative-group.

(1') Conversely, let $\mathcal{A} = (A, \to, \rightsquigarrow, 1)$ be an implicative-group. Define $\beta_{ig}(\mathcal{A}) \stackrel{def.}{=} (A, \leq, \to, \rightsquigarrow, 1)$ by: for any $x, y \in A$, $x \leq y \stackrel{def.}{\Leftrightarrow} x \to y = 1 \stackrel{(IdEq)}{\Longleftrightarrow} x \rightsquigarrow y = 1$.

Then, $\beta_{ig}(\mathcal{A})$ is a p-semisimple left-pBCI algebra.

13.1.2 p-semisimple pBCI(pP) algebras (lattices) and their duals

Recall (Definition 2.5.1) that a pBCI(pP) algebra is a left-pBCI algebra $\mathcal{A}^L = (A^L, \leq, \to^L, \rightsquigarrow^L, 1)$ with the additional property (pP):
(pP) $\exists\, x \odot y \stackrel{notation}{=} \min\{z \mid x \leq y \to^L z\} = \min\{z \mid y \leq x \rightsquigarrow^L z\}$.

We introduce now the following definition.

Definition 13.1.9

(i) Let $\mathcal{A}^L = (A^L, \leq, \to^L, \rightsquigarrow^L, 1)$ be a pBCI(pP) algebra (lattice). We say that $\mathcal{A}^L$ is *p-semisimple*, if (p-s^L) holds.

(i') Dually, let $\mathcal{A}^R = (A^R, \geq, \to^R, \rightsquigarrow^R, 0)$ be a pBCI(pS) algebra (lattice). We say that $\mathcal{A}^R$ is *p-semisimple*, if (p-s^R) holds.

We then prove the following result.

Theorem 13.1.10

(1) Let $\mathcal{A}^L = (A^L, \leq, \to^L, \rightsquigarrow^L, 1)$ be a p-semisimple pBCI(pP) algebra. Define $\alpha_{ig}(\mathcal{A}^L) \stackrel{def.}{=} (A^L, \to^L, \rightsquigarrow^L, 1)$.

Then, $\alpha_{ig}(\mathcal{A}^L)$ is an implicative-group.

(1') Conversely, let $\mathcal{A} = (A, \to, \rightsquigarrow, 1)$ be an implicative-group. Define $\beta_{ig}(\mathcal{A}) \stackrel{def.}{=} (A, \leq, \to, \rightsquigarrow, 1)$ by: for any $x, y \in A$, $x \leq y \stackrel{def.}{\Leftrightarrow} x \to y = 1 \stackrel{(IdEq)}{\Longleftrightarrow} x \rightsquigarrow y = 1$.

Then, $\beta_{ig}(\mathcal{A})$ is a p-semisimple pBCI(pP) algebra.

(2) The above defined mappings α_{ig} and β_{ig} are mutually inverse.

Proof.

(1): Obviously, since any p-semisimple pBCI(pP) algebra is a p-semisimple left-pBCI algebra; then apply Theorem 13.1.6.

(1'): Let $\mathcal{A} = (A, \to, \rightsquigarrow, 1)$ be an implicative-group. Then, $\beta_{ig}(\mathcal{A}) \stackrel{def.}{=} (A, \leq, \to, \rightsquigarrow, 1)$ is a p-semisimple left-pBCI algebra, by Theorem 13.1.6. It remains to prove that (pP) holds.

Indeed, since $(A, \to, \rightsquigarrow, 1)$ is an implicative-group, then, there exists the product $\cdot$, defined by (8.5): $x \cdot y \stackrel{def.}{=} y^{-1} \to x \stackrel{(pDNeg2)}{=} x^{-1} \rightsquigarrow y$,
where the negation $^{-1}$ is defined by: $x^{-1} \stackrel{def.}{=} x \to 1 = x \rightsquigarrow 1$.

Further, by Theorem 8.1.29, the property (pGa$^=$) holds: for all $x, y \in A$,
(pGa$^=$) $x \cdot y = z \Leftrightarrow x = y \to z \Leftrightarrow y = x \rightsquigarrow z$.

Finally note $x \leq y \stackrel{def.}{\Leftrightarrow} x \to y = 1 \stackrel{(IdEq)}{\Longleftrightarrow} x \rightsquigarrow y = 1 \Leftrightarrow x = y$, by (pEq=).

Consequenly, (pGa$^=$) becomes:
$x \cdot y = \min\{z \mid x \leq y \to z\} = \min\{z \mid y \leq x \rightsquigarrow z\}$,
i.e. (pGa$^=$) becomes (pP), thus (pP) holds.

(2): Immediately. □

Remark 13.1.11 (See Theorem 2.11.4, Remarks 2.11.7, 2.11.8 and Remark 8.1.30)
By above proof, we can see better that an implicative-group is an involutive structure that verifies both the property (pP) and the property (pS).

Note that, by above Theorem and above Remark, the p-semisimple pBCI(pP) algebras also coincide with the implicative-groups and, dually, the p-semisimple pBCI(pS) algebras also coincide with the implicative-groups.

Concerning the p-semisimple left-pBCI(pP) lattices, we have only the following result:

Theorem 13.1.12
(1) Let $\mathcal{A}^L = (A^L, \vee, \wedge, \to^L, \rightsquigarrow^L, 1)$ be a p-semisimple pBCI(pP) lattice. Define $\alpha_{ig}(\mathcal{A}^L) \stackrel{def.}{=} (A^L, \to^L, \rightsquigarrow^L, 1)$.

Then, $\alpha_{ig}(\mathcal{A}^L)$ is an implicative-group.

(1') Conversely, let $\mathcal{A} = (A, \to, \rightsquigarrow, 1)$ be an implicative-group. Define $\beta_{ig}(\mathcal{A}) \stackrel{def.}{=} (A, \leq, \to, \rightsquigarrow, 1)$ by: for any $x, y \in A$,
$x \leq y \stackrel{def.}{\Leftrightarrow} x \to y = 1 \stackrel{(IdEq)}{\Longleftrightarrow} x \rightsquigarrow y = 1$.

Then, $\beta_{ig}(\mathcal{A})$ is a p-semisimple pBCI(pP) algebra.

13.1.3 p-semisimple pBCI(pRP) algebras (lattices) and their duals

Recall (Definition 2.6.1) that a *pBCI(pRP) algebra* is a structure $\mathcal{A}^L = (A^L, \leq, \to^L, \rightsquigarrow^L, \odot, 1)$ such that $(A^L, \leq)$ is a poset, $(A^L, \to^L, \rightsquigarrow^L, 1)$ is a left-pME algebra and the pseudo-product $\odot$ is connected to the pseudo-implication $(\to^L, \rightsquigarrow^L)$ by the following property (pRP): for all $x, y, z \in A^L$,
(pRP) $\quad x \leq y \to^L z \Longleftrightarrow y \leq x \rightsquigarrow^L z \Longleftrightarrow x \odot y \leq z$.

We introduce now the following definition.

Definition 13.1.13
(i) Let $\mathcal{A}^L = (A^L, \leq, \rightarrow^L, \rightsquigarrow^L, \odot, 1)$ be a pBCI(pRP) algebra (lattice). We say that $\mathcal{A}^L$ is *p-semisimple*, if (p-s^L) holds.
(i') Dually, let $\mathcal{A}^R = (A^R, \geq, \rightarrow^R, \rightsquigarrow^R, 0)$ be a pBCI(pcoRS) algebra (lattice). We say that $\mathcal{A}^R$ is *p-semisimple*, if (p-s^R) holds.

Recall also (Definition 8.1.34) that an *X-implicative-group* is an algebra $(G, \rightarrow, \rightsquigarrow, \cdot, 1)$ such that the properties (pEx) and (pM) hold and the pseudo-implication $(\rightarrow, \rightsquigarrow)$ and the operation $\cdot$ verify: for all $x, y, z \in G$,
(pGa$^=$) $x = y \rightarrow z \Leftrightarrow y = x \rightsquigarrow z \Leftrightarrow x \cdot y = z$.

We then prove the following Theorem.

Theorem 13.1.14
(1) Let $\mathcal{A}^L = (A^L, \leq, \rightarrow^L, \rightsquigarrow^L, \odot, 1)$ be a p-semisimple pBCI(pRP) algebra. Define $\alpha_{ig}(\mathcal{A}^L) \stackrel{def.}{=} (A^L, \rightarrow^L, \rightsquigarrow^L, \odot, 1)$.
Then, $\alpha_{ig}(\mathcal{A}^L)$ is an X-implicative-group.
(1') Conversely, let $\mathcal{A} = (A, \rightarrow, \rightsquigarrow, \cdot, 1)$ be an X-implicative-group. Define $\beta_{ig}(\mathcal{A}) \stackrel{def.}{=} (A, \leq, \rightarrow, \rightsquigarrow, \odot, 1)$ by: for any $x, y \in A$, $\odot \stackrel{def.}{=} \cdot$ and $x \leq y \stackrel{def.}{\Leftrightarrow} x \rightarrow y = 1 \stackrel{(IdEq)}{\Longleftrightarrow} x \rightsquigarrow y = 1$.
Then, $\beta_{ig}(\mathcal{A})$ is a p-semisimple pBCI(pRP) algebra.
(2) The above defined mappings α_{ig} and β_{ig} are mutually inverse.

Proof. Immediately. □

Note that, by above Theorem, we can say that the p-semisimple pBCI(pRP) algebras coincide with the X-implicative-groups and, dually, the p-semisimple pBCI(pcoRS) algebras coincide with the X-implicative-groups.

Concerning the p-semisimple left-pBCI(pRP) lattices, we have only the following result:

Theorem 13.1.15
(1) Let $\mathcal{A}^L = (A^L, \vee, \wedge, \rightarrow^L, \rightsquigarrow^L, \odot, 1)$ be a p-semisimple pBCI(pRP) lattice. Define $\alpha_{ig}(\mathcal{A}^L) \stackrel{def.}{=} (A^L, \rightarrow^L, \rightsquigarrow^L, \odot, 1)$.
Then, $\alpha_{ig}(\mathcal{A}^L)$ is an X-implicative-group.
(1') Conversely, let $\mathcal{A} = (A, \rightarrow, \rightsquigarrow, \cdot, 1)$ be an X-implicative-group. Define $\beta_{ig}(\mathcal{A}) \stackrel{def.}{=} (A, \leq, \rightarrow, \rightsquigarrow, \odot, 1)$ by: for any $x, y \in A$, $\odot \stackrel{def.}{=} \cdot$ and $x \leq y \stackrel{def.}{\Leftrightarrow} x \rightarrow y = 1 \stackrel{(IdEq)}{\Longleftrightarrow} x \rightsquigarrow y = 1$.
Then, $\beta_{ig}(\mathcal{A})$ is a p-semisimple pBCI(pRP) algebra.

13.1.4 Resuming connections for the family of pBCI algebras

Remark 13.1.16 Note that the p-semisimple pME, pRME=pCI, pre-pBCI, pBCH algebras also coincide with the implicative-groups.

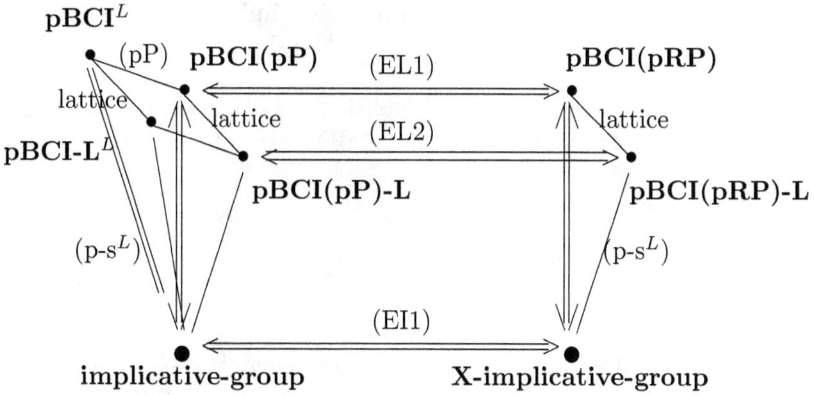

Figure 13.1: The p-semisimple algebras from the family of pBCI algebras

13.1.5 p-semisimple (strong) p*aRM** algebras

Definition 13.1.17

(i) A (strong) left-p*aRM** algebra $\mathcal{A}^L = (A^L, \leq, \rightarrow^L, \rightsquigarrow^L, 1)$ is *p-semisimple*, if (p-s^L) holds.

(i') Dually, a (strong) right-p*aRM** algebra $\mathcal{A}^R = (A^R, \geq, \rightarrow^R, \rightsquigarrow^R, 0)$ is *p-semisimple*, if (p-s^R) holds.

We prove first the following Theorem, establishing the connection between the implicative-hubs and the left- p*aRM** algebras (the dual case is omitted).

Theorem 13.1.18 *(See Theorem 13.1.6 for implicative-groups)*

*(1) Let $\mathcal{A}^L = (A^L, \leq, \rightarrow^L, \rightsquigarrow^L, 1)$ be a p-semisimple left-p*aRM** algebra. Define $\alpha_{ih}(\mathcal{A}^L) \stackrel{def.}{=} (A^L, \rightarrow^L, \rightsquigarrow^L, 1)$.*

Then, $\alpha_{ih}(\mathcal{A}^L)$ is an implicative-hub.

(1') Conversely, let $\mathcal{A} = (A, \rightarrow, \rightsquigarrow, 1)$ be an implicative-hub. Define $\beta_{ih}(\mathcal{A}) \stackrel{def.}{=} (A, \leq, \rightarrow, \rightsquigarrow, 1)$ by: for any $x, y \in A$, $x \leq y \stackrel{def.}{\Leftrightarrow} x \rightarrow y = 1 \stackrel{(IdEq)}{\Leftrightarrow} x \rightsquigarrow y = 1$.

*Then, $\beta_{ih}(\mathcal{A})$ is a p-semisimple left-p*aRM** algebra.*

(2) The above defined mappings α_{ih} and β_{ih} are mutually inverse.

Proof.

(1): Obviously, by definitions of left-p*aRM** algebras and implicative-hubs.

(1'): By Proposition 11.1.4 (2).

(2): Immediately. □

Note that, by above Theorem, we can say that the p-semisimple left-p*aRM** algebras coincide with the implicative-hubs and, dually, the p-semisimple right-p*aRM** algebras also coincide with the implicative-hubs. Hence, we have the equivalences:

(p*aRM**IH) **p*aRM**^{L}** + (p-s^L) $\Leftrightarrow$ **implicative-hub** and, dually,
p*aRM^R** + (p-s^R) $\Leftrightarrow$ **implicative-hub**.

We prove now the following Theorem, establishing the connection between the strong implicative-hubs and the strong left- p*aRM** algebras (the dual case is omitted).

Theorem 13.1.19
(1) Let $\mathcal{A}^L = (A^L, \leq, \rightarrow^L, \rightsquigarrow^L, 1)$ *be a p-semisimple strong left-p*aRM** algebra. Define* $\alpha_{ih}(\mathcal{A}^L) \overset{def.}{=} (A^L, \rightarrow^L, \rightsquigarrow^L, 1)$.
Then, $\alpha_{ih}(\mathcal{A}^L)$ *is a strong implicative-hub.*
(1') Conversely, let $\mathcal{A} = (A, \rightarrow, \rightsquigarrow, 1)$ *be a strong implicative-hub. Define* $\beta_{ih}(\mathcal{A}) \overset{def.}{=} (A, \leq, \rightarrow, \rightsquigarrow, 1)$ *by: for any* $x, y \in A$,
$x \leq y \overset{def.}{\Leftrightarrow} x \rightarrow y = 1 \overset{(IdEq)}{\Leftrightarrow} x \rightsquigarrow y = 1$.
Then, $\beta_{ih}(\mathcal{A})$ *is a p-semisimple strong left-p*aRM** algebra.*
(2) The above defined mappings α_{ih} *and* β_{ih} *are mutually inverse.*

Proof.
(1): Obviously, by definitions of strong left-p*aRM** algebras and strong i-hubs.
(1'): By Proposition 11.1.27.
(2): Immediately. □

Note that, by above Theorem, we can say that the p-semisimple strong left-p*aRM** algebras coincide with the strong implicative-hubs and, dually, the p-semisimple strong right-p*aRM** algebras also coincide with the strong implicative-hubs. Hence, we have the equivalences:
(sp*aRM**IH) **strong p*aRM**^{L}** + (p-s^L) $\Leftrightarrow$ **strong implicative-hub** and,
strong p*aRM^R** + (p-s^R) $\Leftrightarrow$ **strong implicative-hub**.

We have seen in Chapter 11 that the strong implicative-hubs are equivalent with the implicative-goops. Consequently, we have the following:

Corollary 13.1.20 *The p-semisimple strong left-p*aRM** algebras are equivalent with the implicative-goops.*

Remark 13.1.21 Note that the p-semisimple (strong) pM, pRM, paRM, poRM, p*aRM, paRM**, p*aRM** algebras coincide also with the (strong) implicative-hubs.

13.1.6 p-semisimple pML algebras

Definition 13.1.22
(i) A left-pML algebra (lattice) $\mathcal{A}^L = (A^L, \leq, \rightarrow^L, \rightsquigarrow^L, 1)$ is *p-semisimple*, if (p-s^L) holds.

(i') Dually, a right-pML algebra (lattice) $\mathcal{A}^R = (A^R, \geq, \rightarrow^R, \rightsquigarrow^R, 0)$ is *p-semisimple*, if (p-s^R) holds.

If we apply the property (p-s^L) to left-pseudo-ML algebras $(A^L, \leq, \rightarrow^L, \rightsquigarrow^L, 1)$ (which are left-pseudo-M algebras verifying (L) ($x \leq 1$, for all x)), then we obtain that p-semisimple left-pML algebras have $A^L = \{1\}$, hence they are trivial algebras.

Dually, if we apply the property (p-s^R) to right-pseudo-ML algebras $(A^R, \geq, \rightarrow^R, \rightsquigarrow^R, 0)$ (which are right-pseudo-M algebras verifying (F) ($x \geq 0$, for all x)), then we obtain p-semisimple right-pML algebras have $A^R = \{0\}$, hence they are trivial algebras.

Consequently, we immediately obtain the following important generalizing result.

Theorem 13.1.23

(1) The p-semisimple left-pML algebras $(A^L, \leq, \rightarrow^L, \rightsquigarrow^L, 1)$ *have* $A^L = \{1\}$, *i.e. are trivial algebras.*

(1') The p-semisimple right-pML algebras $(A^R, \geq, \rightarrow^R, \rightsquigarrow^R, 0)$ *have* $A^R = \{0\}$, *i.e. are trivial algebras.*

13.2 Quasi-p-semisimple quasi-pseudo-algebras

We have introduced in Chapter 5 the definition of *quasi-p-semisimple* quasi-pseudo-algebra (quasi-pseudo-structure) - Definition 5.1.24.

13.2.1 Quasi-p-semisimple quasi-pBCI algebras

The left-quasi-pBCI algebras were introduced in Chapter 5, as left-quasi-pseudo-structures and also as left-quasi-pseudo-algebras (the dual case is omitted).

We introduce the following definition:

Definition 13.2.1

(i) A left-q-pBCI algebra is *quasi-p-semisimple*, if it verifies: for all x, y,
(q-p-s^L) $x \leq y \iff (1 \rightarrow^L x = 1 \rightarrow^L y$ and $1 \rightsquigarrow x^L = 1 \rightsquigarrow^L y)$.

(i') A right-q-pBCI algebra is *quasi-p-semisimple*, if it verifies: for all x, y,
(q-p-s^R) $x \geq y \iff (0 \rightarrow^R x = 0 \rightarrow^R y$ and $0 \rightsquigarrow x^R = 0 \rightsquigarrow^R y)$.

Remark 13.2.2 Following Definition 1 of left-q-pBCI algebras, a quasi-p-semisimple left-q-pBCI algebra is defined equivalently as an algebra $(A, \rightarrow, \rightsquigarrow, 1)$ verifying: for all $x, y, z \in A$,

(pBB=) $y \rightarrow z = (z \rightarrow x) \rightsquigarrow (y \rightarrow x)$, $y \rightsquigarrow z = (z \rightsquigarrow x) \rightarrow (y \rightsquigarrow x)$,
(q-pD=) $1 \rightarrow y = (y \rightarrow x) \rightsquigarrow x$, $1 \rightsquigarrow y = (y \rightsquigarrow x) \rightarrow x$,
(q-pM) $1 \rightarrow (x \rightsquigarrow y) = x \rightsquigarrow y)$, $1 \rightsquigarrow (x \rightarrow y) = x \rightarrow y$,
(IdEq) $x \rightarrow y = 1 \Longleftrightarrow x \rightsquigarrow y = 1$,
(q-pEq=) $1 \rightarrow x (= 1 \rightsquigarrow x) = 1 \rightarrow y (= 1 \rightsquigarrow y) \Longleftrightarrow x \rightarrow y = 1 \overset{(IdEq)}{\Longleftrightarrow} x \rightsquigarrow y = 1$.

Note that such an algebra is just the *exchanged q-implicative-hub*, as it is defined in Chapter 11, Definition 3.

We shall prove now the following very important result (the dual case is omitted):

Theorem 13.2.3
(1) Let $\mathcal{A} = (A, \leq, \to^L, \rightsquigarrow^L, 1)$ *be a quasi-p-semisimple left-quasi-pBCI algebra. Define* $\alpha_{eqih}(\mathcal{A}) \stackrel{def.}{=} (A, \to^L, \rightsquigarrow^L, 1)$.
Then, $\alpha_{eqih}(\mathcal{A})$ *is an exchanged q-implicative-hub.*
(1') Conversely, let $\mathcal{A} = (A, \to, \rightsquigarrow, 1)$ *be an exchanged q-implicative-hub. Define* $\beta_{eqih}(\mathcal{A}) \stackrel{def.}{=} (A, \leq, \to, \rightsquigarrow, 1)$ *by: for any* $x, y \in A$,
$x \leq y \stackrel{def.}{\Leftrightarrow} x \to y = 1 \stackrel{(IdEq)}{\Longleftrightarrow} x \rightsquigarrow y = 1$.
Then, $\beta_{eqih}(\mathcal{A})$ *is a quasi-p-semisimple left-q-pBCI algebra.*
(2) The above defined mappings α_{eqih} *and* β_{eqih} *are mutually inverse.*

Proof.
(1): Let $\mathcal{A} = (A, \leq, \to^L, \rightsquigarrow^L, 1)$ be a left-quasi-pBCI algebra (i.e. a structure verifying (pBB^L), (pD^L), (Re), (q-pM^L), (q-pAn^L), (IdEq^L), (pEq^L)) verifying also (q-p-s^L). Then, $\mathcal{A}$ verifies also (Id^L), by Proposition 5.1.49, and (pRe'), by (pEq^L). By Proposition 11.3.2 (q17=), (pBB^L) + (q-pAn^L) + (q-pM^L) + (Id^L) $\Longrightarrow$ (pBB=), hence $\mathcal{A}$ verifies (pBB=). Thus, $(A, \to^L, \rightsquigarrow^L, 1)$ is an exchanged q-implicative-hub (Definition 1).
(1'): Let $\mathcal{A} = (A, \to, \rightsquigarrow, 1)$ be an exchanged q-implicative-hub. Then, by Proposition 11.3.37, $(A, \leq, \to, \rightsquigarrow, 1)$ is a left-q-pBCI algebra verifying (q-p-s^L).
(2): Routine. □

By above theorem, we obtain the equivalences:

(qpBCIIH) **q-pBCI**^{L} + (q-p-s^L) $\Longleftrightarrow$ **exchanged q-i-hub** and, dually,
q-pBCI^R + (q-p-s^R) $\Longleftrightarrow$ **exchanged q-i-hub**,

i.e. the quasi-p-semisimple quasi-pBCI algebras coincide with the exchanged q-implicative-hubs.

13.2.2 Quasi-p-semisimple quasi-p*aRM** algebras

The left-quasi-p*aRM** algebras were introduced in Chapter 11, as left-quasi-pseudo-structures and also as left-quasi-pseudo-algebras (the dual case is omitted).
We introduce the following definition.

Definition 13.2.4
(i) A left-q-p*aRM** algebra is *quasi-p-semisimple*, if it verifies (q-p-s^L).
(i') A right-q-p*aRM** algebra is *quasi-p-semisimple*, if it verifies (q-p-s^R).

Remark 13.2.5 Following the definition of left-q-p*aRM** algebras, a quasi-p-semisimple left-q-p*aRM** algebra is defined equivalently as an algebra $(A, \to, \rightsquigarrow, 1)$ verifying (q-pM), (pM1), (IdEq), (q-pEq=), (p*'), (p**').
Note that such an algebra is just the super q-implicative-hub, as it is defined also in Chapter 11.

We shall prove now the following very important result (the dual case is omitted):

Theorem 13.2.6

(1) Let $\mathcal{A} = (A, \leq, \to^L, \leadsto^L, 1)$ *be a quasi-p-semisimple left-q-p*aRM** algebra. Define* $\alpha_{qih}(\mathcal{A}) \stackrel{def.}{=} (A, \to^L, \leadsto^L, 1)$.

Then, $\alpha_{qih}(\mathcal{A})$ *is a super q-implicative-hub.*

(1') Conversely, let $\mathcal{A} = (A, \to, \leadsto, 1)$ *be a super q-implicative-hub. Define* $\beta_{qih}(\mathcal{A}) \stackrel{def.}{=} (A, \leq, \to, \leadsto, 1)$ *by: for any* $x, y \in A$, $x \leq y \stackrel{def.}{\Leftrightarrow} x \to y = 1 \stackrel{(IdEq)}{\Longleftrightarrow} x \leadsto y = 1$.

Then, $\beta_{qih}(\mathcal{A})$ *is a quasi-p-semisimple left-q-p*aRM** algebra.*

(2) The above defined mappings α_{qih} *and* β_{qih} *are mutually inverse.*

Proof.

(1): Let $\mathcal{A} = (A, \leq, \to^L, \leadsto^L, 1)$ be a left-q-p*aRM** algebra (i.e. a structure verifying (IdEq^L), (pEq^L), (q-pAn^L), (Re), (q-pM^L), (p*^{L}), (p**^{L})) verifying also (q-p-s^L). By Proposition 11.3.2 (q9=), (q-pEq=) $\Longleftrightarrow$ (q-p-s^L), hence $\mathcal{A}$ verifies (q-pEq=). (Re) implies (pM1), by (pEq^L); (p*^{L}) and (p**^{L}) imply (p*') and (p**'), respectively, by (pEq^L) also. Thus, $(A, \to^L, \leadsto^L, 1)$ verifies (q-pM^L), (pM1), (IdEq^L), (q-pEq=), (p*'), (p**'), i.e. it is a super q-implicative-hub.

(1'): Let $\mathcal{A} = (A, \to, \leadsto, 1)$ be a super q-implicative-hub. Then, by Proposition 11.3.6 (2), $\mathcal{A}$ is a left-q-p*aRM** algebra verifying (q-p-s^L).

(2): Routine. □

By above theorem, we obtain the equivalences:

(qp*aRM**IH) **q-p*aRM****^{L} + (q-p-s^L) $\Longleftrightarrow$ **super q-i-hub** and, dually,

q-p*aRM**^{R} + (q-p-s^R) $\Longleftrightarrow$ **super q-i-hub**,

i.e. the quasi-p-semisimple quasi-p*aRM** algebras coincide with the super q-implicative-hubs.

13.3 Resuming connections

By the connections presented in the previous sections and the connections presented in Figures 11.6 and 11.11, we obtain the following resuming connections from Figures 13.2 and 13.3.

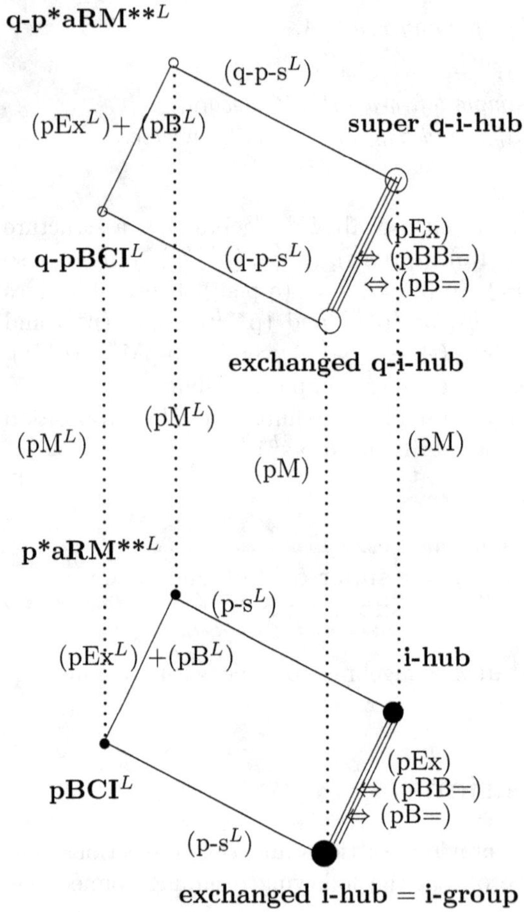

Figure 13.2: The hierarchy No. qi1

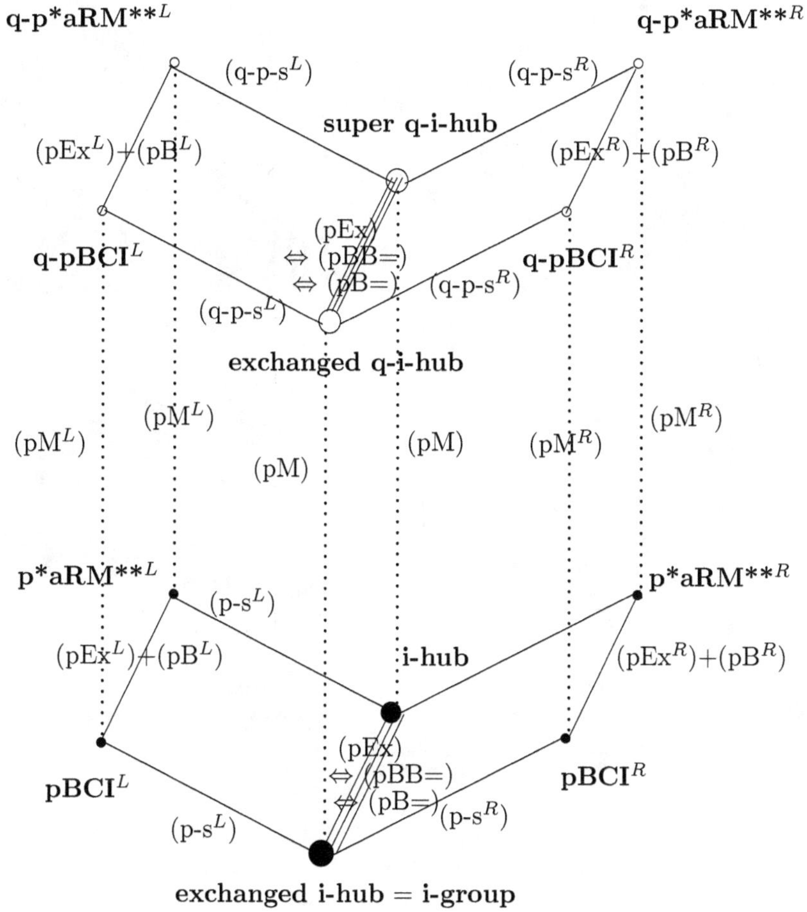

Figure 13.3: The hierarchies No. qi1 and No. qi1'

Chapter 14

Connections at lattice-order level

In this chapter, we present notions and results from [97], [98], [99], [100]. In Section 1, we present connections between the l-implicative-groups and the pseudo-BCK(pP) lattices which provide the pseudo-Wajsberg algebras and the equivalent of pseudo-product algebras. Namely, we prove that the l-implicative-group operations $\vee, \wedge, \rightarrow, \rightsquigarrow, 1$ and $\cdot$, restricted to the negative (positive) cone, determine a structure of left- (right-) pseudo-BCK lattice with pseudo-product (pseudo-sum, respectively) verifying some properties (Theorem 14.1.1 - the main result); we prove, consequently, that by bounding in two different ways the above mentioned left- (right-) pseudo-BCK lattices with pseudo-product (pseudo-sum, respectively) verifying some properties, we obtain a left- (right-) pseudo-Wajsberg algebra or a left-pseudo-Hájek(pP) algebra with the properties (pP1^L), (pP2^L) (right-pseudo-Hájek(pS) algebra with the properties (pP1^R), (pP2^R), dually) (Corollary 14.1.7). In Section 2, we obtain the corresponding connections between the l-groups and the po-ims(pR) which provide the pseudo-MV algebras and the pseudo-product algebras. The analysis of three important examples ends the chapter.

As mentioned in the Introduction (Chapter 1), the pseudo-MV algebras are intervals $[1, u]$ in l-groups [47] and the pseudo-MV algebras are termwise equivalent [24], [25] (see (iE9'), (iE9)) to the pseudo-Wajsberg algebras. Hence, the pseudo-Wajsberg algebras must be connected to a notion that is termwise equivalent to the l-group.

Also the pseudo-product algebras are obtained as bounded sets $\{-\infty\} \cup G^-$ in l-groups and the pseudo-product algebras are equivalent (see (iELM8), (iE8)) to the pseudo-Hájek(pRP) algebras verifying some properties. Hence, the pseudo-Hájek(pRP) algebras verifying some properties (namely (pP1^L) and (pP2^L)) must be connected also to a notion that is termwise equivalent to the l-group.

That notion that is termwise equivalent (see (EIG3), (Eq3)) to the l-group is the *l-implicative-group*, presented in Chapter 8; it is the great piece which missed from the puzzle showing the connections between algebras of logic (pseudo-M algebras, more precisely) and unital magmas:

$$\begin{array}{ccc} Pseudo-M\ algebras & & Unital\ magmas \\ \mathbf{l-implicative-group} & \overset{(EIG3)}{\Leftrightarrow} & \mathbf{l-group} \\ \Downarrow & & \Downarrow \\ \mathbf{pW^L} & \overset{(iE9')}{\Leftrightarrow} & \mathbf{pMV^L} \end{array}$$

and

$$\begin{array}{ccc} \mathbf{l-implicative-group} & \overset{(EIG3)}{\Leftrightarrow} & \mathbf{l-group} \\ \Downarrow & & \Downarrow \\ \mathbf{pHa(pRP)} + (pP1^L) + (pP2^L) & \overset{(iELM8)}{\Leftrightarrow} & \mathbf{pproduct^L} \end{array}$$

The connections are easier to be proved in algebras of logic rather than in unital magmas. Consequently, in this chapter, we present firstly connections between the left-(right-) l-implicative-groups and the pseudo-BCK(pP) lattices (pseudo-BCK(pS) lattices) and then we deduce the connections between the left-(right-) l-groups and the l-ims(pR) (l-ims(pcoR)).

The chapter has three sections.

14.1 Connections between pseudo-M algebras

In this section, we present connections between left-(right-) l-implicative-groups and pseudo-BCK(pP) (pseudo-BCK(pS)) lattices.

We have seen in Proposition 2.4.3 the connection between pBCI lattices and pBCK lattices.

The following theorem shows the deeper connection existing between l-implicative-groups and pBCK lattices; it is the main result of this chapter.

Theorem 14.1.1 *Let* $\mathcal{G} = (G, \vee, \wedge, \rightarrow, \rightsquigarrow, 1)$ *be an l-implicative-group (left-l-implicativ group or right-l-implicative-group). We have, for all* $x, y \in G$:
$x^{-1} \overset{def.}{=} x \rightarrow 1 = x \rightsquigarrow 1$ *and*
$x \cdot y \overset{def.}{=} (x \rightarrow y^{-1})^{-1} = (y \rightsquigarrow x^{-1})^{-1} = y^{-1} \rightarrow x = x^{-1} \rightsquigarrow y.$

(1) Define, for all $x, y \in G^-$:

$$x \rightarrow^L y \overset{def.}{=} (x \rightarrow y) \wedge 1, \quad x \rightsquigarrow^L y \overset{def.}{=} (x \rightsquigarrow y) \wedge 1, \quad x \odot y \overset{def.}{=} x \cdot y. \tag{14.1}$$

Then, $\mathcal{G}^- = (G^-, \wedge, \vee, \rightarrow^L, \rightsquigarrow^L, \mathbf{1} = 1)$ *is a distributive pBCK(pP) lattice (with the pseudo-product* $\odot$*) verifying: for all* $x, y, z \in G^-$,
(pCC^L*)* $x \vee y = (x \rightsquigarrow^L y) \rightarrow^L y = (x \rightarrow^L y) \rightsquigarrow^L y,$
(p@^{L}*)* $x \rightarrow^L y = (x \odot z) \rightarrow^L (y \odot z), \ x \rightsquigarrow^L y = (z \odot x) \rightsquigarrow^L (z \odot y).$

(1') Dually, define for all $x, y \in G^+$:

$$x \rightarrow^R y \overset{def.}{=} (x \rightarrow y) \vee 1, \quad x \rightsquigarrow^R y \overset{def.}{=} (x \rightsquigarrow y) \vee 1, \quad x \oplus y \overset{def.}{=} x \cdot y. \tag{14.2}$$

Then, $\mathcal{G}^+ = (G^+, \vee, \wedge, \to^R, \rightsquigarrow^R, \mathbf{0} = 1)$ *is a distributive pBCK(pS) lattice (with the pseudo-sum* $\oplus$*) verifying: for all* $x, y, z \in G^+$,
(pCC^R) $x \wedge y = (x \to^R y) \rightsquigarrow^R y = (x \rightsquigarrow^R y) \to^R y$;
(p@^{R}) $x \to^R y = (x \oplus z) \to^R (y \oplus z),\ \ x \rightsquigarrow^R y = (z \oplus x) \rightsquigarrow^R (z \oplus y)$.

Proof. (1):

- G^- is closed under the lattice operations $\wedge$ and $\vee$ of $\mathcal{G}$, by Lemma 8.3.3; $(G^-, \wedge, \vee)$ is a distributive lattice, since $(G, \vee, \wedge)$ is a distributive lattice. G^- is closed under $\to^L$ and $\rightsquigarrow^L$, by Lemma 8.3.4.

- We prove now that $\mathcal{G}^-$ is a left-pBCK algebra, i.e. $(G^-, \leq, \mathbf{1} = 1)$ is a poset with greatest element and that properties (pBB^L), (pM^L), (pEq^L) hold, by Definition 2.3.8.

• Obviously, $(G^-, \leq, \mathbf{1} = 1)$ is a poset with greatest element, where $x \leq y$ iff $x \wedge y = x$ iff $x \vee y = y$.

• (pBB^L): We must prove that for all $x, y, z \in G^-$,

$$(14.3)\quad (z \to^L x) \rightsquigarrow^L (y \to^L x) \geq y \to^L z, \quad (z \rightsquigarrow^L x) \to^L (y \rightsquigarrow^L x) \geq y \rightsquigarrow^L z.$$

First, we shall prove that

$$(14.4)\qquad (z \to^L x) \rightsquigarrow^L (y \to^L x) = (y \to^L z) \vee (y \to^L x),$$

$$(z \rightsquigarrow^L x) \to^L (y \rightsquigarrow^L x) = (y \rightsquigarrow^L z) \vee (y \rightsquigarrow^L x).$$

Indeed, denote $A = (z \to^L x) \rightsquigarrow^L (y \to^L x)$. Then,
$A = ([(z \to x) \wedge 1] \rightsquigarrow [(y \to x) \wedge 1]) \wedge 1 \overset{(7.15)}{=}$
$([(z \to x) \wedge 1] \rightsquigarrow (y \to x)) \wedge ([(z \to x) \wedge 1] \rightsquigarrow 1) \wedge 1 \overset{(7.18)}{=}$
$([(z \to x) \wedge 1] \rightsquigarrow (y \to x)) \wedge 1 \overset{(7.14)}{=}$
$([(z \to x) \rightsquigarrow (y \to x)] \vee [1 \rightsquigarrow (y \to x)]) \wedge 1 \overset{(pBB=),(pM)}{=}$
$((y \to z) \vee (y \to x)) \wedge 1 \overset{distributivity}{=}$
$[(y \to z) \wedge 1] \vee [(y \to x) \wedge 1] = (y \to^L z) \vee (y \to^L x)$.

Similarly, denote $B = (z \rightsquigarrow^L x) \to^L (y \rightsquigarrow^L x)$. Then
$B = ([(z \rightsquigarrow x) \wedge 1] \to [(y \rightsquigarrow x) \wedge 1]) \wedge 1 =$
$([(z \rightsquigarrow x) \wedge 1] \to (y \rightsquigarrow x)) \wedge ([(z \rightsquigarrow x) \wedge 1] \to 1) \wedge 1 \overset{(7.18)}{=}$
$([(z \rightsquigarrow x) \wedge 1] \to (y \rightsquigarrow x)) \wedge 1 =$
$([(z \rightsquigarrow x) \to (y \rightsquigarrow x)] \vee [1 \to (y \rightsquigarrow x)]) \wedge 1 =$
$((y \rightsquigarrow z) \vee (y \rightsquigarrow x)) \wedge 1 \overset{distributivity}{=}$
$[(y \rightsquigarrow z) \wedge 1] \vee [(y \rightsquigarrow x) \wedge 1] = (y \rightsquigarrow^L z) \vee (y \rightsquigarrow^L x)$.

Thus, (14.4) holds and therefore (14.3), i.e. (pBB^L), holds.

• (pM^L): We must prove that for all $x \in G^-$ we have

$$(14.5)\qquad \mathbf{1} \to^L x = x = \mathbf{1} \rightsquigarrow^L x.$$

Indeed, $\mathbf{1} \to^L x = (1 \to x) \wedge 1 \overset{(pM)}{=} x \wedge 1 = x$ and $\mathbf{1} \rightsquigarrow^L x = (1 \rightsquigarrow x) \wedge 1 \overset{(pM)}{=} x \wedge 1 = x$. Thus, (14.5), i.e. (pM^L), holds.

• (pEq^L): We must prove that for all $x, y \in G^-$ we have

$$x \leq y \Longleftrightarrow x \to^L y = \mathbf{1} \Longleftrightarrow x \rightsquigarrow^L y = \mathbf{1}. \tag{14.6}$$

But $x \leq y \Longleftrightarrow 1 \leq x \to y \Longleftrightarrow 1 \leq x \rightsquigarrow y$, by Corollary 7.2.10 (ii),
$x \to y \geq 1 \Longleftrightarrow x \to^L y = (x \to y) \wedge 1 = 1$ and
$x \rightsquigarrow y \geq 1 \Longleftrightarrow x \rightsquigarrow^L y = (x \rightsquigarrow y) \wedge 1 = 1$.
Hence, (14.6), i.e. (pEq^L), holds.

Hence, $(G^-, \wedge, \vee, \to^L, \rightsquigarrow^L, \mathbf{1} = 1)$ is a left-pBCK lattice.

- To prove that $\mathcal{G}^-$ is with the pseudo-product $\odot = \cdot$ it is equivalent (see (EL1)) to prove that the property (pRP) holds, i.e. for every $x, y, z \in G^-$:

$$x \odot y \leq z \;\Leftrightarrow\; x \leq y \to^L z = (y \to z) \wedge 1, \tag{14.7}$$

$$x \odot y \leq z \;\Leftrightarrow\; y \leq x \rightsquigarrow^L z = (x \rightsquigarrow z) \wedge 1. \tag{14.8}$$

(14.7): By (pGa$^\leq$), we have $x \odot y = x \cdot y \leq z \Leftrightarrow x \leq y \to z$. But, $x \leq y \to z \Leftrightarrow x \leq y \to^L z$, by (8.10), hence (14.7) holds.
(14.8): Similarly, or by (14.7) and (8.11). Thus, $\mathcal{G}^-$ is a pBCK(pP) lattice.

- To prove that the pBCK(pP) lattice $\mathcal{G}^-$ satisfies the property (pCC^L) means to prove that, for all $x, y \in G^-$, we have:

$$(y \to^L x) \rightsquigarrow^L x = y \vee x = (y \rightsquigarrow^L x) \to^L x. \tag{14.9}$$

Indeed,
$(y \to^L x) \rightsquigarrow^L x = ([(y \to x) \wedge 1] \rightsquigarrow x) \wedge 1 \overset{(7.17)}{=}$
$= ([(y \to x) \rightsquigarrow x] \vee [1 \rightsquigarrow x]) \wedge 1 \overset{(pD=), (pM)}{=} (y \vee x) \wedge 1 = y \vee x$ and
$(y \rightsquigarrow^L x) \to^L x = ([(y \rightsquigarrow x) \wedge 1] \to x) \wedge 1 \overset{(7.17)}{=}$
$= ([(y \rightsquigarrow x) \to x] \vee [1 \to x]) \wedge 1 = (y \vee x) \wedge 1 \overset{(pD=), (pM)}{=} y \vee x$;
thus, (14.9), i.e. (pCC^L), holds.

- The pBCK(pP) lattice $\mathcal{G}^-$ satisfies the property (p@^{L}). Indeed, the l-implicative-group $(G, \vee, \wedge, \to, \rightsquigarrow, 1)$ is termwise-equivalent (see (Eq3)) with the l-group $(G, \vee, \wedge, \cdot, ^{-1}, 1)$, where the property (p@) holds. Hence,
$(x \odot z) \to^L (y \odot z) = [(x \cdot z) \to (y \cdot z)] \wedge 1 \overset{(p@)}{=} (x \to y) \wedge 1 = x \to^L y$ and
$(z \odot x) \rightsquigarrow^L (z \odot y) = [(z \cdot x) \rightsquigarrow (z \cdot y)] \wedge 1 \overset{(p@)}{=} (x \rightsquigarrow y) \wedge 1 = x \rightsquigarrow^L y$,
i.e. (p@^{L}) holds.

Thus, (1) holds.

(1'): has a similar (dual) proof. □

Remark 14.1.2
(i) By Theorem 2.12.4 (1), the properties (pprel^L) and (pdiv^L) are verified by $\mathcal{G}^-$.
(i') By Theorem 2.12.4 (1'), the dual properties (pprel^R) and (pdiv^R) are verified by $\mathcal{G}^+$.

Remarks 14.1.3

(1) An *L-algebra* [158] is an algebra $(X, \to, 1)$ verifying: for all $x, y, z \in X$, $x \to x = x \to 1 = 1$, $1 \to x = x$, $x \to y = y \to x = 1$ imply $x = y$ and the following property (Lalg):
(Lalg) $(x \to y) \to (x \to z) = (y \to x) \to (y \to z)$.

Note that if $(G, \to, \rightsquigarrow, 1)$ is an implicative-group, then $(G, \to, 1)$ and $(G, \rightsquigarrow, 1)$ are not L algebras, since (Lalg) does not hold neither for $\to$ nor for $\rightsquigarrow$, by (pB=).

(i) But $(G^-, \to^L, \mathbf{1} = 1)$ and $(G^-, \rightsquigarrow^L, \mathbf{1} = 1)$ are L algebras, because (Lalg) holds. Indeed,
$(x \to^L y) \to^L (x \to^L z) = ([(x \to y) \wedge 1] \to [(x \to z) \wedge 1]) \wedge 1 \overset{(7.15)}{=}$
$([(x \to y) \wedge 1] \to (x \to z)) \wedge ([(x \to y) \wedge 1] \to 1) \wedge 1 \overset{(7.18)}{=}$
$([(x \to y) \wedge 1] \to (x \to z)) \wedge 1 = ([(x \to y) \to (x \to z)] \vee [1 \to (x \to z)]) \wedge 1 \overset{(pB=)}{=}$
$((y \to z) \vee (x \to z)) \wedge 1 \overset{distrib.}{=} [(y \to z) \wedge 1] \vee [(x \to z) \wedge 1] = (y \to^L z) \vee (x \to^L z)$;
hence, $(y \to^L x) \to^L (y \to^L z) = (x \to^L z) \vee (y \to^L z)$, i.e. (Lalg) holds for $\to^L$.
Similarly, $(x \rightsquigarrow^L y) \rightsquigarrow^L (x \rightsquigarrow^L z) = ([(x \rightsquigarrow y) \wedge 1] \rightsquigarrow [(x \rightsquigarrow z) \wedge 1]) \wedge 1 \overset{(7.15)}{=}$
$([(x \rightsquigarrow y) \wedge 1] \rightsquigarrow (x \rightsquigarrow z)) \wedge ([(x \rightsquigarrow y) \wedge 1] \rightsquigarrow 1) \wedge 1 \overset{(7.18)}{=}$
$([(x \rightsquigarrow y) \wedge 1] \rightsquigarrow (x \rightsquigarrow z)) \wedge 1 = ([(x \rightsquigarrow y) \rightsquigarrow (x \rightsquigarrow z)] \vee [1 \rightsquigarrow (x \rightsquigarrow z)]) \wedge 1 \overset{(pB=)}{=}$
$((y \rightsquigarrow z) \vee (x \rightsquigarrow z)) \wedge 1 \overset{distrib.}{=} [(y \rightsquigarrow z) \wedge 1] \vee [(x \rightsquigarrow z) \wedge 1] = (y \rightsquigarrow^L z) \vee (x \rightsquigarrow^L z)$;
hence, $(y \rightsquigarrow^L x) \rightsquigarrow^L (y \rightsquigarrow^L z) = (x \rightsquigarrow^L z) \vee (y \rightsquigarrow^L z)$, i.e. (Lalg) holds for $\rightsquigarrow^L$ too.

(i') Dually, $(G^+, \to^R, \mathbf{0} = 1)$ and $(G^+, \rightsquigarrow^R, \mathbf{0} = 1)$ are L algebras.

We obtain obviously the following corollary:

Corollary 14.1.4 *If $\mathcal{G}$ is a linearly-ordered l-implicative-group, then:*
(1) the two implications from Theorem 14.1.1 (1) become, for all $x, y \in G^-$:

$$x \to^L y = \begin{cases} 1, & \text{if } x \leq y \\ x \to y, & \text{if } y < x, \end{cases} \qquad x \rightsquigarrow^L y = \begin{cases} 1, & \text{if } x \leq y \\ x \rightsquigarrow y, & \text{if } y < x. \end{cases}$$

(1') the two implications from Theorem 14.1.1 (1') become, for all $x, y \in G^+$:

$$x \to^R y = \begin{cases} 1, & \text{if } x \geq y \\ x \to y, & \text{if } y > x, \end{cases} \qquad x \rightsquigarrow^R y = \begin{cases} 1, & \text{if } x \geq y \\ x \rightsquigarrow y, & \text{if } y > x. \end{cases}$$

Lemma 14.1.5 *Let $(G, \vee, \wedge, \to, \rightsquigarrow, 1)$ be an l-implicative-group. Let $u' < 1$ and $u > 1$ from G. Then,*

(1) the interval $[u', 1] = \{x \in G^- \mid u' \leq x \leq 1\} \subset G^-$ is closed under $\to^L$ and $\rightsquigarrow^L$ in the pBCK(pP) lattice $\mathcal{G}^-$ from Theorem 14.1.1 (1) and

(2) for all $x \in [u', 1]$ we have:

$$x^{-^L} \overset{def.}{=} x \to^L u' = u' \cdot x^{-1}, \quad x^{\sim^L} \overset{def.}{=} x \rightsquigarrow^L u' = x^{-1} \cdot u'.$$

(1') Dually, the interval $[1, u] = \{x \in G^+ \mid 1 \leq x \leq u\} \subset G^+$ is closed under $\to^R$ and $\rightsquigarrow^R$ in the pBCK(pS) lattice $\mathcal{G}^+$ from Theorem 14.1.1 (1') and

(2') for all $x \in [1, u]$ we have:

$$x^{-^R} \stackrel{def.}{=} x \to^R u = u \cdot x^{-1}, \quad x^{\sim^R} \stackrel{def.}{=} x \rightsquigarrow^R u = x^{-1} \cdot u.$$

Proof.

(1): Let $u' \in G^-$ and let $u' \leq x, y \leq 1$. By the properties of the left-pBCK algebra $\mathcal{G}^-$, we have $u' \leq y \leq x \to^L y = (x \to y) \wedge 1 \leq 1$ and $u' \leq y \leq x \rightsquigarrow^L y = (x \rightsquigarrow y) \wedge 1 \leq 1$. Hence, $u' \leq x \to^L y \leq 1$ and $u' \leq x \rightsquigarrow^L y \leq 1$.

(2): $x^{-^L} \stackrel{def.}{=} x \to^L u' = (x \to u') \wedge 1 = (u' \cdot x^{-1}) \wedge 1 = u' \cdot x^{-1}$, since $u' \cdot x^{-1} \leq 1$; indeed, if $x \in [u', 1]$, i.e. $u' \leq x \leq 1$, then $u' \cdot x^{-1} \leq x \cdot x^{-1} = 1$.

$x^{\sim^L} \stackrel{def.}{=} x \rightsquigarrow^L u' = (x \rightsquigarrow u') \wedge 1 = (x^{-1} \cdot u') \wedge 1 = x^{-1} \cdot u'$, since $x^{-1} \cdot u' \leq 1$; indeed, if $x \in [u', 1]$, i.e. $u' \leq x \leq 1$, then $u'^{-1} \geq x^{-1} \geq 1$ and hence $1 = u'^{-1} \cdot u' \geq x^{-1} \cdot u' \geq 1 \cdot u' = u'$.

(1'): Let $u \in G^+$ and let $1 \leq x, y \leq u$. By the properties of the right-pBCK algebra $\mathcal{G}^+$, we have $u \geq y \geq x \to^R y = (x \to y) \vee 1 \geq 1$ and $u \geq y \geq x \rightsquigarrow^R y = (x \rightsquigarrow y) \vee 1 \geq 1$. Hence, $u \geq x \to^R y \geq 1$ and $u \geq x \rightsquigarrow^R y \geq 1$.

(2'): $x^{-^R} \stackrel{def.}{=} x \to^R u = (x \to u) \vee 1 = (u \cdot x^{-1}) \vee 1 = u \cdot x^{-1}$, since $u \cdot x^{-1} \geq 1$; indeed, if $x \in [1, u]$, i.e. $1 \leq x \leq u$, then $u \cdot x^{-1} \geq x \cdot x^{-1} = 1$.

$x^{\sim^R} \stackrel{def.}{=} x \rightsquigarrow^R u = (x \rightsquigarrow u) \vee 1 = (x^{-1} \cdot u) \vee 1 = x^{-1} \cdot u$, since $x^{-1} \cdot u \geq 1$; indeed, if $x \in [1, u]$, i.e. $1 \leq x \leq u$, then $u^{-1} \leq x^{-1} \leq 1$ and hence $1 = u^{-1} \cdot u \leq x^{-1} \cdot u \leq 1 \cdot u = u$. $\square$

Lemma 14.1.6 *Let $(G, \vee, \wedge, \to, \rightsquigarrow, 1)$ be an l-implicative-group. Let $u' < 1$ and $u > 1$ from G. Then,*

(1) the interval $[u', 1] = \{x \in G^- \mid u' \leq x \leq 1\} \subset G^-$ is not closed under $\odot = \cdot$ in the pBCK(pP) lattice $\mathcal{G}^-$ from Theorem 14.1.1 (1), but is closed under the new operation $\odot^L$ defined as follows: for all $x, y \in [u', 1]$,

$$x \odot^L y \stackrel{def.}{=} (x \odot y) \vee u' = (x \cdot y) \vee u'.$$

(2) For all $x, y, z \in [u', 1]$, we have:

$$x \odot^L y \leq z \Leftrightarrow x \cdot y \leq z.$$

(1') Dually, the interval $[1, u] = \{x \in G^+ \mid 1 \leq x \leq u\} \subset G^+$ is not closed under $\oplus = \cdot$ in the pBCK(pS) lattice $\mathcal{G}^+$ from Theorem 14.1.1 (1'), but is closed under the new operation $\oplus^R$ defined as follows: for all $x, y \in [1, u]$,

$$x \oplus^R y \stackrel{def.}{=} (x \oplus y) \wedge u = (x \cdot y) \wedge u.$$

(2') For all $x, y, z \in [1, u]$, we have:

$$x \oplus^R y \geq z \Leftrightarrow x \cdot y \geq z.$$

Proof. (1): $[u', 1]$ is not closed under $\odot = \cdot$. Indeed, let $x, y \in [u', 1]$, i.e. $u' \leq x, y \leq 1$; then, $u' \cdot u' \leq x \cdot y \leq 1$ and $u' \cdot u' \leq 1 \cdot u' = u'$, hence $x \cdot y$ may belong or not to $[u', 1]$.

$[u', 1]$ is closed under $\odot^L$. Indeed, let $u' \leq x, y \leq 1$; since $x \cdot y \leq 1$ and $u' \leq 1$, it follows that $(x \cdot y) \vee u' \leq 1$; we also have that $u' \leq (x \cdot y) \vee u'$; hence,

$$u' \leq x \odot^L y = (x \cdot y) \vee u' \leq 1$$

and thus the interval $[u', 1]$ is closed under $\odot^L$.

(2): If $x \odot^L y \leq z$, i.e. $(x \cdot y) \vee u' \leq z$, then, since $x \cdot y \leq (x \cdot y) \vee u'$, it follows that $x \cdot y \leq z$; conversely, if $x \cdot y \leq z$, then $(x \cdot y) \vee u' \leq z \vee u' = z$, since $u' \leq z \leq 1$, hence $x \odot^L y \leq z$.

(1'), (2'): similarly (dually). □

Let us "bound" the pBCK(pP) lattice $\mathcal{G}^-$ and the pBCK(pS) lattice $\mathcal{G}^+$ from Theorem 14.1.1 in two different ways: first, with an "internal" element, then, with an "external" element. We obtain the equivalent of known results.

Corollary 14.1.7

• *(i) Let* $\mathcal{G}^- = (G^-, \wedge, \vee, \rightarrow^L, \rightsquigarrow^L, \mathbf{1} = 1)$ *be the pBCK(pP) lattice (with the pseudo-product* $\odot = \cdot$*) from Theorem 14.1.1 (1). Let us "bound" this algebra in two different ways:*

1) Let us take $u' < 1$ *from* G^- *and consider the interval* $[u', 1] = \{x \in G^- \mid u' \leq x \leq 1\}$. *Then, the algebra*

$$\mathcal{G}_1^- = ([u', 1], \wedge, \vee, \rightarrow^L, \rightsquigarrow^L, \mathbf{0} = u', \mathbf{1} = 1)$$

is a bounded distributive pBCK(pP) lattice (with the pseudo-product $x \odot^L y \overset{def.}{=} (x \odot y) \vee u' = (x \cdot y) \vee u'$*) with property* (pCC^L)*, i.e. is an equivalent definition (see (iEL10), (iE10)) of the* **left-pseudo-Wajsberg algebra**

$$\mathcal{G}_{1'}^- = ([u', 1], \rightarrow^L, \rightsquigarrow^L, ^{-^L}, ^{\sim^L}, \mathbf{0} = u', \mathbf{1} = 1),$$

where:

$$x^{-^L} \overset{def.}{=} x \rightarrow^L u' = u' \cdot x^{-1}, \quad x^{\sim^L} \overset{def.}{=} x \rightsquigarrow^L u' = x^{-1} \cdot u'.$$

2) Let us consider a symbol $-\infty$ *distinct from the elements of* G. *Define* $G^-_{-\infty} \overset{def.}{=} \{-\infty\} \cup G^-$ *and extend the operations* $\rightarrow^L, \rightsquigarrow^L, \odot$ *from* G^- *to* $G^-_{-\infty}$ *as follows:*

$$x \rightarrow_2^L y \overset{def.}{=} \begin{cases} x \rightarrow^L y, & if \;\; x, y \in G^- \\ -\infty, & if \;\; x \in G^-,\ y = -\infty \\ 1, & if \;\; x = -\infty, \end{cases}$$

$$x \rightsquigarrow_2^L y \overset{def.}{=} \begin{cases} x \rightsquigarrow^L y, & if \;\; x, y \in G^- \\ -\infty, & if \;\; x \in G^-,\ y = -\infty \\ 1, & if \;\; x = -\infty, \end{cases}$$

$$x \odot_2 y \overset{def.}{=} \begin{cases} x \odot y = x \cdot y, & if \;\; x, y \in G^- \\ -\infty, & if \;\; otherwise. \end{cases}$$

We extend the lattice order relation $\leq$ *as follows: we put* $-\infty \leq x$*, for any* $x \in G^-_{-\infty}$*. Then, the algebra*

$$\mathcal{G}_2^- = (G^-_{-\infty}, \wedge, \vee, \to_2^L, \rightsquigarrow_2^L, \mathbf{0} = -\infty, \mathbf{1} = 1)$$

is a **pseudo-Hájek(pP) algebra (with the pseudo-product** $\odot_2$**) verifying properties (pP1^L) and (pP2^L).**

• *(i') Dually, let* $\mathcal{G}^+ = (G^+, \vee, \wedge, \to^R, \rightsquigarrow^R, \mathbf{0} = 1)$ *be the pBCK(pS) lattice from Theorem 14.1.1 (1'). Let us "bound" this algebra in two different ways:*

1') Let us take $u > 1$ *from* G^+ *and consider the interval* $[1, u] = \{x \in G^+ \mid 1 \leq x \leq u\}$*. Then, the algebra*

$$\mathcal{G}_1^+ = ([1, u], \vee, \wedge, \to^R, \rightsquigarrow^R, \mathbf{0} = 1, \mathbf{1} = u)$$

is a bounded pBCK(pS) lattice (with the pseudo-sum $x \oplus^R y \stackrel{def.}{=} (x \oplus y) \wedge u = (x \cdot y) \wedge u$*) with property (pCC^R), i.e. is an equivalent definition (see (iEL10), (iE10)) of the* **right-pseudo-Wajsberg algebra**

$$\mathcal{G}_{1'}^+ = ([1, u], \to^R, \rightsquigarrow^R, {}^{-^R}, {}^{\sim^R}, \mathbf{0} = 1, \mathbf{1} = u),$$

where:

$$x^{-^R} \stackrel{def.}{=} x \to^R u = u \cdot x^{-1}, \quad x^{\sim^R} \stackrel{def.}{=} x \rightsquigarrow^R u = x^{-1} \cdot u.$$

2') Let us consider a symbol $+\infty$ *distinct from the elements of* G*. Define* $G^+_{+\infty} \stackrel{def.}{=} G^+ \cup \{+\infty\}$ *and extend the operations* $\to^R, \rightsquigarrow^R$*,* $\oplus$ *from* G^+ *to* $G^+_{+\infty}$ *as follows:*

$$x \to_2^R y \stackrel{def.}{=} \begin{cases} x \to^R y, & \text{if } x, y \in G^+ \\ +\infty, & \text{if } x \in G^+,\ y = +\infty \\ 1, & \text{if } x = +\infty, \end{cases}$$

$$x \rightsquigarrow_2^R y \stackrel{def.}{=} \begin{cases} x \rightsquigarrow^R, & \text{if } x, y \in G^+ \\ +\infty, & \text{if } x \in G^+,\ y = +\infty \\ 1, & \text{if } x = +\infty, \end{cases}$$

$$x \oplus_2 y \stackrel{def.}{=} \begin{cases} x \oplus y = x \cdot y, & \text{if } x, y \in G^+ \\ +\infty, & \text{if otherwise.} \end{cases}$$

We extend the lattice order relation $\geq$ *as follows: we put* $+\infty \geq x$*, for any* $x \in G^+_{+\infty}$*. Then, the algebra*

$$\mathcal{G}_2^+ = (G^+_{+\infty}, \vee, \wedge, \to_2^R, \rightsquigarrow_2^R, \mathbf{0} = 1, \mathbf{1} = +\infty)$$

is a **pseudo-Hájek(pS) algebra (with the pseudo-sum** $\oplus_2$**) verifying the dual properties (pP1^R) and (pP2^R).**

Proof. (i) 1): By Theorem 14.1.1(1), $\mathcal{G}^- = (G^-, \wedge, \vee, \to^L, \rightsquigarrow^L, \mathbf{1} = 1)$ is a distributive pBCK(pP) lattice (with the pseudo-product $\odot = \cdot$) verifying the properties (pCC^L) and (p@^{L}).

By Lemma 14.1.5 (1), $[u',1]$ is closed under $\to^L$ and $\rightsquigarrow^L$ in $\mathcal{G}^-$; it follows that $\mathcal{G}_1^- = ([u',1], \wedge, \vee, \to^L, \rightsquigarrow^L, \mathbf{0} = u', \mathbf{1} = 1)$ is a bounded distributive pBCK lattice verifying the property (pCC^L).

By Lemma 14.1.5 (2), for all $x \in [u',1]$ we have:

$$x^{-^L} \overset{def.}{=} x \to^L u' = u' \cdot x^{-1}, \quad x^{\sim^L} \overset{def.}{=} x \rightsquigarrow^L u' = x^{-1} \cdot u'.$$

By Lemma 14.1.6 (1), $[u',1]$ is not closed under $\odot = \cdot$ (hence $\mathcal{G}_1^-$ does not verify the property (p@^{L})), but is closed under $\odot^L$ defined as follows: for all $x, y \in [u',1]$,

$$x \odot^L y \overset{def.}{=} (x \odot y) \vee u' = (x \cdot y) \vee u'.$$

We shall prove that the bounded pBCK lattice $\mathcal{G}_1^-$ is with the pseudo-product $\odot^L$. To prove that $\mathcal{G}_1^-$ is with the pseudo-product $\odot^L$, it is equivalent (see (EL1)) to prove that the property (pRP) holds, i.e. for every $x, y, z \in [u',1]$:

$$\text{(14.10)} \qquad x \odot^L y \leq z \Leftrightarrow x \leq y \to^L z = (y \to z) \wedge 1,$$

$$\text{(14.11)} \qquad x \odot^L y \leq z \Leftrightarrow y \leq x \rightsquigarrow^L z = (x \rightsquigarrow z) \wedge 1.$$

(14.10): By (pGa$^{\leq}$), we have $x \odot y = x \cdot y \leq z \Leftrightarrow x \leq y \to z$.
By (8.10), $x \leq y \to z \;\Leftrightarrow\; x \leq y \to^L z$.
By Lemma 14.1.6 (2), $x \odot^L y \leq z \;\Leftrightarrow\; x \cdot y \leq z$.
Thus, (14.10) holds.
(14.11): it follows by (14.10) and (8.11).
Thus, $\mathcal{G}_1^-$ is with the pseudo-product $\odot^L$.

Resuming, $\mathcal{G}_1^-$ is a bounded distributive pBCK(pP) lattice (with the pseudo-product $\odot^L$) verifying the property (pCC^L), i.e. is an equivalent definition (see (iEL10), (iE10)) of the **left-pseudo-Wajsberg algebra**

$$\mathcal{G}_{1'}^- = ([u',1], \to^L, \rightsquigarrow^L, ^{-^L}, ^{\sim^L}, \mathbf{0} = u', \mathbf{1} = 1).$$

(i) 2): By Theorem 14.1.1(1), $\mathcal{G}^- = (G^-, \wedge, \vee, \to^L, \rightsquigarrow^L, \mathbf{1} = 1)$ is a distributive pBCK(pP) lattice (with the pseudo-product $\odot = \cdot$) verifying the properties (pCC^L) (hence the properties (pprel^L) and (pdiv^L), by Theorem 2.12.4) and (p@^{L}).

It is routine to check that $\mathcal{G}_2^- = (G_{-\infty}^-, \wedge, \vee, \to_2^L, \rightsquigarrow_2^L, \mathbf{0} = -\infty, \mathbf{1} = 1)$ is a bounded distributive pBCK(pP) lattice (with the pseudo-product $\odot_2$).

$\mathcal{G}_2^-$ does not verify the property (pCC^L), because for $x \in G^-$ and $y = -\infty$, $x \vee y (= y \vee x) = x$, $(y \to_2^L x) \rightsquigarrow_2^L x = (y \rightsquigarrow_2^L x) \to_2^L x = x$, but $(x \to_2^L y) \rightsquigarrow_2^L y = (x \rightsquigarrow_2^L y) \to_2^L y = 1$. It is routine to check that $\mathcal{G}_2^-$ satisfies the properties (pprel^L) and (pdiv^L).

Hence, $\mathcal{G}_2^-$ is a pseudo-Hájek(pP) algebra (with the pseudo-product $\odot_2$).

$\mathcal{G}_2^-$ does not verify the property (p@^{L}), because for $x \in G^-$, $y = z = -\infty$: $-\infty = x \to_2^L y \neq (x \odot_2 z) \to_2^L (y \odot_2 z) = -\infty \to_2^L -\infty = 1$. It is routine to check that $\mathcal{G}_2^-$ satisfies the properties (pP1^L) and (pP2^L), where:

$$x^- \overset{def.}{=} x \to_2^L -\infty, \qquad x^\sim \overset{def.}{=} x \rightsquigarrow_2^L -\infty.$$

Hence, $\mathcal{G}_2^- = (G_{-\infty}^-, \wedge, \vee, \rightarrow_2^L, \rightsquigarrow_2^L, \mathbf{0} = -\infty, \mathbf{1} = 1)$ is a pseudo-Hájek(pP) algebra verifying properties (pP1^L) and (pP2^L).

(i'): dually. □

14.2 Connections between unital magmas

In this section, we present connections between left- (right-) l-groups and l-ims(pR) (l-ims(pcoR)). We have seen in Proposition 3.4.3 the connection existing between l-ms and l-ims. We shall see now the deeper connections existing between l-groups and l-ims.

First, recall Proposition 7.3.12.

By the equivalences (EIG3), (E2), (iE3), (iE8), (iE9'), we obtain the following analogous of Theorem 14.1.1 and Corollary 14.1.7 for unital magmas.

Theorem 14.2.1 *(See Theorem 14.1.1)*

Let $\mathcal{G} = (G, \vee, \wedge, \cdot, ^{-1}, 1)$ be an l-group (left-one or right-one). We have: for all $x, y \in G$,
$x \rightarrow y \overset{def.}{=} y \cdot x^{-1}$ *and* $x \rightsquigarrow y \overset{def.}{=} x^{-1} \cdot y$.

(1) Define, for all $x, y \in G^-$:

$$x \odot y \overset{def.}{=} x \cdot y, \quad x \rightarrow^L y \overset{def.}{=} (x \rightarrow y) \wedge 1, \quad x \rightsquigarrow^L y \overset{def.}{=} (x \rightsquigarrow y) \wedge 1.$$

Then, $\mathcal{G}_m^- = (G^-, \wedge, \vee, \odot, \mathbf{1} = 1)$ is a distributive l-im(pR) (with the pseudo-residuum $(\rightarrow^L, \rightsquigarrow^L)$) verifying the properties (pCC^L) and (p@^{L}) (while the equivalent algebra, by (EM2), $\mathcal{G}_{m'}^- = (G^-, \wedge, \vee, \odot, \rightarrow^L, \rightsquigarrow^L, \mathbf{1} = 1)$ is a distributive left-pseudo-residuated lattice verifying the properties (pCC^L) and (p@^{L})).

(1') Dually, define for all $x, y \in G^+$:

$$x \oplus y \overset{def.}{=} x \cdot y, \quad x \rightarrow^R y \overset{def.}{=} (x \rightarrow y) \vee 1, \quad x \rightsquigarrow^R y \overset{def.}{=} (x \rightsquigarrow y) \vee 1.$$

Then, $\mathcal{G}_m^+ = (G^+, \vee, \wedge, \oplus, \mathbf{0} = 1)$ is a distributive l-im(pcoR) (with the pseudo-coresiduum $(\rightarrow^R, \rightsquigarrow^R)$) verifying the properties (pCC^R)and (p@^{R}) (while the equivalent algebra, by (EM2), $\mathcal{G}_{m'}^+ = (G^+, \vee, \wedge, \oplus, \rightarrow^R, \rightsquigarrow^R, \mathbf{0} = 1)$ is a distributive right-pseudo-residuated lattice verifying the properties (pCC^R) and (p@^{R})).

Remark 14.2.2 $\mathcal{G}_m^-$ verifies the properties (pprel^L) and (pdiv^L), by Theorem 2.12.4, and $\mathcal{G}_m^+$ verifies the dual properties (pprel^R) and (pdiv^R).

We shall "bound" the above algebras $\mathcal{G}_m^-$ and $\mathcal{G}_m^+$ in two different ways; we obtain known results ((1) see [42], and [150] for the commutative case; (2) see [42], and [72] for the commutative case).

Corollary 14.2.3 *(See Corollary 14.1.7)*

• *(i) Let $\mathcal{G}_m^- = (G^-, \wedge, \vee, \odot, \mathbf{1} = 1)$ be the l-im(pR) from above Theorem 14.2.1 (1). Let us "bound" this algebra in two different ways:*

1) Let us take $u' < 1$ from G^- and consider the interval $[u', 1] = \{x \in G^- \mid u' \leq x \leq 1\}$. Define for all $x, y \in G^-$:

$$x \odot^L y \stackrel{def.}{=} (x \odot y) \vee u' = (x \cdot y) \vee u' \quad (x \oplus^L y \stackrel{def.}{=} (x \cdot u'^{-1} \cdot y) \wedge 1),$$

$$x^{-^L} \stackrel{def.}{=} u' \cdot x^{-1}, \quad x^{\sim^L} \stackrel{def.}{=} x^{-1} \cdot u'.$$

Then, the algebra

$$\mathcal{G}^-_{m_1} = ([u', 1], \wedge, \vee, \odot^L, \mathbf{0} = u', \mathbf{1} = 1)$$

is a bounded l-im(pR) verifying the property (pCC^L), i.e. is an equivalent definition (see (iE10)) of the **left-pseudo-MV algebra**

$$\mathcal{G}^-_{m'_1} = ([u', 1], \odot^L, ^{-^L}, ^{\sim^L}, \mathbf{0} = u', \mathbf{1} = 1).$$

2) Let us consider a symbol $-\infty$ distinct from the elements of G. Define $G^-_{-\infty} \stackrel{def.}{=} \{-\infty\} \cup G^-$ and extend the operations $\odot$, $\to^L, \rightsquigarrow^L$ from G^- to $G^-_{-\infty}$ as follows:

$$x \odot_2 y \stackrel{def.}{=} \begin{cases} x \odot y = x \cdot y, & if \ \ x, y \in G^- \\ -\infty, & if \ \ otherwise. \end{cases}$$

$$x \to^L_2 y \stackrel{def.}{=} \begin{cases} x \to^L y, & if \ \ x, y \in G^- \\ -\infty, & if \ \ x \in G^-, \ y = -\infty \\ 1, & if \ \ x = -\infty, \end{cases}$$

$$x \rightsquigarrow^L_2 y \stackrel{def.}{=} \begin{cases} x \rightsquigarrow^L y, & if \ \ x, y \in G^- \\ -\infty, & if \ \ x \in G^-, \ y = -\infty \\ 1, & if \ \ x = -\infty. \end{cases}$$

We extend the lattice order relation $\leq$ as follows: we put $-\infty \leq x$, for any $x \in G^-_{-\infty}$.
Then, the algebra

$$\mathcal{G}^-_{m_2} = (G^-_{-\infty}, \wedge, \vee, \odot_2, \mathbf{0} = -\infty, \mathbf{1} = 1)$$

is a bounded l-im(pR) verifying the properties (pprel^L), (pdiv^L) and (pP1^L), (pP2^L), while the equivalent algebra, by (iEM8),

$$\mathcal{G}^-_{m'_2} = (G^-_{-\infty}, \wedge, \vee, \odot_2, \to^L_2, \rightsquigarrow^L_2, \mathbf{0} = -\infty, \mathbf{1} = 1)$$

is a left-pseudo-BL algebra verifying (pP1^L), (pP2^L), i.e. is a **left-pseudo-product algebra**.

• *(i') Dually, let $\mathcal{G}^+_m = (G^+, \vee, \wedge, \oplus, \mathbf{0} = 1)$ be the l-im(pcoR) from above Theorem 14.2.1 (1'). Let us "bound" this algebra in two different ways:*

1') Let us take $u > 1$ from G^+ and consider the interval $[1, u] = \{x \in G^+ \mid 1 \leq x \leq u\}$. Define for all $x, y \in G^+$:

$$x \oplus^R y \stackrel{def.}{=} (x \oplus y) \wedge u = (x \cdot y) \wedge u \quad (x \odot^R y \stackrel{def.}{=} (x \cdot u^{-1} \cdot y) \vee 1),$$

$$x^{-^R} \overset{def.}{=} u \cdot x^{-1}, \quad x^{\sim^R} \overset{def.}{=} x^{-1} \cdot u.$$

Then, the algebra

$$\mathcal{G}^+_{m_1} = ([1,u], \vee, \wedge, \oplus^R, \mathbf{0} = 1, \mathbf{1} = u)$$

is a bounded l-im(pcoR) verifying the property (pCC^R), i.e. is an equivalent definition (see (iE10)) of the **right-pseudo-MV algebra**

$$\mathcal{G}^+_{m'_1} = ([1,u], \oplus^R, {}^{-^R}, {}^{\sim^R}, \mathbf{0} = 1, \mathbf{1} = u).$$

2') Let us consider a symbol $+\infty$ distinct from the elements of G. Define $G^+_{+\infty} \overset{def.}{=} G^+ \cup \{+\infty\}$ and extend the operations $\oplus$, $\to^R, \rightsquigarrow^R$ from G^+ to $G^+_{+\infty}$ as follows:

$$x \oplus_2 y \overset{def.}{=} \begin{cases} x \oplus y = x \cdot y, & \text{if } x, y \in G^+ \\ +\infty, & \text{if otherwise.} \end{cases}$$

$$x \to^R_2 y \overset{def.}{=} \begin{cases} x \to^R y, & \text{if } x, y \in G^+ \\ +\infty, & \text{if } x \in G^+,\ y = +\infty \\ 1, & \text{if } x = +\infty, \end{cases}$$

$$x \rightsquigarrow^R_2 y \overset{def.}{=} \begin{cases} x \rightsquigarrow^R y, & \text{if } x, y \in G^+ \\ +\infty, & \text{if } x \in G^+,\ y = +\infty \\ 1, & \text{if } x = +\infty. \end{cases}$$

We extend the lattice order relation $\leq$ as follows: we put $+\infty \geq x$, for any $x \in G^+_{+\infty}$.
Then, the algebra

$$\mathcal{G}^+_{m_2} = (G^+_{+\infty}, \vee, \wedge, \oplus_2, \mathbf{0} = 1, \mathbf{1} = +\infty)$$

is a bounded l-im(pcoR) verifying the properties (pprel^R), (pdiv^R) and (pP1^R), (pP2^R), while the equivalent algebra, by (iEM8),

$$\mathcal{G}^+_{m'_2} = (G^+_{+\infty}, \vee, \wedge, \oplus_2, \to^R_2, \rightsquigarrow^R_2, \mathbf{0} = 1, \mathbf{1} = +\infty)$$

is a right-pseudo-BL algebra verifying (pP1^R), (pP2^R), i.e. is a **right-pseudo-product algebra**.

14.3 The analysis of three commutative examples

Example 1. The additive l-group $\mathcal{Z}_g = (\mathbf{Z}, \vee, \wedge, +, -, 0)$

Note that the additive l-group $\mathcal{Z}_g$ is linearly-ordered, commutative, self-dual. By Theorem 9.2.3, the termwise equivalent l-implicative-group, by (Eq3),

$$\mathcal{Z}_{ig} = (\mathbf{Z}, \vee, \wedge, \to, 0)$$

has the implication $\to$ defined by: $x \to y \overset{def.}{=} y - x$, for all $x, y \in \mathbf{Z}$.

Consider first the commutative l-implicative-group $\mathcal{Z}_{ig} = (\mathbf{Z}, \vee, \wedge, \to, 0)$. Then we have:

By Theorem 14.1.1, taking $Z^- = \{x \in \mathbf{Z} \mid x \leq 0\}$ and $Z^+ = \{x \in \mathbf{Z} \mid x \geq 0\}$, we obtain:

(1) Define, for all $x, y \in Z^-$,

$$x \to^L y \overset{def.}{=} (x \to y) \wedge 0 = \min(0, y - x), \quad x \odot y \overset{def.}{=} x + y,$$

hence the tables of $\to^L$ and $\odot$ are the following:

$\to^L$	...	-3	-2	-1	0
⋮	...	⋮	⋮	⋮	⋮
-3	...	0	0	0	0
-2	...	-1	0	0	0
-1	...	-2	-1	0	0
0	...	-3	-2	-1	0

$\odot$	...	-3	-2	-1	0
⋮	...	⋮	⋮	⋮	⋮
-3	...	-6	-5	-4	-3
-2	...	-5	-4	-3	-2
-1	...	-4	-3	-2	-1
0	...	-3	-2	-1	0

Then, $\mathcal{Z}^- = (Z^-, \wedge, \vee, \to^L, \mathbf{1} = 0)$ is a distributive BCK(P) lattice (with the product $\odot = +$) verifying the properties (CC^L) and (@^{L}).

(1') Define, for all $x, y \in Z^+$,

$$x \to^R y \overset{def.}{=} (x \to y) \vee 0 = \max(0, y - x), \quad x \oplus y \overset{def.}{=} x + y,$$

hence the tables of $\to^R$ and $\oplus$ are the following:

$\to^R$	0	1	2	3	...
0	0	1	2	3	...
1	0	0	1	2	...
2	0	0	0	1	...
3	0	0	0	0	...
⋮	⋮	⋮	⋮	⋮	...

$\oplus$	0	1	2	3	...
0	0	1	2	3	...
1	1	2	3	4	...
2	2	3	4	5	...
3	3	4	5	6	...
⋮	⋮	⋮	⋮	⋮	...

Then, $\mathcal{Z}^+ = (Z^+, \vee, \wedge, \to^R, \mathbf{0} = 0)$ is a distributive BCK(S) lattice (with the sum $\oplus = +$) verifying the properties (CC^R) and (@^{R}).

By Corollary 14.1.7, we obtain:

• (i) Let $\mathcal{Z}^-$ be the above BCK(P) lattice. Let us "bound" this algebra in two different ways:

(1) Let us take $u' = -n$ $(n \geq 1)$ and consider the interval

$$L_{-(n+1)} = [-n, 0] = \{-n, -n+1, \ldots, -2, -1, 0\} \subset Z^-.$$

Then, the algebra

$$\mathcal{Z}_1^L = ([-n, 0], \wedge, \vee, \to^L, \mathbf{0} = -n, \mathbf{1} = 0)$$

is a bounded distributive BCK(P) lattice (with the product $x \odot^L y \stackrel{def.}{=} (x \odot y) \vee -n = \max\{-n, x+y\}$) with the property (CC^L), i.e. is an equivalent definition of the left-Wajsberg algebra

$$\mathcal{Z}^-_{1'} = ([-n, 0], \to^L, ^{-^L}, \mathbf{0} = -n, \mathbf{1} = 0),$$

where:

$$x^{-^L} = x \to^L -n = -n - x.$$

For example, if we take $n = 2$, then we obtain the following tables of $\to^L$ ($x \to^L y = \min(0, y - x)$) and $\odot$ ($x \odot y = \max\{-2, x+y\}$) in $L_{-3} = \{-2, -1, 0\}$:

L_{-3}

$\to^L$	-2	-1	0
-2	0	0	0
-1	-1	0	0
0	-2	-1	0

$\odot$	-2	-1	0
-2	-2	-2	-2
-1	-2	-2	-1
0	-2	-1	0

(2) Let us consider the symbol $-\infty \notin \mathbf{Z}$ and define $Z^-_{-\infty} \stackrel{def.}{=} \{-\infty\} \cup Z^-$. Extend the operations $\to^L$, $\odot$ and the lattice order relation $\leq$ from Z^- to $Z^-_{-\infty}$ as mentioned in Corollary 14.1.7. The new tables are:

$\to^L_2$	$-\infty$	...	-3	-2	-1	0
$-\infty$	0	...	0	0	0	0
$\vdots$	$\vdots$	...	$\vdots$	$\vdots$	$\vdots$	$\vdots$
-3	$-\infty$	...	0	0	0	0
-2	$-\infty$	...	-1	0	0	0
-1	$-\infty$	...	-2	-1	0	0
0	$-\infty$	...	-3	-2	-1	0

$\odot_2$	$-\infty$	...	-3	-2	-1	0
$-\infty$	$-\infty$	...	$-\infty$	$-\infty$	$-\infty$	$-\infty$
$\vdots$	$\vdots$	...	$\vdots$	$\vdots$	$\vdots$	$\vdots$
-3	$-\infty$	...	-6	-5	-4	-3
-2	$-\infty$	...	-5	-4	-3	-2
-1	$-\infty$	...	-4	-3	-2	-1
0	$-\infty$	...	-3	-2	-1	0

Then, the algebra

$$\mathcal{Z}^L_2 = (Z^-_{-\infty}, \wedge, \vee, \to^L_2, \mathbf{0} = -\infty, \mathbf{1} = 0)$$

is a Hájek(P) algebra (with the product $\odot_2$) verifying (P1^L) and (P2^L) (see Example 7 from Chapter 6).

• (i') Let $\mathcal{Z}^+$ be the above BCK(S) lattice. Let us "bound" this algebra in two different ways:

(1') Let us take $u = n$ ($n \geq 1$) and consider the interval

$$L_{n+1} = [0, n] = \{0, 1, 2, \ldots, n\} \subset R^+.$$

Then, the algebra

$$\mathcal{Z}_1^R = (L_{n+1} = [0,n], \vee, \wedge, \to^R, \mathbf{0} = 0, \mathbf{1} = n)$$

is a bounded BCK(S) lattice (with the sum $x \oplus y \overset{def.}{=} (x \oplus y) \wedge u = \min\{n, x+y\}$) with the property (CC^R), i.e. is an equivalent definition of the right-Wajsberg algebra

$$\mathcal{Z}_{1'}^+ = (L_{n+1} = [0,n], \to^R, ^{-^R}, \mathbf{0} = 0, \mathbf{1} = n),$$

where:

$$x^{-^R} = x \to^R n = n - x.$$

For example, we take $n = 2$ and we obtain the following tables for $\to^R$ ($x \to^R y = \max(0, y - x)$) and $\oplus$ ($x \oplus y = \min\{2, x+y\}$):

L_3

$\to^R$	0	1	2
0	0	1	2
1	0	0	1
2	0	0	0

$\oplus$	0	1	2
0	0	1	2
1	1	2	2
2	2	2	2

By Remarks 2.11.8, we introduce the additional "left" operations as follows:

$$x \odot^R y \overset{def.}{=} (x^- \oplus y^-)^-, \quad x \Rightarrow y = (x^- \to^R y^-)^-,$$

namely:

$x \odot^R y \overset{def.}{=} (x^- \oplus y^-)^- = n - [(n-x) \oplus (n-y)] =$
$n - \min\{n, n-x+n-y\} = n - \min\{n, 2n-x-y\} =$
$= n - \begin{cases} n, & \text{if } n \leq 2n-x-y \\ 2n-x-y, & \text{if } n > 2n-x-y \end{cases} = \begin{cases} 0, & \text{if } 0 \leq n-x-y \\ -n+x+y, & \text{if } 0 > n-x-y \end{cases} =$
$\begin{cases} 0, & \text{if } 0 \geq x+y-n \\ x+y-n, & \text{if } 0 < x+y-n \end{cases} = \max\{0, x+y-n\}$ and

$x \Rightarrow y = (x^- \to^R y^-)^- = n - \max\{0, (n-y)-(n-x)\} =$
$n - \max\{0, x-y\} = n - \begin{cases} 0, & \text{if } x-y \leq 0 \\ x-y, & \text{if } x-y > 0 \end{cases} = \begin{cases} n, & \text{if } x \leq y \\ n-x+y, & \text{if } x > y \end{cases} =$
$\min\{n, y-x+n\}$.

We obtain the following tables for $\Rightarrow$ ($x \Rightarrow y = \min\{2, y-x+2\}$) and $\odot^R$ ($x \odot^R y = \max\{0, x+y-2\}$) in L_3 as left-algebra (see [94], page 161):

L_3

$\Rightarrow = \min\{2, y-x+2\}$	0	1	2
0	2	2	2
1	1	2	2
2	0	1	2

$\odot^R$	0	1	2
0	0	0	0
1	0	0	1
2	0	1	2

(2') Let us consider the symbol $+\infty \notin \mathbf{R}$ and define $Z_{+\infty}^+ \overset{def.}{=} Z^+ \cup \{+\infty\}$. Extend the operations $\to^R$, $\oplus$ and the lattice order relation $\geq$ from Z^+ to $Z_{+\infty}^+$ as mentioned in Corollary 14.1.7. Then, the algebra

$$\mathcal{Z}_2^R = (Z_{+\infty}^+, \vee, \wedge, \to_2^R, \mathbf{0} = 0, \mathbf{1} = +\infty)$$

is a Hájek(S) algebra (with the sum $\oplus_2$) verifying ($P1^R$) and ($P2^R$).

Consider now the commutative l-group $\mathcal{Z}_g = (\mathbf{Z}, \vee, \wedge, +, -, 0)$. Then, we have:
By Theorem 14.2.1, we obtain:
(1) Define, for all $x, y \in Z^-$:

$$x \odot y \stackrel{def.}{=} x + y, \quad x \to^L y \stackrel{def.}{=} (x \to y) \wedge 0 = \min\{0, y - x\}.$$

Then, $\mathcal{Z}_m^- = (Z^-, \wedge, \vee, \odot, \mathbf{1} = 0)$ is a distributive l-cim(R) (with the residuum $\to^L$) verifying the properties (CC^L) and ($@^L$).

(1') Dually, define, for all $x, y \in Z^+$:

$$x \oplus y \stackrel{def.}{=} x + y, \quad x \to^R y \stackrel{def.}{=} (x \to y) \vee 0 = \max\{0, y - x\}.$$

Then, $\mathcal{Z}_m^+ = (Z^+, \vee, \wedge, \oplus, \mathbf{0} = 0)$ is a distributive l-cim(coR) (with the coresiduum $\to^R$) verifying the properties (CC^R) and ($@^R$).

By Corollary 14.2.3, we obtain:

• (i) Let $\mathcal{Z}_m^-$ be the l-cim(R) from above. Let us "bound" this algebra in two different ways:

(1) Let us take $-n < 0$ from Z^- and consider the interval $[-n, 0] = \{x \in Z^- \mid -n \leq x \leq 0\}$. Define for all $x, y \in Z^-$:

$$x \odot^L y \stackrel{def.}{=} (x + y) \vee (-n) = \max(-n, x + y), \quad x^- \stackrel{def.}{=} -n - x,$$

$$x \oplus^L y \stackrel{def.}{=} (x - (-n) + y) \wedge 0 = \min(0, x + y + n).$$

Then, the algebra

$$\mathcal{Z}_{m_1}^- = ([-n, 0], \wedge, \vee, \odot^L, \mathbf{0} = -n, \mathbf{1} = 0)$$

is a bounded l-cim(R) verifying the property (CC^L), i.e. is an equivalent definition, by (iE10), of the left-MV algebra

$$\mathcal{Z}_{m_1'}^- = ([-n, 0], \odot^L, {}^{-^L}, \mathbf{0} = -n, \mathbf{1} = 0).$$

(2) Let us consider the symbol $-\infty$ distinct from the elements of $\mathbf{Z}$. Define $Z_{-\infty}^- \stackrel{def.}{=} \{-\infty\} \cup Z^-$ and extend the operations $\odot$, $\to^L$ and the lattice order relation $\leq$ from Z^- to $Z_{-\infty}^-$ as mentioned in Corollary 14.2.3. Then, the algebra

$$\mathcal{Z}_{m_2}^- = (Z_{-\infty}^-, \wedge, \vee, \odot_2, \mathbf{0} = -\infty, \mathbf{1} = 0)$$

is a bounded l-cim(R) verifying the properties ($prel^L$), (div^L) and ($P1^L$), ($P2^L$), while the equivalent algebra, by (iEM8),

$$\mathcal{Z}_{m_2'}^- = (Z_{-\infty}^-, \wedge, \vee, \odot_2, \to_2^L, \mathbf{0} = -\infty, \mathbf{1} = 0)$$

is a left-BL algebra verifying ($P1^L$), ($P2^L$), i.e. is a **left-product algebra** (see [72] for the commutative case) (see Example 7' from Chapter 6).

• (i') Let $\mathcal{Z}_m^+$ be the l-im(coR) from above. Let us "bound" this algebra in two different ways:

(1') Let us take $n > 0$ from Z^+ and consider the interval

$$L_{n+1} = [0, n] = \{x \in Z^+ \mid 0 \le x \le n\} = \{0, 1, 2, \ldots, n\}.$$

Define for all $x, y \in Z^+$:

$$x \oplus^R y \stackrel{def.}{=} (x + y) \wedge n = \min(n, x + y), \quad x^- \stackrel{def.}{=} n - x,$$

$$x \odot^R y \stackrel{def.}{=} (x - n + y) \vee 0 = \max(0, x + y - n).$$

Then, the algebra

$$\mathcal{Z}_{m_1}^+ = (L_{n+1} = [0, n], \vee, \wedge, \oplus^R, \mathbf{0} = 0, \mathbf{1} = n)$$

is a bounded l-cim(coR) verifying the property (CC^R), i.e. is an equivalent definition, by (iE10), of the right-MV algebra

$$\mathcal{Z}_{m_1'}^+ = (L_{n+1} = [0, n], \oplus^R, {}^{-^R}, \mathbf{0} = 0, \mathbf{1} = n).$$

(2') Let us consider the symbol $+\infty$ distinct from the elements of $\mathbf{Z}$. Define $Z_{+\infty}^+ \stackrel{def.}{=} Z^+ \cup \{+\infty\}$ and extend the operations $\oplus$, $\rightarrow^R$ and the lattice order relation $\ge$ from Z^+ to $Z_{-\infty}^+$ as mentioned in Corollary 14.2.3. Then, the algebra

$$\mathcal{Z}_{m_2}^+ = (Z_{+\infty}^+, \vee, \wedge, \oplus_2, \mathbf{0} = 0, \mathbf{1} = +\infty)$$

is a bounded l-cim(coR) verifying the properties (prel^R), (div^R) and (P1^R), (P2^R), while the equivalent algebra, by (iEM8),

$$\mathcal{Z}_{m_2'}^+ = (Z_{+\infty}^+, \vee, \wedge, \oplus_2, \rightarrow_2^R, \mathbf{0} = 0, \mathbf{1} = +\infty)$$

is a right-BL algebra verifying (P1^R), (P2^R), i.e. is a **right-product algebra**.

Example 2. The additive l-group $\mathcal{R}_g = (\mathbf{R}, \vee, \wedge, +, -, 0)$

Note that the additive l-group $\mathcal{R}_g$ is linearly-ordered, commutative, self-dual and has 1 as strong unit.

By Theorem 9.2.3, the termwise equivalent l-implicative-group, by (Eq3),

$$\mathcal{R}_{ig} = (\mathbf{R}, \vee, \wedge, \rightarrow, 0)$$

has the implication $\rightarrow$ defined by: $x \rightarrow y \stackrel{def.}{=} y - x$, for all $x, y \in \mathbf{R}$.

Consider first the commutative l-implicative-group $\mathcal{R}_{ig} = (\mathbf{R}, \vee, \wedge, \rightarrow, 0)$. Then we have:

By Theorem 14.1.1, taking $R^- = \{x \in \mathbf{R} \mid x \le 0\}$ and $R^+ = \{x \in \mathbf{R} \mid x \ge 0\}$, we obtain:

(1) Define, for all $x, y \in R^-$,

$$x \to^L y \stackrel{def.}{=} (x \to y) \wedge 0 = \min(0, y - x);$$

Then, $\mathcal{R}^- = (R^-, \wedge, \vee, \to^L, \mathbf{1} = 0)$ is a distributive BCK(P) lattice (with the product $\odot = +$) verifying the properties (CC^L) and (@^{L}).

(1') Define, for all $x, y \in R^+$,

$$x \to^R y \stackrel{def.}{=} (x \to y) \vee 0 = \max(0, y - x).$$

Then, $\mathcal{R}^+ = (R^+, \vee, \wedge, \to^R, \mathbf{0} = 0)$ is a distributive BCK(S) lattice (with the sum $\oplus = +$) verifying the properties (CC^R) and (@^{R}).

By Corollary 14.1.7, we obtain:

• (i) Let $\mathcal{R}^-$ be the above BCK(P) lattice. Let us "bound" this algebra in two different ways:

(1) Let us take $u' = -1$ and consider the interval $[-1, 0] \subset R^-$. Then, the algebra

$$\mathcal{R}_1^L = ([-1, 0], \wedge, \vee, \to^L, \mathbf{0} = -1, \mathbf{1} = 0)$$

is a bounded BCK(P) lattice (with the product $x \odot^L y \stackrel{def.}{=} (x \odot y) \vee u' = \max\{-1, x + y\}$) with the property (CC^L), i.e. is an equivalent definition of the left-Wajsberg algebra

$$\mathcal{R}_{1'}^- = ([-1, 0], \to^L, ^{-^L}, \mathbf{0} = -1, \mathbf{1} = 0),$$

where:

$$x^{-^L} = x \to^L -1.$$

(2) Let us consider the symbol $-\infty \notin \mathbf{R}$ and define $R^-_{-\infty} \stackrel{def.}{=} \{-\infty\} \cup R^-$. Extend the operations $\to^L$, $\odot$ and the lattice order relation $\leq$ from R^- to $R^-_{-\infty}$ as mentioned in Corollary 14.1.7. Then, the algebra

$$\mathcal{R}_2^L = (R^-_{-\infty}, \wedge, \vee, \to_2^L, \mathbf{0} = -\infty, \mathbf{1} = 0)$$

is a Hájek(P) algebra (with the product $\odot_2$) verifying (P1^L) and (P2^L).

• (i') Let $\mathcal{R}^+$ be the above BCK(S) lattice. Let us "bound" this algebra in two different ways:

(1') Let us take $u = 1$ and consider the interval $[0, 1] \subset R^+$. Then, the algebra

$$\mathcal{R}_1^R = ([0, 1], \vee, \wedge, \to^R, \mathbf{0} = 0, \mathbf{1} = 1)$$

is a bounded BCK(S) lattice (with the sum $x \oplus^L y \stackrel{def.}{=} (x \oplus y) \wedge u = \min\{1, x + y\}$) with the property (CC^R), i.e. is an equivalent definition of the right-Wajsberg algebra

$$\mathcal{R}_{1'}^+ = ([0, 1], \to^R, ^{-^R}, \mathbf{0} = 0, \mathbf{1} = 1).$$

(2') Let us consider the symbol $+\infty \notin \mathbf{R}$ and define $R^+_{+\infty} \stackrel{def.}{=} R^+ \cup \{+\infty\}$. Extend the operations $\to^R$, $\oplus$ and the lattice order relation $\geq$ from R^+ to $R^+_{+\infty}$ as mentioned in Corollary 14.1.7. Then, the algebra

$$\mathcal{R}_2^R = (R^+_{+\infty}, \vee, \wedge, \to_2^R, \mathbf{0} = 0, \mathbf{1} = +\infty)$$

is a Hájek(S) algebra (with the sum $\oplus_2$) verifying ($P1^R$) and ($P2^R$).

Consider now the commutative l-group $\mathcal{R}_g = (\mathbf{R}, \vee, \wedge, +, -, 0)$. Then we have:
By Theorem 14.2.1, we obtain:
(1) Define, for all $x, y \in R^-$:

$$x \odot y \stackrel{def.}{=} x + y, \quad x \to^L y \stackrel{def.}{=} (x \to y) \wedge 0 = \min\{0, y - x\}.$$

Then, $\mathcal{R}_m^- = (R^-, \wedge, \vee, \odot, \mathbf{1} = 0)$ is a distributive l-cim(R) (with the residuum $\to^L$) verifying the properties (CC^L) and ($@^L$).

(1') Dually, define, for all $x, y \in R^+$:

$$x \oplus y \stackrel{def.}{=} x + y, \quad x \to^R y \stackrel{def.}{=} (x \to y) \vee 0 = \max\{0, y - x\}.$$

Then, $\mathcal{R}_m^+ = (R^+, \vee, \wedge, \oplus, \mathbf{0} = 0)$ is a distributive l-cim(coR) (with the coresiduum $\to^R$) verifying the properties (CC^R) and ($@^R$).

By Corollary 14.2.3, we obtain:

• (i) Let $\mathcal{R}_m^-$ be the l-im(R) from above. Let us "bound" this algebra in two different ways:

(1) Let us take $-1 < 0$ from R^- and consider the interval $[-1, 0] = \{x \in R^- \mid -1 \leq x \leq 0\}$. Define for all $x, y \in R^-$:

$$x \odot^L y \stackrel{def.}{=} (x + y) \vee (-1) = \max(-1, x + y), \quad x^- \stackrel{def.}{=} -1 - x,$$

$$x \oplus^L y \stackrel{def.}{=} (x - (-1) + y) \wedge 0 = \min(0, x + y + 1).$$

Then, the algebra

$$\mathcal{R}_{m_1}^- = ([-1, 0], \wedge, \vee, \odot^L, \mathbf{0} = -1, \mathbf{1} = 0)$$

is a bounded l-cim(R) verifying the property $C(C^L)$, i.e. is an equivalent definition, by (iE10), of the left-MV algebra

$$\mathcal{R}_{m_1'}^- = ([-1, 0], \odot^L, {}^{-^L}, \mathbf{0} = -1, \mathbf{1} = 0).$$

(2) Let us consider a symbol $-\infty$ distinct from the elements of $\mathbf{R}$. Define $R_{-\infty}^- \stackrel{def.}{=} \{-\infty\} \cup R^-$ and extend the operations $\odot$, $\to^L$ and the lattice order relation $\leq$ from R^- to $R_{-\infty}^-$ as mentioned in Corollary 14.2.3. Then, the algebra

$$\mathcal{R}_{m_2}^- = (R_{-\infty}^-, \wedge, \vee, \odot_2, \mathbf{0} = -\infty, \mathbf{1} = 0)$$

is a bounded l-cim(R) verifying the properties ($prel^L$), (div^L) and ($P1^L$), ($P2^L$), while the equivalent algebra, by (iEM8),

$$\mathcal{R}_{m_2'}^- = (R_{-\infty}^-, \wedge, \vee, \odot_2, \to_2^L, \mathbf{0} = -\infty, \mathbf{1} = 0)$$

is a left-BL algebra verifying ($P1^L$), ($P2^L$), i.e. is a **left-product algebra** (due to property ($@^L$)) (see [72] for the commutative case).

• (i') Let $\mathcal{R}_m^+$ be the l-im(coR) from above. Let us "bound" this algebra in two different ways:

(1') Let us take $1 > 0$ from R^+ and consider the interval $[0,1] = \{x \in R^+ \mid 0 \leq x \leq 1\}$. Define for all $x, y \in R^+$:

$$x \oplus^R y \stackrel{def.}{=} (x+y) \wedge 1 = \min(1, x+y), \quad x^- \stackrel{def.}{=} 1 - x,$$

$$x \odot^R y \stackrel{def.}{=} (x - 1 + y) \vee 0 = \max(0, x+y-1).$$

Then, the algebra

$$\mathcal{R}_{m_1}^+ = ([0,1], \vee, \wedge, \oplus^R, \mathbf{0} = 0, \mathbf{1} = 1)$$

is a bounded l-cim(coR) verifying the property (CC^R), i.e. is an equivalent definition, by (iE10), of the right-MV algebra

$$\mathcal{R}_{m_1'}^+ = ([0,1], \oplus^R, {}^{-^R}, \mathbf{0} = 0, \mathbf{1} = 1).$$

(2') Let us consider a symbol $+\infty$ distinct from the elements of $\mathbf{R}$. Define $R_{+\infty}^+ \stackrel{def.}{=} R^+ \cup \{+\infty\}$ and extend the operations $\oplus$, $\to^R$ and the lattice order relation $\geq$ from R^+ to $R_{-\infty}^+$ as mentioned in Corollary 14.2.3. Then, the algebra

$$\mathcal{R}_{m_2}^+ = (R_{+\infty}^+, \vee, \wedge, \oplus_2, \mathbf{0} = 0, \mathbf{1} = +\infty)$$

is a bounded l-cim(coR) verifying the properties (prel^R), (div^R) and (P1^R), (P2^R), while the equivalent algebra, by (iEM8),

$$\mathcal{R}_{m_2'}^+ = (R_{+\infty}^+, \vee, \wedge, \oplus_2, \to_2^R, \mathbf{0} = 0, \mathbf{1} = +\infty)$$

is a right-BL algebra verifying (P1^R), (P2^R), i.e. is a **right-product algebra** (due to property (@^{R})).

Consequently,

- in the left-MV algebra $\mathbf{R}_{m_1'}^- = ([-1,0], \odot^L, {}^-, \mathbf{0} = -1, \mathbf{1} = 0)$ we have:

$x \odot^L y = \max(-1, x+y)$ is the basic t-norm, with the associated residuum $x \to^L y = \min(0, y-x)$;

$x \oplus^L y = \min(0, x+y+1)$ is the additional t-conorm ($x \oplus^L y = (x^- \odot^L y^-)^-$), with the associated coresiduum:

$x \Rightarrow^L y = (x^- \to^L y^-)^- = ((-1-x) \to^L (-1-y))^- = [\min(0, (-1-y) - (-1-x))]^- = [\min(0, -y+x)]^- = -1 - \min(0, x-y) = -1 + \max(0, y-x) = \max(-1, y-x-1)$.

- in the right-MV algebra $\mathbf{R}_{m_1'}^+ = ([0,1], \oplus^R, {}^-, \mathbf{0} = 0, \mathbf{1} = 1)$, we have:

$x \oplus^R y = \min(1, x+y)$ is the basic t-conorm, with the associated coresiduum $x \to^R y = \max(0, y-x)$;

$x \odot^R y = \max(0, x-1+y)$ is the additional t-norm ($x \odot^R y = (x^- \oplus^R y^-)^-$), with the associated residuum:

$x \Rightarrow^R y = (x^- \to^R y^-)^- = ((1-x) \to^R (1-y))^- = [\max(0, (1-y) - (1-x))]^- = 1 - \max(0, x-y) = 1 + \min(0, y-x) = \min(1, y-x+1)$, by Remarks 2.11.8.

Note now that, if we make the following translation t of interval $[-1,0]$ to interval $[0,1]$: $t(x) = x' = x+1$, then we obtain that:

• $z' = z+1 = t(z) = t(x \odot^L y) = t(\max(-1, x+y)) = \max(-1, x+y)+1 = \max(0, x+y+1) = \max(0, (x+1)+(y+1)-1) = \max(0, x'+y'-1) = x' \odot^R y'$, i.e. $\odot^L$ from $[-1,0]$ becomes $\odot^R$ from $[0,1]$; **this is the Łukasiewicz t-norm.**

• $z' = z+1 = t(z) = t(x \to^L y) = t(\min(0, y-x)) = \min(0, y-x)+1 = \min(1, y-x+1) = \min(1, (y+1)-(x+1)+1) = \min(1, y'-x'+1) = x' \Rightarrow^R y'$, i.e. $\to^L$ from $[-1,0]$ becomes $\Rightarrow^R$ from $[0,1]$; **this is the associated Łukasiewicz residuum.**

• $z' = z+1 = t(z) = t(x \oplus^L y) = t(\min(0, x+y+1)) = \min(0, x+y+1)+1 = \min(1, x+y+2) = \min(1, (x+1)+(y+1)) = \min(1, x'+y') = x' \oplus^R y'$, i.e. $\oplus^L$ from $[-1,0]$ becomes $\oplus^R$ from $[0,1]$.

• $z' = z+1 = t(z) = t(x \rightsquigarrow^L y) = t(\max(-1, y-x-1) = \max(-1, y-x-1)+1 = \max(0, y-x) = \max(0, (y+1)-(x+1)) = \max(0, y'-x') = x' \to^R y'$, i.e. $\Rightarrow^L$ from $[-1,0]$ becomes $\to^R$ from $[0,1]$.

Example 3. The multiplicative l-group $\mathcal{D}_g = (D = (0, +\infty), \vee, \wedge, \cdot, ^{-1}, 1)$

By Theorem 9.2.3, the termwise equivalent l-implicative-group, by (Eq3),

$$\mathcal{D}_{ig} = (D, \vee, \wedge, \to, 1)$$

has the implication $\to$ defined by: $x \to y \overset{def.}{=} y \cdot x^{-1} = \frac{y}{x}$, for all $x, y \in D$.

We can make the same analysis as for the two examples. We point only that, given the commutative l-group $\mathcal{D}_g = (D = (0, +\infty), \vee, \wedge, \cdot, ^{-1}, 1)$, we obtain:

By Corollary 14.2.3,

• (i) (1) $\mathcal{D}^-_{m'_1} = ([u', 1], \odot^L, ^-, \mathbf{0} = u', \mathbf{1} = 1)$ is a left-MV algebra;

(2) $\mathcal{D}^-_{m'_2} = ([0,1] = \{0\} \cup D^-, \wedge, \vee, \odot_2, \to^L_2, \mathbf{0} = 0, \mathbf{1} = 1)$ is **the standard (left-) product algebra.**

• (i') (1') $\mathcal{D}^+_{m'_1} = ([1, u], \oplus^R, ^-, \mathbf{0} = 1, \mathbf{1} = u)$ is a right-MV algebra;

(2') $\mathcal{D}^+_{m'_2} = (D^+ \cup \{+\infty\}, \vee, \wedge, \oplus_2, \to^R_2, \mathbf{0} = 1, \mathbf{1} = +\infty)$ is a right-product algebra.

Chapter 15

Normal filters/ideals and compatible deductive systems

In this chapter, we present notions and results from [98], [99].

The chapter has six sections.

15.1 Filters/ideals and deductive systems

We recall the following dual definitions from [97].

Definition 15.1.1

(i) Let $\mathcal{A}^L = (A^L, \leq, \odot, 1)$ be a left-po-im. An *($\odot$)-filter*, or simply a *filter* when there is no danger of confusion, of $\mathcal{A}^L$ is a subset $F \subseteq A^L$ which satisfies:
(f1) $1 \in F$,
(pf2) $x, y \in F$ imply $x \odot y \in F$,
(f3) $x \in F$ and $x \leq y$ imply $y \in F$.

(i') Dually, let $\mathcal{A}^R = (A^R, \leq, \oplus, 0)$ be a right-po-im. An *($\oplus$)-ideal* or simply an *ideal*, when there is no danger of confusion, of $\mathcal{A}^R$ is a subset $I \subseteq A^R$ which satisfies:
(i1) $0 \in I$,
(pi2) $x, y \in I$ imply $x \oplus y \in I$,
(i3) $x \in I$ and $x \geq y$ imply $y \in I$.

Definition 15.1.2

(i) Let $\mathcal{A}^L = (A^L, \leq, \to^L, \rightsquigarrow^L, 1)$ be a left-pBCK algebra.
A *($\to^L, \rightsquigarrow^L$)-deductive system*, or simply a *deductive system* when there is no danger

of confusion, of $\mathcal{A}^L$ is a subset $S \subseteq A^L$ which satisfies:

(ds1) $1 \in S$,

(pds2) $x \in S$ and $x \to^L y \in S$ imply $y \in S$
(or $x \in S$ and $x \rightsquigarrow^L y \in S$ imply $y \in S$).

(i') Dually, let $\mathcal{A}^R = (A^R, \leq, \to^R, \rightsquigarrow^R, 0)$ be a right-pBCK algebra. A *$(\to^R, \rightsquigarrow^R)$-deductive system* or simply a *deductive system*, when there is no danger of confusion, of $\mathcal{A}^R$ is a subset $S' \subseteq A^R$ which satisfies:

(ds1') $0 \in S'$,

(pds2') $x \in S'$ and $x \to^R y \in S'$ imply $y \in S'$
(or $x \in S'$ and $x \rightsquigarrow^R y \in S'$ imply $y \in S'$).

Following the ideas from [23], we prove the following general result (the dual case is omitted):

Theorem 15.1.3 *Let $\mathcal{A}^L_t = (A^L, \leq, \to^L, \rightsquigarrow^L, 1)$ be a pBCK(pP) algebra with the pseudo-product $\odot$ and let $\mathcal{A}^L_r = (A^L, \leq, \odot, 1)$ be the equivalent po-im(pR), by (iE1). Then, the deductive systems of $\mathcal{A}^L_t$ coincide with the filters of $\mathcal{A}^L_r$.*

Proof. If $\mathcal{A}^L_t = (A^L, \leq, \to^L, \rightsquigarrow^L, 1)$ is a pBCK(pP) algebra, then the property (pPP) holds, i.e. for all $x, y \in A^L$, $x \leq y \to^L (x \odot y)$ and $y \leq x \rightsquigarrow^L (x \odot y$.
Let S be a deductive system of $\mathcal{A}^L_t$, i.e. (ds1) and (pds2) hold; we must prove that S is an $(\odot)$-filter of $\mathcal{A}^L_r$, i.e. (f1), (pf2), (f3) hold. Indeed, (f1) is (ds1). To prove (pf2), let $x, y \in S$; then $x \to^L (y \to^L (x \odot y)) = 1$, by (pPP); it follows that $y \to^L (x \odot y) \in S$, by (pds2), hence $x \odot y \in S$, by (pds2) again; thus, (pf2) holds. To prove (f3), let $x \in S$, $x \leq y$; then $x \to^L y = 1$, hence $x \to^L y \in S$; it follows by (pds2) that $y \in S$, i.e. (f3) holds.

Conversely, if $\mathcal{A}^L_r = (A^L, \leq, \odot, 1)$ is a po-im(pR), then the property (pRR) holds, i.e. for all $x, y \in A^L$, $(x \to^L y) \odot x \leq y$ and $x \odot (x \rightsquigarrow^L y) \leq y$.
Let F be an $(\odot)$-filter of $\mathcal{A}^L_r$, i.e. (f1), (pf2), (f3) hold; we must prove that F is a deductive system of $\mathcal{A}^L_t$, i.e. (ds1) and (pds2) hold. Indeed, (ds1) is (f1). To prove (pds2), let $x \in F$, $x \to^L y \in F$; then by (pf2), $(x \to^L y) \odot x \in F$; but we have $(x \to^L y) \odot x \leq y$, by (pRR); hence, by (f3), $y \in F$; (or let $x \in F$, $x \rightsquigarrow^L y \in F$; then by (pf2), $x \odot (x \rightsquigarrow^L y) \in F$; but, we have $x \odot (x \rightsquigarrow^L y) \leq y$; hence, by (f3), $y \in F$;) thus, (pds2) holds. □

Note that Theorem 15.1.3 and its dual are induced by Theorems 15.2.3, 15.3.1, 15.4.1.

At lattice-ordered level, we have consequently the following result.

Theorem 15.1.4 *Let $\mathcal{A}^L_t = (A^L, \wedge, \vee, \to^L, \rightsquigarrow^L, 1)$ be a pBCK(pP) lattice with the pseudo-product $\odot$ and let $\mathcal{A}^L_r = (A^L, \wedge, \vee, \odot, 1)$ be the equivalent l-im(pR), by (iE2). Then, the deductive systems of $\mathcal{A}^L_t$ coincide with the filters of $\mathcal{A}^L_r$.*

15.2 Deductive systems of po-implicative-groups

Recall first the definition of a convex po-subgroup of a po-group.

Definition 15.2.1 Let $\mathcal{G}_g = (G, \leq, \cdot, ^{-1}, 1)$ be a po-group.

A *convex po-subgroup* of $\mathcal{G}_g$ is a subset $S \subseteq G$ which satisfies: for all $x, y, a, b \in G$,
(CS1) $1 \in S$,
(pCS2) (i) $x, y \in S$ imply $x \cdot y \in S$ and (ii) $x \in S$ implies $x^{-1} \in S$,
(pCS3) $a, b \in S$ and $a \leq x \leq b$ imply $x \in S$.

Note that a convex po-subgroup is also called a *po-ideal* of the po-group, but we shall see that a better name would be that of *filter-ideal*, because it determines both a filter and an ideal of certain structures built on G^- and G^+ respectively (see Theorem 15.3.1).

We introduce now the notion of "deductive system" of a po-implicative-group.

Definition 15.2.2 Let $\mathcal{G}_{ig} = (G, \leq, \rightarrow, \rightsquigarrow, 1)$ be a po-implicative-group.

A *deductive system* of $\mathcal{G}_{ig}$ is a subset $S \subseteq G$ which satisfies: for all $x, y, a, b \in G$,
(DS1)=(CS1) $1 \in S$,
(pDS2)(i) $x \in S$, $x \rightarrow y \in S$ imply $y \in S$ (or $x \in S$, $x \rightsquigarrow y \in S$ imply $y \in S$);
(ii) $x \in S$ implies $x \rightarrow 1 = x \rightsquigarrow 1 \in S$,
(pDS3)=(pCS3) $a, b \in S$ and $a \leq x \leq b$ imply $x \in S$.

Theorem 15.2.3 *Let $\mathcal{G}_g = (G, \leq, \cdot, ^{-1}, 1)$ be a po-group and let $\mathcal{G}_{ig} = (G, \leq, \rightarrow, \rightsquigarrow, 1)$ be the termwise equivalent po-implicative-group (by (Eq2)).*

Then, the convex po-subgroups of $\mathcal{G}_g$ coincide with the deductive systems of $\mathcal{G}_{ig}$.

Proof. By Corollary 9.2.5, we can use Lemma 7.1.21.

Let S be a convex po-subgroup of $\mathcal{G}_g$, i.e. (CS1), (pCS2), (pCS3) hold. We must prove that S is a deductive system of $\mathcal{G}_{ig}$, i.e. that (DS1), (pDS2), (pDS3) hold. Indeed, (DS1) and (pDS3) hold. (pDS2)(i): Let $x \in S$ and $x \rightarrow y \in S$; then, by (pCS2)(i), $(x \rightarrow y) \cdot x \in S$; but, by ($\alpha$) or by Lemma 7.1.21, $(x \rightarrow y) \cdot x = y$, hence we obtain that $y \in S$; (similarly, let $x \in S$ and $x \rightsquigarrow y \in S$; then, by (pCS2)(i), $x \cdot (x \rightsquigarrow y) \in S$; by ($\alpha$) or by Lemma 7.1.21, $x \cdot (x \rightsquigarrow y) = y$, hence, we obtain that $y \in S$;) thus, (pDS2)(i) holds. (pDS2)(ii) holds by (pCS2)(ii), since $x^{-1} = x \rightarrow 1 = x \rightsquigarrow 1$.

Conversely, let S be a deductive system of $\mathcal{G}_{ig}$, i.e. (DS1), (pDS2), (pDS3) hold. We must prove that S is a convex po-subgroup of $\mathcal{G}_g$, i.e. that (CS1), (pCS2), (pCS3) hold. Indeed, (CS1) and (pCS3) hold by (DS1) and (pDS3). (pCS2)(i): Let $x, y \in S$; by Lemma 7.1.21, $y \rightarrow (x \cdot y) = x$; then, by (pDS2)(i), we obtain that $x \cdot y \in S$. (pCS2)(ii): holds by (pDS2)(ii), since $x^{-1} = x \rightarrow 1 = x \rightsquigarrow 1$. □

Note that this theorem induces Theorem 15.1.3 and its dual.

At lattice-ordered level, we have consequently the following result.

Theorem 15.2.4 *Let $\mathcal{G}_g = (G, \vee, \wedge, \cdot, ^{-1}, 1)$ be an l-group and let $\mathcal{G}_{ig} = (G, \vee, \wedge, \rightarrow, \rightsquigarrow, 1)$ be the termwise equivalent l-implicative-group (by (Eq3)).*

Then, the convex l-subgroups of $\mathcal{G}_g$ coincide with the deductive systems of $\mathcal{G}_{ig}$.

15.3 po-groups (po-implicative groups) on G^-, G^+

In this section, we study the po-groups (po-implicative groups) and the associated algebras on G^-, G^+.

First, we shall analyse the connections between the convex po-subgroups (= filters-ideals) of $\mathcal{G}$, on the one hand, and the filters of $\mathcal{G}^-$ and the ideals of $\mathcal{G}^+$, on the other hand.

Theorem 15.3.1 *Let $\mathcal{G} = (G, \leq, \cdot, ^{-1}, 1)$ be a po-group and S be a convex po-subgroup (= filter-ideal) of $\mathcal{G}$. Then,*

(1) $S^- = S \cap G^-$ is a filter of the left-po-im $\mathcal{G}^- = (G^-, \leq, \odot = \cdot, \mathbf{1} = 1)$ from Proposition 7.2.14 (1).

(1') $S^+ = S \cap G^+$ is an ideal of the right-po-im $\mathcal{G}^+ = (G^+, \leq, \oplus = \cdot, \mathbf{0} = 1)$ from Proposition 7.2.14 (1').

Proof. We prove (1): (f1): holds by (CS1). (pf2): Let $x, y \in S^-$; then, on the one hand, $x, y \in S$ and hence, by (pCS2)(a), $x \cdot y \in S$; on the other hand, $x, y \in G^-$, hence $x \cdot y \in G^-$; consequently, $x \odot y = x \cdot y \in S^-$ and thus, (pf2) holds. (f3): Let $x \in S^-$ and $x \leq y$ ($y \in G^-$); then, $x \in S$, $1 \in S$ and $x \leq y \leq 1$, hence $y \in S$, by (pCS3); consequently, $y \in S^-$ and thus (f3) holds. (1') has a similar proof. □

Recall now the following definition:

Definition 15.3.2 Let $\mathcal{G}_g = (G, \leq, \cdot, ^{-1}, 1)$ be a po-group. A convex po-subgroup S of $\mathcal{G}_g$ is *normal*, if the following condition holds: for all $g \in G$,
(No_g) $\quad S \cdot g = g \cdot S$.

Recall that $S \cdot g = g \cdot S$ means:
(i) for each $h \in S$, there exists $h' \in S$ such that $h \cdot g = g \cdot h'$;
(ii) for each $h' \in S$, there exists $h \in S$ such that $g \cdot h' = h \cdot g$.

We introduce now the following definition:

Definition 15.3.3 Let $\mathcal{G}_{ig} = (G, \leq, \to, \rightsquigarrow, 1)$ be a po-implicative-group. A deductive system S of $\mathcal{G}_{ig}$ is *compatible*, if the following condition holds: for all $x, y \in G$,
(Co_{ig}) $\quad x \to y \in S \Leftrightarrow x \rightsquigarrow y \in S$.

We formulate now the following complex result (which brings together more results):

Theorem 15.3.4 *Let $\mathcal{G}_{ig} = (G, \leq, \to, \rightsquigarrow, 1)$ be a po-implicative-group (or let $\mathcal{G}_g = (G, \leq, \cdot, ^{-1}, 1)$ be a po-group). Let S be a deductive system of $\mathcal{G}_{ig}$ (or, equivalently, a convex po-subgroup of $\mathcal{G}_g$, by Theorem 15.2.3).*

Then, S is compatible if and only if S is normal, i.e. $(Co_{ig}) \Leftrightarrow (No_g)$.

Proof. $(\mathrm{Co}_{ig}) \Longrightarrow (\mathrm{No}_g)$: Suppose that (Co_{ig}) holds and let $g \in G$.
(i) Let $h \in S$; denote $y \stackrel{notation}{=} h \cdot g$ and remark that by (α) we have

$$(g \to y) \cdot g = y = g \cdot (g \rightsquigarrow y).$$

Hence, $y = h \cdot g = (g \to y) \cdot g$, which implies that $h = g \to y$. Hence, $g \to y \in S$. By (Co_{ig}), it follows that $g \rightsquigarrow y \in S$ also. Hence, there exists $h' = g \rightsquigarrow y \in S$ such that $h \cdot g = (g \to y) \cdot g = y = g \cdot (g \to y) = g \cdot h'$.
(ii) Similarly, let $h' \in S$; denote $x \overset{notation}{=} g \cdot h'$ and remark that by (α) we have

$$(g \to x) \cdot g = x = g \cdot (g \rightsquigarrow x).$$

Hence, $x = g \cdot h' = g \cdot (g \rightsquigarrow x)$, which implies $h' = g \rightsquigarrow x$. Hence, $g \rightsquigarrow x \in S$. By ($Co_{ig}$), it follows that $g \rightsquigarrow x \in S$ too. Hence, there exists $h = g \to x \in S$ such that $gh' = g \cdot (g \rightsquigarrow x) = x = (g \to x) \cdot g = h \cdot g$.
By (i) and (ii), we obtain that (No_g) holds.

(No_g) $\Longrightarrow$ (Co_{ig}): Suppose that (No_g) holds and that $x \to y \in S$. By (α), we have

$$(x \to y) \cdot x = y = x \cdot (x \rightsquigarrow y).$$

Put then $h = x \to y \in S$; hence, $h \cdot x = y = x \cdot (x \rightsquigarrow y)$. By ($No_g$), there exists $h' \in S$ such that $h \cdot x = y = x \cdot h'$. Hence, $y = x \cdot (x \rightsquigarrow y) = x \cdot h'$. It follows that $x \rightsquigarrow y = h'$. Thus, $x \rightsquigarrow y \in S$.
Similarly, $x \rightsquigarrow y \in S$ implies that $x \to y \in S$.
Thus, (Co_{ig}) holds. □

We introduce now the following definition:

Definition 15.3.5

(1) Let $\mathcal{M}^L = (M^L, \leq, \odot, 1)$ be a left-po-im. A filter S^L of $\mathcal{M}^L$ is *normal*, if the following condition holds: for all $x \in M^L$,
(No^L) $S^L \odot x = x \odot S^L$.

(1') Let $\mathcal{M}^R = (M^R, \leq, \oplus, 0)$ be a right-po-im. An ideal S^R of $\mathcal{M}^R$ is *normal*, if the following condition holds: for all $x \in M^R$,
(No^R) $S^R \oplus x = x \oplus S^R$.

Then we have the following result:

Proposition 15.3.6 *Let $\mathcal{G} = (G, \leq, \cdot, ^{-1}, 1)$ be a po-group and S be a normal convex po-subgroup of $\mathcal{G}$. Then, we have:*

(1) $S^- = S \cap G^-$ is a normal filter of the left-po-im $\mathcal{G}^- = (G^-, \leq, \odot = \cdot, \mathbf{1} = 1)$ from the Proposition 7.2.14 (1).

(1') $S^+ = S \cap G^+$ is a normal ideal of the right-po-im $\mathcal{G}^+ = (G^+, \leq, \oplus = \cdot, \mathbf{0} = 1)$ from the Proposition 7.2.14 (1').

Proof. (1): By Theorem 15.3.1 (1), S^- is a filter of $\mathcal{G}^-$. It remains to prove that is a normal filter, i.e. (No^L) holds.
(i) Let $h \in S^- = S \cap G^-$, i.e. $h \in S$ and $h \leq 1$. S being normal, by (No_g) it follows that there exists $h' \in S$ such that

$$s \overset{notation}{=} h \cdot x = x \cdot h'.$$

We must prove that $h' \leq 0$. First, notice that $s \leq x$, since $h \leq 1$ implies that $s = h \cdot x \leq 1 \cdot x = x$. Then, $s \leq x$ implies that $s = x \cdot h' \leq x$, and therefore $h' \leq 1$.

Thus, $h' \in S^- = S \cap G^-$ and $h \odot x = x \odot h'$.
(ii) Similarly, for $h' \in S^-$, there exists $h \in S^-$, such that $x \odot h' = h \odot x$.
By (i) and (ii), (No^L) holds.
(1') has a dual proof. □

Recall the following definition.

Definition 15.3.7 (see ([135], Definition 2.2.1)
(1) Let $\mathcal{A}^L = (A^L, \leq, \to^L, \rightsquigarrow^L, 1)$ be a left-pBCK algebra.
We say that a $(\to^L, \rightsquigarrow^L)$-deductive system S^L of $\mathcal{A}^L$ is *compatible*, if the following condition holds: for any $x, y \in A^L$,
(Co^L) $\quad x \to^L y \in S^L \Leftrightarrow x \rightsquigarrow^L y \in S^L$.
(1') Let $\mathcal{A}^R = (A^R, \leq, \to^R, \rightsquigarrow^R, 0)$ be a right-pBCK algebra.
We say that a $(\to^R, \rightsquigarrow^R)$-deductive system S^R of $\mathcal{A}^R$ is *compatible*, if the following condition holds: for any $x, y \in A^R$,
(Co^R) $\quad x \to^R y \in S^R \Leftrightarrow x \rightsquigarrow^R y \in S^R$.

15.4 l-groups (l-implicative groups) on G^-, G^+

In this section, we study the l-groups (l-implicative groups) and the associated algebras on G^-, G^+.

Now we shall analyse the connections between the deductive systems of the l-implicative-group $\mathcal{G}$, on the one hand, and the deductive systems of $\mathcal{G}^-$ and $\mathcal{G}^+$, on the other hand (see Theorem 15.1.3 and Theorems 15.2.3 and 15.3.1).

Theorem 15.4.1 *Let $\mathcal{G} = (G, \vee, \wedge, \to, \rightsquigarrow, 1)$ be an l-implicative-group and S be a deductive system of $\mathcal{G}$. Then,*
(1) $S^- = S \cap G^-$ is a $(\to^L, \rightsquigarrow^L)$-deductive system of the pBCK(pP) lattice

$$\mathcal{G}^- = (G^-, \wedge, \vee, \to^L, \rightsquigarrow^L, \mathbf{1} = 1)$$

from Theorem 14.1.1 (1).
(1') $S^+ = S \cap G^+$ is a $(\to^R, \rightsquigarrow^R)$-deductive system of the pBCK(pS) lattice

$$\mathcal{G}^+ = (G^+, \vee, \wedge, \to^R, \rightsquigarrow^R, \mathbf{0} = 1)$$

from Theorem 14.1.1 (1').

Proof. We prove (1): (ds1) holds by (DS1).
(pds2): Let $x \in S^-$ and $x \to^L y \in S^-$ $(y \in G^-)$.
Then, $x \in S$ and $x \to^L y = \begin{cases} 1, & \text{if } x \leq y \\ (x \to y) \wedge 1, & \text{if } x \not\leq y, \end{cases} \in S$:
- if $x \leq y$, then $x \leq y \leq 1$ and $x, 1 \in S$; then, by (pDS3), we obtain that $y \in S$, hence $y \in S^-$;
- if $x \not\leq y$, then $x \to^L y = (x \to y) \wedge 1 \leq x \to y$. But $y \leq 1$ implies by $(\mathrm{p}{*}^L)$ that $x \to y \leq x \to 1$.
Hence, we have $x \to^L y \leq x \to y \leq x \to 1$ and $x \to^L y, x \to 1 \in S$; hence, by

(pDS3), $x \to y \in S$. Thus, we obtained that $x \in S$ and $x \to y \in S$, hence, by (pDS2)(a), $y \in S$, hence $y \in S^-$.

(1') has a dual proof. □

Note that Theorems 15.2.3, 15.3.1, 15.4.1 imply Theorem 15.1.3 and its dual.

Proposition 15.4.2 *Let $\mathcal{G} = (G, \vee, \wedge, \to, \rightsquigarrow, 1)$ be an l-implicative-group and S be a compatible deductive system of $\mathcal{G}$. Then,*

(1) $S^- = S \cap G^-$ is a compatible $(\to^L, \rightsquigarrow^L)$-deductive system of the pBCK(pP) lattice $\mathcal{G}^- = (G^-, \wedge, \vee, \to^L, \rightsquigarrow^L, \mathbf{1} = 1)$ from Theorem 14.1.1 (1).

(1') $S^+ = S \cap G^+$ is a compatible $(\to^R, \rightsquigarrow^R)$-deductive system of the pBCK(pS) lattice $\mathcal{G}^+ = (G^+, \vee, \wedge, \to^R, \rightsquigarrow^R, \mathbf{0} = 1)$ from Theorem 14.1.1 (1').

Proof. (1): By Theorem 15.4.1 (1), S^- is a $(\to^L, \rightsquigarrow^L)$-deductive system of $\mathcal{G}^-$. It remains to prove that S^- is compatible, i.e. for all $x, y \in G^-$:

$$x \to^L y \in S^L \Leftrightarrow x \rightsquigarrow^L y \in S^-. \tag{15.1}$$

Since $x \to^L y \in G^-$, $x \rightsquigarrow^L y \in G^-$, it remains to prove that

$$x \to^L y \in S \Leftrightarrow x \rightsquigarrow^L y \in S. \tag{15.2}$$

But $x \to^L y = (x \to y) \wedge 1 = (x \to y) \wedge (x \to x) \overset{(7.15)}{=} x \to (y \wedge x)$ and similarly $x \rightsquigarrow^L y = x \rightsquigarrow (y \wedge x)$ and since S is compatible, we have

$$x \to (y \wedge x) \in S \Leftrightarrow x \rightsquigarrow (y \wedge x) \in S. \tag{15.3}$$

It follows that (15.2) holds, hence (15.1) holds.

(1') has a dual proof. □

We then formulate the following complex result:

Theorem 15.4.3

(1) Let $\mathcal{A}^L = (A^L, \wedge, \vee, \to^L, \rightsquigarrow^L, 1)$ be a pBCK(pP) lattice (or let $\mathcal{A}^L_m = (A^L, \wedge, \vee, \odot, 1)$ be a l-im(pR)) verifying (pdiv^L)), where: for all $x, y \in A^L$,
(pdiv^L) $x \wedge y = (x \to^L y) \odot x = x \odot (x \rightsquigarrow^L y)$.

Let S^L be a $(\to^L, \rightsquigarrow^L)$-deductive system of $\mathcal{A}^L$ (or, equivalently, a filter of $\mathcal{A}^L_m$).

Then, S^L is compatible if and only if is normal, i.e. (Co^L) $\Leftrightarrow$ (No^L).

(1') Let $\mathcal{A}^R = (A^R, \vee, \wedge, \to^R, \rightsquigarrow^R, 0)$ be a pBCK(pS) lattice (or let $\mathcal{A}^R_m = (A^R, \vee, \wedge, \oplus, 0)$ be a l-im(pR)) verifying (pdiv^R)), where: for all $x, y \in A^R$,
(pdiv^R) $x \vee y = (x \to^R y) \oplus x = x \oplus (x \rightsquigarrow^R y)$.

Let S^R be a $(\to^R, \rightsquigarrow^R)$-deductive system of $\mathcal{A}^R$ (or, equivalently, an ideal of $\mathcal{A}^R_m$).

Then, S^R is compatible if and only if is normal, i.e. (Co^R) $\Leftrightarrow$ (No^R).

Proof. (1): Suppose that S^L is compatible, i.e. that (Co^L) holds. We must prove that S^L is normal, i.e. that (No^L) holds. Indeed, let $x \in A^L$.

(i) Let $h \in S^L$. We put $s \overset{notation}{=} h \odot x$. First, notice that $s \leq x$, since $h \odot x \leq x$.

Then, by (pdiv^L), we have $s = x \wedge s = (x \to^L s) \odot x = x \odot (x \rightsquigarrow^L s)$.
But $x \to^L s = x \to^L (h \odot x) \geq h$, by ([94] page 354, property (10.3)). Hence $x \to^L s \geq h$ and $h \in S^L$. Since S^L is a filter, it follows that $x \to^L s \in S^L$. Then, by (Co^L), we obtain that $x \rightsquigarrow^L s \in S^L$ too; hence, there exists $h' \overset{notation}{=} x \rightsquigarrow^L s \in S^L$ such that $h \odot x = x \odot h'$.
(ii) Similarly, for $h' \in S^L$, there exists $h \in S^L$ such that $x \odot h' = h \odot x$.
By (i) and (ii), (No^L) holds.

Suppose now that S^L is normal, i.e. that (No^L) holds. We must prove that S^L is compatible, i.e. that (Co^L) holds. Indeed, let $x, y \in A^L$.
Assume $x \to^L y \in S^L$. Then putting $h = x \to^L y$, by (No^L) there exists $h' \in S^L$ such that $h \odot x = x \odot h'$.
But, by (pdiv^L), $x \wedge y = (x \to^L y) \odot x = x \odot (x \rightsquigarrow^L y)$,
i.e. $x \wedge y = h \odot x = x \odot (x \rightsquigarrow^L y)$. Hence, $x \wedge y = x \odot h' = x \odot (x \rightsquigarrow^L y)$.
We must prove that $x \rightsquigarrow^L y \in S^L$. Indeed, by (2.19) and by (pPP),
$x \rightsquigarrow^L y = x \rightsquigarrow^L (x \wedge y) = x \rightsquigarrow^L (x \odot h') \geq h'$.
Hence, $x \rightsquigarrow^L y \geq h'$ and $h' \in S^L$; since S^L is a filter, it follows that $x \rightsquigarrow^L y \in S^L$.
Similarly, assuming $x \rightsquigarrow^L y \in S^L$, we shall obtain that $x \to^L y \in S^L$.
Thus, (Co^L) holds.

(1'): dually. □

Remark 15.4.4 Note that condition (pdiv^L) ((pdiv^R), dually) is a necessary condition in order to have that "*compatible* property is equivalent to *normal* property". In the absence of condition (pdiv^L) ((pdiv^R), dually), we may have any situation.

Open problem 15.4.5 Find an example of pBCK(pP) lattice for instance, not verifying (pdiv^L), which has a filter that is normal but not compatible, or is compatible but not normal.

Finally, we formulate the following complex result:

Theorem 15.4.6 *Let $\mathcal{G}_{ig} = (G, \vee, \wedge, \to, \rightsquigarrow, 1)$ be an l-implicative-group (or let $\mathcal{G}_g = (G, \vee, \wedge, \cdot, ^{-1}, 1)$ be an l-group).*

Let S be a compatible deductive system of $\mathcal{G}_{ig}$ (or, equivalently, a normal convex l-subgroup of $\mathcal{G}_g$).

Then,

(1) $S^- = S \cap G^-$ is a compatible $(\to^L, \rightsquigarrow^L)$-deductive system of the pBCK(pP) lattice $\mathcal{G}^- = (G^-, \wedge, \vee, \to^L, \rightsquigarrow^L, \mathbf{1} = 1)$ (or, equivalently, S^- is a normal filter of the l-im(pR) $\mathcal{G}^-_m = (G^-, \wedge, \vee, \odot = +, \mathbf{1} = 1)$), and S^- is compatible if and only if is normal, i.e. $(Co^L) \Leftrightarrow (No^L)$.

(1') $S^+ = S \cap G^+$ is a compatible $(\to^R, \rightsquigarrow^R)$-deductive system of the pBCK(pS) lattice $\mathcal{G}^+ = (G^+, \vee, \wedge, \to^R, \rightsquigarrow^R, \mathbf{0} = 1)$ (or, equivalently, S^+ is a normal ideal of the l-im(pcoR) $\mathcal{G}^+_m = (G^+, \vee, \wedge, \oplus = +, \mathbf{0} = 1)$), and S^+ is compatible if and only if is normal, i.e. $(Co^R) \Leftrightarrow (No^R)$.

Proof. (1): S^- is a compatible $(\to^L, \rightsquigarrow^L)$-deductive system of $\mathcal{G}^-$ by Proposition 15.4.2 (1). S^- is a normal filter of $\mathcal{G}^-_m$ by Proposition 15.3.6 (1).

$(Co^L) \Leftrightarrow (No^L)$ by Theorem 15.4.3 (1), since $\mathcal{G}^-$ ($\mathcal{G}^-_m$) verifies condition (pCC^L) and (pCC^L) implies (pdiv^L).

(1') has a dual proof. □

In other words, the above Theorem 15.4.6 says that normality (compatibility) at l-group (l-implicative-group) G level is inherited by the algebras obtained by restricting the l-group (l-implicative-group) operations to the negative cone G^- and to the positive cone G^+. Also, it says that the equivalence $(Co_{ig}) \Leftrightarrow (No_g)$ (*compatible* if and only if *normal*), existing at l-group (l-implicative-group) level (Theorem 15.3.4), is preserved by the algebras obtained by restricting the l-group (l-implicative-group) operations to G^- and to G^+.

15.5 l-groups (l-implicative groups) on $[u',1]$, $[1,u]$

In this section, we study the l-groups (l-implicative groups) and the associated algebras on $[u',1]$, $[1,u]$.

Remarks 15.5.1

(a) It is proved, in [58], ([61], Lemma 3.2), that in right-pseudo-MV algebras, an ideal is normal if and only if is compatible. Hence, dually, in left-pseudo-MV algebras, a filter is normal if and only if is compatible.

(b) Note that these dual results follow by Theorem 15.4.3, (1') and (1), respectively, since in a left-pseudo-MV algebra (right-pseudo-MV algebra) condition (pdiv^L) (condition (pdiv^R), respectively) is verified.

We shall clarify in this section how the results from the above Remarks are connected with those at l-group (l-implicative-group) level.

First note that we have the following general theorem, that generalizes the Corollary 14.1.7, (1), (1').

Theorem 15.5.2

(1) Let $\mathcal{A}^L = (A^L, \wedge, \vee, \rightarrow^L, \rightsquigarrow^L, 1)$ be a pBCK(pP) lattice (with the pseudo-product $\odot$) verifying the property (pCC^L). Let us "bound" this algebra with an "internal" element in the following way: let us take $u' < 1$ from A^L and consider the interval $[u',1] = \{x \in A^L \mid u' \leq x \leq 1\} \subset A^L$. Then, the algebra

$$\mathcal{A}^L_1 = ([u',1], \wedge, \vee, \rightarrow^L, \rightsquigarrow^L, \mathbf{0} = u', 1)$$

is a bounded pBCK(pP) lattice (with the pseudo-product $x \odot^L y \overset{def.}{=} (x \odot y) \vee u'$) with property (pCC^L), i.e. is an equivalent definition of the **left-pWajsberg algebra** *(by Theorem 2.12.10 (1)):*

$$\mathcal{A}^L_{1'} = ([u',1], \rightarrow^L, \rightsquigarrow^L, {}^{-^L}, {}^{\sim^L}, \mathbf{0} = u', 1).$$

(1') Let $\mathcal{A}^R = (A^R, \vee, \wedge, \rightarrow^R, \rightsquigarrow^R, 0)$ be a pBCK(pS) lattice (with the pseudo-sum $\oplus$) verifying the property (pCC^R). Let us "bound" this algebra in the following

way: let us take $u > 0$ *from* A^R *and consider the interval* $[0, u] = \{x \in A^R \mid 0 \leq x \leq u\} \subset A^R$. *Then, the algebra*

$$\mathcal{A}_1^R = ([0, u], \vee, \wedge, \to^R, \rightsquigarrow^R, 0, \mathbf{1} = u)$$

is a bounded pBCK(pS) lattice (with the pseudo-sum $x \oplus^R y \stackrel{def.}{=} (x \oplus y) \wedge u$*) with property (pCC*^{R}*), i.e. is an equivalent definition of the* **right-pWajsberg algebra** *(by Theorem 2.12.10 (1')):*

$$\mathcal{A}_{1'}^R = ([0, u], \to^R, \rightsquigarrow^R, {}^{-^R}, {}^{\sim^R}, 0, \mathbf{1} = u).$$

Proof.

(1): $[u', 1]$ is obviously closed under $\wedge$ and $\vee$. $[u', 1]$ is closed under $\to^L$. Indeed, let $x, y \in [u', 1]$; $y \leq 1$ implies $x \to^L y \leq x \to^L 1 = 1$; $u' \leq y$ implies $x \to^L u' \leq x \to^L y$ and $u' \leq x \to^L u'$; hence $u' \leq x \to^L y \leq 1$. Similarly, $[u', 1]$ is closed under $\rightsquigarrow^L$. It follows that $\mathcal{A}_1^L$ is a bounded left-pBCK lattice verifying (pCC^L). Then, it verifies (pDN^L), where $x^- = x^{-^L} \stackrel{def.}{=} x \to^L u'$ and $x^\sim = x^{\sim^L} \stackrel{def.}{=} x \rightsquigarrow^L u'$, for any $x \in [u', 1]$. Consequently, there exists $x \odot^L y = (x \to^L y^-)^\sim = (x \to^L (y \to^L u')) \rightsquigarrow^L u' = ((x \odot y) \to^L u') \rightsquigarrow^L u' \stackrel{(pCC^L)}{=} (x \odot y) \vee u'$.
Thus, $\mathcal{A}_1^L$ is a bounded pBCK(pP) lattice with pseudo-product $\odot^L$ verifying (pCC^L).

(1'): a dual proof. □

Open problem 15.5.3 It remains an open problem if there are other examples than l-implicative-groups producing pseudo-BCK(pP) lattices verifying condition (pCC^L) (pseudo-BCK(pS) lattices verifying condition (pCC^R), respectively).

Note that we can reformulate Theorem 15.5.2, equivalently, by (iE2), (iE4), (iE10), as follows:

Theorem 15.5.4

(1) Let $\mathcal{A}_m^L = (A^L, \wedge, \vee, \odot, 1)$ *be an l-im(pR) (with the pseudo-residuum* $(\to^L, \rightsquigarrow^L)$*) verifying the property (pCC*^{L}*). Let us "bound" this algebra with an "external" element in the following way: let us take* $u' < 1$ *from* A^L *and consider the interval* $[u', 1] \subset A^L$. *Then, the algebra*

$$\mathcal{A}_{m1}^L = ([u', 1], \wedge, \vee, \odot^L, \mathbf{0} = u', 1)$$

is a bounded l-im(pR) (with $x \odot^L y \stackrel{def.}{=} (x \odot y) \vee u'$ *and the pseudo-residuum* $(\to^L, \rightsquigarrow^L)$*) with property (pCC*^{L}*), i.e. is an equivalent definition of the* **left-pMV algebra** *(see [89]):*

$$\mathcal{A}_{m1'}^L = ([u', 1], \odot^L, {}^{-^L}, {}^{\sim^L}, \mathbf{0} = u', 1).$$

(1') Let $\mathcal{A}_m^R = (A^R, \vee, \wedge, \oplus, 0)$ *be a l-im(pcoR) (with the pseudo-coresiduum* $(\to^R, \rightsquigarrow^R)$*) verifying the property (pCC*^{R}*). Let us "bound" this algebra in the*

following way: let us take $u > 0$ from A^R and consider the interval $[0, u] \subset A^R$. Then, the algebra

$$\mathcal{A}^R_{m1} = ([0, u], \vee, \wedge, \oplus^R, 0, \mathbf{1} = u)$$

is a bounded l-im(pcoR) (with $x \oplus^R y \stackrel{def.}{=} (x \oplus y) \wedge u$) with property (pCC^R), i.e. is an equivalent definition of the **right-pMV algebra**

$$\mathcal{A}^R_{m1'} = ([0, u], \oplus^R, ^{-^R}, ^{\sim^R}, 0, \mathbf{1} = u).$$

A filter of the left-pMV algebra $\mathcal{A}^L_{m1'}$ is a filter of the left-po-im $([u', 1], \leq, \odot^L, 1)$ and an ideal of the right-pMV algebra $\mathcal{A}^R_{m1'}$ is an ideal of the right-po-im $([0, u], \leq, \oplus^R, 0)$. Then, we have the following result.

Proposition 15.5.5 *Let $\mathcal{G} = (G, \vee, \wedge, \cdot, ^{-1}, 1)$ be an l-group and S be a convex l-subgroup of $\mathcal{G}$. Then,*

(1) for any $u' < 1$ and $S^- = S \cap G^-$, the set $S^L_{[]} = S \cap [u', 1] = S^- \cap [u', 1]$ is a filter of the left-pMV algebra

$$\mathcal{G}^L_{m1'} = ([u', 1], \odot^L, ^{-^L}, ^{\sim^L}, \mathbf{0} = u', \mathbf{1} = 1).$$

(1') for any $u > 1$ and $S^+ = S \cap G^+$, the set $S^R_{[]} = S \cap [1, u] = S^+ \cap [1, u]$ is an ideal of the right-pMV algebra

$$\mathcal{G}^R_{m1'} = ([1, u], \oplus^R, ^{-^R}, ^{\sim^R}, \mathbf{0} = 1, \mathbf{1} = u).$$

Proof.

(1): First, note that, by Theorem 15.3.1 (1), S^- is a filter of $\mathcal{G}^-$.
We prove that $S^L_{[]} = S^- \cap [u', 1]$ is a filter of $\mathcal{G}^L_{m1'}$:
(F1): $1 \in S^L_{[]}$ since $1 \in S^-$ and $1 \in [u', 1]$.
(F2): Let $x, y \in S^L_{[]}$, i.e. $x, y \in S^-$ and $x, y \in [u', 1]$. We must prove that $x \odot^L y \in S^L_{[]}$. Since S^- is a filter of $\mathcal{G}^-$, it follows that $x \odot y \in S^-$; but $x \odot^L y = (x \odot y) \vee u' \geq x \odot y$, hence $x \odot^L y \in S^-$. On the other hand, $u' \leq x, y \leq 1$ imply that

$$u' \leq x \odot^L y = (x \odot y) \vee u' = (x \cdot y) \vee u' \leq 1,$$

hence $x \odot^L y \in [u', 1]$. Thus, $x \odot^L y \in S^- \cap [u', 1] = S^L_{[]}$.
(F3): Let $x \in S^L_{[]}$, $x \leq y$ ($y \in [u', 1]$). We must prove that $y \in S^L_{[]}$. But S^- is a filter of $\mathcal{G}^-$, hence $y \in S^-$. Since $y \in [u', 1]$, we obtain that $y \in S^- \cap [u', 1] = S^L_{[]}$.

(1'): a dual proof. □

Proposition 15.5.6 *Let $\mathcal{G} = (G, \vee, \wedge, \rightarrow, \rightsquigarrow, 1)$ be an l-implicative-group and S be a compatible deductive system of $\mathcal{G}$. Then,*

(1) for any $u' < 1$ and $S^- = S \cap G^-$, the set $S^L_{[]} = S \cap [u', 1] = S^- \cap [u', 1]$ is a compatible $(\rightarrow^L, \rightsquigarrow^L)$-deductive system of the bounded pBCK(pP) lattice verifying the property (pCC^L)

$$\mathcal{G}^L_1 = ([u', 1], \wedge, \vee, \rightarrow^L, \rightsquigarrow^L, \mathbf{0} = u', \mathbf{1} = 1)$$

from Corollary 14.1.7(i) (i.e. of the left-pWajsberg algebra $\mathcal{G}^L_{1'}$*).*

(1') for any $u > 1$ *and* $S^+ = S \cap G^+$*, the set* $S^R_{[]} = S \cap [1, u] = S^+ \cap [1, u]$ *is a compatible* $(\to^R, \leadsto^R)$*-deductive system of the bounded pBCK(pS) lattice verifying the property* (pCC^R)

$$\mathcal{G}^R_1 = ([1, u], \vee, \wedge, \to^R, \leadsto^R, \mathbf{0} = 1, \mathbf{1} = u)$$

from Corollary 14.1.7(i') (i.e. of the right-pWajsberg algebra $\mathcal{G}^R_{1'}$*).*

Proof. (1): First, note that by Proposition 15.3.6 (1), S^- is a compatible $(\to^L, \leadsto^L)$-deductive system of $\mathcal{G}^-$.
Second, $S^L_{[]}$ is a $(\to^L, \leadsto^L)$-deductive system of $[u', 1]$, since:
(ds1): $\mathbf{1} = 1 \in S^L, [u', 1]$, hence $1 \in S^L_{[]}$.
(ds2): Let $x \in S^L_{[]}$ and $x \leq y$ ($y \in [u', 1]$). It follows that $y \in S^-$. Since we also have that $y \in [u', 1]$, it follows that $y \in S^L \cap [u', 1] = S^L_{[]}$.
Third, we prove that $S^L_{[]}$ is compatible. Indeed, let $x, y \in [u', 1]$.
Assume that $x \to^L y \in S^L_{[]}$, i.e. $x \to^L y \in S^-$ and $x \to^L y \in [u', 1]$. S^- being compatible, it follows that $x \leadsto^L y \in S^-$ too. Since $[u', 1]$ is closed under $\to^L$, $\leadsto^L$ (Lemma 14.1.5), it follows that $x \leadsto^L y \in [u', 1]$. Hence, $x \leadsto^L y \in S^- \cap [u', 1] = S^L_{[]}$.
Assume that $x \leadsto^L y \in S^L_{[]}$. We obtain similarly that $x \to^L y \in S^L_{[]}$. Thus, $S^L_{[]}$ is compatible.

(1'): a dual proof. □

Finally, we formulate the following complex result:

Theorem 15.5.7 *Let* $\mathcal{G}_{ig} = (G, \vee, \wedge, \to, \leadsto, 1)$ *be an l-implicative-group (or let* $\mathcal{G}_g = (G, \vee, \wedge, \cdot, ^{-1}, 1)$ *be an l-group, by (Eq3)). Let S be a compatible deductive system of* $\mathcal{G}_{ig}$ *(or, equivalently, a normal convex l-subgroup of* $\mathcal{G}_g$*). Then,*

(1) for any $u' < 1$ *and* $S^- = S \cap G^-$*, the set* $S^L_{[]} = S \cap [u', 1] = S^- \cap [u', 1]$ *is a compatible* $(\to^L, \leadsto^L)$*-deductive system of the bounded pBCK(pP) lattice verifying the property* (pCC^L)

$$\mathcal{G}^L_1 = ([u', 1], \wedge, \vee, \to^L, \leadsto^L, \mathbf{0} = u', \mathbf{1} = 1)$$

from Corollary 14.1.7 (i), i.e. of the left-pWajsberg algebra $\mathcal{G}^L_{1'}$ *(or, equivalently,* $S^L_{[]}$ *is a normal filter of the bounded l-im(pR) verifying the property* (pCC^L)

$$\mathcal{G}^L_{m1} = ([u', 1], \wedge, \vee, \odot^L, \mathbf{0} = u', \mathbf{1} = 1),$$

i.e. of the left-pMV algebra $\mathcal{G}^L_{m1'}$*), and* $S^L_{[]}$ *is compatible if and only if is normal, i.e.* $(Co^L) \Leftrightarrow (No^L)$*.*

(1') for any $u > 1$ *and* $S^+ = S \cap G^+$*,* $S^R_{[]} = S \cap [1, u] = S^+ \cap [1, u]$ *is a compatible* $(\to^R, \leadsto^R)$*-deductive system of the bounded pBCK(pS) lattice verifying the property* (pCC^R)

$$\mathcal{G}^R_1 = ([1, u], \vee, \wedge, \to^R, \leadsto^R, \mathbf{0} = 1, \mathbf{1} = u)$$

from Corollary 14.1.7 (i'), i.e. of the right-pWajsberg algebra $\mathcal{G}^R_{1'}$ (or, equivalently, $S^R_{[]}$ is a normal ideal of the bounded l-im(pcoR) verifying the property (pCC^R)

$$\mathcal{G}^R_{m1} = ([1,u], \vee, \wedge, \oplus^R, \mathbf{0} = 1, \mathbf{1} = u),$$

i.e. of the right-pMV algebra $\mathcal{G}^R_{m1'}$), and $S^R_{[]}$ is compatible if and only if is normal, i.e. $(Co^R) \Leftrightarrow (No^R)$.

Proof.

(1): $S^L_{[]}$ is a compatible $(\rightarrow^L, \rightsquigarrow^L)$-deductive system of $\mathcal{G}^L_{1'}$, by Proposition 15.5.6 (1). $S^L_{[]}$ is a filter of $\mathcal{G}^L_{m1}$ (i.e. $[u', 1]$), by Proposition 15.5.5(1).
$(Co^L) \Leftrightarrow (No^L)$, by Theorem 15.4.3 (1), since $\mathcal{G}^L_{1'}$ verifies the property (pCC^L) and (pCC^L) implies (pdiv^L); hence $S^L_{[]}$ is a normal filter of $[u', 1]$.

(1'): a dual proof. □

In other words, the above Theorem 15.5.7 says that normality (compatibility) at *l*-group (*l*-implicative-group) G level is inherited by the algebras obtained by restricting the *l*-group (*l*-implicative-group) operations to any segment $[u', 1] \subset G^-$ and to any segment $[1, u] \subset G^+$. Also, it says that the equivalence $(Co_{ig}) \Leftrightarrow (No_g)$ (*compatible* if and only if *normal*), existing at *l*-group (*l*-implicative-group) level (Theorem 15.3.4), is preserved by the algebras obtained by restricting the *l*-group (*l*-implicative-group) operations to $[u', 1]$ and to $[1, u]$ (see also Remarks 15.5.1 (a)).

Open problem 15.5.8 Find a direct proof that a normal convex *l*-group of an *l*-group $\mathcal{G}$ produces a normal filter of $\mathcal{G}^L_{m1'}$ (a normal ideal of $\mathcal{G}^R_{m1'}$) (see Proposition 15.5.5 and Theorem 15.5.7).

15.6 *l*-groups (*l*-implicative groups) on $\{-\infty\} \cup G^-$, $G^+ \cup \{+\infty\}$

In this section, we study the *l*-groups (*l*-implicative groups) and the associated algebras on $\{-\infty\} \cup G^-$, $G^+ \cup \{+\infty\}$.

First note that we have the following general theorem that generalizes the Corollary 14.1.7, (2), (2'):

Theorem 15.6.1

(i) Let $\mathcal{A}^L = (A^L, \wedge, \vee, \rightarrow^L, \rightsquigarrow^L, 1)$ be a pBCK(pP) lattice (with the pseudo-product $\odot$) verifying (pCC^L) and (p@^{L}). Let us "bound" this algebra in the following way (the usual way of "bounding" the Hilbert algebras - in the commutative case): let us consider a symbol $-\infty$ distinct from the elements of A^L. Define $A^L_{-\infty} \stackrel{def.}{=} \{-\infty\} \cup A^L$ and extend the operations $\rightarrow^L, \rightsquigarrow^L, \odot$ from A^L to $A^L_{-\infty}$ as follows:

$$x \rightarrow^L_2 y \stackrel{def.}{=} \begin{cases} x \rightarrow^L y, & \text{if } x, y \in A^L \\ -\infty, & \text{if } x \in A^L,\ y = -\infty \\ 1, & \text{if } x = -\infty, \end{cases}$$

$$x \rightsquigarrow_2^L y \overset{def.}{=} \begin{cases} x \rightsquigarrow^L y, & if \ x, y \in A^L \\ -\infty, & if \ x \in A^L, \ y = -\infty \\ 1, & if \ x = -\infty, \end{cases}$$

$$x \odot_2 y \overset{def.}{=} \begin{cases} x \odot y, & if \ x, y \in A^L \\ -\infty, & if \ otherwise. \end{cases}$$

We extend the lattice order relation $\leq$ as follows: we put $-\infty \leq x$, for any $x \in A^L_{-\infty}$.

Then, the algebra

$$\mathcal{A}_2^L = (A^L_{-\infty}, \wedge, \vee, \rightarrow_2^L, \rightsquigarrow_2^L, \mathbf{0} = -\infty, 1)$$

is a **pseudo-Hájek(pP) algebra (with the pseudo-product $\odot_2$) verifying the properties (pP1^L) and (pP2^L)** *(see [89]).*

(i') Let $\mathcal{A}^R = (A^R, \vee, \wedge, \rightarrow^R, \rightsquigarrow^R, 0)$ be a pBCK(pS) lattice (with the pseudo-sum $\oplus$) verifying the propertiess (pCC^R) and (p@^{R}). Let us "bound" this algebra in the following way: let us consider a symbol $+\infty$ distinct from the elements of A^R. Define $A^R_{+\infty} \overset{def.}{=} A^R \cup \{+\infty\}$ and extend the operations $\rightarrow^R, \rightsquigarrow^R$, $\oplus$ from A^R to $A^R_{+\infty}$ as follows:

$$x \rightarrow_2^R y \overset{def.}{=} \begin{cases} x \rightarrow^R y, & if \ x, y \in A^R \\ +\infty, & if \ x \in A^R, \ y = +\infty \\ 0, & if \ x = +\infty, \end{cases}$$

$$x \rightsquigarrow_2^R y \overset{def.}{=} \begin{cases} x \rightsquigarrow^R, & if \ x, y \in A^R \\ +\infty, & if \ x \in A^R, \ y = +\infty \\ 0, & if \ x = +\infty, \end{cases}$$

$$x \oplus_2 y \overset{def.}{=} \begin{cases} x \oplus y, & if \ x, y \in A^R \\ +\infty, & if \ otherwise. \end{cases}$$

We extend the lattice order relation $\geq$ as follows: we put $+\infty \geq x$, for any $x \in A^R_{+\infty}$.

Then, the algebra

$$\mathcal{A}_2^R = (A^R_{+\infty}, \vee, \wedge, \rightarrow_2^R, \rightsquigarrow_2^R, 0, \mathbf{1} = +\infty)$$

is a **pseudo-Hájek(pS) algebra (with the pseudo-sum $\oplus_2$) verifying the dual properties (pP1^R) and (pP2^R).**

Proof.

(i): Obviously, $\mathcal{A}_2^L$ is a bounded pBCK(pP) lattice, with the pseudo-product $\odot_2$ (by property (pP)):

$x \odot_2 y \overset{def.}{=} \min\{z \in A^L_{-\infty} \mid x \leq y \rightarrow_2^L z\}=$

$$= \min\{z \in A^L_{-\infty} \mid x \leq \begin{cases} y \rightarrow^L z, & \text{if} \ y, z \in A^L \\ -\infty, & \text{if} \ y \in A^L, \ z = -\infty \\ 1, & \text{if} \ y = -\infty \end{cases} \}$$

$$= \begin{cases} x \odot y, & \text{if } x, y \in A^L \\ -\infty, & \text{if } \textit{otherwise}. \end{cases}$$

$\mathcal{A}^L$ verifies the property (pCC^L), hence it verifies the properties (pprel^L) and (pdiv^L), by Theorem 2.12.4 (1). We shall prove that $\mathcal{A}_2^L$ does not verify (pCC^L) anymore, but it satisfies (pprel^L) and (pdiv^L). Indeed,
- $\mathcal{A}_2^L$ does not verify (pCC^L), since for $-\infty$ and any $x \neq 1$, we have:
$x = -\infty \vee x \neq (x \to^L -\infty) \rightsquigarrow^L -\infty = -\infty \rightsquigarrow^L -\infty = 1$.
- $\mathcal{A}_2^L$ verifies (pprel^L): since $\mathcal{A}^L$ verifies (pprel^L), it is sufficient to prove that

$$(x \to_2^L y) \vee (y \to_2^L x) = 1 = (x \rightsquigarrow_2^L y) \vee (y \rightsquigarrow_2^L x), \quad \textit{for } x, y \notin A^L.$$

Indeed, for $x = -\infty$ and $y \in A^L$, we have $(-\infty \to_2^L y) \vee (y \to_2^L -\infty) = 1 \vee -\infty = 1$ and
for $x, y = -\infty$, we have $(-\infty \to_2^L -\infty) \vee (-\infty \to_2^L -\infty) = 1 \vee 1 = 1$,
and similarly for $\rightsquigarrow_2^L$.
- $\mathcal{A}_2^L$ verifies (pdiv^L): since $\mathcal{A}^L$ verifies (pdiv^L), it is sufficient to prove that

$$x \wedge y = (x \to_2^L y) \odot_2 x = x \odot_2 (x \rightsquigarrow_2^L y), \quad \textit{for } x, y \notin A^L.$$

Indeed,
· for $x = -\infty$ and $y \in A^L$, we have $-\infty \wedge y = -\infty$, $(-\infty \to_2^L y) \odot_2 -\infty = 1 \odot_2 -\infty = -\infty$ and $-\infty \odot_2 (-\infty \rightsquigarrow_2^L y) = -\infty \odot_2 1 = -\infty$;
· for $x \in A^L$ and $y = -\infty$, we have $x \wedge -\infty = -\infty$, $(x \to_2^L -\infty) \odot_2 x = -\infty \odot_2 x = -\infty$ and $x \odot_2 (x \rightsquigarrow_2^L -\infty) = x \odot_2 -\infty = -\infty$;
· for $x, y = -\infty$, we have $x \wedge y = -\infty$, $(-\infty \to_2^L -\infty) \odot_2 -\infty = 1 \odot_2 -\infty = -\infty$ and $-\infty \odot_2 (-\infty \rightsquigarrow_2^L -\infty) = -\infty \odot_2 1 = -\infty$.
Thus, $\mathcal{A}_2^L$ is a pseudo-Hájek(pP) algebra.
-$\mathcal{A}_2^L$ verifies (pP1^L). Indeed,

$$x^- = x \to_2^L -\infty = \begin{cases} x \to_2^L -\infty, & \text{if } x \in A^L \\ -\infty \to_2^L -\infty, & \text{if } x = -\infty \end{cases} = \begin{cases} -\infty, & \text{if } x \in A^L \\ 1, & \text{if } x = -\infty. \end{cases}$$

Similarly,

$$x^\sim = x \rightsquigarrow_2^L -\infty = \begin{cases} x \rightsquigarrow_2^L -\infty, & \text{if } x \in A^L \\ -\infty \rightsquigarrow_2^L -\infty, & \text{if } x = -\infty \end{cases} = \begin{cases} -\infty, & \text{if } x \in A^L \\ 1, & \text{if } x = -\infty. \end{cases}$$

Note that $x^- = x^\sim$, for all $x \in A_2^L$.

Then, $$x \wedge x^- = x \wedge \begin{cases} -\infty, & \text{if } x \in A^L \\ 1, & \text{if } x = -\infty. \end{cases} = \begin{cases} x \wedge -\infty, & \text{if } x \in A^L \\ x \wedge 1, & \text{if } x = -\infty. \end{cases} = -\infty$$

and similarly $x \wedge x^\sim = -\infty$.
-$\mathcal{A}_2^L$ verifies (pP2). Denote first

$$F \overset{notation}{=} (x \odot_2 z) \to_2^L (y \odot_2 z) \quad \textit{and} \quad E \overset{notation}{=} (z^-)^- \odot_2 F.$$

We must prove that

$$E \leq x \to_2^L y.$$

There are eight cases:
1. $x \in A^L$, $y \in A^L$, $z \in A^L$,
2. $x \in A^L$, $y \in A^L$, $z = -\infty$,

3. $x \in A^L$, $y = -\infty$, $z \in A^L$,
4. $x \in A^L$, $y = -\infty$, $z = -\infty$,
5. $x = -\infty$, $y \in A^L$, $z \in A^L$,
6. $x = -\infty$, $y \in A^L$, $z = -\infty$,
7. $x = -\infty$, $y = -\infty$, $z \in A^L$,
8. $x = -\infty$, $y = -\infty$, $z = -\infty$.

Recall that $z^- = \begin{cases} -\infty, & \text{if} \quad z \in A^L \\ 1, & \text{if} \quad z = -\infty, \end{cases}$ hence

$(z^-)^- = \begin{cases} -\infty^-, & \text{if} \quad z \in A^L \\ 1^-, & \text{if} \quad z = -\infty \end{cases} = \begin{cases} 1, & \text{if} \quad z \in A^L \\ -\infty, & \text{if} \quad z = -\infty. \end{cases}$

We then obtain:
1: $(z^-)^- = 1$, $E = 1 \odot_2 F = F \overset{(*)}{=} x \to^L y = x \to_2^L y$.
2: $(z^-)^- = -\infty$, $E = -\infty \odot_2 F = -\infty < x \to_2^L y$.
3: $(z^-)^- = 1$, $E = 1 \odot_2 F = F = (x \odot z) \to_2^L (-\infty \odot_2 z) = (x \odot z) \to_2^L -\infty = -\infty = x \to_2^L -\infty$.
4: $(z^-)^- = -\infty$, $E = -\infty \odot_2 F = -\infty = x \to_2^L -\infty$.
5: $(z^-)^- = 1$, $E = 1 \odot_2 F = F = (-\infty \odot_2 z) \to_2^L (y \odot_2 z) = -\infty \to_2^L (y \odot_2 z) = 1 = -\infty \to_2^L y$.
6: $(z^-)^- = -\infty$, $E = -\infty \odot_2 F = -\infty < -\infty \to_2^L y = 1$.
7: $(z^-)^- = 1$, $E = 1 \odot_2 F = F = (-\infty \odot_2 z) \to_2^L (-\infty \odot_2 z) = -\infty \to_2^L -\infty = 1 = x \to_2^L y$.
8: $(z^-)^- = -\infty$, $E = -\infty \odot_2 F = -\infty < x \to_2^L y = 1$.
Similarly one can prove the second inequality.

(i'): a dual proof. □

Open problem 15.6.2 It remains an open problem if there are other examples than l-implicative-groups producing pseudo-BCK(pP) lattices verifying condition (pCC^L) and (p@^{L}) (pseudo-BCK(pS) lattices verifying condition (pCC^R) and (p@^{R}), respectively).

Let now $\mathcal{G}_{ig} = (G, \vee, \wedge, \to, \rightsquigarrow, 1)$ be an l-implicative-group (or let $\mathcal{G}_g = (G, \vee, \wedge, \cdot, ^{-1}, 1)$ be an l-group). Let $S \subseteq G$ be a deductive system of $\mathcal{G}_{ig}$ (or, equivalently, a convex l-subgroup of $\mathcal{G}_g$). Note that:

(1) if $S^- = S \cap G^-$, $-\infty \notin G$ and $G^-_{-\infty} = \{-\infty\} \cup G^-$ (see Corollary 14.1.7(i)(2)), then we have:

$S \cap G^-_{-\infty} = S \cap (\{-\infty\} \cup G^-) \overset{distributivity}{=} (S \cap \{-\infty\}) \cup (S \cap G^-) = \emptyset \cup S^- = S^-$.

(1') similarly, if $S^+ = S \cap G^+$, $-\infty \notin G$ and $G^+_{+\infty} = G^+ \cup \{+\infty\}$ (see Corollary 14.1.7(i')(2')), then we have:
$S \cap G^+_{+\infty} = S^+$.

Hence, we formulate the following complex result:

Theorem 15.6.3 *Let $\mathcal{G}_{ig} = (G, \vee, \wedge, \to, \rightsquigarrow, 1)$ be an l-implicative-group (or let $\mathcal{G}_g = (G, \vee, \wedge, \cdot, ^{-1}, 1)$ be an l-group). Let S be a compatible deductive system of $\mathcal{G}_{ig}$ (or, equivalently, a normal convex l-subgroup of $\mathcal{G}_g$).*
Then, we have:

(1) $S \cap G^-_{-\infty} = S^-$ *is a compatible* $(\rightarrow^L_2, \rightsquigarrow^L_2)$*-deductive system of the pHájek(pP) algebra (with the pseudo-product* $\odot_2$*) verifying the properties* (pP1^L), (pP2^L) *(see Corollary 14.1.7(i)(2))*

$$\mathcal{G}^L_2 = (G^-_{-\infty}, \wedge, \vee, \rightarrow^L_2, \rightsquigarrow^L_2, \mathbf{0} = -\infty, \mathbf{1} = 1)$$

(or, equivalently, S^- *is a normal filter of the left-pproduct algebra (see Corollary 14.2.3(i)(2))*

$$\mathcal{G}^L_{m2'} = (G^-_{-\infty}, \wedge, \vee, \odot_2, \rightarrow^L_2, \rightsquigarrow^L_2, \mathbf{0} = -\infty, \mathbf{1} = 1)),$$

and S^L *is compatible if and only if is normal, i.e.* $(Co^L) \Leftrightarrow (No^L)$.

(1') $S \cap G^+_{+\infty} = S^+$ *is a compatible* $(\rightarrow^R_2, \rightsquigarrow^R_2)$*-deductive system of the pHájek(pS) algebra (with the pseudo-sum* $\oplus_2$*) verifying the properties* (pP1^R), (pP2^R) *(see Corollary 14.1.7(i')(2'))*

$$\mathcal{G}^R_2 = (G^+_{+\infty}, \vee, \wedge, \rightarrow^R_2, \rightsquigarrow^R_2, \mathbf{0} = 1, \mathbf{1} = +\infty)$$

(or, equivalently, S^R *is a normal filter of the right-pproduct algebra (see Corollary 14.2.3(i')(2'))*

$$\mathcal{G}^R_{m2'} = (G^+_{+\infty}, \vee, \wedge, \oplus_2, \rightarrow^R_2, \rightsquigarrow^R_2, \mathbf{0} = 1, \mathbf{1} = +\infty)),$$

and S^R *is compatible if and only if is normal, i.e.* $(Co^R) \Leftrightarrow (No^R)$.

Proof.

(1): $S^- \subseteq G^-$ is a compatible $(\rightarrow^L, \rightsquigarrow^L)$-deductive system of $\mathcal{G}^-$, by Proposition 15.4.2 (1). It follows obviously that S^- is a compatible $(\rightarrow^L_2, \rightsquigarrow^L_2)$-deductive system of $\mathcal{G}^L_2$. S^- is a normal filter of $\mathcal{G}^L_m$, by Proposition 15.3.6 (1). It follows obviously that S^- is a normal filter of $\mathcal{G}^L_{m2'}$. $(Co^L) \Leftrightarrow (No^L)$, by Theorem 15.4.3 (1), since $\mathcal{G}^L_2$ ($\mathcal{G}^L_{m2'}$) verifies the property (pdiv^L).

(1'): a dual proof. □

In other words, the above Theorem 15.6.3 says that normality (compatibility) at l-group (l-implicative-group) G level is inherited by the algebras obtained by restricting the l-group (l-implicative-group) operations to $G^-_{-\infty}$ and to $G^+_{+\infty}$. Also, it says that the equivalence $(Co_{ig}) \Leftrightarrow (No_g)$ (*compatible* if and only if *normal*), existing at l-group (l-implicative-group) level (Theorem 15.3.4), is preserved by the algebras obtained by restricting the l-group (l-implicative-group) operations to $G^-_{-\infty}$ and to $G^+_{+\infty}$.

Finally, we have the following general complex result:

Corollary 15.6.4

(1) Let $\mathcal{A}^L = (A^L, \wedge, \vee, \rightarrow^L, \rightsquigarrow^L, 0, 1)$ *be a pHájek(pP) algebra (with the pseudo-product* $\odot$*) verifying* (pP1^L), (pP2^L)
(or let $\mathcal{A}^L_{m'} = (A^L, \wedge, \vee, \odot, \rightarrow^L, \rightsquigarrow^L, 0, 1)$ *be a left-pproduct algebra).*
Let S^L *be a* $(\rightarrow^L, \rightsquigarrow^L)$*-deductive system of* $\mathcal{A}^L$ *(or, equivalently, a filter of* $\mathcal{A}^L_{m'}$*).*

Then, S^L *is compatible if and only if is normal, i.e.* $(Co^L) \Leftrightarrow (No^L)$.

(1') Let $\mathcal{A}^R = (A^R, \vee, \wedge, \rightarrow^R, \rightsquigarrow^R, 0, 1)$ be a pHájek(pS) algebra (with the pseudo-sum $\oplus$) verifying (pP1^R), (pP2^R)
(or let $\mathcal{A}^R_{m'} = (A^R, \vee, \wedge, \oplus, \rightarrow^R, \rightsquigarrow^R, 0, 1)$ be a right-pproduct algebra).
Let S^R be a $(\rightarrow^R, \rightsquigarrow^R)$-deductive system of $\mathcal{A}^R$ (or, equivalently, an ideal of $\mathcal{A}^R_{m'}$).

Then, S^R is compatible if and only if is normal, i.e. $(Co^R) \Leftrightarrow (No^R)$.

Proof.

(1): It follows by Theorem 15.4.3, because both $\mathcal{A}^L$ and $\mathcal{A}^L_{m'}$ satisfy the property (pdiv^L).

(1'): dually. □

Chapter 16

Representability

In this chapter, we study the representability [98], [99]. First, we present equivalent conditions for an l-implicative-group to be representable. Then, we prove that representability at l-implicative-group G level is inherited by the algebras obtained by restricting the operations from G to G^- and to G^+.

The chapter has two sections.

16.1 Representable l-groups, l-implicative-groups

Recall (see [6], for example) that an l-group is *representable*, if it is a subdirect product of totally-ordered groups. Recall also the following theorem that provides characterizations of representable l-groups, some of them needed in the sequel.

Theorem 16.1.1 *(see [6], Theorem 4.1.1)*

The following are equivalent for an l-group $\mathcal{G} = (G, \vee, \wedge, \cdot, ^{-1}, 1)$*:*
(a) $\mathcal{G}$ *is representable.*
(b) For all $a, b \in G$, $(a \wedge b)^2 = a^2 \wedge b^2$*;*
(b^d*) For all* $a, b \in G$, $(a \vee b)^2 = a^2 \vee b^2$.
(c) For all $a, b \in G$, $a \wedge (b^{-1} \cdot a^{-1} \cdot b) \leq 1$*;*
(c^d*) For all* $a, b \in G$, $a \vee (b^{-1} \cdot a^{-1} \cdot b) \geq 1$.
(d) Each polar subgroup is normal.
(e) Each minimal prime subgroup is normal.
(f) For each $a \in G$, $a > 1$, $a \wedge (b^{-1} \cdot a \cdot b) > 1$*, for all* $b \in G$*;*
(f^d*) For each* $a \in G$, $a < 1$, $a \vee (b^{-1} \cdot a \cdot b) < 1$*, for all* $b \in G$.
Note that ^{d} *means "dual" and* a^2 *means* $a \cdot a$.

Remarks 16.1.2 Note that, in a commutative l-group $\mathcal{G}$, we have: for all $a, b \in G$:

$$(a \wedge b)^2 = a^2 \wedge b^2 \Leftrightarrow (b \to a) \wedge (a \to b) \leq 1.$$

$$(a \vee b)^2 = a^2 \vee b^2 \Leftrightarrow (b \to a) \vee (a \to b) \geq 1.$$

Indeed, for example:
$(a\vee b)^2 = a^2\vee b^2 \Leftrightarrow (a\vee b)\cdot(a\vee b) = a^2\vee b^2 \Leftrightarrow a^2\vee b^2 = [a\cdot(a\vee b)]\vee[b\cdot(a\vee b)] \Leftrightarrow$
$a^2\vee b^2 = a^2\vee(a\cdot b)\vee(b\cdot a)\vee b^2 \Leftrightarrow a^2\vee b^2 = a^2\vee b^2\vee(a\cdot b) \Leftrightarrow a^2\vee b^2 \geq a\cdot b \Leftrightarrow$
$(a^2\vee b^2)\cdot b^{-1} \geq a \Leftrightarrow (a^2\cdot b^{-1})\vee b \geq a \Leftrightarrow [(a^2\cdot b^{-1})\vee b]\cdot a^{-1} \geq 1 \Leftrightarrow$
$(a\cdot b^{-1})\vee(b\cdot a^{-1}) \geq 1 \Leftrightarrow (b\to a)\vee(a\to b) \geq 1.$

We obtain, in the non-commutative case, the following results.

Proposition 16.1.3 *Let $\mathcal{G} = (G,\vee,\wedge,\cdot,^{-1},1)$ be an l-group. Then*

$$(b) \Leftrightarrow (b1) \Leftrightarrow (b2), \qquad (b^d) \Leftrightarrow (b1^d) \Leftrightarrow (b2^d),$$

where:
(b1) *for all* $a,b\in G$, $(b\to a)\wedge(a\rightsquigarrow b) \leq 1\wedge[(b\rightsquigarrow a)\rightsquigarrow(b\to a)]$,
(b2) *for all* $a,b\in G$, $(b\rightsquigarrow a)\wedge(a\to b) \leq 1\wedge[(b\to a)\to(b\rightsquigarrow a)]$;
(b1^d) *for all* $a,b\in G$, $(b\to a)\vee(a\rightsquigarrow b) \geq 1\vee[(b\rightsquigarrow a)\rightsquigarrow(b\to a)]$,
(b2^d) *for all* $a,b\in G$, $(b\rightsquigarrow a)\vee(a\to b) \geq 1\vee[(b\to a)\to(b\rightsquigarrow a)]$.

Proof.
(b^d) $\Leftrightarrow$ (b1^d):
$(a\vee b)^2 = a^2\vee b^2 \Leftrightarrow$
$(a\vee b)\cdot(a\vee b) = a^2\vee b^2 \Leftrightarrow$
$[a\cdot(a\vee b)]\vee[b\cdot(a\vee b)] = a^2\vee b^2 \Leftrightarrow$
$a^2\vee(a\cdot b)\vee(b\cdot a)\vee b^2 = a^2\vee b^2 \Leftrightarrow$
$a^2\vee b^2\vee(a\cdot b)\vee(b\cdot a) = a^2\vee b^2 \Leftrightarrow$
$a^2\vee b^2 \geq (a\cdot b)\vee(b\cdot a) \Leftrightarrow$
$(a^2\vee b^2)\cdot b^{-1} \geq [(a\cdot b)\vee(b\cdot a)]\cdot b^{-1} \Leftrightarrow$
$(a^2\cdot b^{-1})\vee b \geq a\vee(b\cdot a\cdot b^{-1}) \Leftrightarrow$
$a^{-1}\cdot[(a^2\cdot b^{-1})\vee b] \geq a^{-1}\cdot[a\vee(b\cdot a\cdot b^{-1})] \Leftrightarrow$
$(a\cdot b^{-1})\vee(a^{-1}\cdot b) \geq 1\vee(a^{-1}\cdot b\cdot a\cdot b^{-1}) \Leftrightarrow$
$(b\to a)\vee(a\rightsquigarrow b) \geq a^{-1}\cdot b\cdot[(b^{-1}\cdot a)\vee(a\cdot b^{-1})] = (b^{-1}\cdot a)^{-1}\cdot[(b\rightsquigarrow a)\vee(b\to a)] \Leftrightarrow$
$(b\to a)\vee(a\rightsquigarrow b) \geq (b\rightsquigarrow a)\rightsquigarrow[(b\rightsquigarrow a)\vee(b\to a)] \overset{(7.14)}{=} 1\vee[(b\rightsquigarrow a)\rightsquigarrow(b\to a)].$
(b^d) $\Leftrightarrow$ (b2^d):
$(a\vee b)^2 = a^2\vee b^2 \Leftrightarrow \ldots \Leftrightarrow$
$a^2\vee b^2 \geq (b\cdot a)\vee(a\cdot b) \Leftrightarrow$
$[a\vee(b^2\cdot a^{-1})]\cdot a \geq [b\vee(a\cdot b\cdot a^{-1})]\cdot a \Leftrightarrow$
$a\vee(b^2\cdot a^{-1}) \geq b\vee(a\cdot b\cdot a^{-1}) \Leftrightarrow$
$b\cdot[(b^{-1}\cdot a)\vee(b\cdot a^{-1})] \geq b\cdot[1\vee(b^{-1}\cdot a\cdot b\cdot a^{-1})] \Leftrightarrow$
$(b^{-1}\cdot a)\vee(b\cdot a^{-1}) \geq 1\vee(b^{-1}\cdot a\cdot b\cdot a^{-1}) \Leftrightarrow$
$(b\rightsquigarrow a)\vee(a\to b) \geq [(a\cdot b^{-1})\vee(b^{-1}\cdot a)]\cdot b\cdot a^{-1} \Leftrightarrow$
$(b\rightsquigarrow a)\vee(a\to b) \geq [(a\cdot b^{-1})\vee(b^{-1}\cdot a)]\cdot(a\cdot b^{-1})^{-1} \Leftrightarrow$
$(b\rightsquigarrow a)\vee(a\to b) \geq (b\to a)\to[(b\to a)\vee(b\rightsquigarrow a)] = 1\vee[(b\to a)\to(b\rightsquigarrow a)].$
The rest of the proof is similar. □

Remarks 16.1.4 (see Remarks 16.1.2)

Note that

$$(b1) \Longrightarrow (b1"), \quad (b2) \Longrightarrow (b2"); \qquad (b1^d) \Longrightarrow (b1^{d}"), \quad (b2^d) \Longrightarrow (b2^{d}"),$$

where:
(b1") for all $a, b \in G$, $(b \to a) \wedge (a \rightsquigarrow b) \leq 1$,
(b2") for all $a, b \in G$, $(b \rightsquigarrow a) \wedge (a \to b) \leq 1$;
($b1^{d}"$) for all $a, b \in G$, $(b \to a) \vee (a \rightsquigarrow b) \geq 1$,
($b2^{d}"$) for all $a, b \in G$, $(b \rightsquigarrow a) \vee (a \to b) \geq 1$.

Note that the converse implications are not true.

Note also that (b1") and (b2") coincide and that ($b1^{d}"$) and ($b2^{d}"$) coincide.

Proposition 16.1.5 *Let $\mathcal{G} = (G, \vee, \wedge, \cdot, ^{-1}, 1)$ be an l-group. Then*

$$(c) \Leftrightarrow (c1) \Leftrightarrow (c2), \qquad (c^d) \Leftrightarrow (c1^d) \Leftrightarrow (c2^d),$$

where:
(c1) for all $x, y, z, w \in G$, $(x \rightsquigarrow y) \wedge (([((y \rightsquigarrow x) \rightsquigarrow z) \rightsquigarrow z] \to w) \to w) \leq 1$,
(c2) for all $x, y, z, w \in G$, $(x \to y) \wedge (([((y \to x) \to z) \to z] \rightsquigarrow w) \rightsquigarrow w) \leq 1$;
($c1^d$) for all $x, y, z, w \in G$, $(x \rightsquigarrow y) \vee (([((y \rightsquigarrow x) \rightsquigarrow z) \rightsquigarrow z] \to w) \to w) \geq 1$,
($c2^d$) for all $x, y, z, w \in G$, $(x \to y) \vee (([((y \to x) \to z) \to z] \rightsquigarrow w) \rightsquigarrow w) \geq 1$.

Proof.

$(c^d) \Longrightarrow (c1^d)$:
$(x \rightsquigarrow y) \vee (([((y \rightsquigarrow x) \rightsquigarrow z) \rightsquigarrow z] \to w) \to w) =$
$(x^{-1} \cdot y) \vee (([((y^{-1} \cdot x)^{-1} \cdot z)^{-1} \cdot z] \to w) \to w) =$
$(x^{-1} \cdot y) \vee (([(x^{-1} \cdot y \cdot z)^{-1} \cdot z] \to w) \to w) =$
$(x^{-1} \cdot y) \vee (([z^{-1} \cdot y^{-1} \cdot x \cdot z] \to w) \to w) =$
$(x^{-1} \cdot y) \vee ((w \cdot [z^{-1} \cdot y^{-1} \cdot x \cdot z]^{-1}) \to w) =$
$(x^{-1} \cdot y) \vee ((w \cdot z^{-1} \cdot x^{-1} \cdot y \cdot z) \to w) =$
$(x^{-1} \cdot y) \vee (w \cdot (w \cdot z^{-1} \cdot x^{-1} \cdot y \cdot z)^{-1}) =$
$(x^{-1} \cdot y) \vee (w \cdot z^{-1} \cdot y^{-1} \cdot x \cdot z \cdot w^{-1}) =$
$(x^{-1} \cdot y) \vee ((w \cdot z^{-1}) \cdot (x^{-1} \cdot y)^{-1} \cdot (z \cdot w^{-1})) =$
$a \vee (b^{-1} \cdot a^{-1} \cdot b) \geq 1$, by ($c^d$).

$(c1^d) \Longrightarrow (c^d)$:
Take $x = 1$, $y = a$, $z = 1$, $w = b^{-1}$ in ($c1^d$); we obtain:
$(1 \rightsquigarrow a) \vee (([((a \rightsquigarrow 1) \rightsquigarrow 1) \rightsquigarrow 1] \to b^{-1}) \to b^{-1}) \geq 1 \Leftrightarrow$
$a \vee ((a^{-1} \to b^{-1}) \to b^{-1}) \geq 1 \Leftrightarrow$
$a \vee ((b^{-1} \cdot (a^{-1})^{-1}) \to b^{-1}) \geq 1 \Leftrightarrow$
$a \vee ((b^{-1} \cdot a) \to b^{-1}) \geq 1 \Leftrightarrow$
$a \vee (b^{-1} \cdot (b^{-1} \cdot a)^{-1}) \geq 1 \Leftrightarrow$
$a \vee (b^{-1} \cdot a^{-1} \cdot b) \geq 1$. Thus $(c^d) \Leftrightarrow (c1^d)$.

$(c^d) \Longrightarrow (c2^d)$:
$(x \to y) \vee (([((y \to x) \to z) \to z] \rightsquigarrow w) \rightsquigarrow w) =$
$(y \cdot x^{-1}) \vee (([z \cdot (z \cdot (x \cdot y^{-1})^{-1})^{-1}] \rightsquigarrow w) \rightsquigarrow w) =$
$(y \cdot x^{-1}) \vee (([z \cdot (z \cdot y \cdot x^{-1})^{-1}] \rightsquigarrow w) \rightsquigarrow w) =$
$(y \cdot x^{-1}) \vee (([z \cdot x \cdot y^{-1} \cdot z^{-1}] \rightsquigarrow w) \rightsquigarrow w) =$

$(y \cdot x^{-1}) \vee (([z \cdot x \cdot y^{-1} \cdot z^{-1}]^{-1} \cdot w) \rightsquigarrow w) =$
$(y \cdot x^{-1}) \vee ((z \cdot y \cdot x^{-1} \cdot z^{-1} \cdot w) \rightsquigarrow w) =$
$(y \cdot x^{-1}) \vee ((z \cdot y \cdot x^{-1} \cdot z^{-1} \cdot w)^{-1} \cdot w) =$
$(y \cdot x^{-1}) \vee (w^{-1} \cdot z \cdot x \cdot y^{-1} \cdot z^{-1} \cdot w) =$
$a \vee (b^{-1} \cdot a^{-1} \cdot b) \geq 1$, by ($c^d$).

($c2^d$) $\Longrightarrow$ (c^d):
Take $x = 1$, $y = a$, $z = 1$, $w = b$ in ($c2^d$); we obtain:
$(1 \to a) \vee (([((a \to 1) \to 1) \to 1] \rightsquigarrow b) \rightsquigarrow b) \geq 1 \Leftrightarrow$
$a \vee ((a^{-1} \rightsquigarrow b) \rightsquigarrow b) \geq 1 \Leftrightarrow$
$a \vee ((a \cdot b) \rightsquigarrow b) \geq 1 \Leftrightarrow$
$a \vee (b^{-1} \cdot a^{-1} \cdot b) \geq 1$. Thus ($c^d$) $\Leftrightarrow$ ($c2^d$).

The rest of the proof is similar. □

We shall say that an l-implicative-group is *representable*, if it is a subdirect product of totally-ordered implicative-groups. Consequently, an l-implicative-group is representable if and only if its term equivalent l-group is representable. Then, we have the following result, needed in the sequel.

Theorem 16.1.6
The following are equivalent for an l-implicative-group $\mathcal{G} = (G, \vee, \wedge, \to, \rightsquigarrow, 1)$:
(a) $\mathcal{G}$ is representable, (b1), (b2), ($b1^d$), ($b2^d$), (c1), (c2), ($c1^d$), ($c2^d$).

Proof. By Theorem 16.1.1 and Propositions 16.1.3, 16.1.5. □

We can put together Theorems 16.1.1 and 16.1.6 in the following resuming theorem.

Theorem 16.1.7 *Let $\mathcal{G}_g = (G, \vee, \wedge, \cdot, ^{-1}, 1)$ be an l-group or, equivalently, by (Eq3), let $\mathcal{G}_{ig} = (G, \vee, \wedge, \to, \rightsquigarrow, 1)$ be an l-implicative-group. The following are equivalent:*
(a) $\mathcal{G}$ is representable.

(b) For all $a, b \in G$, $(a \wedge b)^2 = a^2 \wedge b^2$,
(b1) For all $a, b \in G$, $(b \to a) \wedge (a \rightsquigarrow b) \leq 1 \wedge [(b \rightsquigarrow a) \rightsquigarrow (b \to a)]$,
(b2) For all $a, b \in G$, $(b \rightsquigarrow a) \wedge (a \to b) \leq 1 \wedge [(b \to a) \to (b \rightsquigarrow a)]$.

(b^d) For all $a, b \in G$, $(a \vee b)^2 = a^2 \vee b^2$,
($b1^d$) For all $a, b \in G$, $(b \to a) \vee (a \rightsquigarrow b) \geq 1 \vee [(b \rightsquigarrow a) \rightsquigarrow (b \to a)]$,
($b2^d$) For all $a, b \in G$, $(b \rightsquigarrow a) \vee (a \to b) \geq 1 \vee [(b \to a) \to (b \rightsquigarrow a)]$.

(c) For all $a, b \in G$, $a \wedge (b^{-1} \cdot a^{-1} \cdot b) \leq 1$,
(c1) For all $x, y, z, w \in G$, $(x \rightsquigarrow y) \wedge (([((y \rightsquigarrow x) \rightsquigarrow z) \rightsquigarrow z] \to w) \to w) \leq 1$,
(c2) For all $x, y, z, w \in G$, $(x \to y) \wedge (([((y \to x) \to z) \to z] \rightsquigarrow w) \rightsquigarrow w) \leq 1$.

(c^d) For all $a, b \in G$, $a \vee (b^{-1} \cdot a^{-1} \cdot b) \geq 1$,
($c1^d$) For all $x, y, z, w \in G$, $(x \rightsquigarrow y) \vee (([((y \rightsquigarrow x) \rightsquigarrow z) \rightsquigarrow z] \to w) \to w) \geq 1$,
($c2^d$) For all $x, y, z, w \in G$, $(x \to y) \vee (([((y \to x) \to z) \to z] \rightsquigarrow w) \rightsquigarrow w) \geq 1$.

(d) Each polar subgroup is normal.
(e) Each minimal prime subgroup is normal.
(f) For each $a \in G$, $a > 1$, $a \wedge (b^{-1} \cdot a \cdot b) > 1$, for all $b \in G$;
(f^d) For each $a \in G$, $a < 1$, $a \vee (b^{-1} \cdot a \cdot b) < 1$, for all $b \in G$.

16.2 Connections at G^-, G^+ level

In this section, we present some connections between the representability at l-implicative-group G level and the representability at G^-, G^+ level.

• **In the commutative case**, recall that we have the following definitions.

Definition 16.2.1
(i) An integral left-residuated lattice $\mathcal{A}^L = (A^L, \wedge, \vee, \odot, \to^L, 1)$ or, equivalently, by (iE2), a BCK(P) lattice $\mathcal{A}^L = (A^L, \wedge, \vee, \to^L, 1)$ with product $\odot$:
(P) there exist $x \odot y \overset{notation}{=} \min\{z \mid x \leq y \to^L z\}$, for all $x, y \in A^L$,
is *representable*, if it is a subdirect product of linearly-ordered ones. It is known that representable such algebras are characterized by the prelinearity condition:

$$(prel^L) \qquad (x \to^L y) \vee (y \to^L x) = 1.$$

(i') Dually, an integral right-residuated lattice $\mathcal{A}^R = (A^R, \vee, \wedge, \oplus, \to^R, 0)$ or, equivalently, by (iE2), a BCK(S) lattice $\mathcal{A}^R = (A^R, \vee, \wedge, \to^R, 0)$ with sum $\oplus$:
(S) there exist $x \oplus y \overset{notation}{=} \max\{z \mid x \geq y \to^R z\}$, for all $x, y \in A^R$,
is *representable*, if it is a subdirect product of linearly-ordered ones; representable such algebras are characterized by the dual prelinearity condition:

$$(prel^R) \qquad (x \to^R y) \wedge (y \to^R x) = 0.$$

Then, we have the following result:

Theorem 16.2.2 *Let $\mathcal{G} = (G, \vee, \wedge, \to, 1)$ be a representable commutative l-implicative-group.*
(1) Define, for all $x, y \in G^-$:

$$x \to^L y \overset{def.}{=} (x \to y) \wedge 1. \tag{16.1}$$

Then, $\mathcal{G}^L = (G^-, \wedge, \vee, \to^L, \mathbf{1} = 1)$ is a representable BCK(P) lattice.
(1') Define, for all $x, y \in G^+$:

$$x \to^R y \overset{def.}{=} (x \to y) \vee 1. \tag{16.2}$$

Then, $\mathcal{G}^R = (G^{\cdot}, \vee, \wedge, \to^R, \mathbf{0} = 1)$ is a representable BCK(S) lattice.

Proof.
(1): By Theorem 14.1.1, $\mathcal{G}^L$ is a BCK(P) lattice. To prove that it is representable, we must prove that (prel^L) holds. Indeed, $(x \to^L y) \vee (y \to^L x) = [(x \to$

$y) \wedge 1] \vee [(y \to x) \wedge 1] = [(x \to y) \vee (y \to x)] \wedge 1 = 1$, by Theorem 16.1.1 and Remarks 16.1.2.

(1'): By Theorem 14.1.1, $\mathcal{G}^R$ is a BCK(S) lattice. To prove that it is representable, we must prove that (prel^R) holds. Indeed, $(x \to^R y) \wedge (y \to^R x) = [(x \to y) \vee 1] \wedge [(y \to x) \vee 1] = [(x \to y) \wedge (y \to x)] \vee 1 = 1$, by Theorem 16.1.1 and Remarks 16.1.2. □

• **In the non-commutative case**, recall that:

(i) a non-commutative integral left-residuated lattice:

$$\mathcal{A}^{\mathcal{L}} = (A^L, \wedge, \vee, \odot, \to^L, \rightsquigarrow^L, 1)$$

or, equivalently, a pBCK(pP) lattice (with the pseudo-product $\odot$):

$$\mathcal{A}^L = (A^L, \wedge, \vee, \to^L, \rightsquigarrow^L, 1)$$

is *representable*, if it is a subdirect product of linearly-ordered ones.
C.J. van Alten [160] proved that such non-commutative algebras are representable if and only if they satisfy the identity:

$$(16.3) \qquad (x \rightsquigarrow^L y) \vee (([((y \rightsquigarrow^L x) \rightsquigarrow^L z) \rightsquigarrow^L z] \to^L w) \to^L w) = 1,$$

or the identity

$$(16.4) \qquad (x \to^L y) \vee (([((y \to^L x) \to^L z) \to^L z] \rightsquigarrow^L w) \rightsquigarrow^L w) = 1.$$

(i') Dually, a non-commutative integral right-residuated lattice:

$$\mathcal{A}^R = (A^R, \vee, \wedge, \oplus, \to^R, \rightsquigarrow^R, 0)$$

or, equivalently, a pBCK(pS) lattice (with the pseudo-sum $\oplus$):

$$\mathcal{A}^R = (A^R, \vee, \wedge, \to^R, \rightsquigarrow^R, 0)$$

is *representable*, if it is a subdirect product of linearly-ordered ones.
Representable such algebras are characterized then by the dual condition:

$$(16.5) \qquad (x \rightsquigarrow^R y) \wedge (([((y \rightsquigarrow^R x) \rightsquigarrow^R z) \rightsquigarrow^R z] \to^R w) \to^R w) = 0,$$

or

$$(16.6) \qquad (x \to^R y) \wedge (([((y \to^R x) \to^R z) \to^R z] \rightsquigarrow^R w) \rightsquigarrow^R w) = 0.$$

We shall prove the following result:

Theorem 16.2.3 *(see Theorem 14.1.1)*

Let $\mathcal{G} = (G, \vee, \wedge, \to, \rightsquigarrow, 1)$ be a representable l-implicative-group. Then,

(1) $\mathcal{G}^L = (G^-, \wedge, \vee, \to^L, \rightsquigarrow^L, \mathbf{1} = 1)$ is a representable pseudo-BCK(pP) lattice (with the pseudo-product $\odot = \cdot$).

(1') $\mathcal{G}^R = (G^{\cdot}, \vee, \wedge, \to^R, \rightsquigarrow^R, \mathbf{0} = 1)$ is a representable pseudo-BCK(pS) lattice (with the pseudo-sum $\oplus = \cdot$).

Proof.

(1): By Theorem 14.1.1, $\mathcal{G}^L$ is a pseudo-BCK(pP) lattice. To prove that $\mathcal{G}^L$ is representable, we must prove that condition (16.3), for example, holds. First denote:

$$A \overset{not.}{=} ((y \rightsquigarrow^L x) \rightsquigarrow^L z) \rightsquigarrow^L z, \quad B \overset{not.}{=} (A \to^L w) \to^L w, \quad C \overset{not.}{=} (x \rightsquigarrow^L y) \vee B.$$

We must prove, by (16.3), that $C = \mathbf{1}$. Indeed,

• **First proof**:

$A = ((y \rightsquigarrow^L x) \rightsquigarrow^L z) \rightsquigarrow^L z = ([(y^{-1} \cdot x) \wedge 1] \rightsquigarrow^L z) \rightsquigarrow^L z =$
$[([(y^{-1} \cdot x) \wedge 1]^{-1} \cdot z) \wedge 1] \rightsquigarrow^L z = [([(x^{-1} \cdot y) \vee 1] \cdot z) \wedge 1] \rightsquigarrow^L z =$
$[[(x^{-1} \cdot y \cdot z) \vee z] \wedge 1] \rightsquigarrow^L z = ([[(x^{-1} \cdot y \cdot z) \vee z] \wedge 1]^{-1} \cdot z) \wedge 1 =$
$([((x^{-1} \cdot y \cdot z) \vee z)^{-1} \vee 1] \cdot z) \wedge 1 = ([[(z^{-1} \cdot y^{-1} \cdot x) \wedge z^{-1}] \vee 1] \cdot z) \wedge 1 =$
$(([(z^{-1} \cdot y^{-1} \cdot x) \wedge z^{-1}] \cdot z) \vee z) \wedge 1 = (((z^{-1} \cdot y^{-1} \cdot x \cdot z) \wedge 1) \vee z) \wedge 1 =$
$[(z^{-1} \cdot y^{-1} \cdot x \cdot z) \wedge 1] \vee z = [(z^{-1} \cdot y^{-1} \cdot x \cdot z) \vee z] \wedge 1.$

$B = (A \to^L w) \to^L w = [(w \cdot A^{-1}) \wedge 1] \to^L w =$
$(w \cdot [(w \cdot A^{-1}) \wedge 1]^{-1}) \wedge 1 = (w \cdot [(A \cdot w^{-1}) \vee 1]) \wedge 1 =$
$((w \cdot A \cdot w^{-1}) \vee w) \wedge 1 =$
$[(w \cdot ([(z^{-1} \cdot y^{-1} \cdot x \cdot z) \vee z] \wedge 1) \cdot w^{-1}) \vee w] \wedge 1 =$
$[([(w \cdot [(z^{-1} \cdot y^{-1} \cdot x \cdot z) \vee z]) \wedge w] \cdot w^{-1}) \vee w] \wedge 1 =$
$[([[(w \cdot z^{-1} \cdot y^{-1} \cdot x \cdot z) \vee (w \cdot z)] \wedge w] \cdot w^{-1}) \vee w] \wedge 1 =$
$[([(w \cdot z^{-1} \cdot y^{-1} \cdot x \cdot z \cdot w^{-1}) \vee (w \cdot z \cdot w^{-1})] \wedge 1) \vee w] \wedge 1 =$
$[[(w \cdot z^{-1} \cdot y^{-1} \cdot x \cdot z \cdot w^{-1}) \wedge 1] \vee [(w \cdot z \cdot w^{-1}) \wedge 1] \vee w] \wedge 1 =$
$[(w \cdot z^{-1} \cdot y^{-1} \cdot x \cdot z \cdot w^{-1}) \wedge 1] \vee [(w \cdot z \cdot w^{-1}) \wedge 1] \vee w \geq$
$(w \cdot z^{-1} \cdot y^{-1} \cdot x \cdot z \cdot w^{-1}) \wedge 1.$

$C = (x \rightsquigarrow^L y) \vee B \geq [(x^{-1} \cdot y) \wedge 1] \vee [(w \cdot z^{-1} \cdot y^{-1} \cdot x \cdot z \cdot w^{-1}) \wedge 1] =$
$[(x^{-1} \cdot y) \vee (w \cdot z^{-1} \cdot y^{-1} \cdot x \cdot z \cdot w^{-1})] \wedge 1 = [a \vee (b^{-1} \cdot a^{-1} \cdot b)] \wedge 1,$
with $a = x^{-1} \cdot y$, $b = z \cdot w^{-1}$.
But $\mathcal{G}$ is representable, hence by Theorem 16.1.1 (c^d), for all $a, b \in G$, $a \vee (b^{-1} \cdot a^{-1} \cdot b) \geq 1$. Hence, $C \geq 1$ and thus $C = 1$, i.e. $C = \mathbf{1}$.

• **Second proof**: Denote

$$D \overset{notation}{=} ((y \rightsquigarrow x) \rightsquigarrow z) \rightsquigarrow z, \qquad E \overset{notation}{=} (D \to w) \to w.$$

By Theorem 16.1.6 ($c1^d$), we have

$$(x \rightsquigarrow y) \vee E \geq 1. \tag{16.7}$$

Then, $A = ((y \rightsquigarrow^L x) \rightsquigarrow^L z) \rightsquigarrow^L z = [([(y \rightsquigarrow x) \wedge 1] \rightsquigarrow z) \wedge 1] \rightsquigarrow^L z \overset{(7.14)}{=}$
$[(((y \rightsquigarrow x) \rightsquigarrow z) \vee (1 \rightsquigarrow z)) \wedge 1] \rightsquigarrow^L z = [(((y \rightsquigarrow x) \rightsquigarrow z) \vee z) \wedge 1] \rightsquigarrow^L z \overset{distrib.}{=}$
$[[((y \rightsquigarrow x) \rightsquigarrow z) \wedge 1] \vee (z \wedge 1)] \rightsquigarrow^L z = ([[((y \rightsquigarrow x) \rightsquigarrow z) \wedge 1] \vee z] \rightsquigarrow z) \wedge 1 \overset{(7.15)}{=}$
$([((y \rightsquigarrow x) \rightsquigarrow z) \wedge 1] \rightsquigarrow z) \wedge (z \rightsquigarrow z) \wedge 1 \overset{(7.14)}{=} ([((y \rightsquigarrow x) \rightsquigarrow z) \rightsquigarrow z] \vee (1 \rightsquigarrow z)) \wedge 1 =$
$(D \vee z) \wedge 1.$

$B = (A \to^L w) \to^L w = ([(D \vee z) \wedge 1] \to w) \wedge 1] \to^L w \overset{(7.14)}{=}$
$[[((D \vee z) \to w) \vee (1 \to w)] \wedge 1] \to^L w = ([[((D \vee z) \to w) \vee w] \wedge 1] \to w) \wedge 1 \overset{(7.14)}{=}$

$(([((D \vee z) \to w) \vee w] \to w) \vee (1 \to w)) \wedge 1 \overset{(7.15)}{=}$
$([(((D \vee z) \to w) \to w) \wedge (w \to w)] \vee w) \wedge 1 \overset{distrib.}{=}$
$[[(((D \vee z) \to w) \to w) \vee w] \wedge (1 \vee w)] \wedge 1 = [(((D \vee z) \to w) \to w) \vee w] \wedge 1 \overset{(7.15)}{=}$
$[([(D \to w) \wedge (z \to w)] \to w) \vee w] \wedge 1 \overset{(7.14)}{=}$
$[[((D \to w) \to w) \vee ((z \to w) \to w))] \vee w] \wedge 1 = [E \vee ((z \to w) \to w) \vee w] \wedge 1.$

$C = (x \rightsquigarrow^L y) \vee B = [(x \rightsquigarrow y) \wedge 1] \vee [(E \vee ((z \to w) \to w) \vee w) \wedge 1] \overset{distrib.}{=}$
$[(x \rightsquigarrow y) \vee E \vee ((z \to w) \to w) \vee w] \wedge 1 = 1,$
since $(x \rightsquigarrow y) \vee E \vee ((z \to w) \to w) \vee w \geq (x \rightsquigarrow y) \vee E \geq 1$, by (16.7), and hence $[(x \rightsquigarrow y) \vee E] \wedge 1 = 1$. Thus, $C = \mathbf{1}$.

(1'): has a dual proof, using Theorem 16.1.1 (c), in the first proof, and Theorem 16.1.6 (c1), in the second proof. □

Finaly, we present some intermediary results and an open problem.

Theorem 16.2.4 *(see Theorem 14.1.1)*

Let $\mathcal{G} = (G, \vee, \wedge, \to, \rightsquigarrow, 1)$ be a representable l-implicative-group. Then,

(1) the pseudo-BCK(pP) lattice $\mathcal{G}^L = (G^-, \wedge, \vee, \to^L, \rightsquigarrow^L, \mathbf{1} = 1)$ (with the pseudo-product $\odot = \cdot$), verifying condition (pCC^L), verifies also the following conditions: for all $a, b \in G^-$,
(i) $(a \vee b)^2 = a^2 \vee b^2$, i.e. $(a \vee b) \odot (a \vee b) = (a \odot a) \vee (b \odot b)$,
(ii) Condition (i) is equivalent with condition

$$[b \to^L (a \rightsquigarrow^L (a \odot a))] \vee [a \rightsquigarrow^L (b \to^L (b \odot b))] = \mathbf{1}. \tag{16.8}$$

(iii) $(b \to^L a) \vee (a \rightsquigarrow^L b) = \mathbf{1}$,
(iv) Condition (iii) implies condition (16.8).

(1') the pseudo-BCK(pS) lattice $\mathcal{G}^R = (G^+, \vee, \wedge, \to^R, \rightsquigarrow^R, \mathbf{0} = 1)$ (with the pseudo-sum $\oplus = \cdot$), verifying the dual condition (pCC^R), verifies also the following conditions: for all $a, b \in G^+$,
(i') $2(a \wedge b) = 2a \wedge 2b$, i.e. $(a \wedge b) \oplus (a \wedge b) = (a \oplus a) \wedge (b \oplus b)$,
(ii') Condition (i') is equivalent with condition

$$[b \to^R (a \rightsquigarrow^R (a \oplus a))] \vee [a \rightsquigarrow^R (b \to^R (b \oplus b))] = \mathbf{0}. \tag{16.9}$$

(iii') $(b \to^R a) \wedge (a \rightsquigarrow^R b) = \mathbf{0}$,
(iv') Condition (iii') implies condition (16.9).

Proof. (1): We denote $\to = \to^L$ and $\rightsquigarrow = \rightsquigarrow^L$.

(i): follows obviously by Theorem 16.1.7 (b^d), since $\mathcal{G}$ is representable.

(ii): We shall prove that (i) $\Leftrightarrow$ (16.8). Indeed,

(i) $\Longrightarrow$ (16.8):

(i) $(a \vee b) \odot (a \vee b) = (a \odot a) \vee (b \odot b) \Leftrightarrow$
$[(a \vee b) \odot a] \vee [(a \vee b) \odot b] = (a \odot a) \vee (b \odot b) \Leftrightarrow$
$a \odot a \vee b \odot a \vee a \odot b \vee b \odot b = a \odot a \vee b \odot b \Leftrightarrow$

$$a \odot b \vee b \odot a \leq a \odot a \vee b \odot b. \tag{16.10}$$

And (16.10) $\Longrightarrow$
$a \odot b \leq a \odot a \vee b \odot b \Longrightarrow$
$b \rightarrow (a \odot b) \leq b \rightarrow (a \odot a \vee b \odot b) \Longrightarrow$

$$a \rightsquigarrow (b \rightarrow (a \odot b)) \leq a \rightsquigarrow (b \rightarrow (a \odot a \vee b \odot b)). \tag{16.11}$$

But $a \rightsquigarrow (b \rightarrow (a \odot b) = b \rightarrow (a \rightsquigarrow (a \odot b)) \leq b \rightarrow b = \mathbf{1}$, since $b \leq a \rightsquigarrow (a \odot b)$.
Hence, (16.11) $\Longrightarrow$
$a \rightsquigarrow (b \rightarrow (a \odot a \vee b \odot b)) = \mathbf{1} \overset{(pprel^L)}{\Leftrightarrow}$
$a \rightsquigarrow [(b \rightarrow a \odot a) \vee (b \rightarrow b \odot b)] = \mathbf{1} \overset{(pprel^L)}{\Leftrightarrow}$
$[a \rightsquigarrow (b \rightarrow a \odot a)] \vee [a \rightsquigarrow (b \rightarrow b \odot b)] = \mathbf{1} \Leftrightarrow$
$[b \rightarrow (a \rightsquigarrow (a \odot a))] \vee [a \rightsquigarrow (b \rightarrow (b \odot b))] = \mathbf{1}$, i.e.(16.8) holds.
Note that we have used an equivalent condition with (pprel^L) denoted (pprel$_{\Rightarrow\vee}$) in [94], pag. 386:
(pprel$_{\Rightarrow\vee}$) $x \rightarrow (y \vee z) = (x \rightarrow y) \vee (x \rightarrow z)$ and $x \rightsquigarrow (y \vee z) = (x \rightsquigarrow y) \vee (x \rightsquigarrow z)$.

(16.8) $\Longrightarrow$ (i):
(16.8) $[b \rightarrow (a \rightsquigarrow (a \odot a))] \vee [a \rightsquigarrow (b \rightarrow (b \odot b))] = \mathbf{1} \Leftrightarrow$
$[a \rightsquigarrow (b \rightarrow (a \odot a))] \vee [a \rightsquigarrow (b \rightarrow (b \odot b))] = \mathbf{1} \overset{(pprel^L)}{\Leftrightarrow}$
$a \rightsquigarrow (b \rightarrow (a \odot a \vee b \odot b)) = \mathbf{1} \Leftrightarrow$
$\mathbf{1} \leq a \rightsquigarrow (b \rightarrow (a \odot a \vee b \odot b)) \Longrightarrow$
$a = a \odot \mathbf{1} \leq a \odot [a \rightsquigarrow (b \rightarrow (a \odot a \vee b \odot b))] \overset{(pdiv^L)}{\Leftrightarrow}$
$a \leq a \wedge (b \rightarrow (a \odot a \vee b \odot b)) \leq a \Longrightarrow$
$a = a \wedge (b \rightarrow (a \odot a \vee b \odot b)) \Leftrightarrow$
$a \leq (b \rightarrow (a \odot a \vee b \odot b)) \Longrightarrow$
$a \odot b \leq (b \rightarrow (a \odot a \vee b \odot b)) \odot b \overset{(pdiv^L)}{\Leftrightarrow}$
$a \odot b \leq b \wedge (a \odot a \vee b \odot b) \leq a \odot a \vee b \odot b \Longrightarrow$
$a \odot b \leq a \odot a \vee b \odot b$.
Similarly,
$b \odot a \leq b \odot b \vee a \odot a$, i.e. $a \odot a \vee b \odot b$ is an upper bound of $a \odot b$ and $b \odot a$.
It follows that $a \odot b \vee b \odot a \leq a \odot a \vee b \odot b$, i.e. (16.10) holds. And we have seen above that (16.10) $\Leftrightarrow$ (i).

(iii): $(b \rightarrow^L a) \vee (a \rightsquigarrow^L b) = [(b \rightarrow a) \wedge 1] \vee [(a \rightsquigarrow b) \wedge 1] =$
$[(b \rightarrow a) \vee (a \rightsquigarrow b)] \wedge 1 \geq (1 \vee [(b \rightsquigarrow a) \rightsquigarrow (b \rightarrow a)]) \wedge 1 = 1 = \mathbf{1}$, by Theorem 16.1.7 ((a) $\Leftrightarrow$ (b1^d).

(iv): Condition (iii) implies condition (16.8). Indeed, since $a \leq a \rightsquigarrow^L (a \odot a)$ and $b \leq b \rightarrow^L (b \odot b)$ by [94], condition (11.3), it follows that
$b \rightarrow^L a \leq b \rightarrow^L [a \rightsquigarrow^L (a \odot a)]$ and $a \rightsquigarrow^L b \leq a \rightsquigarrow^L [b \rightarrow^L (b \odot b)]$, hence
$\mathbf{1} = (b \rightarrow^L a) \vee (a \rightsquigarrow^L b) \leq (b \rightarrow^L [a \rightsquigarrow^L (a \odot a)]) \vee (a \rightsquigarrow^L [b \rightarrow^L (b \odot b)])$, hence
$(b \rightarrow^L [a \rightsquigarrow^L (a \odot a)]) \vee (a \rightsquigarrow^L [b \rightarrow^L (b \odot b)]) = \mathbf{1}$.

(1'): has a dual proof. □

Proposition 16.2.5 *(see Theorem 14.1.1)*

Let $\mathcal{G} = (G, \vee, \wedge, \rightarrow, \rightsquigarrow, 1)$ be an l-implicative-group.

(1) If $\mathcal{G}$ verifies the condition (b1^d") from Remarks 16.1.4:
(b1^d") for all $a, b \in G$, $(b \rightarrow a) \vee (a \rightsquigarrow b) \geq 1$,

then the pseudo-BCK(pP) lattice $\mathcal{G}^L = (G^-, \wedge, \vee, \rightarrow^L, \rightsquigarrow^L, \mathbf{1} = 1)$ *verifies the condition (iii) from Theorem 16.2.4 (1):*
(iii) for all $a, b \in G^-$, $(b \rightarrow^L a) \vee (a \rightsquigarrow^L b) = \mathbf{1} = 1$.

(1') If $\mathcal{G}$ *verifies the condition (b1") from Remarks 16.1.4:*
(b1") for all $a, b \in G$, $(b \rightarrow a) \wedge (a \rightsquigarrow b) \leq 1$,
then the pseudo-BCK(pS) lattice $\mathcal{G}^R = (G^+, \vee, \wedge, \rightarrow^R, \rightsquigarrow^R, \mathbf{0} = 1)$ *verifies the condition (iii') from Theorem 16.2.4 (1'):*
(iii') for all $a, b \in G^+$, $(b \rightarrow^R a) \wedge (a \rightsquigarrow^R b) = \mathbf{0} = 1$.

Proof.

(1): $(b \rightarrow^L a) \vee (a \rightsquigarrow^L b) = [(b \rightarrow a) \wedge 1] \vee [(a \rightsquigarrow b) \wedge 1] \stackrel{distrib.}{=}$
$[(b \rightarrow a) \vee (a \rightsquigarrow b)] \wedge 1 \stackrel{(b1^{d}")}{=} 1 = \mathbf{1}$.

(1'): $(b \rightarrow^R a) \wedge (a \rightsquigarrow^R b) = [(b \rightarrow a) \vee 1] \wedge [(a \rightsquigarrow b) \vee 1] =$
$[(b \rightarrow a) \wedge (a \rightsquigarrow b)] \vee 1 \stackrel{(b1")}{=} 1 = \mathbf{0}$. □

Open problems 16.2.6

(1) Find if there are connections between the representability of $\mathcal{G}^L = (G^-, \wedge, \vee, \rightarrow^L, \rightsquigarrow^L, \mathbf{1} = 1)$ (or of the left-pseudo-MV algebra $[u', 1]$) and the conditions (i) $\Leftrightarrow$ (16.8), (iii).

(1') Find if there are connections between the representability of $\mathcal{G}^R = (G^+, \vee, \wedge, \rightarrow^R, \rightsquigarrow^R, \mathbf{0} = 1)$ (or of the right-pseudo-MV algebra $[1, u]$) and the conditions (i') $\Leftrightarrow$ (16.9), (iii').

Open problem 16.2.7 Find connections between the representability at l-group (l-implicative-group) G level and the representability at $[u', 1] \subset G^-$, $[1, u] \subset G^+$ level and at $G^-_{-\infty}$, $G^+_{+\infty}$ level.

Chapter 17

States and implicative-states

In this chapter, we present the states [98], [99]. We introduce the notions of *additive-state*, or *state* for short, on a po-group with strong unit and *implicative-state* on a po-implicative-group with strong unit and prove they coincide. Next, we introduce the notions of *state morphism* on an l-group with strong unit and *implicative-state morphism* on an l-implicative-group with strong unit and prove they coincide. Finally, we define the distance functions d_1^L, d_2^L and d_1^R, d_2^R on an l-group and prove some properties, following the ideas in the pseudo-BL algebras case. Then, we introduce the notion of *Bosbach-state* on an l-group with strong unit, prove some properties and prove that any state is a Bosbach-state, following the ideas from [57].

The chapter has three sections.

17.1 Additive-states and implicative-states

At poset level, we generalize to arbitrary po-groups with strong unit the definition of states for the abelian po-groups from [67]. We also introduce the notion of *implicative-state* on the term equivalent po-implicative-group and we prove that the implicative-states coincide with the states.

Let $\mathcal{G}_g = (G, \leq, \cdot, ^{-1}, 1)$ be a po-group with strong unit u, i.e. $u \geq 1$ and for every $x \in G$, there exists some positive integer n such that $x \leq \underbrace{u \cdot u \cdot \ldots \cdot u}_{n \text{ times}}$. Denote it by $(G, u)_g$.

Also, let $\mathcal{G}_{ig} = (G, \leq, \rightarrow, \rightsquigarrow, 1)$ be the term equivalent po-implicative-group with strong unit u. Denote it by $(G, u)_{ig}$.

Let $\mathcal{R}_g = (\mathbf{R}, \leq, +, -, 0)$ be the additive abelian po-group of real numbers with strong unit 1. Denote it by $(\mathbf{R}, 1)_g$.

Also, let $\mathcal{R}_{ig} = (\mathbf{R}, \leq, \rightarrow = \rightsquigarrow, 0)$ be the term equivalent commutative po-implicative-group of real numbers with strong unit 1. Denote it by $(\mathbf{R}, 1)_{ig}$.

Definition 17.1.1 (See also [48])

Let $(G,u)_g$ and $(\mathbf{R},1)_g$. An *additive-state*, or a *state* for short, on $(G,u)_g$ is any positive (or, equivalently, order preserving) po-group homomorphism $s : G \longrightarrow \mathbf{R}$ verifying $s(u) = 1$, i.e. s is a state if and only if the following properties hold: for all $x, y \in G$,
(s1) $s(x \cdot y) = s(x) + s(y)$,
(s2) $x \geq 1$ implies $s(x) \geq 0$,
(s3) $s(u) = 1$.

Proposition 17.1.2 *Let s be a state on $(G,u)_g$. Then, the following properties hold: for all $x, y \in G$,*
(s4) $s(1) = 0$,
(s5) $s(x^{-1}) = -s(x)$,
(s6) $x \leq y$ *implies* $s(x) \leq s(y)$,
(s7) $s(x \to y) = s(x) \to s(y) = s(x \rightsquigarrow y)$.

Proof.
(s4): $s(x) = s(x \cdot 1) = s(x) + s(1)$ implies $s(1) = 0$.
(s5): $0 = s(1) = s(x \cdot x^{-1}) = s(x) + s(x^{-1})$ implies $s(x^{-1}) = -s(x)$.
(s6): $x \leq y$ implies $y \cdot x^{-1} \geq 1$, hence $s(y \cdot x^{-1}) \geq s(1)$, i.e. $s(y) + s(x^{-1}) \geq 0$, hence $s(y) - s(x) \geq 0$, hence $s(x) \leq s(y)$.
(s7): $s(x \to y) = s(y \cdot x^{-1}) = s(y) + s(x^{-1}) = s(y) - s(x) = s(x) \to s(y)$ and $s(x \rightsquigarrow y) = s(x^{-1} \cdot y) = s(x^{-1}) + s(y) = -s(x) + s(y) = s(x) \rightsquigarrow s(y) = s(x) \to s(y)$.
□

We introduce the following definition.

Definition 17.1.3 Let $(G,u)_{ig}$ and $(\mathbf{R},1)_{ig}$. An *implicative-state* on $(G,u)_{ig}$ is any map $s : G \longrightarrow \mathbf{R}$ verifying: for all $x, y \in G$,
(s7) $s(x \to y) = s(x) \to s(y) = s(x \rightsquigarrow y)$,
(s2) $x \geq 1$ implies $s(x) \geq 0$,
(s3) $s(u) = 1$.

Proposition 17.1.4 *Let s be an implicative-state on $(G,u)_{ig}$. Then, the following properties hold: for all $x, y \in G$,*
(s4) $s(1) = 0$,
(s5) $s(x^{-1}) = -s(x)$.

Proof.
(s4): $s(1) \overset{(pRe')}{=} s(x \to x) = s(x) \to s(x) \overset{(pRe')}{=} 0$,
(s5): $s(x^{-1}) = s(x \to 1) = s(x) \to 0 = -s(x)$. □

Then, we have the following theorem.

Theorem 17.1.5 *The states on the po-group $\mathcal{G}_g = (G, \leq, \cdot, ^{-1}, 1)$ with strong unit u coincide with the implicative-states on the term equivalent po-implicative-group $\mathcal{G}_{ig} = (G, \leq, \to, \rightsquigarrow, 1)$.*

Proof.

Let s be a state on $\mathcal{G}_g$. To prove that s is an implicative-state on $\mathcal{G}_{ig}$, it is sufficient to prove that (s7) holds for all $x, y \in G$. Indeed,
$s(x \to y) = s(y \cdot x^{-1}) = s(y) + s(x^{-1}) = s(y) + (-s(x)) = s(x) \to s(y)$ and
$s(x \rightsquigarrow y) = s(x^{-1} \cdot y) = s(x^{-1}) + s(y) = -s(x) + s(y) = s(x) \rightsquigarrow s(y) = s(x) \to s(y)$.

Let s be an implicative-state on $\mathcal{G}_{ig}$. To prove that s is a state on $\mathcal{G}_g$, it is sufficient to prove that (s1) holds for all $x, y \in G$. Indeed,
$s(x \cdot y) = s((x \to y^{-1})^{-1}) = -s(x \to y^{-1}) = -(s(x) \to s(y^{-1})) =$
$-(s(x) \to (-s(y))) = s(x) + s(y)$. □

17.2 State morphisms and implicative-state morphisms

Al lattice level, we recall the definition of *state morphism*, we introduce the notion of *implicative-state morphism* and prove that they coincide.

Consider the following two dual properties:
(s0) $s(x \wedge y) = s(x) \wedge s(y) = \min(s(x), s(y))$, for all $x, y \in G$,
(s0') $s(x \vee y) = s(x) \vee s(y) = \max(s(x), s(y))$, for all $x, y \in G$.

Recall first the following definition.

Definition 17.2.1 Let $(G, u)_g$ be an l-group $\mathcal{G}_g = (G, \vee, \wedge, \cdot, ^{-1}, 1)$ with strong unit u and let $(\mathbf{R}, 1)_g$ be the additive abelian l-group of real numbers $\mathcal{R}_g = (\mathbf{R}, \max, \min, +, -, 0)$ with strong unit 1.

A *state morphism* on $(G, u)_g$ is a state s on $\mathcal{G}_g$ verifying the property (s0).

Note that (s0) can be replaced by the weaker condition

$$s(x) \wedge s(y) \leq s(x \wedge y),$$

since $s(x \wedge y) \leq s(x) \wedge s(y)$ always holds (indeed, $x \wedge y \leq x$, y implies $s(x \wedge y) \leq s(x)$, $s(y)$, i.e. $s(x \wedge y)$ is a lower bound of $s(x)$, $s(y)$; hence, $s(x \wedge y) \leq s(x) \wedge s(y)$).

Proposition 17.2.2 *Let s be a state-morphism on $(G, u)_g$. Then, (s0') holds too.*

Proof. $s(x \vee y) = s(x \cdot (x \wedge y)^{-1} \cdot y) = s(x) - s(x \wedge y) + s(y) = s(x) - s(x) \wedge s(y) + s(y) = s(x) \vee s(y)$, by Proposition 7.3.2 (3). □

Note that an equivalent definition of a state-morphism on (G, u) would be, by Proposition 17.2.2 and by Proposition 7.3.2 (3), the following:
A *state-morphism* on $\mathcal{G}_g$ is any state s on $\mathcal{G}_g$ verifying the property (s0'), and then (s0) will follow.

We introduce now the following definition.

Definition 17.2.3 Let $(G, u)_{ig}$ be an l-implicative-group $\mathcal{G}_{ig} = (G, \vee, \wedge, \to, \rightsquigarrow, 1)$ with strong unit u and let $(\mathbf{R}, 1)_{ig}$ be the additive abelian l-implicative-group of real numbers $\mathcal{R}_{ig} = (\mathbf{R}, \max, \min, \to = \rightsquigarrow, 0)$ with strong unit 1.

An *implicative-state morphism* on $\mathcal{G}_{ig}$ is an implicative-state s on $\mathcal{G}_{ig}$ verifying (s0).

By Theorem 17.1.5, we immediately obtain the following theorem.

Theorem 17.2.4 *The states morphisms on the l-group $\mathcal{G}_g = (G, \vee, \wedge, \cdot, ^{-1}, 1)$ with strong unit u coincide with the implicative-state morphisms on the term equivalent l-implicative-group $\mathcal{G}_{ig} = (G, \vee, \wedge, \rightarrow, \rightsquigarrow, 1)$.*

17.3 Bosbach-states

17.3.1 Distance functions

Following the ideas from [42], we introduce the following dual "distances".

Definition 17.3.1 Let $(G, \vee, \wedge, \cdot, ^{-1}, 1)$ be an l-group. We define the following dual *distance functions*, by (7.20), (7.21) and (7.22), (7.23): for all $x, y \in G$,

$$d_1^L(x,y) \overset{def.}{=} (x \vee y) \rightarrow (x \wedge y) \in G^-, \quad d_2^L(x,y) \overset{def.}{=} (x \vee y) \rightsquigarrow (x \wedge y) \in G^-,$$

$$d_1^R(x,y) \overset{def.}{=} (x \wedge y) \rightarrow (x \vee y) \in G^+, \quad d_2^R(x,y) \overset{def.}{=} (x \wedge y) \rightsquigarrow (x \vee y) \in G^+.$$

Proposition 17.3.2 *Let $(G, \vee, \wedge, \cdot, ^{-1}, 1)$ be an l-group. Then, the above defined distance functions verify the following properties (see [42]): for all $x, y, z \in G$,*
(1) $d_1^L(x,y) = d_1^L(y,x), \quad d_2^L(x,y) = d_2^L(y,x)$,
(1') $d_1^R(x,y) = d_1^R(y,x), \quad d_2^R(x,y) = d_2^R(y,x)$,
(2) $d_1^L(x,y) = 1 \Leftrightarrow x = y \Leftrightarrow d_2^L(x,y) = 1$,
(2') $d_1^R(x,y) = 1 \Leftrightarrow x = y \Leftrightarrow d_2^R(x,y) = 1$,
(3) $d_1^L(x,1) = d_2^L(x,1) = \begin{cases} x^{-1} & , \textit{if } \; x \geq 1 \\ x & , \textit{if } \; x < 1, \end{cases}$
(3') $d_1^R(x,1) = d_2^R(x,1) = \begin{cases} x^{-1} & , \textit{if } \; x \leq 1 \\ x & , \textit{if } \; x > 1, \end{cases}$
(4) $d_1^L(x,y) = d_2^L(x^{-1}, y^{-1}), \quad d_2^L(x,y) = d_1^L(x^{-1}, y^{-1})$,
(4') $d_1^R(x,y) = d_2^R(x^{-1}, y^{-1}), \quad d_2^R(x,y) = d_1^R(x^{-1}, y^{-1})$,
(5) $d_2^L(x,y) \cdot d_2^L(y,z) \cdot d_2^L(x,y) \leq d_2^L(x,z)$,
$\quad d_2^L(y,z) \cdot d_2^L(x,y) \cdot d_2^L(y,z) \leq d_2^L(x,z)$,
(6) $d_1^L(x,y) \cdot d_1^L(y,z) \cdot d_1^L(x,y) \leq d_1^L(x,z)$,
$\quad d_1^L(y,z) \cdot d_1^L(x,y) \cdot d_1^L(y,z) \leq d_1^L(x,z)$,
(5') $d_2^R(x,y) \cdot d_2^R(y,z) \cdot d_2^R(x,y) \geq d_2^R(x,z)$,
$\quad d_2^R(y,z) \cdot d_2^R(x,y) \cdot d_2^R(y,z) \geq d_2^L(x,z)$,
(6') $d_1^R(x,y) \cdot d_1^R(y,z) \cdot d_1^R(x,y) \geq d_1^R(x,z)$,
$\quad d_1^R(y,z) \cdot d_1^R(x,y) \cdot d_1^R(y,z) \geq d_1^R(x,z)$.

Proof.

(1), (1'): Obvious, by (7.20) and (7.22).

(2): $d_1^L(x,y) = 1 \Leftrightarrow (x \rightarrow y) \wedge (y \rightarrow x) \wedge 1 = 1 \Leftrightarrow (x \rightarrow y) \wedge (y \rightarrow x) \geq 1 \Leftrightarrow x \rightarrow y \geq 1,\ y \rightarrow x \geq 1 \Leftrightarrow y \cdot x^{-1} \geq 1,\ x \cdot y^{-1} \geq 1 \Leftrightarrow y \geq x,\ x \geq y \Leftrightarrow x = y$. The other equivalence has a similar proof.

(2'): has a similar proof.

(3): By (7.20), $d_1^L(x,1) = (x \to 1) \wedge (1 \to x) \wedge 1 = x^{-1} \wedge x \wedge 1$:
- if $x \geq 1$, then $x^{-1} \leq 1$, hence $d_1^L(x,1) = 1 \wedge x^{-1} = x^{-1}$;
- if $x \leq 1$, then $x^{-1} \geq 1$, hence $d_1^L(x,1) = x \wedge 1 = x$.
$d_2^L(x,1) = (x \rightsquigarrow 1) \wedge (1 \rightsquigarrow x) \wedge 1 = x^{-1} \wedge x \wedge 1 = d_1^L(x,1)$.

(3'): has a similar proof.

(4): $d_1^L(x,y) = (x \to y) \wedge (y \to x) \wedge 1 \overset{(pDNeg1)}{=} [y^{-1} \rightsquigarrow x^{-1}] \wedge [x^{-1} \rightsquigarrow y^{-1}] \wedge 1 = d_2^L(x^{-1}, y^{-1})$.
$d_2^L(x,y) = (x \rightsquigarrow y) \wedge (y \rightsquigarrow x) \wedge 1 \overset{(pDNeg1)}{=} [y^{-1} \to x^{-1}] \wedge [x^{-1} \to y^{-1}] \wedge 1 = d_1^L(x^{-1}, y^{-1})$.

(4'): has a similar proof.

(5): by Proposition 7.3.2 (1') and Proposition 7.1.18 (a), we obtain:
$d_2^L(x,y) \cdot d_2^L(y,z) \cdot d_2^L(x,y) =$
$[(x \rightsquigarrow y) \wedge (y \rightsquigarrow x) \wedge 1] \wedge [(y \rightsquigarrow z) \wedge (z \rightsquigarrow y) \wedge 1] \wedge [(x \rightsquigarrow y) \wedge (y \rightsquigarrow x) \wedge 1] =$
$a \wedge b \wedge c \wedge d \wedge e \wedge f \wedge m \wedge n \wedge p \wedge$
$a' \wedge b' \wedge c' \wedge d' \wedge e' \wedge f' \wedge m' \wedge n' \wedge p' \wedge$
$a'' \wedge b'' \wedge c'' \wedge d'' \wedge e'' \wedge f'' \wedge m'' \wedge n'' \wedge p''$,
where:
$a = (x \rightsquigarrow y) \cdot (y \rightsquigarrow z) \cdot (x \rightsquigarrow y)$, $b = (x \rightsquigarrow y) \cdot (y \rightsquigarrow z) \cdot (y \rightsquigarrow x)$,
$\mathbf{c} = (x \rightsquigarrow y) \cdot (y \rightsquigarrow z) \cdot 1 = x \rightsquigarrow z$,
$d = (x \rightsquigarrow y) \cdot (z \rightsquigarrow y) \cdot (x \rightsquigarrow y)$, $e = (x \rightsquigarrow y) \cdot (z \rightsquigarrow y) \cdot (y \rightsquigarrow x)$,
$f = (x \rightsquigarrow y) \cdot (z \rightsquigarrow y) \cdot 1$,
$m = (x \rightsquigarrow y) \cdot 1 \cdot (x \rightsquigarrow y)$, $\mathbf{n} = (x \rightsquigarrow y) \cdot 1 \cdot (y \rightsquigarrow x) = \mathbf{1}$, $p = (x \rightsquigarrow y) \cdot 1 \cdot 1$,

$a' = (y \rightsquigarrow x) \cdot (y \rightsquigarrow z) \cdot (x \rightsquigarrow y)$, $b' = (y \rightsquigarrow x) \cdot (y \rightsquigarrow z) \cdot (y \rightsquigarrow x)$,
$c' = (y \rightsquigarrow x) \cdot (y \rightsquigarrow z) \cdot 1$,
$d' = (y \rightsquigarrow x) \cdot (z \rightsquigarrow y) \cdot (x \rightsquigarrow y)$, $e' = (y \rightsquigarrow x) \cdot (z \rightsquigarrow y) \cdot (y \rightsquigarrow x)$,
$f' = (y \rightsquigarrow x) \cdot (z \rightsquigarrow y) \cdot 1$,
$\mathbf{m}' = (y \rightsquigarrow x) \cdot 1 \cdot (x \rightsquigarrow y) = \mathbf{1}$, $n' = (y \rightsquigarrow x) \cdot 1 \cdot (y \rightsquigarrow x)$, $p' = (y \rightsquigarrow x) \cdot 1 \cdot 1$,

$a'' = 1 \cdot (y \rightsquigarrow z) \cdot (x \rightsquigarrow y)$, $b'' = 1 \cdot (y \rightsquigarrow z) \cdot (y \rightsquigarrow x)$, $c'' = 1 \cdot (y \rightsquigarrow z) \cdot 1$,
$d'' = 1 \cdot (z \rightsquigarrow y) \cdot (x \rightsquigarrow y)$, $\mathbf{e}'' = 1 \cdot (z \rightsquigarrow y) \cdot (y \rightsquigarrow x) = z \rightsquigarrow x$, $f'' = 1 \cdot (z \rightsquigarrow y) \cdot 1$,
$\mathbf{m}'' = 1 \cdot 1 \cdot (x \rightsquigarrow y)$, $n'' = 1 \cdot 1 \cdot (y \rightsquigarrow x)$, $\mathbf{p}'' = 1 \cdot 1 \cdot 1 = \mathbf{1}$.
But, $d_2^L(x,y) \cdot d_2^L(y,z) \cdot d_2^L(x,y) \leq \mathbf{c} \wedge \mathbf{e}'' \wedge \mathbf{p}'' = (x \rightsquigarrow z) \wedge (z \rightsquigarrow x) \wedge 1 = d_2^L(x,z)$.
The second inequality of (5) has a similar proof.

(6), (5'), (6'): have similar proofs. □

17.3.2 Bosbach-states on l-groups with strong unit

In this subsection, $\mathcal{G}_g = (G, \vee, \wedge, \cdot, ^{-1}, 1)$ is an l-group with strong unit u, denoted by $(G,u)_g$, and $\mathcal{R}_g = (\mathbf{R}, \max, \min, +, -, 0)$ is the additive abelian l-group of real numbers with the strong unit 1, denoted by $(\mathbf{R}, 1)_g$.

Proposition 17.3.3 *(See [57], Proposition 2.1)*

Let $s : G \longrightarrow \mathbf{R}$ such that $s(1) = 0$. Then, the following are equivalent: for all $x, y \in G$,

(i) $s(x \vee y) + s(d_1^L(x, y)) = s(x \wedge y)$,
(ii) $s(y) + s((y \to x) \wedge 1) = s(x \wedge y)$,
(iii) $s(x) + s((x \to y) \wedge 1) = s(y) + s((y \to x) \wedge 1)$.

Proof.

(i) $\Longrightarrow$ (ii): Let us consider $a \leq b$ in G; then $a \wedge b = a$ and $a \vee b = b$, hence $d_1^L(a, b) = b \to a$. It follows, by (i), that

$$s(b) + s(b \to a) = s(a). \tag{17.1}$$

Let us take $a = x \wedge y$ and $b = y$. By (17.1), we obtain:
$s(y) + s((y \to x) \wedge 1) = s(y) + s(y \to (x \wedge y)) = s(x \wedge y)$, i.e. (ii) holds.

(ii) $\Longrightarrow$ (iii): $s(x) + s((x \to y) \wedge 1) \overset{(ii)}{=} s(y \wedge x) = s(x \wedge y) = s(y) + s((y \to x) \wedge 1)$, i.e. (iii) holds.

(iii) $\Longrightarrow$ (i): $s(x \vee y) + s(d_1^L(x, y)) = s(x \vee y) + s((x \vee y) \to (x \wedge y)) \overset{(7.20)}{=}$
$s(x \vee y) + s([(x \vee y) \to (x \wedge y)] \wedge 1) \overset{(iii)}{=} s(x \wedge y) + s([(x \wedge y) \to (x \vee y)] \wedge 1) =$
$s(x \wedge y) + s(1) = s(x \wedge y)$, since
$(x \vee y) \to (x \wedge y) \leq 1$ and hence $(x \vee y) \to (x \wedge y) = [(x \vee y) \to (x \wedge y)] \wedge 1$ and since
$x \wedge y \leq x \vee y$ and hence $[(x \wedge y) \to (x \vee y)] \geq 1$, by Corollary 7.2.12. □

The following proposition has a similar proof.

Proposition 17.3.4 *(See [57], Proposition 2.2)*
Let $s : G \longrightarrow \mathbf{R}$ such that $s(1) = 0$. Then, the following are equivalent: for all $x, y \in G$,
(i) $s(x \vee y) + s(d_2^L(x, y)) = s(x \wedge y)$,
(ii) $s(y) + s((y \rightsquigarrow x) \wedge 1) = s(x \wedge y)$,
(iii) $s(x) + s((x \rightsquigarrow y) \wedge 1) = s(y) + s((y \rightsquigarrow x) \wedge 1)$.

Dually, we have the next Propositions 17.3.5 and 17.3.6.

Proposition 17.3.5 *Let $s : G \longrightarrow \mathbf{R}$ such that $s(1) = 0$. Then, the following are equivalent: for all $x, y \in G$,*
(i') $s(x \wedge y) + s(d_1^R(x, y)) = s(x \vee y)$,
(ii') $s(y) + s((y \to x) \vee 1) = s(x \vee y)$,
(iii') $s(x) + s((x \to y) \vee 1) = s(y) + s((y \to x) \vee 1)$.

Proposition 17.3.6 *Let $s : G \longrightarrow \mathbf{R}$ such that $s(1) = 0$. Then, the following are equivalent: for all $x, y \in G$,*
(i') $s(x \wedge y) + s(d_2^R(x, y)) = s(x \vee y)$,
(ii') $s(y) + s((y \rightsquigarrow x) \vee 1) = s(x \vee y)$,
(iii') $s(x) + s((x \rightsquigarrow y) \vee 1) = s(y) + s((y \rightsquigarrow x) \vee 1)$.

We can now define the Bosbach-state on l-groups with strong unit.

Definition 17.3.7 A *Bosbach-state* on $(G,u)_g$ is a function $s : G \longrightarrow \mathbf{R}$ such that: for all $x, y \in G$,

(S1) $s(x) + s((x \to y) \wedge 1) = s(y) + s((y \to x) \wedge 1)$,

(S2) $s(x) + s((x \rightsquigarrow y) \wedge 1) = s(y) + s((y \rightsquigarrow x) \wedge 1)$,

(S1') $s(x) + s((x \to y) \vee 1) = s(y) + s((y \to x) \vee 1)$,

(S2') $s(x) + s((x \rightsquigarrow y) \vee 1) = s(y) + s((y \rightsquigarrow x) \vee 1)$,

(S3) $s(1) = 0$, $s(u) = 1$,

(S3') $x \geq 1$ implies $s(x) \geq 0$.

Note that Propositions 17.3.3, 17.3.4 (17.3.5, 17.3.6) provide equivalent conditions to (S1), (S2) ((S1'), (S2'), respectively).

Proposition 17.3.8 *(See [57], Proposition 2.7)*

Let s be a Bosbach-state on $(G,u)_g$. *Then, we have: for all* $x, y \in G$,

(S4) $s((x \to y) \wedge 1) = s((x \rightsquigarrow y) \wedge 1)$,

(S4') $s((x \to y) \vee 1) = s((x \rightsquigarrow y) \vee 1)$,

(S5) $s(d_1^L(x,y)) = s(d_2^L(x,y))$,

(S5') $s(d_1^R(x,y)) = s(d_2^R(x,y))$,

(S6) $s(x) + s(x^{-1} \wedge 1) = s(x \wedge 1)$,

(S6') $s(x) + s(x^{-1} \vee 1) = s(x \vee 1)$,

(S7) $s(x^{-1}) = -s(x)$,

(S8) $x \leq y$ *implies* $s(y \to x) = s(y \rightsquigarrow x) = s(y) \to s(x)$,

(S8') $x \leq y$ *implies* $s(x \to y) = s(x \rightsquigarrow y) = s(x) \to s(y)$,

(S9) $x \leq y$ *implies* $s(x) \leq s(y)$.

Proof.

(S4): $s(x) + s((x \to y) \wedge 1) = s(y \wedge x) = s(x) + s((x \rightsquigarrow y) \wedge 1)$, by Propositions 17.3.3 (ii) and 17.3.4 (ii). Hence, $s((x \to y) \wedge 1) = s((x \rightsquigarrow y) \wedge 1)$.

(S4'): $s(x) + s((x \to y) \vee 1) = s(y \vee x) = s(x) + s((x \rightsquigarrow y) \vee 1)$, by Propositions 17.3.5 (ii') and 17.3.6 (ii'). Hence, $s((x \to y) \vee 1) = s((x \rightsquigarrow y) \vee 1)$.

(S5): $s(x \vee y) + s(d_1^L(x,y)) = s(x \wedge y) = s(x \vee y) + s(d_2^L(x,y))$, by Propositions 17.3.3 (i) and 17.3.4 (i). Hence, $s(d_1^L(x,y)) = s(d_2^L(x,y))$.

(S5'): dually, by Propositions 17.3.5 (i') and 17.3.6 (i').

(S6): $s(x) + s(x^{-1} \wedge 1) \overset{(7.9)}{=} s(x) + s((x \to 1) \wedge 1) \overset{(S1)}{=} s(1) + s((1 \to x) \wedge 1) \overset{(pM)}{=} 0 + s(x \wedge 1) = s(x \wedge 1)$.

(S6'): $s(x) + s(x^{-1} \vee 1) \overset{(7.9)}{=} s(x) + s((x \to 1) \vee 1) \overset{(S1')}{=} s(1) + s((1 \to x) \vee 1) \overset{(pM)}{=} 0 + s(x \vee 1) = s(x \vee 1)$.

(S7): First, we prove that: $s(x^{-1}) + s(x \wedge 1) = s(x^{-1} \wedge 1)$.

Indeed, $s(x^{-1}) + s(x \wedge 1) = s(x^{-1}) + s((x^{-1})^{-1} \wedge 1) \overset{(S6)}{=} s(x^{-1} \wedge 1)$.

Now, $s(x \wedge 1) \overset{(S6)}{=} s(x) + s(x^{-1} \wedge 1) = s(x) + s(x^{-1}) + s(x \wedge 1)$. It follows that $s(x) + s(x^{-1}) = 0$, i.e. $s(x^{-1}) = -s(x)$.

(S8): Let $x \leq y$; then $x \to y \geq 1$, $x \rightsquigarrow y \geq 1$ and $y \to x \leq 1$, $y \rightsquigarrow x \leq 1$, by Corollary 7.2.12. Then:

$s(y) + s(y \to x) = s(y) + s((y \to x) \wedge 1) \overset{(S1)}{=} s(x) + s((x \to y) \wedge 1) = s(x) + s(1) = s(x)$; hence, $s(y \to x) = s(x) - s(y) = s(y) \to s(x)$. Similarly,

$s(y)+s(y \rightsquigarrow x) = s(y)+s((y \rightsquigarrow x)\wedge 1) \overset{(S2)}{=} s(x)+s((x \rightsquigarrow y)\wedge 1) = s(x)+s(1) = s(x)$; hence, $s(y \rightsquigarrow x) = s(x) - s(y) = s(y) \to s(x)$.

(S8'): Let $x \leq y$; then $x \to y \geq 1$, $x \rightsquigarrow y \geq 1$ and $y \to x \leq 1$, $y \rightsquigarrow x \leq 1$, by Corollary 7.2.12. Then:

$s(y) = s(y) + 0 = s(y) + s(1) = s(y) + s((y \to x) \vee 1) \overset{(S1')}{=} s(x) + s((x \to y) \vee 1) = s(x) + s(x \to y)$; hence, $s(x \to y) = s(y) - s(x) = s(x) \to s(y)$. Similarly, $s(x \rightsquigarrow y) = s(y) - s(x) = s(x) \to s(y)$.

(S9): Let $x \leq y$; hence $x \to y \geq 1$. By (S3'), $s(x \to y) \geq 1$; but, by (S8'), $s(x \to y) = s(y) - s(x)$; it follows that $s(y) - s(x) \geq 1$, hence $s(x) \leq s(y)$. □

Theorem 17.3.9 *Any state is a Bosbach-state.*

Proof. By definitions of states and Bosbach-states, it remains to prove (S1)-(S2'). Indeed,

(S1): $s(x) + s((x \to y) \wedge 1) = s((x \to y) \wedge 1) + s(x) \overset{s\ state}{=} s([(x \to y) \wedge 1] \cdot x) = s([(x \to y) \cdot x] \wedge x) = s(y \wedge x) = s(x \wedge y) = \ldots = s(y) + s((y \to x) \wedge 1)$.

(S2): $s(x) + s((x \rightsquigarrow y) \wedge 1) \overset{s\ state}{=} s(x \cdot [(x \rightsquigarrow y) \wedge 1]) = s([x \cdot (x \rightsquigarrow y)] \wedge x) = s(y \wedge x) = s(x \wedge y) = \ldots = s(y) + s((y \rightsquigarrow x) \wedge 1)$.

(S1'): $s(x) + s((x \to y) \vee 1) = s((x \to y) \vee 1) + s(x) \overset{s\ state}{=} s([(x \to y) \vee 1] \cdot x) = s([(x \to y) \cdot x] \vee x) = s(y \vee x) = s(x \vee y) = \ldots = s(y) + s((y \to x) \vee 1)$.

(S2'): $s(x) + s((x \rightsquigarrow y) \vee 1) \overset{s\ state}{=} s(x \cdot [(x \rightsquigarrow y) \vee 1]) = s([x \cdot (x \rightsquigarrow y)] \vee x) = s(y \vee x) = s(x \vee y) = \ldots = s(y) + s((y \rightsquigarrow x) \vee 1)$. □

Open problem 17.3.10 Study the restrictions of the various kinds of states from the l-group level to the G^-, G^+ level, the $[u', 1]$, $[1, u]$ level, the $G^-_{-\infty}$, $G^+_{+\infty}$ level.

Bibliography

[1] J.C. Abbott, *Implicational algebras*, Bull. Math. Soc. Sci. Math. Roumanie, 11 (59), 1967, pp. 3–23.

[2] —*Semi-Boolean algebra*, Mat. Vesnik, 4 (19), 1967, pp. 177–198.

[3] S.S. Ahn, K.S. So, *On ideals and uppers in BE-algebras*, Sci. Mat. Japonica, Online e-2008, pp. 351–357.

[4] C.J. van Alten, *Representable biresiduated lattices*, J. Algebra 247, 2002, pp. 672–691.

[5] M. Anderson, P. Conrad, J. Martinez, *The lattice of convex l-subgroups of a lattice-ordered group*, in: A.M.W. Glass, W.C. Holland (Eds.), Lattice-Ordered Groups, D. Reidel, Dordrecht, 1989, pp. 105–127.

[6] M. Andersen, T. Feil, Lattice-Ordered Groups - An Introduction -, D. Reidel Publishing Company, 1988.

[7] E.G. Beltrametti, G. Cassinelli, The Logic of Quantum Mechanics, Addison-Wesley, Reading, 1981.

[8] G. Birkhoff, Lattice Theory, 1st ed., *American Mathematical Society Colloquium Publications*, American Mathematical Society, Providence, R.I., 1948.

[9] G. Birkhoff, J. von Neumann, *The logic of quantum mechanics*, Ann. Math., 37, 1936, pp. 823–834.

[10] K. Blount, C. Tsinakis, *The structure of residuated lattices*, Internat. J. Algebra Comput. 13 (4), 2003, pp. 437–461.

[11] V. Boicescu, A. Filipoiu, G. Georgescu, S. Rudeanu, Łukasiewicz-Moisil algebras, *Annals of Discrete Mathematics*, 49, North-Holland, 1991.

[12] G. Boole, An investigation of the laws of thought, Cambridge 1854, reprinted by Dover Press, New York, 1967.

[13] R.A. Borzooei, A. Borumand Saeid, A. Rezaei, A. Radfar, R. Ameri, *On pseudo-BE algebras*, Discussiones Mathematicae, General Algebra and Applications, 33, 2013, pp. 95–108.

[14] B. Bosbach, *Concerning bricks*, Acta Math. Acad. Sci. Hungar., 38, 1981, pp. 89–104.

[15] —*Concerning semiclans*, Arch. Math., 37, 1981, pp. 316–324.

[16] —*Concerning cone algebras*, Algebra Universalis, 15, 1982, pp. 58–66.

[17] —Residuationsstrukturen, Univ. Kassel, 1997.

[18] M. Botur, J. Kühr, L. Liu, C. Tsinakis, *The Conrad program: from l-groups to algebras of logic*, J. Algebra 450, 2016, pp. 173–203.

[19] M. Botur, I. Chajda, R. Halaš, J. Kühr, J. Paseka, Algebraic Methods in Quantum Logic, http://www.researchgate.net/publication/274319102

[20] F. Bou, F. Paoli, A. Ledda, H. Freytes, *On some properties of quasi-MV algebras and* $\sqrt{'}$*quasi-MV algebras. Part II*, Soft Comput., 12 (4), 2008, pp. 341–352.

[21] F. Bou, F. Paoli, A. Ledda, M. Spinks, R. Giuntini, *The Logic of Quasi-MV algebras*, Journal of Logic and Computation, 20, 2, 2010, pp. 619–643.

[22] R.H. Bruck, A Survey of Binary Systems, Springer-Verlag, 1971.

[23] D. Buşneag, S. Rudeanu, *A glimpse of deductive systems in algebra*, Cent. Eur. J. Math., 8 (4), 2010, pp. 688–705.

[24] R. Ceterchi, *On Algebras with Implications, categorically equivalent to Pseudo-MV Algebras*, The Proceedings of the Fourth International Symposium on Economic Informatics, INFOREC Printing House, Bucharest, Romania, May 1999, pp. 912–916.

[25] —*Pseudo-Wajsberg Algebras*, Mult. Val. Logic (A special issue dedicated to the memory of Gr. C. Moisil), 2001, Vol. 6, No. 1-2, pp. 67–88.

[26] C. C. Chang, *Algebraic analysis of many valued logics*, Trans. Amer. Math. Soc. 88, 1958, pp. 467–490.

[27] M.A. Chaudhry, *On BCH-algebras*, Math. Japonica, 36, 1991, pp. 665–676.

[28] W.J. Chen, W.A. Dudek, *Quantum computational algebra with a non-commutative generalization*, Mathematica Slovaca, 2016, DOI: https://doi.org/10.1515/ms-2015-0112.

[29] —*Ideals and congruences in quasi-pseudo-MV algebras*, Soft Computing, 2017.

[30] W.J. Chen, Bijan Davvaz, *Some classes of quasi-pseudo-MV algebras*, Logic Journal of the IGPL, Volume 24, Issue 5, 2016, pp. 655–672, https://doi.org/10.1093/jigpal/jzw034

[31] R. Cignoli, A. Torrens, *An algebraic analysis of product logic*, Centre de Recerca Matematica, Preprint no. 363, Barcelona, 1997.

[32] R. Cignoli, I.M.L. D'Ottaviano, D. Mundici Algebraic Foundations of many-valued Reasoning, Kluwer, 2000, Volume 7.

[33] J. Cirulis, *Implication in sectionally pseudocomplemented posets*, Acta Sci. Math. (Szeged) 74, 2008, pp. 477–491.

[34] —*Subtraction-like operations in nearsemilattices*, Dem. Math., 43, 2010, pp. 725–738.

[35] —*Quasi-orthomodular posets and weak BCK-algebras*, Order, DOI 10.1007/s11083-013-9309-1.

[36] L.C. Ciungu, Non-commutative Multiple-Valued Logic Algebras, Springer, 2014.

[37] P.M. Cohn, Universal Algebra, 2nd ed., *Mathematics and Its Applications*, vol. 6, D. Reidel Publishing Co., Dordrecht–Boston, Mass., 1981.

[38] P.F. Conrad, Lattice Ordered Groups, An Introduction, *Tulane Lecture Notes*, New Orleans, 1970.

[39] M.R. Darnel, Theory of Lattice-Ordered Groups, Marcel Dekker, 1995.

[40] A. Diego, Sur les algèbres de Hilbert, Ph. D. Thesis, *Collection de Logique math.*, Serie A, XXI, Gauthier-Villars, 1966 (traduction faites d'après la Thèse édi'ee en langue espagnole sous le titre: Sobre Algebras de Hilbert, parue dans la Collection *notas de Logica Matematica*, Univ. Nacional del Sur, Bahia Blanca, 1961).

[41] R. P. Dilworth, *Non-commutative residuated lattices*, Trans. of the American Math. Soc. 46, 1939, pp. 426–444.

[42] A. Di Nola, G. Georgescu, A. Iorgulescu, *Pseudo-BL algebras: Part I*, Multi. Val. Logic, 2002, Vol. 8 (5-6), pp. 673–714.

[43] —*Pseudo-BL algebras: Part II*, Multi. Val. Logic, 2002, Vol. 8 (5-6), pp. 717–750.

[44] W.A. Dudek, Y.B. Jun, *Pseudo-BCI algebras*, East Asian Math. J., 24, 2008, pp. 187–190.

[45] A. Dvurečenskij, *On partial addition in pseudo-MV algebras*, In: Proc. Fourth Inter. Symp. on Econ. Inform., May 6-9, 1999, Bucharest, Eds I. Smeureanu et al., INFOREC Printing House, Bucharest, 1999, pp. 952–960.

[46] —*States on pseudo MV-algebras*, Studia Logica 68, 2001, pp. 301–327.

[47] —*Pseudo MV-algebras are intervals in l-groups*, Journal of Australian Mathematical Society, 70, 2002, pp. 427–445.

[48] —*States on Unital Partially-Ordered Groups*, Kybernetika, 38, 2002, pp. 297–318.

[49] A. Dvurečenskij, S. Pulmanova, New Trends in Quantum Structures, Kluwer Academic Publishers, Ister Science, Dordrecht, Bratislava, 2000.

[50] G. Dymek, *p-semisimple pseudo-BCI-algebras*, J. of Multiple-Valued Logic and Soft Computing, 19, 2012, pp. 461–474.

[51] —*On compatible deductive systems of pseudo-BCI-algebras*, J. of Multiple-Valued Logic and Soft Computing, 22, 2014, pp. 167–187.

[52] P. Flondor, G. Georgescu, A. Iorgulescu, *Pseudo-t-norms and pseudo-BL algebras*, Soft Computing, 5, No 5, 2001, pp. 355–371.

[53] J.M. Font, A. J. Rodriguez, A. Torrens, *Wajsberg algebras*, Stochastica Vol. VIII, No. 1, 1984, pp. 5–31.

[54] H. Furstenberg, *The inverse operation in groups*, Proc. Amer. Math. Soc., 6, 1955, pp. 991–997.

[55] N. Galatos, C. Tsinakis, *Generalized MV-algebras*, J. Algebra, 283, 2005, pp. 254–291.

[56] N. Galatos, P. Jipsen, T. Kowalski, H. Ono, Residuated lattices: An Algebraic Glimpse at Substrucural Logics, *Studies in Logic and the Foundation of Mathematics*, Vol. 151, Elsevier, 2007.

[57] G. Georgescu, *Bosbach states on fuzzy structures*, Soft Computing no. 8, 2004, pp. 217–230.

[58] G. Georgescu, A. Iorgulescu, *Pseudo-MV Algebras: a non-commutative Extension of MV Algebras*, The Proceedings of the Fourth International Symposium on Economic Informatics, INFOREC Printing House, Bucharest, Romania, May 1999, pp. 961–968.

[59] —*Pseudo-BL algebras: A non-commutative extension of BL algebras*, Abstracts of The Fifth International Conference FSTA 2000, Slovakia, February 2000, pp. 90–92.

[60] —*Pseudo-BCK algebras: An extension of BCK algebras*, Proceedings of DMTCS'01: Combinatorics, Computability and Logic (C.S. Calude, M.J. Dinneen, S. Sburlan (Eds)), Springer, London, 2001, pp. 97–114.

[61] —*Pseudo-MV algebras*, Mult. Val. Logic (A special issue dedicated to the memory of Gr. C. Moisil), Vol. 6, Nr. 1-2, 2001, pp. 95–135.

[62] —Logică matematică (Romanian), Editura ASE (Academy of Economic Studies Publishing House), 2010.

[63] J. Gil-Férez, A. Ledda, F. Paoli, C. Tsinakis, *Projectable l-groups and algebras of logic: Categorical and algebraic connections*, Journal of Pure and Applied Algebra 220, 2016, pp. 3514–3532.

[64] J. Gil-Férez, A. Ledda, C. Tsinakis, *Hulls of ordered algebras: Projectability, strong projectability and lateral completeness*, Journal of algebra, 483, 2017, pp. 429–474.

[65] R. Giuntini, *Weakly linear quantum MV algebras*, Algebra Universalis, 53, 1, 2005, pp. 45–72.

[66] A.M.W. Glass, Partially Ordered Groups, *Series in Algebra*, World Scientific, 1999.

[67] K.R. Goodearl, Partially ordered commutative groups with interpolation, *Mathematical Surveys and Monographs*, no. 20, A.M.S, Providence, Rhode Island, 1986.

[68] S. Gottwald, A Treatise on Many-Valued Logics, *Studies in Logic and Computation*, vol. 9, Research Studies Press: Baldock, Hertfordshire, England, 2001.

[69] G. Grätzer, Universal Algebra (paperback), 2nd ed., Springer, 2008.

[70] A. Grzaślewicz, *On some problem on BCK-algebras*, Math. Japonica 25, No. 4, 1980, pp. 497–500.

[71] P. Hájek, Metamathematics of fuzzy logic, *Inst. of Comp. Science, Academy of Science of Czech Rep., Technical report* 682, 1996.

[72] —Metamathematics of fuzzy logic, Kluwer Acad. Publ., Dordrecht, 1998.

[73] P. Hájek, L. Godo, F. Esteva, *A Complete Many-Valued Logic with Product-Conjunction*, Archive for Mathematical Logic, Vol. 35, No. 3, 1996, pp. 191–208.

[74] Paul R. Halmos, Algebraic Logic, AMS Chelsea Publishing 1962.

[75] L. Henkin, *An Algebraic Characterization of Quantifiers*, Fundamenta Mathematicae, vol. 37, 1950, pp. 63–74.

[76] G. Higman, B.H. Neumann, *Groups as grupoids with one law*, Publicationes Mathematicae, Debrezin, 2, 1952, pp. 215–221.

[77] D. Hilbert, *Die Logischen Grundlagen der Mathematik*, Mathematischen Annalen, 88 Band, 1923, pp. 151–165.

[78] D. Hilbert, W. Ackermann, Principles of Mathematical Logic, AMS Chelsea Publishing, 1958.

[79] D. Hilbert, P. Bernays, Gründlagen der Math., Erster Band, Berlin, 1934.

[80] A. Horn, *Logic with truth values in a linearly ordered Heyting algebra*, J. Symbolic Logic, 34, 1969, pp. 395–408.

[81] Q.P. Hu, X. Li, *On BCH-algebras*, Math. Sem. notes, 11, 1983, pp. 313–320.

[82] —*On proper BCH-algebras*, Math. Japon., 30, 1985, pp. 659–661.

[83] Y.S. Huang, BCI-algebras, Science Press, Beijing, China, 2006.

[84] Y. Imai, K. Iséki, *On axiom systems of propositional calculi XIV*, Proc. Japan Academy, 42, 1966, pp. 19–22.

[85] C. Ioniţă, *Pseudo-MV algebras as semigroups*, The Proceedings of the Fourth International Symposium on Economic Informatics, INFOREC Printing House, Bucharest, Romania, May 1999, pp. 983–987.

[86] A. Iorgulescu, *Some direct ascendents of Wajsberg and MV algebras*, Scientiae Mathematicae Japonicae, Vol. 57, No. 3, 2003, pp. 583–647.

[87] — *On pseudo-BCK algebras and porims*, Scientiae Mathematicae Japonicae, Vol. 10, Nr. 16, 2004, pp. 293–305.

[88] — *Pseudo-Iséki algebras. Connection with pseudo-BL algebras*, J. of Multiple-Valued Logic and Soft Computing, Vol. 11 (3-4), 2005, pp. 263–308.

[89] — *Classes of pseudo-BCK algebras - Part I*, J. of Multiple-Valued Logic and Soft Computing, Vol. 12, No. 1-2, 2006, pp. 71–130.

[90] — *Classes of pseudo-BCK algebras - Part II*, J. of Multiple-Valued Logic and Soft Computing, Vol. 12, No. 5-6, 2006, pp. 575–629.

[91] — *On BCK algebras - Part I.a: An attempt to treat unitarily the algebras of logic. New algebras*, J. of Universal Computer Science, Vol. 13, no. 11, 2007, pp. 1628–1654.

[92] — *On BCK algebras - Part I.b: An attempt to treat unitarily the algebras of logic. New algebras*, J. of Universal Computer Science, Vol. 14, no. 22, 2008, pp. 3686–3715.

[93] — *Monadic Involutive Pseudo-BCK algebras*, Acta Universitatis Apulensis, No. 15, 2008, pp. 159–178.

[94] — Algebras of logic as BCK algebras, Academy of Economic Studies Press (Editura ASE), Bucharest, 2008.

[95] — *Asupra algebrelor Booleene* (Romanian), Revista de logică http://egovbus.net/rdl, 25.01.2009, pp. 1–25.

[96] — *Classes of examples of pseudo-MV algebras, pseudo-BL algebras and divisible bounded non-commutative residuated lattices*, Soft Computing, Vol. 14, no. 4, 2010, pp. 313–327.

[97] — *The implicative-group - a term equivalent definition of the group coming from algebras of logic - Part I*, Preprint nr. 11/2011, Institutul de Matematică "Simion Stoilow" al Academiei Române, Bucureşti, România.

[98] — *The implicative-group - a term equivalent definition of the group coming from algebras of logic - Part II*, Preprint nr. 12/2011, Institutul de Matematică "Simion Stoilow" al Academiei Române, Bucureşti, România.

[99] — *On l-implicative-groups and associated algebras of logic*, presented at Congmatro 2011-Braşov, Bulletin of the Transilvania University of Brasov # Series III: Mathematics, Informatics, Physics, Vol 5(54) 2012, Special Issue: Proceedings of the Seventh Congress of Romanian Mathematicians, 179-194, published by Transilvania University Press, Brasov and Publishing House of the Romanian Academy.

[100] — *New algebras and new connections between/in the algebras of logic and the monoidal algebras*, Libertas Mathematica, Vol. 35 (Dedicated to Nicolae Dinculeanu and Solomon Marcus in celebration of their 90th Birthday) No. 1, 2015, pp. 13–55.

[101] — *Quasi-algebras vs. regular algebras - Part I*, (Dedicated to Sergiu Rudeanu) Scientific Annals of Computer Science vol. 25 (1), 2015, pp. 1–43. doi: 10.7561/SACS.2015.1.ppp

[102] — *Quasi-algebras vs. regular algebras - Part II*, 2015, Preprint nr. 3/2017 (*http : //imar.ro/ increst*/2017/3_2017.*pdf*).

[103] — *Quasi-algebras vs. regular algebras - Part III*, 2015, Preprint nr. 4/2017 (*http : //imar.ro/ increst*/2017/4_2017.*pdf*).

[104] — *New generalizations of BCI, BCK and Hilbert algebras - Parts I, II*, (Dedicated to Dragoş Vaida) J. of Multiple-Valued Logic and Soft Computing, Vol. 27, No. 4, 2016, pp. 353–406, pp. 407–456 (a previous version available from December 6, 2013, at http://arxiv.org/abs/1312.2494).

[105] — *Implicative-groups*, Colocviu de matematică-informatică, Curtea de Argeş, Romania, 10-11 octombrie 2016.

[106] — Implicative-groups vs. groups and generalizations, 1st ed., Matrix Rom, Bucharest, 2018.

[107] — *Generalizations of MV algebras, ortholattices and Boolean algebras*, ManyVal 2019, Bucharest, Romania, November 1-3, 2019.

[108] — *Quasi-i-Boolean algebras vs. quasi-m Boolean algebras*, 11th Conference of the European Society for Fuzzy Logic and Technology (EUSFLAT 2019), Atlantis Studies in Uncertainty Modelling, Vol. 1, 2019, pp. 334–341.

[109] — *Algebras of logic vs. Algebras*, Landscapes in Logic, Vol. 1 Contemporary Logic and Computing, Editor Adrian Rezuş, College Publications, 2020, pp. 157–258.

[110] — *On Quantum-MV algebras - Part I: Orthomodular algebras*, Scientific Annals of Computer Science, Vol. 31(2), 2021, pp. 163–221.

[111] — *On Quantum-MV algebras - Part II: Orthomodular lattices, softlattices and widelattices*, Trans. Fuzzy Sets Syst., Vol. 1, No. 1, 2022, pp. 1–41.

[112] — *On quantum-MV algebras - Part III: The properties (m-Pabs-i) and* (WNM_m), Scientiae Mathematicae Japonicae, Vol. 86, No. 1, 2023, pp. 49–81.
Online version at https://www.jams.or.jp/notice/scmjol/2022.html, 2022, pp. 1–33.

[113] — BCK algebras versus m-BCK algebras. Foundations, *Studies in Logic 96*, College Publications, 2022.

[114] A. Iorgulescu, M. Kinyon, *Putting bounded involutive lattices, De Morgan algebras, ortholattices and Boolean algebras on the "map"*, Journal of Applied Logics (JALs-ifCoLog), Special Issue on Multiple-Valued Logic and Applications, Vol. 8, No. 5, 2021, pp. 1169–1213.

[115] — *Two generalizations of bounded involutive lattices and of ortholattices*, Journal of Applied Logics (JALs-ifCoLog), Vol. 8, No. 7, 2021, pp. 2173–2218.

[116] — *Putting Quantum MV algebras on the "map"*, Scientiae Mathematicae Japonicae, Vol. 85, No. 2, 2022, pp. 89–115.
Online version at https://www.jams.or.jp/notice/scmjol/2021.html, 2021, pp. 1–27.

[117] — Non-commutative algebras. Pseudo-BCK algebras versus m-pseudo-BCK algebras, *Studies in Logic 107*, College Publications, 2024.

[118] K. Iséki, *An algebra related with a propositional calculus*, Proc. Japan Acad., 42, 1966, pp. 26–29.

[119] —*BCK-Algebras with condition (S)*, Math. Japonica, 24, No. 1, 1979, pp. 107–119.

[120] —*On BCK-Algebras with condition (S)*, Math. Japonica, 24, No. 6, 1980, pp. 625–626.

[121] K. Iséki, S. Tanaka, *An introduction to the theory of BCK-algebras*, Math. Japonica 23, No.1, 1978, pp. 1–26.

[122] P. Jipsen, C. Tsinakis, *A Survey of Residuated Lattices*, in: Jorge Martinez (Ed.), Ordered Algebraic Structures, Kluwer, Dordrecht, 2002, pp. 19–56.

[123] P. Jipsen, A. Ledda, F. Paoli, *On some properties of quasi-MV algebras and $\sqrt{'}$quasi-MV algebras. Part IV*, Rep. Math. Logic, 48, 2013, pp. 3–36.

[124] Y.B. Jun, M.S. Kang, *Fuzzifications of generalized Tarski filters in Tarski algebras*, Computers and Mathematics with Applications, 61, 2011, pp. 1–7.

[125] Y.B. Jun, E.H. Roh, H.S. Kim, *On BH-algebras*, Scientiae Mathematicae Japonicae, 1, 1998, pp. 347–354.

[126] H.S. Kim, Y.H. Kim, *On BE-algebras*, Sci. Math. Jpn., online e-2006, 2006, pp. 1192–1202.

[127] J.D. Kim, Y.M. Kim, E.H. Roh, *A note on GT-algebras*, J. Korea Soc. Math. Educ. Ser. B Pure Appl. Math., 16, 2009, pp. 59–68.

[128] E.P. Klement, R. Mesiar, E. Pap, Triangular norms, Kluwer Academic Publishers, 2000.

[129] Y. Komori, *The variety generated by BCC-algebras is finitely based*, Reports Fac. Sci. Shizuoka Univ., 17, 1983, pp. 13–16.

[130] —*The class of BCC-algebras is not a variety*, Math. Japonica, 29, 1984, pp. 391–394.

[131] T. Kowalski, F. Paoli, *On some properties of quasi-MV algebras and $\sqrt{'}$quasi-MV algebras. Part III*, Rep. Math. Logic, 45, 2010, pp. 161–1999.

[132] W. Krull, *Axiomatische Begründung der allgemeinen Idealtheorie*, Sitzungsberichte der physikalisch medizinischen Societät der Erlangen 56, 1924, pp. 47–63.

[133] J. Kühr, *Representable dually residuated lattice-ordered monoids*, Discuss. Math. Gen. Algebra Appl. 23, 2003, pp. 115–123.

[134] —*Pseudo BL-algebras and DRl-monoids*, Math. Bohem. 128, 2003, pp. 199–208.

[135] —Pseudo-BCK-algebras and related structures, Universita Palackého v Olomouci, 2007.

[136] A. Ledda, M. Konig, F. Paoli, R. Giuntini, *MV Algebras and Quantum Computation*, Studia Logica, 82, 2, 2006, pp. 245–270.

[137] A. Ledda, F. Paoli, C. Tsinakis, *Lattice-theoretic properties of algebras of logic*, J. Pure Appl. Algebra 218 (10), 2014, pp. 1932–1952, http://dx.doi.org/10.1016/j.jpaa.2014.02.015.

[138] T.D. Lei, C.C. Xi, *p-radical in BCI-algebras*, Math. Japonica, 30 (4), 1985, pp. 511–517.

[139] L. Leuştean, Representations of Many-valued Algebras, Editura Universitară, Bucharest, 2010.

[140] Y. H. Lin, *Some properties on commutative BCK-algebras*, Scientiae Mathematicae Japonicae, 55, No. 1, 2002, pp. 129–133, :e5, 235-239.

[141] P. Mangani, *On certain algebras related to many-valued logics (Italian)*, Boll. Un. Mat. Ital. (4) **8**, 1973, pp. 68–78.

[142] D.J. Meng, *BCI-algebras and abelian groups*, Math. Japonica, 32, 1987, pp. 693–696.

[143] J. Meng, Y.B. Jun, BCK-algebras, Kyung Moon SA CO., Seoul, 1994.

[144] B.L. Meng, *CI-algebras*, Sci. Math. Jpn., 71, 1, 2010, pp. 11–17.

[145] —*Atoms in CI-algebras and singular CI-algebras*, Sci. Math. Jpn., 72, 1, 2010, pp. 319–324.

[146] G. Metcalfe, F. Paoli, C. Tsinakis, *Ordered algebras and logic*, in: H. Hosni, F. Montagna (Eds.), Uncertainty and Rationality, in: Publications of the Scuola Normale Superiore di Pisa, vol. 10, 2010, pp. 1–85.

[147] Gr.C. Moisil, Essais sur les logiques non-chrysippiennes, Editions de l'Academie de la Republique Socialiste de Roumanie, 1972.

[148] F. Montagna, *Subreducts of MV algebras with product and product residuation*, Algebra Universalis, 53, 2005, pp. 109–137.

[149] L. Monteiro, *Les algébres de Heyting et de Łukasiewicz trivalentes*, Notre Dame J. Formal Logic, 11, 1970, pp. 453–466.

[150] D. Mundici, *MV-algebras are categorically equivalent to bounded commutative BCK-algebras*, Math. Japonica 31, No. 6, 1986, pp. 889–894.

[151] —*Interpretation of AF C^*-algebras in Łukasiewicz sentential calculus*, J. Funct. Anal., 65, 1986, pp. 15–63.

[152] —Advanced Łukasiewicz calculus and MV-algebras, *Trends in Logic 35*, Springer, 2011.

[153] F. Paoli, Substructural Logics, *Trends in Logic*, vol. 13, Kluwer, Dordrecht, 2002.

[154] F. Paoli, A. Ledda, R. Giuntini, H. Freytes, *On some properties of quasi-MV algebras and $\sqrt{'}$quasi-MV algebras. Part I*, Rep. Math. Logic, 44, 2009, pp. 31–63.

[155] D. Piciu, Algebras of Fuzzy Logic, Editura Universitaria Craiova, Craiova, 2007.

[156] J. Rachunek, *A non-commutative generalization of MV-algebras*, Czechoslovak Math. J., 52 (127), 2002, pp. 255–273.

[157] A. Rezaei, A. Borumand Saeid, K. Yousefi Sikari Saber, *On pseudo-CI algebras*, Soft Computing, 23, 2019, pp. 4643–4654.

[158] W. Rump, *L-algebras, self-similarity, and l-groups*, J. of Algebra, 320, no. 6, 2008, pp. 2328–2348.

[159] M. Spinks, Contributions to the theory of Pre-BCK-Algebras, PhD Thesis, Monash University, 2003.

[160] C.J. van Alten, *Representable biresiduated lattices*, J. Algebra, 247, 2002, pp. 672–691.

[161] A. Walendziak, *Pseudo-BCH algebras*, Discussiones Mathematicae, General algebra and Applications, 35, 2015, pp. 5–19.

[162] A.M. Wille, Residuated Structures with Involution, Shaker Verlag, Aachen, 2006.

[163] X.H. Zhang, *BIK^+-logic and non-commutative fuzzy logics*, Fuzzy Systems Math., 21, 2007, pp. 31–36.

[164] X.H. Zhang, R. Ye, *BZ-algebras and groups*, J. Math. Phys. Sci., 29, 1995, pp. 223–233.

Index

www.ingramcontent.com/pod-product-compliance
Lightning Source LLC
Chambersburg PA
CBHW061231220326
41599CB00028B/5392